Methods in Enzymology

Volume XXXIV
AFFINITY TECHNIQUES
Enzyme Purification: Part B

METHODS IN ENZYMOLOGY

EDITORS-IN-CHIEF

Sidney P. Colowick   Nathan O. Kaplan

Methods in Enzymology

Volume XXXIV

Affinity Techniques

Enzyme Purification: Part B

EDITED BY

William B. Jakoby

SECTION ON ENZYMES AND CELLULAR BIOCHEMISTRY
NATIONAL INSTITUTE OF ARTHRITIS, METABOLISM, AND DIGESTIVE DISEASES
NATIONAL INSTITUTES OF HEALTH
BETHESDA, MARYLAND

Meir Wilchek

DEPARTMENT OF BIOPHYSICS
THE WEIZMANN INSTITUTE OF SCIENCE
REHOVOT, ISRAEL

ACADEMIC PRESS   New York   San Francisco   London   1974
A Subsidiary of Harcourt Brace Jovanovich, Publishers

COPYRIGHT © 1974, BY ACADEMIC PRESS, INC.
ALL RIGHTS RESERVED.
NO PART OF THIS PUBLICATION MAY BE REPRODUCED OR
TRANSMITTED IN ANY FORM OR BY ANY MEANS, ELECTRONIC
OR MECHANICAL, INCLUDING PHOTOCOPY, RECORDING, OR ANY
INFORMATION STORAGE AND RETRIEVAL SYSTEM, WITHOUT
PERMISSION IN WRITING FROM THE PUBLISHER.

ACADEMIC PRESS, INC.
111 Fifth Avenue, New York, New York 10003

United Kingdom Edition published by
ACADEMIC PRESS, INC. (LONDON) LTD.
24/28 Oval Road, London NW1

Library of Congress Cataloging in Publication Data
Main entry under title:

Enzyme purification and related techniques.

(Methods in enzymology, v. 22, 34)
Vol. 2, edited by W. B. Jakoby and M. Wilchek, has
title: Enzyme purification part B, with special title:
Affinity methods.
Includes bibliographical references.
1.   Enzymes–Purification.   I.   Jakoby, William B.,
Date   ed.   II.   Wilchek, Meir, ed.   III.   Series:
Methods in enzymology, v. 22 [etc.]   [DNLM;   1.   Enzymes
–Isolation–Purification.   W1ME9615K v. 22 etc. / QU135
E604]
QP601.C733 vol. 22   547'.758   79-26902
ISBN 0–12–181897–7 (v. 34)

PRINTED IN THE UNITED STATES OF AMERICA

Table of Contents

Section I. Introduction

Section II. Coupling Reactions and General Methodology

Section III. Attachment of Specific Ligands and Methods for Specific Proteins

A. Coenzymes and Cofactors

B. Sugars and Derivatives

C. Amino Acids and Peptides

D. Nucleic Acids, Nucleotides, and Derivatives

E. Other Systems

Section IV. Purification of Synthetic Macromolecules

Section V. Purification of Receptors

Section VI. Immunological Approaches

Contributors to Volume XXXIV

Article numbers are in parentheses following the names of contributors.
Affiliations listed are current.

K. L. AGARWAL (84), *Departments of Biology and Chemistry, Massachusetts Institute of Technology, Cambridge, Massachusetts*

YASUO AKANUMA (93), *3rd Department of Internal Medicine, University of Tokyo, Tokyo, Japan*

ROBERT H. ALLEN (28), *Department of Internal Medicine, Washington University School of Medicine, St. Louis, Missouri*

CHARALAMPOS ARSENIS (26), *Department of Biological Chemistry, University of Illinois College of Medicine, Chicago, Illinois*

Y. ASHANI (74), *Israel Institute for Biological Research, Ness-Ziona, Israel*

GILBERT ASHWELL (87), *National Institute of Arthritis, Metabolism, and Digestive Diseases, National Institutes of Health, Bethesda, Maryland*

BERNARD R. BAKER* (65), *Department of Chemistry, University of California, Santa Barbara, California*

ATARA BAR-ELI (42), *Department of Organic Chemistry, The Weizmann Institute of Science, Rehovot, Israel*

ROBERT BARKER (29, 35, 56), *Department of Biochemistry, University of Iowa, Iowa City, Iowa*

EVELYN R. BARRACK (71), *Department of Pharmacology and Experimental Therapeutics, The Johns Hopkins University School of Medicine, Baltimore, Maryland*

STANDISH BARRY (8, 77) *Department of Biochemistry, University College, Galway, Ireland*

EUGENE A. BAUER (47), *Division of Dermatology, Washington University School of Medicine, St. Louis, Missouri*

ED BAYER (20), *Department of Bio-physics, The Weizmann Institute of Science, Rehovot, Israel*

HELGA BEIKIRCH (60), *Abteilung Chemie, Max-Planck-Institut für Experimentelle Medizin, Göttingen, Germany*

ANN M. BENSON (71), *Department of Pharmacology and Experimental Therapeutics, The Johns Hopkins University School of Medicine, Baltimore, Maryland*

O. BERGLUND (18), *Medicinka Nobelinstitutet, Biochemiska Avdelningen, Karolinska Institutet, Stockholm, Sweden*

DAVID H. BING (92), *Center for Blood Research, Boston, Massachusetts*

PETER M. BLUMBERG (43), *Department of Biology, Massachusetts Institute of Technology, Cambridge, Massachusetts*

KEITH BROCKLEHURST (66), *Department of Biochemistry and Chemistry, The Medical College, St. Bartholomew's Hospital, London, England*

WILLIAM H. BROWN (83), *Department of Chemistry, Beloit College, Beloit, Wisconsin*

ROBERT L. BURGER (28), *Department of Internal Medicine, Washington University School of Medicine, St. Louis, Missouri*

Y. BURSTEIN (50), *Department of Organic Chemistry, The Weizmann Institute of Science, Rehovot, Israel*

MICHAEL BUSTIN (94), *Department of Chemical Immunology, The Weizmann Institute of Science, Rehovot, Israel*

JAN CARLSSON (66), *Department of Biochemistry and Chemistry, The Medical College, St. Bartholomew's Hospital, London, England*

P. J. CASHION (84), *Department of Biology, University of New Brunswick, Fredricton, New Brunswick, Canada*

IRWIN M. CHAIKEN (82), *National Institute of Arthritis, Metabolism, and*

* Deceased.

xiii

Digestive Diseases, National Institutes of Health, Bethesda, Maryland

WILLIAM W.-C. CHAN (41), *Department of Biochemistry, McMaster University, Hamilton, Ontario, Canada*

CHAO-KUO CHIANG (29), *Department of Biochemistry, University of Iowa, Iowa City, Iowa*

ICHIRO CHIBATA (21, 44), *Department of Biochemistry, Research Laboratory of Applied Biochemistry, Tanabe Seiyaku Co., Ltd., Osaka, Japan*

B. A. K. CHIBBER (48), *Department of Biochemistry, Purdue University, West Lafayette, Indiana*

KENNETH J. CLEMETSON (37), *Theodor-Kocher Institut, Bern, Switzerland*

LOUIS A. COHEN (7), *National Institute of Arthritis, Metabolism, and Digestive Diseases, National Institutes of Health, Bethesda, Maryland*

DENNIS A. CORNELIUS (83), *Department of Chemistry, University of Arizona, Tucson, Arizona*

FRIEDRICH CRAMER (60), *Abteilung Chemie, Max-Plack-Institut für Experimentelle Medizin, Göttingen, Germany*

DAVID B. CRAVEN (17), *Biochemistry Department, Liverpool University, Liverpool, England*

ERIC M. CROOK (66), *Department of Biochemistry and Chemistry, The Medical College, St. Bartholomew's Hospital, London, England*

PEDRO CUATRECASAS (6, 34, 79, 85, 86), *Department of Pharmacology and Experimental Therapeutics, The Johns Hopkins University School of Medicine, Baltimore, Maryland*

PETER V. DANENBERG (64), *McArdle Laboratory for Cancer Research, Madison, Wisconsin*

PETER D. G. DEAN (17), *Biochemistry Department, Liverpool University, Liverpool, England*

D. G. DEUTSCH (48), *Department of Medicine, The University of Chicago, Medical School, Chicago, Illinois*

YADIN DUDAI (73), *Department of Bio-*

physics, The Weizmann Institute of Science, Rehovot, Israel

F. ECKSTEIN (18, 78), *Abteilung Chemie, Max-Planck-Institut für Experimentelle Medizin, Göttingen, Germany*

G. M. EDELMAN (15), *The Rockefeller University, New York, New York*

MARVIN EDELMAN (59), *Department of Plant Genetics, The Weizmann Institute of Science, Rehovot, Israel*

ARTHUR Z. EISEN (47), *Department of Medicine, Washington University School of Medicine, St. Louis, Missouri*

ZELIG ESHHAR (94), *Department of Pathology, Harvard Medical School, Boston, Massachusetts*

A. M. FILBERT (4), *Biomaterials Research and Development, Corning Glass Works, Corning, New York*

HAROLD M. FLOWERS (33), *Department of Biophysics, The Weizmann Institute of Science, Rehovot, Israel*

M. FRIDKIN (84), *Department of Organic Chemistry, The Weizmann Institute of Science, Rehovot, Israel*

LUCIANO FRIGERI (72), *Istituto di Patologia Generale, Università di Padova, Padua, Italy*

A. M. FRISCHAUF (78), *Abteilung Molekular Biologie, Max-Planck-Institut für Biophysikalische Chemie, Göttingen, Germany*

BARBARA C. FURIE (75), *American Red Cross Blood Research Laboratory, Bethesda, Maryland*

BRUCE FURIE (75), *Laboratory of Chemical Biology, National Institute of Arthritis, Metabolism, and Digestive Diseases, National Institutes of Health, Bethesda, Maryland*

MARIAN GORECKI (42, 57), *Department of Organic Chemistry, The Weizmann Institute of Science, Rehovot, Israel*

DAVID J. GRAVES (10), *Department of Chemical and Biochemical Engineering, University of Pennsylvania, Philadelphia, Pennsylvania*

TADHG GRIFFIN (8), *Department of Biochemistry, University College, Galway, Ireland*

NOAM HARPAZ (33), *Department of Biophysics, The Weizmann Institute of Science, Rehovot, Israel*

MICHAEL J. HARVEY (17), *Biochemistry Department, Liverpool University, Liverpool, England*

MASAKI HAYASHI (93), *3rd Department of Internal Medicine, University of Tokyo, Tokyo, Japan*

CHARLES HEIDELBERGER (64), *McArdle Laboratory for Cancer Research, Madison, Wisconsin*

ROBERT L. HILL (29, 35, 56), *Department of Biochemistry, Duke University, Durham, North Carolina*

H. F. HIXSON, JR. (51), *Abbott Diagnostics Division, Abbott Laboratories, Inc., North Chicago, Illinois*

VÁCLAV HOŘEJŠÍ (13, 36), *Department of Biochemistry, Charles University, Prague, Czechoslovakia*

ROGER L. HUDGIN (87), *National Institute of Arthritis, Metabolism, and Digestive Diseases, National Institutes of Health, Bethesda, Maryland*

JOHN K. INMAN (3), *Laboratory of Immunology, National Institute of Allergy and Infectious Diseases, National Institutes of Health, Bethesda, Maryland*

AKIO IWASHIMA (27), *Department of Biochemistry, Kyoto Prefectural University of Medicine, Kyoto, Japan*

WILLIAM B. JAKOBY (1), *Section on Enzymes and Cellular Biochemistry, National Institute of Arthritis, Metabolism, and Digestive Diseases, National Institutes of Health, Bethesda, Maryland*

E. JAY (84), *Department of Biochemistry and Molecular Biology, Cornell University, Ithaca, New York*

JOHN J. JEFFREY (47), *Division of Dermatology, Washington University School of Medicine, St. Louis, Missouri*

BENGT JERGIL (19), *Biochemical Division, Chemical Center, University of Lund, Lund, Sweden*

BERNARD T. KAUFMAN (22), *National Institute of Arthritis, Metabolism, and Digestive Diseases, National Institutes of Health, Bethesda, Maryland*

MICHAEL N. KAZARINOFF (26), *Section of Biochemistry, Molecular and Cell Biology, Cornell University, Ithaca, New York*

H. G. KHORANA (84), *Departments of Biology and Chemistry, Massachusetts Institute of Technology, Cambridge, Massachusetts*

MAREK P. J. KIERSTAN (66), *Department of Biochemistry and Chemistry, The Medical College, St. Bartholomew's Hospital, London, England*

ROY L. KISLIUK (23), *Departments of Biochemistry and Pharmacology, Tufts University School of Medicine, Boston, Massachusetts*

JAN KOCOUREK (13, 36) *Department of Biochemistry, Charles University, Prague, Czechoslovakia*

LEONARD D. KOHN (88), *National Institute of Arthritis, Metabolism, and Digestive Diseases, National Institutes of Health, Bethesda, Maryland*

TORE KRISTIANSEN (31), *Institute of Biochemistry, University of Uppsala, Uppsala, Sweden*

RAPHAEL LAMED (55), *Department of Biophysics, The Weizmann Institute of Science, Rehovot, Israel*

ROBERT J. LEFKOWITZ (89), *Departments of Medicine and Biochemistry, Duke University Medical Center, Durham, North Carolina*

JURIS LIEPNIEKS (52), *Department of Chemistry, Purdue University, West Lafayette, Indiana*

ALBERT LIGHT (52), *Department of Chemistry, Purdue University, West Lafayette, Indiana*

UNO LINDBERG (58, 63), *Department of Microbiology, The Wallenberg Laboratory, Uppsala University, Uppsala, Sweden*

HALINA LIS (32), *Department of Biophysics, The Weizmann Institute of Science, Rehovot, Israel*

U. Z. LITTAUER (81), *Department of Biochemistry, The Weizmann Institute of Science, Rehovot, Israel*

DAVID M. LIVINGSTON (91), *Department of Medicine, The Childrens Cancer Research Foundation and Harvard Medical School, Boston, Massachusetts*

REUBEN LOTAN (32), *Department of Biophysics, The Weizmann Institute of Science, Rehovot, Israel*

CHRISTOPHER R. LOWE (17), *Biochemical Division, Chemical Center, University of Lund, Lund, Sweden*

DONALD B. McCORMICK (26), *Section of Biochemistry, Molecular and Cell Biology, Cornell University, Ithaca, New York*

THOMAS MACIAG (53), *Department of Biochemistry, University of Pennsylvania School of Medicine, Philadelphia, Pennsylvania*

PHILIP W. MAJERUS (28), *Departments of Internal Medicine and Biological Chemistry, Washington University School of Medicine, St. Louis, Missouri*

STEVEN MARCH (6), *Department of Pharmacology and Experimental Therapeutics, The Johns Hopkins University School of Medicine, Baltimore, Maryland*

STUART L. MARCUS (38, 40), *Walker Laboratory, Sloan-Kettering Institute for Cancer Research, Rye, New York*

JAMES S. MARSHALL (39), *Department of Medicine, University Hospitals of Cleveland, Cleveland, Ohio*

ISAMU MATSUMOTO (30), *Division of Chemical Toxicology and Immunochemistry, Faculty of Pharmaceutical Sciences, University of Tokyo, Tokyo, Japan*

ATSUKO MATSUURA (27), *Department of Biochemistry, Kyoto Prefectural University of Medicine, Kyoto, Japan*

YUHSI MATUO (21, 44), *Department of Biochemistry, Research Laboratory of Applied Biochemistry, Tanabe Seiyaku Co., Ltd., Osaka, Japan*

CAROL S. MEHLMAN (28), *Department of Internal Medicine, Washington University School of Medicine, St. Louis, Missouri*

E. T. MERTZ (48), *Department of Biochemistry, Purdue University, West Lafayette, Indiana*

TALIA MIRON (5), *Department of Biophysics, The Weizmann Institute of Science, Rehovot, Israel*

ANATOL G. MORELL (87), *Division of Genetic Medicine, Department of Medicine, Albert Einstein College of Medicine, Bronx, New York*

KLAUS MOSBACH (16, 19, 76), *Biochemical Division, Chemical Center, University of Lund, Lund, Sweden*

J. C. NICOLAS (69, 70), *Groupe des Recherches sur la Biochimie des Steroids, Institut de Biologie, Montpelier, France*

A. H. NISHIKAWA (51), *Chemical Research Division, Hoffman-La Roche, Inc., Nutley, New Jersey*

ERNESTO NOLA (86), *Patologia Generale, Universita di Napoli, Naples, Italy*

YOSHITSUGU NOSE (27), *Department of Biochemistry, Kyoto Prefectural University of Medicine, Kyoto, Japan*

PÁDRAIG O'CARRA (8, 77), *Department of Biochemistry, University College, Galway, Ireland*

DONALD S. O'HARA (89), *Departments of Medicine and Physiological Chemistry, Massachusetts General Hospital and Harvard Medical School, Boston, Massachusetts*

TOSHIAKI OSAWA (30), *Division of Chemical Toxicology and Immunochemistry, Faculty of Pharmaceutical Sciences, University of Tokyo, Tokyo, Japan*

M. K. PANGBURN (50), *Aerospace Medical Research Laboratory/THP Wright-Patterson Air Force Base, Dayton, Ohio*

INDU PARIKH (6, 79, 85, 86) *Department of Pharmacology and Experimental Therapeutics, The Johns Hop-*

kins University School of Medicine, Baltimore, Maryland

EDWARD J. PASTORE (23), Department of Chemistry, University of California, San Diego, La Jolla, California

AVRAHAM PATCHORNIK (42), Department of Organic Chemistry, The Weizmann Institute of Science, Rehovot, Israel

JACK PENSKY (39), Department of Medicine, University Hospitals of Cleveland, Cleveland, Ohio

TORGNY PERSSON (58), Department of Microbiology, The Wallenberg Laboratory, Uppsala University, Uppsala, Sweden

LAURENCE T. PLANTE (23), Department of Biology, University of California, San Diego, La Jolla, California

MOHINDAR POONIAN (54), Roche Research Division, Hoffman-La Roche, Inc., Nutley, New Jersey

JERKER PORATH (2), Institute of Biochemistry, University of Uppsala, Uppsala, Sweden

WILLIAM E. PRICER, JR. (87), National Institute of Arthritis, Metabolism, and Digestive Diseases, National Institutes of Health, Bethesda, Maryland

GIOVANNI A. PUCA (86), Patologia Generale, Universita di Napoli, Naples, Italy

E. KENDALL PYE (53), Department of Biochemistry, University of Pennsylvania School of Medicine, Philadelphia, Pennsylvania

JOHN B. ROBBINS (90), National Institute of Child Health and Human Development, National Institutes of Health, Bethesda, Maryland

MALKA ROBERT-GERO (61), Laboratoire d'Enzymologie, Centre National de la Recherche Scientifique, Gif-sur-Yvette, France

ADOLFO RUIZ-CARRILLO (68), The Rockefeller University, New York, New York

JOHN A. RUPLEY (83), Department of Chemistry, University of Arizona, Tucson, Arizona

U. RUTISHAUSER (15), The Rockefeller University, New York, New York

RYUJIRO SANO (44), Department of Biochemistry, Research Laboratory of Applied Biochemistry, Tanabe Seiyaku Co., Ltd., Osaka, Japan

TODASHI SATO (44), Department of Biochemistry, Research Laboratory of Applied Biochemistry, Tanabe Seiyaku Co., Ltd., Osaka, Japan

ALAN N. SCHECHTER (62), National Institute of Arthritis, Metabolism, and Digestive Diseases, National Institutes of Health, Bethesda, Maryland

PETER W. SCHILLER (62), National Institute of Arthritis, Metabolism, and Digestive Diseases, National Institutes of Health, Bethesda, Maryland

H. SCHMITT (81), Department of Biochemistry, The Weizmann Institute of Science, Rehovot, Israel

RACHEL SCHNEERSON (90), National Institute of Child Health and Human Development, National Institutes of Health, Bethesda, Maryland

WILLIAM H. SCOUTEN (24), Chemistry Department, Bucknell University, Lewisburg, Pennsylvania

SHMUEL SHALTIEL (9), Department of Chemical Immunology, The Weizmann Institute of Science, Rehovot, Israel

NATHAN SHARON (32), Department of Biophysics, The Weizmann Institute of Science, Rehovot, Israel

M. C. SHAW (46), Science and Technology, Ministry of State, Ottawa, Ontario, Canada

ANDREW F. SHRAKE (83), Department of Chemistry, University of Arizona, Tucson, Arizona

VINCENZO SICA (86), Patologia Generale, Universita di Napoli, Naples, Italy

HANS-ULRICH SIEBENEICK (65), Knoll AG, Ludwigshafen am Rhein, Germany

ISRAEL SILMAN (73), Department of Biophysics, The Weizmann Institute of Science, Rehovot, Israel

ERIC J. SIMON (80), *Department of Medicine, New York University Medical Center, New York, New York*

L. A. Æ. SLUYTERMAN (67), *Philips Research Laboratories, Eindhoven, The Netherlands*

MORDECHAI SOKOLOVSKY (45), *Department of Biochemistry, The George S. Wise Center for Life Sciences, Tel Aviv University, Tel Aviv, Israel*

EDWARD STEERS, JR. (34), *National Institute of Arthritis, Metabolism, and Digestive Diseases, National Institutes of Health, Bethesda, Maryland*

RICHARD J. STOCKERT (87), *Division of Genetic Medicine, Department of Medicine, Albert Einstein College of Medicine, Bronx, New York*

GEORGE P. STRICKLIN (47), *Division of Dermatology, Washington University School of Medicine, St. Louis, Missouri*

JACK L. STROMINGER (43), *Departments of Biochemistry and Molecular Biology, Harvard University, Cambridge, Massachusetts*

ANTHONY J. SURUDA (71), *Department of Medicine, University of Connecticut Health Center, Farmington, Connecticut*

TOMOJI SUZUKI (49), *Division of Plasma Proteins, Institute for Protein Research, Osaka University, Osaka, Japan*

PATRIK SWANLJUNG (72), *Corporate Laboratory, Imperial Chemical Industries, Ltd., Runcorn, Cheshire, England*

HIDENOBU TAKAHASHI (49), *Division of Plasma Proteins, Institute for Protein Research, Osaka University, Osaka, Japan*

MIHO TAKAHASHI (41), *Mitsubishi-kasei Institute of Life Sciences, Tokyo, Japan*

PAUL TALALAY (71), *Department of Pharmacology and Experimental Therapeutics, The Johns Hopkins University School of Medicine, Baltimore, Maryland*

RAMON L. TATE (88), *National Institute of Arthritis, Metabolism, and Diges-tive Diseases, National Institutes of Health, Bethesda, Maryland*

E. BRAD THOMPSON (25), *National Cancer Institute, National Institutes of Health, Bethesda, Maryland*

G. TOMLINSON (46), *Institute of Molecular Biology, University of Oregon, Eugene, Oregon*

TETSUYA TOSA (21, 44), *Department of Biochemistry, Research Laboratory of Applied Biochemistry, Tanabe Seiyaku Co., Ltd., Osaka, Japan*

IAN P. TRAYER (29, 35, 56), *Department of Biochemistry, University of Birmingham, Birmingham, England*

T. VISWANATHA (47), *Department of Chemistry, University of Waterloo, Waterloo, Ontario, Canada*

FRIEDRICH VON DER HAAR (11, 60), *Abteilung Chemie, Max-Planck-Institut für Experimentell Medizin, Göttingen, Germany*

HOUSTON F. VOSS (74), *Department of Chemistry, University of Colorado, Boulder, Colorado*

TOVA WAKS (94), *Department of Chemical Immunology, The Weizmann Institute of Science, Rehovot, Israel*

DONALD F. H. WALLACH (12), *Departments of Therapeutic Radiology and Physiology, Tufts-New England Medical Center, Boston, Massachusetts*

JEAN-PIERRE WALLER (61), *Laboratoire d'Enzymologie, Centre National de la Recherche Scientifique, Gif-sur-Yvette, France*

K. A. WALSH (50), *Department of Biochemistry, University of Washington, Seattle, Washington*

H. H. WEETALL (4), *Molecular Biology and Immunology, Corning Glass Works, Corning, New York*

MICHAEL K. WEIBEL (53), *Department of Biochemistry, University of Pennsylvania School of Medicine, Philadelphia, Pennsylvania*

ARTHUR WEISSBACH (54), *Department of Cell Biology, Roche Institute of Molecular Biology, Nutley, New Jersey*

J. WIJDENES (67), *Philips Research*

Laboratories, Eindhoven, The Nether-
lands

MEIR WILCHEK (1, 5, 14, 20, 55, 57), *Department of Biophysics, The Weizmann Institute of Science, Rehovot, Israel*

GARY WILCOX (37), *Department of Biological Sciences, University of California, Santa Barbara, California*

IRWIN B. WILSON (74), *Department of Chemistry, University of Colorado, Boulder, Colorado*

ROGER J. WINAND (88), *Département de Clinique et de Sémiologie Médicale, Institute de Médicine, Université de Liège, Liège, Belgium*

YUN-TAI WU (10), *Department of Chemical and Biochemical Engineering, University of Pennsylvania, Philadelphia, Pennsylvania*

KOZO YAMAMOTO (44), *Department of Biochemistry, Research Laboratory of Applied Biochemistry, Tanabe Seiyaku Co., Ltd., Osaka, Japan*

Preface

The last few years have brought progress in affinity chromatography to what must surely be the log phase of development. The interest and enthusiasm for affinity methods is understandable since all of us faced with the problem of purifying macromolecules would prefer an "easy way." The way of affinity chromatography literally seizes on the specificity of the macromolecule to bring about adsorption of only that population with such specificity. Ideally, specificity in the form of a competing ligand is again used to desorb the macromolecule. With good fortune, a tissue extract may be purified to a degree approaching homogeneity for one protein by a single pass through an appropriate affinity column. That the goal is seldom attained does not preclude aiming for it; the fact that it is occasionally reached serves as a spur.

We would have been delighted to be able to present definitive studies which could be used as a clear outline for the design of new systems. Unfortunately, this is not possible. In this explosive phase of development our choice has been one of waiting until the point of definition arrives or of presenting the developments of affinity methodology, as applied to purification, in its present imperfect form. Obviously, we have chosen the latter course. Despite the limitations, we believe that we are now at a stage in which the simplistic notions of affinity chromatography can be examined; there is just sufficient experience to allow beginning guidelines to be formed and suggestions for correction to be advanced.

The result of assembling such experience is a volume in which the organization is imperfect and in which there is much repetition. Methods of purifying dehydrogenases, for example, are presented in several separate sections, and certain individual enzymes claim similar distribution. We chose not to mention the number of descriptions for the best means of adding cyanogen bromide. Yet we included all this material and allowed the repetition so that the investigator contemplating the affinity approach can both obtain an idea of what may be best for the individual system and what the expectation may be as to the limits to which a method can be stretched. Since the titles of the articles cannot be completely informative, we have added separate lists of ligands and of the macromolecules which have been purified with them. In addition, much of the literature on purification methods which was not included is referred to in the first article.

Although our approach and concentration have been oriented toward enzyme purification, affinity methods are presented for such diverse systems as cells, antibodies, and specifically modified proteins. The material on antibody serves mainly as a guide to the enzymologist since methods specific to working with these species of protein have been examined in great detail elsewhere.

WILLIAM B. JAKOBY
MEIR WILCHEK

METHODS IN ENZYMOLOGY

EDITED BY

Sidney P. Colowick and Nathan O. Kaplan

VANDERBILT UNIVERSITY
SCHOOL OF MEDICINE
NASHVILLE, TENNESSEE

DEPARTMENT OF CHEMISTRY
UNIVERSITY OF CALIFORNIA
AT SAN DIEGO
LA JOLLA, CALIFORNIA

METHODS IN ENZYMOLOGY

EDITORS-IN-CHIEF

Sidney P. Colowick   Nathan O. Kaplan

List of Isolated Substances

Article numbers follow each entry.

A

Acetylcholinesterase, 73, 74
Adrenergic receptor, 89
D-Alanine carboxypeptidase, 42, 43
Alcohol dehydrogenase, 8, 17
Aminoacyl-tRNA synthetase, 11
Aminotransferases, 25
Anthranilate phosphoribosyl transferase, 38
Anthranilate synthetase complex, 40
Antibody, 15, 90
araC Protein, 37
Arginyl-tRNA synthesis, 11
Asialoglycoproteins, 87
Asparaginase, 44
Aspartase, 44
Aspartate β-decarboxylase, 44
ATPase, 72
Avidin, 20
Azophenylglycoside haptens, 3

B

B_{12}-binding proteins, 28
Biotin-binding proteins, 20

C

Carboxypeptidase B, 48
Cells, 15, 90, 94
Cholera toxin, 79
Chymotrypsin, 46, 74
Coagulant protein, 75
Coenzyme A, 21
Collagenase, 47
Complement, 92
Creative kinase, 17
Cysteine peptides, 14
Cytochrome c reductase, 26

D

Dehydrogenases, 9, 16, 17, 56
3-Deoxy-D-*arabino*-heptulosonate 7-phosphate synthetase, 41
Diaphorase, 24
Dihydrofolate reductase, 3, 22, 23

Dihydroneopterin triphosphate synthetase, 56
DNA, 59
DNA polymerase, 54
DNase inhibitor proteins, 63

E

Erythrocytes, 15
Estradiol dehydrogenase, 69, 70
Estrogen receptor, 86

F

L-Fucose-binding protein, 29

G

α-Galactosidase, 33
β-Galactosidase, 8, 34
Galactosyl transferase, 29, 35, 56
Glucokinase, 29, 56
Glucose-6-phosphate dehydrogenase, 17, 55
Glutamate dehydrogenase, 9
Glutamate synthetase, 9
Glutamate synthetase adenyltransferase, 9
Glutamine synthetase, 9
Glutathione reductase, 17
Glyceraldehyde-3-phosphate dehydrogenase, 8, 9, 16, 17
Glycerokinase, 17
Glycogen phosphorylase, 9
Glycogen synthetase, 9
Glycolate oxidase, 26
Glycoproteins, 31, 32, 36, 87
Guanine deaminase, 65

H

Hemagglutinins, 29, 30, 36
Heme peptide, 14
Hexokinase, 17, 56
Histidine-binding protein, 9
Histidinyl-tRNA synthetase, 11
Histones, 68

List of Ligands

Article numbers follow each entry.

Methods in Enzymology

Volume XXXIV
AFFINITY TECHNIQUES
Enzyme Purification: Part B

Section I

Introduction

[1] The Literature on Affinity Chromatography

By MEIR WILCHEK and WILLIAM B. JAKOBY

Shortly after the introduction, over twenty years ago, of a method for affinity chromatography of anti-hapten antibodies,[1] a system was developed by Lerman[2] for the purification of tyrosinase, based on the same principles that we accept today. However, widespread application of the affinity approach awaited development of more suitable matrices and recognition of the variety of reactions available for the attachment of ligands to them. Once this stage was accomplished, the ensuing rapid growth is reflected in this volume which, despite its size, is insufficient for coverage of all or even most of the methods available in the literature.

In order to partially correct this limitation, many of the relevant isolation methods which are not included in detail, are referred to in the table. Oversights will be noted in this compilation; in particular, only a sampling of the methods using specific antibody are included. Nevertheless, the table represents a large fraction of the available references[3-122] to affinity methods for the isolation of specific proteins or groups of proteins.

PROTEINS PURIFIED BY AFFINITY CHROMATOGRAPHY

Protein	Carrier	Ligand	Reference footnote
I. Enzymes			
β-N-Acetylhexosaminidase	Agarose	Glycopeptide	3
Acrosin	Cellulose	Benzamidine	4
	Cellulose	Soybean trypsin inhibitor	5
Adenosine deaminase	Agarose	Adenosine	6
Alcohol dehydrogenase	Glass	NAD	7
Aldehyde oxidase	Agarose	N-Benzyl-6-methyl nicotinamide	8
Alginase	Acrylamide	Alginic acid	9

(Continued)

[1] D. H. Campbell, E. Leuscher, and L. S. Lerman, *Proc. Nat. Acad. Sci. U.S.* **37**, 575 (1951); L. S. Lerman, *Nature* **172**, 635 (1953).
[2] L. S. Lerman, *Proc. Nat. Acad. Sci. U.S.* **39**, 232 (1953).

TABLE (*Continued*)

Protein	Carrier	Ligand	Reference footnote
Aminopeptidase	Agarose	Hexamethylenediamine	10
Anhydrochymotrypsin	Agarose	Lima bean inhibitor	11
Arylsulfatase A	Agarose	Psychosine sulfate	12
Aspartate aminotransferase	Agarose	Pyridoxal 5′-phosphate	13
	Agarose	Pyridoxamine 5′-phosphate	14
(Na$^+$ + K$^+$)-ATPase	Agarose	Digoxin	15
	Agarose	ATP analog	16
Barstar	Agarose	Barnase	17
Bromelain	Agarose	ε-Aminocaproyl-D-tryptophan methyl ester	18
Carbonic anhydrase	Agarose	Sulfanilamide	19
Choline acetyltransferase	Agarose	Bromostyrylpyridinium salt	21
Cholismate mutase	Agarose	Tryptophan	20
Carboxypeptidase A	Agarose	L-Tyrosyl-D-tryptophan	22
	Agarose	D-Tryptophan	23
	Cellulose	Glycyl-D-phenylalanine	24
Deoxyribonucleases	Agarose	DNA	25
DNA-polymerase	Acrylamide	DNA	26
Dihydroneopterin triphosphate synthetase	Agarose	GTP	27
Elastase	Agarose	ε-NH$_2$-caproyl-L-Ala-L-Ala-L-Ala	28
Estradiol-17β dehydrogenase	Agarose	Estriol 16-hemisuccinate	29
Galactosyltransferase	Agarose	α-Lactalbumin	30
(Gentamicin)-acetyltransferase	Agarose	Gentamicin C 1	31
β-Glucuronidase	Agarose	Saccharo-1,4-lactone	32
Glutathione reductase	Agarose	Oxidized glutathione	33
Glyceraldehyde-3-phosphate dehydrogenase	Agarose	NAD	34
Glycerol-3-phosphate dehydrogenase	Agarose	Glycerol-3-phosphate	35
Glycosidase	Agarose	Glycosides	36
Isoleucyl-tRNA synthetase	Agarose	Aminoacyl-tRNA	37
α-Isopropylmalate synthetase	Agarose	Leucine	38
β-Lactamase (see also Penicillinase)	Agarose	Ampicilline	39
Lactose synthetase	Agarose	α-Lactalbumin	40
Lipoprotein lipase	Agarose	Heparin	41
Lipoxygenase	Agarose	Lipoic acid	42
Lysozyme	Deaminated chitin	Deaminated chitin	43

TABLE (*Continued*)

Protein	Carrier	Ligand	Reference footnote
Lysozyme	Cellulose	Chitin	44
Kallikrein	Ethylenemaleic anhydride	Trypsin-kallikrein inhibitor	45
	Cellulose	Kunitz soybean inhibitor	46
Neuraminidase	Agarose	N-(4-Aminophenyl)oxamic acid	47
Papain	Agarose	Glycylglycyl-(O-benzyl)-L-tyrosylarginine	48
Penicillinase	Agarose	Cephalosporin C	49
Pepsin	Agarose	Poly-L-lysine	50
Phenylalanine: tRNA ligase	Agarose	Phenylalanine	51
Phosphofructokinase		Cibacron blue FsG-A	52
Plasmin	Ethylenemaleic anhydride	Trypsin-kallikrein inhibitor	45
Polynucleotide phosphorylase	Agarose	Poly(A)	53
Protease, insulin-specific	Agarose	Insulin	54
Proteases, alkaline	Agarose	Ovoinhibitor	55
Protease, neutral	Agarose	ϵ-NH$_2$-caproyl-glycyl-leucine	56
Protease, wheat	Agarose	Hemoglobin	57
Protein kinase	Agarose	N^6-Aminocaproyl-3',5'-cAMP	58
Pteroyl oligo-γ-L-glutamyl endopeptidase	Agarose	γ-Oligo-L-glutamate	59
Pyridoxal kinase	Agarose	ADP	60
Pyruvate kinase	Sephadex	Cibacron blue	61
Renin	Agarose	Pepstatin	62
Ribonuclease, 3 amino-tyrosine (pancreatic)	Agarose	5'-(4-Aminophenylphosphoryl)uridine (2')3'-phosphate	63
Ribonuclease (*E. coli*)	Agarose	DNA	64
RNA-dependent DNA polymerase	Cellulose	Poly(deoxythymidine)$_{12-18}$	65
RNA polymerase	Agarose	DNA	66
	Sephadex	DNA	67
Threonine deaminase	Agarose	Isoleucine	68
Thrombin	Agarose	4-Chlorobenzylamine	69
	Agarose	Benzamidine	70
Thymidine kinase	Agarose	5'-Amino-5'-deoxythymidine	71
Trypsin	Agarose	Ovomucoid	72, 73
	Cellulose	4-(4'-Aminophenoxypropoxy)-benzamidine	74

(*Continued*)

TABLE (*Continued*)

Protein	Carrier	Ligand	Reference footnote
	Copolymer ethyl maleic anhydride	Trypsin-kallikrein inhibitor	75
	Agarose	Antidinitrophenyl antibody	76
Tyrosinase	Cellulose	Phenol	2
	Agarose	Dihydroxyphenyl derivative	77
Tyrosine hydroxylase	Agarose	3-Iodotyrosine	78
Xanthine oxidase	Agarose	Allopurinol	79
Xylosyltransferase	Agarose	Proteoglycan (Smith-degraded)	80

System	Carrier	Ligand	Reference footnote

II. Other Systems

System	Carrier	Ligand	Reference footnote
Adipose cells	Agarose	Insulin	81
Albumin	Agarose	Fatty acids	82
Aldosterone-binding	Agarose	Deoxycorticosterone hemi-succinate	83, 84
Antitrypsin	Agarose	Concanavalin A	85, 86
Antitrypsin	Agarose	Trypsin	87
Bilirubin	Agarose	Albumin	88, 89
Cholinergic receptor	Agarose	Quaternary ammonium salts	90, 91
Cholinergic receptor	Agarose	*Naga naga* neurotoxin	92
Chorionic gonadotropin	Agarose	Concanavalin A	93
Chorionic somatomammotropin	Agarose	Antibody	94
Coagulation factor	Agarose	Heparin	95
Corticosteroid-binding globulin	Agarose	Cortisol hemisuccinate	96
DNA-binding protein	Agarose	Single-stranded DNA	97
DNA, defined sequences	Cellulose	RNA	98
Fat cells	Agarose	Insulin	81
α-Fetoprotein	Agarose	Concanavalin A	99
Follicle-stimulating hormone	Agarose	Concanavalin A	93
Gal-Repressor	Agarose	p-Aminophenyl-α-thio-galactoside	100
Glycoproteins	Agarose	Concanavalin A	101
Hemaglobin	Agarose	Haptoglobin	102
Heme peptides	Agarose	Albumin	89
Hepatitis β-antigen	Agarose	Concanavalin A	103

TABLE (*Continued*)

System	Carrier	Ligand	Reference footnote
Influenza virus	Agarose	*N*-(4-Aminophenyl)oxamic acid	47
Interferon	Agarose	Antibody	105
Lac operon polysomes	Agarose	β-Galactosidose inhibitor	106
Lectins	Various	Sugars	107
Lipids	Agarose	Dodecylamine	108
Liver cell membranes	Agarose	Glucagon	109
Luteinizing hormone	Agarose	Concanavalin A	93
Luteinizing hormone	Agarose	Antibody	110
Lymphocytes, antihapten specific	Acrylamide	β-Lactoside haptens	111
	Agarose	Phytomitogens	112
Ovoinhibitor	Agarose	Chymotrypsin	113
Phage, T4	Agarose	Polylysine	104
Polysomes	Agarose	Antibody to specific protein	114
Properdin	Agarose	Antibody	115
Retinal-binding protein	Agarose	Prealbumin	116
Arg-tRNA	Agarose	Antibody	117
tRNA	Agarose	Amino acid-tRNA synthetase	118
tRNA	Agarose	Antidinitrophenyl antibody	119
β-Thiogalactoside-binding protein	γ-Globulin	*p*-Aminophenyl-β-thiogalactoside	120
Teichoic acid	Concanavalin	Concanavalin A	121
8-*lysine*-Vasopressin	Agarose	Neurophysin	122

[3] G. Dawson, R. L. Propper, and A. Dorfman, *Biochem. Biophys. Res. Commun.* **54**, 1102 (1973).

[4] W. D. Schleuni, H. Schiessler, and H. Fritz, *Hoppe-Seyler's Z. Physiol. Chem.* **354**, 550 (1973).

[5] E. Fink, H. Schliesser, M. Arnhold, and H. Fritz, *Hoppe-Seyler's Z. Physiol. Chem.* **353**, 1633 (1972).

[6] S. Barry and P. O'Carra, *FEBS Lett.* **37**, 134 (1973).

[7] M. K. Weibel, E. R. Doyle, A. E. Humphrey, and H. J. Bright, *Biotech. Bioeng.* Suppl. 3, 167 (1972).

[8] A. E. Chu and S. Chaykin, *Fed. Proc., Fed. Amer. Soc. Exp. Biol.* **31**, 468 (1972).

[9] V. V. Favorov, *Int. J. Biochem.* **4**, 107 (1973).

[10] K. Vosbeck, K. F. Chow, and W. M. Awad, *Fed. Proc., Fed. Amer. Soc. Exp. Biol.* **32**, 504 (1973).

[11] H. Ako, C. A. Ryan, and R. J. Foster, *Biochem. Biophys. Res. Commun.* **46**, 1639 (1972).

[12] J. L. Breslow and H. R. Sloan, *Biochem. Biophys. Res. Commun.* **46**, 919 (1972).

[13] E. Ryan and P. F. Fottrell, *FEBS Lett.* **23**, 73 (1972).

[14] R. Collier and G. Kohlhaw, *Anal. Biochem.* **42**, 48 (1971).
[15] T. B. Okarma, P. Tramell, and S. M. Kalman, *Mol. Pharmacol.* **8**, 476 (1972).
[16] B. H. Anderson, F. W. Hulla, H. Fasold, and H. A. White, *FEBS Lett.* **37**, 338 (1973).
[17] R. W. Hartley, D. L. Rogerson, and J. R. Smeaton, *Prep. Biochem.* **2**, 243 (1972).
[18] D. Bobb, *Prep. Biochem.* **2**, 347 (1972).
[19] S. O. Falkbring, P. O. Gothe, P. O. Nyman, L. Sundberg, and J. Porath, *FEBS Lett.* **24**, 229 (1972).
[20] S. S. Husain and H. G. Mautner, *Fed. Proc., Fed. Amer. Soc. Exp. Biol.* **32**, 564 (1973).
[21] B. Sprossler and F. Lingens, *FEBS Lett.* **6**, 232 (1970).
[22] P. Cuatrecasas, M. Wilchek, and C. B. Anfinsen, *Proc. Nat. Acad. Sci. U.S.* **61**, 636 (1968).
[23] G. R. Reeck, K. A. Walsh, M. A. Hermodson, and H. Neurath, *Proc. Nat. Acad. Sci. U.S.* **68**, 1226 (1971).
[24] J. R. Uren, *Biochim. Biophys. Acta* **236**, 67 (1971).
[25] J. C. Schabort, *J. Chromatogr.* **73**, 253 (1972).
[26] L. F. Cavalieri and E. Carroll, *Proc. Nat. Acad. Sci. U.S.* **67**, 807 (1970).
[27] R. J. Jackson, R. M. Wolcott, and T. Shiota, *Biochem. Biophys. Res. Commun.* **51**, 428 (1973).
[28] A. Janoff, *Fed. Proc., Fed. Amer. Soc. Exp. Biol.* **32**, 292 (1973).
[29] C. C. Chin and J. C. Warren, *Steroids* **22**, 373 (1973).
[30] R. Mawal, J. F. Morrison, and K. E. Ebner, *J. Biol. Chem.* **246**, 7106 (1971).
[31] F. Legoffic and N. Moreau, *FEBS Lett.* **29**, 289 (1973).
[32] R. G. Harris, J. J. M. Rowe, P. S. Stewart, and D. C. Williams, *FEBS Lett.* **29**, 189 (1973).
[33] J. J. Harding, *J. Chromatogr.* **77**, 191 (1973).
[34] J. D. Hocking and J. I. Harris, *FEBS Lett.* **34**, 280 (1973).
[35] P. D. Holohan, K. Mahajan, and T. P. Fondy, *Fed. Proc., Fed. Amer. Soc. Exp. Biol.* **32**, 476 (1973).
[36] E. Junowicz and J. E. Paris, *Biochim. Biophys. Acta* **321**, 234 (1973).
[37] S. Bartkowiak and J. Pawelkiewicz, *Biochim. Biophys. Acta* **272**, 137 (1972).
[38] G. Doellgas and G. Kohlhaw, *Fed. Proc., Fed. Amer. Soc. Exp. Biol.* **31**, 424 (1972).
[39] F. Legoffic, R. Labia, and J. Andrillo, *Biochim. Biophys. Acta* **315**, 439 (1973).
[40] P. Andrews, *FEBS Lett.* **9**, 297 (1970).
[41] T. Olivecrona, T. Egelrud, P. H. Iverius, and U. Lindahl, *Biochem. Biophys. Res. Commun.* **43**, 524 (1971).
[42] S. Grossman, M. Trop, S. Yaroni, and M. Wilchek, *Biochim. Biophys. Acta* **289**, 77 (1972).
[43] F. D. Katz, *Abstr. Pap. ACS* **164**, 14 (1972).
[44] T. Imoto and K. Yagishita, *Agr. Biol. Chem.* **37**, 1191 (1973).
[45] H. Fritz, H. Schultz, M. Neudecker, and E. Werle, *Angew. Chem.* **78**, 775 (1966).
[46] H. Fritz, G. Wunderer, and B. Dittmann, *Hoppe-Seyler's Z. Physiol. Chem.* **353**, 893 (1972).
[47] P. Cuatrecasas and G. Illiano, *Biochem. Biophys. Res. Commun.* **44**, 178 (1971).
[48] S. Blumberg, I. Schechter, and A. Berger, *Eur. J. Biochem.* **15**, 97 (1970).
[49] L. U. Crane, G. E. Bettinge, and J. O. Lampen, *Biochem. Biophys. Res. Commun.* **50**, 220 (1973).
[50] B. Nevaldine and B. Kassell, *Biochim. Biophys. Acta* **250**, 207 (1971).

[51] P. I. Forrester and R. L. Hancock, *Can. J. Biochem.* **51**, 231 (1973).

[52] H. J. Bohme, S. C. Kopper, J. Schultz, and E. Hofmann, *J. Chromatogr.* **69**, 209 (1972).

[53] H. Lehrach and K. H. Scheit, *Hoppe-Seyler's Z. Physiol. Chem.* **353**, 731 (1972).

[54] W. C. Duckworth, M. A. Heinemann, and A. E. Kitabchi, *Proc. Nat. Acad. Sci. U.S.* **69**, 3698 (1972).

[55] G. Feinstein and A. Gertler, *Biochim. Biophys. Acta* **309**, 196 (1973).

[56] L. G. Sparrow and A. B. McQuade, *Biochim. Biophys. Acta* **302**, 90 (1973).

[57] G. K. Chua and W. Bushuk, *Biochem. Biophys. Res. Commun.* **37**, 545 (1969).

[58] M. Wilchek, Y. Salomon, M. Lowe, and Z. Selinger, *Biochem. Biophys. Res. Commun.* **45**, 1177 (1971).

[59] P. K. Saini and I. H. Rosenberg, *Fed. Proc., Fed. Amer. Soc. Exp. Biol.* **32**, 928 (1973).

[60] J. T. Neary and W. F. Diven, *J. Biol. Chem.* **245**, 5585 (1970).

[61] P. Roschlau and B. Hess, *Hoppe-Seyler's Z. Physiol. Chem.* **353**, 441 (1972).

[62] P. Corvol, C. Devaux, and J. Menard, *FEBS Lett.* **34**, 189 (1973).

[63] M. Gorecki, M. Wilchek, and A. Patchornik, *Biochim. Biophys. Acta* **229**, 590 (1971).

[64] S. C. Weatherford, L. S. Weisberg, D. T. Achord, and D. Apirion, *Biochem. Biophys. Res. Commun.* **49**, 1307 (1972).

[65] B. I. Gerwin and J. B. Milstein, *Proc. Nat. Acad. Sci. U.S.* **69**, 2599 (1972).

[66] C. Nüsslein and B. Heyden, *Biochem. Biophys. Res. Commun.* **47**, 282 (1972).

[67] D. Rickwood, *Biochem. Biophys. Acta* **269**, 47 (1972).

[68] I. Rahimi-Laridjani, H. Grimminger, and F. Lingens, *FEBS Lett.* **30**, 185 (1973).

[09] A. R. Thompson and E. W. Davie, *Biochim. Biophys. Acta* **250**, 210 (1971).

[70] G. Schmer, *Hoppe-Seyler's Z. Physiol. Chem.* **353**, 810 (1972).

[71] W. Rohde and A. G. Lezius, *Hoppe-Seyler's Z. Physiol. Chem.* **352**, 1507 (1971).

[72] G. Feinstein, *FEBS Lett.* **7**, 353 (1970).

[73] N. C. Robinson, R. W. Tye, H. Neurath, and K. A. Walsh, *Biochemistry* **10**, 2743 (1971).

[74] G. W. Jameson and D. T. Elmore, *Biochem. J.* **124**, 66p (1971).

[75] H. Fritz, B. Brey, A. Schmal, and E. Werle, *Hoppe-Seyler's Z. Physiol. Chem.* **350**, 617 (1969).

[76] M. Wilchek and M. Gorecki, *FEBS Lett.* **31**, 149 (1973).

[77] S. P. O'Neil, D. J. Graves, and J. J. Ferguson, *J. Macromol. Sci. Chem.* **7**, 1159 (1973).

[78] R. Shiman, M. Akino, and S. Kaufman, *J. Biol. Chem.* **246**, 1330 (1971).

[79] D. Edmondson, V. Massey, G. Palmer, L. M. Beachan, and G. B. Elion, *J. Biol. Chem.* **247**, 1597 (1972).

[80] N. B. Schwartz and L. Roden, *Fed. Proc., Fed. Amer. Soc. Exp. Biol.* **32**, 560 (1973).

[81] D. D. Soderman, J. Germershausen, and H. M. Katzen, *Proc. Nat. Acad. Sci. U.S.* **70**, 792 (1973).

[82] T. Peters, H. Taniuchi, and C. B. Anfinsen, *J. Biol. Chem.* **248**, 2447 (1973).

[83] J. H. Ludens, J. R. DeVries, and D. D. Fanestil, *J. Steroid Biochem.* **3**, 193 (1972).

[84] J. H. Ludens, J. R. DeVries, and D. D. Fanestil, *J. Biol. Chem.* **247**, 7533 (1972).

[85] I. E. Liener, O. R. Garrison, and Z. Pravda, *Biochem. Biophys. Res. Commun.* **51**, 436 (1973).

[86] R. J. Murthy and A. Hercz, *FEBS Lett.* **32**, 243 (1973).

[87] J. Lebas, A. Hayem, A. Laine, and R. Havez, *C. R. Acad. Sci. Ser. D* **276**, 2741 (1973).

[88] P. H. Plotz, P. Berk, B. F. Scharschmidt, J. K. Gordon, and J. Vergalla, *J. Clin. Invest.* **53**, 778 (1974).

[89] M. Wilchek, *Anal. Biochem.* **49**, 572 (1972).

[90] J. Schmidt and M. A. Raftery, *Biochem. Biophys. Res. Commun.* **49**, 572 (1972).

[91] R. W. Olsen, J. C. Meunier, and J. P. Changeux, *FEBS Lett.* **28**, 96 (1972).

[92] E. Karlsson, E. Heilbronn, and L. Widlund, *FEBS Lett.* **28**, 107 (1972).

[93] M. L. Dufau, T. Tswinhara, and K. J. Catt, *Biochim. Biophys. Acta* **278**, 281 (1972).

[94] B. D. Weintraub, *Biochem. Biophys. Res. Commun.* **39**, 83 (1970).

[95] P. W. Gentry and B. Alexander, *Biochem. Biophys. Res. Commun.* **50**, 500 (1973).

[96] W. Rosner and H. L. Bradlow, *J. Clin. Endocrinol.* **33**, 193 (1971).

[97] H. Schaller, C. Nüsslein, F. J. Bonhoeffer, C. Kurz, and I. Nietzschmann, *Eur. J. Biochem.* **26**, 474 (1972).

[98] H. Y. Shih and M. A. Martin, *Proc. Nat. Acad. Sci. U.S.* **70**, 1697 (1973).

[99] M. Pagé, *Can. J. Biochem.* **51**, 1213 (1973).

[100] J. S. Parks, M. Gottesman, K. Shimada, R. A. Weisberg, R. L. Perlman, and I. Pastan, *Proc. Nat. Acad. Sci.* **68**, 1891 (1971).

[101] D. Allan, J. Auger, and M. J. Crumpton, *Nature (London) New Biol.* **236**, 23 (1972).

[102] M. Klein and C. Mihaesco, *Biochem. Biophys. Res. Commun.* **52**, 774 (1973).

[103] A. R. Neurath, A. M. Prince, and A. Lippin, *J. Gen. Virol.* **19**, 391 (1973).

[104] L. Sundberg and S. Hoglund, *FEBS Lett.* **37**, 70 (1973).

[105] J. D. Sipe, J. de Maeyer-Guignard, B. Fauconnier, and E. De Maeyer, *Proc. Nat. Acad. Sci. U.S.* **70**, 1037 (1973).

[106] U. Melcher and K. A. Marcker, *Plant Physiol.* **49**, 7 (1972).

[107] H. Lis and N. Sharon, *Annu. Rev. Biochem.* **42**, 541 (1973).

[108] D. G. Deutsch, D. J. Fogleman, K. N. Von Kaulla, *Biochem. Biophys. Res. Commun.* **50**, 758 (1973).

[109] F. Krug, B. Desbuquois, and P. Cuatrecasas, *Nature (London) New Biol.* **234**, 268 (1971).

[110] D. Gospodarowicz, *J. Biol. Chem.* **247**, 6491 (1972).

[111] L. Wofsy, J. Kimura, and P. Truffa-Bachi, *J. Immunol.* **107**, 725 (1971).

[112] M. F. Greaves and S. Bauminger, *Nature (London) New Biol.* **235**, 67 (1972).

[113] G. Feinstein, *Biochim. Biophys. Acta* **236**, 73 (1971).

[114] C. H. Faust, P. Vassalli, and B. Mach, *Experientia* **28**, 745 (1972).

[115] J. O. Minta and I. H. Lepow, *J. Immunol.* **111**, 286 (1973).

[116] A. Vahlquist, S. F. Nilsson, and P. A. Peterson, *Eur. J. Biochem.* **20**, 160 (1971).

[117] H. Inouye, S. Fuchs, M. Sela, and U. Z. Littauer, *J. Biol. Chem.* **248**, 8125 (1973).

[118] J. Denburg and M. Deluca, *Proc. Nat. Acad. Sci. U.S.* **67**, 1057 (1970).

[119] S. Fuchs, B. Bonavida, M. Sela, E. Ziv, S. Rappaport, and Y. Lapidot, *Isr. J. Med. Sci.* **7**, 636 (1971).

[120] S. Tomino and K. Faigen, "The Lactose Operon," p. 233. Cold Spring Harbor Laboratory, Cold Spring Harbor, New York, 1970.

[121] R. J. Doyle, D. C. Birdsell, and F. E. Young, *Prep. Biochem.* **3**, 13 (1973).

[122] P. Pradelle and S. Jard, *FEBS Lett.* **26**, 189 (1972).

Section II

Coupling Reactions and General Methodology

[2] General Methods and Coupling Procedures

By Jerker Porath

Principle

In nature, reactions are catalyzed and regulated by stereospecific inter-actions between molecular species present in the solid and liquid phases of living systems. The specificity of the interactions may be high as in the formation of certain enzyme–substrate and antigen–antibody com-plexes or low as in albumin–lipid and some carbohydrate–lectin com-plexes. The degree of complementarity between interacting molecular surfaces or regions of the constituents of a complex determines the mag-nitude of attraction, and thus the strength of a complex. Weak complexes dissociate spontaneously and reform with ease, usually at a high rate; strong complexes form easily, but their dissociation may require more or less drastic changes in the composition of the medium.

In the simplest possible case, two components, A and B, will form a binary complex $A \cdots B$. The properties of the complex differ from those of either component. This is evidently true for the molecular size. Molecular sieving in beds of macroreticular hydrophilic gels may there-fore be used to accomplish a separation of A or B from similar com-pounds not able to form the AB complex.

However, it is clear that if B is covalently attached to a high polymer or matrix substance, (M), to form immobilized B, i.e., (M)B, the separa-tion of A from substances of identical size or charge is greatly facilitated. When (M) is so large that it forms an insoluble, macroreticular network or gel, the desired material can be transported into and out from the interstitial space, and the solid phase, i.e., the gel, can easily be separated from the solution. Furthermore, and most important when a weak com-plex is formed, fractionation can be made very much more effective by application of chromatography. Chromatography rather than batch pro-cedures may be conveniently used also when A is almost completely trans-ferred from the solution to the solid phase by the formation of a stable complex $(M)B \cdots A$.

The component B attached to the solid *matrix* support is called a *ligand*. The attachment may be made by physical adsorption or by a chemical reaction that links the ligand to the matrix by a covalent bond. The latter method is far superior and should be used whenever possible. It is often desirable or even necessary to introduce an interconnecting link between the ligand and the matrix to serve as a *spacer* to allow or facilitate specific interaction between sterically fitting surfaces on the adsorbate molecule and the ligand.

A number of review articles have been published in the rapidly expanding field of biospecific adsorption and bioaffinity chromatography.[1-13]

Theoretical Aspects

Development of a mathematical theory for the mechanisms of biospecific adsorption is likely to entail almost insurmountable difficulties. Only general theories for adsorption chromatography can be applied, i.e., theories valid without respect to the mechanisms responsible for the adsorbate–adsorbent interactions.

It has been found experimentally that complex formation between interacting species, i.e., ligand and solutes, is sometimes sterically hindered by the carrier matrix.[14-16] A semiquantitative account of the influence of the matrix can be obtained by comparing the association constants in free solution and in the two-phase systems used in biospecific affinity chromatography. The latter association constants may be found by determination of adsorption isotherms or by the retention volumes in frontal chromatography.

Suppose that a zone of a certain substance A travels through a bed of a permeable gel with the void volume of V_0 and a total volume of V_t. Depending on the extent of permeation, the substance will be held up in the bed and will be eluted at a volume V_e so that $V_0 < V_e < V_t$. The

[1] J. Porath and T. Kristiansen, in "The Proteins" (H. Neurath and R. Hill, eds.), 3rd ed., Vol. I. Academic Press, New York, 1975, in press.

[2] P. Cuatrecasas and C. B. Anfinsen, this series, Vol. 22, p. 345.

[3] P. Cuatrecasas, Advan. Enzymol. Relat. Areas Mol. Biol. 36, 29 (1972).

[4] P. Cuatrecasas and C. B. Anfinsen, Amer. Rev. Biochem. 40, 259 (1971).

[5] J. Porath and L. Sundberg, "The Chemistry of Biosurfaces" (M. Hair, ed.), Vol. 2, p. 633. Dekker, New York, 1972.

[6] G. Feinstein, Naturwissenschaften 58, 389 (1971).

[7] F. Friedborg, Chromatogr. Rev. 14, 121 (1971).

[8] O. Pihar, Chem. Listy 65 (7), 713 (1971).

[9] R. H. Reiner and A. Walch, Chromatographia 4, 578.

[10] J. Chudzik and A. Koj, Post. Biochem. 18, 73 (1972).

[11] V. Kocemba-Sliwowska, Post. Biochem. 18, 59 (1972).

[12] H. Lang, Nachr. Chem. Tech. 20 (4), 71 (1972).

[13] H. O. Orth and W. Brümmer, Angew. Chem. 84, 319 (1972).

[14] P. Cuatrecasas, M. Wilchek, and C. B. Anfinsen, Proc. Nat. Acad. Sci. U.S. 61, 636 (1968).

[15] P. Cuatrecasas, J. Biol. Chem. 245, 3059 (1970).

[16] P. O'Carra, S. Barry, and T. Griffin, Biochem. Soc. Trans. 1, 289 (1973).

retention due to the molecular sieving capacity of the gel may be described by the distribution coefficient K_{av}[17]:

$$K_{av} = (V_e - V_0)/(V_t - V_0) \tag{1}$$

If the substance A is adsorbed at local points or regions occupied by certain groups evenly distributed in the gel, and provided that the adsorption is not dependent on concentration (linear adsorption), then additional retention will occur and the center of the migrating solute zone will appear after passage of a volume V'_e. If A is adsorbed in a nonlinear fashion, the retention volume should be determined by frontal chromatography; i.e., the substance A at a certain concentration is continuously introduced into the top of the bed and the effluent is analyzed for detection of the breakthrough of the substance A. The position of the front in such a chromatogram, V'_e, depends on the concentration and the affinity (association) constant. If equilibrium is attained under the prevailing chromatographic conditions, adsorption isotherms may be established based on the retention volumes. If we further assume that all adsorption centers in the gel are equally strong and that interaction will occur only with one single group or region in the solute molecule, the following expression for the relative retention S can be deduced:

$$S = (V'_e - V_e)/(V_t - V_0) = (K_{av} \cdot kr)/(1 + kC_A) \tag{2}$$

where k is the association constant for the formation of the adsorbate (M)BA, r is the concentration of active centers (functional ligands), and C_A the total concentration of A. S may be considered as a chromatographic measure of affinity. S can be decreased by including in the eluent a substance D that will compete with A for the adsorption sites. With k' as the association constant for D, the following expression for relative retention of A is valid:

$$S = K_{av} \cdot kr/(1 + kC_A + k'C_D) \tag{3}$$

At high concentration of the displacer substance, D, S approaches zero and A will be eluted at or very close to V_e, which is the volume determined by molecular sieving in the absence of adsorption. It is also possible to choose $k'C_D$ so that $kC_A \ll 1 + k'C_D$. Under these conditions, S will be independent of the concentration of A, and the latter substance will travel through the bed as a symmetrical zone with the relative retention due to adsorption equal to $K_{av} kr/1 + k'C_D$. For good performance of linear chromatography, r should be large, i.e., a high degree of substitution of ligands is desirable. However, for high resolution, the adsorp-

[17] T. C. Laurent and J. Killander, *J. Chromatogr.* 14, 317 (1964).

tion centers should be uniform in structure and distribution, a requirement that can be only partially fulfilled in practice.

Techniques of Displacement and Elution

Nonspecific Displacement (One-Step Elution) and Gradient Elution

Biospecific affinity chromatography is most frequently executed in two steps: (1) adsorption by contacting the sample with the adsorbent in a suitable medium and (2) desorption by a single-step change in the solvent medium. If the substance to be purified is virtually completely adsorbed, batch procedures may be preferable to chromatography, especially if in the latter case the adsorbent is not easily packed in fast-flowing beds. However, beds of good flow characteristics may permit faster operation.

Desorption may be accomplished by alteration in pH or in ionic strength of the medium (Fig. 1). Owing to the nonideal adsorption and desorption conditions under these circumstances, incomplete separation from contaminants is likely to result. In other words, the selectivity will be less than expected from ideal behavior. Improved separation may be obtained by gradient elution (Fig. 2).

Complete elution of substrates, inhibitors, or effectors can usually be effected by change in pH and/or salt concentration. However, in the case of immunosorption the binding of the enzyme to its antibody may be so strong that drastic elution conditions will be necessary. Chaotropic

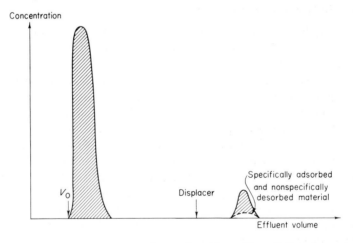

FIG. 1. Model chromatogram illustrating biospecific affinity chromatography with one-step elution (displacement) using a nonspecific displacer. Striped areas represent impurities.

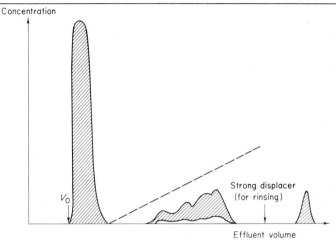

FIG. 2. Model chromatogram illustrating fractionation and incomplete purification of specifically adsorbed substances by means of a concentration gradient of a nonspecific eluting agent (- - -).

ions such as trichloroacetate or perchlorate, can be tried. Lowering the polarity of the solvent, e.g., by including glycerol, may have a beneficial effect when hydrophobic bonding is contributing significantly to the formation of the adsorption complex or when an excess of a hydrophobic spacer substituent is likely to obscure the specific character of the adsorption.

Specific Displacement and Gradient Elution

Selectivity in purification can be increased by the use of biospecifically mediated desorption as well as adsorption. The desorption can obviously be accomplished by inclusion in the eluent of species competing with the ligand. This situation prevails when k' in Eq. (3) is large. If we are dealing with enzymes capable of undergoing conformational changes in the presence of an allosteric effector, while adsorption depends on inhibitor or substrate analog interaction, we are in a favorable position to make use of two different kinds of specificity to effectuate purification. Ternary complex formation and subsequent dissociation has been used for purification of NAD- and NADP-dependent enzymes (see this volume [8], [16] and [77]) and has been shown to be a highly selective method for purification of these enzymes. A gradient of a specific displacer should, in principle, increase the separation power and make possible the isolation of isoenzymes or enzymes with qualitatively similar or identical but quantitatively different specificity (see Fig. 3). This technique has but barely come into practice.

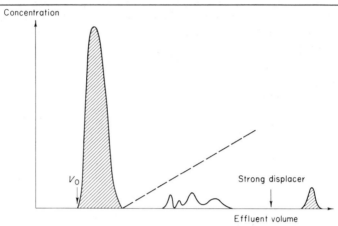

FIG. 3. Model chromatogram illustrating an ideal gradient (- - -) elution of specifically adsorbed substances of different affinities toward the ligands in the adsorbent. The gradient substance competes with the adsorbed substances for the adsorption sites.

Matrix Selection

Proper selection of a matrix or carrier for the ligands is of decisive importance for the successful application of stereospecific adsorption. The ideal matrix should meet the following demands:

1. Insolubility
2. Sufficient permeability
3. High rigidity and suitable particle form
4. Zero adsorption capacity
5. Chemical stability under the conditions required for adsorption, desorption, and regeneration
6. Chemical reactivity allowing ligands and spacers to be introduced
7. Resistance toward microbial and enzymatic attack
8. Hydrophilic character

No matrix support is ideal in all these respects. Porous glass, for example, has excellent rigidity and permeability but lacks chemical stability in alkaline solution and is indeed not devoid of nonspecific adsorption centers. Cellulose, frequently used to prepare immunosorbents, has a heterogeneous microstructure which strongly detracts from its suitability as a matrix, and, in addition, the nonspecific adsorption capacity of commercial cellulose is far from negligible. Polyacrylamide meets many of the listed requirements (see this volume [3]), but ligands or spacers may not be easily introduced without simultaneous introduction of amino or carboxylic groups, which certainly will act as adsorption centers for proteins. Cross-linked dextran, Sephadex, in its most permeable form is a good ligand carrier but is superseded in all respects by agarose. Agarose, particularly in cross-linked bead form, is at present the matrix substance of choice for preparation of stereospecific adsorbents.[18]

The cross-linking of agarose decreases the number of hydroxyl groups and consequently lowers the capacity for introducing ligands. This is of only secondary importance in view of the advantages gained by the increase in alkaline and heat stability and the improvement in several other properties. As will be described later, in some activation procedures cross-linking takes place simultaneously with the introduction of reactive groups for subsequent binding of ligands to the matrix. Among the excellent properties of agarose should be mentioned the resistance to bacteria and enzymes likely to be present in the laboratory. No "agarase" has yet been found to cause an observable attack on cross-linked agarose. Acid sensitivity as manifested at low pH (<2.5) is a weak point. Rigidity is higher than for many other carrier materials, but an even greater rigidity would be desirable. Much harder gels can be obtained by cross-linking with divinylsulfone. Agarose cross-linked by this reagent retains high permeability, but unfortunately—owing to alkaline instability of the ether linkages—the working range for the divinylsulfone cross-linked agarose is limited to a pH below 9. It is likely that the tertiary structure of the agar polysaccharides with helical regions alternating with single chains in random coil conformation confers upon the gels a unique constellation of properties: high permeability associated with high rigidity and a capability of retaining exchangeable water. Therefore, at the same matrix content, an agarose gel has an exclusion limit far above that of Sephadex and probably other gels formed from synthetic polymers. The present view is that agarose gels are composed of open hollow regions capable of accommodating solutes of large molecular size solutes, these regions being interlinked by dense supporting helical matrix structures.[19]

Methods of Attachment

Ligands are covalently linked to polymers containing hydroxylic or amino groups by means of bifunctional reagents, X-R-Y, where X and Y are "connector" groups that react with the functional groups in the polymer and ligand, respectively X and Y may be identical. The attachment is usually made in two steps:

[18] J. Porath, J.-C. Janson, and T. Låås, *J. Chromatogr.* **60**, 167 (1971).
[19] D. A. Rees, *Chem. Ind.* **16**, 630 (1972).

Activation step:

$$\text{(M)}{-}OH + X{-}R{-}Y \longrightarrow \text{(M)}{-}O{-}R{-}Y + HX$$

Coupling step:

$$\text{(M)}{-}O{-}R{-}Y + H_2N\text{-Ligand} \longrightarrow \text{(M)}{-}O{-}R{-}Y'\text{-Ligand} + NH_3$$

Activation may also include several steps, as in the case of diazo coupling:

$$\text{(M)}{-}OH + ClCH_2{-}\langle\bigcirc\rangle{-}NO_2 \longrightarrow \text{(M)}{-}O{-}CH_2{-}\langle\bigcirc\rangle{-}NO_2$$

$$\text{(M)}{-}OCH_2{-}\langle\bigcirc\rangle\, NO_2 \xrightarrow{\text{reduction}} \text{(M)}{-}O{-}CH_2{-}\langle\bigcirc\rangle{-}NH_2$$

$$\text{(M)}{-}O{-}CH_2{-}\langle\bigcirc\rangle{-}NH_2 \xrightarrow{\text{diazotization}} \text{(M)}{-}O{-}CH_2{-}\langle\bigcirc\rangle{-}N_2^+$$

It is important to mention that, after coupling of the desired ligand, reactive Y groups may still be present. Deactivation should be carried out so as to avoid chemisorption of proteins and other substances upon contact with a sample to be fractionated. Deactivation may occur by spontaneous hydrolysis but, if this is not the case, coupling with a low molecular weight substance should be made without converting the matrix into an ion exchanger. Glycine, neutral dipeptides, and ethanolamine are deactivating substances that should be considered.

Only a few methods will be described in detail. They have been extensively studied and optimized to suit the purpose of covalently attaching ligands to agarose (and other hydroxylic polymers). The bisoxirane method is particularly attractive since here a long-chain reagent is chosen as spacer and introduced simultaneously with the activation reaction. The methods all have the merit of avoiding the introduction of hydrophobic substituents. In the case of bisoxirane and divinylsulfone, no additional charges are introduced in the matrix, and the ligand will be fixed onto a flexible hydrophilic spacer.

Most methods used to immobilize proteins are based on the coupling of amino groups. Those described here in some detail are useful primarily for coupling of amino groups. Other methods have been described in review articles[1–13,20–23] (see also this volume [3] and [4]). To be mentioned is the isocyanide reaction, thoroughly studied by Ugi and collaborators and introduced for ligand coupling by Axén et al.[24,25] Its versatility is clear from the fact that four different functional groups, COOH, NH$_2$, CO, and NC, take part in the reaction, and any one of them may be the reacting group in the ligand-forming reagent. Unfortunately, the efficiency of coupling amino acids and proteins to neutral carbonyl-containing polymers has not been very impressive thus far.

Attachment of certain ligands may present great difficulties. Reactive groups may be absent from the ligand substance in question and synthesis of a suitable derivative may be necessary. This situation is exemplified by aminoalkylation of adenine nucleotides (see this volume [16]). Similarly, when the substance to be bound to the matrix is insoluble in water, derivatization may be necessary, e.g., by conjugating a steroid with a hydrophilic substituent. Another possibility would be to perform the fixation reaction in an organic or mixed solvent. Unfortunately, other complications are then likely to arise. Many of the commonly used reactions often proceed at a much slower rate in such solvents, and the gel may shrink and become impermeable to the reactants. Thus far, no systematic work has been carried out to elaborate general methods for producing biospecific adsorbents with ligands of this nature.

1. The Cyanogen Bromide Method of Coupling

Cyanogen bromide in alkaline solution will form cyanamides with amines. Solid polymers containing amino groups can therefore be activated by cyanogen bromide since the cyanamide groups may be utilized as intermediates for coupling with amines which react to form substituted guanidines. In attempts to utilize cyanogen bromide in this fashion, we discovered that the hydroxylic polymers themselves can be activated,[26–29] and that the reactive polymers contained imino carbonate ester groups.

[20] I. H. Silman and E. Katchalski, Annu. Rev. Biochem. 35, 873 (1966).
[21] J. K. Inman and H. M. Dintzis, Biochemistry 8, 4074 (1969).
[22] R. Goldman, L. Goldstein, and E. Katchalski, "Biochemical Aspects of Reactions on Solid Supports" (G. R. Stark, ed.), p. 1. Academic Press, New York, 1971.
[23] P. Cuatrecasas, "Biochemical Aspects of Reactions on Solid Supports" (G. R. Stark, ed.), p. 79. Academic Press, New York, 1971.
[24] R. Axén, P. Vretblad, and J. Porath, Acta Chem. Scand. 25, 1129 (1971).
[25] R. Axén and P. Vretblad, Acta Chem. Scand. 25, 2711 (1971).
[26] R. Axén, J. Porath, and S. Ernback, Nature (London) 214, 1302 (1967).
[27] J. Porath, R. Axén, and S. Ernback, Nature (London) 215, 1491 (1967).
[28] R. Axén and S. Ernback, Eur. J. Biochem. 18, 351 (1971).
[29] J. Porath, Nature 218, 834 (1968).

1,2-Diols are especially liable to react with cyanogen halides to form cyclic imino carbonates. Cross-linking between polymer chains will take place to a certain extent, and nonreactive carbonate is also formed. In the coupling step a substance containing amino groups will form at least three products, presumably through isocyanate as an unstable intermediate. These include the imido carbonate derivative, substituted carbamate, and isourea.

Activation step:

$$\text{Matrix}\!\!\begin{array}{c}\text{OH}\\\text{OH}\end{array} + \text{BrCN} \longrightarrow \begin{cases} \text{Matrix}\!\!\begin{array}{c}\text{O}\\\text{O}\end{array}\!\!\text{C=NH} \quad \text{Reactive} \\[2ex] \text{Matrix}\!\!\begin{array}{c}\text{OCONH}_2\\\text{OH}\end{array} \quad \text{Non-reactive} \end{cases}$$

Coupling step:

$$\text{Matrix}\!\!\begin{array}{c}\text{O}\\\text{O}\end{array}\!\!\text{C=NH} + \text{H}_2\text{NR} \longrightarrow \begin{cases} \text{Matrix}\!\!\begin{array}{c}\text{O}\\\text{O}\end{array}\!\!\text{C=NR} \\[2ex] \text{Matrix}\!\!\begin{array}{c}\text{OCONHR}\\\text{OH}\end{array} \\[2ex] \text{Matrix}\!\!\begin{array}{c}\text{O—C}\!\!\begin{array}{c}\text{NH}\\\text{NHR}\end{array}\\\text{OH}\end{array} \end{cases}$$

The reactive species formed in the activation step have different degrees of stability. After acid treatment a product is formed that can be freeze-dried and stored for long periods of time. Agarose (and cross-linked agarose) activated and thus stabilized is commercially available (Pharmacia).

A number of variations differing slightly in detail have been published and are claimed to represent improvements. Under certain conditions such as high concentration of cyanogen bromide, a high degree of substitution may be reached. At the same time the cross-linking of the gel matrix may become extensive. Therefore, the coupling product may not possess the high adsorption capacity expected from the ligand content. This is not at all a serious problem. In fact, a lower degree of substitution may be preferable, since under these circumstances a larger fraction of the ligands will be capable of adsorption.

A large part of the cyanogen bromide will be hydrolyzed to form bromide, cyanide, and cyanate in alkaline solution. Continuous removal of hydroxyl ions therefore occurs simultaneously with imido carbonate formation. Usually the consumption of hydroxylic ions can be compensated for by a corresponding addition of alkali. Autotitration can be circumvented by the use of a high concentration of buffer.[30] The simplified procedure is as follows.

Spherical beads of agarose (whether or not cross-linked) are washed with a 2 M potassium phosphate buffer, prepared by mixing and diluting 2 parts of 5 M K_3PO_4 and one part of 5 M K_2HPO_4 with a pH of about 12.1. The interstitial water is removed by filtration on a Büchner funnel. Ten grams of gel, p percent with respect to dry agarose,[31] are suspended in 2.5 p ml cold 5 M potassium phosphate buffer prepared as above. The gel suspension is then diluted with distilled water to a volume of 20 ml. A solution, p ml, containing 100 mg of cyanogen bromide per milliliter, is added in portions during a period of 2 minutes. The temperature is maintained between 5 and 10° under gentle stirring during a reaction time of 10 minutes. The product is washed on a glass filter with cold distilled water to neutral reaction.

More weakly activated gels can be conveniently obtained by decreasing the amount of cyanogen bromide and buffer, e.g., to 0.12 p ml of the 5 M buffer solution and 0.05 p ml of the cyanogen bromide solution, respectively.

The coupling may be illustrated as follows: 20 ml of the activated product is washed with 0.25 M sodium bicarbonate at pH 9 and transferred to a suitable cylindrical reaction vessel. Twenty milliliters of 0.25 M sodium bicarbonate is added followed by 50–500 mg of the substance to be coupled. The reaction vessel is slowly rotated end-over-end at room temperature or in the cold for 24 hours. The reaction may also be made in a suitable beaker or conical flask with magnetic stirrer. The product is washed on a Büchner funnel or directly in the column to be used for chromatography.

Variations in the coupling conditions can be introduced within certain limits of pH, temperature, and so on. If necessary, the pH may be decreased. This will result in a decrease in the amount of ligand coupled but may be compensated for by a better survival of biological activity. Below pH 6 the coupling efficiency is usually too low to be practical. At a higher pH, there may be an increase in coupling yield, but also in damage to the ligand and extent of cross-linking. Indeed, the latter may be so large that sterical hindrance will cause a substantial amount of the ligands to be inaccessible to solutes otherwise adsorbed.

[30] J. Porath, K. Aspberg, H. Drevin, and R. Axén, *J. Chromatogr.* 86, 53 (1973).
[31] p is the percentage of matrix substance (w/v) in swollen gel and is equal to 4 and 6 for Sepharose 4B and 6B, respectively.

Although most applications of cyanogen bromide coupling have involved agarose and cross-linked agarose, other hydroxyl-containing polymers may also be converted to biospecific adsorbents by coupling of suitable ligands in the same manner.

The cyanogen bromide method may be used for coupling two compounds containing hydroxylic groups, e.g., two carbohydrates. For this purpose a diamine is used to accomplish the connection:

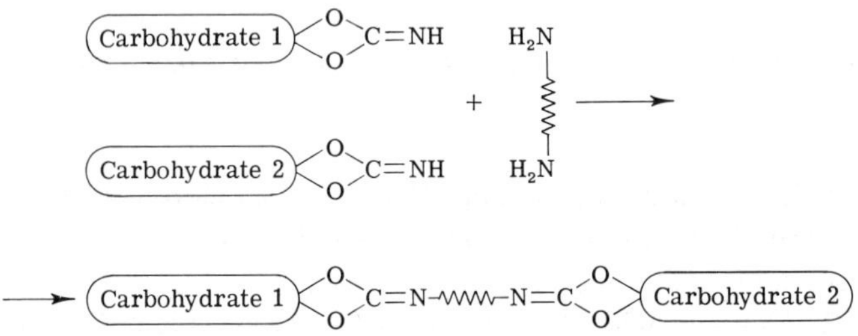

The carbohydrates may be monosaccharides or polysaccharides. For example, glucose,[32] galactose, or lactose may be linked to agarose.

The stability of the coupled products has been questioned.[33] In a series of experiments in the author's laboratory, Kristiansen et al. coupled ^{14}C-labeled glucose to agarose[32] in the manner just described. It is possible that certain coupled products formed by reactions with groups other than aliphatic and aromatic amino groups may be less stable than those tested in the author's laboratory. It was found that radioactivity was not released within a wide range of pH. At very low pH, 1–2.5, leakage was observed. Very likely the activity found in the washings is due to a progressive hydrolysis of the gel-forming agarose itself. Presence of esterases may also account for the release of ligand substance under physiological conditions. At the other end of the pH scale—above about pH 12—hydrolysis of the connecting linkages apparently occurs.

2. Bisoxirane Coupling

Bisoxiranes (bisepoxides) are particularly useful reagents for introducing low molecular weight ligands through amino or hydroxyl groups.[34,35] As in the case described above for the cyanogen bromide method, the introduction takes place in two steps: (1) the activation in which one of the oxirane groups of the reagent is allowed to react with

[32] T. Kristiansen, L. Sundberg, and J. Porath, *Biochim. Biophys. Acta* **184**, 93 (1969).
[33] G. I. Tesser, H.-U. Fisch, and R. Schwyzer, *FEBS Lett.* **13**, 56 (1972).
[34] J. Porath and L. Sundberg, *Protides Biol. Fluids, Proc. Colloq.* **18**, 401 (1970).
[35] L. Sundberg and J. Porath. To be published.

hydroxyl in the polymer matrix, leaving the other group free; and (2) the coupling wherein the oxirane ether is allowed to react with the ligand-forming substance. With bisglycidyl ether of 1,4-dihydroxy-*n*-butane, the following reactions take place:

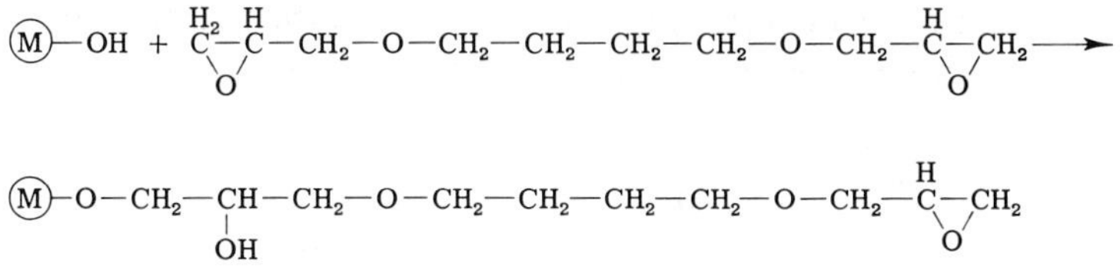

$$\text{M}-\text{OH} + \overset{H_2}{\underset{\diagdown O \diagup}{C}}-\overset{H}{C}-CH_2-O-CH_2-CH_2-CH_2-CH_2-O-CH_2-\overset{H}{\underset{\diagdown O \diagup}{C}}-CH_2 \longrightarrow$$

$$\text{M}-O-CH_2-\underset{OH}{CH}-CH_2-O-CH_2-CH_2-CH_2-CH_2-O-CH_2-\overset{H}{\underset{\diagdown O \diagup}{C}}-CH_2$$

The intermediate oxirane-containing polymer is attacked by nucleophiles, and by the coupling with R-NH$_2$ an alkylated amine is formed:

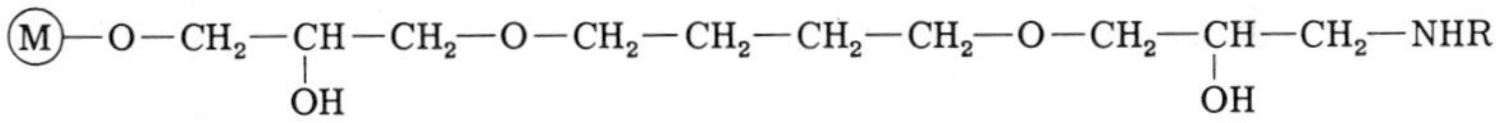

$$\text{M}-O-CH_2-\underset{OH}{CH}-CH_2-O-CH_2-CH_2-CH_2-CH_2-O-CH_2-\underset{OH}{CH}-CH_2-NHR$$

An alcohol, R—OH, reacts to give an ether:

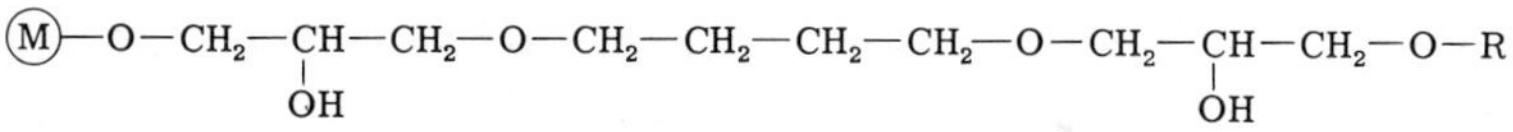

$$\text{M}-O-CH_2-\underset{OH}{CH}-CH_2-O-CH_2-CH_2-CH_2-CH_2-O-CH_2-\underset{OH}{CH}-CH_2-O-R$$

Optimal or nearly optimal conditions for activation have been worked out for agarose.[35] The following procedures are recommended as exemplified by 6% agarose gel (Sepharose 6B).

Synthesis of a highly activated gel on a small scale can be accomplished in the following manner:

About 7 ml of sedimented Sepharose 6B gel are washed with distilled water on a glass filter, and excess water is removed by suction. The gel is suspended in 5 ml of 1 *M* NaOH containing 2 mg of NaBH$_4$ per milliliter and 5 ml of 1,4-butanediol-diglycidylether. The reaction is allowed to proceed at 23° for 10 hours under slow stirring. The activated gel is thoroughly washed with distilled water on a glass filter and stored until used in distilled water at 5°. The gel may be kept at least 1 week without any significant decrease in the number of oxirane groups, which should constitute about 1000 μmoles per gram of dry gel.

The synthesis of a gel with lower oxirane content can be performed as above except for a shorter reaction time or a lower concentration of the bisoxirane. For example, at a reaction time of 5 hours with 1 ml of reagent, the concentration of the oxirane groups in the activated ether will be about 200 μmoles per gram of dry gel.

The coupling should be performed in an alkaline solution at a pH between 11 and 12, if possible. At a lower pH the reaction proceeds more slowly and in an extremely alkaline medium, >pH 13, cross-linking is likely to lower the yield. The coupling of peptides may be illustrated as follows: 3 ml of highly activated Sepharose 6B, dried by suction, is suspended in 2 ml of 0.2 M sodium carbonate at pH 11, containing 300 mg of glycylleucine. The suspension is kept at 40° for 15 hours with gentle stirring. The product is sequentially washed with water, 1 M NaCl solution, and the buffer to be used. The gel will contain about 600 μmoles of bound peptide per gram of dry gel.

In those cases in which the amine to be coupled is only slightly soluble at room temperature, the reaction may be carried out at elevated temperature. Aromatic amines, for example, may be introduced into the gel at 40–80° or at even higher temperatures. Proteins may also be coupled. The lower pH necessary to avoid destruction of the protein can partly be compensated for by prolonging the contact time. Attachment of soybean trypsin inhibitor to agarose may be used as an example of protein coupling: 3 ml of highly activated Sepharose 6B is treated with 50 ml of soybean inhibitor dissolved in 2 ml of 0.2 M sodium carbonate at pH 9.0. After 2 days at 23° with gentle stirring, the reaction is terminated. The product has been found to contain about 6 μmoles of protein (144 mg) per gram of dry gel.

The optimum conditions for coupling hydroxyl compounds have not yet been worked out. Preliminary studies have shown that high temperature (>40°) and high pH (12–13) will be necessary to achieve a high degree of substitution. Moreover, the hydroxylic compounds differ greatly in their reactivity toward the oxirane ether gels.

Oxirane groups are quite stable in the media used for chromatography. Removal of excess active groups therefore may present a problem. In the case of stable ligands, storage at pH 11–12 for 1 week or longer can be recommended to remove residual reactive groups. Gels containing proteins may be deactivated by long-time treatment with a concentrated glycine or hydroxylamine solution. Some groups, inaccessible to peptides and proteins, do remain in the latter case, as found[36] by titration with $Na_2S_2O_3$, but this fact is of little or no importance for the chromatography of proteins.

The extremely high stability of the linkage between ligand and matrix is a great advantage whenever treatment of the adsorbent under drastic alkaline conditions is necessary. Another advantage of the method, as exemplified here with a long-chain bisoxirane reagent, is the introduction of a hydrophilic spacer group that places the ligand at a considerable

---

[36] L. Sundberg and T. Wåhlstrand. To be published.

distance, on the average, from the polysaccharide chains. The importance of introduction of such a group in the reagent has recently been demonstrated by Sundberg and Wåhlstrand.[36] Sulfanilamide coupled to Sepharose 6B with 1,4-n-butane-bisglycidylether yields an adsorbent of high capacity for human carbonic anhydrase whereas cyanogen bromide coupling gives a heavily substituted product of only low adsorption capacity.

3. Divinylsulfone Coupling

Other bifunctional reagents, for example, the bisaziridines, may be used for attachment of compounds containing amino or hydroxyl groups to hydroxylic matrices. For the preparation of biospecific adsorbents it is desirable to select reagents that do not introduce ionogenic groups (bisaziridines will form amines and convert the carrier to an anion exchanger). Divinylsulfone, occasionally used for cross-linking synthetic polymers, may be used about as are bisoxiranes.[35,37,38] The first step will lead to the vinylsulfonyl ethyl ether of the hydroxylic polymer:

$$\textcircled{M}-OH + CH_2=CH-SO_2-CH=CH_2 \longrightarrow \textcircled{M}-O-CH_2-CH_2-SO_2-CH=CH_2$$

with cross-linking as a side reaction. The vinyl groups introduced into the matrix are more reactive than are the oxirane groups. They will thus couple to amines, alcohols, and phenols at lower temperatures and at lower pH than the oxirane intermediates described in the preceding section. The divinylsulfone coupling method is well suited for the attachment of hydroxylic compounds, R—OH, to form products presumably containing the following connector-spacer structure:

$$\textcircled{M}-O-CH_2-CH_2-SO_2-CH_2-CH_2-OR$$

Unfortunately, the products are unstable in alkaline solution (the amino link above about pH 8 and the hydroxyl link at about pH 9 or 10). Since a pH of 11–12 may be necessary for the reaction to proceed smoothly, studies must often be made to optimize the coupling in each particular case. Very satisfactory adsorbents have been produced in this manner with galactose-Sepharose 6B.[39]

Activation. Ten milliliters of sedimented Sepharose 6B is gently suspended in 10 ml of 1 M sodium carbonate at pH 11, and 2 ml of divinylsulfone is added. The mixture is carefully stirred for 70 minutes at room temperature. The reaction is stopped by lowering the pH to 7.0 and wash-

---

[37] J. Porath and L. Sundberg, *Nature (London)* **238**, 261 (1972).
[38] J. Porath and L. Sundberg. To be published.
[39] J. Porath and B. Ersson. To be published.

ing the gel on a glass filter funnel with small portions of water (total volume of 500 ml). Preparation of adsorbents containing low molecular weight ligands may be exemplified by sulfanilamide-Sepharose 6B.[37]

Coupling. Ten milliliters of activated gel (as above) are suspended in 10 ml of 1 M sodium carbonate at pH 11.0, at 50°, containing 500 mg of sulfanilamide. The mixture is gently stirred at 50° for 2 hours. The product is washed on a glass filter funnel with small portions of the following solutions: 500 μl of sodium carbonate at pH 11.0 containing 1 M NaCl, 500 ml of 0.2 M glycine at pH 3.0 containing 1 M NaCl, and 500 ml of 1 M NaCl at 50°. The coupled sulfanilamide is 550 μmoles per gram of dry gel according to the nitrogen content as determined by the Kjeldahl method.

Alkali-stable proteins, such as soybean trypsin inhibitor, may be coupled as follows: Ten milliliters of activated Sepharose 6B is mixed with 10 ml of 0.3 M sodium carbonate at pH 10 containing 200 mg of protein. The suspension is gently stirred with a magnetic stirrer at room temperature for 2 hours. After the coupling is completed, the product is washed with the following solutions: 500 ml of 0.3 M sodium carbonate at pH 10.0 containing 1 M NaCl, 500 ml of 0.3 M glycine at pH 3.0 containing 1 M NaCl, and finally with 500 ml of 0.05 M Tris chloride at pH 7.5 containing 0.5 M NaCl in which the gel is stored at 5°. In the case of soybean trypsin inhibitor, the gel contains 12 μmoles per gram of dry conjugate, and the gel is able to adsorb a stoichiometric amount of trypsin which, in a subsequent step, can be quantitatively desorbed.

The washing procedures can be varied widely. To ensure essentially complete blocking, treatment with 1 M glycine of pH 7–8 can be recommended.

Extensive cross-linking occurs with divinylsulfone. As with bisoxirane, the permeability of agarose gels is often slightly decreased, but this can be compensated for by the use of a starting gel of lower matrix content. Divinylsulfone cross-linked gel of 4% agarose usually exhibits a higher permeability than 6% non-cross-linked gel and, surprisingly, sometimes even higher than the non-cross-linked starting gel. A definite advantage gained by the cross-linking with divinylsulfone has been discovered: the gels are considerably more rigid and will sediment to form beds of excellent flow properties. Divinylsulfone cross-linked agarose gels, therefore, are very well suited for chromatography in the pH range of 3 to 8 and for shorter periods, may be left in contact with solutions of higher alkalinity provided the temperature is kept below 10° and contact time at only two or so hours at a pH of 10. Increase in rigidity has also been observed upon substitution of benzyl groups into cross-linked agarose. Such gels show strong hydrophobic adsorption. It remains to be seen whether and to what extent divinylsulfone-treated, cross-linked agarose will exhibit similar adsorption properties.

Concluding Remarks

Bioaffinity chromatography and other related methods are likely to become increasingly important. Purification of receptor-containing organelles and fragments, isolation of enzyme complexes of higher order with immobilized effectors and coenzymes, isolation of enzyme precursors and genetically defective, inactive enzymes with immunoadsorbents are only a few examples of application. Unfortunately, the methods, so convenient and intellectually attractive, have not yet advanced to the stage where they are able to fulfill all expectations. Thus far we have been concerned mostly with the problems involved in limiting nonspecific ionic adsorption by selecting matrices of the lowest possible content of fixed ionogenic substituents. It now seems evident that the matrix substances themselves may be involved in other kinds of interactions, especially hydrophobic and hydrogen bonding. For this reason, more attention might be paid to the solvent composition which presumably ought to be selected to minimize nonspecific interactions but still retain a maximum of specific adsorption. Mixed solvents systems containing, in addition to water, polyhydric alcohols, such as glycols, glycerol, or hexitols, and strong breakers of hydrogen bonds at carefully controlled concentrations may be used to modify and improve the bioaffinity methods.

Attention should also be paid to the properties of the spacer groups chosen. Introduction of spacers containing amino groups without subsequent complete blocking of the residual connector, will lower the degree of adsorption specificity whereas the additional complication of a hydrophobic spacer will further enhance the nonspecific adsorption capacity as a consequence of hydrophobic interaction.

The density distribution of adsorption centers, specific as well as nonspecific, will be progressively more important as the size of solute molecules increases. Desorption requires simultaneous detachment from all interacting areas of the adsorbate–adsorbent association complex, the likelihood of which will be small at high density of adsorption centers in both solute and adsorbent. Furthermore, for large molecules and organelles, the diffusion rate may be too slow to make chromatography meaningful. For this reason it is likely that batch procedures will be more effective once bioadsorbents of desirable properties have been developed.

Undoubtedly the biospecific adsorbents provide tools both for fractionation and for model studies of various enzyme interactions. For fractionation work, it is not likely that extreme specificity will be critically important. On the contrary, it may be anticipated that adsorbents showing a high degree of group specificity will be generally more useful and that their lack in individual specificity will be compensated for by the use of specific elution or displacement procedures. The group-specific adsorbents constitute a type of adsorbent between those of unique specificity and those of lowest degree of specificity, i.e., adsorbents of the classical types, such as the commonly used ion exchangers, lipophilic, amphiphilic, and "aromatophilic" adsorbents. The classical type of adsorbents will not be outdated by the bioadsorbents but should be further developed so that the initial stages become more efficient in the purification schemes for mixtures of natural and synthetic compounds.

[3] Covalent Linkage of Functional Groups, Ligands, and Proteins to Polyacrylamide Beads

By JOHN K. INMAN

New methods for preparing chemical derivatives of cross-linked polyacrylamide are presented here. They have been designed to overcome several problems encountered when the derivatives earlier described by Inman and Dintzis[1] were used in affinity chromatography. Emphasis has now been placed on (1) providing spaced-out functional groups for attaching specific ligands, and (2) selecting chemical reactions and conditions which tend to minimize formation of additional cross-links in the gel structure. Thus, ligands can be placed in orientations favorable for specific binding with proteins or macromolecules which is necessary for their selective retention. Large moieties should be able to enter the pores of the polymer as readily after chemical derivatization as before.

In general, polyacrylamide supports provide advantages, such as very low nonspecific adsorption of biological macromolecules, wide limits for chemical and thermal stability, freedom from attack by enzymes (especially those that attack the polysaccharide structure of certain other supports). Ligands can be firmly and permanently attached to cross-linked polyacrylamide since they are part of a completely covalent network of stable bonds. A wide selection and concentration range of useful chemical groupings can be placed on the framework structure in a controlled fashion. An especially important consideration is that the distribution of functional groups and support substructures are readily randomized so that extensive clustering of groups and peculiar microenvironments are avoided.

The structural network of cross-linked polyacrylamide gels, a part of

[1] J. K. Inman and H. M. Dintzis, *Biochemistry* **8**, 4074 (1969).

which is shown below, consists mainly of segments of linear polyethylene with alternate backbone carbon atoms bearing primary amide groups (carbamyl groups, —$CONH_2$). The great abundance of these amide groups[2] is responsible for the marked hydrophilic character of polyacrylamide and for its low adsorption to macromolecules. Linear polyacrylamide itself is water soluble. Insoluble gel networks are formed by including some bifunctional monomer, e.g., N,N'-methylenebisacrylamide, in the polymerization reaction to produce cross-linkages. One such cross-link is illustrated (running horizontally) in the first structural diagram.

Chemical derivatives of polyacrylamide can be prepared by (1) copolymerization of acrylamide and cross-linker with a functional group-bearing acrylic or vinyl monomer,[3] or (2) derivatization of plain, preformed polyacrylamide gels.[1] The latter approach will be followed in the work presented here because of advantages gained from the commercial availability of various grades of polyacrylamide particles in spherical bead form[4] which are ideally suited for column adsorption experiments. It is possible, therefore, to prepare adsorbents with independent selection of functional group, degree of substitution (group density), porosity, and particle size without having to cope, in each and every instance, with a bead-forming copolymerization process and sizing operation.

Commercial polyacryamide beads[4] are sold in a dry state. Upon being mixed with water or aqueous solutions, the particles gradually swell and, after a period of 4–48 hours, reach a limiting size. The solvent-filled pores inside the gel matrix are of macromolecular dimensions and will exclude large molecules depending on their size and on the degree of cross-linkage in the polymer. While this property is essential for the gel filtration process, it may impose a limitation on the usefulness of these particles in affinity chromatography by virtue of the fact that excluded molecules have relatively little gel surface with which to interact. The most porous bead presently available (Bio-Gel P-300[4]) can include molecules with molecular weights up to about 400,000. Thus, many enzymes and biological

---

[2] About 14 mmoles per gram dry weight, or 62% of the polymer mass.

[3] Polyacrylamide derivatives (as nonbeaded fragments) bearing useful functional groups are made in this way and sold by Koch-Light Laboratories Ltd., Colnbrook, Bucks, England and by Aldrich Chemical Co. (USA) under the trade name Enzacryl. A copolymerization approach was recently described by Y. Ohno and M. A. Stahmann, *Immunochemistry* 9, 1077 and 1087 (1972).

[4] The polyacrylamide gel beads are purchased under the trade name, Bio-Gel P, from Bio-Rad Laboratories, 32nd and Griffin Avenue, Richmond, California 94804 and their distributors. The beads are offered in various size ranges and porosities as described in their catalog. Type P-2, for example, is highly cross-linked and, when fully swollen, has a molecular weight exclusion limit of 1800; higher P values (P-4 through P-300) have progressively fewer cross-links and higher molecular weight exclusion limits.

macromolecules can be separated effectively on polyacrylamide adsorbents. Because of the softness of low cross-linked beads and consequent resistance to flow in columns, one should select bead porosities no greater than that needed to obtain an effective interactive surface. Flow rates can be improved by using only larger-size beads.[5] Some successful affinity separations, however, have been achieved with large moieties interacting only at the outer surfaces of the polyacrylamide particles, notably in separations of whole, viable cells according to their membrane receptor specificities.[6-10]

The following sections give procedures and examples that represent only a small sampling from a great many possibilities. It is hoped that each interested investigator will freely vary them or devise new derivatives as special problems or needs arise.

General Procedures

Handling and Washing of Gels. Polyacrylamide gel particles adhere somewhat to glass surfaces. In order to facilitate transfer and mixing operations, one should siliconize all glassware that the gels contact.[11] Polyethylene, polypropylene, or, ideally, the clear TPX™ (Nalgene) labware may be used without prior treatment.

Gels are washed free of excess reactants and by-products with 0.1–0.2 M NaCl (saline) by (1) suction filtration on porcelain Büchner funnels using Whatman No. 1 or No. 5 paper circles, and (2) repeated batch washing: stirring with 5–25 volumes of saline, allowing settling by gravity, and siphoning off the supernatant. It is advantageous to limit the period of settling in order to remove fines. The two methods of washing are

---

[5] One may, for example, discard the slowest sedimenting one-third of the beads in the 50 to 100-mesh range. Certain derivatives of very large beads are mechanically fragile, so that agitation methods may have to be changed after making low-power microscopic checks. Also, the manufacturers recommend mixing P-300 derivatives 1:2 with Bio-Gel P-30 (underivatized) for faster flow rates in columns.

[6] P. Truffa-Bachi and L. Wofsy, *Proc. Nat. Acad. Sci. U.S.* 66, 685, (1970).

[7] L. Wofsy, J. Kimura, and P. Truffa-Bachi, *J. Immunol.* 107, 725 (1971); C. Henry, J. Kimura, and L. Wofsy, *Proc. Nat. Acad. Sci. U.S.* 69, 34 (1972).

[8] J. K. Inman, in "Mechanisms of Cell-Mediated Immunity" (R. T. McCluskey and S. Cohen, eds.), p. 375. Wiley (Interscience), New York, 1974.

[9] S. Sell and T. An, *J. Immunol.* 107, 1302 (1971); J. D. Stobo, A. S. Rosenthal, and W. E. Paul, *J. Immunol.* 108, 1 (1972).

[10] P. G. H. Gell and H. P. Godfrey, in "Contact Hypersensitivity in Experimental Animals—Collegium Internationale Allergologicum." Karger, Basel, in press.

[11] Clean, *dry* glassware may be siliconized by being filled with a 2% v/v solution of dichlorodimethylsilane in benzene. After a standing period of 0.5 hour (or longer), the solution is returned to a storage bottle for reuse. Treated vessels are then rinsed with acetone, air-dried, and heated briefly in a drying oven.

usually used in the above sequence.[12] In cases of the most porous gels or those with extremely fine particles, batch washing may have to be used throughout because of slowness of flow on the suction filter. For substances that are difficult to remove completely by washing, alternate funnel and batch washes, or column washing can be effective. Some derivatives with poorly water-soluble ligands may have to be washed with solvents and/or solvent–saline mixtures. Formamide, in which polyacrylamide remains swollen, has been an effective washing solvent in some cases.

Gel derivatives should be stored in the wet, swollen state under refrigeration. Freezing and thawing may cause some fragmentation of the beads. The gels can be stored in neutral solutions containing 0.02–0.1% sodium azide as preservative, or in a "storage buffer" made up as follows: Dissolve in water and make up to 1 liter, 11.69 g of NaCl, 0.74 g of disodium ethylenediaminetetraacetate ($2H_2O$), 6.18 g of boric acid, and 5 ml of 1 N NaOH. Then add with mixing 0.3 ml of a 0.5% solution of pentachlorophenol in ethanol. The pH should be in the range of 7.0 to 7.5.

TNBS Tests. The formation of trinitrophenyl derivatives of primary amines, hydrazides [see Eq. (21)] and hydrazines provides the basis for an extremely useful color test for these groups either when bound to gels or in free solution. In the former case, a small amount of saline-washed gel is suspended in saturated sodium borate in a spotplate well (or test tube). Three to 5 drops of a 3% w/v solution of 2,4,6-trinitrobenzene-sulfonic acid (TNBS) in water are added with mixing. Primary amines give an orange color, after several minutes, ranging from yellow to orange-red depending on the concentration of groups in the gel. Primary hydrazides give a rust color which deepens to a magenta on standing. No color within 30 seconds can be considered a negative test. Washings are tested by mixing samples (4 or 5 ml) with one-half their volume of saturated sodium borate, adding TNBS solution as above, and comparing with a saline control after 5 minutes.

Functional Group Density.[13] One gram of dry polyacrylamide gel will

---

[12] When resuspending gels from suction filter cakes, one should transfer the material to a beaker and add fluid at first in small amounts and mix with a wide-bladed spatula after each addition. Otherwise, small clumps of gel remain which are difficult to disperse even with prolonged stirring.

[13] The term "functional group density" (or "group density") will be used to express the degree of substitution of amide groups in terms of millimoles of introduced group per dry gram of original polyacrylamide. The term "specific capacity,"[1] or simply "capacity," has often been used to mean the same thing. However, it would seem preferable to reserve "capacity" as an expression of the maximum amount of a moiety that the adsorbent is able to bind in its eventual application.

swell to a settled or packed bed volume that ranges from 4 ml to 40 ml or so depending on the degree of cross-linkage.[14] Up to several days may be required to reach full swelling from the dried state at room temperature. The functional group density obtained under a given set of conditions will be essentially independent of degree of swelling or cross-linkage if expressed in terms of millimoles or micromoles per dry gram. Therefore, rules for controlling group densities will be expressed in these terms.

In applications with specific gels, it is more convenient to quantitate functional groups as milli- or micromoles per milliliter of packed bed volume. This quantity will be referred to as functional group *concentration* and can be obtained simply by dividing the functional group density by the bed volume per dry gram. An estimate of the latter value is given by the manufacturer,[4] or it can be found by allowing a weighed sample of the original gel to swell for a day or two in 0.2 M NaCl and then measuring the volume of settled bed in a TPX graduated cylinder. Strictly, dry weights of gels as obtained from the manufacturer should be corrected for adsorbed atmospheric moisture (usually amounting to 5–10% of the weight) by drying a sample in vacuum over anhydrous calcium chloride (see *Analytical Procedures*).

Primary Derivatives

Equations (1)–(3) indicate how normally neutral and inert side-chain amide moieties can undergo reactions which yield more reactive functional groups.[15] Any one of these three reactions may serve as the first step in a derivatization sequence. The extent of the reaction, that is, the proportion of amide converted, can be easily controlled as described below or as shown graphically by Inman and Dintzis.[1] The degree of substitution in the final derivative is usually determined at this stage.

The simplest way to modify the amide groups is through acid- or base-catalyzed hydrolysis to yield side-chain carboxyl groups (structure I). However, relatively few opportunities exist for subsequent reactions in aqueous media.[16] So far, water-soluble carbodiimides have been the reagents of choice in effecting subsequent reactions with the carboxyl function. The acyl hydrazide derivative (II) is also easy to prepare but provides many opportunities for the next stage of derivatization. It can be

---

[14] The degree of swelling is reproducible only if the polymer bears very few charged groups, or if it swells in solutions possessing enough electrolyte to effectively shield bound charges. Charged polyacrylamide beads can swell to several times their usual hydrated volume (in salt and buffer solution) when washed with distilled water because of internal electrostatic repulsions.

[15] The primary amide groups are stable with respect to hydrolysis in aqueous media at room temperature or below in the pH range 2 to 9 and for limited time at pH 0 to 10. In a slightly narrower pH range, the gels can withstand temperatures of up to 100° or more.

PA (Polyacrylamide)

$$
\begin{array}{ll}
\text{H}_2\text{C} \quad \text{O} & \text{H}_2\text{C} \quad \text{O} \\
\text{HC—C—NH}_2 & \text{HC—C—NH}_2 \\
\text{H}_2\text{C} \quad \text{O} & \text{H}_2\text{C} \quad \text{O} \\
\text{HC—C—NH}_2 & \text{HC—C—NH}_2 \\
\text{H}_2\text{C} \quad \text{O} & \text{O} \quad \text{CH}_2 \\
\text{HC—C—N—CH}_2\text{—N—C—CH} & \\
\end{array}
$$

$$
\begin{array}{ll}
\text{PA} & \text{PA} \\
\text{H}_2\text{C} \quad \text{O} & \text{H}_2\text{C} \quad \text{O} \\
\text{HC—C—NH}_2 + \text{OH}^- \longrightarrow & \text{HC—C—O}^- + \text{NH}_3 \quad (1) \\
& (\text{I})
\end{array}
$$

$$
\begin{array}{ll}
\text{PA} & \text{PA} \\
\text{H}_2\text{C} \quad \text{O} & \text{H}_2\text{C} \quad \text{O} \\
\text{HC—C—NH}_2 + \text{H}_2\text{NNH}_2 \longrightarrow & \text{HC—C—NNH}_2 + \text{NH}_3 \quad (2) \\
\qquad\qquad \text{Hydrazine} & \qquad\quad \text{H} \\
& (\text{II})
\end{array}
$$

$$
\begin{array}{ll}
\text{PA} & \text{PA} \\
\text{H}_2\text{C} \quad \text{O} & \text{H}_2\text{C} \quad \text{O} \\
\text{HC—C—NH}_2 + \text{H}_2\text{NCH}_2\text{CH}_2\text{NH}_2 \longrightarrow & \text{HC—C—NCH}_2\text{CH}_2\text{NH}_2 \quad (3) \\
\qquad\qquad \text{Ethylenediamine} & \qquad\quad \text{H} \\
& (\text{III}) \quad + \text{NH}_3
\end{array}
$$

acylated, alkylated, or converted directly to an acylating function, the acyl azide (XXIII) according to Eq. (16). The aminoethyl derivative (III) can be prepared directly and provides a convenient primary aliphatic amino group. A few select preparations of the hydrazide and aminoethyl derivatives are available commercially.[4]

Preparation of Carboxyl Derivatives (I) by Deamidation [Eq. (1)]. The dry beads as they come from the manufacturer[4] are added gradually, with efficient magnetic stirring, to a mixture of 0.50 M sodium carbonate and 0.50 M sodium bicarbonate (pH 10.0 or 10.5, see below). The volume of carbonate solution is chosen to be several times the expected

---

[10] Cross-linked polyacrylamide will not swell in most organic solvents. A few exceptions among the latter are glacial acetic acid, formic acid, formamide, and ethylenediamine (at 90°). In solvent–water solutions the degree of swelling diminishes rapidly as the proportion of solvent increases and is usually quite small above 50% v/v. Some degree of swelling is needed for reactions to occur inside the beads.

bed volume of the swollen gel (see above section on *Functional Group Density*). After about 0.5 hour of stirring, the gel suspension is heated quickly to 60° (hot water or steam bath), and the container is placed in a water bath controlled at 60°. During the ensuing heating period, the reaction mixture should either be stirred slowly or occasionally swirled. At the end of the desired heating period the suspension is cooled in ice-water, and the gel is washed on a funnel with large volumes of 0.2 M NaCl.

The functional group density can be controlled by the pH using a 3-hour heating period at 60°,[1] or simply by limiting the time of heating at a given pH. A useful rule is the relationship D (millimoles/dry gram) = 0.36t (at pH 10.0) or 0.58t (at pH 10.5), where t is the heating time in hours at 60°. The reactions in carbonate buffer can be carried out at room temperature instead of 60° for longer periods of time (roughly one day in place of each hour at 60°). The carboxyl group density may be measured by titration as described under *Analytical Procedures*.

Preparation of Hydrazide Derivatives (II). The dry beads are added gradually with stirring to an aqueous solution of hydrazine (1 M to 6 M).[17] The suspension is stirred slowly or swirled occasionally while being held at a constant temperature. The reaction is stopped after the desired time by funnel washing of the gel. Large volumes of 0.2 M NaCl are used if flow rates permit. Washing is completed by a series of batch treatments until supernatants (free of beads) become negative or show only pale violet color in the TNBS test. The gel should give a positive color reaction. The washed gel is kept in storage buffer (see above under *Handling and Washing of Gels*).

The functional group density (D) is regulated by the concentration of hydrazine and the time and temperature of the reaction. Derivatization may be carried out at room temperature or at higher temperatures. A useful guide for arriving at a desired group density is the relation, D (millimoles/g) = $h \cdot C \cdot t$, where C is the molar concentration of hydrazine, t is the reaction time in hours, and h is a coefficient which is a function of temperature. At 25.0°, $h = 0.0130$ for group densities of 0 to 0.2 mmole/g and 0.0112 for densities of about 1.0. At 50.0°, $h = 0.123$ at low densities and 0.111 for densities of near 1.0 mmole/g. At other temperatures h is related to the above values by the expression, $\log_{10} h_2/h_1 =$

---

[17] Aqueous solutions of hydrazine can be prepared by diluting reagent grade hydrazine hydrate (20.4 M) or anhydrous 97 + % hydrazine (30.4 M). A great deal of heat is generated when the latter is mixed with water. Therefore, it should be added carefully *to* water. Dilutions should be carried out in a hood. The concentration of a hydrazine solution can be checked by diluting a sample to about 0.1 M and titrating with standard 1 or 2 N HCl to a methyl red end point (pH 5). One mole of hydrazine is equivalent to 1 mole of HCl since the alterate nitrogen atom of a hydrazinium ion is protonated at a very much lower pH.

$0.0395(T_2 - T_1)$. Values of h_1 and h_2 correspond, respectively, to temperatures T_1 and T_2. The above expression is based on observations that an increase in temperature of 10° raises reaction rates by a factor of 2.48. Hydrazide group densities may be measured by the method given in *Analytical Procedures.*

Hydrazine is a base, so consequently some carboxyl group is generated via Eq. (1) during the formation of hydrazide. The amount of carboxyl formed depends very little on the hydrazine concentration and increases at rates of about 0.0013 and 0.010 mmole/g per hour at 25° and 50°, respectively. Thus, the amount of carboxyl in relation to hydrazide formed is small being about 10 mole% of the hydrazide in 1 M hydrazine reactions and only 1.7 mole% in 6 M reactions.

Preparation of Aminoethyl Derivatives (III). The dry beads are added in very small portions (to lessen lump formation) to anhydrous (98%) ethylenediamine preheated to 90°[18] and stirred efficiently. The operations are carried out in a ventilated hood. About 1 g of dry gel is added to each 20 ml of ethylenediamine. Stirring is continued and temperature of the suspension is maintained at 90°. At the end of the reaction period the mixture is cooled to near room temperature, then the contents are mixed with an equal volume of crushed ice. The gel is promptly washed with 0.2 M NaCl, 0.001 N HCl on a Büchner funnel and then by batch washes until supernatants are negative by the TNBS test. The gel should give a positive test at this point.

The density of aminoethyl groups increases approximately linearly with the time of heating at 90°.[1] Up to a density of 1.2 mmoles/g, about 0.30 mmole/g of amide are converted per hour according to Eq. (3). Beyond this point the rate falls off progressively. The reaction rate depends somewhat on water content of the ethylenediamine, increasing with increasing concentration of water. However, carboxyl groups are formed by alkaline hydrolysis in increasing amounts *in relation to* aminoethyl groups at high moisture contents. Therefore, ethylenediamine is used as it comes from the supplier. Aminoethyl and carboxyl group densities may be measured by titration as described under *Analytical Procedures.*

Spaced-out Functional Groups

Ligands that are attached closely to the support structure of an adsorbent may interact very weakly with the specific binding protein. By attaching the same ligand to the same support through a flexible "bridge" or "arm," consisting of a chain of 5 or more atoms, a dramatic increase in binding may occur. Cuatrecasas[19-21] has reported a number of such

---

[18] Glycerol is used in the heating bath instead of water in order to prevent excessive absorption of water vapor by the reaction mixture.

[19] P. Cuatrecasas, *J. Biol. Chem.* **245**, 3059 (1970).

observations with affinity adsorbents based on agarose. Presumably, these findings reflect the importance of correct spacial orientation and freedom from steric hindrance in the binding process. The same phenomenon should apply equally well to polyacrylamide supports. Hence, a selected group of polyacrylamide derivatives that provide useful functional groups spaced out from the polymer backbone are described below.

Succinylhydrazide (IV) and Glutarylhydrazide (VI) Derivatives. The primary hydrazide group can be acylated readily and completely with an excess of a carboxylic acid anhydride. The reactions can be carried out in aqueous media to yield N,N'-diacylhydrazides. Examples employing succinic anhydride and glutaric anhydride (V) are shown in Eqs. (4)

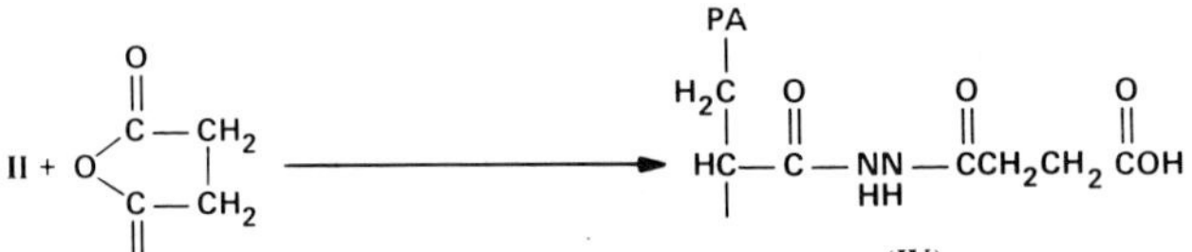

Succinic anhydride

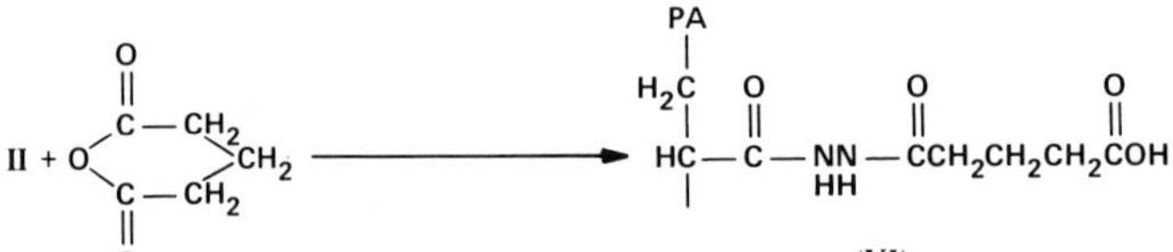

Glutaric anhydride

(V)

and (5). These cyclic anhydrides open up on reaction to leave a free terminal carboxyl group. The resulting succinylhydrazide (IV) or glutaryl-hydrazide (VI) structures provide carboxyls spaced out 6 or 7 atoms, respectively, from the polymer chain.

The preparation of succinylhydrazide gels was described by Inman and Dintzis.[1] The glutarylhydrazide derivative is prepared in a similar way as follows[22]: The hydrazide derivative is washed and resuspended[12] in 0.1 M NaCl to a volume several times the packed volume. A magnet bar and pH electrode are introduced. Under efficient stirring, solid glutaric

[20] P. Cuatrecasas and C. B. Anfinsen, this series, Vol. 22, p. 345.
[21] P. Cuatrecasas, *Advan. Enzymol.* 36, 29 (1972).
[22] J. K. Inman, unpublished work, 1973.

anhydride[23,24] (V) is added in an amount equal to 3 mmoles per milli-mole of hydrazide group estimated to be present. A minimum of 3 mg of anhydride should be added per milliliter of suspension. The pH is allowed to drop to 4.0 and is held in the range of 3.7–4.3 by additions of 1 N (or 2 N) NaOH. Reaction is complete in 15–20 minutes. A small sample of gel may be washed on a funnel with saline and given the TNBS test. If the test is positive, showing unreacted hydrazide, then more glutaric anhydride is added and the 15-minute, pH 4 reaction is repeated. The gel is finally washed alternately by funnel and batch tech-niques for several cycles.

Preparation of 6-Bromoacetamidocaproic Acid (BACA, VII). BACA is a useful intermediate for introducing spaced-out bromoacetyl and α-amino groups. It can be prepared in the following way[25]: Fourteen ml (0.10 mole) of triethylamine are added to a solution of 13.9 g (0.10 mole) of bromoacetic acid in 330 ml of ethyl acetate. The resulting solu-tion is transferred to a 1000-ml separatory funnel, and 8.8 ml (0.10 mole) of bromoacetyl bromide is added.[26] The funnel is shaken for about 1 minute (releasing excess vapor pressure) and allowed to stand for 10–15 minutes. A solution of 19.7 g (0.15 mole) of 6-aminocaproic acid and 29.8 g (0.21 mole) of Na_2HPO_4 (dibasic, anhydrous) is pre-pared in 280 ml of water (by warming, stirring, and cooling back to room temperature). This solution is then added to the mixture in the separatory funnel, and the latter is immediately shaken. The heavy precipitate of triethylammonium bromide formed in the previous step quickly dissolves. Shaking of the funnel is continued vigorously for 3 minutes then less so for 7 more minutes. The lower layer is separated

[23] Glutaric anhydride (not to be confused with glutaraldehyde) can be purchased from Aldrich Chemical Co. (Milwaukee, Wisconsin) or Eastman Organic Chem-icals (from their various distributors). Darkly colored material should be purified as follows: Dissolve 25 g of the crude anhydride in 125 ml of benzene. Add 2.5 g of activated charcoal and stir for 1 hour at room temperature. Filter and reduce the volume of filtrate to about 60 ml by boiling (in a hood). Cool the solution to 20° and add gradually, with stirring, 180 ml of ethyl ether. Cover and allow to stand until crystallization starts. Place the well-covered container in a refrigerator for 1 hour. Collect crystals by suction filtration, wash with about 30 ml of ether and dry under vacuum. A white crystalline material should be obtained which melts at 53–55°. Store in a desiccator.

[24] Glutaric anhydride, unlike succinic anhydride, is readily dissolved in water and aqueous solutions. For this reason it should be a more satisfactory reagent for modifying amino and other nucleophilic groups on adsorbents and proteins. Once dissolved in water it hydrolyzes rapidly (10–15 minutes at pH 4).

[25] J. K. Inman, unpublished work, 1972.

[26] Bromoacetic acid and bromoacetyl bromide are highly injurious to skin and mu-cous membranes on brief contact and should be handled with care.

and 21.1 g (0.153 mole) of $NaH_2PO_4 \cdot H_2O$ (monobasic) and 19.3 g (0.33 mole) of NaCl are stirred with it until dissolved. The resulting solution is adjusted to pH 5.50 with $10\,N$ NaOH, added back to the upper layer in the separatory funnel, and shaken for several minutes. The upper layer is set aside and the lower layer is shaken twice with fresh 300-ml volumes of ethyl acetate. The three upper layers are combined and shaken three times with 45-ml volumes of a solution having the composition, $2.30\,M$ NaCl, $0.200\,M$ Na_2HPO_4, $0.279\,M$ NaH_2PO_4 (pH 5.6). The lower layers are discarded. The upper layer is stirred with 60 g of anhydrous Na_2SO_4, filtered and flash evaporated to remove solvent. The oily residue is treated under reduced pressure until it solidifies.

The residue is refluxed and stirred with 1.0 g of activated charcoal (Darco G 60) in 500 ml of benzene for 20 minutes. The mixture is filtered while hot (through fluted Whatman No. 1 paper). The filtrate is allowed to cool to about 40°; the bottom and walls of the beaker are rubbed with a glass rod to induce crystallization in the oily precipitate. After an overnight stand at room temperature, the solid is collected on a suction filter (fritted glass) and recrystallized from 500 ml of hot benzene. The white crystals are collected the next day, washed with a small volume of benzene, and dried in air and under reduced pressure. The following results were obtained on a typical trial: yield 7.0 g (28% theory), mp 74–77°. Found (%): C 38.1, H 5.9, N 5.4, Br 31.5. Calculated (%): C 38.1, H 5.6, N 5.6, Br 31.7. Calc. MW = 252.12.

Bromoacetamidocaproyl-Hydrazide Derivative (IX).[25] Hydrazide groups are acylated as shown in Eq. (7) by a mixed anhydride (VIII) formed from BACA (VII) and isobutyl chloroformate [Eq. (6)]. N-Methylmorpholine removes HCl produced by the reaction. There is no need to isolate the anhydride, and so the reaction mixture is added directly to the hydrazide gel suspension. An example of such a preparation is given below.

Six millimoles of mixed anhydride are prepared: In 30 ml of N,N-dimethylformamide are dissolved 1.60 g (6.35 mmoles) of 6-bromo-acetamidocaproic acid (BACA) and 0.70 ml (6.3 mmoles) of N-methyl-morpholine. The solution is cooled to $-15°$ in an alcohol bath (by adding limited amounts of dry ice to the bath), and 0.80 ml (6.0 mmoles) of isobutyl chloroformate[27] are added under magnetic stirring. Some precipitate of methylmorpholinium chloride forms. The mixture is stirred for 20 minutes at $-15°$ and kept at that temperature until being added to the gel.

About 120-ml bed volume of hydrazide derivative on Bio-Gel P-200

---

[27] Ethyl chloroformate (0.57 ml) could probably be used instead with equally satisfactory results.

$$BrCH_2\overset{O}{\underset{H}{\overset{\|}{C}}}N(CH_2)_5\overset{O}{\overset{\|}{C}}OH \;+\; Cl\overset{O}{\overset{\|}{C}}OCH_2CH(CH_3)_2 \;+\; O{\langle}{\rangle}NCH_3$$

(VII)   Isobutyl chloroformate   *N*-Methylmorpholine

(6)

$$BrCH_2\overset{O}{\underset{H}{\overset{\|}{C}}}N(CH_2)_5\overset{O}{\overset{\|}{C}} - O - \overset{O}{\overset{\|}{C}}OCH_2CH(CH_3)_2 \;+\; O{\langle}{\rangle}\overset{+}{\underset{H}{N}}CH_3 \quad Cl^-$$

(VIII)

+ II

(7)

PA
|
$$\underset{|}{H_2C} \quad O$$
$$HC - \overset{O}{\overset{\|}{C}} - \underset{HH}{NN} - \overset{O}{\overset{\|}{C}}CH_2CH_2CH_2CH_2CH_2\overset{O}{\underset{H}{N}}\overset{\|}{C}CH_2Br$$
|

(IX)

+ 2NH₃

(8)

PA
|
$$\underset{|}{H_2C} \quad O$$
$$HC - \overset{O}{\overset{\|}{C}} - \underset{HH}{NN} - \overset{O}{\overset{\|}{C}}CH_2CH_2CH_2CH_2CH_2\overset{O}{\underset{H}{N}}\overset{\|}{C}CH_2NH_2 \;+\; NH_4Br$$
|

(X)

beads (having a group density of 0.6 mmole/g) is washed and suspended[12] to 200 ml volume with a buffer of the composition, $0.1\,M$
KH_2PO_4, $0.1\,M$ K_2HPO_4. The suspension is stirred in an ice bath until
its temperature is below 3°. A pH electrode is inserted. The mixed anhydride reaction mixture is now added in portions under efficient stirring.
The pH is maintained all along near 6.8 by adding $1\,N$ NaOH. The
mixture is stirred for 30 more minutes in the ice bath while the pH is
kept at 6.7–6.8. A sample of the gel is then washed and tested with
TNBS in borate solution. If the test is negative, the entire preparation
can be washed; if the test is slightly positive, 0.6 ml of acetic anhydride
may be added to acetylate (block) unreacted hydrazide; if it is strongly
positive, more mixed anhydride should be prepared and added as described above. The finished gel is washed with $0.2\,M$ NaCl.

Glycinamidocaproyl-Hydrazide Derivative (X).[25] Terminal bromo-
acetamido groups of (IX) are readily converted to glycinamido groups
of (X) in ammonium chloride–ammonia buffer, pH 10, according to
Eq. (8). The resulting α-amino groups are spaced out from the polymer
backbone with a 12-atom chain. They show a relatively low $pK_a(8.1)$

expected of a glycinamido structure and should exhibit increased reactivity with reagents that hydrolyze easily.

The bromoacetamidocaproyl-hydrazide gel (as prepared above) is resuspended in 0.2 M NaCl to 200 ml. Thirteen grams of NH_4Cl are added and dissolved, and the pH is adjusted to 10.0 with conc. NH_4OH (roughly 45 ml being required). The suspension is stirred at room temperature for 1 hour. The gel is then washed with 0.2 M NaCl until washings are TNBS negative. The gel should then give a positive (orange) TNBS test.

Aminoethylglutaramyl-Hydrazide Derivative (XIII).[22] Ethylenediamine can be coupled to glutarylhydrazide gel (VI) with a water-soluble coupling agent such as, 1-ethyl-3(3-dimethylaminopropyl) carbodiimide (XI), as shown in Eq. (9). The by-product (XII) is also water-soluble

$$
\begin{array}{c}
\text{PA} \\
|\\
\text{H}_2\text{C} \quad\quad \text{O} \quad\quad\quad \text{O} \quad\quad\quad \text{O}\\
|\quad\quad\; \|\quad\quad\quad\; \|\quad\quad\quad\; \|\\
\text{HC}-\text{C}-\text{NN}-\text{CCH}_2\text{CH}_2\text{CH}_2\text{COH} + \text{H}_2\text{NCH}_2\text{CH}_2\overset{+}{\text{N}}\text{H}_3 + \text{C}_2\text{H}_5\text{N}=\text{C}=\overset{+}{\underset{\text{H}}{\text{N}}}\text{(CH}_2\text{)}_3\text{N(CH}_3\text{)}_2 \quad^+\text{Cl}^-\\
|\quad\quad\; \text{HH}\\
\end{array}
$$

(VI)   (XI)

(9)

$$
\begin{array}{c}
\text{PA} \\
|\\
\text{H}_2\text{C} \quad\quad \text{O} \quad\quad\quad \text{O} \quad\quad\quad \text{O}\\
|\quad\quad\; \|\quad\quad\quad\; \|\quad\quad\quad\; \|\\
\text{HC}-\text{C}-\text{NN}-\text{CCH}_2\text{CH}_2\text{CH}_2\text{CNCH}_2\text{CH}_2\overset{+}{\text{N}}\text{H}_3 + \text{C}_2\text{H}_5\overset{}{\underset{\text{H}}{\text{N}}}\overset{\text{O}}{\overset{\|}{\text{C}}}\text{N(CH}_2\text{)}_3\overset{+}{\underset{\text{H}}{\text{N}}}\text{(CH}_3\text{)}_2 \quad ^+\text{Cl}^-\\
|\quad\quad\; \text{HH}\quad\quad\quad\quad\quad\quad \text{H}\\
\end{array}
$$

(XIII)   (XII)

and easily removed. A large excess of ethylenediamine is used in order to minimize cross-linking and maximize the amount of free terminal amino group. The resulting derivative (XIII) can thus be prepared with available chemicals although its amino groups ionize with a pK_a slightly more than 1 unit higher than the α-amino groups of (X).

Derivative (XIII) can be prepared as follows: Fifty milliliters of packed glutarylhydrazide gel (VI) is resuspended to 100 ml with 0.1 M NaCl, and 13.3 g of ethylenediamine dihydrochloride (Eastman Organic Chemicals) is added. The pH is adjusted to 4.7. Over a 15-minute period, at room temperature, 1.25 g of EDC-HCl[28] is added in small portions while the suspension is stirred and held at pH 4.7 with addition of 1.0 N

---

[28] EDC·HCl refers to 1-ethyl-3-(3-dimethylaminopropyl) carbodiimide hydrochloride, MW = 191.71. It may be obtained from Cyclo Chemical, Sigma Chemical Co., Calbiochem, or Pierce Chemical Co.

HCl as required. Stirring is continued and pH 4.7 is maintained for 3 hours longer. The gel is washed with large volumes of 0.2 M NaCl.

Aminoethyldithio and Sulfhydryl Derivatives (XV, XVI, and XVII).[22] Derivatives (XV) and (XVI) provide a means for dissociating subsequently bound ligands from the gel matrix by mild reduction of the disulfide bond. Reduction of (XV) or (XVI) with excess mercaptoethanol or dithiothreitol yields a spaced-out sulfhydryl derivative, as, for example, (XVII) via Eq. (12).

$$
\begin{array}{c}
\text{PA} \\
| \\
\text{H}_2\text{C} \quad \text{O} \\
| \quad \parallel \\
\text{HC}-\text{COH} + \text{H}_2\text{NCH}_2\text{CH}_2\text{S}-\text{SCH}_2\text{CH}_2\overset{+}{\text{N}}\text{H}_3 + \text{XI} \\
| \\
\text{(I)} \qquad \text{(XIV)} \qquad\qquad\qquad \text{(10)}
\end{array}
$$

$$
\begin{array}{c}
\text{PA} \\
| \\
\text{H}_2\text{C} \quad \text{O} \\
| \quad \parallel \\
\text{HC}-\text{C}-\underset{\text{H}}{\text{N}}\text{CH}_2\text{CH}_2\text{S}-\text{SCH}_2\text{CH}_2\overset{+}{\text{N}}\text{H}_3 + \text{XII} \\
| \\
\text{(XV)}
\end{array}
$$

$$
\text{VI} + \text{XIV} + \text{XI} \longrightarrow
\begin{array}{c}
\text{PA} \qquad\qquad\qquad\qquad\qquad\qquad\qquad (11)\\
| \\
\text{H}_2\text{C} \quad \text{O} \qquad\quad \text{O} \qquad\quad \text{O} \\
| \quad \parallel \qquad\quad \parallel \qquad\quad \parallel \\
\text{HC}-\text{C}-\underset{\text{HH}}{\text{NN}}-\text{CCH}_2\text{CH}_2\text{CH}_2\text{C}\underset{\text{H}}{\text{N}}\text{CH}_2\text{CH}_2\text{S}-\text{SCH}_2\text{CH}_2\overset{+}{\text{N}}\text{H}_3 \\
| \\
\text{(XVI)} \qquad\qquad + \text{(XII)}
\end{array}
$$

+ Excess
RSH

(12)

$$
\begin{array}{c}
\text{PA} \\
| \\
\text{H}_2\text{C} \quad \text{O} \qquad\quad \text{O} \qquad\quad \text{O} \\
| \quad \parallel \qquad\quad \parallel \qquad\quad \parallel \\
\text{HC}-\text{C}-\underset{\text{HH}}{\text{NN}}-\text{CCH}_2\text{CH}_2\text{CH}_2\text{C}\underset{\text{H}}{\text{N}}\text{CH}_2\text{CH}_2\text{SH} + \text{RS}-\text{SCH}_2\text{CH}_2\overset{+}{\text{N}}\text{H}_3 \\
| \\
\text{(XVII)}
\end{array}
$$

Cystamine (XIV), present in excess, can be coupled efficiently to gels bearing carboxyl groups by means of water-soluble carbodiimides as shown in Eqs. (10) and (11). A carboxyl-bearing gel, such as (I), (IV), and (VI), is washed in 0.1 M NaCl and allowed to settle; the supernatant fluid is siphoned off. An equal volume of a filtered, aqueous, 1.0 M solution of cystamine dihydrochloride (Aldrich Chemical Co.) is added, and the pH of the suspension is adjusted to 4.7. Under efficient stirring at room temperature, 12.0 mg of EDC-HCl[28] per milliliter of suspension is added in small portions over 10–15 minutes. The suspension is stirred for 4 hours longer. The pH is maintained at 4.6–4.8 during the addition of carbodiimide and throughout the subsequent reaction period. The gel is then washed with 0.2 M NaCl until washings are TNBS negative.

Ligands may be attached to the terminal amino groups then later dissociated with 0.05–0.10 M mercaptoethanol or 0.005–0.02 M dithiothreitol at pH 8–9. In the preparation of sulfhydryl derivatives, (XV) and (XVI) are treated with 0.2 M mercaptoethanol or 0.02 M dithiothreitol in borate buffer (pH 9.0) for 10 minutes at room temperature shortly before use. They are washed just before use with N_2-aerated, pH 6.8, phosphate-buffered saline.

Spaced-Out Hydrazide Derivatives (XX) and (XXII).[22] Extended carboxyl derivatives of polyacrylamide, e.g., (VI), can be coupled with *tert*-butyl carbazate (XVIII) by water-soluble carbodiimide (XI) to give spaced-out hydrazide derivatives blocked with *tert*-butyloxycarbonyl groups [e.g., (XIX); see Eq. (13)]. This approach is less likely to produce unwanted cross-links than could occur if unblocked hydrazine were used instead of the carbazate. Compound (XIX) offers the additional advantage of a stable storage form of the hydrazide gel since during storage it is possible that oxidation processes might convert some acyl hydrazide to diimide which in turn could acylate remaining free hydrazide groups. Resulting internal cross-links would cause loss of gel porosity.

Just prior to the next stage of derivatization, (XIX) is converted to free hydrazide (XX) by standing in 1.0 N HCl for 7 hours at room temperature [Eq. (14)]. Only 0.02 mmole/g of carboxyl group is introduced by acid hydrolysis of the polymer amide during this period. For conversion to reactive azide, the suspension in 1.0 N HCl is then cooled to 0° and one-tenth its volume of cold, 1.0 M sodium nitrite is added to convert the hydrazide to the active azide derivative for coupling to amines or proteins (see next two sections below).

The preparation of derivative (XIX) is carried out as follows: Glutarylhydrazide gel (VI) is thoroughly washed and resuspended in 0.1 M NaCl to 1.5 times its bed volume. One-half the bed volume of a filtered, aqueous, 0.6 M solution of *tert*-butyl carbazate (Aldrich Chemical Co.) is added. The pH is adjusted to 4.7, EDC-HCl[28] (12.0 mg/ml of total volume) is added in portions to the stirred suspension at room temperature. The pH is maintained at 4.6–4.8 during the addition of carbodiimide (10–15 minutes) and during the next 3–4 hours. Stirring is maintained throughout. The gel is then washed with 0.2 M NaCl until washings are TNBS negative.

In a more direct approach, a spaced-out hydrazide, such as (XXII), may be prepared according to Eq. (15) by coupling adipic acid dihydrazide (XXI) to the carboxyl derivative (I): About 25 ml (bed volume) of (I) is suspended to 50 ml in 0.1 M NaCl, and 4.36 g of adipic acid dihydrazide (Eastman Organic Chemicals) is added; the mixture is stirred for 5 minutes. The pH is then adjusted to 4.6, and 750 mg of EDC-HCl[28] are added, in portions, with efficient stirring. The pH is kept at 4.6 at all times by adding 1.0 N HCl as required. Stirring and pH control are continued for the next 3–4 hours. The gel is washed with 0.2 M NaCl until washings are TNBS negative. The product should be kept in storage buffer.

Attachment of Specific Ligands

The final stage of derivatization requires special knowledge of the binding moiety to be separated. Some ingenuity is needed in selecting or preparing a ligand having a suitable functional group for attachment to the support. With polyacrylamide one is essentially constrained to working in aqueous or semiaqueous media. Yet many reactions forming stable bonds can be conducted in the presence of water with adequate efficiency. Some useful reactions have been discussed above in connection with spaced-out functional groups. In general, one can successfully join ligands to the support by one of the following types of linkage: amide (including acyl hydrazide), secondary amine (by alkylation of amino, hydrazide, and hydrazino groups), sulfide (by alkylation of thiols), or azo (through diazonium salts).

The activated species may be on the support and the receptor (nucleophilic) group on the ligand or vice versa. A principal consideration in selecting a method of attachment is to avoid the need for protecting or blocking groups on the ligand. This strategy involves much extra labor. Furthermore, subsequent removal of protecting groups may result in undesirable side reactions with the polyacrylamide structure, such as hydrolysis of its carbamyl groups. It is helpful to keep in mind that the acyl azide group, as in (XXIII), will react selectively with primary aliphatic or aryl amines or with thiols in the presence of carboxyl and various kinds of hydroxyl groups. Examples are given below of polyacrylamide derivatives that have been prepared and used for affinity separations.

Reaction of Ligand Amino Groups with Polyacrylamide Acyl Azide Derivatives. The conversion of gel-bound hydrazide to acyl azide is carried out at 0° as indicated in Eq. (16). The azide is unstable and should be used immediately after it is formed. Coupling reactions should be performed also near 0°.

Rickli and Cuendet[29] prepared a lysine-coupled polyacrylamide P-300 (XXIV) by allowing the azide derivative to react with lysine as shown in Eq. (17). Reaction was expected mainly at the α-amino group because its pK_a is 1.4 units lower than that of the ϵ-amino group. Lysine bound in this manner served as an analog of ϵ-aminocaproic acid, an inhibitor of plasminogen activation to plasmin, as was demonstrated earlier by Deutsch and Mertz[30] with lysine bound to agarose. Lysine–polyacrylamide proved to be an extremely effective affinity adsorbent for the isolation of plasminogen directly from fresh human blood plasma.[29] Elution was carried out with a buffered solution of ϵ-aminocaproic acid. A great improvement in performance was noted by comparison with an agarose–lysine adsorbent. Polyacrylamide, in contrast with agarose, allowed higher levels of substitution with lysine and showed no nonspecific retention of plasma proteins, thus affording high yield and purity of plasminogen. Potential capacity of the adsorbent was found to be quite high so that plasminogen molecules, which have a molecular weight of 90,000, must have effectively penetrated the gel pores.

Truffa-Bachi and Wofsy[6] reacted histamine with the polymer azide

[29] E. E. Rickli and P. A. Cuendet, *Biochim. Biophys. Acta* **250**, 447 (1971).
[30] D. G. Deutsch and E. T. Mertz, *Science* **170**, 1095 (1970).

$$
\begin{array}{c}
\underset{\text{(II)}}{\overset{\displaystyle\underset{|}{\text{PA}}}{\underset{|}{\overset{\displaystyle\underset{|}{\text{H}_2\text{C}}}{\text{HC}}}} - \underset{\text{Nitrous acid}}{\overset{\text{O}}{\underset{\text{H}}{\text{C}}} - \text{NNH}_2 + \text{HNO}_2} \xrightarrow[\text{dil. HCl}]{0^\circ} \underset{\text{(XXIII)}}{\overset{\displaystyle\underset{|}{\text{PA}}}{\underset{|}{\overset{\displaystyle\underset{|}{\text{H}_2\text{C}}}{\text{HC}}}} - \overset{\text{O}}{\text{C}} - \text{N}_3 + 2\text{H}_2\text{O}}
\end{array}
\quad (16)
$$

$$\text{(XXIII)} + \underset{\text{Lysine}}{\text{H}_2\text{N} - \overset{\overset{+}{\text{NH}_3}}{\underset{|}{\underset{\overset{|}{(\text{CH}_2)_4}}{\text{CH}}}}} \underset{\text{O}=\text{C}-\text{O}^-}{} \longrightarrow \underset{\text{(XXIV)}}{\overset{\text{PA}}{\text{HC}} - \overset{\text{O}}{\text{C}} - \underset{\text{H}}{\text{N}} - \text{CH} + \text{HN}_3}
\quad (17)$$

$$\text{(XXIII)} + \underset{\text{Histamine}}{\text{H}_2\text{NCH}_2\text{CH}_2 - \text{C}=\text{CH}} \longrightarrow \underset{\text{(XXV)}}{\text{HC} - \text{C} - \underset{\text{H}}{\text{N}}\text{CH}_2\text{CH}_2 - \text{C}=\text{CH} + \text{HN}_3}
\quad (18)$$

$$\text{(XXIII)} + \underset{\substack{m\text{ - Aminobenzeneboronic} \\ \text{acid}}}{\text{H}_2\text{N} - \bigcirc - \text{B} \overset{\text{OH}}{\underset{\text{OH}}{}}} \longrightarrow \underset{\text{(XXVI)}}{\text{HC} - \text{C} - \underset{\text{H}}{\text{N}} - \bigcirc - \text{B} \overset{\text{OH}}{\underset{\text{OH}}{}} + \text{HN}_3}
\quad (19)$$

according to Eq. (18) to obtain the slightly spaced-out imidazole derivative (**XXV**). A solution of histamine dihydrochloride was added to the azide-forming reaction mixture, and the pH was adjusted to 9.5 with triethylamine. Coupling was allowed to proceed for 1 hour at 0°. Finally, ethanolamine was added to a concentration of 1.7 M to react with any remaining azide groups.

Aryl amine groups of m-aminobenzeneboronic acid were allowed to react with polyacrylamide azide by Eq. (19) as described by Weith et al.,[31] who reported also the preparation of a corresponding cellulose

[31] H. L. Weith, J. L. Wiebers, and P. T. Gilham, *Biochemistry* 9, 4396 (1970).

derivative and its use in chromatographic separations of nucleic acids, sugars, and other polyols through specific complex formation with bound boronic acid groups. The azide derivative (XXIII)[1] was prepared from the hydrazide form of Bio-Gel P-60 (group density 1 mmole/g), washed with water at 0°, and added to a cold, 0.2 M solution of m-aminobenzeneboronic acid at pH 8.5. Coupling proceeded for 24 hours at 0°. Remaining azide groups were removed by adding an equal volume of concentrated ammonium hydroxide to the reaction mixture. Boronic acid groups were bound to the extent of 0.3 mmole/g.[31]

Reaction of Ligand Carboxyl Groups with Aminoethyl Polyacrylamide via Carbodiimide. Erickson and Mathews[32] prepared an affinity adsorbent by coupling N^{10}-formylaminopterin to the aminoethyl derivative of Bio-Gel P-150 (group density of about 1 mmole/g) by means of the carbodiimide EDC-HCl.[28] The reaction was carried out in aqueous dimethylformamide solution at pH 4.7 and room temperature for 18–24 hours. Stable amide bonds were established between glutamate carboxyls of the ligand and amino groups on the support [Eq. (20)]. The resulting

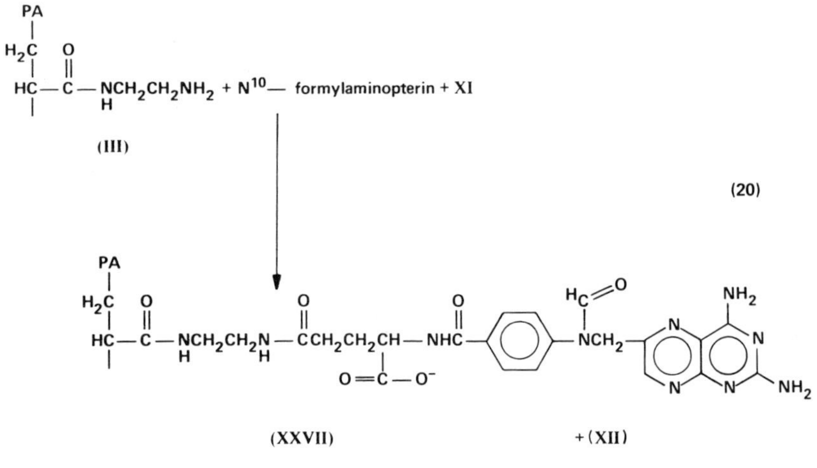

PA
|
H₂C   O
| ‖
HC — C — NCH₂CH₂NH₂ + N¹⁰— formylaminopterin + XI
| H

(III)

(20)

PA
|
H₂C   O
| ‖
HC — C — NCH₂CH₂N — CCH₂CH₂CH — NHC —〈 〉— NCH₂ — [pteridine]
| H   H
        O=C — O⁻

(XXVII)   +(XII)

derivative (XXVII) was used to purify dihydrofolate reductase (specified by T4 bacteriophage). A single adsorption to (XXVII) followed by specific elution with soluble ligand (at 0.2 mM) in a column gave a 3000-fold purification. The purified enzyme (MW 29,000) appeared to be a homogeneous protein. The adsorbent was reused several times without change in its performance.

Dinitro- and Trinitrophenylation of Amino and Hydrazide Gels. Polyacrylamide hydrazide gels react rapidly with 2,4,6-trinitrobenzenesulfonate

[32] J. S. Erickson and C. K. Mathews, *Biochem. Biophys. Res. Commun.* 43, 1164 (1971).

(TNBS) at pH 9 [Eq. (21)] to form an N'-trinitrophenylhydrazide (XXVIII).[1] The dinitrophenylhydrazide has been prepared with the analogous reagent 2,4-dinitrobenzenesulfonate (DNBS).[6] The preceding two derivatives have also been formed with picryl chloride and dinitrofluorobenzene, respectively.[10] Aminoethyl polyacrylamide reacts in borate buffer (pH 9.3, 20 hours) with DNBS to form the dinitrophenylaminoethyl derivative (XXIX) by Eq. (22).[1] Analogously, the trinitrophenylaminoethyl derivative has been prepared with TNBS.[6]

Coupling of Diazonium Salts to Polyacrylamide Derivatives (see also this Volume [6]). Diazonium salts are prepared from aryl (aromatic) amine derivatives of ligands with cold, aqueous nitrous acid [Eq. (23)]. Truffa-Bachi and Wofsy[6] coupled diazotized *p*-aminophenylglycosides to the histamine derivative (XXV) of Bio-Gel P-6 beads to form a stable, azo-linked derivative (XXX). Reaction was carried out in borate buffer, pH 9.0, at 4° for several hours [Eq. (24)]. The modified beads were used for affinity chromatography of cells bearing receptors specific for azophenylglycoside haptens.[6,7]

Phenolic structures such as the *p*-hydroxyphenethyl[1] or glycylglycyltyrosine[19] derivatives also may be coupled with diazonium salts under the above conditions. Structure (XXXI) is an example of a ligand attached to the latter through an azo link.[19]

Alkylation of a Ligand by a Bromoacetyl Polyacrylamide. Aminoethyl-Bio-Gel P-300 (III) was bromoacetylated by Cuatrecasas[19] with the active ester, *O*-bromoacetyl-*N*-hydroxysuccinimide, to give structure (XXXII). The above reagent can be prepared in dioxane by reacting together *N*-hydroxysuccinimide, bromoacetic acid, and dicyclohexyl carbo-

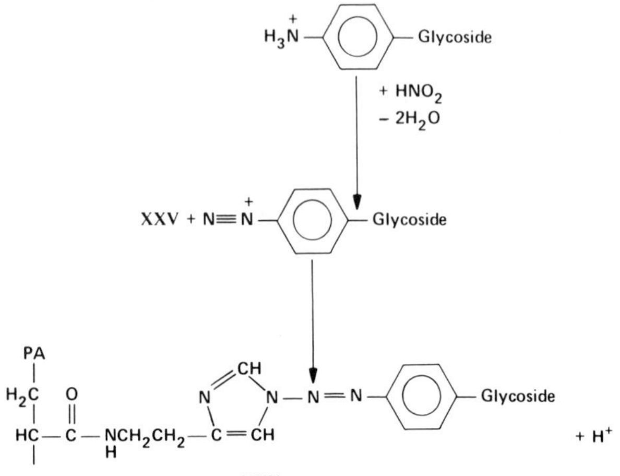

$$H_3\overset{+}{N}-\bigcirc-\text{Glycoside}$$

$$+ \text{HNO}_2 \qquad (23)$$
$$- 2\text{H}_2\text{O}$$

$$\text{XXV} + \overset{+}{N}\equiv N-\bigcirc-\text{Glycoside} \qquad (24)$$

(XXX)

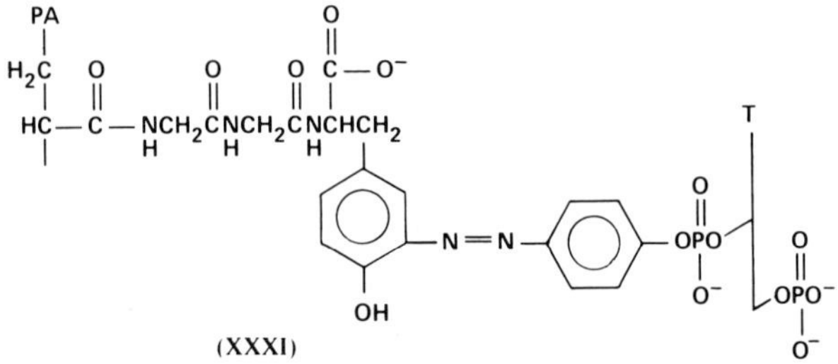

(XXXI)

diimide in nearly equimolar proportions. Dicyclohexylurea is removed by filtration, and the filtrate is added directly to the suspension of aminoethyl gel in pH 7.5 phosphate buffer.[19] Alkylation of an aryl amine-bearing ligand is shown in Eq. (25). The resulting structure (XXXIII) (as well as XXXI), can be used for the purification of staphylococcal nuclease by affinity chromatography.[19] Bromoacetyl derivatives of polyacrylamide, such as (XXXII) and (IX), will react readily with sulfhydryl groups of ligands at pH 8–9.5 and, under appropriate conditions, also with phenolic and imidazolic compounds.

Special Uses of the Hydrazide Derivative. Vasilenko *et al.*[33] were able to bind tRNA through its dialdehyde derivative to polyacrylamide hydrazide (II). The interaction proved stable over a wide range of temperature and pH and served as the basis for the purification of tRNA. The authors described the complex as a hydrazone, but its stability suggests rather the carbinol hydrazide structure (XXXIV). Hydrazones could

[33] S. K. Vasilenko, L. V. Obukhova, and V. I. Yamkovy, *Biokhimiya* 36, 1288 (1971).

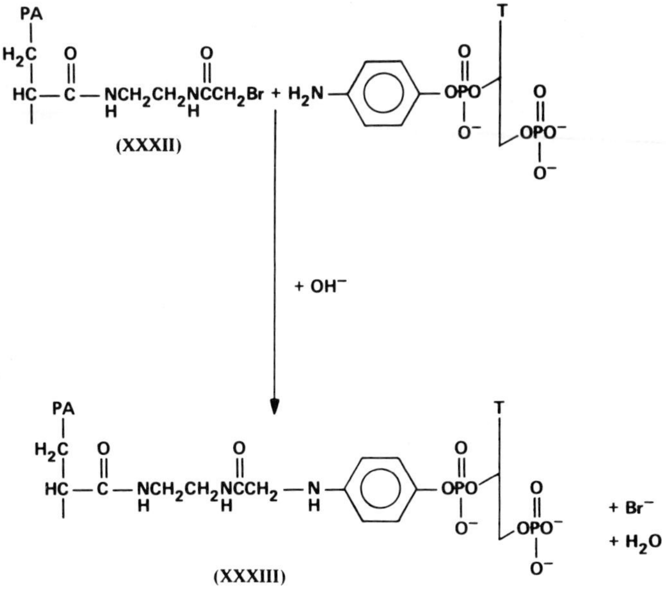

(XXXII)

+ OH⁻   (25)

(XXXIII)   + Br⁻   + H₂O

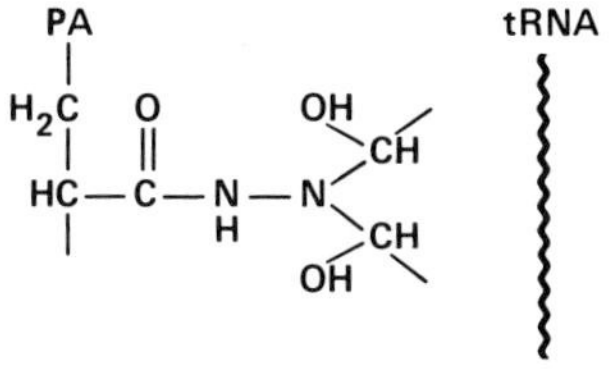

(XXXIV)

be formed between ketones or aldehydes and hydrazino derivatives of polyacrylamide. The latter can be prepared from bromoacetyl derivatives, (XXXII) or (IX), by reaction with hydrazine in aqueous solution.

Attachment of Proteins

In some affinity separations proteins serve as the immobilized ligand. Due consideration should be given to the combined size of the interacting components before selecting a highly porous polyacrylamide support. If the combined molecular weight of the components exceeds several hundred thousand and the capacity demand is not high, then beads with small-sized pores and small diameters may be tried with the view of using only their outer surfaces.

A variety of methods are possible for covalently linking proteins to polyacrylamide. Considerable control may be attained over the type of amino acid side chain involved and the number of side chains ligated.

Specific activity of the bound protein can be adversely affected by establishing too many bonds to each protein molecule. Therefore, functional group densities should not be higher than necessary. Furthermore, high group densities increase the opportunity for undesired cross-linkages to occur within the gel structure since, in protein attachment, the activated group is usually placed on the support. This problem may be important in the formation of acyl azide from hydrazide derivatives.[19-21] Accordingly, the use of spaced-out functional groups could be advantageous by increasing the efficiency of attachment and thereby decreasing the required group density. Spacers may also impart greater rotational freedom to the bound protein molecules.

A number of active functional groups have been employed for binding proteins to polyacrylamide. Several of them are quite specific for the primary amino groups of lysyl and N-terminal residues, such as acyl azide[1,34] and isothiocyanato groups.[35] The diazonium salts[1,34] react with tyrosyl, histidyl, lysyl, and possibly other residues. Some activated or activatable polyacrylamide derivatives are available commercially.[3,4] Polyacrylamide Enzacryls™ are available as hydrazide, aryl amine, polyacetol, polythiol, and polythiolactone derivatives, but they come as nonbeaded fragments and may have a rather high content of carboxylate groups. The latter can increase nonspecific retention of proteins by ion exchange adsorption. Bio-Gel beads are sold in several forms as aminoethyl and hydrazide derivatives.[4] The bromoacetyl derivatives, (IX) and (XXXII), should react slowly with amino groups of proteins at pH values greater than 8.5 to form stable, secondary amine links. The reaction rate with free sulfhydryl groups should be somewhat higher.

Coupling by Reaction with the Acyl Azide Derivative. A useful approach to covalent binding of proteins is shown by Eq. (26).[1] The hydrazide derivative (II) or a spaced-out hydrazide form (XX or XXII) is suspended in 1.0 N HCl. The suspension is cooled to 0°, and one-tenth its volume of cold, 1.0 M sodium nitrite is added with stirring. After 2–3 minutes, the gel is washed briefly on a funnel with 0.2 M NaCl (cooled to 0°). The drained gel is immediately added to a cold, buffered solution of protein. Borate buffers in the pH range 8–9 may be used. The suspension is stirred in an ice bath overnight. Unreacted azide groups should be converted back to carbamyl groups by adding one-fifth volume of 3 M NH₄Cl, 1 M NH₄OH, and stirring several hours before final washing.

Coupling with Water-Soluble Carbodiimide. Carboxyl derivatives (I),

[34] S. A. Barker, P. T. Somers, R. Epton, and J. V. McLaren, *Carbohyd. Res.* 14, 287 (1970); S. A. Barker and R. Epton, *Process Biochem.* 14 (1970).

[35] R. Epton and T. H. Thomas, *in* "An Introduction to Water-Insolubilized Enzymes." Koch-Light Laboratories Ltd., Colnbrook, Bucks, England, 1971.

$$
\begin{array}{c}
\text{PA} \\
|\\
\text{H}_2\text{C} \quad \text{O} \\
| \quad || \\
\text{HC} - \text{C} - \text{N}_3 + \text{H}_2\text{N} - \text{Protein} \longrightarrow \\
|
\end{array}
\qquad
\begin{array}{c}
\text{PA} \\
|\\
\text{H}_2\text{C} \quad \text{O} \\
| \quad || \\
\text{HC} - \text{C} - \text{N} - \text{Protein} + \text{HN}_3 \\
| \qquad\quad \text{H}
\end{array}
\qquad (26)
$$

(XXIII)

or HNO$_2$ + (XX) or (XXII)

$$
\begin{array}{c}
\text{PA} \\
|\\
\text{H}_2\text{C} \quad \text{O} \\
| \quad || \\
\text{HC} - \text{C OH} + \text{H}_2\text{N} - \text{Protein} +(\text{XI}) \longrightarrow \\
|
\end{array}
\qquad
\begin{array}{c}
\text{PA} \\
|\\
\text{H}_2\text{C} \quad \text{O} \\
| \quad || \\
\text{HC} - \text{C} - \text{N} - \text{Protein} +(\text{XII}) \\
| \qquad\quad \text{H}
\end{array}
\qquad (27)
$$

(I)

or (IV or VI)

(IV), or (VI) can be coupled to protein by means of a water-soluble carbodiimide [Eq. (27)]. Although several investigators[36,37] have chosen this approach, in principle, amino derivatives of polyacrylamide should also bind protein under the same conditions through activation of carboxyl groups on the latter. In either case there exists the problem of cross-linkage between protein molecules themselves which can decrease their desired activity. The cross-linking may be lessened by adding carbodiimide to the carboxylated polymer prior to adding protein as was done in the following procedure by Mezzasoma and Turano.[37]

Bio-Gel P-150 (carboxylate (I), 50 mg dry weight) was suspended in 3 ml of water and the pH was adjusted to 4.7–5.4 (with dilute HCl). Fifteen to 30 mg of EDC-HCl[28] was added. After 3 minutes 20 mg of protein were added, and the mixture was stirred at room temperature for 1–2 hours. All during the above procedure the pH was kept in the range of 4.7–5.4. Washing was performed by centrifugation. Chymotrypsin (MW 24,000) was bound more efficiently than glutamic dehydrogenase (MW 2,000,000) because of its ability to penetrate the gel pores.

Diazonium Salt and Isothiocyanato Forms from Aryl Amine Derivatives. Inman and Dintzis[1] described the preparation of an aryl amine (*p*-aminobenzamidoethyl) derivative of polyacrylamide (XXXVI) starting with an aminoethyl gel as shown in Eq. (28). An improved procedure for preparing (XXXVI) is given below.[22]

Fifty milliliters of bed volume of Bio-Gel P-200 aminoethyl derivative (group density 0.5 mmoles/g; concentration 17 μmoles/ml of bed) is

[36] J. D. Stobo, A. S. Rosenthal, and W. E. Paul, *J. Immunol.* **108**, 1 (1972).
[37] I. Mezzasoma and C. Turano, *Boll. Soc. Ital. Biol. Sper.* **47**, 407 (1971).

$$\text{(III)} + N_3\overset{O}{\overset{\|}{C}}-\underset{}{\boxed{}}-NO_2 \quad \xrightarrow{+ \text{Et}_3N \;|\; + Na_2S_2O_4} \; | \; \longrightarrow \; \underset{\text{(XXXVI)}}{HC-\overset{\overset{PA}{|}}{\underset{|}{\overset{H_2C}{C}}}-\overset{O}{\overset{\|}{C}}-NCH_2CH_2NC-\boxed{}-NH_2} \qquad (28)$$

$$\text{(XXXV)}$$

$$\text{(XXXVI)} + HNO_2 \longrightarrow \underset{\text{(XXXVII)}}{HC-\overset{\overset{PA}{|}}{\underset{|}{\overset{H_2C}{C}}}-\overset{O}{\overset{\|}{C}}-\underset{H}{N}CH_2CH_2\underset{H}{N}\overset{O}{\overset{\|}{C}}-\boxed{}-\overset{+}{N}\equiv N} \qquad (29)$$

$$+ \text{ Protein}$$

$$\underset{\text{(XXXVIII)}}{HC-\overset{\overset{PA}{|}}{\underset{|}{\overset{H_2C}{C}}}-\overset{O}{\overset{\|}{C}}-\underset{H}{N}CH_2CH_2\underset{H}{N}\overset{O}{\overset{\|}{C}}-\boxed{}-N=N-\text{Protein}} \qquad (30)$$

suspended in 0.2 M NaCl to a volume of 100 ml. A solution of 1.15 g (6 mmoles) of p-nitrobenzoyl azide (Eastman Organic Chemicals) in 100 ml of tetrahydrofuran (THF) is added with stirring followed by 1.0 ml of triethylamine (Et_3N). Stirring is continued at room temperature for 45 minutes. Throughout this period an additional 1.0 ml of Et_3N is added in portions. The gel is then washed with THF–saline 1:1 v/v on a funnel, batch-washed in 100 ml of formamide, and funnel-washed with saline (0.2 M NaCl). The TNBS test on the gel should be negative or show light yellow. The gel is resuspended in saline to 90 ml and 1.0 ml of acetic anhydride is added dropwise with efficient stirring. The pH is kept between 6 and 7 by addition of 2 N NaOH as required. Acetic anhydride blocks any unreacted aliphatic amino groups which may later survive cold nitrous acid treatment and then cross-link with diazonium groups during the coupling reaction. At the same time, acetic anhydride provides a source of acetate buffer for the next step. After 30 minutes the pH is adjusted to 6.5. The suspension is warmed to 50° in a water bath and 4.5 g of sodium hydrosulfite (sodium dithionite, $Na_2S_2O_4$) is added. The mixture is stirred at 50–55° for 40 minutes, then the gel, now as (XXXVI), is washed thoroughly in turn with 0.2 M, 2.0 M, and 0.2 M NaCl. A TNBS test on the gel should show a bright orange color after 5 minutes of reaction in the spotplate.

The diazonium form, (XXXVII) via Eq. (29), should be prepared just prior to use as it has limited stability. The conditions may be the same as used for converting hydrazide to azide (see above) except that about 15–20 minutes in cold, nitrous acid, rather than 2–3 minutes, should be allowed. The gel is washed[38] and coupled with protein in the same manner as with the acyl azide. Borate buffers, ca. pH 8.5, are preferred. Coupling reactions proceed from 2 to 24 hours at 0° to 4°.

Derivative (XXXVI) can be converted to the active isothiocyanato derivative (XXXIX), with thiophosgene [Eq. (31)] and coupled to

amino groups of proteins [Eq. (32)] resulting in thiourea links as shown by (XL).

Conversion with thiophosgene can be carried out as follows[22]: Six milliliters of bed volume of the p-aminobenzamidoethyl gel (XXXVI) are washed and suspended in 0.50 M K_2HPO_4 up to 12 ml. A solution of 0.2 ml of thiophosgene in 6 ml of tetrahydrofuran (THF) is added and the mixture is stirred vigorously at room temperature for 60 minutes.[39] The vessel is covered to retard evaporation of solvent. The mixture is

[38] Washing with cold saline should be thorough enough to obviate the need for sulfamic acid or sulfamate treatment, which tends to fill the gel pores with nitrogen gas. Gas blocking (floating beads) occurs sometimes when dealing with hydrazide, azide, or diazonium forms of the gel. If this occurs the suspension is placed in a beaker (less than half filled) in a desiccator and vacuum-treated to produce gentle boiling for a minute or two.

[39] Thiophosgene ($CSCl_2$) is volatile, and its vapors are extremely poisonous. All transfers, reactions, and washings must be conducted in a well-ventilated hood. Residues contaminated with $CSCl_2$ are treated with concentrated ammonium hydroxide, allowed to stand several hours, and washed away with plenty of water.

washed on a funnel carefully with 200 ml of THF–0.1 M NaCl 1 : 1 v/v and then with 0.1 M NaCl. The activated gel is stirred with protein or ligand dissolved in pH 9 borate buffer. Coupling is allowed to proceed 2–6 hours at room temperature or overnight at 4°.

Coupling of Protein to Polyacrylamide with Glutaraldehyde.[40] Underivatized polyacrylamide beads can be activated by exposure to buffered, aqueous solutions of glutaraldehyde. After being washed, the beads retain the ability to covalently bind proteins for many days. A considerable fraction of the original biological activity (enzymological, immunochemical) remains after binding. The coupled products appear to be stable for long periods of time when stored at pH 7.4 and 4°. The nature of the chemical reaction between glutaraldehyde and polyacrylamide beads is not clear. Variables involved in activation and coupling were carefully studied by Weston and Avrameas.[40] The following procedure is planned according to the optimum conditions they found: Bio-Gel P-300 beads are activated by incubation at 37° for 16–41 hours in a 5% solution of glutaraldehyde (one-fifth of the volume as 25% v/v) in 0.1 M potassium phosphate buffer, pH 6.9. The beads are washed by centrifugation with pH 6.9 and 7.7 phosphate buffers. Coupling with proteins is carried out in 0.1 M potassium phosphate, pH 7.7, for 18 hours at 4°.

Analytical Procedures

Functional group densities[13] of primary derivatives can be estimated fairly reliably from the guiding rules given for their preparation. However, it may be of interest to determine group densities of primary, secondary, or final derivatives following certain variations of procedure. Precise methods for determination of carboxyl, hydrazide, and amino group densities ("specific capacities") based on acid-base titration have been given in detail by Inman and Dintzis.[1] Simpler procedures that are adequate for most purposes are described below.

Dry Weight Determination. For many experiments the functional group content of a sample need only be related to its settled or packed bed volume. However, better reproducibility is possible when functional groups are related to dry weight. A dry weight is determined in the following manner: A titrated sample is poured into a tared, 30- or 50-ml Gooch crucible (with coarse sintered disk). It is washed by suction with 0.2 N HCl (for carboxyl) or 0.2 M sodium carbonate (for amino groups) to render the groups uncharged.[14] The sample is next washed with distilled water to remove electrolyte and finally with methanol to shrink the gel and remove water. The material is allowed to dry thoroughly

[40] P. D. Weston and S. Avrameas, *Biochem. Biophys. Res. Commun.* **45**, 1574 (1971); T. Ternynck and S. Avrameas, *FEBS Lett.* **23**, 24 (1972).

in air and then under vacuum over anhydrous calcium chloride. The crucible is weighed immediately after being removed from the desiccator. Vacuum-dried polyacrylamide will absorb moisture from room air and gain weight initially at about 2% per hour.

Determination of Carboxyl Groups. A sample of gel is washed and suspended in 0.2 M NaCl (to not more than twice its bed volume) and the pH is adjusted to 6.5–7.5 with dilute NaOH. A pH electrode is inserted after calibration against 50–100 ml of 0.0025 N HCl, 0.2 M NaCl (as pH = 2.60). Standard HCl (v, ml) of normality N is added with magnetic stirring until the pH reads (steadily) 2.60. The volume of suspension (V, ml) is measured. The carboxyl group content of the sample is measured by the milliequivalents of bound hydrogen ion = milliequivalents of HCl added (vN) minus milliequivalents of free H$^+$ (0.0025V). The carboxyl group density is obtained after dividing by the dry weight of the sample.

Determination of Hydrazide Groups. The pK_a of acyl hydrazide groups is close to 2.6. Direct titration is therefore very difficult on account of the high free hydrogen ion correction. The groups can be converted quantitatively to the succinylhydrazide form (IV) with excess succinic anhydride (0.6 g/25 ml of suspension) at pH 4.0 (otherwise, follow procedure for glutarylhydrazide derivative). The gel sample is then washed thoroughly with 0.2 M NaCl, and the carboxyl content is determined as described above. The hydrazide group density, D, obtained with the subsequently determined dry weight, relates to the succinylhydrazide derivative with its incremental mass of succinyl groups. The group density in terms of millimoles of hydrazide, if sample is structure (II), per gram of original polyacrylamide is $D' = D/(1 - 0.115D)$. The density of carboxyl groups produced during reaction of the gel with hydrazine can be estimated as indicated in the last paragraph under *Preparation of Hydrazide Derivatives* and subtracted from D'.

Determination of Aliphatic Amino Groups.[41] The gel sample is washed and resuspended in 0.2 M KCl (to not more than twice its bed volume). The pH is adjusted to 11.00 with dilute KOH after calibrating the electrode with standard pH 10 buffer. Standard HCl (v, ml) of normality N is added to bring the pH to about 7 (or pH 6 for α-amino groups). The volume of suspension (V, ml) is measured. The amino group content is measured by the milliequivalents of bound hydrogen ion = milliequivalents of HCl added (vN) minus milliequivalents of free OH$^-$ titrated

---

[41] Amino groups cannot be titrated by the method given here if they are connected to the support through a diacylhydrazide link, as in derivatives (X), (XIII), and (XVI), because of the overlapping ionization in the link structure, PA-CO-NHNH-COR' = PA-CO-NHN$^-$-COR' + H$^+$. The pK_a for this system is about 11.0.

$(0.001(V - v))$. The group density in terms of the mass of original polyacrylamide, D', is obtained from the density with respect to the dry weight of derivative by means of the expression, $D' = D/(1 - D \cdot \Delta r)$. The quantity Δr must be calculated for the particular derivative and is equal to 0.001 times the molecular weight of all atoms in a residue of the derivative that occur in addition to those in the original carbamyl side-chain. Thus, for the aminoethyl derivative (III), $\Delta r = 0.043$.

Concluding Remarks

A principal virtue of polyacrylamide supports in affinity chromatography is low nonspecific adsorption of proteins.[42] In cases where efficient, specific elution is not feasible, this fact becomes of prime importance for achieving a high separation factor. In order to maintain this favorable property, one should avoid loading the support unnecessarily with hydrophobic structures or charged groups. The most likely source of the latter would be carboxylate groups resulting from hydrolysis of the polyacrylamide itself. The derivatization procedures have been designed to keep hydrolysis at a low level. Carboxylate groups can be decreased by treating a gel, such as (X) and presumably other amino forms, with water-soluble carbodiimide (XI) in the presence of $1 M$ NH_4Cl at pH 5.4.[22] A similar procedure has been used for this purpose in the case of Sephadex beads and cellulose membranes.[43]

Spacers may be lengthened by combinations of methods presented above. Carboxyl groups can be spaced out farther by glutarylation of the aminoethyl rather than the hydrazide derivative. The reaction with glutaric anhydride, in this instance, is most efficiently accomplished at pH 9 rather than 4. Cases may arise where specific binding is too strong, and effective elution becomes impossible. One way to solve such a problem may be to shorten or eliminate the spacer. Finally, it might be pointed out that small changes in the structure or spacing of a ligand can have large effects on binding and separation properties, and that it may prove worthwhile to experiment with several forms of the bound ligand.

[42] Low nonspecific adsorption of protein to Bio-Gel beads was demonstrated by sensitive mixed agglutination reaction with sensitized red cells by C. J. Sanderson [*Immunology* 21, 719 (1971)].
[43] H.-C. Chen, L. C. Craig, and E. Stoner, *Biochemistry* 11, 3559 (1972).

[4] Porous Glass for Affinity Chromatography Applications

By H. H. WEETALL and A. M. FILBERT

This report deals with "methods" of attaching ligands to inorganic carriers for affinity chromatography.

An ideal carrier for the affinity chromatography of enzymes should possess a number of desirable characteristics. It should exist as a porous network to allow rapid movement of large macromolecules throughout its entire structure and be uniform in size and shape to exhibit good flow properties. The surface area of the carrier would be necessarily large to allow attainment of a high effective concentration of coupled inhibitor, and the support surface should exhibit little specific attraction for proteins. The material should be mechanically and chemically stable to various conditions of pH, ionic strength, temperature, and the presence of denaturants. The availability of controlled-pore glass provides a material that fulfills many of the above-mentioned "ideal, support criteria," and presents a new and unique carrier for use in affinity chromatography applications.

Preparation of Controlled-Pore Diameter Glasses

Certain sodium borosilicate glass compositions exist (see Fig. 1), which, after heat treatment, can be leached to form a porous glass framework.[1] During heat treatment the base, glass separates into two intermingled and continuous glassy phases. One phase, rich in boric oxide, is soluble in acids; the other phase is high in silica and is stable toward acid solutions. The boric acid-rich phase may be leached out of the glass, leaving a porous structure of very high silica content. Porous diameters for these glasses are in the range of 30 to 60 Å, and the pore volume is approximately 28% of the total sample volume.

Larger-pore glasses can be prepared from the same sodium borosilicate glass compositions. After the heat treatment and acid leaching steps, a mild caustic treatment enlarges pore diameters by removing siliceous residue from pore interiors.[2] By carefully controlling the various physical and chemical treatment parameters, glasses are produced which exhibit extremely narrow pore size distributions, as shown in Fig. 2. The con-

---

[1] M. E. Nordberg, *J. Amer. Ceram. Soc.* **27**, 299 (1944).
[2] H. P. Hood and M. E. Nordberg (1940). U.S. Patent No. 2,221,709, "Borosilicate Glass."

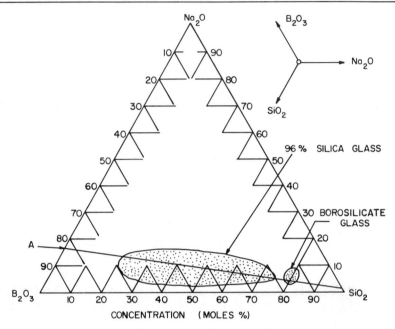

FIG. 1. Phase diagram of sodium borosilicate system.

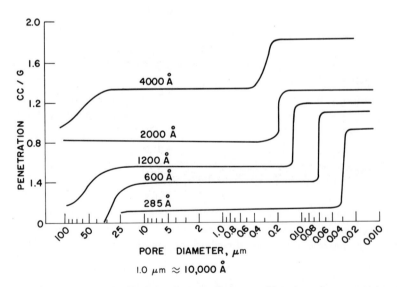

FIG. 2. Pore size distribution of controlled-pore diameter glass supports. All values were determined by the mercury intrusion method.

trolled-pore glass, thus produced, is a 96% silica glass, with 3–5% B_2O_3 content, and trace quantities of several metal oxides.

Surface Area of Controlled-Pore Glass

The surface area of controlled-pore glasses is a function of pore diameter and pore volume. Surface area increases with increased pore volume and decreases with increasing pore diameter, as is shown by the data in the table.

Surface Properties

The surface properties of controlled-pore glass are greatly affected by the thermal history of the glass, heat treatment playing a significant role in determining the ratio of adjacent to single hydroxyl groups which exist on the glass surface. Infrared spectra of the glass change as a function of temperature. On a wet silica surface three bands can be identified as 3747, 3660, 3450 cm^{-1}, and 1650 cm^{-1}. The bands at 3450 and 1650 cm^{-1} are typical of the vibrations associated with molecular water. When temperature is raised to about 150°, these two bands disappear, but the bands at 3747 and 3660 cm^{-1} become more evident. As the temperature is raised further, the band at 3660 cm^{-1} slowly disappears. At 500° it is observed only as a tail on the low frequency side of the 3747 cm^{-1} band. At 800°, only the band at 3747 cm^{-1} remains. These bands are attributed to the vibrations of, respectively, surface hydroxyl groups spaced sufficiently far apart that they do not interact with each other, surface hydroxyl groups so close together they are hydrogen-bonded to each other, and molecular water physically adsorbed on the surface of the silica (Fig. 3).

The rehydration of a silica surface depends on its previous thermal

CONTROLLED-PORE GLASS PHYSICAL PROPERTIES

Pore diameter (Å)	Pore volume SA (M²/g) at PV = 0.70	cc/g 1.00
75	249	356
125	149	214
175	107	153
240	78	111
370	50	72
700	27	38
1250	15	21
2000	9	13

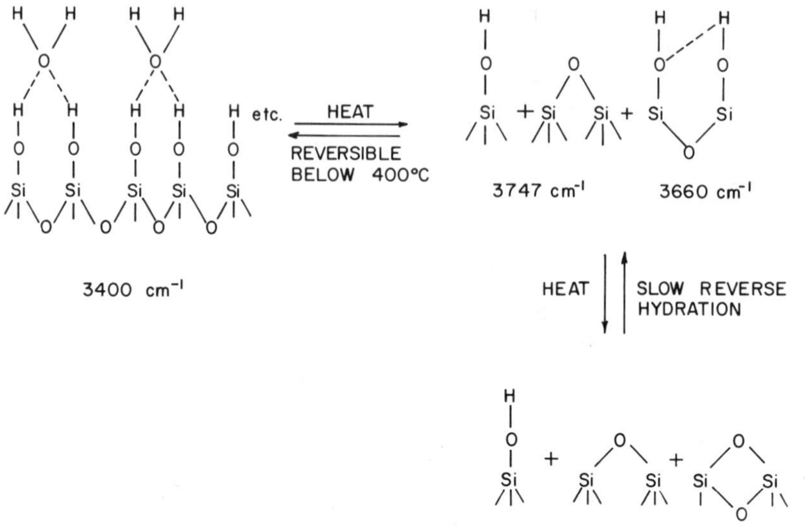

Fig. 3. Dehydration of a silica surface.

history. Up to about 400°, the dehydroxylation of the surface is reversible, but above this temperature the removal of the adjacent hydroxyl groups from the surface causes the silica to become hydrophobic. Indeed, after extensive heating at 850°, a controlled-pore glass surface will not readily adsorb water unless the hydroxylated surface is reconstituted by a drastic treatment such as boiling in concentrated nitric acid. In this case, the water adsorption is associated with the hydrogen-bonded hydroxyl groups and occurs on these groups in preference to the freely vibrating group. In fact, removal of the freely vibrating group from the surface by silanization techniques has little effect on the water adsorption that occurs at low partial pressures.

In addition to the surface silanols, the presence of boron in the glass results in the formation of surface Lewis acid sites as shown in Fig. 4. Although B_2O_3 analysis for the bulk glass is only in the order of 3% with a B:Si ratio of 1:20, infrared studies indicate a much higher surface concentration of boron. Thus, the heat treatment and leaching procedures not only increase surface area of the glass, but also change its surface composition. One now finds a surface boron concentration of 2.4×10^{20} boron atoms per gram of glass and 8 surface silanol groups per square millimicron. The resulting porous glass, with its surface hydroxyls and Lewis acid sites, exhibits a slight negative charge in aqueous media, and this in large part explains the strong adsorption of basic materials.

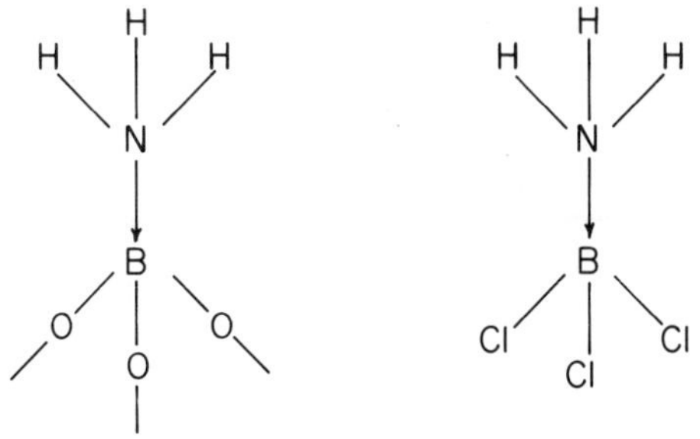

LEWIS ACID SITE   LEWIS ACID - BASE
     SURFACE   COMPLEX

FIG. 4. Concept of Lewis acid site on the surface of glass developed by analogy to inorganic Lewis acid halides.

Durability of Controlled-Pore Glass

The degree of solubility or the corrosion rate of controlled-pore glass (CPG) is a function of temperature, time, solution content, volume, pH, glass composition and surface area. Since CPG is essentially pure silica, it generally exhibits a high degree of durability, but its high surface area contributes to a higher than normal solubility.

Generally, one can expect solubility to increase by a factor of 1.5 for each 10° increase in temperature. Solubility of the glass is extremely pH dependent and is most stable under strong acid conditions. Controlled-pore glass which has been coupled to hydrophilic silanes shows improved durability, and its solubility rate in aqueous systems is decreased substantially. These derivatives (CPG silanes) are excellent starting points for use of the glass in affinity chromatography applications.

Preparation of the Silanized Carrier

The inorganic carrier is treated with an organosilane containing an organic functional group at one end and a silylalkoxy group at the other. Coupling of the silane to the carrier is through the surface silanol or oxide groups to the silylalkoxy groups.[3] In the case of the trialkoxy silanes, polymerization most likely occurs between adjacent silanes. The resulting product is an inorganic carrier having available organic functional groups. The reaction sequence may be schematically represented as in Fig. 5 with a triethoxysilane.

Typical organosilanes commercially available include epoxy-, vinyl-, sulfhydryl-, alkylamine-, alkylchloro-, and phenylsilanes.

Silanes can be attached to inorganic carriers either in organic sol-

[3] W. Haller, *Nature* (*London*) **206**, 693 (1965).

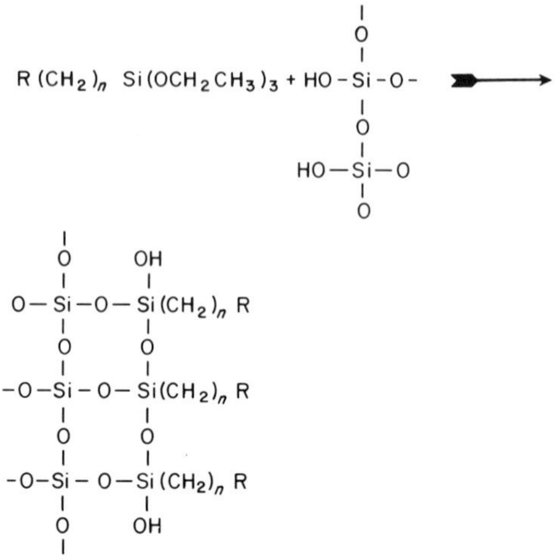

$$R(CH_2)_n \; Si(OCH_2CH_3)_3 + HO-\overset{\displaystyle \overset{|}{O}}{\underset{\displaystyle \underset{|}{O}}{Si}}-O-$$

$$HO-\overset{|}{\underset{|}{Si}}-O$$

$$\begin{array}{c} \overset{|}{O} \quad\; OH \\ | \qquad | \\ O-Si-O-Si(CH_2)_n\,R \\ | \qquad\; | \\ O \qquad O \\ | \qquad\; | \\ -O-Si-O-Si(CH_2)_n\,R \\ | \qquad\; | \\ O \qquad O \\ | \qquad\; | \\ -O-Si-O-Si(CH_2)_n\,R \\ | \qquad\; | \\ O \qquad OH \\ | \end{array}$$

FIG. 5. Silanization of the surface of porous glass. R represents an organic functional group.

vents[4-7] or aqueously.[8] In the case of some organosilanes, organic solvents must be used for reasons of solubility.

Organic Silanization of an Inorganic Surface

The glass or inorganic carrier is first cleaned by boiling in 5% nitric acid solution for 45 minutes and then washed and dried 24 hours at 115°.

To 1 g of dry, clean carrier is added 75 ml of 10% γ-aminopropyl-triethoxysilane (Union Carbide, A-1100) dissolved in toluene. The preparation is refluxed for 12–24 hours, washed thoroughly with toluene, followed by acetone, and air-dried. The resulting product is the alkylamine carrier.

Aqueous Silanization of an Inorganic Carrier

The aqueous procedure, although it does not add as great a quantity of amine groups to the carrier, appears to produce a dynamically more stable product when exposed to continuous use over long periods of time.

The carrier is first cleaned as described for the organic procedure but is not dried. The washed carrier is placed in a 10% solution of

[4] M. L. Hair and I. D. Chapman, *J. Amer. Ceram. Soc.* 49, 651 (1966).
[5] H. H. Weetall, *Nature (London)* 223, 959 (1969).
[6] H. H. Weetall, *Science* 166, 615 (1969).
[7] H. H. Weetall and L. S. Hersh, *Biochim. Biophys. Acta* 185, 464 (1969).
[8] H. H. Weetall and N. B. Havewala, *Biotech. Bioeng. Symp.* 3, 241 (1972).

γ-aminopropyltriethoxysilane dissolved in distilled water and the pH is immediately adjusted with 6 N HCl to pH 3.45. The suspension is placed in a water bath for 2.75 hours at 75°, removed and washed on a Büchner funnel with an equal volume of water, i.e., the same volume as used for dilution of the silane. The product is dried overnight at 100°. The resulting material has an available alkylamine.

Methods of Activating the Silanized Inorganic Carrier

Most of our studies have been carried out using an alkylaminosilane derivatized inorganic carrier. This derivative is prepared using commercially available γ-aminopropyltriethoxysilane and aqueous silanization.[8]

Preparation of an Arylamine

The arylamine is prepared from the alkylaminosilane carrier by reaction with p-nitrobenzoylchloride, followed by reduction with sodium dithionite. The reaction is represented schematically as shown in Fig. 6.

To prepare the arylamine, the alkylamine derivative is refluxed overnight in chloroform containing 10% triethylamine (v/v) and p-nitrobenzoylchloride (5 g/100 g of carrier). The product is washed with chloroform and allowed to air-dry. Reduction of the nitro groups is accomplished by boiling in 10% sodium dithionite solution in water in a volume large enough to exceed 1 g of sodium dithionite per gram of carrier. The carrier is boiled for 30 minutes and filtered while hot. The carrier is washed with slightly acidified water until the odor of sulfur

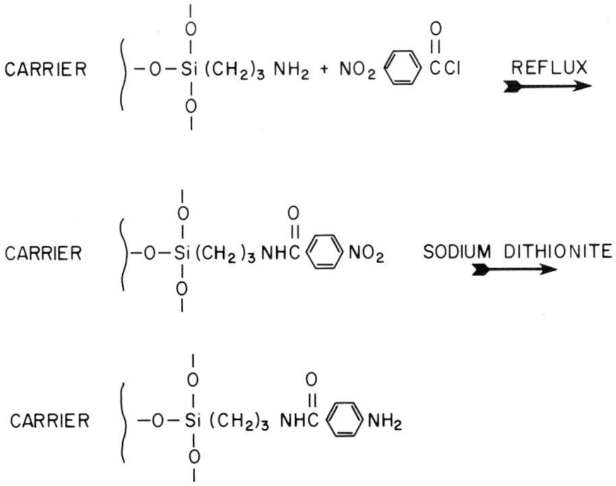

Fig. 6. Preparation of the arylamine derivative from the alkylamine derivative.

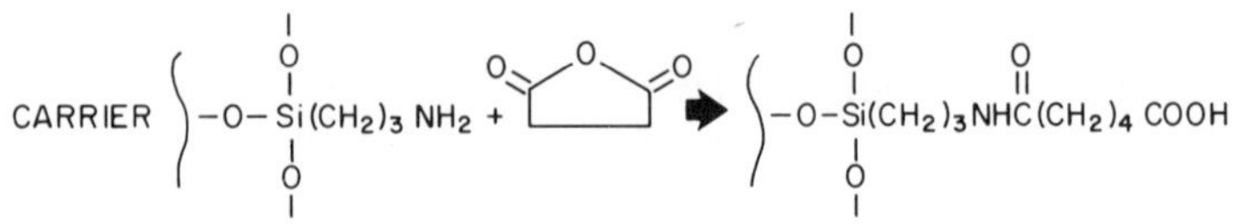

$$\text{CARRIER} \left(\begin{array}{c} | \\ O \\ | \\ -O-Si(CH_2)_3\,NH_2 \\ | \\ O \\ | \end{array} \right. + \underset{O}{\overset{O}{\triangle}} \longrightarrow \left(\begin{array}{c} | \\ O \\ | \\ -O-Si(CH_2)_3NHC(CH_2)_4\,COOH \\ | \\ O \\ | \end{array} \right.$$

Fig. 7. Preparation of the carboxyl derivative from the alkylamine derivative.

is no longer detectable. The final washes are with distilled water, and the product is dried at between 60° and 80°.

The carrier can be reused by burning off the organic material at 625° for 12 hours under O_2 if possible. The repeat procedure must include cleansing and rehydration of the surface with boiling nitric acid.

Preparation of the Carboxyl Derivative

The carboxylated derivative is prepared by reaction of the alkylamine carrier with succinic anhydride.

To each gram of alkylamine derivative is added 0.3 g of succinic anhydride dissolved in 50 mM sodium phosphate at pH 6.0. The reaction is continued for 12 hours and is followed by exhaustive washing with distilled water. The final product, the carboxyl derivative, may be dried and stored. The reaction is schematically represented in Fig. 7.

Preparation of the Aldehyde Derivative

The active aldehyde of the alkylamine carrier is prepared by reaction with glutaraldehyde. The product readily reacts with amines.

The reaction is carried out initially at room temperature. To 1 g of alkylamine glass, add enough 2.5% glutaraldehyde solution (25% glutaraldehyde diluted 1 to 10 in 0.1 M sodium phosphate at pH 7) to cover the glass. The entire reaction mixture is placed in a desiccator and attached to an aspirator to remove air and gas bubbles from within the particles. Reaction is continued in the aspirator for 60 minutes.

The product is removed and filtered on a Büchner funnel and washed with distilled water. The resulting derivative is the activated aldehyde. Schematically the reaction is represented in its simple form in Fig. 8.

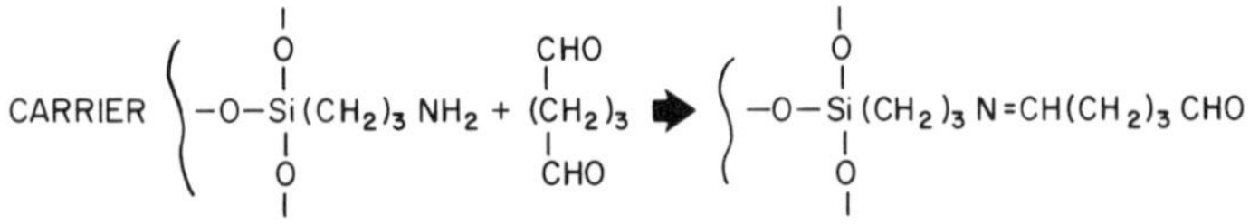

$$\text{CARRIER} \left(\begin{array}{c} | \\ O \\ | \\ -O-Si(CH_2)_3\,NH_2 \\ | \\ O \\ | \end{array} \right. + \begin{array}{c} CHO \\ | \\ (CH_2)_3 \\ | \\ CHO \end{array} \longrightarrow \left(\begin{array}{c} | \\ O \\ | \\ -O-Si(CH_2)_3\,N{=}CH(CH_2)_3\,CHO \\ | \\ O \\ | \end{array} \right.$$

Fig. 8. Preparation of the aldehyde derivative from the alkylamine derivative.

Methods of Covalently Attaching Proteins and Other Organic
Molecules to the Silanized Derivative

The three derivatives previously described may be utilized for the
covalent attachment of any organic molecule of choice. In addition, these
three derivatives can easily be converted to other derivatives having functional groups of different characteristics. Several methods for coupling an
organic moiety are presented here although others may be found throughout this volume, particularly in Articles [1] and [2].

Alkylamine Carrier

Amide Linkage. The alkylamine may be covalently coupled to any
free carboxyl group using a carbodiimide.[9-11] In the case of proteins or
other water-soluble materials, one may utilize a soluble carbodiimide. In
the case of organic soluble materials, an organic solvent may be suitable.
In this case N,N'-dicyclohexyl carbodiimide would be useful.

WATER-SOLUBLE CARBODIIMIDES. The coupling is carried out at room
temperature or at 0°. With an enzyme, the 0° approach may be preferred. To 1 g of alkylamine-glass is added 50 mg of protein and 10 ml
of distilled water to which 40 mg of 1-cyclohexyl-3(2-morpholinoethyl)
carbodiimide metho-p-toluenesulfonate had been added and adjusted to
pH 4.0. After addition of the diimide solution, the mixture is adjusted
to pH 4.0 and monitored for 20–30 minutes. Continuous agitation is
recommended for maximum coupling. After the first 30 minutes, the
suspension is placed in an aspirator for 20 minutes to ensure complete
wetting and is refrigerated overnight. The following morning, the solution
is decanted and the derivative is washed several times with distilled water.
The product may be stored at 6° as a wet cake or, in some instances,
dried. With some derivatives activation is best achieved at pH 10–11.

If the carrier is a carboxy derivative, one can form a pseudourea of
the carboxyl group, remove excess carbodiimide, and in this way prevent
possible cross-linking wherever both carboxyl and amine groups are
available.

To form the pseudourea of a carboxyl derivative, one simply adds a
large excess of carbodiimide to the derivative, adjusts to pH 4.0, and
allows the reaction to take place for several hours. The excess is removed
by filtering with distilled water slightly acidified. The product should be

[9] N. Weliky, H. H. Weetall, R. V. Gilden, and D. H. Campbell, *Immunochemistry*
1, 219 (1964).
[10] D. G. Hoove and D. E. Koshland, Jr., *J. Biol. Chem.* **242**, 2447 (1967).
[11] N. Weliky, F. S. Brown, and E. L. Dale, *Arch. Biochem. Biophys.* **131**, 1 (1969).

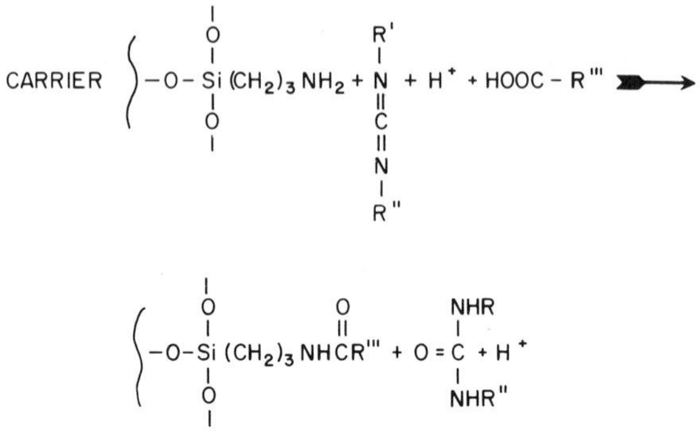

$$\left(\text{CARRIER}\right) - O - \underset{\underset{O}{|}}{\overset{\overset{O}{|}}{Si}} (CH_2)_3\,NH_2 + \underset{\underset{R''}{\overset{\|}{N}}}{\overset{R'}{\underset{\|}{N}}} + H^+ + HOOC - R''' \blacktriangleright\!\!\longrightarrow$$

$$\left(- O - \underset{\underset{O}{|}}{\overset{\overset{O}{|}}{Si}} (CH_2)_3\,NH\overset{\overset{O}{\|}}{C}R''' + O = \underset{\underset{NHR''}{|}}{\overset{NHR}{|}}C + H^+ \right.$$

Fig. 9. Preparation of an amide-linked ligand with the alkylamine derivative and a carbodiimide. R' and R'' represent groups on the carbodiimide, e.g., cyclohexyl groups; R''' represents the ligand having a free carboxyl group.

capable of reacting with amine group or a ligand. With some derivatives, best results are achieved at pH 10–11.

WATER-INSOLUBLE CARBODIIMIDES. Coupling between a derivative having free carboxyl or amine groups is carried out in a manner similar to that described above. However, the reaction is carried out in ethanol. As an example, N,N'-dicyclohexyl carbodiimide is soluble in ethanol and may be used to link amines to carboxyls as described for the water-soluble derivatives. The carbodiimide reaction is schematically represented in Fig. 9.

Thiourea Linkage. An alkylamine carrier may be converted to an isothiocyanate and covalently attached to a free amine group.[12,13] The isothiocyanate will react with free amine groups. The reaction is schematically represented in Fig. 10.

To 1 g of alkylamine-glass is added enough 10% thiophosgene in CHCl$_3$ to reflux for at least 4 hours; usually 75–125 ml is adequate. The carrier is washed with CHCl$_3$, quickly air-dried, and used immediately. Fifty milligrams of protein or ligand dissolved in 50 mM NaHCO$_3$,

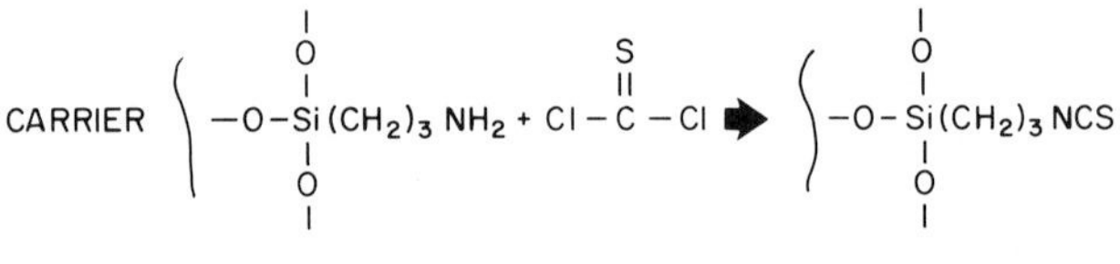

$$\left(\text{CARRIER}\right) \; \left(- O - \underset{\underset{O}{|}}{\overset{\overset{O}{|}}{Si}}(CH_2)_3\,NH_2 + Cl - \overset{\overset{S}{\|}}{C} - Cl \; \blacktriangleright \; \left(- O - \underset{\underset{O}{|}}{\overset{\overset{O}{|}}{Si}}(CH_2)_3\,NCS\right.$$

Fig. 10. Preparation of the isothiocyanate from the alkylamine derivative.

[12] L. Wide, R. Axén, and J. Porath, Immunochem. 4, 381 (1967).
[13] L. Wide and J. Porath, Biochim. Biophys. Acta 130, 257 (1967).

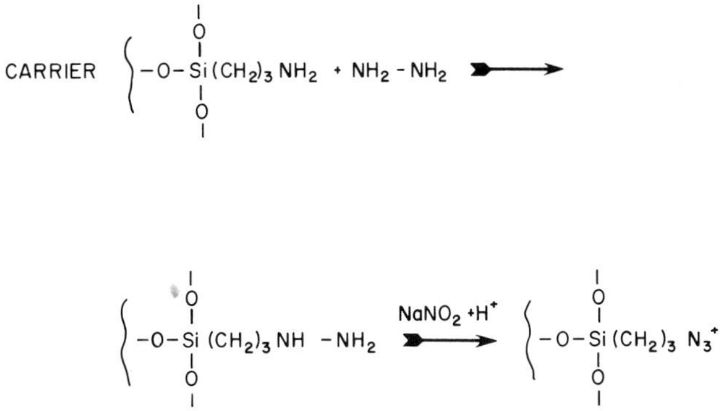

$$\text{CARRIER} \left(-O-\overset{\overset{\displaystyle |}{O}}{\underset{\underset{\displaystyle |}{O}}{Si}}(CH_2)_3\,NH_2 \;+\; NH_2-NH_2 \;\blacktriangleright\!\!-\!\!\!\longrightarrow\right.$$

$$\left(-O-\overset{\overset{\displaystyle |}{\overset{\bullet}{O}}}{\underset{\underset{\displaystyle |}{O}}{Si}}(CH_2)_3\,NH\;-NH_2 \;\xrightarrow{\;NaNO_2\,+H^+\;}\; \left(-O-\overset{\overset{\displaystyle |}{O}}{\underset{\underset{\displaystyle |}{O}}{Si}}(CH_2)_3\;N_3^+\right.\right.$$

FIG. 11. Preparation of the hydrazide from the alkylamine derivative.

buffer pH 9–10, is added to the isothiocyanate derivative. It is continuously agitated with the use of an aspirator for 30 minutes and for an additional 2 hours at room temperature. The product is washed with distilled water and stored as a wet cake at 6°.

Hydrazide Linkage. This reaction can be used for the covalent attachment of an alkylamine carrier to a compound containing an available primary aliphatic amine.[14] The reaction is schematically represented in Fig. 11.

This activated carrier should react with free primary aliphatic amines under slightly alkaline conditions. However, in our laboratories we have not attempted this method of activating and coupling compounds to inorganic carriers.

The Aldehyde Carrier

The aldehyde carrier will react directly at neutral pH with primary amines.[7] The reaction is gentle, but should be carried out in ice. However, it is imperative that excess glutaraldehyde be washed out before the coupling step. The preparation of the aldehyde derivative has been described above.

Place the derivative in the solution containing the protein or ligand to be coupled. As little liquid as possible should be used, and the reactants should be maintained at pH 7–9. Continue the reaction for at least 2 hours. Wash and store. This compound may be reduced with agents such as 1.0% sodium borohydride at acid pH. The reaction sequence in its simplest form is represented in Fig. 12.

[14] J. K. Inman and H. M. Dentzis, *Biochemistry* 8, 4074 (1969).

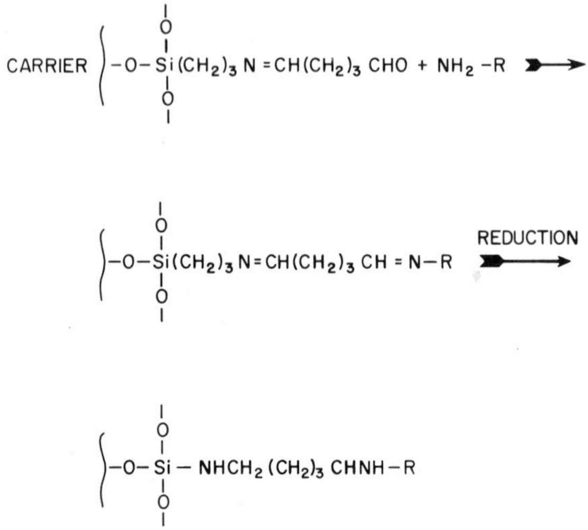

$$\text{CARRIER} \left\{ -O-\overset{\overset{|}{O}}{\underset{\underset{|}{O}}{Si}}(CH_2)_3\,N = CH(CH_2)_3\,CHO \;+\; NH_2-R \;\blacktriangleright\!\!\longrightarrow \right.$$

$$\left\{ -O-\overset{\overset{|}{O}}{\underset{\underset{|}{O}}{Si}}(CH_2)_3\,N = CH(CH_2)_3\,CH = N-R \quad \overset{\text{REDUCTION}}{\blacktriangleright\!\!\longrightarrow} \right.$$

$$\left\{ -O-\overset{\overset{|}{O}}{\underset{\underset{|}{O}}{Si}} - NHCH_2\,(CH_2)_3\,CHNH-R \right.$$

FIG. 12. The covalent coupling of an amine functional ligand with the alde-hyde derivative followed by reduction with sodium borohydride.

Carboxylated Carriers

The carboxylated carriers may be used to couple both aryl and alkyl amines.

Amide Linkage. The mechanism is similar to that shown for the alkylamine-carriers using the carbodiimide. A major advantage of using the carboxyl derivative is that the carboxyl group can be activated and the excess carbodiimide removed before coupling to the amine (Fig. 13).

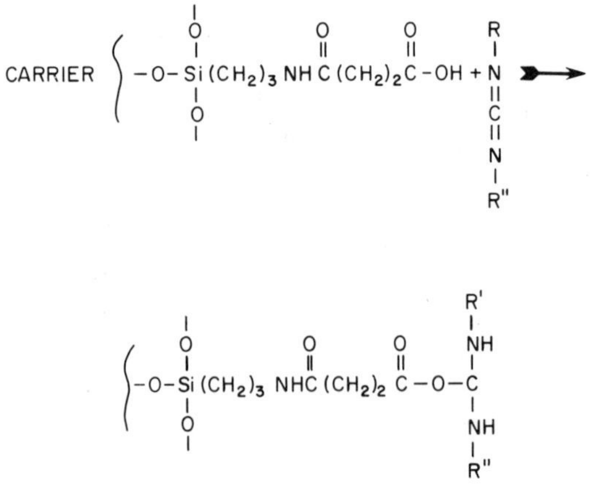

$$\text{CARRIER} \left\{ -O-\overset{\overset{|}{O}}{\underset{\underset{|}{O}}{Si}}(CH_2)_3\,NH\,\overset{O}{\overset{\|}{C}}(CH_2)_2\overset{O}{\overset{\|}{C}}-OH \;+\; \underset{\underset{|}{R''}}{\overset{\overset{R}{|}}{N}}\!\!=\!\!\underset{}{\overset{}{\underset{\|}{C}}}\!\!=\!\!N \;\blacktriangleright\!\!\longrightarrow \right.$$

$$\left\{ -O-\overset{\overset{|}{O}}{\underset{\underset{|}{O}}{Si}}(CH_2)_3\,NHC\,(CH_2)_2\,\overset{O}{\overset{\|}{C}}-O-\underset{\underset{\underset{|}{R''}}{NH}}{\overset{\overset{\overset{|}{R'}}{NH}}{C}} \right.$$

FIG. 13. The preparation of the pseudourea of a carboxyl derivative by reac-tion with a carbodiimide.

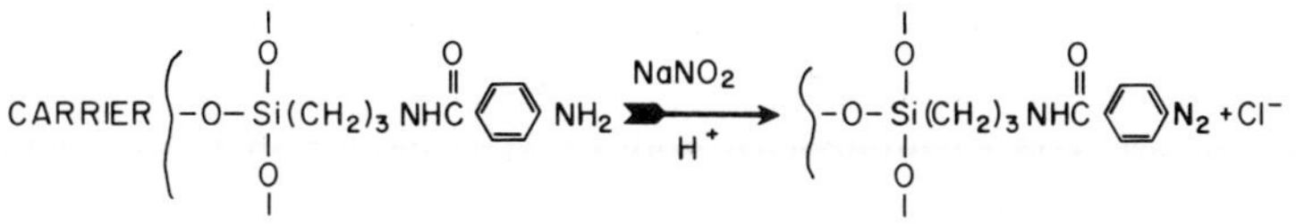

FIG. 14. The preparation of the diazonium chloride from the arylamine derivative.

The pseudourea formed in this reaction will react with amines forming an amide linkage. The reaction can be carried out under slightly acid conditions by addition of the ligand dissolved in either water or organic solvent to the activated carrier at pH 4.0.

Arylamine Derivative

Diazonium salts will react with phenolic compounds and other heterocyclics.[5,14–18] These salts will also react with alkylamines to form triazines. However, the triazine is not a preferred linkage because of rather poor stability. The reaction is generally conducted under slightly alkaline conditions.

The diazotization procedure is carried out in an ice bath. To 1 g of arylamine glass add 10 ml of 2 N HCl and 250 mg of solid NaNO$_2$. Place the entire reaction mixture in a desiccator and attach to an aspirator so as to remove air and gas bubbles in the particles. Continue diazotization in aspirator for 20 minutes, maintaining the suspension packed in ice. The reaction is represented schematically in Fig. 14.

Remove and filter on a Büchner funnel and wash with ice-cold water containing 1% sulfamic acid if desired. Place the diazotized glass in the ligand solution previously adjusted to pH 8–9, using as little liquid as possible; a slurry is the preferred state. Maintain pH by the addition of 2 N NaOH. The use of a buffer such as 50 mM NaHCO$_3$ or 50 mM Tris will aid in pH control. If possible, use 50–100 mg of protein or other ligand per gram of glass. Allow the reaction to continue for at least 60 minutes. Samples of the solution may be withdrawn at intervals for quantitative determinations in order to establish when the maximum coupling rate decreases; usually this occurs within 60 minutes. Continuing the reaction may add an additional 15–20% of bound ligand. Wash in

---

[15] R. Goldman, L. Goldstein, and E. Katchalski, in "Biochemical Aspects of Reactions on Solid Supports" (G. R. Stark, ed.), pp. 1–72. Academic Press, New York, 1971.
[16] H. H. Weetall, Res./Develop. 22, 18 (1971).
[17] N. Weliky and H. H. Weetall, Immunochemistry 2, 293 (1965).
[18] D. H. Campbell and N. Weliky, in "Methods in Immunology and Immunochemistry" (C. A. Williams and W. Chase, eds.), Vol. I. Academic Press, New York, 1967.

distilled water and store in closed container immediately after filtration; the product usually contains 50–70% water by weight.

There are a large number of additional reactions that can be utilized for coupling of enzymes, antigens, antibodies, and ligands to inorganic supports. Such methods are reviewed in more detail elsewhere.[15,18,19]

[19] A. Bar-Eli and E. J. Katchalski, *Biol. Chem.* **238**, 1690 (1963).

[5] Polymers Coupled to Agarose as Stable and High Capacity Spacers

By MEIR WILCHEK and TALIA MIRON

For successful use of affinity chromatography, the polymer-bound ligand must be sufficiently distant from the polymer surface to minimize steric interference. This is achieved by introducing a spacer between the solid matrix and the ligand.[1]

Since the only effective method available for binding various molecules to Sepharose is by its activation with cyanogen bromide,[2] the spacer must contain a free amino group through which the binding can take place.

Two different approaches for introducing the spacer may be used. First, the ligand can be prepared with a long hydrocarbon chain containing an amino group, and this derivative is coupled to the solid matrix.[1] The long chain serves as the spacer. The second approach is to bind a spacer, such as ε-aminocaproic acid, hexamethylenediamine, cystamine, *p*-aminobenzoic acid, tyramine, or *p*-hydroxymercuribenzoate, directly to Sepharose so that a ligand can be attached to these derivatives by various methods.[3,4]

In both methods the ligand is coupled monovalently to Sepharose through N-substituted isourea bonds, which are not completely stable.

[1] P. Cuatrecasas, M. Wilchek, and C. B. Anfinsen, *Proc. Nat. Acad. Sci. U.S.* **61**, 636 (1968).
[2] J. Porath, R. Axén, and S. Ernback, *Nature (London)* **215**, 1491 (1967).
[3] P. Cuatrecasas, *J. Biol. Chem.* **245**, 3059 (1970).
[4] M. Wilchek and M. Rotman, *Isr. J. Chem.* **8** (1970).

Therefore these conjugates exhibit a small but constant "leakage" of the ligands from the solid matrix. This phenomenon could limit the use of affinity chromatography to the isolation of very small amounts of proteins and would render the interaction between Sepharose-bound hormones and intact cells questionable.

Principle

We describe here the preparation of stable and high capacity agarose derivatives achieved by coupling polylysine, or multipoly-DL-alaninepolylysine to Sepharose.[5] Poly-DL-alanine was introduced to achieve structureless straight chains which serve as very long spacers between the solid matrix and the ligand. An additional advantage in using poly-DL-alanine is that its amino group has a lower pK than polylysine and can, therefore, be used more effectively in coupling reactions. The high stability of the product is the result of the numerous positions at which polylysine is coupled to the agarose.

Polylysyl- and poly-DL-alanyllysyl-Sepharose were used for additional ligand substitution in a variety of ways (Fig. 1). The methods of prepara-

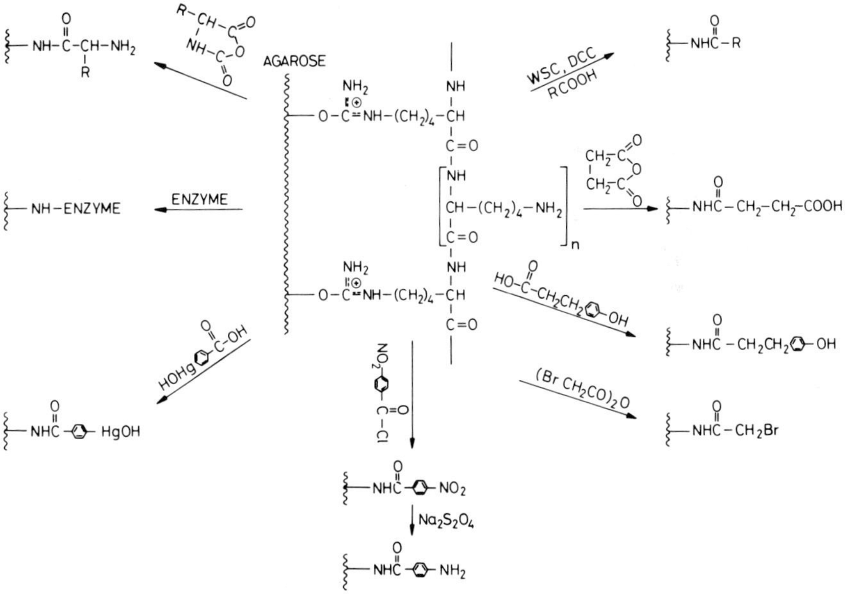

FIG. 1. Procedures for preparing derivatives of agarose-polylysine or agarose-multi-poly-DL-alaninepolylysine.

[5] M. Wilchek, *FEBS Lett.* 33, 70 (1973).

tion of the various derivatives are commonplace and are found in this volume[6] as well as elsewhere.[3,7,8]

Each of the derivatives has strong ion exchange properties due to incomplete coupling of the ligands to the amino or carboxyl groups, and also because the N-substituted isourea formed on coupling of amines to cyanogen bromide-activated Sepharose retains the basicity of the amino group. (A pK of 10.4 was found by potentiometric titration of the isourea derivative formed on coupling butylamine to Sepharose.)

Adipic acid dihydrazide[9,10] polyglutamic acid hydrazide,[11] and linear polyacrylic hydrazide[12] have been coupled to cyanogen bromide-activated Sepharose. These conjugates were prepared in order to obtain stable, high capacity agarose derivatives, free of ion exchange properties.

The polyacrylic hydrazide-Sepharose derivatives are easy to prepare and behave as ideal insoluble carriers for affinity chromatography. They retain the properties of Sepharose; namely, minimal nonspecific interaction with proteins, good flow rate, and a very loose porous network which permits entry and exit of large molecules. In addition, they are mechanically and chemically stable to the conditions of coupling and elution. However, these derivatives also have some of the properties of poly-acrylamide gels, in that they lack charged groups even after coupling to cyanogen bromide-activated Sepharose; a pK of 4.2 was found on poten-tiometric titration of semicarbazide coupled to Sepharose, and in that they have a very large number of modifiable groups. These hydrazide-agarose derivatives, which can be considered "universal columns," can either be utilized directly, e.g., coupling of RNA or nucleotides, or can be further derivatized, as shown in Fig. 2. The methods of preparing the derivatives and others, can be found in this volume[6] and elsewhere.[3,7-9]

Preparations

Preparation of Polylysyl-Sepharose[5]

Sepharose, 25 ml, is activated with 2.5 g of cyanogen bromide as previously described.[1,2] The activated gel is washed with 0.1 M sodium

[6] See the article by Porath [2], Inman [3], Weetall [4], and Cuatrecasas [6] among others.
[7] P. Cuatrecasas and C. B. Anfinsen, this series, Vol. 22, p. 345.
[8] J. K. Inman and H. M. Dintzis, *Biochemistry* 8, 4074 (1969).
[9] R. Lamed, Y. Levin, and M. Wilchek, *Biochim. Biophys. Acta* 304, 231 (1973).
[10] M. Wilchek and R. Lamed, this volume [55].
[11] J. Kovacs, V. Bruckner, and K. Kovacs, *J. Chem. Soc.* 145 (1953).
[12] V. W. Kern, T. Hucke, R. Hollander, and R. Schneider, *Makromol. Chem.* 22, 31 (1957).

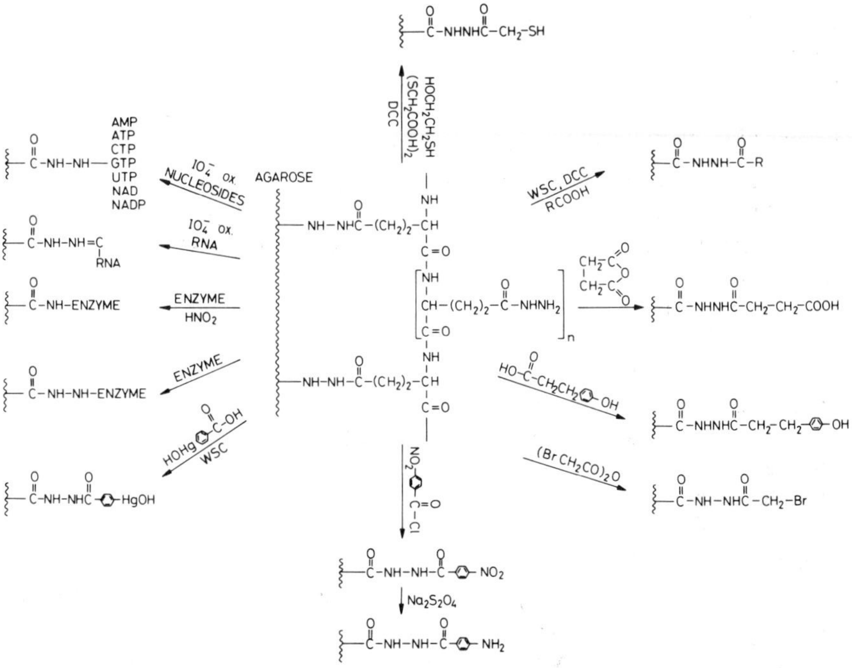

FIG. 2. Procedures for preparing derivatives of agarose-polyhydrazides.

bicarbonate and added to an equal volume of polylysine (1 g) or multi-poly-DL-Ala-Lys (Miles-Yeda) in the same solvent. After 16 hours at 4° the gel is washed sequentially with water, 0.1 N hydrochloric acid, 0.1 M sodium bicarbonate and water. Polylysyl-Sepharose derivatives, containing up to 30 mg of polylysine per milliliter of Sepharose, can be obtained; this value corresponds to about 250 μmoles of lysine per milliliter. More than 80% of the lysine residues are available for further derivatization, as determined by reaction with fluoridinitrobenzene, followed by estimation of the ε-DNP-lysine formed after total hydrolysis.

Coupling of DL-*Alanine-N-Carboxyanhydride to*
  Polylysyl-Sepharose[5]

Polylysyl-Sepharose, 10 ml, is washed with an aqueous solution of triethylamine or with 0.1 N NaOH, water, methanol, and, finally, with dioxane. To the dioxane suspension, 4 g of DL-Ala-N-carboxyanhydride in 10 ml of dioxane are added and the reaction mixture is stirred gently at room temperature. After 24 hours, the substituted Sepharose is washed with a large volume of water until samples of the washings show no color on reaction with 2,4,6-trinitrobenzene sulfonate. The average length of the spacer is between 5 and 10 alanyl residues per lysine according to the ratio of lysine to alanine after total hydrolysis.

Preparation of Polymethyl Acrylate[12,13]

Sodium laurylsulfate, 250 mg, 5 ml of freshly redistilled methylacrylate, 1 ml of 1% thioglycolic acid, and 125 mg of ammonium persulfate are added successively to 50 ml of distilled water with magnetic stirring in a fume hood. The emulsion is refluxed for 2.5 hours in a water bath at 80° until the odor of methylacrylate has almost completely disappeared.

After polymerization the emulsion is poured, with stirring, into 100 ml of ice-cold water to which 100 ml of cold 2 N HCl are added. The product, which coagulates in the form of white flakes, is kept for 30 minutes in the cold and is then washed by decantation with large amounts of cold water.

Preparation of Polyacrylic Hydrazide

Polymethylacrylate, 5 g, is dispersed into 70 ml of 98% hydrazine hydrate. The reaction mixture is stirred vigorously and heated to 100° in a boiling water bath. After 3 hours, a clear viscous solution is obtained. The reaction mixture is cooled to room temperature, the insoluble particles are filtered through gauze, and the filtrate is poured with constant stirring into 500 ml of ice-cold methanol containing 1 ml of glacial acetic acid. The precipitate is filtered, washed with cold methanol–acetic acid (500:1), and redissolved in 150 ml of water. The insoluble particles are removed by filtration. This solution may be used immediately for coupling to Sepharose, or may be lyophilized, taking care to remove any trace of moisture so as to prevent cross-linking of the product.

Preparation of Polyacrylic Hydrazide-Sepharose

Sepharose 4B, activated with cyanogen bromide,[1,2] is suspended in 3 volumes of the above cold polyacryl hydrazide solution in either water or 0.1 N NaHCO$_3$. The reaction is allowed to proceed overnight at 4° with slow stirring. The conjugate is washed with 0.1 M sodium chloride until the washings are free of color on reaction with 2,4,6-trinitrobenzene sulfonate. An average capacity of 20–25 μmoles of available hydrazide per milliliter of Sepharose is obtained.

[13] D. G. Knorr, S. D. Misina, and L. C. Sandachtchiev, Izv. Sib. Otd. Akad. Nauk SSSR Ser. Khim. Nauk 11, 134 (1964).

[6] Topics in the Methodology of Substitution Reactions with Agarose

By INDU PARIKH, STEVEN MARCH, and PEDRO CUATRECASAS

Agarose beads have been one of the most useful solid supports in affinity chromatography. This chapter describes various alternative methods that can be used to activate agarose for the covalent attachment of small ligands and proteins. In addition, some general problems encountered in the use of such derivatives are presented and possible solutions are suggested.

Cyanogen Bromide Method for Activating Agarose

Cyanogen halides react rapidly with the hydroxyl groups of carbohydrates to form cyanate esters.[1,2] The reaction occurs most rapidly at a pH between 10 and 12. The cyanate may react further with hydroxyl groups to form an imidocarbonate intermediate. It has been postulated, from studies of low-molecular-weight model compounds, that a cyclic imidocarbonate is formed as a major product following the reaction of cyanogen halides with polymeric carbohydrates (Fig. 1A). The highly reactive intermediate can be isolated, and agarose beads which contain this reactive group are commercially available (Pharmacia). Imidocarbonates react with amines to form N-imidocarbonates (Fig. 1B), isoureas (Fig. 1C), or N-carbamates (Fig. 1D). However, the actual products formed in the reaction between cyanogen halide-activated agarose and the amino group of alanine, for example, are difficult to study owing to the insolubility of the carbohydrate and the relatively low concentrations of the products. Recent studies on the isoelectric point of products of the reaction between CNBr-activated polysaccharides and a primary amine indicate that an isourea is the major product.[3]

Titration Method of Activation

The agarose beads, washed thoroughly with large volumes of 0.1 M NaCl and water, are suspended in water to achieve a total volume equal to twice the packed volume of the gel. A volume of the 1:1 slurry is added to a beaker containing a magnetic stirrer and a few chunks of ice.

[1] R. Axén, J. Porath, and S. Ernbäck, *Nature* (*London*) **214**, 1302 (1967). See also J. Porath, this volume [2].
[2] P. Cuatrecasas, *J. Biol. Chem.* **245**, 3059 (1970).
[3] B. Svensson, *FEBS Lett.* **29**, 167 (1973).

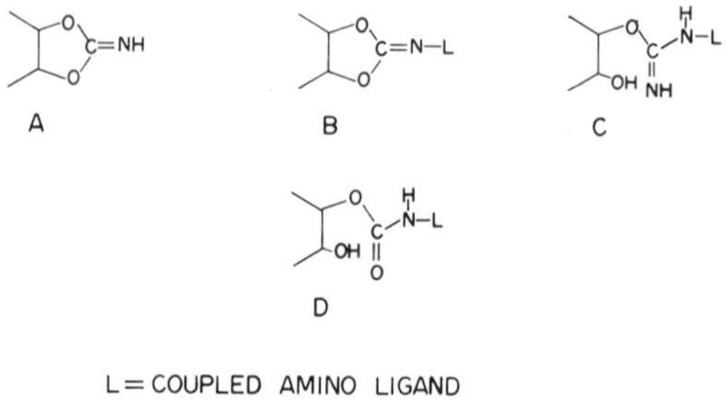

A   B   C

D

L = COUPLED AMINO LIGAND

FIG. 1. Proposed chemical structures for the intermediates and products formed by the reaction of CNBr with agarose and of the activated agarose with amino ligands. See R. Axén, J. Porath, and S. Ernbäck, *Nature* (*London*) **214**, 1302 (1967). See also this volume [2].

The CNBr is weighed into a glass bottle in a fume hood and capped until used. The base used for titration is poured into a 50-ml burette with a 2-mm bore stopcock. For amounts in the range of 50 mg of CNBr per milliliter of packed beads, 2 M NaOH is adequate. When higher degrees of activation are desired, 300 mg of CNBr per milliliter and 8 M NaOH should be used. All the CNBr is added at once, and the rate of flow of the burette is adjusted to maintain the pH near 11. Ice is occasionally added to the slurry to maintain the temperature near 30°. In 10–12 minutes the CNBr dissolves and the rate of base consumption decreases. At this time the slurry is poured into a chilled sintered-glass funnel (containing chunks of ice), and the gel is rapidly (2–3 minutes) filtered with suction and washed with about 5–10 volumes of ice-cold distilled water. The gel is filtered to a moist cake, the outlet of the funnel is covered with Parafilm, and the buffer containing the material to be coupled is added to the funnel and mixed immediately with a glass rod. The slurry is poured into a plastic bottle, and the suspension is gently shaken on a mechanical shaker (preferred) or stirred for at least 12 hours at 4°. Agitation of the slurry should be by gentle means since shattering of the beads may occur upon vigorous stirring. After completion of the coupling reaction, the gel is washed briefly and suspended in a 4-fold volume of 1 M glycine in order to promote masking of activated groups that may still remain on the agarose. The suspension is shaken for 4 hours at room temperature, or for 12 hours at 4° if higher temperatures are not well tolerated by the ligand.

Buffer Method of Activation

Despite the relative simplicity of the titration method, a faster and safer alternative method has been developed.[4] In this modification, the

[4] S. C. March, I. Parikh, and P. Cuatrecasas, *Anal. Biochem.*, in press.

reaction is performed in carbonate buffer. The coupling efficiencies obtained are comparable to those observed after activation by the titration method. A typical reaction involves adding a volume of 1:1 washed agarose to an equal volume of 2 M sodium carbonate. The slurry is cooled in an ice bath with slow stirring. After cooling to 5°, the rate of stirring is increased and the desired amount of CNBr, dissolved in acetonitrile (2 g of CNBr per milliliter of CH_3CN) is added. After 2 minutes the beaker is decanted into a coarse disk sintered-glass funnel, and the gel is washed and coupled to a ligand as described above. The results are reproducible; the mean deviation in the quantity of ligand or protein coupled in a group of individually activated gels has been as low as 4%. Cooling the reaction before and during activation is important. For example, coupling may be reduced by as much as 35% if the reaction is performed at room temperature.

General Considerations in the Use of CNBr-Activated Gels

Modification of the Conditions of Reaction and Coupling. The amount of ligand coupled to the gel depends on the amount of CNBr added. Typically this varies between 50 and 300 mg of CNBr per milliliter of packed beads. For example, with 200 mg of CNBr per milliliter of agarose and 0.5% albumin in 0.2 M $NaHCO_3$ at pH 9.5, approximately 5 mg of protein is bound per milliliter of packed gel. Similarly, if the concentration of low molecular weight ligand, e.g., alanine, is 0.1 M, the amount coupled is about 10 μmoles per milliliter of gel. The actual coupling efficiency will depend on the specific ligand used.

The quantity of CNBr and the exact composition of the buffer used in the coupling reaction should be adapted to the specific system under study. These conditions have been described in detail elsewhere.[2] A standard condition is the use of 200 mg of CNBr per milliliter of agarose, and of 0.2 M sodium bicarbonate at pH 9.5 as the buffer for the coupling reaction. Smaller quantities of CNBr, lower pH values, and high concentrations of ligand will decrease the probability of multipoint attachments of proteins (especially those of high molecular weight) to the gel, a condition that may lead to decreased or altered biological activity. In coupling macromolecules, e.g., concanavalin A or enzymes, it may be desirable to add ligands, e.g., competitive inhibitors, which bind to the active site of the protein. This may protect and perhaps prevent coupling through essential residues of the active site and, in addition, may stabilize the conformation of the protein so that the coupled protein will be more likely to retain its native structure.

In many cases, the interposition of spacers or "arms" between the matrix and the ligand greatly increases the effectiveness of the adsorbent. A variety of spacer molecules can be attached to agarose, and many chemical reactions exist that can be used to couple ligands and proteins to these derivatized agarose gels.[2] Diaminodipropylamine (Eastman) has been one of the most useful spacer molecules because it is relatively long and because it exhibits very minimal hydrophobic properties as compared to strictly methylenic diamine compounds such as hexamethylenediamine. Whenever possible, it is advantageous to first attach such spacers to the ligand rather than to the gel since the adsorbents prepared in this way are less likely to exhibit nonspecific or ionic properties that can interfere in subsequent affinity chromatography experiments.

Hazards of CNBr. CNBr is a toxic chemical, and awareness of the hazards associated with CNBr activation reactions is important. CNBr, which sublimes rather rapidly at room temperature, is a powerful lachrimator, and is highly toxic. Allowing the chemical to stand for months at room temperature can result in the formation of explosive compounds. Filtrates collected after activation contain large amounts of cyanide ion and should never be allowed to come to neutral or acidic pH. Disposal of the filtrates should be performed with care and only after the gaseous products are dissipated from the filter flask in a fume hood for several hours. The importance of adequate ventilation during the activation procedures is emphasized.

Sodium Periodate Method of Activation

Nearly all previously described methods for coupling ligands or proteins to agarose depend on the initial modification of the gel with CNBr.[1,2] An alternative method which is rapid, simple, and safe and which promises to result in chemically stable ligand-agarose bonds has been developed. The method depends on the oxidation of cis-vicinal hydroxyl groups of agarose by sodium metaperiodate ($NaIO_4$) to generate aldehyde functions[5] (Fig. 2). These aldehydic functions react at pH 4–6 with primary amines to form Schiff bases; these are reduced with sodium borohydride ($NaBH_4$) to form stable secondary amines. The reductive stabilization of the intermediate Schiff base is best achieved with sodium cyanoborohydride ($NaBH_3CN$) since this reagent, at a pH between 6 and 6.5, preferentially reduces Schiff bases without reducing the aldehyde functions of the agarose.[6] Since sodium cyanoborohydride selectively reduces only the Schiff bases, it shifts the equilibrium of the reaction (Fig. 2) to the right and thereby drives the overall reaction to completion.

Very useful derivatives can be prepared by allowing the periodate-oxidized agarose to react with a bifunctional, symmetrical dihydrazide

[5] C. J. Sanderson and D. V. Wilson, *Immunology* **20**, 1061 (1971).
[6] R. F. Borch, M. D. Bernstein, and H. D. Durst, *J. Amer. Chem. Soc.* **93**, 2897 (1971).

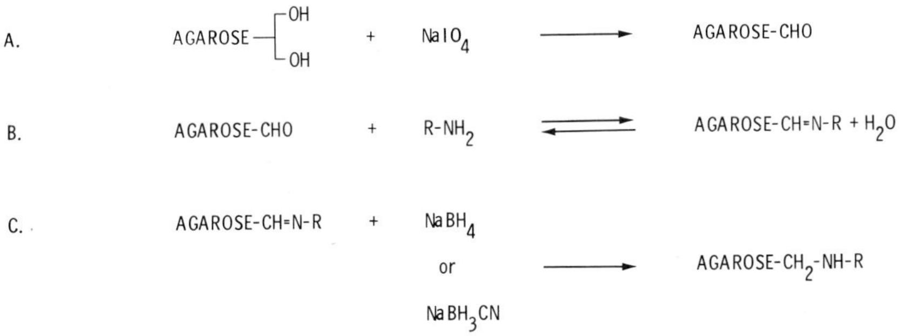

A. $\quad$ AGAROSE $-\begin{bmatrix} -OH \\ -OH \end{bmatrix}$ $\quad + \quad$ NaIO$_4$ $\quad\longrightarrow\quad$ AGAROSE-CHO

B. $\quad$ AGAROSE-CHO $\quad + \quad$ R-NH$_2$ $\quad\rightleftharpoons\quad$ AGAROSE-CH=N-R + H$_2$O

C. $\quad$ AGAROSE-CH=N-R $\quad + \quad$ NaBH$_4$

$\qquad\qquad\qquad$ or $\qquad\qquad\longrightarrow\quad$ AGAROSE-CH$_2$-NH-R

$\qquad\qquad\qquad$ NaBH$_3$CN

FIG. 2. Reductive amination of periodate-oxidized agarose.

(Fig. 3). This reaction is quite favorable since carbonyl groups react with hydrazides more completely than with primary amines. The matrix is thus converted to a hydrazido/hydrazone form and can be reduced with NaBH$_4$ to produce unsymmetric hydrazides while the unreacted hydrazido groups are unaffected. The stable hydrazido-agarose derivatives can be stored and used at will for coupling proteins or ligands that contain carboxyl or amino groups with carbodiimide reagents or by conversion of the acyl hydrazide to the acyl azide form (Fig. 3).

Although the periodate oxidation activation method for agarose generally results in a lower degree of ligand substitution than can be achieved with the CNBr method, the facility and safety of the method, and greater stability of the ligand-agarose bonds formed, may offer advantages. In many cases enzyme purification by affinity chromatography is achieved satisfactorily with ligand substitution of 0.05–1.0 μmole per milliliter of agarose, which can readily be achieved with these procedures. The "ac-

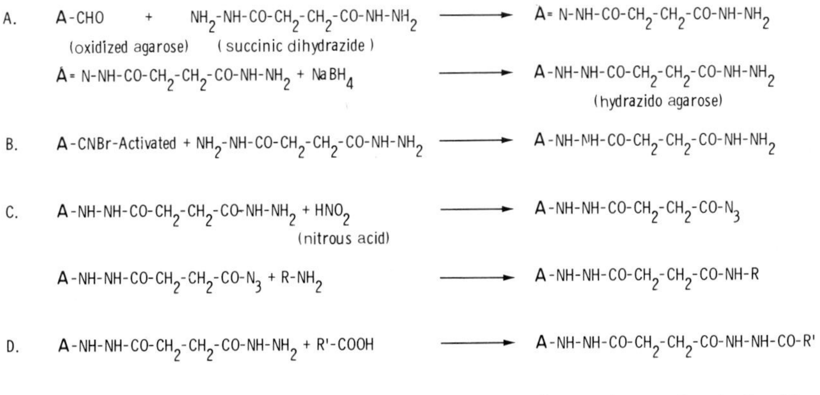

A. A-CHO $\quad + \quad$ NH$_2$-NH-CO-CH$_2$-CH$_2$-CO-NH-NH$_2$ $\quad\longrightarrow\quad$ A= N-NH-CO-CH$_2$-CH$_2$-CO-NH-NH$_2$
    (oxidized agarose)   (succinic dihydrazide)

$\quad$ A= N-NH-CO-CH$_2$-CH$_2$-CO-NH-NH$_2$ + NaBH$_4$ $\quad\longrightarrow\quad$ A-NH-NH-CO-CH$_2$-CH$_2$-CO-NH-NH$_2$
$\qquad\qquad\qquad\qquad\qquad\qquad\qquad\qquad\qquad$ (hydrazido agarose)

B. A-CNBr-Activated + NH$_2$-NH-CO-CH$_2$-CH$_2$-CO-NH-NH$_2$ $\quad\longrightarrow\quad$ A-NH-NH-CO-CH$_2$-CH$_2$-CO-NH-NH$_2$

C. A-NH-NH-CO-CH$_2$-CH$_2$-CO-NH-NH$_2$ + HNO$_2$ $\quad\longrightarrow\quad$ A-NH-NH-CO-CH$_2$-CH$_2$-CO-N$_3$
$\qquad\qquad\qquad\qquad$ (nitrous acid)

$\quad$ A-NH-NH-CO-CH$_2$-CH$_2$-CO-N$_3$ + R-NH$_2$ $\quad\longrightarrow\quad$ A-NH-NH-CO-CH$_2$-CH$_2$-CO-NH-R

D. A-NH-NH-CO-CH$_2$-CH$_2$-CO-NH-NH$_2$ + R'-COOH $\quad\longrightarrow\quad$ A-NH-NH-CO-CH$_2$-CH$_2$-CO-NH-NH-CO-R'

FIG. 3. Chemical reactions involved in the preparation and use of a hydrazido-agarose.

TABLE I
INCORPORATION OF RADIOACTIVE LIGANDS INTO GELS ACTIVATED
BY THE PERIODATE/HYDRAZIDE PROCEDURES[a]

| | Incorporation | | |
Matrix	SDH (μmoles/ml)	Alanine (μmoles/ml)	Albumin (mg/ml)
Agarose	1.2	0.25	0.20
Particulate cellulose	210	1.1	0.21
Beaded, porous cellulose	500	1.1	0.16

[a] Gels were oxidized with 0.5 M sodium metaperiodate for 2 hours and allowed to react with excess succinic dihydrazide (SDH) as described in the text. SDH incorporation was calculated from elemental analysis.

tivated" hydrazido-agarose derivatives, which are readily formed by these procedures, may be especially useful in the coupling of proteins to agarose through primary amino groups of the protein.[7]

Other solid supports, such as cross-linked dextrans and cellulose, unlike agarose contain very large numbers of cis vicinal hydroxyl groups and are thus very well suited for activation and coupling by the periodate oxidation procedures. Cellulose, because of its stability to the conditions of oxidation, can be substituted to a very high degree with the periodate method (Table I). The more porous Sephadex gels (G-75 and G-200) shatter after periodate treatment (5 mM, 2 hours) and therefore are not recommended.

Procedures for Sodium Periodate Oxidation of Agarose

To a suspension of 100 ml of agarose in 80 ml of water is added 20 ml of 1 M NaIO$_4$. The suspension, placed in a 500-ml tightly closed polyethylene bottle, is gently shaken on a mechanical shaker for 2 hours at room temperature. The oxidized agarose is filtered and washed on a coarse sintered-glass funnel with 2 liters of water.

Reductive Amination of Oxidized Agarose

Use of Sodium Borohydride. The washed, oxidized agarose is allowed to react with a solution of the ligand which contains an amino group. For example, an ω-aminoalkyl agarose derivative can be prepared by adding the oxidized gel, 100 ml, to 100 ml of 2 M aqueous diaminodipropylamine (Eastman Chemicals) at pH 5.0. After 6–10 hours of gentle shaking at room temperature. The pH is raised to 9 with solid Na$_2$CO$_3$. Ten ml of freshly prepared 5 M NaBH$_4$ are added in small aliquots to the mag-

---

[7] I. Parikh and P. Cuatrecasas, this volume [79].

netically stirred suspension and kept at 4° over a 12-hour period with precautions to avoid excessive foaming. The reduced agarose derivative is washed in a 250-ml sintered-glass funnel with 2 liters of 1 M NaCl without suction over a 4-hour period. After stopping the outflow (with a rubber stopper), the agarose derivative is incubated in an equal volume of 1 M NaCl for 15 hours while on the funnel, and washed with an additional 2 liters of 1 M NaCl over a 2-hour period. The filtrates of the wash are occasionally checked for the presence of free diamine by the TNBS[2,8] test or with ninhydrin. The extent of substitution of diamine, as determined by reaction with ^{14}C-labeled acetic anhydride, is 2–3 μmoles per milliliter of agarose.

Use of Sodium Cyanoborohydride. The product of reaction of the NaIO$_4$-oxidized agarose with the amino ligand can also be subjected to reductive amination with sodium cyanoborohydride (NaBH$_3$CN). Although sodium borohydride is efficient, the use of NaBH$_3$CN is in some cases advantageous, e.g., when the quantity of the amino ligand is limited. Since the latter reducing agent drives the reaction toward completion, relatively more efficient use of the amino ligand will result. This method is also preferable when the ligand to be coupled is sensitive to the higher pH values (pH 9–10) necessary with NaBH$_4$.

A 1 to 50 mM solution of the amino ligand, a designation which includes proteins as well as smaller molecules, in 0.5 M phosphate buffer at pH 6, containing 0.5 mM sodium cyanoborohydride (Alfa Chemicals) is prepared at room temperature and centrifuged at 3000 g for 10 minutes to remove insoluble material. The pH of the solution is adjusted, if necessary, to 6. The solution is then added to an equal volume of periodate-oxidized agarose which has previously been washed with 1–2 volumes of 0.5 M phosphate buffer at pH 6. The suspension is gently shaken for 3 days in a closed, capped polyethylene bottle at room temperature with a mechanical shaker. The gel is extensively washed as described above. The unreacted aldehyde groups on the agarose matrix are then reduced with a solution 1 M NaBH$_4$ (1 ml per each milliliter of agarose gel) for 15 hours at 4°. The substituted agarose is washed extensively again. The substitution of ligand is about 2 μmoles per milliliter of agarose when 50 mM diaminodipropylamine is used.

Hydrazidosuccinyl-Agarose

Preparation of Succinic Dihydrazide (SDH). Sixty-four grams (2 moles) of hydrazine (99%) is stirred magnetically in a 1-liter Erlenmeyer

---

[8] 2,4,6-Trinitrobenzene sulfonic acid (TNBS); 2–3 drops of 1.5% ethanolic solution of TNBS are added to a mixture of 0.5 ml of filtrate and 0.5 ml of saturated sodium borate. The presence of diamine is indicated by formation of intense yellow color (420 nm) and of hydrazide by brick red color (500 nm).

flask with 100 ml of absolute ethanol at room temperature. Diethyl succinate (35 g; 0.2 mole) is added dropwise over 3–4 hours from a dropping funnel. The temperature of the reaction mixture rises to 45–50°. The clear reaction mixture is allowed to stand overnight at room temperature, and the resultant crystalline succinic dihydrazide (SDH) is filtered, washed with a liter of ice-cold absolute ethanol, and dried. The dihydrazide (mp 168–170 with decomposition), obtained in 90% yield (26.5 g), is used without further purification.

Use of Periodate-Oxidized Agarose. A suspension containing 100 ml of oxidized agarose and 100 ml of 0.1 M SDH (pH adjusted to 5.0) is gently shaken for 1–2 hours at room temperature and the pH is raised to 9 with solid Na_2CO_3. Ten milliliters of freshly prepared aqueous $NaBH_4$ (5 M) are added in small aliquots, with gentle stirring at 4°, over a 12-hour period (avoiding excessive foaming). The reduced agarose derivative is washed on a 250-ml sintered-glass funnel with 2 liters of 1 M NaCl without suction over a 3- to 4-hr period. The agarose derivative is incubated with an equal volume of 1 M NaCl for 15 hours at 24° in the same funnel, by stopping the outflow with a rubber stopper. It is washed with an additional 2 liters of 1 M NaCl over 2–3 hours. The substitution of SDH, calculated from the total nitrogen content of the hydrazidoagarose, varies between 1 and 2 μmoles per milliliter of packed agarose.

Use of Cyanogen Bromide-Activated Agarose. Agarose, activated with CNBr is treated with an equal volume of 0.5 M SDH in 0.2 M $NaHCO_3$ at pH 9.0. The suspension is gently shaken for 15 hours at 4° and an additional 4–6 hours at room temperature. The hydrazido-agarose is washed with about 10 volumes of 1 M NaCl over a coarse sintered-glass funnel and incubated with 1 M glycine at pH 9.0 for 4 hours at room temperature. The substituted agarose is extensively washed with 1 M NaCl until SDH is no longer detected in the washings.

Activation and Use of Hydrazido-Agarose. The hydrazido-agarose derivative is stable for months if stored at 4° in the presence of 0.02 M NaN_3. Carboxylic group-containing ligands can be coupled to the hydrazido-agarose with the use of carbodiimide reagents by established procedures.[2,9] Primary or secondary amines can also be coupled readily to the hydrazido-agarose after formation of agarose-azide derivatives with $NaNO_2$ and dilute hydrochloric acid, as described below.

Preparation and Use of Hydrazido-Cellulose

Preparation. Cellulose fibers or beaded cellulose,[10] 10 g, are suspended in 1 liter of 0.5 M sodium metaperiodate. The slurry is shaken at

[9] D. G. Hoare and D. E. Koshland, *J. Biol. Chem.* **242**, 2447 (1967).
[10] The experimental, beaded, porous cellulose was a gift from Pharmacia.

room temperature. After 2 hours, 30 ml (0.54 mole) of ethylene glycol are added and the slurry is shaken for an additional hour. The oxidized beads are washed on a coarse sintered-glass funnel with ten portions of 200 ml each of water. The washed gel is adjusted to a 1:1 slurry with water. SDH (4 g, 27 mmoles, in 29 ml of water) is added, the slurry is stirred and the pH is adjusted to 5 with concentrated HCl. After 1 hour the pH is adjusted to 9 with solid sodium carbonate. Sodium borohydride (1 g, 27 mmoles) is added, and the slurry is shaken overnight in a vented container. The beads are washed with 200 ml of 0.1 M acetic acid followed by five washes of 400 ml each of water and stored as a 1:1 slurry in water. Table I shows the incorporation of SDH.

Activation and Reaction. The derivatized beads are activated by a modification of the method used by Inman and Dintzis.[11] One volume of a mixture of concentrated HCl and concentrated H_3PO_4 (90:10, v:v) is added to 10 volumes of a chilled 1:1 slurry of SDH beads. One volume of chilled 1 M $NaNO_2$ is added, and the suspension is stirred in an ice bath for 1–2 minutes. One volume of about 12 M sodium hydroxide is added. The concentration of the base is such that equal volumes of the mixed acid and the base, diluted in water, produces a pH near 6. The neutralized suspension is added directly to the coupling solution, which consists of 10 volumes of 0.2 M sodium bicarbonate at pH 9.5, containing an appropriate concentration of ligand. The slurry is shaken at 4° for 2 hours and then washed extensively with the appropriate buffers. The amount of alanine and albumin coupled to the gels by this method is indicated in Table I. The use of SDH in this method provides an intrinsic, "active" spacer on the gel. Furthermore, the hydrazido-agarose derivatives may be especially useful for immobilizing glycoproteins, oligosaccharides, or glycolipids which can be oxidized by $NaIO_4$.[12]

Stable Activated Derivatives of Agarose

Agarose derivatives containing stable "activated" functional groups, like N-hydroxysuccinimide esters,[13] are convenient for routine applications. The N-hydroxysuccinimide ester and hydrazido derivatives of agarose can be used, by simple and mild procedures, to immobilize proteins and complex group-containing ligands to agarose, without interfering or complicating reactions. For example, it is very difficult to couple amino acids through their amino group to carboxylic-agarose derivatives without selectively blocking other functional groups of the amino acid. The principal alternative procedures for coupling amino group containing

---

[11] J. K. Inman and H. M. Dintzis, *Biochemistry* **8**, 4074 (1969). See also J. K. Inman, this volume [3].
[12] I. Parikh and P. Cuatrecasas, this volume [79].
[13] P. Cuatrecasas and I. Parikh, *Biochemistry* **11**, 2291 (1972).

ligands to derivatized agarose involve the use of water-soluble carbodi-imide reagents or alkylating groups[2,14] and require much longer reaction periods, are much less specific for amino groups (especially alkylating agarose derivatives), result in complicating side reactions if the ligand contains other functional groups, and are less likely to proceed to completion.

If it is desired to place the protein at a distance from the matrix backbone by coupling to hydrocarbon extensions attached to the polymer, the current alternative procedures that can be used will (a) result in reaction of tyrosyl or histidyl residues of the protein (diazonium derivatives); (b) lead to molecular cross-linking and polymerization reactions of the protein (carbodiimide reagents); and (c) result in slow reactions which do not discriminate between amino, imidazole, phenolic, or sulfhydryl groups[2] of the protein (bromoacetyl derivatives). The activated agarose derivatives, on the other hand, react rapidly and almost exclusively with amino groups of the protein under very mild conditions.

N-Hydroxysuccinimide Ester of Agarose[13]

This activated agarose derivative is easily prepared and is stable when stored in anhydrous dioxane or as a dry powder. Exposure of this derivatized agarose to dioxane over extended periods of time (over 6 months) does not appreciably change the chemical and physical properties of the agarose beads. When coupling is performed with buffers at a pH in the range of 5 to 8.5, the unprotonated form of aliphatic or aromatic amino groups reacts very rapidly (5–30 minutes) under very mild (4°) conditions to form stable amide bonds (Fig. 4). Above pH 8.5, however, there is rapid hydrolysis of the active ester. Of the amino acid functional groups tested, only SH-groups compete effectively with free amino groups for reaction with the activated ester of agarose.

Preparation of N-Hydroxysuccinimide Ester of Agarose[13]

Aminoalkyl agarose is prepared by reaction of 3,3'-diaminodipropylamine, or other suitable diamine compound with CNBr-activated agarose as described previously.[2] Succinylamino agarose is prepared as described earlier[2,14] by allowing succinic anhydride (100 mmoles) to react with 100 ml of diaminodipropylaminoagarose and 100 ml of water in a 500-ml beaker. The pH is maintained at 6 by the addition of 5 M NaOH. Ice is added periodically to maintain the temperature below 30°. When no further change in pH occurs, the suspension is maintained for 5 more hours

---

[14] P. Cuatrecasas and C. B. Anfinsen, this series, Vol. 22 [31], 1972.

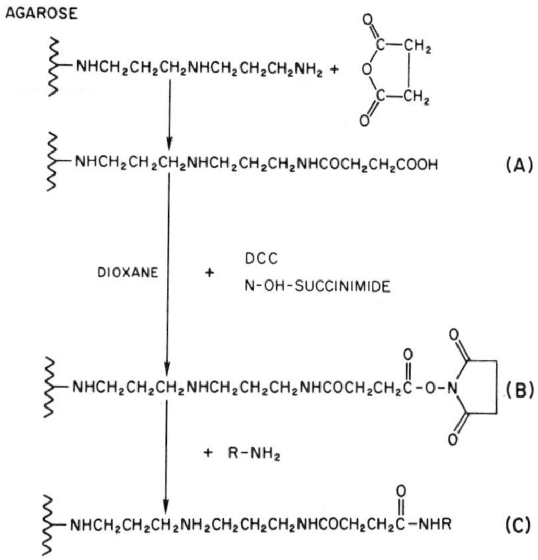

Fig. 4. Scheme of reactions involved in the preparation and use of N-hydroxy-succinide esters of agarose.

at 4°. The derivatized agarose is washed with about 2 liters of water at room temperature and incubated with 0.1 N NaOH for 30–40 minutes at 24°. This step is necessary to cleave labile carboxylic groups (hemisuccinates of agarose hydroxyl groups) formed during the succinylation reaction. If these labile groups are not hydrolyzed at this stage, slow release of the substituted ligand may occur at neutral pH during subsequent steps.

The succinylated agarose is then washed with 2 liters of water and 1 liter of anhydrous dioxane,[15] the latter to obtain anhydrous conditions, and the gel is suspended in 200 ml of dioxane. Solid N-hydroxysuccinimide and N',N'-dicyclohexyl carbodiimide (DCC) are added to achieve a final concentration of 0.1 M each. The suspension is gently shaken for 90 minutes at room temperature. The N-hydroxysuccinimide ester of agarose is washed under suction with 800 ml of dioxane (10 minutes). This is followed immediately with 300–400 ml of methanol (with frequent stirring with a glass rod) to remove the precipitated dicyclohexylurea, and the gel is washed again with 300 ml of dioxane. The N-hydroxysuccinimide ester of agarose is stored, protected from light and at 24° as a 1:1 suspension in dioxane under anhydrous conditions in a tightly capped glass container

---

[15] Dioxane is dried by passage through a neutral alumina column.

TABLE II
EFFECT OF pH ON THE COUPLING OF AMINO ACIDS THROUGH
α-AMINO AND THROUGH ε-AMINO GROUPS[a]

	μMoles per milliliter of packed gel	
pH of coupling	L-Alanine	N-α-Ac-L-Lys
5.0	0.24	0.06
6.0	0.55	0.21
6.5	0.63	—
7.0	0.70	0.36
8.0	0.81	0.44
8.5	0.85	0.56
9.0	0.77	0.61
10.0	0.60	0.48

[a] N-Hydroxysuccinimide ester of agarose (Fig. 4, B), prepared as described in the text, is treated with 10 mM L-[³H]alanine or with N-α-acetyl-L-lysine in buffers of varying pH for 6 hours at 4°. The degree of lysine substitution is determined by amino acid analysis. Data from P. Cuatrecasas and I. Parikh, *Biochemistry* **11**, 2291 (1972).

over molecular sieves.[16] For storage in a dry powder form, the dioxane suspension is filtered and the gel is dried briefly (30–60 seconds) on a sintered-glass funnel followed by lyophilization.

Coupling of Ligand to Activated Agarose. The dioxane suspension of the activated agarose is briefly (30–60 seconds) dried under vacuum over a sintered-glass funnel. The slightly moist cake is added, with continuous stirring, to the buffer (4°) containing the aminoligand in a volume equal to twice that of packed agarose. The pH of the buffer can be varied between 5 and 8.5 (Table II) and the time of the reaction at 4° can vary from 10 minutes to 6 hours. Some buffers cannot be used in the coupling reaction: those containing Tris-, ammonium-, or guanidinium salts, glycine, and others which carry a nucleophilic group capable of reacting with the agarose active ester. Buffers of citrate, phosphate, acetate, and bicarbonate are adequate. At the termination of the coupling reaction the gel should be incubated with 1 M glycine at pH 9 for 2 hours at room temperature to mask residual active ester groups. The adsorbent is finally washed extensively with the desired buffer.

Other condensing agents, such as those described in Table III, at equimolar concentrations, are more effective than DCC in the formation of the N-hydroxysuccinimide ester of agarose. When agarose adsorbents

[16] Molecular sieves (Fisher), types 4A and 5A in the form of ⅛-inch granules or pellets, are useful in maintaining anhydrous conditions. See also, L. F. Fieser and M. Fieser, "Reagents for Organic Synthesis." Wiley, New York, 1967.

TABLE III

USE OF VARIOUS CONDENSING REAGENTS IN THE PREPARATION OF
N-HYDROXYSUCCINIMIDE ESTER OF AGAROSE[a]

Coupling reagent	L-Alanine coupled (μmoles/ml of packed agarose)
N,N'-Dicyclohexyl carbodiimide	1.5
1-Cyclohexyl-3-(2-morpholinethyl) carbodiimide metho-p-toluenesulfonate	4.2
N-Ethoxycarbonyl-2-ethoxyl-1,2-dihydroquinoline	3.5
1-Ethyl-3-(3-dimethylaminopropyl) carbodiimide	4.0

[a] Succinylated agarose (Fig. 4, A) is incubated for 70 minutes at room temperature in dioxane with 0.1 M N-hydroxysuccinimide and 0.1 M of the indicated condensing reagent as described in the text. The active agarose ester is reacted for 8 hours (4°) in 0.1 M NaHCO$_3$ buffer, pH 8.5, containing 50 mM L-[^{3}H]alanine (8 Ci/mole). The coupled derivatives are washed with bicarbonate buffer as described in the text before determination of radioactive content. Data from P. Cuatrecasas and I. Parikh, *Biochemistry* 11, 2291 (1972).

with a higher degree of substitution are desired, it may be advantageous to use one of these reagents.

Selectivity of the Coupling Reaction. It is possible under certain conditions to couple a ligand through α-NH$_2$ groups in preference to ϵ-NH$_2$ or δ-NH$_2$ groups. The conditions described in Table II, however, do not allow discrimination between α- and ϵ-NH$_2$ groups even at the lowest pH, i.e., pH 5.0, tested. Selectivity for reaction with α-NH$_2$ groups can be enhanced (with an approximately 50% fall in the total degree of substitution) by shortening the reaction time (10–20 minutes) and by performing the coupling reaction at a low pH (6.0 or below). In this way, it is possible to take advantage of the large differences that exist in the rates of coupling of these two types of groups (Fig. 5) which in turn reflect differences in pK. During coupling of a peptide or protein discrimination may be even more favorable than for simple amino acids because the α-NH$_2$ group of a peptide is a weaker base than the α-NH$_2$ group of a simple amino acid.

Other Properties of N-Hydroxysuccinimide Ester of Agarose. The activated ester of agarose is very stable when stored in dioxane under anhydrous conditions. After storage for nearly 4 months at room temperature, 80% of the capacity to react with L-alanine is retained. In aqueous buffers at 4°, the activated ester is considerably more stable at pH 6.3 than at pH 8.6. Almost 80% of the capacity to react with L-alanine is retained after exposure of the gel to 0.1 M sodium acetate at pH 6.5 for 40 minutes at 4°. In contrast, only 20% of the coupling

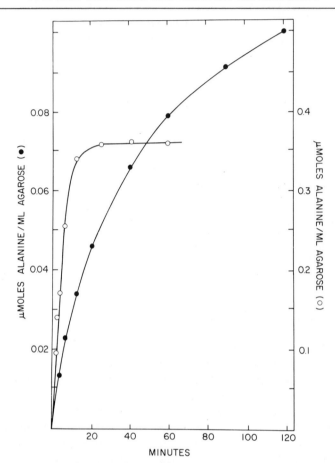

Fig. 5. Rate of reaction of N-hydroxysuccinimide ester of agarose with L-[³H]alanine at 4° in 0.1 M NaHCO₃ buffer, pH 8.6 (○—○), and in 0.1 M sodium acetate buffer, pH 6.3 (●—●). The reaction of N-α-acetyl-L-lysine in 0.1 M sodium acetate buffer, pH 6.3, yields 3 nmoles at 20 minutes and 22 nmoles at 120 minutes (not shown in the figure). Data from Cuatrecasas and Parikh (13).

capacity is retained with buffer at pH 8.6. Increasing the concentration of amino ligand proportionally increases the amount of ligand coupled at pH 6.4 but not at pH 8.4.

Hydrazido Derivatives of Agarose

This type of "activated" agarose has general properties comparable to those of the N-hydroxysuccinimide ester derivatives. Immediately before use, the acylhydrazide groups of these derivatives are converted to the corresponding azide form by a simple reaction with $NaNO_2$ in dilute acid. The acyl azide agarose reacts rapidly with amino group-containing ligands. By selecting a pH near 6 for the coupling reaction, substitution can be directed to occur relatively specifically with α-NH_2 groups relative to ϵ-NH_2 groups of peptides or proteins.

The hydrazido-agarose derivatives are prepared by reaction of succinic dihydrazide with either CNBr activated agarose (page 83) or with $NaIO_4$ activated agarose (page 80). Hydrazido derivatives of agarose substituted with macromolecular spacer arms are prepared by hydrazinolysis of the ester functions which are formed on reaction of bromoacetic ester with derivatives such as poly-L-lysine-agarose or poly-(L-lysyl-DL-alanine)-agarose.

Preparation of Hydrazido Derivatives of Agarose

Hydrazido derivatives of agarose, prepared by the reaction of succinic dihydrazide with sodium periodate-oxidized agarose or with CNBr-activated agarose have been described above. The hydrazido derivatives of agarose gels which contain macromolecular arms offer many special advantages over the succinic dihydrazido-agarose. Depending on the size and molecular weight of the polyfunctional macromolecule used, the ligand or protein to be coupled can be very conveniently separated from the agarose matrix by distances of varying length; approximately 150 Å with poly-(L-lysyl-DL-alanine) of 37,500 daltons.

Poly-(L-*lysyl-*DL-*alanyl-hydrazido*)-*Agarose.* The agarose derivative of poly-(L-lysyl-DL-alanine), prepared by the CNBr method, is converted to a poly-N-carboxymethyl ester by reaction with bromoacetic ester (Eastman). The polyester is then converted to the hydrazide form by treatment with aqueous hydrazine.

Poly-(L-lysyl-DL-alanine)-agarose, 10 ml, is suspended in 10 ml of saturated sodium borate and 1.5 ml of 2-bromoacetic ester is added. The suspension is gently shaken overnight in a tightly closed polyethylene bottle at 24°. The agarose derivative is washed over a sintered-glass funnel with 100 ml of water and 100 ml of dioxane. The agarose cake is suspended in 10 ml of 5 M aqueous hydrazine solution. The suspension, after shaking gently for an additional 8–10 hours at 24°, is filtered and washed extensively with 1 M NaCl solution until the TNBS[2,8] test for hydrazine in the wash is negative. The reaction should be performed in a well ventilated hood.

Coupling Ligands to Hydrazido-Agarose. The coupling of amino group-containing ligands and proteins to hydrazido-agarose is performed by essentially the same method as described earlier for polyacrylamide-hydrazide.[11] Because of the very limited stability of the intermediate acyl azide formed by nitrous acid, the time and temperature of the activation and coupling reactions must be carefully controlled.

Poly-(L-lysyl-DL-alanyl-hydrazido)-agarose, 10 ml, is suspended in 8 ml of water, and 2 ml of 1 M HCl are added. The suspension is cooled in an ice bath for 30 minutes and, while stirring, 2 ml of an ice-cold solution of 1 M NaNO$_2$ is added dropwise over 1-minute period. The suspension is stirred for an additional 2–3 minutes, then rapidly (1–2 minutes) filtered with suction on a coarse sintered-glass funnel (previously cooled) and washed with 20–30 ml of cold 5 mM HCl. The outlet of the sintered-glass funnel is covered with Parafilm, and a 50 mM solution of [^{14}C]-L-alanine (0.1 μCi/μmole) in 0.2 M sodium bicarbonate at pH 8, is added while being stirred with a glass rod. The suspension is transferred to a polyethylene vial and shaken gently for 15 hours at 4°. The substituted agarose is washed extensively with 1 M NaCl. Substitution of [^{14}C]alanine is about 1 μmole per milliliter of agarose.

Preparation of Hydrazido-Albumin

Another approach to the preparation of hydrazido macromolecular derivatives is to synthesize hydrazido-albumin, which can then be coupled to agarose, stored in buffer, and activated for use when needed. Hydrazido-albumin is prepared by esterifying the aspartate and glutamate carboxyl groups with methanol, HCl being used as the catalyst. The esterified albumin is then treated with hydrazine to convert the esters to acyl hydrazides. The hydrazido-albumin is coupled to agarose in denaturing solvents (urea or guanidine·HCl) with CNBr. These agarose beads are then activated by reaction with dilute nitrous acid to convert the hydrazido groups to acyl azide groups. The latter can react rapidly with amines to form amides.

Hydrazido-albumin is prepared by dissolving 5 g of bovine albumin in 100 ml of anhydrous methanol and adding 72 mg (20 mmoles) of anhydrous HCl gas. The solution is stirred overnight at room temperature. A white precipitate forms which is collected by filtration, washed once with ethanol, and suspended in 100 ml of anhydrous ethanol. Hydrazine (3.2 g, 300 mmoles) is added and allowed to react for 20 hours at room temperature while the suspension is stirred. The solvents are removed on a vacuum evaporator, and the residue is suspended in ethanol and dried again. The residue is dissolved in 6 M guanidine·HCl to give a final concentration of 10 mg/ml. One volume of the protein solution is added to 1 volume of 0.2 M NaHCO$_3$ at pH 9.5, and allowed to react with 2 volumes of CNBr-activated agarose by the buffer activation method (page 78). The hydrazido-albumin-agarose is activated with nitrous acid as described above. Preliminary studies with insulin as a test ligand indicate that the coupling efficiency of this macromolecular gel is about 10% of that of CNBr-activated agarose.

Macromolecular Spacer Arms

It has been recognized that ligands attached to agarose by the CNBr activation reaction (directly or through a spacer arm) are slowly released from the matrix into the buffer medium.[17-19] The quantity of ligand released (leakage or bleeding) depends on the temperature and on the specific buffer. The general aspects of this problem are discussed below in the section on stability of gel-bound ligands.

An important approach in the resolution of this problem involves the use, as "anchoring" reagents, of a variety of water-soluble polyfunctional polymers which contain large numbers of primary amino groups. By the proper selection of reaction conditions it is possible to couple the polyfunctional polymers so as to achieve multipoint attachment of the polymer to the backbone of the agarose matrix with resultant increased chemical stability. The probability of spontaneous release of the ligand bound to the CNBr-activated agarose decreases geometrically with the number of attachment points. For example, if alanine is released with a probability of one part per thousand (0.1%), a ligand attached (by the same chemical means) at two points might be expected to be released with a probability of one part per million (0.0001%).

In addition to increasing the chemical stability of the ligand–agarose complex, these polyfunctional spacer arms offer other advantages in affinity chromatography. The macromolecular spacers provide appreciably greater separations of ligand from the matrix backbone than can be achieved with the conventional spacers available. Furthermore, these polymers exhibit minimal hydrophobicity, and the immediate microenvironment created in the vicinity of the ligands or proteins substituted on such polyfunctional spacer arms appear to be especially favorable in many types of interactions encountered in affinity chromatography. For example, estradiol derivatives of albumin-agarose display much greater affinity for uterine estrogen receptors than do comparable derivatives containing conventional spacers (15–25 Å in length).[18,19] Similarly, leukocytes interact much better with agarose beads containing histamine and catecholamines attached to the beads through albumin spacers.[20]

[17] J. H. Ludens, J. R. DeVries, and D. D. Fanestil, *J. Biol. Chem.* **247**, 7533 (1972).
[18] V. Sica, E. Nola, I. Parikh, G. A. Puca, and P. Cuatrecasas, *Nature* (*London*) New Biol. **244**, 36 (1973).
[19] I. Parikh, V. Sica, E. Nola, G. A. Puca, and P. Cuatrecasas, this volume [86].
[20] Y. Weinstein, K. L. Melmon, H. R. Bourne, and M. Sela, *J. Clin. Invest.* **52**, 1349 (1973).

Some of the biospecific adsorbents prepared with the use of polyfunctional spacer arms are so effective that significant extraction of cholera toxin occurs even when the gels are diluted 200- to 600-fold with unsubstituted agarose.[12,21] Adsorbents diluted in this way can often be used routinely for the purification of receptors by affinity chromatography since selective adsorption of receptors in excess of 90% can still be achieved in the cases tested. An important consequence of procedures which use such diluted adsorbents is that the leakage of ligand from the gel is drastically reduced,[19] thus minimizing the complications that can result from such leakage during chromatography of the sample. Furthermore, by diluting the agarose adsorbent with unsubstituted agarose, the possible complications or difficulties encountered in the thorough washing of the gel before use are reduced, and the nonspecific adsorption of proteins is minimized. Finally, elution of the specifically adsorbed proteins is greatly facilitated if such elution is to be performed by an exchange reaction which utilizes buffers containing specific competitive ligands. The preparation and properties of four different polyfunctional agarose derivatives which have been found useful in the isolation and purification of estrogen,[19] insulin receptors,[7] and cholera toxin[12,21] are described here. The four macromolecules used as polyfunctional spacer arms are (a) poly-L-lysine: average MW 160,000 (Schwarz-Mann); (b) poly-(L-lysyl-DL-alanine): average MW 37,500 (Miles); this is a branched copolymer of poly-L-lysine (backbone) and poly-DL-alanine (side chains); the lysine to alanine ratio is 1:15; (c) native albumin: bovine albumin, grade A (Pentex); (d) denatured albumin: urea-denatured bovine albumin, grade A (Pentex).

Poly-(L-lysyl-DL-alanine)-Agarose and Poly-L-lysine-Agarose[22]

The branched-chain copolymer of L-lysine (backbone) and DL-alanine (side chains) is coupled to agarose by the previously described CNBr activation method. An aqueous suspension of 100 ml of packed agarose, in a total volume of 200 ml, is activated with 30 g of finely divided CNBr. The activated agarose cake is immediately added to a solution containing 150 mg of the poly-(L-lysyl-DL-alanine) copolymer in 75 ml of 0.2 M NaHCO$_3$ at pH 9; the polymer is dissolved in 5 ml of water and added to 70 ml of NaHCO$_3$ buffer. The coupling reaction is carried out by gentle agitation on a mechanical shaker at 4° for 15 hours, and for an additional 6 hours at room temperature. The substi-

[21] P. Cuatrecasas, I. Parikh, and M. D. Hollenberg, *Biochemistry* **12**, 4253 (1973).
[22] See also M. Wilchek and T. Miron, this volume [5].

tuted agarose is filtered through a coarse sintered-glass funnel and washed on the funnel without suction with 1 liter of 1 M NaCl over 30–60 minutes. The gel is incubated in 100 ml of 1 M glycine at pH 9 for 4 hours at 24° to mask any remaining activated agarose groups.

The agarose is washed again without suction with 2 liters of 1 M NaCl over 2–3 hours and with 1 liter of water. The substitution of the copolymer on the agarose is 1.2 mg per milliliter of packed gel as judged by the recovery of unreacted polymer in the filtrates. Poly-L-lysine-agarose is prepared by the same procedures.

Albumin-Agarose and Denatured Albumin-Agarose

Albumin is coupled to CNBr-activated agarose in the presence of a high concentration of urea to promote the coupling of this protein in its unfolded state, thus increasing the likelihood of multipoint attachment of the protein to the solid support. The presence of high concentrations of urea does not significantly interfere with the coupling reaction. One hundred milliliters of agarose, activated with 30 g of CNBr, is treated with 400 mg of bovine serum albumin (Pentex) dissolved in 100 ml of 0.2 M NaHCO$_3$ at pH 9, containing 10 M urea. After shaking gently for 15 hours at 4°, followed by reaction for 6 hours at 24°, the adsorbent is washed at 24° with 1 liter of 1 M NaCl (30–60 minutes) and incubated at 24° in 100 ml of 1 M glycine at pH 9, for 4 hours. The albumin-agarose is washed again with 2 liters of 1 M NaCl (2–3 hours) and with 1 liter of water. The substitution of albumin, as judged by recovery in the filtrates, is 2.4 mg per milliliter of packed gel. Native albumin is coupled to agarose (agarose-albumin) as described for denatured albumin except that the albumin is dissolved in 0.2 M NaHCO$_3$ at pH 9 in the absence of urea. The substitution of albumin is 3.5 mg per milliliter of gel.

The Use of Macromolecular Agarose Derivatives

Agarose derivatives containing these polyfunctional macromolecules may be used for coupling ligands containing carboxyl or amino groups as described in this chapter and elsewhere in this volume. For example, the primary amino groups of such agarose derivatives are readily converted to contain a carboxylic acid function by reaction with succinic anhydride. The macromolecular derivatives containing amino or carboxyl groups can be derivatized in the several ways outlined. It may be especially advantageous to use amino acid polymers consisting entirely of D-stereo-isomers since such polymers would be expected to be resistant to hydrolysis by proteases present in crude tissue preparations.

Stability of Gel-Bound Ligands—The Problem of Leakage

Nature of the Problem and Its Control by Manipulating the Experimental Conditions

Early reports on "insolubilized" enzymes attested to the long-term retention of activity of the covalently attached proteins.[23] Enzymically active gels have been used continuously for months with little apparent diminution of activity.[24] Affinity columns have also been reused repeatedly without appreciable loss of binding potency.[25] Nonetheless, recent applications of affinity chromatography to special systems of very high affinity have revealed that in virtually all systems which have been examined carefully, including ligands coupled to agarose by the CNBr procedure, significant and continuous "leakage," i.e., solubilization, of the coupled ligand occurs.[19]

Such leakage of the bound ligand from the gel may interfere seriously with affinity chromatographic procedures. The presence of a small amount of free, solubilized ligand in the gel buffer will not seriously interfere with the interaction of the specific protein with the gel-bound ligand provided that the affinity of the complex is not high, i.e., less than 10 μM, and that the affinity of the protein for the gel-bound ligand is similar to that for the free ligand. In such cases, the rates of dissociation of the complexes are fast and an equilibrium state will be achieved on the column; concentrations of free ligand as high as 1–5% of that present in the bound form will not significantly compete with the much higher concentration of bound ligand. However, if the affinity of the complex under study is very high, e.g., K_d of 1 nM, as is frequently observed in hormone–receptor interactions, and if the free ligand which leaks from the gel is appreciably more effective in complex formation than the gel-bound ligand, the free ligand may bind effectively to the protein in an essentially irreversible manner. In such cases, the protein may not adsorb to the column, and unless special procedures are used to detect the presence of the soluble ligand–protein complex in the gel effluents, the "inactivated" protein may not be detected and may lead to the erroneous conclusion that irreversible adsorption to gel has occurred when, in fact, no protein was bound to the gel. These problems have been studied in detail in the purification of estrogen receptors with estradiol-containing gels.[18,19] In certain cases, the release of estradiol in amounts as low as

[23] H. H. Weetall, *Biochim. Biophys. Acta* **212**, 1 (1970).
[24] H. P. J. Bennett, D. F. Elliott, B. E. Evans, P. J. Lowry, and C. McMartin, *Biochem. J.* **129**, 695 (1972).
[25] P. J. Robinson, P. Dunnill, and M. D. Lilly, *Biochim. Biophys. Acta* **285**, 28 (1972).

0.01% of the total steroid bound can completely "inactivate" the receptor protein applied and render the column ineffective. Since one of the critical factors is the relative affinity of the specific protein for the gel-bound ligand compared to the gel-free ligand, it is important to select chemical conditions for coupling of the ligand to the gel such that the least stable bond in the series of coupling reactions be the initial bond formed by the CNBr reaction. By this means the probability will be increased that the free ligand which is solubilized will not differ greatly from the gel-bound ligand since the chemical composition of the ligand itself will be similar in both cases.

In systems of very high affinity it is therefore important that the affinity adsorbents used contain the least quantity of ligand substitution compatible with effective, specific adsorption. Even if the free, released ligand is not much more effective than that which is still coupled to the gel, the larger concentration of gel-bound ligand will result in the release of proportionately greater quantities of free ligand. This free ligand can bind to the protein in solution, and the stable complex will pass through the column undissociated. If the chromatographic procedure involves a long period of time, as is usually the case in receptor purification where large quantities of sample must be applied and where the columns must be washed extensively before specific elution procedures are instituted, release of ligand will lead to a progressive and continual loss of the specific protein.

Because most specific adsorbents used in high-affinity systems require very small quantities (in some cases approaching stoichiometry) of ligand substitution, it is generally wise to serially dilute the adsorbent with unsubstituted agarose to determine the maximal degree of dilution which is required. An alternative but less convenient approach is to compare a series of different gels, separately prepared, which contain progressively less substitution. Dilution is readily achieved by mixing with untreated agarose. Columns containing such mixtures have the additional advantage of exhibiting less nonspecific adsorption of proteins than gels which have been activated or otherwise treated chemically, even if the degree of substitution is low. Estradiol-agarose derivatives have been used effectively to extract estrogen receptors even when diluted 1:200 with unsubstituted agarose.[19]

Interestingly, the obvious and in some cases dramatic heterogeneity of bead composition does not appear to be a handicap in practice. In fact, there are theoretical reasons why such affinity adsorbents may be more effective than those which contain an equal degree of ligand substitution but which, rather than having the ligand distributed over a few beads, are uniformly substituted. The relatively high local concentration of ligand within the interstices of the matrix of a given bead may aid in retaining the protein within that bead once penetration has occurred. Multiple interactions and collisions can occur within a given bead since the diffusion distances are short. The principal problem which is encountered in using gels which contain relatively few specifically derivatized beads is that the protein entering the gel must find and interact with those few beads. Since this problem deals only with considerations of diffusion, the flow rates of such columns must be appropriately decreased.

Other experimental conditions may be manipulated to decrease the problems arising from ligand leakage. For example, lowering the temperature to 4° can be expected to decrease the rate of the leakage. Rapid flow rates or short periods of time will prevent the accumulation of significant concentrations of free ligand. Batchwise methods of purification may be inferior since the released ligand would progressively accumulate.

The continuous release of a ligand or protein from a gel may represent predominantly the release of material that is nonspecifically *adsorbed* rather than the release of covalently bound material by cleavage of chemical bonds. A number of highly hydrophobic or water-insoluble ligands, such as estradiol, cholesterol, and certain dyes, as well as certain polypeptides, e.g., insulin, glucagon, may adsorb tenaciously to agarose and other gels. Removal of adsorbed estradiol from agarose derivatives requires many days of continuous washing with organic solvents, such as methanol or dioxane. Sometimes adsorbed material can be removed more effectively with solutions that contain high concentrations of protein, e.g., 5% albumin for estradiol gels, which exhibit affinity for the ligand.[18] Dextran Blue may require weeks of continuous washing for removal after it is applied on certain agarose derivatives, despite the high degree of water-solubility of this dye. Similarly, insulin and glucagon adsorbed on previously derivatized (CNBr) and masked (glycine or alanine) gels may require weeks of washing as well as the use of strong denaturants. Thus, when an adsorbent is found to continuously release coupled ligand, the principal initial concern must be directed toward the use of exhaustive washing procedures. Since the adsorbed ligand, when released, will in most cases possess higher affinity for the specific protein than does the coupled ligand released by cleavage of the bond formed by CNBr, especially if spacer arms are used, the problems that can result from the desorption of such ligands are likely to be the most serious.

In addition to the general class of high-affinity systems described above, other interactions that can be studied with insolubilized ligands and proteins may be complicated by the spontaneous release of bound material. For example, examination of the biological activity of insolubilized hormones or other biological mediators may be complicated by the release of significant quantities of the bound substance. In addition to exhaustive washing of the gel, such studies must be carefully controlled to determine whether significant activity has been released into the medium. Frequently the concentration of the gel or of the quantity of ligand on the gel must be decreased progressively until the quantity of material released cannot account for the biological effects observed. The conditions, i.e., time, temperature, and nature of buffer, must be carefully regulated to minimize leakage.

Stability of CNBr-Coupled Ligands

Extensive washing and a variety of treatments of the derivatized agarose, including the use of acids, bases or organic solvents, have been tried to eliminate leakage with only limited success.

In order to obtain some quantitative estimate of the leakage phenomenon of ligands attached to agarose by the CNBr procedure, the process has been studied with agarose gels to which a simple ligand (L-alanine), a protein (albumin), or a polypeptide (insulin) have been coupled. Tritiated alanine is a simple and stable ligand. ^{125}I-Labeled albumin contains several amino groups and thus has the potential for multipoint attachment. ^{125}I-Labeled insulin contains three amino groups for potential attachment, but coupling occurs by only a single bond.[26] Comparisons of the leakage of alanine with insulin may thus primarily reflect differences in nonspecific adsorption. To define "covalently bound," a standard wash sequence is used. The washes are chosen such that they tend to reduce or disrupt known kinds of noncovalent interactions but can still be considered compatible with many macromolecules. The material that is not removed by this washing procedure is operationally defined as that bound to the matrix.

The agarose matrix is activated by the buffer-CNBr reaction described elsewhere in this chapter. The following standard conditions are employed for coupling: protein or peptide, 5 mg/ml; amino acid, 0.1 M; buffer, 0.1 M sodium bicarbonate at pH 9.5; coupling reaction time, not less than 12 hours. The reaction is performed on a mechanical shaker at 4°. After coupling, the beads are transferred to disposable chromatography columns of 2 ml volume and washed at room temperature with 20 ml of (a) 0.1 M sodium bicarbonate at pH 10.0, (b) 2 M urea, and (c) 0.1 M sodium acetate at pH 4; each of these buffers also contains 0.5 M NaCl. Each washing step is done over a 2–4 hour period. The columns are washed with 10 ml of 0.05 M sodium phosphate at pH 7.0 and stored

[26] P. Cuatrecasas, *Proc. Nat. Acad. Sci. U.S.* **69**, 1277 (1972).

TABLE IV

EFFECTS OF VARIOUS TREATMENTS OF CNBr-ACTIVATED AGAROSE ON RATES OF LIGAND RELEASE

Treatment	Amount incorporated			Rate on day 0[a]			Rate on day 25[a]		
	Alanine (μmole/ml)	Albumin (mg/ml)	Insulin (mg/ml)	Alanine	Albumin	Insulin	Alanine	Albumin	Insulin
None	8	7.5	3.3	0.2	0.03	1	0.03	0.006	0.2
Glycine quench[b]	6	5.5	—	1	0.05	—	0.05	0.002	—
NaBH$_4$ reduction[c]	9	4.6	3.2	1	1	2	0.03	0.1	0.1

[a] Rates are expressed as percent of total bound ligand released per day at 4°, as described in the text and in Fig. 2.
[b] Incubation of the various substituted agarose with 1 M glycine for 4 hours at 4°.
[c] Incubation of the various substituted agarose in the presence of 2 ml of 0.1 M sodium borohydride and 0.2 M NaHCO$_3$, pH 9.5.

at 4°. At various times the columns are washed at room temperature with 0.05 M sodium phosphate at pH 7.0, and the washings are assayed for radioactivity. In an effort to modify the rate of leakage the coupled ligands are subjected to various treatments as noted in Table IV.

The rate of leakage from substituted agarose does not follow first-order kinetics over a 30-day period, tending to slow more than would be expected from a first-order extrapolation of the rates observed during the first few days. All the experiments with CNBr-activated materials gave similar rates. In general, the rate of release of albumin is less than that of alanine. Insulin is released more rapidly than alanine, probably reflecting differences in the effectiveness of the original washing procedure or, possibly, disulfide bond reduction and elimination of one of peptide chains of insulin. No chemical treatment of the coupled materials has been successful in significantly decreasing the rate of leakage, although reduction increased the rate (Table IV). From studies such as these, the following rule of thumb is suggested. Simple monoamines will be lost at an initial rate of about 0.1% per day whereas proteins may be lost at a rate about five times slower. The rates of ligand release may decrease by about 6-fold after 30 days.

Control of the Leakage Problem

Currently the leakage problem must be primarily controlled by exhaustive and prolonged washing and by manipulation of the experimental conditions (chemical method of coupling, degree of ligand substitution, flow rates, temperature, etc.).

Another major approach involves the use of soluble polyfunctional spacers to separate the gel matrix from the specific ligand or protein. Since polymers such as the branched copolymer of lysine (backbone) and alanine (side chains) can attach to agarose at multiple points, the overall chemical stability of the bonding of this polymer to the gel, and therefore of other substituents coupled to the polymer, will be greatly enhanced. The use of such macromolecular spacers has been described above as well as in another article.[22]

The leakage of ligands from adsorbents prepared with CNBr probably reflects primarily the fission of the isourea (or imidocarbonate) bonds formed. Unfortunately, attempts to stabilize these bonds by reduction of the isourea bonds with $NaBH_4$ have not yielded more stable derivatives (Table IV). Although slow solubilization of the carbohydrate matrix of untreated agarose is known to occur, it is unlikely that this is a major reason for the leakage since CNBr-treated gels undergo marked stabilization due to cross-linking. Such gels, for example, in contrast to unsubstituted agarose, withstand 1 N HCl and 1 N NaOH. Furthermore, the proportion of total ligand that can be lost over prolonged time periods is generally much greater than any change that can be measured in the packed volumes of the gels. Nevertheless, a greater degree of cross-linking of agarose, which may for example be achieved by treating with epichlorohydrin,[27] may be useful in diminishing gel solubilization.

Methods of coupling ligands to agarose other than that which use CNBr may be necessary to achieve greater stability of the ligand–matrix bond. The periodate oxidation method of coupling offers such an advantage in principle since those bonds are inherently of greater stability.

Alternative kinds of solid supports, polymers, or gels may be helpful in the search for adsorbent–ligand bonds of greater stability. Glass beads containing ligands coupled by silane methods appear to exhibit rates of leakage that are far greater than those observed with agarose. Leakage, is such that it has not been possible to use glass or polyacrylamide derivatives in the purification of estrogen[19] or insulin[26] receptors.

Preliminary studies on cellulosic matrices activated by periodate oxidation and alkylation indicate that this activation method results in about 10-fold higher initial rates of leakage, but that the rates decrease much more rapidly with time. It may be that less drastic oxidation of these matrices will result in lower subsequent leakage rates. Present data suggest that the coupling efficiency of cellulosic matrices "armed" with succinate hydrazides is 20- to 100-fold lower than that found with agarose and CNBr.

[27] J. Porath, J. C. Janson, and T. Laas, *J. Chromatogr.* **60**, 167 (1971).

[7] Ligand Coupling via the Azo Linkage

By LOUIS A. COHEN

General Considerations

In the earliest studies on the covalent coupling of proteins to insoluble supports,[1] the high reactivities of phenols (tyrosine) and of imidazoles (histidine) toward aryldiazonium ions were recognized and were used to advantage. Since 1951, coupling via the azo linkage has been applied widely to the preparation of immunoadsorbents[2] and, to a lesser degree,

[1] D. H. Campbell, E. Luescher, and L. S. Lerman, *Proc. Nat. Acad. Sci. U.S.* **37**, 575 (1951).

[2] D. H. Campbell and N. Weliky, *Methods Immunol. Immunochem.* **1**, 365 (1967).

to the immobilization of enzymes and polypeptides.[3] Yet more recently, this technique has been explored for the coupling of smaller molecules—substrates, inhibitors, and hormones—in the hope of fractionating and purifying proteins and active membrane fragments via affinity chromatography.[4]

Of the ligand coupling methods considered in any new venture into affinity chromatography, the diazonium technique often falls into last place, possibly because three or four steps are required in the standard synthetic sequence.[5] The method offers some advantages, however, which should not be overlooked: (1) the reaction is rapid and usually complete, even at $0°$; (2) coupling may be achieved in aqueous buffer, at pH 5–8; (3) unused diazonium groups are readily blocked with simple phenols; (4) the amine precursor to the diazonium ion can be prepared in large quantity and stored for later use; (5) no excess charge is introduced, nor are the ionic properties of the ligand perturbed significantly; (6) the length of the phenylazo (or polyphenylazo) linkage can provide the often needed "long arm" spacer; (7) the rigidity of the linkage can help prevent a small ligand from folding back on itself; (8) the intense color of the azo chromophore provides a rapid signal of degradation or solubilization of the support; and (9) should the adsorbed material resist elution, the azo linkage can be broken readily by reduction, releasing both ligand and "antiligand."

For individual cases, there are occasional limitations: (1) the phenol or imidazole group of a given ligand or polypeptide sequence may be required for binding; (2) arylazo functions may have strong nonspecific adsorptive properties of their own; and (3) the site of coupling may not be known with certainty unless degradative experiments are done. In coupling reactions above pH 7, phenol and imidazole (in a polypeptide) often compete for the diazonium ion, provided they are equally accessible. Although indole (tryptophan), the amino group, and the guanidino group are also known to undergo coupling,[6] their reactions may be slower than

[3] E. Katchalski, I. Silman, and R. Goldman, *Advan. Enzymol.* **34**, 445 (1971); I. H. Silman and E. Katchalski, *Annu. Rev. Biochem.* **35**, 873 (1966); R. Goldman, L. Goldstein, and E. Katchalski, *in* "Biochemical Aspects of Reactions on Solid Supports" (G. R. Stark, ed.), Chap. 1. Academic Press, New York, 1971.
[4] (a) P. Cuatrecasas, *in* "Biochemical Aspects of Reactions on Solid Supports" (G. R. Stark, ed.), Chap. 2. Academic Press, New York, 1971; (b) P. Cuatrecasas and C. B. Anfinsen, this series, Vol. 22, p. 345; (c) P. Cuatrecasas and C. B. Anfinsen, *Annu. Rev. Biochem.* **40**, 259 (1971).
[5] More direct approaches are currently under investigation in the author's laboratory.
[6] (a) A. N. Howard and F. Wild, *Biochem. J.* **65**, 651 (1957); (b) T. F. Spande and G. G. Glenner, *J. Amer. Chem. Soc.* **95**, 3400 (1973).

those of phenol or imidazole; in effect, few data are available on the extent of coupling to these five functional groups in proteins, either as a function of pH or concentration of diazonium ion. At pH 5–6, imidazole couples in preference to phenol, on the basis of limited studies. In acidic media, tryptophan is probably the only species capable of coupling at a reasonable rate.[6b] The possibility of effecting specific coupling to tryptophan at low pH has yet to be explored, and may prove to be of value.

The site of coupling is most readily determined by dithionite reduction of the azo bond to an aminopolymer and an amino derivative of the ligand. If the liberated species is simple and stable, its structure can often be ascertained by chromatographic or spectroscopic comparison with known material (3-aminotyrosine,[7a] 2-aminohistidine[7b]). In the case of a released polypeptide, acid hydrolysis and amino acid analysis may be necessary. Alternatively, the support may be solubilized by degradation (periodate cleavage or acid hydrolysis[8] for agarose or cellulose) and the azo compound identified by its characteristic visible spectrum. Agarose and some of its derivatives have been found readily soluble in 50% aqueous glycerol or 1,2-propanediol, permitting direct spectroscopic assay[9]; unfortunately, azo derivatives appear to be insoluble in these media.

Comments

Diazonium coupling has been utilized with the four major support systems—cellulose, glass beads, agarose, and polyacrylamide beads. Coupling products derived from p-aminobenzylcellulose[2,10] and related compounds[11] were first utilized as immunoadsorbents.[2] Lerman[12] introduced the concept of affinity chromatography by concentrating tyrosinase with phenolic ligands which had been bound to p-diazobenzyl cellulose. Soybean or potato trypsin inhibitor, azo linked to the same backbone, was effective in purifying trypsin,[13a] while azo-coupled estradiol proved to be capable of binding soluble uterine estradiol receptor protein.[13b]

Although most applications of coupling with diazotized micro glass

---

[7] (a) M. Gorecki, M. Wilchek, and A. Patchornik, *Biochim. Biophys. Acta* **229**, 590 (1971); (b) W. Nagai, K. L. Kirk, and L. A. Cohen, *J. Org. Chem.* **38**, 1971 (1973).

[8] D. Failla and D. V. Santi, *Anal. Biochem.* **52**, 363 (1973).

[9] Peter A. Cohen, unpublished observations (1972).

[10] N. Weliky and H. H. Weetall, *Immunochemistry* **2**, 293 (1965).

[11] M. M. Behrens, J. K. Inman, and W. E. Vannier, *Arch. Biochem. Biophys.* **119**, 411 (1967).

[12] L. S. Lerman, *Proc. Nat. Acad. Sci. U.S.* **39**, 232 (1953).

[13] (a) V. V. Mosolov and E. V. Lushnikova, *Biokhimiya* **35**, 440 (1970); (b) B. Vonderhaar and G. C. Mueller, *Biochim. Biophys. Acta* **176**, 626 (1969).

beads have involved the immobilization of enzymes (acylase,[14a] alkaline phosphatase,[14b] trypsin or papain,[14c] urease,[14d] and β-galactosidase[14e]), this technique has also been effective in binding NAD^+ as an active coenzyme.[15] Epinephrine, norepinephrine, and several of their analogs were found to retain normal physiological activities after being bound to "diazonium glass."[16]

Essentially the same sequence of reactions used to prepare "diazonium glass" has been applied to agarose as a support[4b,17]: (1) cyanogen bromide-activated agarose (cyanoagarose) is allowed to react with an α,ω-diamine of desired chain length; (2) the terminal primary amine is acylated with p-nitrobenzoyl azide; (3) the terminal nitro group is reduced to an amino group with sodium dithionite; (4) the anilino group is diazotized and the diazonium ion (diazonium agarose) is coupled to the ligand.

Diazonium agarose has been linked to the 2- or 4-carbon of estradiol[17]; the product was effective in removing estradiol binding protein from human serum and calf uterine extracts. Since the protein could not be displaced from the gel readily, the azo linkage was reduced with dithionite, releasing the complete aminoestradiol–protein complex. Diazonium agarose is also effective in coupling phenol- and catecholamines.[18a] While the coupling of phenols, such as tyramine, octopamine, or serotonin, pose no special problems, catecholamines or their monoethers are best coupled at pH 5–6; this procedure avoids oxidation of the catechol to an o-quinone and formation of an adrenochrome. Routine use of the low pH for the coupling of all phenolic amines is recommended, since the amino group of the side chain is deactivated by protonation. Even at pH 5, coupling of phenolic ligands is surprisingly rapid (2–5 minutes); enhanced reactivity may be due to preliminary adsorption of the ligand to the agarose (as in an enzyme–substrate complex). In preliminary studies, agarose-bound metaraminol was found effective in blocking 80% of the uptake of norepinephrine by intact rat iris or atrium membranes.[18b]

Modifications of diazonium agarose, containing much longer spacers

[14] (a) H. H. Weetall, *Abstr. 158th Meeting, Amer. Chem. Soc. Contrib.* 153, *Biochem. Sect.;* (b) *Nature (London)* **223**, 959 (1969); (c) *Science* **116**, 615 (1969); (d) H. H. Weetall and L. S. Hersh, *Biochim. Biophys. Acta* **185**, 464 (1969); (e) J. H. Woychik and M. V. Wondilowski, *ibid.* **289**, 347 (1972).

[15] M. K. Weibel, H. H. Weetall, and H. J. Bright, *Biochem. Biophys. Res. Commun.* **44**, 347 (1971).

[16] J. C. Venter, J. E. Dixon, P. R. Maroko, and N. O. Kaplan, *Proc. Nat. Acad. Sci. U.S.* **69**, 1141 (1972).

[17] P. Cuatrecasas, *J. Biol. Chem.* **245**, 3059 (1970).

[18] (a) Louis A. Cohen, to be published; (b) Ingeborg Hanbauer, to be published.

than ethylenediamine, have been used for coupling to glucagon[19a] and to insulin.[19b] At pH 5, coupling is strongly selective for histidine in both polypeptides; at higher pH, both tyrosine and histidine form azo linkages. Although azo-linked glucagon is effective in binding its receptor protein in rat liver membranes, the analogous insulin derivative is ineffective; affinity chromatography was successful, however, with insulin linked to agarose by other methods. In a similar manner, the phenolic alkaloid d-tubocurarine has been coupled to agarose[20]; this ligand was used for the purification of acetylcholinesterase.

In some instances, the ligand itself can be equipped with an amino group suitable for diazotization, requiring that a phenolic recipient be bound to the agarose. Thus, N-p-aminophenyloxamic acid was diazotized and coupled to agarosyl-Gly-Gly-Tyr[21]; this adduct is effective in binding neuraminidase, as well as influenza virus particles which contain this enzyme in an accessible site.[22] Staphylococcal nuclease has been purified by use of the adduct of 3'-(4-diazophenylphosphoryl)deoxythymidine 5'-phosphate and agarosyl-Gly-Gly-Tyr.[23] The p-aminophenyl ester of phosphorylcholine was diazotized and bound to agarosyl-Gly-Tyr, the adduct being used to isolate a myeloma protein which binds phosphorylcholine.[24] Fractionation of nucleoside deoxyribosyl transferases was achieved on a support prepared by allowing m-phenylene diamine to react with cyanoagarose, and by coupling diazotized 6-(p-aminobenzylamino)purine to the immobilized aniline.[25]

Preparation of ω-Aminoalkylagarose[26]

Procedure. A cold, aqueous solution of the diamine (ethylenediamine, 3,3'-diaminodipropylamine, etc.) is adjusted to pH 10 and is added to a suspension of cyanoagarose[27] in an equal volume of cold water. For each milliliter of packed gel (as measured by gentle centrifugation), 2–3

[19] (a) F. Krug, B. Desbuquois, and P. Cuatrecasas, *Nature (London) New Biol.* **234**, 268 (1971); (b) P. Cuatrecasas, *Proc. Nat. Acad. Sci. U.S.* **69**, 1277 (1972).

[20] M. J. Jung and B. Belleau, *Mol. Pharmacol.* **8**, 589 (1972).

[21] This nomenclature was introduced by R. Schwyzer and J. Frank, *Helv. Chim. Acta* **55**, 2678 (1972).

[22] (a) P. Cuatrecasas and G. Illiano, *Biochem. Biophys. Res. Commun.* **44**, 178 (1971); (b) The ion-exchange properties of agarosyl-Gly-Gly-Tyr may be eliminated by use of agarosyl-Gly-Gly-tyramine (M. Wilchek, personal communication).

[23] P. Cuatrecasas, M. Wilchek, and C. B. Anfinsen, *Proc. Nat. Acad. Sci. U.S.* **61**, 636 (1968).

[24] B. Chesebro and H. Metzger, *Biochemistry* **11**, 766 (1972).

[25] R. Cardinaud and J. Holguin, *Biochem. J.* **127**, 30P (1972).

[26] This procedure is a slight modification of that given in references 4b and 17.

[27] Prepared according to references 4b and 17, or purchased from Pharmacia Chemicals, Inc.

mmoles of diamine are used. Should the diamine be insufficiently soluble in water, dimethylformamide or tetramethylurea may be added to increase solubility. The mixture is stirred at 0 to 5° for 15–18 hours and is then filtered through a coarse sintered-glass funnel with gentle suction. The gel is washed with sufficient portions of distilled water to render the filtrate neutral. In the case of ethylenediamine, the product contains about 12 μmoles of aminoethyl group per milliliter of settled gel. For longer or bulkier diamines, the degree of coupling may be somewhat less. The gel should not be permitted to dry out during suction or storage. Amino-alkylagarose appears to survive storage at low temperature for at least 4–6 months.

Preparation of p-Aminobenzamidoalkylagarose[26]

Procedure. Sodium borate (0.2 *M*, pH 9.3) is diluted with dimethyl-formamide (6:4). The aminoalkylagarose is suspended in a volume of this solvent mixture twice its own settled volume, and the mixture is cooled to 0–5°. A solution of *p*-nitrobenzoyl azide in 0.1 *M* dimethylformamide is added gradually to the gel suspension, with stirring. The total amount of azide added is 80–100 μmoles per milliliter of settled gel. The reaction mixture is stirred at 5° for 1 hour and at ambient temperature for an additional 3–4 hours. Completion of the acylation reaction can be checked by loss of a color reaction with trinitrobenzenesulfonic acid.[17,28] The gel is washed thoroughly with 50% dimethylformamide and with water.

The *p*-nitrobenzamidoalkylagarose is suspended in a solution of 0.2 *M* sodium dithionite in 0.5 *M* sodium bicarbonate at pH 8.5, and the mixture is maintained at 40° for 1–2 hours with stirring. The gel is filtered, then washed thoroughly with water. This product may be stored in distilled water at low temperature for later use.

Coupling of p-Aminobenzamidoalkylagarose to Ligands[18a]

The previous product is suspended in an equal volume of ice-cold 0.5 *N* hydrochloric acid. A solution of sodium nitrite (0.1 *M*, 100 μmoles per milliliter of gel) is added, and the mixture is stirred for 7 minutes in an ice bath. The gel is filtered through a coarse sintered-glass funnel, is washed thoroughly with cold water and is resuspended in an equal volume of cold acetate (0.2 *M*, pH 5.0–5.5) or phosphate (0.2 *M*, pH 6–7) buffer.[29] The ligand (phenol or imidazole compound) is dissolved in a similar volume of buffer, or in buffer–dimethylformamide (2:1) if necessary, and this solution is added in one portion to the suspension of diazo-

---

[28] Impure batches of dimethylformamide may give a positive test.
[29] The diazonium agarose should be kept cold and used immediately after generation. It cannot be stored without decomposition.

nium agarose, stirred in an ice bath. Formation of the colored azo gel begins immediately, and coupling is usually complete in 10–30 minutes. During this time, the mixture is allowed to warm to room temperature. If desired, any unreacted diazonium groups[30] may be blocked at this point by addition of a dilute solution of β-naphthol or of 2,6-di-*tert*-butylphenol in 50% dimethylformamide. The gel is filtered and washed thoroughly with water, with 50% dimethylformamide, and again with water, and is stored (cold) as a suspension in water or buffer.

[30] This procedure is usually unnecessary for small ligands, especially when used in moderate excess in the coupling reaction.

[8] Interfering and Complicating Adsorption Effects in Bioaffinity Chromatography[1]

By Pádraig O'Carra, Standish Barry, and Tadhg Griffin

The term affinity chromatography was originally introduced to denote adsorption chromatography based on enzymatically specific interactions,[2] as distinct from adsorption dependent on gross physicochemical properties, which in this context has generally been referred to as nonspecific adsorption. This terminology is not entirely satisfactory since the term affinity could be validly applied to any form of adsorption, specific or otherwise, while the so-called nonspecific adsorption effects often have quite a high degree of apparent specificity, as evidenced by the enzyme purifications achievable by ion-exchange chromatography.

Discussion is facilitated by the use of less ambiguous terms, and for this reason a modified terminology proposed elsewhere[3] will be followed here. Adsorption dependent on biologically specific interactions (as illustrated diagrammatically in Fig. 1A) is termed *bioaffinity* or *biospecific adsorption*. Adsorption which does not fall into this category is termed *nonbiospecific adsorption* rather than merely nonspecific adsorption. A similar distinction is applied in regard to the elution process, the term *bioelution* being applied to elution promoted by a soluble, biospecific

[1] Based in part on research supported by grants from the Medical Research Council and National Science Council of Ireland.
[2] P. Cuatrecasas, M. Wilchek, and C. B. Anfinsen, *Proc. Nat. Acad. Sci. U.S.* **61**, 636 (1968).
[3] P. O'Carra, in FEBS Symposium on "*Industrial Aspects of Biochemistry*" (B. Spenser, ed.), p. 107. North-Holland Publ., Amsterdam (1974).

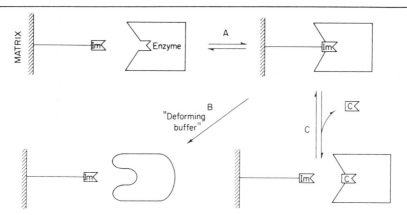

FIG. 1. Diagrammatic representations of (A) biospecific adsorption; (B) elution by a "deforming buffer"; (C) bioelution with a soluble, competitive counterligand. Im, immobilized ligand; C, competitive counterligand.

counterligand competing with the immobilized one, as visualized in Fig. 1C. By contrast, nonbiospecific elution may involve gross physicochemical parameters (e.g., ionic strength-dependent elution from ion-exchange resins) or gross conformational changes in the enzyme caused by so-called "deforming buffers" as illustrated diagrammatically in Fig. 1B.

Although many workers have referred to the possibility of complicating nonbiospecific adsorption effects in affinity chromatographic systems, they have not always rigorously eliminated such interference. As a result, much of the affinity chromatography so far reported may be complicated by such phenomena. As shown below, it now appears that some important model studies probably depend largely, if not entirely, on nonbiospecific effects.

The confusion of nonbiospecific with biospecific effects in the adsorption process is of more than academic interest and has considerable practical consequences. First, such confusion has resulted in the formulation of incorrect generalizations or rules-of-thumb which have to some extent adversely affected the design of new affinity chromatographic systems.[3,4] Apart from this, the failure to recognize or eliminate nonbiospecific complications early in the development of a particular affinity chromatographic system interferes directly with the main advantages that bioaffinity offers over conventional methods of purification. These anticipated advantages are: (a) rationality of design and operation, following a logic

[4] P. O'Carra, submitted to *Biochem. J.*

based on the known biological specificity of the enzyme to be purified; (b) ultraspecificity, deriving from the specificity of biological interactions.

Interfering nonbiospecific adsorption may invalidate the rationale and affect the specificity in a largely unpredictable manner. Even though the operation of a system dominated by nonbiospecific adsorption may be adjustable to yield a reasonably good purification, the optimization of its efficiency will not be conducive to logic based on biospecificity. Even when the system is optimized by a trial-and-error approach, attempts to extend its use—for example to the purification of the same enzyme from another source—will have no guarantee of success owing to the fact that the gross physicochemical properties of biologically homologous enzymes do not necessarily parallel their biospecificity. Such practical consequences are evident, for example, in the case of the system for ribonuclease A discussed below.

It is even more crucial to distinguish rigorously between biospecific and nonbiospecific effects when an affinity chromatographic system is used as a tool for specificity studies or other enzyme-ligand binding studies in which the chromatographic behavior is interpreted in terms of active-site binding.[5,6] The care which must be exercised in this regard may be illustrated by some secondary applications of the affinity system developed for the purification of β-galactosidase.[7] As discussed under β-galactosidase in the section on examples, the evidence now indicates that this system is largely, if not entirely, dependent on nonbiospecific interactions rather than on specific active-site binding.[8] This fact does not necessarily affect the value of this system in the original limited application for which it was developed (as a step in the purification of β-galactosidase from *Escherichia coli*), but it does seriously undermine any conclusions drawn from the chromatographic results on the assumption that these reflect active-site-specific binding—as, for example, the interpretation of the chromatographic behavior of various mutant and fragmentary forms of β-galactosidase in terms of comparative active-site competence.[9]

This β-galactosidase system has also played an important role in the development of the view that immobilized ligands with dissociation constants in the millimolar range are capable of promoting very strong specific adsorption,[10,11] and this view in turn has led to the development of the

[5] P. O'Carra and S. Barry, *FEBS Lett.* **21**, 281 (1972).
[6] S. Barry and P. O'Carra, *Biochem. J.* **135**, 595 (1973).
[7] E. Steers, Jr., P. Cuatrecasas, and H. B. Pollard, *J. Biol. Chem.* **246**, 196 (1971).
[8] P. O'Carra, S. Barry, and T. Griffin, *Biochem. Soc. Trans.* **1**, 289–290 (1973).
[9] M. R. Villarejo and I. Zabin, *Nature (London) New Biol.* **242**, 50.52 (1973).
[10] P. Cuatrecasas and C. B. Anfinsen, this series, Vol. 22, p. 345.
[11] P. Cuatrecasas and C. B. Anfinsen, *Annu. Rev. Biochem.* **40**, 259 (1971).

concept of "progressively perpetuating effectiveness."[12] This concept is not consistent with much other evidence, or with the theoretical treatment summarized in the next section, and it is difficult to reconcile with kinetic theory in general.[3,4] Again, the observations that largely gave rise to this concept are explainable in terms of nonbiospecific rather than biospecific adsorption.

Detection of Nonbiospecific Interference

A recently developed theoretical treatment of affinity chromatography has facilitated rigorous analysis.[3,4] The chief value of such analysis to date has been the uncovering of discrepancies that have been traced to the forms of interference discussed below.

Some previous attempts at theoretical treatment failed to adequately reveal such discrepancies, either because the treatment[12] consisted of the formulation of generalizations based on observations that were themselves complicated with nonbiospecific effects or because the treatment[13,14] failed to relate the enzyme–ligand binding in a useful or quantified manner with chromatographic parameters.

Rigorous and useful analysis becomes possible only if the retardation of the enzyme by the column of adsorbent is quantified in units directly relevant to practical chromatographic operations, but free from dependence on such arbitrary and variable factors as the volume of the column and the rate of chromatographic irrigation. A simple and effective method of quantification is achieved by expressing the elution volume of the enzyme as a ratio relative to the operating volume of the column, after subtraction of the breakthrough volume. In other words, retardation (R) is made equivalent to the elution volume of the enzyme expressed in column-volume units, minus one unit representing the initial breakthrough volume through which even a completely unretarded substance must pass.

When expressed thus, the retardation due solely to bioaffinity, R_{bio}, may be readily shown[4] to be related to the molar dissociation constant, K_{im}, and the effective molar concentration, [Im], of the immobilized ligand by the following equation.

$$R_{bio} = [Im]/K_{im} \qquad (1)$$

Addition of a competitive counterligand to the irrigant will result in a decrease in the biospecific retardation depending on the relative concentrations and dissociation constants of the immobilized and soluble ligands.

[12] P. Cuatrecasas, *Advan. Enzymol. Relat. Areas Mol. Biol.* **36**, 29 (1972).
[13] R. H. Reiner and A. Walch, *Chromatographia* **4**, 578 (1971).
[14] H. F. Hixson, Jr. and A. H. Nishikawa, *Fed. Proc., Fed. Amer. Soc. Exp. Biol.* **30**, 1078 (Abstract) (1971).

The resultant retardation, again assuming dependence only on pure bioaffinity, may be shown[4] to be described by the equation

$$R_{bio} = \frac{[Im]}{K_{im}}\left(\frac{K_c}{K_c + [C]}\right) \tag{2}$$

where K_c and $[C]$ are the dissociation constant and the concentration respectively of C, the soluble, competitive counterligand.

Other theoretical considerations indicate that for multisubunit enzymes the immobilized state of the ligand should result effectively in a decrease in the dissociation constant, relative to that of an exactly equivalent soluble ligand, by a factor equal to the number of equivalent binding sites per enzyme molecule.[4] This appears to be the only complication introduced intrinsically by the immobilized state as such, and any other discrepancies between theoretical prediction and experimental observation are regarded as possible indications of interference.

Negative discrepancies, i.e., retardations lower than predicted, are in general readily attributable either to occlusion of the ligand, usually by the matrix, resulting in a decrease in the accessibility (and hence in the effective concentration) of the ligand, or to steric hindrance by the immobilization linkage at the point of attachment of the ligand, resulting in an increase in the effective dissociation constant.

Retardations which are considerably greater than predicted (after due allowance for any "multiple binding-site effect") have invariably been traced to interference from nonbiospecific adsorption effects. Such interference also results in anomalous bioelution behavior; in cases of severe interference, bioelution is usually not demonstrable at all (see below for examples).

The principal reason for the initial lack of awareness of such interference in affinity chromatographic studies is probably attributable to the most commonly practiced methods of elution. These have most commonly employed so-called "deforming buffers," that is buffers of extreme pH, ionic strength, or other physicochemical property, which elute the macromolecule by deforming or distorting its conformation, thereby weakening or abolishing the chromatographic adsorption, supposedly as visualized in Fig. 1B.

Such methods of elution must be regarded as largely nonbiospecific in character, and they are much less logical (and less effective) than biospecific methods of elution using biospecific counterligands; a deforming buffer is just as likely to exert an effect on gross physicochemical interactions between enzyme and adsorbent as on biospecific ones. Indeed, in many cases, the eluting parameters bear no demonstrable relationship whatever to any known biospecific interactions of the enzyme being purified, and the entire chromatographic process may easily depend on non-biospecific effects rather than on bioaffinity.

Thus in many studies the nonbiospecific nature of the elution process may hide a similar lack of biospecificity in the adsorption process. Examples of this effect are to be found in the first three examples in the section on examples, below.

Sources and Nature of the Interference

Until recently, ion-exchange effects seem to have been the only type of positive interference seriously considered by many workers (see, e.g., references 15 and 16), and it appears to have been widely believed that interference could be adequately controlled by eliminating ionic groups in the matrix material and in the spacer arms used. In many cases, however, the ligands themselves are ionic and may promote ion-exchange binding. Interference from this source seems possible in the affinity system for ribonuclease A, as outlined below in the section giving examples and in the chromatography of certain NAD-linked dehydrogenases described by Lowe and Dean[17] as outlined in the examples section. However, studies with "mock affinity systems" indicate that the most common source of interference stems from the spacer arms in common use, in spite of their largely nonionic character. "Mock affinity systems" consist of analogs of the affinity gels with the biospecific ligand either absent or replaced by a close chemical analog having little or no biospecific affinity for the enzyme under consideration—see, for example, the investigation of the system for β-galactosidase (section on examples, below) and for NAD-linked dehydrogenases (section on examples, below).

Such studies have revealed widespread interference from the hydrocarbon spacer arms commonly in use.[3,8,18] The nature of the interfering adsorption has not as yet been fully elucidated, but the extent and magnitude of the interference correlates closely with the extent of the hydrophobic regions in the spacer arms. This suggests that the interference results largely from what are loosely termed hydrophobic interactions, and the elimination of much of the nonbiospecific binding by replacement of the hydrophobic spacer arms by polar or hydrophilic ones (as described below in the sections on control of interference and on examples) seems to confirm this interpretation.

Hydrocarbon-substituted agarose gels have recently received attention

[15] P. D. G. Dean and C. R. Lowe, *Biochem. J.* **127**, 11–12p (1972).
[16] H. F. Hixson, Jr. and A. H. Nishikawa, *Arch. Biochem. Biophys.* **154**, 501 (1973).
[17] C. R. Lowe and P. D. G. Dean, *FEBS Lett.* **14**, 313 (1971).
[18] P. O'Carra, S. Barry, and T. Griffin, unpublished work (1973).

as adsorbents in their own right for chromatography of proteins.[19-22] Such "hydrophobic chromatography" has been regarded by some authors as a form of affinity chromatography,[21,22] but the adsorption involved has been recognized in most cases as being basically nonbiospecific in character, and the development of such hydrophobic chromatography has independently drawn attention to the possible complications deriving from hydrocarbon spacer arms in bioaffinity chromatography.[23]

In a number of cases, the nonbiospecific adsorption caused by predominantly hydrophobic spacer arms could be eliminated by raising the ionic strength[3,5,6,18] as well as by replacing the arm by a polar one. The azo-NAD system for lactate dehydrogenase described below under examples typifies such behavior. It is generally considered that hydrophobic interactions should be strengthened rather than weakened by raising the ionic strength of the medium. Therefore the salt-sensitive nonbiospecific adsorption in such systems may involve other interactions, such as van der Waals and possibly electrostatic interactions of a nonionic nature.

A further, and seemingly paradoxical, complication appears to be explainable only on the basis of nonbiospecific adsorption which does not become obtrusive unless the enzyme is brought into close contact with the spacer arm by prior biospecific adsorption on an attached ligand. The type of behavior which suggests such a complicated type of interference is exemplified by the case of glyceraldehyde-3-phosphate dehydrogenase discussed below. This phenomenon has been termed *ligand-dependent nonbiospecific adsorption*[3] and is characterized by the proportion of the enzyme which is bioelutable becoming progressively smaller the longer the enzyme is left adsorbed on the ligand-arm assembly. The enzyme seems to become progressively more enmeshed in the hydrophobic spacer arm the longer it is left in close proximity to it.[3,6] This effect also is abolished by replacing the arm with a hydrophilic counterpart as described below.

One further type of interference may become operative where the spacer arm is long and flexible and capable of interacting noncovalently with the attached ligand. Molecular models have suggested in a number of such cases that the entire ligand-arm assembly might be capable of adopting a folded conformation in which the ligand "curls back" and interacts with the spacer arm, e.g., by hydrophobic interactions, in such a

[19] R. J. Yon, *Biochem. J.* **126**, 765 (1972).

[20] Z. Er-el, Y. Zaidenzaig, and S. Shaltiel, *Biochem. Biophys. Res. Commun.* **49**, 383 (1972). See also this volume [9].

[21] B. H. J. Hofstee, *Anal. Biochem.* **52**, 430 (1973).

[22] P. Roschlau and B. Hess, *Hoppe-Seyler's Z. Physiol. Chem.* **353**, 441 (1972).

[23] S. Shaltiel and Z. Er-el, *Proc. Nat. Acad. Sci. U.S.* **70**, 778 (1973). See also this volume [9].

way that it becomes masked or occluded. Such "conformational occlusion" has been suggested in explanation of the unexpected chromatographic ineffectiveness of certain affinity gels bearing hydrophobic ligands attached through hydrophobic spacer arms.[3,6]

Compound Affinity

Although nonbiospecific adsorption is, in general, an undesirable complication in affinity chromatography, a number of cases have now been documented in which nonbiospecific interactions could be judiciously controlled in such a way as to usefully reinforce weak bioaffinities.[3,6,18] In such cases the operating conditions are carefully adjusted or balanced so as to allow a residual level of nonbiospecific interaction which is just sufficient to reinforce the weak bioaffinity synergistically without dominating or obscuring it. The reinforced bioaffinity behaves very much like strong bioaffinity, and the enzyme may be cleanly eluted with biospecific counterligands.

This effect has been termed *compound affinity*.[3,6] Examples are described below in the last section. Another example of such an effect is the reinforcement of the weak "abortive complex affinity" of the H-type isoenzyme of lactate dehydrogenase described elsewhere in the volume.[24] In such compound affinity it appears that the two types of weak interaction working in conjunction somehow cause strong adsorption of the enzyme, while either alone is incapable of causing more than weak retardation. Thus, when either is abolished, the enzyme is eluted. Such compound affinity is therefore readily distinguishable from pure bioaffinity by observing the effect on the chromatographic behavior produced by eliminating the nonbiospecific interactions completely.

In the examples described in the last section, the reinforcing interactions seem to derive from the hydrophobic spacer arms as evidenced by the sensitivity of the effect to changes in the character, as distinct from the length, of these arms. It has been suggested[6] that in such systems a positive reinforcing effect may constitute a more important element in the role of the arm than the passive spacing effect.

Although such reinforcement of weak bioaffinity can be very useful, the critical involvement of nonbiospecific interactions introduces some of the complications already discussed in the first section in relation to nonbiospecific adsorption effects generally. Since compound affinity involves a certain dependence on the gross physicochemical properties of the enzyme as well as on its bioaffinity, a method based on such compound

---

[24] P. O'Carra and S. Barry, this volume [77].

affinity may not be automatically transferable to an homologous enzyme from another source.

Control and Elimination of Interference

Approaches to the control or elimination of nonbiospecific adsorption fall into two categories: (a) adjustment of operating conditions, (b) modification of the adsorbent.

Only the former approach seems applicable where the interfering adsorption, as well as the bioaffinity, derives from the ligand itself, but in the majority of cases, where the interference derives from the spacer arms, both approaches are possible.

The ionic strength of the irrigant has been found the most useful adjustable parameter in the operating conditions to date.[3,5,6] A salt, e.g., NaCl or KCl, is simply added to the irrigant in increasing concentrations until satisfactory control of the nonbiospecific adsorption is achieved. Interfering ion-exchange adsorption is most obviously controllable by such means, but as mentioned in the preceding section, the interaction of some enzymes with spacer arms of apparently hydrophobic (and certainly nonionic) character may also be eliminated by increasing the ionic strength sufficiently. Examples of such control may be found in the next section and elsewhere in this volume.[24]

Where the bioaffinity of the system is weak, it may be more advantageous to allow the operation of compound affinity as discussed in the preceding section. This requires careful, and largely trial-and-error, balancing of the ionic strength rather than arbitrarily increasing it to a high level. Examples of such control, as distinct from elimination, of nonbiospecific interactions may also be found in the next section and elsewhere in this volume.[24]

While in certain cases the bioaffinity of enzymes for active-site ligands may be affected by changes in ionic strength, available kinetic evidence suggests that such cases are exceptional. Relatively high concentrations of NaCl or KCl in particular have little effect on the specific ligand binding of many enzymes in spite of the seemingly widespread assumption to the contrary in the field of affinity chromatography—see, for example, the case of the adsorption of lactate dehydrogenase on ribosyl-linked NAD^+ discussed in the next section.

By contrast, many other possible types of adjustment of the operating conditions—including, for example, pH and temperature changes—are more likely to disturb the bioaffinity and are less generally useful. Nevertheless, adjustment of the operating temperature has been found to be important in the perfection of at least one affinity chromatographic system.[24] In view of the temperature-dependence of at least some nonbiospecific

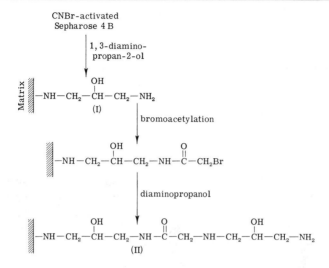

Fig. 2. Structures and method of preparation of typical polar or hydrophilic spacer arms. Beaded agarose gel (Sepharose 4B, Pharmacia) is activated with CNBr in the usual manner [P. Cuatrecasas and C. B. Anfinsen, this series, Vol. 22, p. 345; P. Cuatrecasas, *J. Biol. Chem.* **245**, 3059 (1970)]. (See also this volume [2] and [6].) The washed, activated gel is quickly mixed with an equal volume of an ice-cold 3 M solution of 1,3-diaminopropan-2-ol adjusted to pH 10, and the mixture is stirred gently at 4° for about 20 hours. The resulting derivatized gel (I) is washed with about 40 volumes of 0.2 M saline followed by water, and it is then bromoacetylated as described by Cuatrecasas (*loc. cit*). The bromoacetylated gel is mixed with an equal volume of 0.4 M 1,3-diaminopropan-2-ol in 0.1 M NaHCO₃ buffer adjusted to pH 9, and the mixture is stirred gently at room temperature (ca. 15°) for 3 days. The gel is washed with 0.1 M saline (about 40 volumes) and with distilled water (about 10 volumes), and any unsubstituted bromoacetyl groups remaining are blocked by incubation of the gel with 0.2 M mercaptoethanol in 0.1 M NaHCO₃ buffer, pH 9 at 15° for 24 hours followed by washing as before. The process of bromoacetylation followed by addition of diaminopropanol may be repeated to yield even longer arms. The gels may be further derivatized by a number of the general methods described throughout this volume so as to provide suitable termini for attachment of the ligands. Data from P. O'Carra, S. Barry, and T. Griffin, unpublished.

adsorption effects and probably of many bioaffinities, it should be borne in mind that an affinity system which works well at one temperature may be much less successful, or may require further operational adjustments, if the operating temperature is changed.[3,24]

    In several cases of adsorption of enzymes on hydrophobic spacer arms, salt concentrations even above 1 M did nothing to ameliorate the interference (see e.g., the examples of β-galactosidase and alcohol dehydrogenase in the next section). Attempts to counteract such severe hydrophobic interference by addition of organic solvents or urea to the irrigant have met with little success, mainly because adequate concentrations of these additives are also capable of denaturing the enzymes.

However, the alternative general approach—modification of the adsorbent, particularly the spacer arms—has shown promise with a number of such "sticky" enzymes. The most successful tactic has been to incorporate polar groups, e.g., carbinol, amido and secondary amino groups, at regular intervals along the length of the spacer arms so that no extensive hydrophobic region remains. Prototype examples of such hydrophilic arms are shown in Fig. 2.

Evidence of the effectiveness of such modified spacer arms in eliminating nonbiospecific adsorption is to be found in examples devoted to β-galactosidase, glyceraldehyde-3-phosphate dehydrogenase, and yeast alcohol dehydrogenase in the next section. Such hydrophilic spacer arms appear to eliminate nonbiospecific adsorption so completely that even the marginal level of such adsorption required as a reinforcing factor in compound affinity is also eliminated (cf., in next section, the examples in subsections on azo-linked NAD^+ and on the 6-linked analogs of NAD^+ and AMP).

Some Examples

β-Galactosidase

The affinity chromatographic system developed by Steers et al.[7] for β-galactosidase employs a β-thiogalactoside ligand attached to agarose via a spacer arm, as represented by structure (III). The method of elution is of the "deforming buffer" type and employs 0.1 M borate at pH 10. This affinity system promoted very strong adsorption of the enzyme even when the substitution level of the gel was as low as 0.6 mM, i.e., 0.6 μmole of ligand per milliliter of packed gel.[7] The long spacer arm of derivative (III) was found to be necessary for this strong adsorption, derivatives with shorter arms being ineffective.[7] This was interpreted purely in terms of elimination of steric hindrance, and these results have been widely quoted in support of the view that really long spacer arms may be necessary to eliminate steric hindrance.[7,10–13,25]

However, since the K_i value for the β-thiogalactoside ligand is as high as 5 mM,[7,11] there is an apparent wide discrepancy between the very weak retardation predicted by Eq. 1 (above) and the very strong adsorption observed with derivative (III).[3,9] Furthermore, satisfactory bioelution was not observed with this system.[3,7]

[25] P. Cuatrecasas, J. Biol. Chem. 245, 3059 (1970).

$$—NH—CH_2—CH_2—CH_2—NH—CH_2—CH_2—CH_2—NH—\overset{O}{\overset{||}{C}}—CH_2—CH_2—\overset{O}{\overset{||}{C}}—HN—\langle\rangle—S-\beta\text{-Galactoside}$$

(III)

$$—NH—CH_2—CH_2—CH_2—NH—CH_2—CH_2—CH_2—NH—\overset{O}{\overset{||}{C}}—CH_2—CH_2—\overset{O}{\overset{||}{C}}—HN—\langle\rangle—O-\alpha\text{-Glucoside}$$

(IV)

$$—NH—CH_2—CH_2—CH_2—CH_2—CH_2—CH_2—NH—\overset{O}{\overset{||}{C}}—CH_2—HN—\langle\rangle$$

(V)

Nonbiospecific analogs of derivative (III) were therefore prepared and studied. Derivative (IV) is an analog which differs from (III) in that the β-galactoside ligand is replaced by an α-glucoside residue (for which β-galactosidase should show little or no bioaffinity). Derivative (V) is a representative example of a number of "mock affinity gel" carrying no recognizable ligand at all. The chromatographic behavior of β-galactosidase on derivative (IV) was indistinguishable from its behavior on derivative (III)[3,18] and the enzyme was even more strongly adsorbed on the ligandless derivative (V) and on a number of similar analogs. Elution could still be achieved with 0.1 M borate at pH 10, but only slowly and with much tailing.[3,9] The enzyme was not eluted from any of the gels by biospecific counterligands, i.e., soluble β-galactosides, even when high salt concentrations were applied at the same time.[3,8,18]

The adsorption involved in this system therefore seems to be predominantly, attributable to nonbiospecific interactions involving the spacer arms. The hydrocarbon nature of these arms and the apparent ineffectiveness of high salt concentrations in counteracting the adsorption suggest that the interactions are of a hydrophobic nature. This seems to be confirmed by the fact that β-galactosidase was not strongly adsorbed on gels carrying polar or hydrophilic spacer arms such as that of derivative (II) (Fig. 2).

Ribonuclease A

The affinity gel developed by Wilchek and Gorecki[26] for ribonuclease A utilizes as adsorbent agarose-APUP, a uridine 2'(3'),5'-diphosphate ligand immobilized by attachment to agarose via an aminophenyl spacer group. The enzyme is satisfactorily adsorbed on this gel only when applied

[26] M. Wilchek and M. Gorecki, *Eur. J. Biochem.* 11, 491 (1969). See also this volume [57].

in dilute buffer at pH 5.2. In the original work elution was achieved with 0.2 M acetic acid, but elution can also be achieved by raising the pH to 7.5 at moderate ionic strength (0.1 to 0.2 M KCl) or by raising the ionic strength to a high level (0.5 to 1 M KCl) at pH 5.2.[27] The pH optimum of the enzyme is in the region of pH 7.0, and salt concentrations up to 0.2 M are not notably inhibitory.[28] It is therefore difficult to rationalize the adsorption and elution behavior satisfactorily in terms of pure bioaffinity. In view of the anionic nature of the ligand, due to its phosphate groups, the possibility that the adsorption may be complicated by nonbiospecific ion-exchange effects must be considered. This possibility is also suggested by the fact that the chromatographic behavior of bovine pancreatic ribonuclease A on simple anionic ion-exchange resins such as IRC-50 and CM-cellulose is in many respects similar to its behavior on the affinity gel; adsorption of the enzyme is strong at or near pH 5.2 and is weakened by raising the pH or by increasing the ionic strength.[27,29] The enzyme may also be eluted—though much less satisfactorily—with aqueous acetic acid.[27]

Difficulties have been reported by several workers who have used this affinity system in the purification of ribonucleases from sources other than beef pancreas. A ribonuclease from a myxomycete[30] and one of broad specificity from a shellfish source[31] behaved quite differently from the bovine pancreatic enzyme; the purification achievable in these cases was not considered useful. Even pancreatic ribonuclease from other mammalian species behaved differently and modifications of the operating conditions had to be developed by a trial-and-error approach rather than by logic based on biospecificity.[32]

These difficulties are understandable if the adsorption and elution depend largely on gross electrostatic properties since these bear no necessary relationship to biospecificity.

Adsorption of Lactate Dehydrogenase on Ribosyl-Linked NAD+

Lowe and Dean[17,33] have described model affinity chromatographic studies of lactate dehydrogenase and other dehydrogenases on a

[27] C. Mahon and P. O'Carra, unpublished work.

[28] F. M. Richards and H. W. Wyckoff, in "The Enzymes" (P. D. Boyer, ed.), 3rd Ed., Vol. IV, p. 647. Academic Press, New York (1971).

[29] C. H. W. Hirs, S. Moore, and W. H. Stein, J. Biol. Chem. 200, 493 (1953).

[30] D. R. Farr, H. Amster, and M. Horisberger, Arch. Microbiol. 85, 249 (1972).

[31] W. Murphy and D. B. Johnson, Comp. Biochem. Physiol. in press; S. Bree, D. B. Johnson, and M. P. Coughlan, Comp. Biochem. Physiol. in press.

[32] R. K. Wierenga, J. D. Huizinga, W. Gaastra, G. W. Welling, and J. J. Beintema, FEBS Lett. 31, 181 (1973).

[33] C. R. Lowe and P. D. G. Dean, Biochem. J. 133, 515 (1973).

number of immobilized NAD[+] preparations. One of these was prepared by direct coupling of NAD[+] to a CNBr-activated polysaccharide matrix and another by carbodiimide-promoted attachment of NAD[+] via the terminal carboxyl group of a caprylic acid-substituted polysaccharide matrix. It was initially assumed that the point of attachment of the NAD[+] in these derivatives was the 6-amino group of the adenine residue,[17] but it has since been established that attachment is mainly, if not entirely, through the ribosyl hydroxyl groups of the NAD[+].[6,34]

In the model studies of Lowe and Dean,[17,33] enzymes are applied in a buffer of low ionic strength and elution is achieved by a concentration gradient of KCl. These authors quantify affinity chromatographic effectiveness in terms of "binding" which is defined as the millimolar salt concentration needed to elute an enzyme. On the basis of this criterion both of the above-mentioned immobilized NAD[+] derivatives were considered to be effective affinity systems for lactate dehydrogenase.

However, kinetic studies show that the KCl concentrations used for elution of this enzyme have little or no effect on the biospecific binding of soluble NAD[+] to lactate dehydrogenase.[6] Moreover, at the low ionic strength of the starting buffer the adsorbed enzyme is not susceptible to bioelution by soluble NAD[+] added to the irrigant. This evidence suggests that the adsorption-elution behavior observed with these derivatives is dominated by salt-sensitive, nonbiospecific adsorption.

Dean and Lowe[15] and Lowe et al.[35] have suggested that the CNBr activation step introduces ionic groups into the matrix which must be neutralized or eliminated before satisfactory results are obtainable. However the NAD[+] residue itself is ionic and would be expected to promote some ion-exchange adsorption at low ionic strength. This presumably dominates the chromatography, since at most only very weak bioaffinity can be detected with the ribosyl-linked NAD[+] derivatives under discussion.[6] This is consistent with the detailed picture of the NAD[+]-binding site of lactate dehydrogenase which has recently emerged from the X-ray diffraction studies of Rossman and co-workers[36,37]: both ribosyl residues of the NAD[+] molecule are buried in the binding crevice and are involved in the binding interactions of NAD[+] with the enzyme. Attachment of the NAD[+] molecule

[34] K. Mosbach, H. Guilford, R. Ohlsson, and M. Scott, *Biochem. J.* **127**, 625 (1972).
[35] C. R. Lowe, M. J. Harvey, D. B. Craven, and P. D. G. Dean, *Biochem. J.* **133**, 499 (1973).
[36] M. G. Rossman, M. J. Adams, M. Buehner, G. C. Ford, M. L. Hackert, P. J. Lentz, Jr., A. McPherson, Jr., R. W. Schevitz, and I. E. Smiley, *Cold Spring Harbor Symp. Quant. Biol.* **36**, 179 (1972).
[37] M. J. Adams, M. Buchner, K. Chandrasekhar, G. C. Ford, M. L. Hackert, A. Liljas, M. G. Rossman, I. E. Smiley, W. S. Allison, J. Everse, N. O. Kaplan, and S. S. Taylor, *Proc. Nat. Acad. Sci. U.S.* **70**, 1968 (1973).

through either of these residues would therefore be expected to drastically decrease or abolish the bioaffinity of the resulting derivative.

Azo-Linked NAD[+]

Immobilized NAD[+] derivatives may also be prepared by incubation of NAD[+] with diazotized derivatives of matrix–spacer arm assemblies such as derivative (VI).[6] The NAD[+] is thought to attach through the 8-position of the adenine residue.[6,38] Immobilized NADP[+] analogs can be similarly prepared,[6] and NAD[+] has been similarly attached via the hydrophilic spacer arms described above in the section on control of interference.[18] A group of such derivatives, (VI) to (IX), have been used for affinity chromatographic studies of a range of NAD[+]-linked dehydrogenases, and these studies have provided useful insights into some of the complications that can be encountered in affinity chromatography.[6,18]

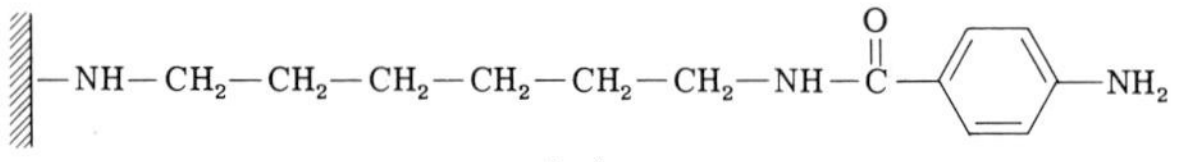

(VI)

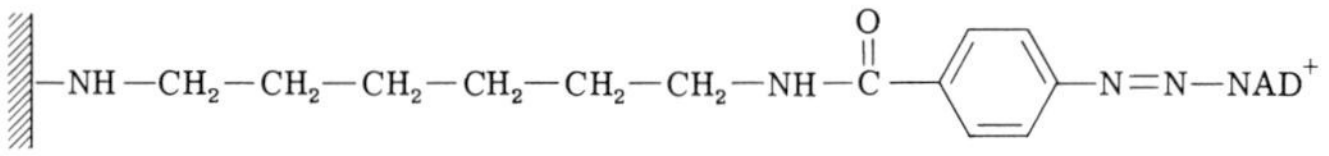

(VII)

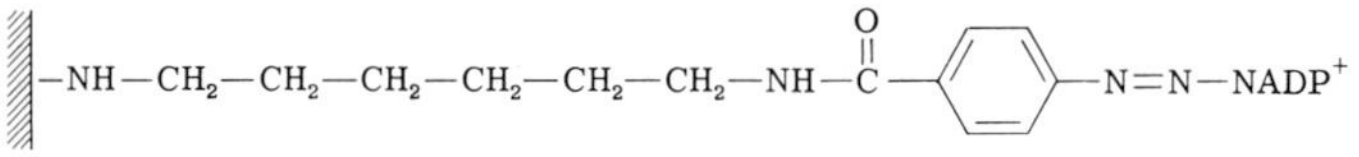

(VIII)

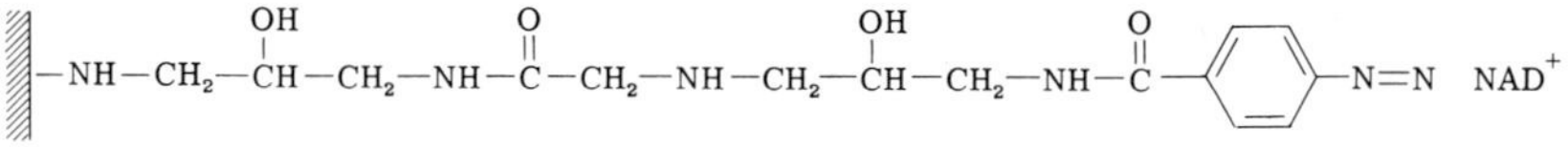

(IX)

Yeast Alcohol Dehydrogenase.[6,18] This enzyme was firmly adsorbed on derivative (VII), but it could neither be eluted by addition of soluble NAD[+] to the irrigant nor by KCl concentrations as high as 1 M. Addition of both NAD[+] and high salt concentrations to the irrigant simultaneously also failed to elute the enzyme. On the other hand, the enzyme showed no detectable affinity for the hydrophilic derivative (IX). It thus seems likely that the adsorption on (VII) is predominantly or entirely due to interac-

[38] M. K. Weibel, H. H. Weetall, and H. J. Bright, *Biochem. Biophys. Res. Commun.* **44,** 347 (1971).

tions with the hydrophobic spacer arm of this derivative. The enzyme is also adsorbed on a number of agarose derivatives carrying hydrophobic groups and no biospecific ligand.[18]

Glyceraldehyde-3-phosphate Dehydrogenase.[6,18] The enzyme from rabbit muscle is adsorbed strongly on derivative (VII). The adsorption is not noticeably affected by high concentrations of KCl, and the enzyme is cleanly eluted by 2 mM NAD$^+$ if a salt concentration of at least 0.2 M is maintained in the irrigant. Under such conditions the enzyme is not adsorbed on either the ligandless analog (VI) or on the NADP$^+$ analog (VIII).

These results indicate clearly the biospecific nature of the adsorption of this enzyme on derivative (VII). However, the yield of the enzyme elutable with NAD$^+$ becomes progressively lower the longer the enzyme is left adsorbed on the gel before elution. This loss is far in excess of the loss of activity in control experiments when samples of the enzyme are put through equivalent procedures on a "blank gel" such as derivative (VI).

The hypothesis that this time-dependent loss of activity is due to slow nonbiospecific adsorption onto the hydrophobic spacer arm, after the enzyme has first bound biospecifically to the immobilized NAD$^+$ is supported by the results obtained when the spacer arm is replaced by a hydrophilic one as in derivative (IX). The enzyme is just as strongly adsorbed on this derivative as on derivative (VII), but when it is eluted with NAD$^+$, the yield is greater than 90% and does not decrease significantly even when the enzyme is left on the column for 1.5 hours, by which time the yield of NAD$^+$-elutable enzyme from derivative (VII) has decreased to about 10%.

Lactate Dehydrogenase.[6,18] When lactate dehydrogenase is chromatographed at low ionic strength, e.g., in 0.05 M potassium phosphate at pH 7.4, it is strongly adsorbed on derivative (VII), but is not bioelutable; it is similarly adsorbed onto the ligandless analog (VI). At low ionic strength, therefore, the adsorption is dominated by nonbiospecific adsorption attributable to the matrix–spacer arm assembly. Inclusion of 0.5 M KCl abolishes such interfering adsorption, as evidenced by the lack of adsorption of the enzyme on derivatives (VI) and (VIII) under such conditions. However, at such a high salt concentration the enzyme remains only weakly retarded by derivative (VII). Since such salt concentrations do not interfere significantly with the binding of soluble NAD$^+$ to lactate dehydrogenase,[6] this result suggests that, although the azo-linked NAD$^+$ derivative retains some bioaffinity for this enzyme, it is rather weak, presumably as a result of partial steric hindrance by the immobilization linkage. This seems to be confirmed by the fact that lactate dehydrogenase

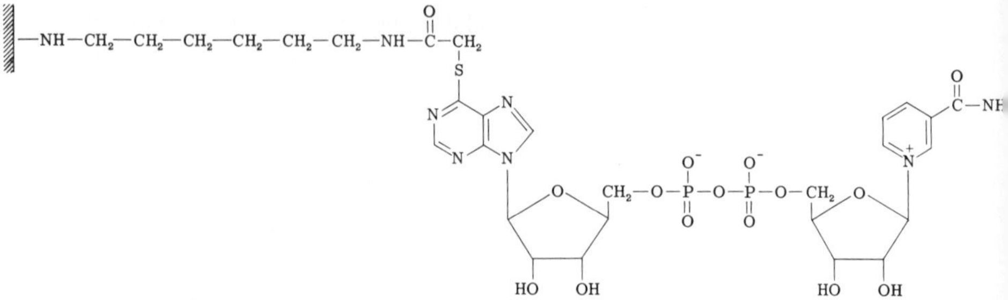

(X)

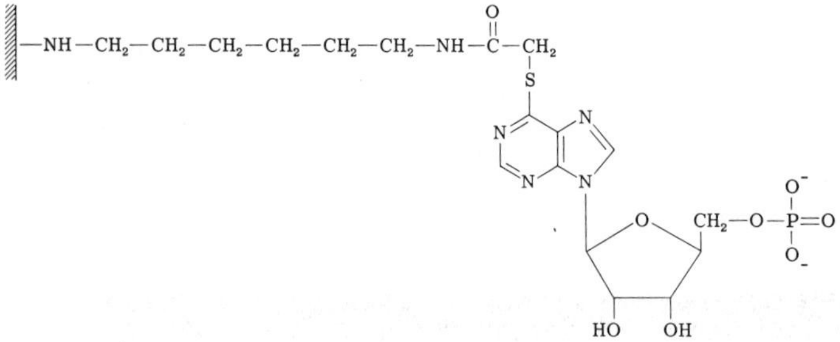

(XI)

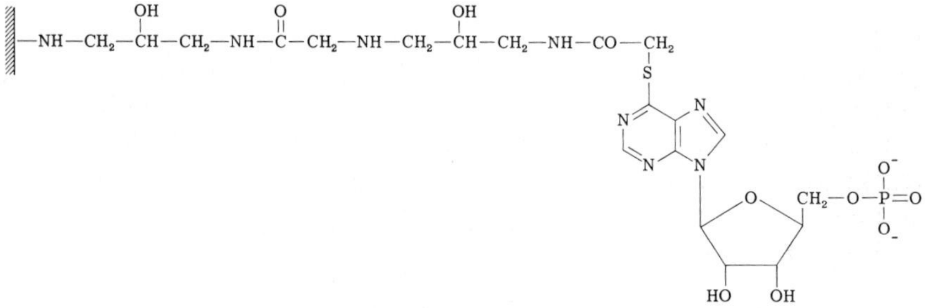

(XII)

shows only very weak affinity for derivative (IX) at salt concentrations as low as 0.1 M. The inclusion of 0.2 M KCl in the irrigating buffer was sufficient to largely eliminate the nonbiospecific adsorption of this enzyme on derivatives (VI) and (VIII). At this intermediate ionic strength, lactate dehydrogenase was strongly adsorbed on derivative (VII), but not on derivative (IX), and could be cleanly eluted with the biospecific counterligands NADH and 5'-AMP (a weak competitive inhibitor against NAD+). This seemingly clear-cut bioaffinity is of a much stronger nature than could be accounted for by bioaffinity alone, in view of the results at high ionic strength and those with the hydrophilic spacer arm. The discrepancy is most plausibly explained on the basis of the compound affinity concept discussed in a previous section.

Analogs of NAD⁺ and AMP Linked through the 6 Position of
the Purine Residue[18,39]

Derivatives (X) to (XII) may be regarded as immobilized thio analogs of NAD⁺ and 5′-AMP linked through the 6-position of the purine residue. They are easily prepared by reacting the 6-mercaptopurine analogs of the nucleotides with bromoacetyl-terminating matrix-arm assemblies.[39] This linkage procedure was considered promising for affinity chromatography of certain NAD-linked dehydrogenases, including lactate dehydrogenase, because it was known from kinetic studies that derivatization at the 6-position does not drastically affect binding of NAD⁺ to these enzymes. Mosbach *et al.*[34,40,41] have synthesized 6-linked derivatives by another method on the basis of the same rationale.

Lactate dehydrogenase was strongly adsorbed on the 6-immobilized NAD⁺ analog (X).[39] Bioelution with soluble NAD⁺ was clear-cut if the salt concentration in the irrigant was maintained above 0.1 M and the affinity remained strong at high ionic strengths (e.g., 0.5 M KCl) indicating that the strong affinity of derivative (X) for this enzyme is probably pure bioaffinity. This contrasts with the behavior of the azo-linked derivative (VII) which binds lactate dehydrogenase strongly only at moderate ionic strengths, apparently with the help of nonbiospecific reinforcement (see above).

5′-AMP is a weak competitive inhibitor of lactate dehydrogenase (K_i = ca. 2 mM). At the substitution level of the batches of derivative (XI) prepared to date (<1 mM), only very weak biospecific retardation would be predicted from Eq. 1. In agreement with this, the enzyme was only marginally adsorbed on derivative (XI) at high ionic strength (0.5 M KCl). However, at low ionic strength the enzyme was strongly adsorbed on this derivative and the ionic strength could be adjusted to intermediate values (e.g., 0.1 M) such that the enzyme remained strongly adsorbed but could be cleanly eluted with biospecific counterligands, such as 5′-AMP and NADH. On the hydrophilic derivative (XII) the enzyme remained only marginally adsorbed at such intermediate ionic strength demonstrating the positive contribution of the hydrophobic arm of derivative (XI) to the compound affinity.

Malate dehydrogenase is the least "sticky" enzyme so far encountered in our work. The enzyme from beef heart is only marginally retarded by hydrophobic "mock affinity gels," such as derivative (VI), even at low ionic strengths. This enzyme displays no detectable affinity for the azo-

[39] S. Barry and P. O'Carra, *FEBS Lett.* **37**, 134 (1973).
[40] R. Ohlsson, P. Brodelius, and K. Mosbach, *FEBS Lett.* **25**, 234 (1972).
[41] K. Mosbach, this volume [16].

linked NAD⁺ derivative (VII) but is strongly adsorbed on the 6-linked NAD⁺ analog (X) at moderate ionic strengths (0.1–0.2 M) and is cleanly bioelutable with NADH. Malate dehydrogenase shows only very weak affinity for the 6-linked AMP analog (XI) and, in contrast to lactate dehydrogenase, this affinity is not reinforced at moderate ionic strengths, presumably owing to the general lack of nonbiospecific interaction of this enzyme with hydrophobic spacer arms.[42]

[42] P. O'Carra and S. Barry, unpublished work (1973).

[9] Hydrophobic Chromatography

By SHMUEL SHALTIEL

Development of the Basic Principle

In the course of our studies on enzymes involved in glycogen metabolism, we attempted to coat beaded agarose with glycogen and use it for affinity chromatography[1] of these enzymes. The procedure used for the preparation of the columns consisted of: (a) activation of the agarose with CNBr[2] and reaction with an α,ω-diaminoalkane to obtain an ω-aminoalkyl agarose, (b) binding of CNBr-activated glycogen to the ω-aminoalkyl agarose. Two glycogen-coated agarose preparations which differed only in the length of the hydrocarbon chains bridging the ligand and the bead (Seph-C_8-NH-glycogen[3] and Seph-C_4-NH-glycogen) displayed unexpected differences in the retention of glycogen phosphorylase b. Whereas the column with 8-carbon-atom bridges adsorbed the enzyme, the column with 4-carbon-atom bridges did not even retard it.[4] Considering the large molecular dimensions of glycogen (Fig. 1), it did not seem likely that changing the length of the extensions from 8 to 4 carbon atoms would have such a dramatic effect on the retention power of the column. Rather, we were initially inclined to think that the column with the shorter side chains had failed to bind glycogen because of a reduced accessibility of the amino groups. This possibility was excluded by repeating the syn-

[1] P. Cuatrecasas, M. Wilchek, and C. B. Anfinsen, *Proc. Nat. Acad. Sci. U.S.* **61**, 636 (1968).

[2] R. Axén, J. Porath, and S. Ernbäck, *Nature (London)* **214**, 1302 (1967).

[3] Seph-C_n-NH-glycogen represents Sepharose 4B activated with CNBr, reacted with an α,ω-diaminoalkane n-carbon atoms long, then coupled with CNBr-activated glycogen.

[4] Z. Er-el, Y. Zaidenzaig, and S. Shaltiel, *Biochem. Biophys. Res. Commun.* **49**, 383 (1972).

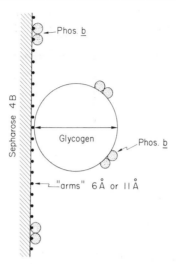

FIG. 1. Schematic representation of the Seph-C_n-NH-glycogen columns, showing the approximate molecular dimensions of phosphorylase b, glycogen and the hydrocarbon "arms." It is apparent from this scheme why changing the length of the hydrocarbon "arms" from ~6 Å to ~11 Å is not likely to affect the accessibility of the ligand in this case. The scheme also depicts the relatively large number of free "arms" that may interact directly with the enzyme. From S. Shaltiel, *in* "Metabolic Interconversion of Enzymes, 1973" (E. H. Fischer, E. G. Krebs, H. Neurath, and E. R. Stadtman, eds.), p. 379. Springer-Verlag, Berlin.

thesis of the two columns with radioactively labeled glycogen. The resulting columns were shown to contain very similar amounts of glycogen, yet only the column with the 8-carbon-atom bridges retained phosphorylase b.

From the way the columns were prepared (first coating the beads with the ω-aminoalkyl "arms," then attaching the preactivated glycogen), it was possible to visualize that some of the ω-aminoalkyl groups did not react with glycogen and remained free and accessible to interaction with the phosphorylase b molecules (Fig. 1). We therefore synthesized two control agarose derivatives, Seph-C_8-NH$_2$ and Seph-C_4-NH$_2$ (Fig. 2). Surprisingly, these control columns were identical in their elution pattern with the appropriate glycogen-coated columns, i.e., Seph-C_8-NH$_2$ retained phosphorylase b, whereas Seph-C_4-NH$_2$ did not. Had Seph-C_8-NH$_2$ been acting merely as an ion-exchange column, one would expect Seph-C_4-NH$_2$ to have a similar elution pattern, as the two columns contained the same type and number of charges (arising from the activation of the agarose with CNBr and from protonated ω-amino groups) per unit weight. Since the two columns did differ in the length of their side chains, the

Abbreviation	Structure
Seph-C$_1$	‖-NH-CH$_3$
Seph-C$_2$	‖-NH-CH$_2$-CH$_3$
Seph-C$_3$	‖-NH-CH$_2$-CH$_2$-CH$_3$
Seph-C$_4$	‖-NH-CH$_2$-CH$_2$-CH$_2$-CH$_3$
Seph-C$_5$	‖-NH-CH$_2$-CH$_2$-CH$_2$-CH$_2$-CH$_3$
Seph-C$_6$	‖-NH-CH$_2$-CH$_2$-CH$_2$-CH$_2$-CH$_2$-CH$_3$
Seph-C$_8$	‖-NH-CH$_2$-CH$_2$-CH$_2$-CH$_2$-CH$_2$-CH$_2$-CH$_2$-CH$_3$

Abbreviation	Structure
Seph-C$_2$-NH$_2$	‖-NH-CH$_2$-CH$_2$-NH$_2$
Seph-C$_3$-NH$_2$	‖-NH-CH$_2$-CH$_2$-CH$_2$-NH$_2$
Seph-C$_4$-NH$_2$	‖-NH-CH$_2$-CH$_2$-CH$_2$-CH$_2$-NH$_2$
Seph-C$_5$-NH$_2$	‖-NH-CH$_2$-CH$_2$-CH$_2$-CH$_2$-CH$_2$-NH$_2$
Seph-C$_6$-NH$_2$	‖-NH-CH$_2$-CH$_2$-CH$_2$-CH$_2$-CH$_2$-CH$_2$-NH$_2$
Seph-C$_8$-NH$_2$	‖-NH-CH$_2$-CH$_2$-CH$_2$-CH$_2$-CH$_2$-CH$_2$-CH$_2$-CH$_2$-NH$_2$

FIG. 2. Structure of alkyl agaroses (left panel) and ω-aminoalkyl agaroses (right panel) used for hydrophobic chromatography. From Z. Er-el, Y. Zaidenzaig, and S. Shaltiel, *Biochem. Biophys. Res. Commun.* 49, 383 (1972) and S. Shaltiel and Z. Er-el, *Proc. Nat. Acad. Sci. U.S.* 70, 778 (1973).

possibility was considered that it is the length of the hydrocarbon extensions that determines the retention power of these columns.

It could be argued that the column with longer "arms" was more effective, because the amino groups at the tip of these "arms" could reach out farther to interact with less accessible charged groups in the protein molecule. In an attempt to check this possibility, we synthesized a homologous series of hydrocarbon-coated agarose (Fig. 2, left panel) with no charged groups at the tip of the "arms" and studied their ability to retard or retain glycogen phosphorylase *b*. When mixtures of D-glyceraldehyde-3-phosphate dehydrogenase and phosphorylase *b* were passed through columns of these agaroses, it was found that under the conditions of our experiment the dehydrogenase is not adsorbed or retarded by any of the columns (Seph-C$_1$ through Seph-C$_6$). However, whereas Seph-C$_1$ and Seph-C$_2$ excluded phosphorylase *b*, Seph-C$_3$ retarded the enzyme and higher alkyl agaroses adsorbed it. Elution of phosphorylase *b* from Seph-C$_4$ was possible by using a "deforming buffer," which was previously shown to bring about reversible structural changes in the enzyme.[5] The binding of phosphorylase *b* to Seph-C$_6$ is apparently very tight, since the enzyme could not be eluted from the column even when the pH of the deforming buffer was lowered to 5.8, which increases its deforming power.[5,6] Recovery of phosphorylase *b* from Seph-C$_6$ was so far possible only in an inactive form, by washing the column with 0.2 M CH$_3$COOH (Fig. 3).

[5] S. Shaltiel, J. L. Hedrick, and E. H. Fischer, *Biochemistry* 5, 2108 (1966).
[6] J. L. Hedrick, S. Shaltiel, and E. H. Fischer, *Biochemistry* 8, 2422 (1969).

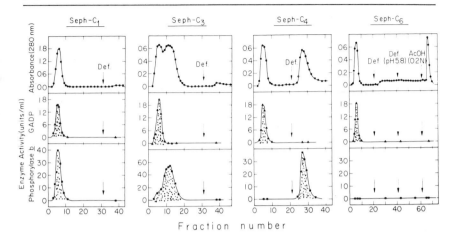

Fraction number

FIG. 3. Preferential adsorption of glycogen phosphorylase *b* on hydrocarbon-coated agaroses varying in the length of their alkyl side chains. Samples containing D-glyceraldehyde-3-phosphate dehydrogenase (GADP) (—▲—) and glycogen phosphorylase (—●—) were applied to the columns which were equilibrated at 22° with a buffer composed of 50 m*M* sodium β-glycerophosphate, 50 m*M* 2-mercaptoethanol, and 1 m*M* EDTA at pH 7.0. Nonadsorbed protein was washed off with the same buffer and elution was initiated (arrow) with a deforming buffer (Def.) (0.4 *M* imidazole, 50 m*M* 2-mercaptoethanol adjusted to pH 7.0 with citric acid). In the case of Seph-C₆, attempts were also made to elute phosphorylase with a more acidic deformer (pH 5.8) and subsequently with 0.2 *M* CH₃COOH. Data from Z. Er-el, Y. Zaidenzaig, and S. Shaltiel, *Biochem. Biophys. Res. Commun.* **49**, 383 (1972).

These observations led to the development of a new approach to the purification of proteins and other biomolecules. This approach, hydrophobic chromatography,[4,7] makes use of homologous series of hydrocarbon-coated agaroses (Seph-C$_n$-X; where X = H, NH₂, COOH, OH, C₆H₅, etc.) to achieve separation of proteins. Each member in the series offers hydrophobic "arms" or "yardsticks" of a different size which interact with accessible hydrophobic pockets in the various proteins, retaining only some proteins out of a mixture. Further resolution is then achieved by gradually changing the nature of the eluting solvent. Work done in our laboratory and elsewhere[8–16] has already demonstrated the wide applicability of this type of chromatography in protein purification.

[7] S. Shaltiel and Z. Er-el, *Proc. Nat. Acad. Sci. U.S.* **70**, 778 (1973).

[8] S. Shaltiel, *in* "Metabolic Interconversion of Enzymes, 1973" (E. H. Fischer, E. G. Krebs, H. Neurath, and E. R. Stadtman, eds.), p. 379. Springer-Verlag, Berlin.

[9] S. Shaltiel, G. F. Ames, and K. D. Noel, *Arch. Biochem. Biophys.* **159**, 174 (1973).

[10] R. J. Yon, *Biochem. J.* **126**, 765 (1972).

[11] B. H. J. Hofstee and N. F. Otillio, *Biochem. Biophys. Res. Commun.* **50**, 751 (1973).

Preparation of Hydrocarbon-Coated Agaroses

Alkyl Agaroses (Seph-C_n; n = 1–10). Sepharose 4B (Pharmacia) is activated by reaction with CNBr at pH 10.5–11 and 22°. One gram of CNBr (dissolved in 1 ml of dioxane) is added to each 10 g (wet weight[17]) of agarose suspended in 20 ml of water. Activation is initiated by raising the pH of the mixture to 11 with 5 N NaOH. The reaction is allowed to proceed for 8 minutes with gentle swirling, while maintaining the pH between 10.5 and 11 with 5 N NaOH. Activation is terminated by addition of crushed ice, filtration, and washing the gel with ice-cold deionized water (100 ml).

The appropriate *n*-alkylamine (4 moles per mole of CNBr used for the activation of the agarose) is dissolved in 20 ml of a solvent composed of equal volumes of *N,N*-dimethylformamide and 0.1 M NaHCO$_3$ (pH 9.5) and the pH of the solution is readjusted if necessary to 9.5. With the higher members in the series ($n \geqslant 8$) a gel or a precipitate may form. In such cases the mixture is heated to 60° to obtain a clear solution, is cooled rapidly and is mixed with the activated agarose before precipitation recurs.

The activated agarose is suspended in 20 ml of 0.1 M NaHCO$_3$ at pH 9.5 and mixed with the alkylamine solution. The reaction is allowed to proceed at room temperature (20–25°) for 10 hours with gentle swirling. Subsequently the alkyl agarose is filtered; washed with five times its volume of each of the following: (a) water, (b) 0.2 M CH$_3$COOH, (c) water, (d) 50 mM NaOH, (e) water, (f) dioxane–water (1:1), (g) 0.2 M CH$_3$COOH; then washed with 20 volumes of water. Finally each column is washed and equilibrated with the buffer chosen for chromatography. The same procedure should be used throughout the series in order to make the columns identical in all respects except for the length of their alkyl side chains.

Alkyl agaroses can be used repeatedly and stored for months at 4° in aqueous suspension. Bacteriostatic agents (e.g., toluene) may be

---

[12] H. Jakubowski and J. Pawelkiewicz, *FEBS Lett.* **34**, 150 (1973).

[13] T. Peters, H. Taniuchi, and C. B. Anfinsen, *J. Biol. Chem.* **248**, 2447 (1973).

[14] S. Shaltiel, S. P. Adler, D. Purich, C. Kaban, P. Senior, and E. R. Stadtman, in preparation.

[15] G. B. Henderson, S. Shaltiel, and E. E. Snell, *Biochemistry,* submitted for publication (1973).

[16] The use of resin-linked aliphatic chains for the separation of membrane proteins, which have a biospecific affinity for such chains in lipids, has been proposed by H. Weiss and T. Bucher, *Eur. J. Biochem.* **17**, 561 (1970).

[17] Before weighing, the agarose is placed on a Büchner funnel to drain off excess water until the very first signs of dryness appear.

added to prolong the shelf life of the columns. Between runs it is recommended that the column be washed with $1 M$ NaCl, water and the buffer to be used in the next run. Additional washing with any one of the washing solvents mentioned above should be attempted (preferably on a Büchner funnel) if clogging causes a considerable decrease in the capacity or flow rate of the column.

ω-*Aminoalkyl Agaroses* ($Seph\text{-}C_n\text{-}NH_2$; $n = 1\text{--}12$). This homologous series is synthesized by the procedure described for the alkyl agarose series, using 4 moles of the α,ω-diaminoalkane per mole of CNBr used for the activation of the agarose. The large excess of the diamine ensures a high yield of coupling and lowers the probability of cross-linking between the beads.

Selected Examples of Protein Purification

Example 1. Use of an Alkyl Agarose in the Purification of Glycogen Phosphorylase b.[4] Under the conditions used in the experiment described in Fig. 3, the reversible binding of glycogen phosphorylase b to Seph-C$_4$ is quite specific. It is therefore possible to use this column for the preparative purification of the enzyme. When crude muscle extract, 0.7 units/mg, is subjected to chromatography on a Seph-C$_4$ column, over 95% of the protein in the sample is excluded. The excluded fractions have essentially no phosphorylase b activity. Upon elution with the deforming buffer, a small amount of protein emerges, with very high phosphorylase activity. The specific activity of the eluted enzyme, after dialysis to remove the deforming buffer, is as high as 68 units/mg, approaching that of the crystalline enzyme[18] (~ 80 units/mg). The yield of the process is over 95%, based on activity measurements.

Example 2. Use of ω-Aminoalkyl Agaroses in the Purification of Glycogen Synthetase.[7] When an extract of rabbit muscle is subjected to chromatography on ω-aminoalkyl agaroses, the columns with short "arms" ($n = 2$ and $n = 3$) exclude glycogen synthetase I, but the enzyme is retained on Seph-C$_4$-NH$_2$, from which it can be eluted with a linear NaCl gradient (Fig. 4). A column of δ-aminobutyl agarose, which retains the synthetase, excludes glycogen phosphorylase b, which in this column series and under the same conditions requires side chains 5 (or 6) carbon atoms long for retention (Fig. 4). Therefore, it is possible to isolate glycogen synthetase by passage of muscle extract through δ-aminobutyl agarose, then to extract phosphorylase b by subjecting the excluded proteins to chromatography on ω-aminohexyl agarose.

In a representative experiment,[7] 70 ml of the extract containing 1750

---

[18] E. H. Fischer, E. G. Krebs, and A. B. Kent, *Biochem. Prep.* 6, 68 (1958).

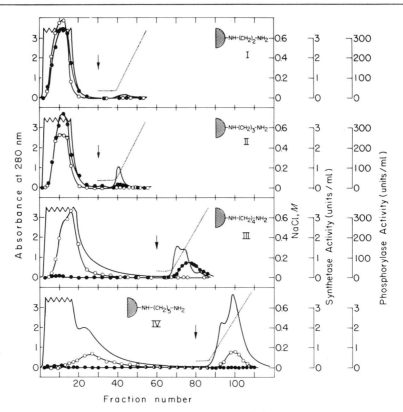

FIG. 4. Preferential adsorption of glycogen synthetase I and glycogen phosphorylase b on Seph-C_n-NH$_2$. Muscle extract was applied to each of the columns ($n = 2$–5) equilibrated at 22° with a buffer composed of 50 mM sodium β-glycerophosphate, 50 mM 2-mercaptoethanol, and 1 mM EDTA at pH 7.0. Unadsorbed protein was washed off until the absorbance at 280 nm (——) decreased below 0.05. Thereafter a linear NaCl gradient in the same buffer was applied ($\cdots$). Synthetase I (—●—) and phosphorylase b (—○—) activities of the fractions were monitored. Data from S. Shaltiel and Z. Er-el, *Proc. Nat. Acad. Sci. U.S.* **70**, 778 (1973).

mg of protein and a synthetase activity of 0.2 unit per milligram of protein were applied to a Seph-C_4-NH$_2$ column (10 × 2.4 cm) equilibrated at 22° with the β-glycerophosphate buffer described in the legend to Fig. 4. After the unadsorbed protein (monitored by the absorbance at 280 nm) was washed off, a linear NaCl gradient in the same buffer was applied. The I form of the synthetase was purified 25- to 50-fold in one step. This experiment also illustrates the high capacity of hydrocarbon-coated agaroses.

Example 3. Purification by Selective Exclusion.[9] In principle, it should

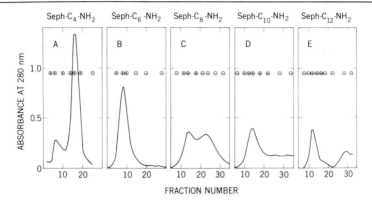

FIG. 5. Exclusion of the histidine-binding protein, J, by ω-aminoalkyl agaroses. Samples of *Salmonella typhimurium* shock fluid were applied to each of the columns, which were equilibrated (at 22°) with 10 mM sodium phosphate at pH 7.2. Protein J was detected (+ or −) by immunodiffusion. Data from S. Shaltiel, G. F. Ames, and K. D. Noel, *Arch. Biochem. Biophys.* **159**, 174 (1973).

be possible to achieve purification of proteins by hydrophobic chromatography not only by selective retention of a given protein, but also by exclusion of a desired protein, when most of the other proteins in the mixture are retained by the column. This principle is illustrated in the purifaction of the histidine-binding protein J from *Salmonella typhimurium* using ω-aminododecyl agarose.

Upon subjection of *S. typhimurium* to osmotic shock, numerous proteins are released into the shock fluid. Chromatography of this fluid on Seph-C_n-NH$_2$ columns shows that protein J is excluded from all the columns tested (Fig. 5). However, as the length of the hydrocarbon side chains is gradually increased, the excluded fractions contain less and less protein (see decrease in absorbance at 280 nm). Furthermore, there is a remarkable decrease in the number of protein bands found in the peak fractions, as judged by polyacrylamide gel electrophoresis in the presence of sodium dodecyl sulfate. In the peak fractions excluded from Seph-C_8-NH$_2$ and Seph-C_{12}-NH$_2$ columns, protein J is clearly the major component in the mixture. On a preparative scale, a purification of 7- to 10-fold in one step is achieved.[9]

Example 4. Resolution of the Metabolically Interconvertible Forms of Glycogen Phosphorylase.[19] Regulatory proteins, which are designed to assume more than one stable conformation and thus regulate the rate of metabolic processes, seem to be good candidates for separation by hydrophobic chromatography. This is illustrated in the separation of the *a* from the *b* form of glycogen phosphorylase. These two forms are iden-

[19] Z. Er-el and S. Shaltiel, *FEBS Lett.* **40**, 142 (1974).

tical in their amino acid sequence except for a unique serine residue in each of the enzyme protomers (MW 100,000)[20] which becomes O-phosphorylated upon conversion[21] of the b form (a dimer) to the a form (a tetramer). Phosphorylase a is retained by Seph-C_1 whereas, under the same conditions, hydrocarbon chains 4-carbon-atoms long are necessary to retain the b form of the enzyme (Fig. 3). It thus becomes possible to fish out phosphorylase a by passage through Seph-C_1 and then to extract the b form of the enzyme by chromatography on Seph-C_4.

An Exploratory Kit for Choosing the Most Appropriate Column and Optimizing Resolution[8,22]

One of the advantages of the above-mentioned families of columns is that they constitute homologous series, in which each member has hydrophobic side chains one-carbon-atom longer than the preceding one. In fact, they contain "hydrophobic yardsticks" of different lengths. It thus becomes possible to optimize the isolation of a protein by choosing the member in the series that will be the most efficient for the purpose. The separation possibilities are greatly amplified by the synthesis of additional families of hydrocarbon-coated agaroses that have, on their hydrocarbon side chains, one or more functional groups. The resulting columns will derive their discriminating power from hydrophobic, ionic, and perhaps other types of interactions. However, it will be possible to systematically increase the contribution of hydrophobic interactions within the series so as to achieve optimal resolution of a desired protein.

Since we have a variety of column-series at our disposal, it becomes necessary to determine, in a fast and simple manner, which family of adsorbents and which member within that family is the most appropriate for adsorbing a desired protein. For that purpose, a method was developed,[22] in which analytical kits of small columns (5 × 0.5 cm) are used. Each kit contains a homologous series of columns, e.g., Seph-C_n-NH_2; $n = 1$–12, as well as two control columns, one of unmodified agarose and the other of agarose activated by CNBr, then reacted with ammonia. Small aliquots, 50–200 μl, of the protein mixture are applied on each of the columns. The amount of excluded enzyme is determined for each column and plotted as percent activity against the number of carbon atoms (n) in the alkyl side chains. The column with the smallest n which is capable of retaining the desired enzyme is the column of choice for purification. An example of the use of such a kit is given in Fig. 6, which depicts the retention by Seph-C_n-NH_2 of several enzymes involved in the regulation of glutamine metabolism in $E.$ $coli.$[14] As seen in Fig. 6,

[20] P. Cohen, T. Duewer, and E. H. Fischer, *Biochemistry* 10, 2683 (1971).
[21] E. H. Fischer, A. Pocker, and J. C. Saari, *Essays Biochem.* 6, 23 (1970).
[22] S. Shaltiel, in preparation.

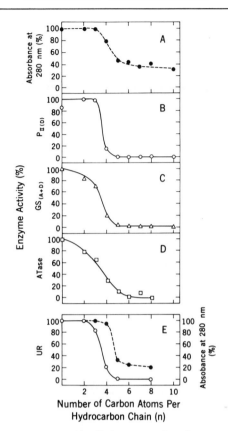

FIG. 6. Selection of an ω-aminoalkyl agarose for chromatography of enzymes active in the regulation of glutamine metabolism. Each of a series of Seph-C$_n$-NH$_2$ columns ($n = 0$–12) was equilibrated at 4° with a buffer composed of 10 mM imidazole·HCl, 0.1 mM EDTA, and 10 mM 2-mercaptoethanol at pH 7.0. Aliquots of 0.2 ml of an extract of *E. coli* were applied to each column, and the first 2.5 ml of effluent were collected and assayed (A–D). For the uridyl-removing enzyme (E) a partially purified preparation was used for which the buffer included 20 mM 2-methylimidazole·HCl, 10 mM 2-mercaptoethanol, 0.1 mM EDTA, and 100 mM KCl at pH 7.6. Absorption at 280 nm (----). Enzymes assayed: ATase, ATP: glutamine synthetase adenylyltransferase; GS, glutamine synthetase; P$_{IID}$, uridylylated protein stimulating deadenylylation; UR, uridylyl-removing enzyme. Data from S. Shaltiel, S. P. Adler, D. Purich, C. Kaban, P. Senior, and E. R. Stadtman, in preparation.

Seph-C$_5$-NH$_2$ seemed to be the column of choice for retention of these enzymes. This column still excluded about 50% of the proteins in the *E. coli* extract.

The resolution potential of hydrophobic chromatography columns is not exhausted by choosing the column with the most appropriate "arm." Further resolution can be achieved during elution. Thus, in spite of the fact that Seph-C$_5$-NH$_2$ retains several enzymes involved in glutamine metabolism (Fig. 6), they can be made to detach from the column one after the other, by gradually changing the nature of the eluting solvent, its ionic strength (Fig. 7), ionic composition, pH, hydrophobicity, denaturing power, or temperature.

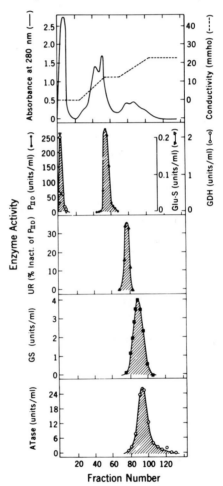

FIG. 7. Chromatography of an extract of *Escherichia coli* on Seph-C$_5$-NH$_2$. The column was equilibrated at 4° with the imidazole·HCl buffer (pH 7.0) described in the legend to Fig. 6. Unadsorbed protein was removed by washing before a linear KCl gradient in the same buffer was applied with a step at 0.25 M KCl. Enzymes assayed: Glu-S, glutamate synthetase; GDH, glutamate dehydrogenase as well as the enzymes listed in the legend to Fig. 6. Data from S. Shaltiel, S. P. Adler, D. Purich, C. Kaban, P. Senior, and E. R. Stadtman, in preparation.

Reporting the Use of Hydrophobic Chromatography

When describing the use of hydrophobic chromatography columns, it is important to specify the exact loading and eluting conditions (pH, ionic strength, buffer composition, temperature). A change in any one of these variables may determine whether a given protein will be retained or excluded by a specific member of the homologous series.

On the Mechanism of Action of Hydrocarbon-Coated Agaroses

It is suggested that the resolving power of hydrocarbon-coated agaroses is derived mainly from differences in the size and distribution of available hydrophobic pockets or regions in a specific protein. Retardation or retention is achieved through lipophilic interactions between these pockets and appropriately sized hydrocarbon side chains on the agarose. This suggestion is based on the following observations:

1. The members of the homologous series of alkyl agaroses (Seph-C_n) used, for example in the experiment depicted in Fig. 3, were all prepared in an identical procedure. They all contain a similar number of alkyl side chains per bead, and these chains are attached in all cases through the same type of linkage. Yet, under identical conditions of pH, ionic strength, buffer composition and temperature, their ability to retain phosphorylase *b* depends on the length of the hydrocarbon side chains, passing from no retention, through retardation, to reversible binding, up to very tight binding as the hydrocarbon "arms" are gradually lengthened. Purified agaroses have been reported to contain negative charges,[23] and it has been proposed[24] that positive charges may be introduced as a result of the activation with CNBr. However, the agarose beads do not bind phosphorylase *b* until alkyl side chains have been attached to them. Mere activation of the beads with CNBr and subsequent reaction with ammonia, i.e., simulating the functional groups linking the "arms," does not suffice. Furthermore, the above-mentioned charges should be equally present in all the members of the homologous series and therefore cannot account for the results described in Fig. 3. It would seem that it is the length of the hydrocarbon chains which is mainly responsible for the gradation observed in the retention power of the columns.

2. In view of the fact that a 4-carbon-atom alkyl group does not bear any special resemblance to the substrates of phosphorylase *b* nor to any of its known effectors, and since it is possible to design conditions

[23] J. Porath, J.-C. Janson, and T. Låås, *J. Chromatogr.* **60**, 167 (1971).
[24] J. Porath, *Nature (London)* **218**, 834 (1968).

for binding different proteins to this type of column, it does not appear likely that we are dealing here with interactions involving exclusively catalytic or regulatory sites of the enzyme (as in affinity chromatography[25-29]). This is in agreement with the observation that proteins bound to such columns remain, at least in some cases, catalytically active.[8,30]

3. Proteins with similar charge and molecular weight require different lengths of "arms" for retention on alkyl agarose columns.

4. The tightness of binding of a given protein increases with the lengthening of the hydrocarbon "arm" (up to a limit, of course) as judged by how drastic a procedure must be applied in order to detach it from the column.

5. Proteins bound to alkyl agarose columns can be detached from them by increasing the hydrophobicity of the eluting solvent.[7,11]

It should be emphasized that ionic interactions may affect the elution pattern of alkyl agaroses. In the case of the ω-aminoalkyl agaroses (Seph-C_n-NH_2) which have amino groups at the tip of their hydrocarbon "arms," ionic interactions are most likely to occur. However, this does not imply that these ionic interactions will necessarily support the hydrophobic interactions and increase the affinity of the column for a given protein. In fact, they may often decrease it. For example, side chains 4-carbon-atoms long suffice to retain glycogen phosphorylase b in the alkyl agarose series whereas 6-carbon-atom side chains are needed for retention of this enzyme under the same conditions in the ω-aminoalkyl series.[4,7] Similarly, while Seph-C_1 binds glycogen synthetase I, Seph-C_4-NH_2 is needed in the ω-aminoalkyl series.[7] These results are not surprising if we keep in mind that by implanting a charged group at the tip of the hydrocarbon chains we not only allow the column to participate in ionic interactions, but also reduce the hydrophobic character of the column. In some cases retained proteins can be detached from the Seph-C_n column by increasing the ionic strength or altering the pH of the eluting solvent,[19] raising the possibility that in these cases ionic interactions are directly involved in the binding of the macromolecule to the modified agarose. However, when dealing with proteins whose conformation and aggregation state are strongly dependent on intramolecular ionic interactions, it is quite conceivable that ionic strength and pH affect the struc-

[25] P. Cuatrecasas and C. B. Anfinsen, this series, Vol. 22, p. 345.

[26] L. S. Lerman, Proc. Nat. Acad. Sci. U.S. 39, 232 (1953).

[27] C. Arsenis and D. B. McCormick, J. Biol. Chem. 239, 3093 (1964).

[28] M. Wilchek and M. Gorecki, Eur. J. Biochem. 11, 491 (1969).

[29] G. Feinstein, Naturwissenschaften 58, 389 (1971).

[30] B. H. J. Hofstee and N. F. Otillio, Biochem. Biophys. Res. Commun. 53, 1137 (1973).

ture of the protein molecule so that there is a change in the size and availability of its hydrophobic pockets.

The suggestion that hydrophobic interactions play a key role in the resolutions achieved with columns from the ω-aminoalkyl series can account for several observations. For example, minimizing the ionic interactions by means of high-salt eluting solvents is not enough for detaching some of the proteins retained on these columns. Similarly, protein J from S. typhimurium, which should be negatively charged at pH 7.0, is excluded at low ionic strength from all the ω-aminoalkyl agaroses tested (Fig. 5) despite being positively charged at this pH. Furthermore, comparison of ω-aminoalkyl agaroses with DEAE-cellulose, the latter generally assumed to separate proteins by ionic interactions, shows differences in elution patterns.[14] Finally, the Seph-C_n-NH_2 series exhibits a gradation in the retention of a specific protein as the length of the ω-aminoalkyl chain increases, a gradation similar to that observed with the Seph-C_n series.

Relevance to Affinity Chromatography and Ion Exchange Chromatography

Introduction of hydrocarbon spacers between ligand molecules and agarose beads has been recommended recently[1,31] and extensively used in affinity chromatography.[1,25-29] The usefulness of these spacers was attributed to a relief of steric restrictions imposed by the matrix backbone and to an increased flexibility and mobility of the ligand as it protrudes farther into the solvent.[31] In some instances, this may indeed be the case. However, our results suggest that hydrocarbon extensions, in and of themselves, contribute to the binding of proteins through hydrophobic interactions, i.e., through a mechanism that does not involve specific recognition of the ligand. Such hydrophobic interactions are more liable to occur when the columns are prepared by first coating the beads with hydrocarbon arms and then attaching the specific ligand to those extensions.[31] In this case, some of the arms may remain unattached to the ligand and may bind other proteins through hydrophobic interactions, thereby interfering with the specificity of the system.

It seems advisable, therefore, to use the alternative method of column preparation, which involves synthesizing first a ligand with a side chain and then attaching the elongated ligand to the agarose beads.[31] This procedure will minimize the probability of performing unintentional hydrophobic chromatography when affinity chromatography is intended. In order also to maintain specificity at the elution step, elution with a sub-

[31] P. Cuatrecasas, *J. Biol. Chem.* **245**, 3059 (1970).

strate or inhibitor is preferred to elution with other solvents. In each case, it is desirable that a control column with ligand-free arms be used so as to evaluate the contribution of hydrophobic interactions to the retention capacity of affinity chromatography columns.

The results obtained with the ω-aminoalkyl agarose series raise the possibility that the commonly used ion exchangers[32] which contain branched hydrocarbon chains, e.g., diethylaminoethyl groups, are in fact special cases of mixed ionic and hydrophobic adsorbents, and could be regarded as isolated members of larger families that have not yet been systematically investigated. This suggestion could account for some abnormalities in the binding of proteins to ion exchangers on the "wrong" side of their isoelectric point.

Note Added in Proof

Since this review article was prepared for publication, several additional papers dealing with the subject have appeared. These include: (a) B. H. J. Hofstee, *Anal. Biochem.* **52**, 430 (1973); (b) J. Porath, L. Sundberg, N. Fornstedt, and J. Olsen, *Nature (London)* **245**, 465 (1973); (c) S. Hjertén, *J. Chromatogr.* **87**, 325 (1973); (d) R. A. Rimerman and G. W. Hatfield, *Science* **182**, 383 (1973).

[32] S. R. Himmelhoch, this series, Vol. 22, p. 273.

[10] On Predicting the Results of Affinity Procedures

By DAVID J. GRAVES and YUN-TAI WU

Although a good number of empirical guidelines have been developed for affinity chromatography, including the use of spacer molecules and the preference for Sepharose 4B as a support phase, theoretical guidelines based on physicochemical properties and relationships have been almost nonexistent. Furthermore, although affinity separations can be spectacularly successful, many initial attempts can lead to failure, and the investigator has to base his next try for success on a rather hazy interpretation of why the failure occurred. It was our feeling that simple kinetic and equilibrium models of the affinity adsorption and desorption events would lead to a better understanding of the results and a more rapid and positive success in applying affinity methods to a particular problem.

Description of the Models

It seemed most appropriate to analyze the adsorptive and desorptive phases of affinity chromatography individually, and within each of these phases first to consider limitations based only on equilibrium relation-ships. Models that take rate processes (diffusion and reaction) into ac-count are not covered in the analysis presented here. The entire column chromatographic process is quite complicated if one considers all the events that take place when a sample is added to a feed stream (which usually changes in composition during the elution process) and the column effluent is monitored to detect emerging species. Except in certain idealized situations, the mathematics become quite involved and almost impossible to handle except on a computer. The individual steps in a batch experiment, on the other hand, can be analyzed readily. It might be added that these statements can be applied with equal conviction to the experiment itself. It is far easier to analyze the results of a batch ad-sorption or desorption than those of a column effluent profile (although analysis of moments and other techniques can be applied to column results).[1]

The Equilibrium Model for Adsorption with a Fixed Binding Constant

The first and simplest question is how an enzyme (or other species)[2] is distributed between the solution phase and the solid ligand-containing phase for a fixed value of the equilibrium constant and when no com-peting ligands are present in solution. If changes in pH, ionic strength, temperature, etc., manifest themselves simply as a change in this binding constant, then the distribution can be calculated both during equilibrium adsorption and desorption conditions. If the concentration of ligand is denoted by (L) and that of the enzyme by (E) then

$$E + L \underset{k_{-1}}{\overset{k_1}{\rightleftharpoons}} EL \tag{1}$$

and

$$K_i = \frac{(E)(L)}{(EL)} \tag{2}$$

Equation (2) is written in the customary manner so that K_i is the disso-ciation rate constant k_{-1} divided by k_1. To use this basic relationship in a gel phase containing immobile ligand groups, it must be modified appro-priately. Let us assume that we have a volume of agarose or similar gel v' which consists of a somewhat smaller volume v of solution entrapped

---

[1] D. Grubner and D. Underhill, *J. Chromatogr.* **73**, 1 (1972).

[2] For simplicity, we will henceforth use "enzyme" as a generic term denoting the molecule of interest which is to be recovered. Molecules which are to be removed and discarded will be called "contaminant protein."

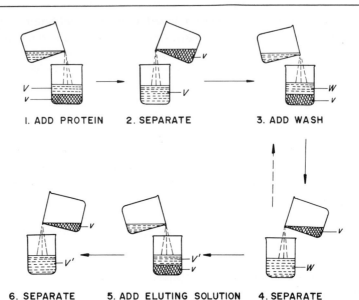

1. ADD PROTEIN   2. SEPARATE   3. ADD WASH

6. SEPARATE   5. ADD ELUTING SOLUTION   4. SEPARATE

FIG. 1. Schematic representation of the batch adsorption, washing, and elution model which is used to represent an affinity purification procedure. Equilibrium is assumed to be achieved in each step.

within the loose network of molecular strands. (We could base our calculations on either v' or v, but for simplicity and clarity in the resulting equations we will use the latter.)

Let us further assume that within the gel a bound ligand is present at concentration L_0 (moles per volume v) and that no enzyme is present in v. Next, a volume V of enzyme at concentration E_0 is added to the gel, and the system is allowed to come to equilibrium. The situation is depicted schematically as step 1 in Fig. 1. Using Eq. (2) and equations describing the conservation of enzyme and ligand molecules, an expression can be derived for the concentration of bound enzyme at equilibrium (EL):

$$(\text{EL}) = \frac{(L_0)}{2} [B][1 - \sqrt{1 - A}] \tag{3}$$

where

$$B = \frac{K_i(V + v)}{L_0 v} + \frac{E_0 V}{L_0 v} + 1 \tag{4}$$

and

$$A = \frac{4E_0 V}{B^2 L_0 v} \tag{5}$$

Fortunately, this rather cumbersome expression can be simplified considerably in the case where A is much less than 1. For such a situation

$$\sqrt{1 - A} \doteq 1 - (A/2) \tag{6}$$

and Eq. (3) reduces to the following form

$$(EL) \doteq \frac{L_0 E_0 V}{K_i(V + v) + L_0 v + E_0 V} \tag{7}$$

The error involved in substituting approximation (6) into Eq. (3) is about 2.6%, 5.2%, and 17.1% when A is 0.1, 0.2, and 0.5, respectively. The quantity B defined in Eq. (4) usually has a value close to unity, and, assuming it is exactly equal to 1.0, Eq. (5) states that the value of A is 4 times the ratio of the number of moles of enzyme to the number of moles of ligand. Under normal circumstances, enzyme concentrations in solution are 10^{-5} M or less, and a typical ligand concentration in the gel phase is 10^{-2} M. Therefore, at most reasonable ratios of solution volume V to gel volume v, the approximation will introduce very little error into whatever equation it is used with.

It is of interest to present a few more relationships that can be derived using the same approximation. A most useful question, which can be answered, is: What fraction of the total enzyme will be bound to the ligand at equilibrium? This quantity is easily shown to be

$$\frac{\text{bound enzyme}}{\text{total enzyme}} = \frac{(EL)v}{E_0 V} \doteq \frac{L_0 v}{K_i(V + v) + L_0 v + E_0 V} \tag{8}$$

In most cases of practical interest, $E_0 V$ is much smaller than the other two terms and may be dropped from Eq. (8). We will omit the $E_0 V$ term in all later derivations.

Equation (8) is most useful for determining the effectiveness of binding for a given set of the parameters K_i, L_0, E_0, V, and v. A second useful equation is that giving the distribution between the gel and external solution phases and includes the enzyme present in solution within the gel (we will refer to this portion as "gel-entrapped" enzyme).

$$\frac{\text{bound + entrapped enzyme}}{\text{total enzyme}} = \frac{V(E_0 - E)}{V E_0} \doteq \frac{L_0 v}{K_i(V + v) + L_0 v}$$
$$+ \left(\frac{v}{V + v}\right)\left[1 - \frac{L_0 v}{K_i(V + v) + L_0 v}\right] \tag{9}$$

This fraction represents the separation achieved simply by separating the two phases, and the difference between (9) and (8) represents that enzyme which is simply entrapped. Another equation is a linear form that can be used to find the values of K_i and L_0 from the experimental variables E_0, E, V, and v.

$$\frac{\text{free} + \text{entrapped enzyme}}{\text{bound enzyme}} = \frac{(V + v)E}{(E_0)V - E(V + v)}$$

$$\doteq \frac{V}{L_0 v}E_0 + \frac{K_i}{L_0}\left(\frac{V + v}{v}\right) \quad (10)$$

Data points representing the left-hand side of Eq. (10) can be plotted on the ordinate and their corresponding E_0 values on the abscissa of a graph in the classical Lineweaver-Burk fashion. The slope of the straight line through these points is then $V/(L_0 v)$ (from which L_0 can be found) and the intercept on the y axis is $K_i(V + v)/(L_0 v)$. The value of K_i can be found from this intercept. Alternately, a least-squares regression analysis can be used to find these quantities. An additional expression is a form of Eq. (3) which has been manipulated to give the ratio of bound to total enzyme for those cases where the approximate expression (8) lacks accuracy.

$$\frac{\text{bound enzyme}}{\text{total enzyme}} = \frac{(EL)v}{(E_0)V} = \frac{v(L_0)}{2V(E_0)}(B)[1 - \sqrt{1 - A}] \quad (11)$$

Similarly, the more exact analog of equation (9) can be written as follows:

$$\frac{\text{bound} + \text{entrapped enzyme}}{\text{total enzyme}} = \frac{L_0 v(B)(1 - \sqrt{1 - A})}{2E_0 V}$$

$$+ \left(\frac{v}{V + v}\right)\left[1 - \frac{L_0 v(B)(1 - \sqrt{1 - A})}{2E_0 V}\right] \quad (12)$$

Any number of additional expressions could be given, but these are perhaps the most useful. For those desiring to derive different relationships, the following identities may be of aid:

Total moles of enzyme present $\equiv (E_0)V$ (13)

Moles of ligand-bound enzyme $\equiv (EL)v \equiv V(E_0 - E) - v(E)$ (14)

Moles of enzyme in solution within gel $\equiv (E)v$ (15)

Moles of enzyme in solution outside gel $\equiv (E)V$ (16)

However, these equations fail to provide an intuitive feel for how the parameters involved will affect the separation. If we return to Eq. (8) and again apply the restriction that the total number of moles of enzyme is much smaller than the number of moles of ligand and add a restriction that the volume V is much smaller than the volume v, we eliminate E_0, V,

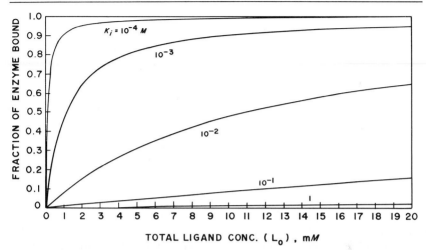

FIG. 2. Fractional enzyme binding for low enzyme concentrations as represented by Eq. (17). This general hyperbolic graph also can be used with the more correct Eq. (8) and even to estimate the fractional ligand saturation (see text).

and v and can clearly see the effects of different values of K_i and L_0:

$$\frac{\text{bound enzyme}}{\text{total enzyme}} \doteq \frac{L_0}{K_i + L_0} \tag{17}$$

This relationship is plotted in Fig. 2. As might be expected, the resulting curve for any value of K_i is a rectangular hyperbola similar to the Langmuir adsorption isotherm or the Michaelis-Menten equation. As an alternative approach, Fig. 2 may be used as a plot of Eq. (8) simply by substituting $L_0 v$ for L_0 on the abscissa and $K_i(V + v) + (E_0)V$ for K_i as the variable parameter.

The significance of the fractional adsorption values given in Fig. 2 is best appreciated by reference to realistic experimental values. Typical ligand concentrations which presently are achieved experimentally are 10 mM. At this ligand concentration, Eq. (17) predicts that K_i values of 10^{-3} M or less would result in effective enzyme binding, whereas higher values would result in rather ineffective retention. Interestingly enough, experimental work by several different investigators has established a similar empirical rule of thumb.[3] It also should be added parenthetically that simply because a soluble inhibitor has a particular K_i value does not imply that this same value will hold when the inhibitor becomes a ligand covalently coupled to a support phase. We and others have observed substantial increases in K_i (by three orders of magnitude in one of our cases).

[3] E. Steers, P. Cuatrecasas, and H. Pollard, *J. Biol. Chem.* **246**, 196 (1971).

Saturation of Gel Phase

Before leaving the binding analysis and considering elution, it is of some interest to look at the other side of the picture, to see what fraction of the ligand groups become saturated with enzyme under any given set of conditions. This fraction is simply EL/L_0 and the approximate expression for it can be found very easily simply by dividing both sides of Eq. (7) by L_0. A hyperbolic equation in $E_0 V$ results, and Fig. 2 again can be used to evaluate the fractional saturation simply by substituting $E_0 V$ for the quantity L_0 on the abscissa and using $K_i (V + v) + L_0 v$ instead of K_i as the variable parameter. The fraction EL/L_0 then is simply read on the ordinate scale.

For convenience, this calculation has been carried out for several representative values of K_i, E_0, and ω, where ω is the ratio of V to v. These are presented in Table I as percentages. The most striking feature shown by these results is that in most cases there is very little saturation of the gel phase. Chromatographers should be able to see clearly that enzyme can begin to "leak out" of a column during an adsorption experiment long before all of the reactive groups in the gel are saturated. For example, if $E_0 = 10^{-6} M$, $L_0 = 10^{-2} M$, $K_i = 10^{-3} M$ and $V = v$, about 17% of the

TABLE I

PERCENTAGE LIGAND SATURATION UNDER VARIOUS CONDITIONS

(FOR $L_0 = 0.01 M$)

E_0 (M)	K_i (M)	EL/L_0 (%)		
		$\omega = 1$	$\omega = 10$	$\omega = 100$
10^{-4}	10^{-3}	0.8264	4.5455	8.2645
	10^{-4}	0.9709	8.2645	33.2226
	10^{-5}	0.9881	9.0009	47.5964
	10^{-6}	0.9899	9.0818	49.7488
10^{-5}	10^{-3}	0.0833	0.4739	0.8929
	10^{-4}	0.0979	0.8929	4.7393
	10^{-5}	0.0997	0.9794	8.3264
	10^{-6}	0.0999	0.9890	9.0082
10^{-6}	10^{-3}	0.0083	0.0476	0.0900
	10^{-4}	0.0098	0.0900	0.4950
	10^{-5}	0.0100	0.0988	0.9001
	10^{-6}	0.0100	0.0998	0.9803
10^{-7}	10^{-3}	0.0008	0.0048	0.0090
	10^{-4}	0.0010	0.0090	0.0497
	10^{-5}	0.0010	0.0099	0.0907
	10^{-6}	0.0010	0.0100	0.0989

enzyme in solution will fail to adsorb although less than 0.1% of the gel capacity is exhausted. Several authors have mentioned that the experimental "capacity" of a column frequently is far lower than the amount of titratable ligand present. Although this result usually is attributed to steric exclusion or ionic repulsion, equilibrium effects alone can be a sufficient explanation.

An implicit assumption inherent in all the foregoing analysis is that no species are competing with the enzyme of interest for ligand binding. In certain affinity methods, such as the separation of a mixture of dehydrogenases on a relatively nonspecific coenzyme column,[4] the simple relationships just presented might not apply too well. An important advantage of affinity methods over others such as ion exchange, though, is that since the total number of moles of enzyme is much less than that of ligand, one enzyme would be unlikely to interfere with the adsorption of a second enzyme. There are plenty of ligand sites for both species and each should act relatively independently. Furthermore, one can determine whether the simple theory presented here is justified in a specific case by comparing the behavior of a pure enzyme with that of the same enzyme when it is part of a crude mixture. The purification achieved during elution, of course, would be highly dependent on the presence of similar enzymes.

The Equilibrium Model for Elution by a Change in K_i

After the degree of enzyme binding has been established, the next question that arises is what fraction can be removed under conditions of elution. Although elution can be achieved by adding a competing soluble inhibitor, a simple and common technique is simply to change the pH or the ionic strength of the buffer solution passing through the column. Although the effects of such manipulations can be complicated, as a first approximation we will assume that the only result is an increase in the value of K_i when the ionic environment is changed. Again, we will examine only the batch process rather than the continuous column washing. The overall affinity procedure can be broken down into a number of steps shown in Fig. 1: (1) volume v of gel is equilibrated with volume V of enzyme with an equilibrium constant K_i; (2) V is separated from v; (3) volume w of washing solution is added to v and equilibrated (K_i is unchanged); (4) w is separated from v, removing a fraction of the contaminant proteins which are assumed to have been entrapped in the gel (some bound and entrapped enzyme is lost too); (4A) steps 3 and 4 are repeated as often as desired to reduce contaminant to a low level; (5)

[4] C. R. Lowe and P. D. G. Dean, *FEBS Lett.* 14, 313 (1971).

TABLE II
Variation in Properties during a Successful Multistep Affinity Procedure

Property	Step in procedure[a]					
	1	2	3	4	5	6
Enzyme conc. (solution)	Low	Low	Very low	Very low	High	High
Enzyme conc. (gel)	High	High	High	High	Low	Low
Equilibrium constant	K_i	K_i	K_i	K_i	K'_i	K'_i
Retained fraction	$V + v$	v	$w + v$	v	$V' + v$	V'
Discarded fraction	—	V	—	w	—	v
Contaminant protein level	High	High	Low	Low	Very low	Very low

[a] Steps shown in Fig. 1.

volume V' of eluting solution is added to v and equilibrated at a new higher equilibrium constant K'_i; (6) V' is separated from v and the purified enzyme is recovered. If desired, steps 5 and 6 could be repeated to recover more enzyme, but in practice such multiple elution is uncommon. In fact, it is rare to find several sequential applications of steps 3 and 4. Table II indicates how several variables change during these steps.

Enzyme Elution without Washing

The derivation of the equations describing this process is almost identical to that used in the binding analysis, but because of the added complexity of the process, we will consider only the forms based on approximation (6). In fact, all we have to do is use an equation such as (9) and apply it twice, the second time substituting for the initial amount of enzyme present, (E_0V), the moles of bound plus gel-entrapped enzyme calculated from the first application of Eq. (9). (As far as the system is concerned, it does not know whether the "initial" enzyme was supplied from an external volume of solution or came from the gel phase itself.) The final result of this substitution is an expression for the fraction of initial enzyme which is bound to ligand at the end of the process. (Let us temporarily assume that no washing step is included.) Subtracting this fraction from unity gives a fraction which is the sum of that enzyme lost in V, that enzyme recovered in V' and that enzyme entrapped in v:

$$\frac{\text{entrapped} + V \text{ and } V' \text{ fractions of enzyme}}{\text{total enzyme}}$$

$$\doteq 1 - \frac{L_0v(K_iv + L_0v)}{(K_i(V + v) + L_0v)(K'_i(V' + v) + L_0v)} \quad (18)$$

However, the enzyme removed in V can be calculated readily from Eq.

(8) by subtracting from unity and multiplying by the appropriate volume fraction:

$$\frac{\text{enzyme in } V}{\text{total enzyme}} \doteq \left(\frac{V}{V+v}\right)\left(1 - \frac{L_0 v}{K_i(V+v) + L_0 v}\right) \qquad (19)$$

The difference between (18) and (19), then, when multiplied by the correct volume fraction, gives the enzyme recovered.

$$\frac{\text{enzyme in } V'}{\text{total enzyme}} \doteq \left(\frac{V'}{V'+v}\right)\left\{1 - \frac{L_0 v(K_i v + L_0 v)}{(K_i(V+v) + L_0 v)(K'_i(V'+v) + L_0 v)}\right.$$
$$\left. - \left(\frac{V}{V+v}\right)\left[1 - \frac{L_0 v}{K_i(V+v) + L_0 v}\right]\right\} \qquad (20)$$

Even this approximate expression is useful in understanding the overall process. Note that recovery is strongly influenced by ratio of V' to v and less by that of V to v. The final term in Eq. (20) sets an asymptotic limit to the recovery of enzyme even with a very high value of K'_i. A graphical example, which Fig. 3 provides, gives us a clearer picture of what Eq. (20) means. A favorably small sample size ($V/v = 1$) and a large elution volume ($V'/v = 10$) have been chosen to simulate a reasonable experimental procedure. The unfavorable nature of K_i values greater than $10^{-3} M$ is again evident, but we also can see that when K_i is too low, a second type of difficulty can arise. Unless K_i can be raised to a K'_i of at least $10^{-2} M$, a good fraction of the enzyme initially present will remain bound

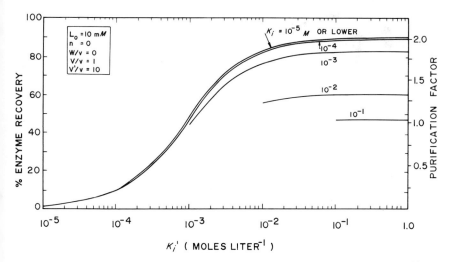

FIG. 3. Fractional recovery and purification factor for the case of no washing steps, a 1:1 sample-to-gel volume ratio, and a 10:1 eluting solution-to-gel volume ratio.

to the gel phase following the elution. Although the reasoning is somewhat intuitive, it stands to reason that manipulation of K_i from a value of 10^{-4} to 10^{-2} M would be easier than from 10^{-8} M to this same value.

Contaminant Recovery

In connection with Fig. 3 and Eq. (20), it is interesting to calculate what recovery would be achieved if there were no binding groups, and entrapment within the gel were the only mechanism operating. In this case, the recovery can be calculated simply from volume ratios:

$$\frac{\text{protein recovered}}{\text{initial protein}} = \frac{vV'}{(V + v)(V' + v)} \qquad (21)$$

In the example given in Fig. 3, the fraction calculated from Eq. (21) is 0.45. Comparing this value with Eq. (20) as plotted in Fig. 3, it is clear that the ligand is contributing very little to the recovery of enzyme when K_i is 10^{-2} M and practically nothing when K_i is as high as 10^{-1} M. The right-hand ordinate indicates what purification factor (P_f) may be achieved for any graphically determined value of enzyme recovery, assuming that contaminant protein recovery is specified by Eq. (21). The purification factor is defined as the final enzyme to total protein ratio divided by the initial ratio. In a usual case of a crude enzyme mixture, total protein is approximately equal to the contaminant protein level, and P_f is equal to the ratio of Eq. (20): Eq. (21). It is quite clear that without a washing step, the purification that can be achieved is severely restricted, even with favorably low values of K_i and favorably high values of K'_i.

The calculation of how effective a sequence of washing steps would be on the reduction of contaminant is quite simple. The relative distribution between a wash of volume w and the gel is simply given by a volume ratio, and n successive washes raise this fraction to the power n:

$$\frac{\text{contaminant protein in elutant}}{\text{total contaminant protein}} = \left[\frac{vV'}{(V + v)(V' + v)}\right]\left(\frac{v}{w + v}\right)^n \qquad (22)$$

If w/v is 10, then the concentration ratio 0.45 given previously is reduced to 0.041 by one washing step and to 0.004 by two washes. Also, it should be noted that if a given volume of washing solution is available, it is more efficient to split this volume up into a number of small washes rather than use it all in one wash. If the wash solution specified previously were used as 10 washes of 1 volume each, rather than one of 10 volumes, Eq. (22) predicts that the concentration ratio would be 0.0004, almost 10 times lower than it would be with 2 washes of 10 volumes each. This result points out the advantage that column techniques can have over batch adsorption and elution procedures, since flowing wash solution through a fixed bed of material approximates a large number of small washing steps.

Enzyme Elution following Multiple Washes

The calculation of enzyme recovery is a more complicated matter because of the simultaneous binding by ligand and physical entrapment within the gel. The derivation, however, follows that just given for no washing steps. It can be shown that the fraction of total enzyme which remains bound after n washing steps is given by

$$\frac{\text{fraction enzyme bound}}{\text{total enzyme}} = f_{Bn} \doteq \frac{L_0 v (K_i v + L_0 v)^n}{(K_i(V + v) + L_0 v)(K_i(w + v) + L_0 v)^n} \tag{23}$$

and similarly that the enzyme remaining bound following the elution step is given by

$$f_{BE} \doteq \frac{L_0 v (K_i v + L_0 v)^{n+1}}{(K_i(V + v) + L_0 v)(K'_i(V' + v) + L_0 v)(K_i(w + v) + L_0 v)^n} \tag{24}$$

Equations (8) and (18) contain similar terms, as would be expected. The fraction recovered in the elutant can be found by multiplying the appropriate volume fraction by the fraction of enzyme present in V' and in v after equilibration as before. This fraction contained in V' and v is simply unity minus the fraction bound minus the sum of each fraction (f_i) removed in wash i:

$$\frac{\text{recovered enzyme}}{\text{total enzyme}} = f_R \doteq \left(\frac{V'}{V' + v}\right)\left[1 - f_{BE} - \sum_{i=0}^{n} f_i\right] \tag{25}$$

The fraction f_0 represents the loss during binding (step 2 in Fig. 1). The fractions (f_i) are related through a relatively simple recursion relationship:

$$f_0 \doteq \left(\frac{V}{V + v}\right)\left[1 - \frac{L_0 v}{K_i(V + v) + L_0 v}\right] \tag{26}$$

$$f_1 \doteq \left(\frac{w}{w + v}\right)[1 - f_{B1} - f_0] \tag{27}$$

$$f_n \doteq \left(\frac{w}{w + v}\right)\left[1 - f_{Bn} - \sum_{i=0}^{n-1} f_i\right] \tag{28}$$

The fractions quickly become quite complex because each term contains all of the previous fractions. For convenience in computation, the fraction removed in the nth wash can be expressed in an alternate way as a finite series containing only f_0 and the bound fractions f_{Bi}:

$$f_n \doteq \frac{w}{w+v} \left\{ -f_{Bn} + (1 - f_0) \left[\sum_{i=0}^{n-1} \frac{(n-1)!}{(n-1-i)!\,i!} \left(\frac{-w}{w+v}\right)^i \right] \right.$$
$$\left. - \sum_{i=1}^{n-1} f_{Bi} \left[\sum_{j=0}^{n-1-i} \frac{(n-1-i)!}{(n-1-i-j)!\,j!} \left(\frac{-w}{w+v}\right)^{j+1} \right] \right\} \quad (29)$$

Although Eq. (29) appears somewhat complex, it can be evaluated readily by a simple computer program, and each value f_n can be substituted in Eq. (25) to provide an estimate of enzyme recovery and the purification factor. For ony one or two washes, Eq. (25) is not overly complex, and it is easier to use an explicit expression rather than a combination of Eqs. (29) and (25). With only one washing step, the following expression applies:

$$f_R \doteq \left(\frac{V'}{V'+v}\right)$$
$$\times \left\{ 1 - \frac{L_0 v (L_0 v + K_i v)^2}{(K_i(V+v) + L_0 v)(K'_i(V'+v) + L_0 v)(K_i(w+v) + L_0 v)} \right.$$
$$\left. - \left(\frac{w}{w+v}\right) [1 - f_{B1} - f_0] - f_0 \right\} \quad (30)$$

and for two washes, the appropriate expression is

$$f_R \doteq \left(\frac{V'}{V'+v}\right)$$
$$\times \left\{ 1 - \frac{L_0 v (L_0 v + K_i v)^3}{(K_i(V+v) + L_0 v)(K'_i(V'+v) + L_0 v)(K_i(w+v) + L_0 v)^2} \right.$$
$$\left. - \left(\frac{w}{w+v}\right) [2 - f_{B2} - f_{B1} - f_0] + \left(\frac{w}{w+v}\right)^2 [1 - f_{B1} - f_0] - f_0 \right\} \quad (31)$$

Although these expressions are useful for specific calculations under any desired conditions, they fail to provide any insight on the influence of the various parameters involved. Figure 4 represents a case where one wash of relatively large volume ($w/v = 10$) is added to the situation previously described and plotted in Fig. 3 ($V/v = 1$, $V'/v = 10$, $L_0 = 0.01$). The most spectacular effect of adding the wash is that the maximum purification factor which can be achieved has jumped from 2 to 22. In fact, even for K_i values as high as $10^{-2}\,M$, the purification has been increased significantly. When K_i is $10^{-5}\,M$ or below, adding this washing step causes no appreciable decrease in the amount of enzyme that can be

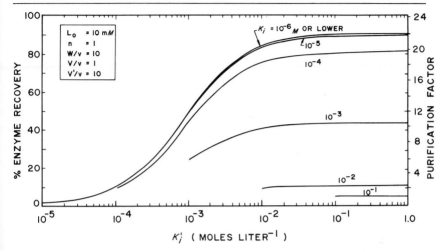

FIG. 4. Recovery and purification factor for the case of a single washing step where the wash is 10 times the gel volume. Other parameters are the same as in Fig. 3.

recovered. For $K_i = 10^{-4} M$ the recovery at high K'_i values drops from about 90 to 82%, but when K_i is $10^{-3} M$ a rather drastic drop from 83 to 44% recovery takes place. Such a loss may be tolerable since the purification factor jumps from less than 2 to more than 10, but the performance under these conditions is far from ideal.

We should again point out, however, that the situation will be considerably better when such a material is used in a continuous column process rather than in a batch adsorption and elution. In fact, Figs. 3 and 4 provide a good deal of guidance in deciding whether a particular affinity system will be successful as a batch process or must be used in a column. Figure 5 demonstrates that two washes at a 10:1 volume ratio produce purifications in the neighborhood of 200 for sufficiently low values of K_i. As is evident, K_i values of greater than $10^{-4} M$ now are quite poor with respect to the amount of enzyme which can be recovered. Figures 6 and 7 illustrate two final cases, those of two and of ten washes of one gel volume each. Very high purifications can be achieved in the last case provided that K_i is low enough to ensure that the enzyme remains "stuck" to the gel while impurity is rinsed away.

Multiple Washes and Multiple Elutions

The mathematics involved in expanding the multiple-wash single-elution case to multiple-wash multiple-elution is straightforward. Instead of the fraction f_R we will use the nomenclature $f_{R(j)}$ to denote the frac-

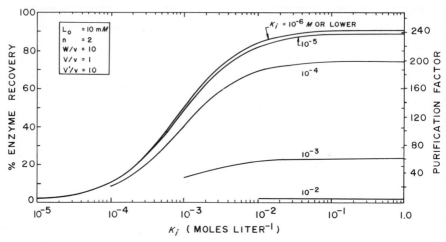

FIG. 5. Recovery and purification factor for the case of two washing steps. Otherwise the conditions are those of Fig. 4.

tional recovery of enzyme in elution j. The $f_{R(j)}$ are then given by the following expressions:

$$f_{R(1)} = f_R \tag{32}$$

$$f_{R(2)} = \left(\frac{V'}{V' + v}\right)\left[1 - f_{BE(2)} - f_{R(1)} - \sum_{i=0}^{n} f_i\right] \tag{33}$$

$$f_{R(m)} = \left(\frac{V'}{V' + v}\right)\left[1 - f_{BE(m)} - \sum_{j=1}^{m-1} f_{R(j)} - \sum_{i=0}^{n} f_i\right] \tag{34}$$

where m represents the result of the mth elution. The fraction f_R is defined by Eq. (25), and the new quantity $f_{BE(j)}$ is the fraction remaining bound to the gel phase following the jth elution. It can be found from the equation

$$f_{BE(j)} = f_{Bn}\left[\frac{K_i v + L_0 v}{K'_i(V' + v) + L_0 v}\right]^j \tag{35}$$

where the value f_{Bn} has been defined previously by Eq. (23). Multiple elutions may prove to be attractive (or in some cases the only feasible method) if K_i is very low and cannot be manipulated to a high enough K'_i value to give good recovery with a single elution. Although we are not considering rate effects in this treatment, it should be mentioned that multiple batch elutions in such cases might offer significant advantages

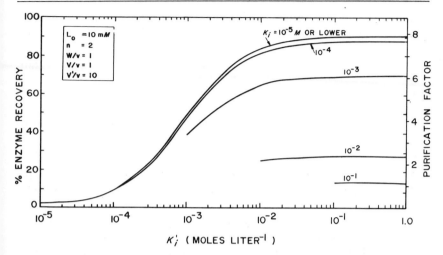

FIG. 6. Recovery and purification factor for the case of two washing steps with a 1:1 wash-to-gel volume ratio. Other conditions are those of Fig. 4.

over column elution. Equations (32)–(34) therefore can have a very practical use.

The Equilibrium Model for Elution by a Competitive Inhibitor

A second common technique for recovering bound materials from affinity gels is the addition of a soluble inhibitor solution. In such a case we will assume that K_i remains unchanged during the elution process. The

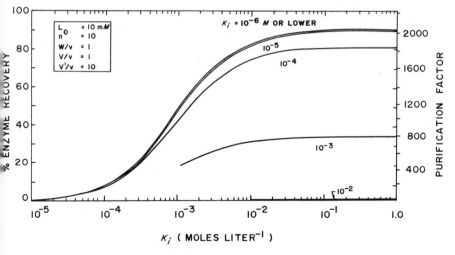

FIG. 7. Recovery and purification factor for the case of 10 washing steps. Other conditions are those of Fig. 6.

derivation of an appropriate expression includes Eqs. (1) and (2), and we add the relationships

$$E + I \underset{k_{-2}}{\overset{k_2}{\rightleftharpoons}} EI \tag{36}$$

and

$$K_s = \frac{k_{-2}}{k_2} = \frac{(E)(I)}{(EI)} \tag{37}$$

All the symbols previously defined have their original significance (V' now is the volume of inhibitor solution) and we need add only I_0 to represent the initial inhibitor concentration, (I) to represent the final equilibrium inhibitor concentration, (EI) the equilibrium level of complexed enzyme in solution, and vE_i to represent the moles of enzyme present in the gel prior to addition of the inhibitor solution. This latter term most generally will signify both bound and gel-entrapped enzyme, so that E_i is a pseudo-concentration rather than either a bound or entrapped concentration alone. Furthermore, all the calculations just given on the effects of washing can still be applied prior to this elution analysis. The equation representing the bound plus entrapped enzyme remaining in the gel after n washes is

$$\frac{vE_i}{VE_0} \doteq \left(\frac{v}{w+v}\right)\left[1 - f_{Bn} - \sum_{i=0}^{n-1} f_i\right] + f_{Bn} \tag{38}$$

where f_{Bn} can be taken from Eq. (23) and the f_i from Eqs. (26) through (28) and (29).

Now, if we combine Eqs. (2) and (37) along with the appropriate material balances, as will obtain a quite complex expression containing ten terms and cubic in the concentration of (EI). Fortunately an asymptotic analysis reveals that we may neglect the second and third powers of (EI) and a simpler expression results:

$$(EI) \doteq \frac{vV'I_0E_i}{v(V'+v)(2 - K_s/K_i)E_i + V'(V'+v)I_0 + K_s(V'+v)^2 + K_sL_0v(V'+v)/K_i} \tag{39}$$

With two inhibitors present, one in solution and one as an immobilized ligand, it can be shown that there will be very little free enzyme (E) present after equilibration, although there may be a considerable amount in solution as (EI). In most cases, then, there would be little error involved in stating that the total enzyme in solution is $V'(EI)$ rather than $V'[(EI) + (E)]$. With this simplification, Eq. (39) can be rearranged to provide an expression for the fractional recovery of enzyme. A new quantity $\rho = V'/v$ has been introduced to provide a more convenient form.

$$\frac{\text{recovered enzyme}}{\text{total enzyme in washed gel}} = \frac{V'(\text{EI})}{v\text{E}_i}$$

$$\doteq \left(\frac{\rho}{\rho + 1}\right) \left[\frac{\rho \text{I}_0}{2\text{E}_i + \rho \text{I}_0 + K_s(1 + \rho) + \dfrac{K_s}{K_i}(\text{L}_0 - \text{E}_i)}\right] \quad (40)$$

As we have shown previously, L_0 normally is far larger than E_i. In most cases then, an equivalent expression is

$$\frac{V'(\text{EI})}{v\text{E}_i} \doteq \left(\frac{\rho}{\rho + 1}\right) \left[\frac{\rho \text{I}_0}{\rho \text{I}_0 + K_s(1 + \rho + \text{L}_0/K_i)}\right] \quad (41)$$

Although this equation could be plotted, the parameters include in addition to L_0, which we have always set at 10 mM, the quantities ρ, I_0, K_s, and K_i. To simplify the representation slightly, it is useful first to consider the case where L_0/K_i is much greater than ρ so that the $K_s(1 + \rho)$ term in the denominator can be neglected. A realistic maximum value for ρ is roughly 10, since larger values would result in large dilutions of the recovered enzyme. With L_0 equal to 10 mM, K_i values of less than $10^{-4}\ M$ will cause the ratio of these two parameters to dominate the others. Even when $K_i = 10^{-3}\ M$, only about a 10% error will be produced by neglecting $1 + \rho$. Then instead of K_s and K_i, we can consider only their ratio:

$$\frac{V'(\text{EI})}{v\text{E}_i} \doteq \left(\frac{\rho}{\rho + 1}\right) \left[\frac{\rho \text{I}_0}{\rho \text{I}_0 + K_s\text{L}_0/K_i}\right] \quad (42)$$

Equation (42) has been represented graphically in Fig. 8 for $\text{L}_0 = 10$ mM. Two representative ρ values equal to 1 and 10 are shown. The influence of the inhibitor concentration in causing enzyme elution now is quite clearly evident as is that of the inhibitor to ligand binding constant ratio. Notice that if the constants are about equal, 10 to 100 mM inhibitor (depending on ρ) will do an effective job of elution. If the inhibitor is ten times better than the ligand, these concentrations can be decreased by about an order of magnitude, while if it is ten times poorer, 0.1 to 1 M concentrations will be needed.

Just as we did with Fig. 2, we can use Fig. 8 as a representation of the more complete solution, which in this case is given by Eq. (41). All that is necessary to use the quantity $(K_s/0.01)(1 + \rho + \text{L}_0/K_i)$ instead of K_s/K_i as the parameter shown in this figure.

Either Eq. (41) or (42) also can be used to predict the recovery on successive elutions. The only modification needed is to multiply $v\text{E}_i$ by the

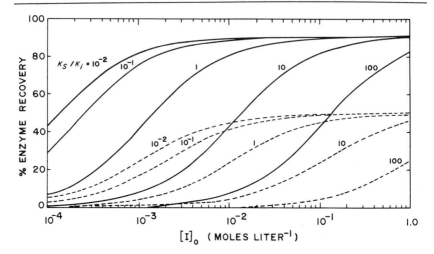

FIG. 8. Enzyme recovery for elution by the addition of soluble inhibitor. The solid curves represent V' to v ratios of 10 and the dashed curves V' to v ratios of 1. No washing steps are assumed. Note that the elution volume ratio strongly influences the overall recovery which is possible.

quantity $(1 - R)$ where R is the ratio calculated from the first application of one of these equations. Any number of additional elutions can be handled in a similar fashion. Purification factors have not been shown in Fig. 8 because they depend on how much impurity has been removed by prior washing steps. The overall purification factor can easily be found from the overall enzyme recovery and protein recovery as we have done previously.

Simulation of Column Chromatography Results

As was mentioned at the outset, the complete analysis of column chromatography is quite complicated. However, one way in which columns have been modeled successfully in the past is as a series of well-mixed compartments. Usually the fluid phase is considered to move continuously from one compartment to the next, but models based on intermittent flow and equilibration can be equally successful. In the extreme case of an infinite number of compartments, the two models become identical.[5]

The equations just presented are directly useful in simulating column behavior if the column is assumed to consist of only one well-stirred compartment. Such a model is physically not realistic, but in behavior the single and multicompartment cases are remarkably similar. Therefore, to

[5] C. J. King, "Separation Processes," p. 410. McGraw-Hill, New York, 1971.

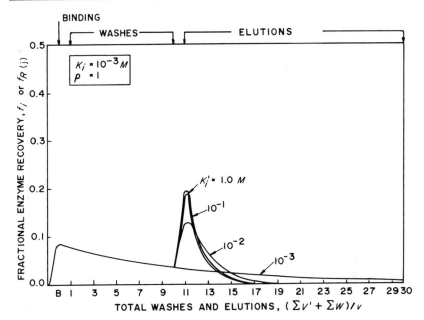

FIG. 9. Simulated column results for a poor binding case ($K_i = 10^{-3} M$) where elution is achieved by a change in K_i to K'_i. Considerable enzyme loss occurs during washing.

obtain a realistic estimate for the shape of an emerging peak, one only need apply equations such as (29) to calculate the fraction of enzyme removed in each wash or (34) to calculate the fraction removed in each elution. As an example of the technique involved, Figs. 9 through 11 have been included. These cases assume that elution is achieved through a change in K_i rather than through the addition of inhibitor. The ratios of wash volume w and of elution volume V' to the gel volume v have been set at unity, which is a reasonable approximation to void fractions which might be encountered experimentally. We have arbitrarily chosen to equilibrate 10 wash volumes and then 20 elution volumes with the gel phase. Smooth curves have been drawn through the data points to eliminate the sharp discontinuities predicted by the simplified mathematical representation. Figure 9 for $K_i = 10^{-3} M$, shows that a great deal of enzyme (61.6%) is lost during the washing peak. Elution at K'_i values of 1.0, 0.1, and 0.01 M then remove essentially all (38.4%) of the remaining enzyme, and do so within 10 elution volumes. Elution at $K'_i = K_i$ removes only about 31.7% even after 20 washes.

With a K_i value of $10^{-4} M$, Fig. 10 shows that the behavior is quite different. Now, only 10.3% of the enzyme is lost during washing steps.

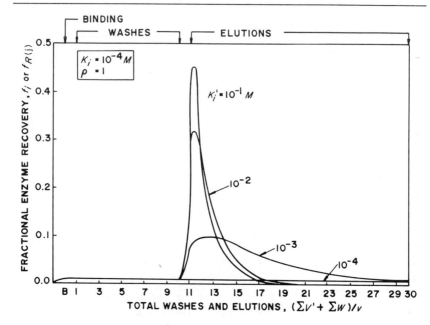

FIG. 10. Simulated column results for moderately good binding ($K_i = 10^{-4}\,M$). Little enzyme is lost during washing.

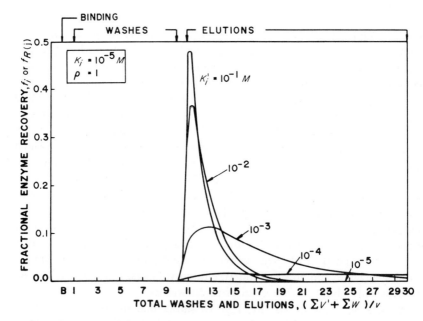

FIG. 11. Simulated column results for very good binding ($K_i \leq 10^{-5}$). Practically no enzyme is lost by washing.

Elution at $K'_i = 0.1$ or 0.01 removes about 89.7% of the enzyme. For $K'_i = 10^{-3} M$ the recovery is 86.2% and for $K'_i = 10^{-4} M$ it drops to only 16%. Figure 11 illustrates the very favorable case of $K_i = 10^{-5} M$ or lower. Now, essentially, no enzyme is lost during washing. Elution at K'_i values from 10^{-1} to $10^{-5} M$ give respective recoveries of 98.9%, 98.9%, 95.6%, 29.7%, and 1.9% of the initial enzyme after 20 washes. High values of K'_i clearly are desirable since they lead to more concentrated solutions of the recovered enzyme.

Conclusions

It is worth reiterating a few of the points that have been made throughout this analysis. First, the fractional binding of enzyme by the ligand can be approximated by an equation which is hyperbolic in the ligand concentration (L_0). The practical limit for L_0 at the present time is about 10 mM, and at this concentration it can be predicted that the critical K_i for producing an effective affinity material is about $10^{-3} M$. Lower values than this will produce good results and higher values poor results. If, however, L_0 could be increased to 50 mM then it should be possible to use ligands whose K_i values are also about 5 times larger. A linear equation was found to determine K_i from experimental data.

The ligand saturation level has been shown to be approximately hyperbolic in the concentration of enzyme. Consequently, to achieve a relatively high degree of gel saturation, the enzyme solution should be as concentrated as possible. Some sacrifice in the *fraction* of enzyme removed from solution may result, but a larger total number of moles of enzyme can be loaded on the column with a concentrated enzyme solution. Depending on one's goal, such manipulation may or may not be desirable. An interesting and important result is that, in most cases, very little of the gel can be saturated with enzyme (typically less than 1% of L_0). One should be aware of this equilibrium-based limitation, and not immediately assume that steric factors are responsible for poor utilization of ligand.

Washing and elution by changing K_i to a new value were studied and at least two important facts emerged. First, K_i values of $10^{-3} M$ look much less attractive than they do when binding alone is considered because so much enzyme is lost through washing. If K_i is $10^{-5} M$ or less, batch adsorption and elution may be as good or better than column chromatography but for values of $10^{-3} M$, only column methods look attractive. It was found that whatever the value of K_i, the increased value K'_i during the elution step had to be at least $10^{-2} M$ in order to remove enzyme from the gel effectively. Achieving such a high value can be a serious limitation if the initial K_i is very low. The number of washing steps and the volume of each wash have as their primary effect an influence on the removal of contaminant. The purification factor therefore is strongly influenced by the thoroughness of washing. If K_i is less than $10^{-4} M$, very little enzyme will be lost with a relatively large amount of washing.

Elution by the addition of a soluble inhibitor was examined and the important parameter was shown to be the relative ratio of the inhibitor to the ligand equilibrium constants. If the ratio is unity, then 0.01 to 0.1 M inhibitor concentrations are needed. The concentration needed and ratio of equilibrium constants are related inversely, and as a practical limit, inhibitors 100 times poorer than the ligand might be used with success at very high concentrations (1 M).

A final interesting result was that the batch multiple elution equations were quite effective in simulating column effluent profiles. Several representative cases were shown and if desired the crude model can be refined easily.

Nomenclature

English Symbols

A	A parameter defined by Eq. (5)
B	A parameter defined by Eq. (4)
E	Enzyme concentration at equilibrium
E_i	Pseudo enzyme concentration representing ligand-bound and gel-entrapped enzyme following washing but prior to elution [see Eq. (38)]
E_0	Initial enzyme concentration in solution phase before contact with gel phase
(EL)	Concentration of enzyme–ligand complex at equilibrium
f_{BE}	Fraction of initial enzyme remaining bound to ligand after n washes and one elution
f_{Bn}	Fraction of initial enzyme remaining bound to ligand after n washing steps
f_n	Fraction of initial enzyme lost when volume w of wash n is discarded
f_0	Fraction of initial enzyme lost when V is discarded
f_R	Fraction of initial enzyme recovered in V' after n washes and one elution
$f_{R(m)}$	Fraction of initial enzyme recovered in V' with elution m after n washes
I_0	Initial concentration of inhibitor
K_i	Ligand–enzyme equilibrium constant during adsorption steps; defined by Eq. (2)
K'_i	Ligand–enzyme equilibrium constant during elution
K_s	Equilibrium constant for enzyme–soluble inhibitor complex
k_1	Rate constant for enzyme–ligand combination
k_{-1}	Rate constant for enzyme–ligand dissociation
k_2	Rate constant for the combination of enzyme with soluble inhibitor
k_{-2}	Rate constant for the dissociation of enzyme–soluble inhibitor complex
L	Ligand concentration at equilibrium
L_0	Initial concentration of ligand in gel
m	Number of elution steps
n	Number of washing steps
P_f	Purification factor: final enzyme to total protein ratio divided by the initial ratio; approximately equal to the ratio of Eq. (21) to f_R or to Eq. (20)
R	Fraction enzyme recovery defined by Eq. (41) or (42)
v	Volume of gel phase which is accessible to penetrating protein molecules

v' Actual volume of gel phase
V Volume of enzyme solution phase initially added to the gel
V' Volume of each eluting solution
w Volume of each wash solution

Greek Letters

ρ Ratio of the elution volume V' to the gel volume v
ω Ratio of the sample volume V to the gel volume v

[11] Affinity Elution: Principles and Applications to Purification of Aminoacyl-tRNA Synthetases

By Friedrich von der Haar

The most effective methods of enzyme purification are those that make use of specific interactions between enzymes and their specific ligands. In affinity chromatography, for example, the ligand is bound to an insoluble support. Depending on the specificity of ligand enzyme interaction, a single enzyme or a group of enzymes could be bound to such an affinity polymer. Ligands interacting with a group of related enzymes have been named "general ligands."[1]

Another approach to enzyme purification is that of elution with a specific ligand.[2] In this case, proteins are bound without particular specificity to a polymer, but the binding is specifically overcome by complex formation of the enzyme with ligand. In addition to general cation exchangers,[3-16] very specific polymers have been used: chitin for the puri-

---

[1] K. Mosbach, H. Guilford, R. Ohlsson, and M. Scott, *Biochem. J.* **127**, 625 (1972).
[2] This series: L. H. Heppel, Vol. 2, p. 570; B. M. Pogell, Vol. 9, p. 9; S. Pontremoli, Vol. 9, p. 625; O. M. Rosen, S. M. Rosen, and B. L. Horecker, Vol. 9, p. 632.
[3] B. M. Pogell, *Biochem. Biophys. Res. Commun.* **7**, 225 (1962).
[4] This series: B. M. Pogell and M. G. Sarngadharan, Vol. 22, p. 379.
[5] M. G. Sarngadharan, A. Watanabe, and B. M. Pogell, *J. Biol. Chem.* **245**, 1926 (1970).
[6] O. M. Rosen, S. M. Rosen, and B. L. Horecker, *Arch. Biochem. Biophys.* **112**, 411 (1965).
[7] O. M. Rosen, *Arch. Biochem. Biophys.* **114**, 31 (1966).
[8] J. Fernando, M. Enser, S. Pontremoli, and B. L. Horecker, *Arch. Biochem. Biophys.* **126**, 599 (1968).
[9] S. Pontremoli, S. Traniello, B. Luppis, and W. A. Wood, *J. Biol. Chem.* **240**, 3459 (1965).
[10] H. Carminatti, E. Rozengurt, and L. Jiménez de Asúa, *FEBS Lett.* **4**, 307 (1969).
[11] J. A. Illingworth and K. F. Tipton, *Biochem. J.* **118**, 253 (1970).

fication of lysozyme[17,18] or insoluble starch for the purification of glycogen synthetase[19] and α-amylase.[20] Since this last approach makes use of the interaction between groups with an affinity for each other, the phrase "affinity elution" may be a useful one in analogy to affinity labeling and affinity chromatography.

The principles of affinity elution from ion exchangers depend upon the rationalization that any ion exchanger can be taken as an affinity polymer containing "general ligands"[1] if binding occurs by reason of groups located at the binding site. Charged groups at the binding sites are shielded against interaction with the ion exchanger as a result of complex formation of the protein with its ligand. This may be due to steric factors or, in the most favorable cases, to neutralization by opposite charges on the ligand. If the difference in accessible charged groups between enzyme and enzyme–ligand complex is significant, the enzyme may be bound to an ion exchanger under certain conditions of buffer, pH, and ionic strength whereas the enzyme–ligand complex is not bound. At present it is not possible to predict the extent to which the charge of an enzyme has to be altered in order to become unbound. The purification of fructose-1,6-diphosphatase from different sources with fructose 1,6-diphosphate,[3–10] isocitrate dehydrogenase with sodium isocitrate[11] or sodium citrate,[12] and glucose-6-phosphate dehydrogenase with glucose 6-phosphate[13,14] show that a maximum of 4, 3, and 2 charges, respectively, on the ligand are able to overcome binding of the enzyme to an ion exchanger under a given set of conditions.

It would appear that many of the advantages of affinity chromatography may be achieved relatively simply by taking advantage of the technique of affinity elution. Affinity elution system can be treated semiquantitatively. There exist interdependent equilibria between the enzyme (E) bound to the ion exchanger (IE), Eq. (1), and the enzyme bound to the ligand (lig) in solution, Eq. (2).

---

[12] L. D. Barnes, G. D. Kuehn, and D. E. Atkinson, *Biochemistry* **10**, 3939 (1971).

[13] T. Matsuda and Y. Yugari, *J. Biochem.* **61**, 535 (1967).

[14] A. Watanabe and K. Takeda, *J. Biochem.* **72**, 1277 (1972).

[15] J. Eley, *Biochemistry* **8**, 1502 (1969).

[16] F. von der Haar, *Eur. J. Biochem.* **34**, 84 (1973).

[17] I. F. Pryme, P. E. Joner, and H. B. Jensen, *Biochem. Biophys. Res. Commun.* **36**, 676 (1969).

[18] I. F. Pryme, P. E. Joner, and H. B. Jensen, *FEBS Lett.* **4**, 50 (1969).

[19] This series: L. F. Leloir and S. H. Goldemberg, Vol. 5, p. 145.

[20] P. S. Thayer, *J. Bacteriol.* **66**, 656 (1953).

$$K_1 = \frac{[\text{E} \times \text{IE}]}{[\text{E}] \times [\text{IE}]} \tag{1}$$

$$K_2 = \frac{[\text{E} \times \text{lig}]}{[\text{E}] \times [\text{lig}]} \tag{2}$$

$$[\text{E} \times \text{lig}] = \frac{K_2 \times [\text{lig}] \times [\text{E} \times \text{IE}]}{K_1 \times [\text{IE}]} \tag{3}$$

The ion exchanger is taken as a macromolecule possessing a finite number of binding sites, an assumption that holds for a given set of conditions, e.g., pH, ionic strength, and a particular quantity of enzyme bound to the polymer.[21] Equation (3) is derived if one assumes that free enzyme concentration, [E], is equal for Eqs. (1) and (2). In fact, this is the limiting case for efficient elution since [E] in Eq. (2) should not exceed [E] in Eq. (1). Otherwise, the enzyme would remain bound to the polymer. Consequently, one can estimate the concentration of ligand required to achieve efficient elution.

The purification of a given enzyme in this system will be due, in general, to the specificity of the enzyme–ligand interaction as is the case for any other affinity system. However, if several complexes, for example $(\text{E}_1 \times \text{lig})$ and $(\text{E}_2 \times \text{lig})$, can be formed it is difficult to predict the order of elution of the enzymes because of the difficulty in determining K_1. If K_1^1, the association constant for E_1 with the polymer and K_1^2, the association constant for E_2 with the polymer, are significantly different, then these enzyme species will easily be separated by simple salt gradient chromatography, and no obstacle exists to their separation. If K_1^1 and K_1^2, the association constant for E_1 with the ligand, is about the same as K_2^2, the association constant for E_2 with the ligand, both enzymes are eluted together. If K_2^1 is bigger than K_2^2 then E_1 will be eluted first followed by E_2. Thus, it is possible to separate glyceraldehyde phosphate dehydrogenase ($K_{m\text{NAD}} = 5 \times 10^{-5}$) from lactate dehydrogenase ($K_{m\text{NAD}} = 1 \times 10^{-4}$) with 0.15 m$M$ NAD$^+$ as eluting solvent and an affinity polymer carrying NAD$^+$ as a general ligand.[1] In this case K_2^1 is only one half of K_2^2. In a similar manner, fructose-1,6-diphosphatase could be separated from aldolase[3] as well as from pyruvate kinase[10] by use of a low concentration of fructose 1,6-diphosphate as the eluting agent and cation exchangers as polymers.

A general approach to the purification of aminoacyl-tRNA synthetases is presented below. This has been successful in the purification of various synthetases from different species in the author's laboratory and appears

[21] A. Kočent, in "Säulenchromatographie an Cellulose-Ionenaustauschern" (M. Rybák, Z. Brada, and J. M. Hais, eds.), p. 105. Fischer, Jena, DDR, 1966.

to be applicable to any aminoacyl-tRNA synthetase. As an illustration, data are presented for the purification of phenylalanyl-tRNA synthetase from baker's yeast. There is one other report for use of affinity elution of a polynucleotide-dependent enzyme, the purification of ribonuclease from chicken pancreas by elution with polynucleotides from phosphocellulose.[15]

Affinity Elution of Aminoacyl-tRNA Synthetases

Starting Material

Any enzyme mixture containing aminoacyl-tRNA synthetase may be used for affinity elution provided that it is freed from polynucleotides either by RNase and DNase digestion[22] or by treatment with an anion exchanger.[16]

If nothing is known of a certain aminoacyl-tRNA synthetase to be purified, the following procedure has been found useful: Cells are broken by any convenient method, and cell debris is moved by centrifugation. Ammonium sulfate is added (30% saturation; 176 g per liter of supernatant), and the precipitate is discarded. An additional 273 g of ammonium sulfate (70% saturation) are added per liter of solution, and the precipitate is suspended in 5 ml of buffer A for each 20 g of protein. The turbid slurry is dialyzed against buffer A. The dialyzate is percolated through a column of DEAE-cellulose [about 300 g of DE-52 cellulose (Whatman) wet weight per 100 g of ammonium sulfate pellet]. The protein solution is treated with 472 g of ammonium sulfate per liter of filtrate. The precipitate is collected by centrifugation and dialyzed against buffer B. This dialyzate may be stored for several months at $-18°$.

Buffers

A: Potassium phosphate, 0.2 M, at pH 7.2 containing 1 mM EDTA and 1 mM dithioerythritol

B: Potassium phosphate, 30 mM, at pH 7.2 containing 50% glycerol (v/v), 1 mM EDTA and 1 mM dithioerythritol

C: Potassium phosphate, 30 mM, at pH 7.2 containing 10% glycerol (v/v), 1 mM EDTA and 1 mM dithioerythritol

Enzyme Assay

The following assay system has been found generally useful for testing aminoacyl-tRNA synthetases. It should be noted that the components may be at suboptimal concentrations for any specific aminoacyl-tRNA

[22] H. Heider, E. Gottschalk, and F. Cramer, *Eur. J. Biochem.* **20**, 144 (1971).

synthetase; nevertheless, the system gives good results if a whole range of different aminoacyl-tRNA synthetases must be tested. Several modifications described previously[23] have been found to be unnecessary. A stock solution that can be stored for several days at 4° contains the following reagents: 300 mM Tris chloride at pH 7.6, 100 mM KCl, 20 mM magnesium sulfate, 0.4 mM ATP, 40 μM [^{14}C]amino acids, and 50 mM mercaptoethanol. Bovine serum albumin is added at a final concentration of 1 mg per milliliter of stock solution. To 50 μl of this solution are added 50 μl of aqueous solution containing 400 A_{260} units of unfractionated tRNA from brewer's yeast. The solution is warmed to 37° and the reaction is started by addition of an appropriate amount of enzyme. Aliquots of 10 μl are withdrawn at intervals, spotted on Whatman 3 MM filter paper and prepared for liquid scintillation counting as described.[23] The data given for phenylalanyl-tRNA synthetase have been obtained by use of the assay optimized for this particular enzyme, using 20 μM [^{14}C]phenylalanine concentration instead of 0.06 mM.[16] One unit of aminoacyl-tRNA synthetase is defined as the amount of enzyme which incorporates 1 nmole of amino acid into tRNA in 1 minute.

Affinity Elution

Determination of Maximum Capacity of Phosphocellulose

A phosphocellulose column[24] of 5-ml bed volume is prepared and equilibrated with buffer C. To this column are applied 5 A_{280} units of protein dissolved in buffer C. Protein is washed through the column with buffer C, and the eluted protein is tested for the specific aminoacyl-tRNA synthetase to be purified. The procedure is repeated until 10–20% of the activity applied to the column at a given step is washed through with buffer C. The amount of specific enzyme protein that can be adsorbed quantitatively on phosphocellulose is defined as the "maximum capacity" of the exchanger for that enzyme. Since the capacity of commercially available phosphocellulose varies among batches, this determination should be repeated whenever a new phosphocellulose batch is obtained. Maximum capacity for different synthetases varies over a wide range. For instance, for one enzyme mixture prepared from baker's yeast, the maximum capacity was 4 A_{280} units for seryl-tRNA synthetase, 6 A_{280} units for phenylalanine-tRNA synthetase, and 40 A_{280} units for tyrosyl-tRNA synthetase.

Determination of maximum capacity must be performed on a column,

[23] E. Schlimme, F. von der Haar, and F. Cramer, *Z. Naturforsch. B* **24**, 631 (1969).
[24] Phosphocellulose P-11 (Whatman) was used. However, CM-cellulose and CM-Sephadex give similar results and may be used in a completely analogous way.

not batchwise, since a significant portion of the enzyme may be adsorbed on fines, which could be transferred as a suspension to the assay mixture.

Determination of KCl Concentration Needed for Aminoacyl-tRNA Synthetase Desorption

A column with 5-ml bed volume is prepared in the same way as described for determination of maximum capacity. It is loaded with enzyme up to 70% of its maximum capacity and is washed thereafter with buffer C containing increasing amounts of KCl; usually increments of 30 mM KCl have been used. The protein eluted from the column with this buffer is tested for enzyme activity. The amount of KCl needed to remove enzyme from the column is called the "minimum salt concentration."

Preparative Affinity Elution

The bed volume of the column is calculated according to the maximum capacity determined as A_{280} units of protein times 70% of maximum capacity. A short column of large diameter may be used. The protein is applied to the column with buffer C, the column is washed with buffer C containing 60% of the minimum salt concentration in order to remove as much unrelated protein as possible. The column is then reequilibrated with buffer C containing 30% of minimum salt. If this value is below 50 mM KCl, then buffer C without any added salt can be used for reequilibration as well as for affinity elution. Finally, the enzyme is eluted with buffer C containing a specific tRNA plus 30% of minimum salt. Concentrations of 0.1 to 2.0 A_{260} units of tRNA per milliliter or buffer C have been found useful. Since it has been shown that other proteins can be eluted with tRNA besides the cognate aminoacyl-tRNA synthetase,[16] the actual tRNA concentration should be chosen so that the enzyme is washed out with a significant amount of solution containing tRNA. The eluate is collected in and is tested for enzyme activity. In the system described, we did not find that the purification of aminoacyl-tRNA synthetase was significantly influenced by magnesium. However, since the general effect of magnesium with respect to tRNA structure is known, 10 mM magnesium sulfate was often added during affinity elution.

Separation of Enzyme and tRNA

The mixture of tRNA and enzyme is adsorbed into a small DEAE-cellulose column (1-ml bed volume per 20 A_{260} units of tRNA). If the salt concentration in buffer C is too high for adsorption of enzyme to DEAE-cellulose, the solution may be diluted appropriately. Enzyme can be desorbed with buffer A, and the tRNA can be desorbed and reused after elution with 1 M NaCl solution. If the enzyme is not adsorbed to DEAE-cellulose, then the mixture of tRNA and enzyme is filtered over DEAE-cellulose for removal of tRNA and the enzyme is concentrated on a small PC-cellulose column.

Purification of Phenylalanyl-tRNA Synthetase

The purification of phenylalanyl-tRNA synthetase from baker's yeast is compared for different starting materials. In one case (Experiment I in Table I), the protein solution had been prepared according to the procedure given in this article. In Experiment II, a partially purified phenylalanyl-tRNA synthetase fraction was used which had been obtained by ammonium sulfate precipitation and subsequent adsorption and elution from DEAE-cellulose.[16] In both cases maximum capacity was found to be 6 A_{280} units of protein per milliliter of phosphocellulose bed volume. Minimum salt was 0.13 M KCl. Consequently, buffer C, without additional salt, was used for affinity elution. In the first experiment, 100 ml of tRNA[Phe] from baker's yeast[25] was used for elution at a concentration of 0.6 A_{260} unit per milliliter of buffer C. In the second experiment, the concentration was 2 A_{260} units tRNA[Phe] per milliliter of buffer C, and 150 ml were used for elution.

TABLE I

AFFINITY ELUTION OF PHENYLALANYL-tRNA SYNTHETASE FROM DIFFERENT STARTING MATERIALS[a]

	Protein (A_{280} units)[b]	Specific activity (units/A_{280} unit)	Total activity (units)	Yield (%)	Purification factor
Experiment I					
Starting material	4500	1.4	6,220	100	1
Affinity eluate	3.7	620	2,280	34	450
Experiment II					
Starting material	3250	20.5	66,500	100	1
Affinity eluate	50	1200	61,500	93	60

[a] Column size was 900-ml bed (6.5 × 27 cm) in the first experiment; 700-ml bed in the second experiment (6.5 × 21 cm).

[b] One A_{280} unit is the quantity of material contained in 1 ml of buffer C which has an absorbance of 1 at 280 nm, when measured in a 1-cm pathlength cell.

[25] D. Schneider, R. Solfert, and F. von der Haar, *Hoppe-Seyler's Z. Physiol. Chem.* **353**, 1330 (1972).

The amount of enzyme that may be handled in a single experiment is drastically reduced in the experiment in which an unfractionated enzyme preparation was used because of the limited capacity of phosphocellulose. Moreover, the yield in this case is much lower than that in which fractionated aminoacyl-tRNA synthetase was used. The main reason for this is that during adsorption of the crude enzyme to phosphocellulose slight irregularities are produced in the column and the enzyme is eluted within a broader volume; only the first fractions show high specific activity. The specific activity is lower in the first of the two experiments, although direct conclusions with respect to purity cannot be drawn from these values. Homogeneous phenylalanyl-tRNA synthetase, purified by gel electrophoresis from identical starting material, showed specific activities in a range from 800 to 1200 when tested in the same assay system as that used here. These preparations have been found to undergo a slow renaturation during storage, reaching a plateau at a specific activity of 1200 after several months.[26]

Examination by SDS gel electrophoresis of the purified product from both experiments of Table I reveals the presence of only the *a* and *b* band of phenylalanyl-tRNA synthetase as the major components. In

TABLE II

AFFINITY ELUTION OF AMINOACYL-tRNA SYNTHETASES
WITH UNFRACTIONATED tRNA

Amino acid tested	Specific activity starting material (units/A_{280} unit)	Specific activity affinity eluate (units/A_{280} unit)	Purification factor
Serine	0.16	0.68	4
Valine	0.74	14.10	19
Isoleucine	0.53	4.52	9
Tyrosine	0.70	2.21	3
Lysine	0.80	2.82	4
Proline	<0.001	<0.001	—
Aspartic acid	0.05	<0.001	—
Alanine	0.02	<0.001	—
Leucine	0.48	2.26	5
Glutamic acid	0.004	<0.001	—
Threonine	0.19	5.63	29
Histidine	0.38	1.27	3
Glycine	<0.001	<0.001	—
Arginine	0.51	18.21	35

[26] F. Fasiolo, N. Befort, Y. Boulanger, and J. P. Ebel, *Biochim. Biophys. Acta* **217**, 305 (1970).

contrast, an unspecific background is present in Experiment II consisting of many minor bands, each representing less than 0.5% of the phenylalanyl-tRNA synthetase.

Other Amino Acid tRNA Synthetases

In Table II are shown the results obtained by washing the column after elution of phenylalanyl-tRNA synthetase from Experiment I with 100 ml of buffer C containing 10 A_{260} units per milliliter of unfractionated tRNA from baker's yeast. Specific activities of starting material and eluate for 15 amino acids are compared. Several enzymes are present only at very low levels in the starting material, and significant activity is therefore not found for these amino acids in the product. For the enzymes present in the starting material, purification factors of from 3 to 35 were obtained. The reason for low purification will vary. For example, maximum capacity for seryl-tRNA synthetase was 4 A_{280} units per milliliter of bed volume; the column was probably overloaded for seryl-tRNA synthetase. On the other hand, tyrosyl-tRNA synthetase is very tightly bound to phosphocellulose, and efficient elution is obtained only if additional salt is added to buffer C. The data presented here indicate that affinity elution may be applied to many, if not all, aminoacyl-tRNA synthetases once appropriate conditions are found.

[12] Affinity Density Perturbation

By Donald F. H. Wallach

Increasing information indicates that the surface—and cytoplasmic—membranes of animal cells contain topologically extended arrays of diverse, but genetically regulated specific functional systems, involving defined receptors as well as ancillary machinery. A thorough understanding of the structure and function of these systems requires the development of specific means by which a given receptor domain can be extracted from other membrane areas. One approach to such specific membrane subfractionation, termed *affinity density perturbation,* appears to offer considerable promise,[1,2] and this report presents the principles involved, a model system, and pertinent pilot studies.

[1] D. F. H. Wallach, B. Kranz, E. Ferber, and H. Fischer, *FEBS Lett.* **21**, 29 (1972).
[2] D. F. H. Wallach and P. S. Lin, *Biochim. Biophys. Acta Rev. Biomembranes,* **300**, 211 (1973).

Fractionation Principle

The fractionation principle is schematized in Fig. 1. Membranes are first physically sheared into minute vesicles. The membrane fragments bearing a given receptor are rapidly separated ultracentrifugally according to the density-increase caused when the receptor combines with a specific ligand which is coupled covalently to a particle of very high density. Membranes and ligand are radioactively labeled with different isotopes, providing the method with unusual quantitative discrimination. The formation of membrane–ligand complexes can be blocked or reversed when desired by addition of reagents of higher affinity for the ligand than the receptor, or an excess of receptor analogs with similar affinity. The density-perturbing particles can be visualized electron microscopically,

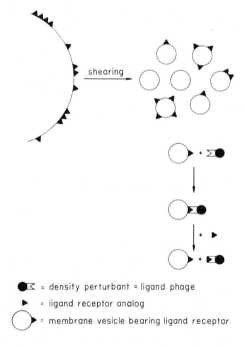

FIG. 1. Schematic of affinity density perturbation applied to membrane subfractionation: A plasma membrane bearing multiple receptors is sheared into membrane fragments carrying different numbers of receptors in varying distributions. These are allowed to react with the ligand (⊐) coupled to the density perturbant, i.e., K29 phage (●), producing a membrane–receptor–ligand–phage complex of higher density than the membrane itself and of lower density than the density perturbant. Addition of a low molecular weight dissociating agent (▲) returns the membrane and density perturbant to their original densities. From D. F. H. Wallach, B. Kranz, E. Ferber, and H. Fischer, *FEBS Lett.* **21**, 29 (1972).

creating the opportunity to map receptor topology. Many samples can be run rapidly and concurrently under highly controlled conditions. Although the first application, illustration, and perhaps major interest lie in the membrane field, this system has as wide a range of application as affinity chromatography and, in addition, offers several advantages.

Model System

In the model system, homogeneous concanavalin A (Con A), radioactively labeled with ^{125}I is used as ligand. This is converted into a density perturbant by glutaraldehyde-coupling it to purified coliphage K29.[3] α,α-Trehalose, which has a modest affinity ($K_4 = 5.38 \times 10^{-3}$ liters/mole[4]) for Con A, is used to dissociate membrane–ligand complexes. Ligands, other than Con A, (e.g., antibodies, hybrid antibodies, peptide hormones) can be employed, as can other coupling methods.[5] Other density perturbants can be envisaged, but small size and stability of K29 offers an unusual advantage. (K29 is an icosahedron about 450 Å in diameter with a density of 1.495 g/ml.)

The membrane fragments are pig lymphocyte plasma membrane vesicles, prepared by surface shearing of the intact cells.[6] About 10^7 Con A receptors are expected per cell,[7] i.e., one receptor per 100 Å^2 surface area, approximately 100 times the number of vesicles produced by shearing. This large number of receptors has been advantageous in developing the system, but future studies must take into account possible large variation in the affinity constants of diverse receptor sites as well as possible steric interference due to the density perturbant.

Methods

Preparation of K29 Coliphage

K29 coliphage, serotype 0.9:K29 (A): H⁻, is propagated on the host organism, *Escherichia coli* B161 using Merck (7882) Standard Nutrient Broth[8] and an initial multiplicity of 4–5. The bacterial lysate produced by disruption of the host cells is first freed of debris by centrifugation at 8000–12,000 g_{av} for 20 minutes. The supernatant fluid is concentrated 100-fold or more by vacuum dialysis and recentrifuged at 8000–12,000 g_{av}

---

[3] F. Wold, this series, Vol. 11, p. 638.
[4] L. L. So and F. F. Goldstein, *Biochim. Biophys. Acta* **165**, 398 (1968).
[5] A. F. Schick and S. J. Singer, *J. Biol. Chem.* **236**, 2471 (1961).
[6] E. Ferber, K. Resch, D. F. H. Wallach, and W. Imm, *Biochim. Biophys. Acta* **266**, 494 (1972).
[7] G. Edelman and C. F. Milette, *Proc. Nat. Acad. Sci. U.S.* **68**, 2436 (1971).
[8] S. Stirm and E. Freund-Mölbert, *J. Virol.* **8**, 330 (1971).

for 20 minutes. The supernatant liquid is then layered upon a CsCl step gradient (density 25° = 1.14–1.55 g/ml), 0.1 M Tris chloride at pH 7.2. After centrifugation for 30 minutes at 28,000 rpm (Spinco Rotor SW 39 or SW 56), the phages form a narrow band at density 25° = 1.45–1.51 g/ml, below the remaining bacterial debris. The phage layer is aspirated and dialyzed exhaustively against 0.15 M NaCl. The phage preparations contain between 1 and 2 × 10^{14} plaque-forming units (PFU) per milliliter.

Membrane Preparation

Preparation of Lymphocyte Suspension. Pig mesenteric lymph nodes removed within 10–20 minutes after slaughtering, are put into cold iso-tonic, phosphate-buffered saline (pH 7.2), freed of adherent connective tissue, cut into small pieces, and macerated gently in a glass TenBroek tissue grinder (Bellco, clearance of 0.004–0.006 inch), to yield the lymphocytes. These are filtered through a small column of Perlon wool (Farbwerke Hoechst, Frankfurt, Germany) to remove debris. The cells are pelleted (5 minutes at 300 g; International PR 6 centrifuge), washed in phosphate-buffered saline, resuspended, and incubated for 30 minutes at 37° in a column of Perlon wool. The lymphocytes are eluted from the column using about four times the volume of phosphate-buffered saline applied. The column treatment eliminates macrophages and dead lymphocytes (~20%) in the original cell suspension. Except for the column incubation, all steps are at 0°.

Cell Disruption. The lymphocytes are suspended at a concentration of 5 × 10^7 to 1 × 10^8 cells per ml in 20 mM 4-(hydroxyethyl)-1-piperazinylethane-2-sulfonic acid (HEPES) chloride at pH 7.2, 0.13 M NaCl, 0.5 mM MgCl$_2$. An equal volume of 0.5 M sucrose is gently added and mixed. The cell suspension is equilibrated in a "bomb" (Artisan Metal Products, Waltham, Massachusetts) with 50 atm of N$_2$ for 15 minutes or 30 atm for 20 minutes, with constant stirring. Cell disruption occurs after dropwise release of the suspension from the "bomb." Shortly after the release, EDTA is added to a final concentration of 1 mM.

Removal of Nuclei and "Large Granules" (Mitochondria and Lyso-somes). Nuclei can be individually pelleted at 450 g for 15 minutes. After removal of nuclei, "large granules" are pelleted at 20,000 g for 15 minutes (Sorvall, Centrifuge RC 2B, Rotor 34).

Isolation of the Microsomal Fraction. These small particles are sedi-mented at 100,000 g for 50 minutes at 0° (Spinco L2 Ultracentrifuge, Rotor 50.1) washed once with 0.01 M HEPES at pH 7.5 and subse-quently with 1 mM HEPES chloride at pH 7.5. If the procedure is stopped at this stage, the particles are suspended in 0.25 M sucrose, 1 mM HEPES chloride at pH 7.5, and frozen at −70°.

Separation of Plasma Membrane and Endoplasmic Reticulum. The washed microsomes are dialyzed for 1 hour against 1 mM HEPES chloride–1 mM MgCl$_2$ at pH 8.2. The microsomal suspension, 1 ml, containing about 1 mg of protein is layered on top of 3 ml of 26% dextran (Dextran-150, Pharmacia) in 1 mM HEPES chloride at pH 8.2, 1 mM MgCl$_2$, density 25° = 1.088–1.092 g/ml, in an SW 56 Spinco rotor tube. The interface between microsomal suspension and dextran is disturbed by gentle stirring with a glass rod. Centrifugation at 400,000 g for 3 hours separates two major fractions: the upper, at the buffer-dextran interface, contains primarily plasma membrane vesicles[6]; the pellet derives primarily from endoplasmic reticulum.[6] For storage, fractions are mixed with equal volumes of 0.25 M sucrose, 20 mM HEPES chloride at pH 7.5, 5 mM EDTA and are maintained at −70°.

Radioiodination of Membranes

Plasma membrane vesicles in 10 mM phosphate, pH 7.0 are labeled with ^{131}I as follows: 3.0 ml of membrane (∼11.2 mg of protein per milliliter) is allowed to react with 0.02 ml of Na ^{131}I (0.2 mCi) and 0.5 ml of chloramine T (0.1 mg) added dropwise. The mixture is stirred 7 minutes at 0°, 0.5 ml (0.1 mg) of Na$_2$S$_2$O$_5$ in 10 mM potassium phosphate at pH 7.0 is added, and the membranes are freed of the reagents by dialysis against the same phosphate buffer and/or by repeated washing in this buffer at 55,000 rpm for 30 minutes.

Radioiodination of Con A

Con A (Miles-Seravac), about 1 mg/ml in 1 M NaCl, 10 mM potassium phosphate at pH 7.0 is first clarified by centrifugation (55,000 rpm, 45 minutes, Spinco Rotor SW 56). The preparation, 3.0 ml, is allowed to react with 0.02 ml of Na ^{125}I (0.2 mCi) at 0°. Chloramine T (0.1 mg) is added dropwise, and the mixture is stirred for 7 minutes Na$_2$S$_2$O$_5$, 0.1 mg in 0.5 ml of 10 mM potassium phosphate is added, the mixture is dialyzed for 24 hours against 5 changes of 2 liters each of 1 M NaCl and centrifuged (55,000 rpm, 45 minutes rotor SW 56). The supernatant liquid, containing about 3 mg of protein and 3×10^7 cpm/ml, is stored at −20° until used.

Coupling of Con A to K29

K29, 0.1 ml (1.5 × 10^{13} PFU), 0.1 ml of ^{125}I–Con A, and 0.025 ml of glutaraldehyde (0.05 mg), all in 0.15 M NaCl at pH 7.25 are allowed to react with gentle agitation for 1 hour at 4°. The labeled phages are freed of large aggregates by low speed centrifugation, concentrated by pelleting (55,000 rpm, 45 minutes, Rotor SW 56) and gently resuspended in 0.15 M NaCl. They are then purified by repeated banding in continuous CsCl gradients of density 1.10–1.70 (30,000 rpm, 60 minutes, Rotor SW 56) and dialyzed exhaustively against 0.15 M NaCl.

Density Perturbation

[131]I labeled membranes and ^{125}I–Con A–K29, containing 3×10^{-10} moles of Con A, both in 0.15 M NaCl are allowed to react for 2 hours at 20° with and without 0.1 mM α,α-trehalose, in a total volume of 0.5 ml. Then 0.4 ml of the mixture is layered on a continuous 4.0 ml CsCl gradient (density = 1.10–1.60) and immediately centrifuged to isopycnic equilibrium (40,000 rpm, 45 minutes Rotor SW 56). Unreacted membranes and phage conjugates are similarly treated. Thereafter, the gradients are cut into 0.4-ml samples. After density determination, these are freed of CsCl by dialysis against 0.15 M NaCl and the distribution of ^{125}I and ^{131}I is determined.

Results

Salt density gradients can be used effectively for certain membrane fractionations. Aggregation does not occur if the concentration of multivalent ions is kept low and there is no dissociation of protein from lipid.

There is 99% recovery of the Con A–K29 label and membrane label. Untreated membranes (^{131}I) equilibrate sharply and rather symmetrically about a density of 1.18. In contrast, pure Con A–K29 phage conjugate pellets; its density, when iodinated, approaches 1.6.

Membrane (~1 mg protein), reacted with Con A–K29, containing 3×10^{-10} mole Con A (2×10^{13} molecules), distributes in a broad, irregular layer, with maximal density about 1.30–1.34, overlapping only slightly (<10%) with the normal membrane distribution and exhibiting a distinct shoulder at density 1.40. Of the reacted membrane material, 15% reaches densities greater than 1.45, but none is trapped in the pellet. Concurrently, a peak for the Con A–K29 conjugate (^{125}I) appears at the same density as the new membrane layer, and ~20% of the conjugate label moves to lower densities. Since no membrane label is associated with the remaining high density conjugate, the latter, presumably, represents unreacted or unreactive material.

The different sedimentation rates and densities of the Con A–K29 and its membrane complexes would foster some dissociation of the latter which might skew the distribution of the complexes to lower densities. However, the rapidity of the centrifugal separation minimizes this possibility and the irregularity and width of the membrane distribution, as well as the shoulders in both membrane and Con A–K29 distributions suggest a heterogeneity in the number and/or affinities of the Con A binding sites.

Excess α,α-trehalose (2×10^4 moles of sugar per mole of conjugated Con A) partially dissociates the membrane Con A–K29 complex. Accordingly, distribution of membrane label shifts to lower density and splits into two components, one coincident with the density of unreacted membranes and one overlapping the low density region of the membrane Con A–K29 complex. It appears that all membranes have lost some ligand, and some membrane fragments have lost all the ligand. This phenomenon, particularly the emergence of the low density component, most likely reflects the distribution of binding sites, particularly those of low affinity. α,α-Trehalose should effect more than 90% dissociation of Con A binding; that this is not seen here is attributed to the fact that the sugar remains at the top of the gradient, while the particulate components quickly sediment. This process favors persistence of the complexes and allows one to detect the presence of low-affinity binding sites.

Comments

Affinity density perturbation cannot yet be considered a standard method. However, pilot studies indicate that the fractionation principle is valid and that the approach represents an urgently needed innovation of particular relevance to membrane biology. It allows both elucidation of membrane topology at the biochemical, as well as micromorphologic levels and isolation of receptor domains which can "trigger" a cell upon reaction with a specific ligand. Obvious examples of the latter are (a) the loci of membrane interaction with peptide hormones or neurotransmitters, (b) receptors for antigens and lectins, (c) membrane-bound antibody. For these examples, we would require the density perturbant coupled to (a) hormones or transmitters, (b) specific antigen or lectin, and (c) specific anti-immunoglobulins. In such applications, the interest lies, not merely in the isolation of the receptor, but in the membrane machinery which translates membrane-ligand binding into biological action.

Affinity density perturbation was developed as a technique for specific membrane subfractionation. However, there is no intrinsic reason why this approach should not form a valuable adjunct to affinity techniques in general.

[13] Affinity Electrophoresis: Separation of Phytohemagglutinins on O-Glycosyl Polyacrylamide Gels

By Václav Hořejší *and* Jan Kocourek

Affinity electrophoresis on polyacrylamide gel is a separation technique that combines the advantages of polyacrylamide gel electrophoresis and affinity chromatography. It allows a quick microscale analysis of protein mixtures with selective separation of those components that have combining sites complementary to a specific ligand. This ligand is covalently attached to a part of the polyacrylamide matrix of the gel column, forming a layer of "affinity gel." Whereas the proteins with combining sites for the particular ligand interact with the affinity gel and remain adsorbed to it, other proteins of the mixture undergo regular separation on the small-pore gel layer. A synchronously run standard electrophoresis of an identical sample serves as control. The method permits one both to detect intact protein molecules and to decide whether their subunits and fragments still contain combining sites for the given ligand. It also renders possible a quick decision as to which type of specific carrier would be suitable for preparative affinity chromatography.

In the present article, affinity electrophoresis of the phytohemagglutinins is described.[1] These are plant proteins which contain combining sites for carbohydrates.[2,3] For this purpose, advantage has been taken of the O-glycosyl polyacrylamide derivatives,[4] which are useful as matrices in the preparative affinity chromatography of phytohemagglutinins (this volume [36]).

O-Glycosyl polyacrylamide gels are prepared by copolymerization of alkenyl O-glycosides with acrylamide and the cross-linking agent, N,N'-methylene bisacrylamide. In the copolymer the O-glycosides are attached through their alkyl chains to the polyacrylamide matrix as shown in Fig. 1 of Article [36] in this volume. For affinity electrophoresis of phytohemagglutinins, copolymers prepared from the readily accessible allyl glycosides can be used.

Preparation of Allyl Glycosides

General directions for the preparation of allyl glycosides are given in Article [36] of this volume.

---

[1] The method reported here has been published in V. Hořejší and J. Kocourek, *Biochim. Biophys. Acta* **336**, 338 (1974).

[2] N. Sharon and H. Lis, *Science* **177**, 949 (1972).

Preparation of Gels for Affinity Electrophoresis

Solutions

Stock solutions[5]

A: 1 N HCl, 48 ml; Tris, 36.6 g; N,N,N',N'-tetramethylene-diamine (TEMED), 0.23 ml; water to 100 ml (pH 8.9)

B: 1 N HCl, approximately 48 ml; Tris, 5.98 g; TEMED, 0.46 ml; water to 100 ml (pH adjusted to 6.7 with 1 N HCl)

C: Acrylamide, 28.0 g; N,N'-methylene bisacrylamide (Bis), 0.735 g; water to 100 ml

D: Acrylamide, 10.0 g; Bis 2.5 g; water to 100 ml

E: Riboflavin, 4 mg; water to 100 ml

Working solutions

Small-pore solution (1): 1 part A, 2 parts C, 1 part water (pH 8.9 ± 0.1)

Small-pore solution (2): ammonium persulfate 0.14 g, water to 100 ml

Large-pore solution: 1 part B, 2 parts D, 1 part E, 4 parts water (pH 6.7 ± 0.1)

Affinity solution: alkenyl glycoside 0.5–1.0 g; mixture of equal parts of small-pore solutions (1) and (2) to 100 ml

Stock buffer solutions for reservoirs: Tris, 6.0 g; glycine, 28.8 g; water to 1 liter (pH 8.3)

Wash-solution for destaining gel: glacial acetic acid 70 ml; water to 1 liter

Fixative-stain solution: Amido Black 10 B, 1.0 g; 7% acetic acid, 100 ml

Procedure. For affinity electrophoresis the polyacrylamide gel electrophoresis system of Davis[5] has been adopted, but any other modification of the disc polyacrylamide electrophoresis is suitable. As shown in Fig. 1, (3), the complete gel column is composed of three layers: (a) large-pore gel, (b) affinity gel, and (c) small-pore gel. In the preparation of the large-pore gel and for the application of the sample, water is used instead of the 40% sucrose solution employed by Davis.[5] The affinity gel is prepared in a separate glass tube of the same diameter as the tubes used for electrophoresis. The affinity solution is poured into the tube and

---

[3] H. Lis and N. Sharon, *Annu. Rev. Biochem.* **42**, 541 (1973).
[4] V. Hořejší and J. Kocourek, *Biochim. Biophys. Acta* **297**, 346 (1973).
[5] B. J. Davis, *Ann. N.Y. Acad. Sci.* **121**, 404 (1964).

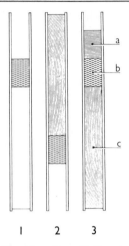

1   2   3

FIG. 1. Procedure for gel setting for affinity electrophoresis. (a) Large-pore gel; (b) affinity gel; (c) small-pore gel.

allowed to polymerize. After the polymerization is completed, the gel is taken out of the tube and thoroughly washed with running water overnight (12 hours) in order to remove residual nonpolymerized glycoside. The gel is then allowed to equilibrate (8 hours) in the Tris chloride buffer (solution A diluted 1:8). From the gel cylinder a piece of about 5 mm is cut and inserted into the electrophoretic tube as depicted in Fig. 1(1). The tube is turned upside down, filled with a mixture of equal parts of the small-pore solutions 1 and 2 [Fig. 1(2)], allowed to polymerize and turned back. The same operation is carried out with the large-pore solution [Fig. 1(3)].

General Conditions

Affinity electrophoresis may be performed in any standard apparatus for disc polyacrylamide gel electrophoresis. Polymerization of the large-pore solution is effected by UV irradiation; polymerization of both the affinity solution and the small-pore solution occurs 30 minutes after mixing of the small-pore solutions 1 and 2. Samples are usually run at a current of 4 mA per tube with bromophenol blue as tracing dye.

The concentration limit of the alkenyl glycoside in the affinity solution is about 0.1%. Below this limit the number of the incorporated glycoside molecules becomes too low to bring about the affinity phenomenon. Hence the mobility of the active protein, in comparison with the mobility of the same protein on the control gel, is not influenced. If residual nonpolymerized glycoside has not been completely washed out

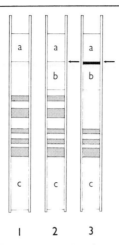

FIG. 2. Electrophoresis of the protein fraction with hemagglutinating activity from pea seeds. (a) Large-pore gel; (b) affinity gel; (c) small-pore gel. (1) Control standard gel. (2) Affinity gel (b), formed by O-α-D-galactopyranosyl polyacrylamide. (3) Affinity gel (b), formed by O-α-D-mannosyl polyacrylamide. Arrows in (2) and (3) indicate boundary between large-pore gel and affinity gel. The large-pore gel layer, corresponding to part a in Fig. 1, has been removed.

from the affinity gel, the active proteins will not adsorb firmly to the affinity gel but will migrate at a reduced rate in comparison with the same proteins on the control gel.

Affinity Electrophoresis of Phytohemagglutinins

Figure 2 shows a typical example of active-protein separation by affinity electrophoresis. A protein fraction from pea seeds (*Pisum sativum* L.)[6] with hemagglutinating activity contains five major protein components [control gel, Fig. 2(1)]; two of them are phytohemagglutinins (isophytohemagglutinins) interacting with α-D-mannosyl and α-D-glucosyl, but not with α-D-galactosyl structures. Therefore, identical electrophoretic patterns are obtained on the control gel (1) and on the gel prepared from O-α-D-galactosyl polyacrylamide (2). In the third tube (3) the affinity layer is formed by an α-D-mannosyl derivative that brings about strong adsorption of the two phytohemagglutinins in the form of a narrow band at the top of the affinity layer. The mobility of the three other proteins of the mixture remain unchanged.

[6] G. Entlicher, J. V. Koštíř, and J. Kocourek, *Biochim. Biophys. Acta* **221**, 272 (1970).

[14] Isolation of Specific and Modified Peptides Derived from Proteins

By MEIR WILCHEK

The chemical modification of proteins is an important tool in the study of their structure–function relationship. A prerequisite for the assessment of this relationship is the precise localization of the modified residue in the amino acid sequence. The major difficulty in such localization is that of isolation of the peptide which contains the modified residue. The modifying reagent often reacts with several residues and to a varying extent so that a protein digest will contain several modified peptides, each present in less than molar amounts. Conventional methods for isolation of peptides require laborious and time-consuming steps, each of which can lead to reduction in the final yield of the desired peptide. Therefore, it is desirable to have methods for one-step isolation of the modified peptide from a peptide mixture derived from a modified protein. Such methods must rely on a reagent which possesses affinity only for the group introduced during the modification procedure. For the present purpose, chemical modification of proteins is divided into three groups: (a) site-directed modification or affinity labeling[1] of residues at the active site of the protein; (b) selective modification of one or a few residues, due to their hyperreactivity or localization at the surface of the protein[2,3]; and (c) general modification of all side chains of a certain amino acid by a group-specific reagent.[4]

Purification of Affinity-Labeled Peptides

Site-specific reagents are used extensively for affinity labeling of the active site of proteins.[1] In such studies, analogs of substrates (in the case of enzymes) or haptens (in the case of antibodies) which possess a chemically reactive group are allowed to react with the protein. The specific and reversible binding of the analog at the combining site of the protein results in the formation of a covalent bond, so that a residue at or near that site is modified. Subsequent enzymatic or hydrolytic hydrolysis of the protein results in a series of peptides, one of which contains the affinity label.

[1] S. J. Singer, *Advan. Protein Chem.* **22**, 1 (1967).
[2] B. L. Vallee and J. F. Riordan, *Annu. Rev. Biochem.* **38**, 733 (1969).
[3] E. Shaw, *Physiol. Rev.* **50**, 244 (1970).
[4] L. Cohen, *Annu. Rev. Biochem.* **37**, 695 (1968).

A general method for the purification of such modified peptides by affinity chromatography has been developed wherein the labeled peptide can be isolated in essentially one step on an agarose column to which the native protein itself is bound as ligand.[5] Selective adsorption of the active site peptide follows by virtue of the interaction between the affinity label (bound to the peptide) and the active site of the protein. After the column has been washed, the modified peptide may be eluted under conditions that dissociate the protein ligand bond. This method has been successfully applied to isolation of labeled peptides from affinity-labeled bovine pancreatic ribonuclease with the use of a ribonuclease-Sepharose column[6] and for the isolation of two different affinity-labeled peptides from staphylococcal nuclease[7] which had been allowed to react with bromoacetyl[8] or diazonium salt of deoxythymidine 3',5'-diphosphate.[7,9] A third labeled peptide, resulting from a much less effective affinity label, was not retained by the Sepharose-nuclease column, demonstrating the important point that only peptide-bound ligands having a high affinity will be adsorbed. Similarly, peptides from affinity-labeled anti-DNP antibodies[10] and from dinitrophenol-binding mouse myeloma protein (MOPC 315) have been purified on an anti-DNP-Sepharose column.[11,12]

Purification of Selective Modified Peptides on Antibody Columns

The method described above can be extended to many other chemical modifications of proteins if specific antibodies are used as the isolating agents. Methods are available for obtaining specific antibodies against small molecules[13,14]; they are suitable reagents for the isolating of modified peptides which have the respective hapten attached to them. In fact, any chemical modification creates a new antigenic determinant (hapten) which can elicit antibodies specific to this hapten. Peptides which con-

[5] D. Givol, Y. Weinstein, M. Gorecki, and M. Wilchek, *Biochem. Biophys. Res. Commun.* 38, 827 (1970).

[6] M. Gorecki, M. Wilchek, and A. Patchornik, *Isr. J. Chem.* 8, 173 pp. (1970).

[7] M. Wilchek, *FEBS Lett.* 7, 161 (1970).

[8] P. Cuatrecasas, M. Wilchek, and C. B. Anfinsen, *J. Biol. Chem.* 244, 4316 (1969).

[9] P. Cuatrecasas, *J. Biol. Chem.* 245, 574 (1970).

[10] Y. Weinstein, M. Wilchek, and D. Givol, *Biochem. Biophys. Res. Commun.* 35, 694 (1969).

[11] J. Haimovich, H. N. Eisen, E. Hurwitz, and D. Givol, *Biochemistry* 11, 2389 (1972).

[12] D. Givol, P. H. Strausbauch, E. Hurwitz, M. Wilchek, J. Haimovich, and H. N. Eisen, *Biochemistry* 10, 3461 (1971).

[13] K. Landsteiner (1936), "The Specificity of Serological Reactions." Dover, New York, 1962.

[14] D. Pressman and A. L. Grossberg, "The Structural Bases of Antibody Specificity." Benjamin, New York, 1968.

tain such small molecules as a result of the same chemical modification of the protein may then be isolated with the antibodies.

Two examples will illustrate the use of this approach: the isolation of nitrotyrosyl containing peptides from nitrotyrosyl lysozyme and the isolation of arsanilazotyrosyl (Ars) containing peptides from carboxypeptidase A modified with *p*-azobenzene arsonate.

Isolation of Group-Specific Modified Peptides—for Sequence Studies

Improvements in amino acid and sequence analysis have made peptide purification the rate-limiting step in the sequencing of proteins. Specific, single-step purification of peptide groups would greatly refine this aspect of protein sequencing. A powerful method of sequence analysis consists in labeling proteins with a suitable group prior to degradation. The location and isolation of peptides containing specific amino acids could be achieved by subsequent affinity chromatography. If a specific modification of all the residues of the same kind, e.g., tyrosine or lysine or tryptophan, were performed, one can foresee the use of insoluble antibodies for the isolation of all peptides containing these residues, i.e., all tyrosyl peptides, lysyl peptides, or tryptophyl peptides. This approach would enable isolation of peptides from enzymatic digest of protein by reason of their amino acid constitution, in contrast to the classical approach of isolation according to size and change.

The methodology for isolation of peptides from protein digests for sequencing must be even more sensitive than methods afforded by today's techniques. The specific modification of the different residues with fluorescence or radioactive haptenic groups should enable isolation of very small quantities for this purpose.

In the following sections, we describe the use of anti-DNP antibodies for the isolation of peptides containing residues to which a DNP group has been covalently linked by various means. This group has the advantage of having a high molar extinction and can, therefore, be easily monitored. Antibodies to it can easily be prepared, and reagents containing radioactive groups can also be made available. Figure 1 illustrates some of the possibilities and conditions for reaction currently under study for modifying specifically different residues with the introduction of a DNP group. Of particular interest are modifications of methionine residues for the purpose of obtaining overlapping peptides to establish positions of cyanogen bromide cleavage. Cyanogen bromide would provide all the peptides containing C-terminal methionine whereas alkylation with N^a-bromoacetyl-N^ϵ-DNP-lysine would yield peptides with methionine in the mid portion of the peptide chain.

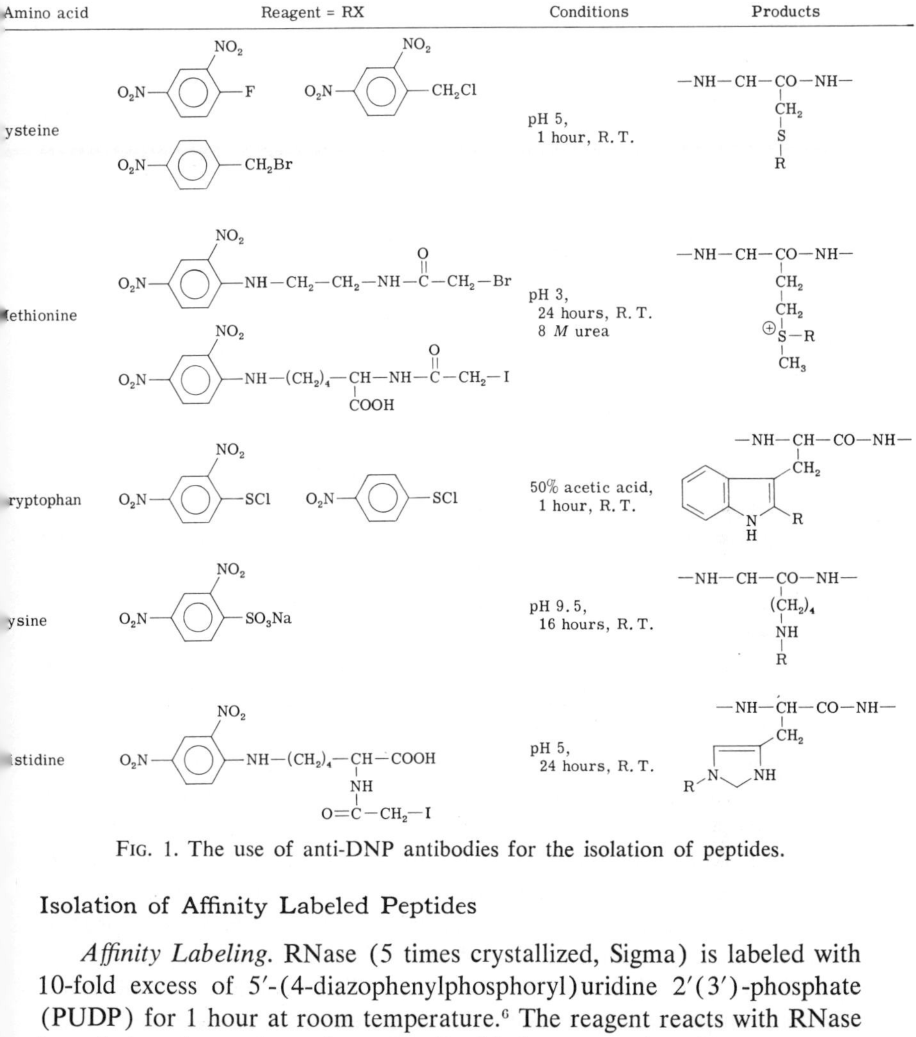

Amino acid	Reagent = RX	Conditions	Products

Cysteine — reagents: $O_2N-C_6H_3(NO_2)-F$, $O_2N-C_6H_3(NO_2)-CH_2Cl$, $O_2N-C_6H_4-CH_2Br$ — Conditions: pH 5, 1 hour, R.T. — Products: $-NH-CH-CO-NH-$, CH_2, S, R

Methionine — reagents: $O_2N-C_6H_3(NO_2)-NH-CH_2-CH_2-NH-CO-CH_2-Br$, $O_2N-C_6H_3(NO_2)-NH-(CH_2)_4-CH(COOH)-NH-CO-CH_2-I$ — Conditions: pH 3, 24 hours, R.T., 8 M urea — Products: $-NH-CH-CO-NH-$, CH_2, CH_2, $\overset{\oplus}{S}-R$, CH_3

Tryptophan — reagents: $O_2N-C_6H_3(NO_2)-SCl$, $O_2N-C_6H_4-SCl$ — Conditions: 50% acetic acid, 1 hour, R.T. — Products: indole ring with $-NH-CH-CO-NH-$, CH_2, N–H, R

Lysine — reagent: $O_2N-C_6H_3(NO_2)-SO_3Na$ — Conditions: pH 9.5, 16 hours, R.T. — Products: $-NH-CH-CO-NH-$, $(CH_2)_4$, NH, R

Histidine — reagent: $O_2N-C_6H_3(NO_2)-NH-(CH_2)_4-CH(COOH)-NH-CO-CH_2-I$ — Conditions: pH 5, 24 hours, R.T. — Products: imidazole ring with $-NH-CH-CO-NH-$, CH_2, R

FIG. 1. The use of anti-DNP antibodies for the isolation of peptides.

Isolation of Affinity Labeled Peptides

Affinity Labeling. RNase (5 times crystallized, Sigma) is labeled with 10-fold excess of 5'-(4-diazophenylphosphoryl)uridine 2'(3')-phosphate (PUDP) for 1 hour at room temperature.[6] The reagent reacts with RNase in a 1:1 molar ratio and specifically labels a tyrosyl residue.

Staphylococcal nuclease is affinity labeled with 3'-(4-bromoacetyl-amidophenylphosphoryl)deoxythymidine 5'-phosphate (pdTp) or with 3'-(4-diazophenylphosphoryl)deoxythymidine 5'-phosphate.[7-9] The reagents react with the nuclease in a 1:1 molar ration and label tyrosyl and lysyl residues.

Affinity labeling of protein 315 (isolated from sera of mice bearing tumor MOPC 315 and DNP antibodies[10] is performed by reaction of the proteins (1 μM) with radioactive 1-bromoacetyl,-1'-DNP-ethylenediamine (BADE) (2 μM) in 0.1 M NaHCO$_3$ at pH 9.0 and 37°. The same conditions are also used for reaction of the protein with N^α-bromoacetyl-N^ϵ-DNP-L-lysine (BADL).[10,12,15]

Preparation of Protein-Sepharose Conjugates.[5] The conjugates of RNase, staphylococcal nuclease, or anti-DNP antibody with agarose are prepared from cyanogen bromide-activated Sepharose.[16] Sepharose 4B is washed on a sintered-glass funnel with water. The washed Sepharose, 100 g (wet weight), is suspended in water (300 ml) and solid CNBr (10 g) is added to the suspension. The solution is brought to pH 11 with 5 N NaOH and maintained between pH 10.8 and 11.2 for 8 minutes by the addition of NaOH. Continuous stirring during the reaction assures complete dissolution of the CNBr during the first 5 minutes. The reaction is terminated by filtration and washing with water. The activated Sepharose is added to the respective protein solution in 0.1 M NaHCO₃ at pH 8. Protein concentration may vary from 0.5 to 5 mg per milliliter. Sepharose, 30 g (wet weight) is used per gram of antibody or enzyme, whereas 15 g of Sepharose are used per gram of antigen. The suspension is stirred slowly at 4° for 16 hours and then washed until absorbance is no longer detected in the filtrate. The yield of binding, as estimated from the absorbance in the solutions before and after coupling to Sepharose, is usually 95%.

The antibody–Sepharose conjugate containing 15–20 mg of antibodies per milliliter of packed Sepharose is washed with 10 mM sodium phosphate–0.15 M NaCl (PBS), suspended in a known volume of the same buffer, and stored at 4°. The capacity of these conjugates is tested in a small column, prepared in Pasteur pipettes with known amounts of antibody-Sepharose conjugate, using [¹⁴C]ε-N-DNP-lysine for anti-DNP-Sepharose. Anti-DNP-Sepharose can adsorb 1.3 moles of [¹⁴C]ε-N-DNP-lysine per mole of bound antibody. These results are reproducible and demonstrate, in this case, that the coupling of the antibody to Sepharose does not result in significant loss of activity. In the same way, the conditions for elution of bound haptens or modified peptides, as well as the regeneration of binding capacity after each elution, can be tested.

An amount of antibody-Sepharose conjugate, predetermined to bind all the desired peptides present in a digest, is packed into a small column and the column is washed with 0.1 M NH₄HCO₃. The enzymatic digest, after inactivation of the proteolytic enzyme, is applied to the column. The column is washed with 0.1 M NH₄HCO₃ until the absorbance of the fractions is less than 0.02 at 280 nm. Elution of the adsorbed peptides

[15] P. H. Strausbauch, Y. Weinstein, M. Wilchek, S. Shaltiel, and D. Givol, *Biochemistry* **10**, 4342 (1971).
[16] R. Axén, J. Porath, and S. Ernbäck, *Nature* (*London*) **214**, 1302 (1967).

is performed with 1 M NH$_4$OH, or 6 M guanidine·HCl, depending on previous experience with text samples.

The capacity of RNase Sepharose column is checked with uridine 2'(3'),5'-diphosphate[5] and of the staphylococcal nuclease column with deoxythymidine 3',5'-diphosphate.[7]

Isolation of PUDP-peptide on RNase-Sepharose Column.[5] PUDP-RNase (15 mg) is reduced and carboxymethylated in 8 M urea–1 M Tris chloride at pH 8.6. The alkylated protein is digested with trypsin at pH 8.5, and the reagents are removed on a Sephadex G-25 column. The digest is brought to pH 5 (maximal conditions for the reversible binding of the inhibitor to RNase) and passed through an RNase-Sepharose column (1 × 4 cm) that contains 50 mg of bound RNase at a flow rate of 80 ml/hr. Peptides labeled with PUDP are not desorbed by this means as determined by A_{342}. The column is washed with water followed by 0.1 M sodium acetate. Bound PUDP-peptide is eluted with 0.8 M NH$_4$OH as a sharp yellow band in 60% yield. Paper electrophoresis at pH 3.5 frees the acidic PUDP-peptide from minor impurities. Upon amino acid analysis, this peptide has been shown to correspond to the region between 67 and 85 in the sequence of RNase. After elution, the RNase-Sepharose column is washed immediately with water and may be reused several times thereafter.

Isolation of Affinity-Labeled Peptides from Staphylococcal Nuclease.[7] Staphylococcal nuclease is affinity-labeled with 3'(4-[^{14}C]bromoacetamido-phenolphosphoryl)deoxythymidine 5'-phosphate (pdTp) as described previously.[8] The alkylated protein (1.7 mg) is digested with trypsin at 37° in 50 mM potassium borate at pH 8. After 3 hours, CaCl$_2$ is added to the solution to a final concentration of 10 mM. The digest is then applied to a nuclease-Sepharose column (0.5 × 2 cm) containing 5 mg (0.3 μmole) of bound enzyme. Binding capacity may be determined with pdTp and is usually about 0.1 μmole, 30% of the theoretical value. No peptide labeled with pdTp emerges from the column but peptide can be eluted with NH$_4$OH at pH 11. After elution, the column is washed immediately with 50 mM potassium borate at pH 8 and may be used again. The yield of radioactive peptide is usually greater than 80% of the amount applied to the column. Amino acid analysis of the radioactive peptide has shown that lysine 48 and lysine 49 are alkylated.[8]

On the same column a peptide labeled with 3'-(4-diazophenylphos-phoryl)deoxythymidine 5'-phosphate is isolated.[7] After further purification by one paper electrophoresis at pH 3.6, the yellow peptide is hydrolyzed, and its amino acid composition is consistent with residues 114–127 in the amino acid sequence of nuclease.

Preparation and Isolation of Affinity-Labeled Peptides from BADE and BADL Labeled Protein-315.[11] Labeled protein-315 is completely reduced and aminoethylated in 7 M guanidine·HCl, and light and heavy chains are separated on Sephadex G-100 in 6 M urea–1 M acetic acid as described.[17] The labeled chains are digested with trypsin for 2 hours at pH 8.6. After digestion the labeled heavy chain (285 mg in 39 ml) containing 2.4 μmoles of dinitrophenyl (DNP) label is passed through a Sepharose column (2 × 10 cm) containing 400 mg of goat anti-DNP antibodies.[18] The column is equilibrated with 0.1 M NH$_4$HCO$_3$. In an experiment, the fraction unadsorbed by the column contained 88% of the absorbance at 280 nm and only 14% of the radioactivity. After the unadsorbed fraction emerges, the specifically adsorbed radioactive peptides are eluted with 7 M guanidine·HCl. More than 95% of the adsorbed radioactive material is desorbed from the antibody column and is applied to a Sephadex G-50 column (2 × 120 cm) to remove guanidine·HCl as well as certain antibodies which may also be eluted. The eluted antibodies are probably due to the guanidine·HCl treatment.[18] Amino acid analysis of the radioactive material has shown that it is a peptide of 20 residues in which lysine is modified.[11,19]

The same procedure has been used for the isolation of modified peptides from the light chain of protein 315[11] and from affinity labeled anti-DNP antibodies.[5]

Much milder conditions are needed for elution when the anti-DNP antibody column is used for isolation of labeled peptides from porcine anti-DNP antibodies labeled with a *m*-nitrobenzenediazonium fluoroborate.[20] Peptides containing tyrosine labeled by *m*-nitrobenzene diazonium reagents are markedly retarded on the column. However, they are released from the column by elution with water alone because of the lower affinity of mononitrobenzenes with the antibodies.

Purification of Selective Modified Peptides on Antibody Columns

Isolation of Nitrotyrosine-Containing Peptides

Preparation of Antigens and Antibodies.[21] A conjugate of bovine serum albumin and nitrotyrosine is prepared by coupling 3-NO$_2$-tyrosine (325 mg) to albumin (200 mg) with 1-ethyl-3(3-dimethylaminopropyl) carbodiimide (300 mg) in 16 ml of 0.15 M NaCl, 10 mM sodium phos-

[17] E. J. Goetzl and M. Metzger, *Biochemistry* 9, 3862 (1970).
[18] M. Wilchek, V. Bocchini, M. Becker, and D. Givol, *Biochemistry* 10, 2828 (1971).
[19] D. Givol, M. Wilchek, H. N. Eisen, and J. Haimovich, *Immunoglobulins* 27, 77 (1972).
[20] F. Franek, *Eur. J. Biochem.* 19, 176 (1971).
[21] M. Helman and D. Givol, *Biochem. J.* 125, 971 (1971).

phate at pH 7.4 (PBS) according to Halloram and Parker.[22] The material is dialyzed against 0.1 M NH$_4$HCO$_3$ and lyophilized. In one trial, 29 moles of nitrotyrosine were found per mole of BSA. Rabbit γ-globulin (RGG) may be nitrated with tetranitromethane according to Sokolovsky *et al.*[23] and the resulting NO$_2$-γ-globulin has been found to bear 13 moles of nitrotyrosine and 30 moles of tyrosine per mole of globulin. Antisera are produced by injecting rabbits or goats with 1 mg of antigen (BSA derivatives), emulsified with complete Freund's adjuvant in multiple intradermal sites. The serum is tested for antibodies with nitro-RGG. Antisera prepared in our laboratory contained between 0.5 and 1.0 mg of antinitrotyrosyl antibodies as determined by precipitin reaction with NO$_2$-RGG. Antibodies are isolated by adsorption of pooled sera on NO$_2$-RGG-Sepharose conjugate followed by elution with 0.1 M acetic acid. The average yield of eluted antibodies was 0.8 mg/ml of serum. Purified antibodies are dialyzed against PBS and stored at $-20°$. Precipitin analysis of these antibodies show that they are 70% precipitable by NO$_2$-RGG. Coupling of antinitrotyrosyl antibodies to Sepharose is performed as described above.[5]

Isolation of Nitrotyrosine Peptides from Lysozyme.[21] Lysozyme is nitrated with tetranitromethane (TNM) using 10-fold molar excess of TNM per lysozyme as described by Atassi and Habeeb.[24] The NO$_2$-lysozyme (2 mg) is reduced in 8 M urea, 0.2 M Tris chloride at pH 8.2 and 0.1 M β-mercaptoethanol for 1 hour at 37°, and is alkylated with 0.15 M iodoacetate for 30 minutes. The reduced and alkylated protein is dialyzed against 0.1 M NH$_4$HCO$_3$ and digested with trypsin (1:50 w/w ratio of enzyme to protein) for 3 hours at 37°. The tryptic digest is applied to a column (1 $\times$ 6 cm) of antinitrotyrosyl antibody-Sepharose conjugate which contains 30 mg of antibodies. Most of the absorbance at 280 nm (80%) emerges unretarded from the column. The column is washed with 0.1 M NH$_4$HCO$_3$ and the adsorbed yellow peptides are eluted with 1 M NH$_4$OH at a yield of about 55% as determined by absorbance at 381 nm. The column is immediately washed with 0.1 M NH$_4$HCO$_3$, and may be used again at least 10 times. Paper electrophoresis reveals three yellow peptides free of other contaminants. Amino acid analysis of these peptides shows that both tyrosine 20 and 23 are nitrated.

The same procedure has been used for the isolation of nitrotyrosine peptides from carboxypeptidase.[25]

[22] M. J. Halloram and C. W. Parker, *J. Immunol.* **96**, 373 (1966).
[23] M. Sokolovsky, J. F. Riordan, and B. L. Vallee, *Biochemistry* **5**, 3582 (1966).
[24] M. Z. Atassi and A. F. S. A. Habeeb, *Biochemistry* **8**, 1385 (1969).
[25] M. Sokolovsky, *Eur. J. Biochem.* **25**, 267 (1972).

Isolation of Arsanilazotyrosyl-Containing Peptides[18]

Preparation of Antigens and Antibodies. Arsanilazotyrosyl-BSA and -RSA are prepared by allowing the proteins to react with 50- to 100-fold molar excess of a freshly prepared solution of diazotized *p*-arsanilic acid for 2 hours at 0°. The reaction with protein is stopped with an aqueous solution of phenol and then extensively dialyzed.

Antisera are produced in goats by injecting 2 mg of the antigen, arsanilazotyrosyl-BSA in complete Freund's adjuvant, at multiple intradermal sites. Further immunizations are performed with the same antigen when necessary. The antibody content of the serum is determined by a precipitin reaction with arsanilazotyrosyl-RSA. Antiarsanilic sera generally contain 0.9–1.2 mg of antibody per milliliter.

The specific antiarsanilazo antibodies are purified by adsorption on a Sepharose immunoadsorbent conjugated with 4-(*p*-aminophenylazo)-phenylarsonic acid (Aldrich). After 1 hour of incubation, the immunoadsorbent is washed with PBS on a sintered-glass funnel until the absorbance at 280 nm is less than 0.05. The antibodies are eluted by incubation of the Sepharose conjugate for 1 hour with 0.3 *M* *p*-arsanilic acid (brought to pH 8 with NaOH) in PBS. The eluted antibodies are dialyzed for 3 days against PBS to remove the hapten until the ratio of absorbance at 280/250 nm is 2.8. The yield of eluted antibodies is 100%.

Isolation of Arsanilazotyrosyl Peptide from Carboxypeptidase A. Coupling of antiarsanilazo antibodies to the activated Sepharose is performed as described earlier.[18] The capacity of the conjugate is tested with *N*-benzyloxycarbonyl (*p*-azobenzenearsonate)-L-tyrosine (ϵ_{325}^M — 21.600).

p-Arsanilazo-*N*-succinylcarboxypeptidase A is prepared by the method of Kagan and Vallee.[26] The protein (0.5 μmole in 5 ml of 0.1 *M* NH_4-HCO_3) is digested with subtilisin for 3 hours at 37°. The reaction mixture is acidified with 2 *N* HCl to pH 2.5 to inactivate subtilisin, and, after 30 minutes, is neutralized with 2 *N* NaOH. This digest is applied to an anti-Ars-Sepharose column (1 × 3.4 cm containing 0.44 μmole of bound antibody). The column is washed with 0.1 *M* NH_4CO_3 until the absorbance of the effluent at 280 nm is lower than 0.03. About 95% of the yellow arsanilazo peptides are adsorbed to the column, while most of the material absorbing at 280 nm emerges unretarded. The adsorbed azopeptides are eluted with 1 *M* ammonium hydroxide. The movement of an orange boundary serves as a visible check on elution, since the yellow color of the

---

[26] H. M. Kagan and B. L. Vallee, *Biochemistry* 8, 4223 (1969).

azophenol changes to orange at alkaline pH. After elution the column is washed immediately with PBS until the pH is neutral. The capacity of the used column, when tested, is usually found to be between 80 and 85% of the initial capacity. No further change in capacity is observed even after six cycles of adsorption and elution. The total yield of peptides is over 90% of the amount applied to the column, as estimated from the absorbance at 325 nm. The yellow eluent is separated by paper electrophoresis at pH 6.5 and is found to have 4 spots. From a chymotryptic digest of the protein a peptide compatible with the sequence of residues 247–257 is isolated. Tyrosine 248 is, therefore, implicated as the major arsanilated tyrosine.[18,27]

Isolation of Group-Specific Modified Peptides—for Sequence Studies

Preparation of Anti-DNP Antibodies. DNP-bovine serum albumin and DNP-ovalbumin are prepared according to the method of Eisen *et al.*[28] using 2,4-dinitrobenzene sulfonate in a weight ratio of 1:10 to protein. Antisera are produced in goats by injecting 2 mg of the antigen (DNP-bovine serum albumin) in complete Freund's adjuvant at multiple intradermal sites (see this volume [92]). The antibody content of the serum, determined by the precipitin reaction with the cross-reacting modified protein, DNP-ovalbumin, is between 2.5 and 3 mg antibody per milliliter.

Anti-DNP antibodies are purified using DNP-ovalbumin-Sepharose conjugate as the immunoabsorbent. The adsorbed antibodies are eluted by 0.1 M acetic acid. The eluate is brought to pH 5.0 and dialyzed against phosphate-buffered saline (10 mM sodium phosphate, 0.15 M NaCl, pH 7.4). About 85% of the anti-DNP-activity applied to the immunoadsorbent is recovered.

Isolation of the Modified Peptides on Antibody-Sepharose Columns. Anti-DNP-Sepharose[5] containing a sufficient amount of antibodies to bind all the modified peptides present in the digest are packed into a column and washed with 0.1 M NH$_4$HCO$_3$. The enzyme digest is applied to this column and the unmodified peptides are eluted with 0.1 M NH$_4$HCO$_3$ until the absorbance at 280 nm is below 0.02. The column is washed with water and the adsorbed peptides are eluted with 2 M propionic acid, 20% formic acid or 6 M guanidine·HCl.

Isolation of Cysteine-Containing Peptides.[29] Reaction of cysteine-containing peptides and proteins with N^α-bromoacetyl-N^ϵ-DNP-L-lysine

[27] J. T. Johansen, D. M. Livingston, and B. L. Vallee, *Biochemistry* 11, 2584 (1972).
[28] H. N. Eisen, S. Belman, and M. E. Carstem, *J. Amer. Chem. Soc.* 75, 4583 (1953).
[29] T. Miron and M. Wilchek, unpublished observations.

(BADL) results in the alkylation of the SH group of these compounds. The DNP-containing derivatives are specifically adsorbed on anti-DNP-Sepharose column.

Papain may be alkylated with BADL at pH 8.5. The modified protein is reduced, carboxymethylated, and digested with trypsin and chymotrypsin. The digest is applied to the anti-DNP-column in 0.1 M NH$_4$HCO$_3$. After extensive washing, a yellow peptide can be eluted with 6 M guanidine·HCl. To remove the guanidine·HCl from the DNP peptide the eluent is passed through a short column (1 × 5 cm) of Poropak Q (Waters Associates, Inc.). DNP-containing peptides are adsorbed to this column whereas guanidine·HCl emerges unretarded. The DNP-peptide is eluted with a 50% solution of ethyl alcohol in water. Amino acid analysis of the isolated peptide from papain shows that cysteine-25 is modified as expected.

The same procedure has been used also for the isolation of a 40 amino acid residue peptide containing cysteine from tobacco mosaic virus protein. Other reagents useful for the isolation of cysteine peptides are described in Fig. 1.

Isolation of Methionine-Containing Peptides.[29] BADL selectively alkylates methionine residues at an acid pH to give the positively charged sulfonium salt possessing a DNP group. Peptides containing this modified residue are adsorbed on an anti-DNP column. A solution of 50 mM glycyl-L-methionylglycine in 50 mM sodium acetate at pH 4 is made 0.1 M in BADL. The reaction is allowed to proceed at 30° for 15 hours; at this time better than 95% alkylation has been achieved. Excess BADL is removed by extraction with ether. The aqueous solution is brought to pH 7.4 and applied to the anti-DNP column. The DNP-containing peptide is adsorbed to the column and eluted with 10% formic acid. The yield is quantitative.

The two methionine residues of lysozyme have been alkylated with BADL at pH 3.5. The labeled peptides are being isolated.

Isolation of Tryptophan-Containing Peptides.[30] Tryptophan residues in proteins are modified with the highly specific reagent, dinitrophenylsulfenyl chloride (DNPS-Cl).[31] The tryptophan residues are converted by this reaction into a DNP-thioether function in the 2-positions of the indole nucleus, and can be selectively adsorbed and purified on an anti-DNP antibody column. The method has been used for the isolation of the tryptophan containing peptides from human serum albumin and cytochrome *c*.

[30] M. Wilchek and T. Miron, *Biochim. Biophys. Acta* **278**, 1 (1972).
[31] E. Scoffone, A. Fontana, and R. Rocchi, *Biochemistry* **7**, 971 (1968).

Sulfenylation of Human Serum Albumin. A solution of human serum albumin (0.5 μmole/ml) in 30% acetic acid is allowed to react with an excess of DNPS-Cl in glacial acetic acid (usually 20 μmoles of DNPS-Cl/ μmole of albumin). The final concentration of acetic acid is 50%. Sulfenylation is performed in the dark at room temperature with constant stirring for 1 hour. The excess of DNPS-Cl is removed by centrifugation (10,000 rpm, 15 minutes) at 5°. The modified protein is precipitated by a cold solution of 1 N HCl in acetone. The precipitate is washed with cold ether, dissolved in water, and lyophilized.

Isolation of DNPS-Tryptophan Peptide from DNPS-Human Serum Albumin. DNPS-albumin (8 mg/ml) in 0.1 N NH₄HCO₃ (0.13 μmole/ ml) is digested twice with 1% (w/w) trypsin for 24 hours at 37°. The digest is applied to an anti-DNP-Sepharose column (2 × 15 cm) containing 0.5 μmole of bound antibody. Most of the yellow material as measured at 330 nm, is adsorbed to the column and can be eluted with 10% formic acid. The overall yield of the eluted peptide is over 60% as judged from the adsorbance at 330 nm. The column is immediately washed with water and 0.1 M NH₄HCO₃.

The yellow formic acid eluate from human serum albumin was chromatographed on a preparative thin-layer chromatograph (silica gel) and found to contain only one peptide. The amino acid composition of this peptide is Ala₃, Arg, Val, and DNPS-Trp. The DNPS-Trp content was determined spectroscopically since this residue is not stable under the conditions commonly used for the hydrolysis of proteins. This peptide is identical in its amino acid analysis and N-terminal residue to the peptide isolated by Swaney and Klotz[32] (Ala-Trp-Ala-Val-Ala-Arg) from the same protein.

The same procedure has been used for the isolation of a tryptophan containing peptide from cytochrome c.[30]

Isolation of Tyrosine-Containing Peptides.[33] Tyrosine-containing peptides can be isolated after nitration on antinitrotyrosine antibody columns or, after arsanylation, on anti-Ars-antibodies as described above. A more complicated method for isolation of tyrosine peptide with anti-DNP antibodies was recently described.[34] The protein is first nitrated with tetranitromethane and reduced with sodium hydrosulfite to the aminotyrosine.[35]

---

[32] J. B. Swaney and I. M. Klotz, *Biochemistry* **9**, 2570 (1970).

[33] M. Wilchek and D. Givol, "Peptides 1971," p. 203. North-Holland Publ., Amsterdam, 1973.

[34] M. Bustin and D. Givol, *Biochim. Biophys. Acta* **263**, 459 (1972).

[35] M. Sokolovsky, J. F. Riordan, and B. L. Vallee, *Biochim. Biophys. Res. Commun.* **27**, 20 (1967).

After dinitrophenylation at pH 4.8 and enzymatic digestion, the DNP-aminotyrosine peptides are specifically separated from the peptide mixture on an anti-DNP-Sepharose column. The procedure can be especially useful in isolating very small amounts of radioactively labeled tyrosine containing peptides.

Miscellaneous

Isolation of the Heme Peptide of Cytochrome c.[36] The ferriporphyrin c-peptide obtained from horse heart cytochrome c by digestion with pepsin can be purified by affinity chromatography on bovine or human serum albumin coupled to Sepharose.

The bovine serum albumin-Sepharose and human serum albumin-Sepharose matrices are prepared by coupling the respective proteins to cyanogen bromide-activated Sepharose 4B.

Cytochrome c (2 mg) is digested with 0.05 mg of pepsin for 14 hours at pH 1.5 and at room temperature. The digest is brought to pH 8 with ammonium bicarbonate and is applied to a bovine serum albumin-Sepharose column (0.5 × 4 cm) that contains 18 mg of bound albumin per milliliter of Sepharose. No peptide-containing heme (absorbance at 410 nm) emerges from the column, whereas most of the digest passes through the column. Since heme is visible to the unaided eye, retention of the heme peptide is detected as a red band. After the column is washed extensively with 0.1 M ammonium bicarbonate, the red heme peptide which had been adsorbed is eluted with 0.8 M ammonium hydroxide as a sharp red band. Recovery of the heme peptide is more than 90% of the amount applied to the column as judged by the adsorbance at 410 nm. After paper electrophoresis, only one peptide has been detected. Amino acid analysis of this peptide has shown that it is identical to the heme undecapeptide isolated by Tsou.[37] After elution, the bovine serum albumin Sepharose column is washed immediately with water and with 0.1 M ammonium bicarbonate and is ready for use. The capacity of the column remains constant for several cycles of operation. Identical results have been obtained with human serum albumin.

Concluding Remarks

The introduction of Sepharose as a convenient support for insolubilized antibodies and the improvement of techniques for affinity chromatography provide rapid and simple procedures for the isolation of modified peptides. The high capacity of CNBr-activated Sepharose and the used of purified antibodies allow the employment of very small columns for the iso-

---

[36] M. Wilchek, *Anal. Biochem.* **49**, 572 (1972).
[37] C. L. Tsou, *Biochem. J.* **49**, 362 (1951).

lation of micromole quantities of modified peptides. The specific isolation of modified peptides by adsorption to and elution from an antibody column is a procedure in the time scale of 1–2 hours.

The adsorption of most (about 95%) modified peptides from a peptide mixture can be accomplished under mild conditions, i.e., neutral pH and a range of ionic strengths. However, elution of modified peptides from antibody-Sepharose requires different conditions which depend on the specific ligand as well as on the specific antibody. It has been found that 1 M NH$_4$OH, if it dissociates the antibody-hapten bonds, is the best solvent for elution of adsorbed peptides. After such treatment, the matrix can be regenerated to almost full capacity and can be re-used many times. Less than 0.1% of the bound antibodies are released from the Sepharose; and the eluted peptides are obtained in a highly purified state.

On the other hand, the use of 6 M guanidine·HCl required for elution of DNP-peptides causes substantial inactivation of antibodies and releases 2–3% of the antibody protein which had been bound covalently to Sepharose. Attempts should be made to prepare antibodies to colored, charged molecules that can be eluted under mild conditions.

Obviously, if the eluate contains several modified peptides, conventional methods should be used to further separate and purify them.

[15] Specific Fractionation and Manipulation of Cells with Chemically Derivatized Fibers and Surfaces

By G. M. EDELMAN and U. RUTISHAUSER

The specific fractionation of cells has not assumed the importance in cytology that fractionation of molecules has in chemistry, perhaps because of its inherent difficulty, and because other methods of obtaining certain pure lines of cells are available. It is obvious that extended development of specific methods of cell separation and fractionation would have enormous value in cell biology, virology, and immunology. Besides providing the opportunity to obtain "pure" cell populations, such methods would, for example, be valuable in isolation of cell surface markers, selection of cells at a time when such markers appeared, determination of the influence of chemically specific interactions of cell surfaces, and determination of the action of molecules that are specifically prevented from entering the cell.

Although a number of physical methods for the fractionation of eukaryotic tissues and cells are in current use, there are few methods utilizing the chemical properties and specificities of the cell as a means for both cell fractionation and cell manipulation. The main requirements of a chemical method for the manipulation of cell populations are specificity, general applicability, high yield and maintenance of cell viability. Two main approaches to a chemically specific method of cell fractionation suggest themselves. The first, which to the authors' knowledge has not yet been implemented, would rely upon the development of a spectroscopically detectable marker that interacted with an appropriately specific reagent *within* a living cell followed by physical separation of the marked cell. This faces a number of difficulties and therefore several workers have resorted to a second approach utilizing specific interaction with cell *surface* components.

Whatever the fractionation procedure, it is necessary to define a criterion of purity. For molecules, purity can be assessed by an analysis of the elemental composition, physical and optical properties, or the covalent and three-dimensional structure. Unfortunately, because of the dynamic state of the cell and the extreme complexity of a cell phenotype, these procedures are not generally applicable to cell fractionation. As a result it is necessary to define purity in terms of a single phenotypic feature that can be assayed by specific methods. Even if the isolated cells are homogeneous with respect to one property, they usually are heterogeneous for many others. Repeated selection of the same cell population by many different criteria would therefore be necessary to obtain cells that were closely similar. Considering the difficulties in handling and maintaining the viability of cells *in vitro,* the realization of such an extensive procedure is not likely. In any case, the isolation of cells all having even one property in common greatly facilitates the characterization of that property and its relationship to other aspects of the cell phenotype.

Although the relative fragility of cells as compared with molecules limits the conditions and extent of fractionation procedures, the size and metabolic capabilities of a cell can be distinct advantages. The ability to visualize directly and count individual cells is also a decided advantage in evaluating the quantitative results of a fractionation procedure and in excluding artifacts caused by nonspecific adsorption, cell damage, and aggregation.

Affinity Fractionation of Cells

Most of the parameters used in the fractionation of molecules have also been used for cells. These include (1) physical properties,[1] such as

[1] K. Shortman, *Annu. Rev. Biophys. Bioeng.* 1, 93 (1972).

size, density, shape, adsorption or adherence, and charge, (2) selection in biological systems,[2] and (3) chemical methods based on the presence of specific receptors or binding sites on the cell surface. As in the case of the fractionation of molecules, many of these procedures complement each other and may be used successively in the separation of complete mixtures.

The chemical separation of cells according to their affinity for a solid support, usually beads coated with a biospecific reagent, has been achieved in a variety of systems. Plastics,[3] glass,[4] polyacrylamide,[5] and agarose[6] have been used as solid supports with adsorbed or covalently attached substances, such as antigens, antibodies, lectins, or hormones, providing the requisite specificity. In most instances, the solid is packed into a column through which the cell suspension is passed. The properties of the retained and effluent cells are then compared to those of the unfractionated cells.

The most extensive application of affinity cell fractionation has been in the analysis of lymphoid cell populations. The removal of specific antigen-binding cells from an immunized lymphoid cell population was first accomplished by Wigzell and co-workers[3,7,8] using columns of antigen-coated plastic or glass beads. Unfortunately, a large number of cells were also retained nonspecifically and because elution by antigen could not be achieved, it was not possible to study the properties of the specifically bound cells or to isolate enriched populations of lymphocytes with specific receptors. Modifications of the antigen column procedure by Wofsy and co-workers[9-11] permitted the isolation of purified antigen-binding and plaque-forming cells specific for one hapten. The use of large polyacrylamide beads derivatized with hapten resulted in considerably less non-specific retention, and specific elution of bound cells in 10–90% yield was achieved by washing the column with solutions of free hapten.

[2] J. F. A. P. Miller, A. Basten, J. Sprent, and C. Cheers, *Cell. Immunol.* **2**, 469 (1971).
[3] H. Wigzell and B. Anderson, *J. Exp. Med.* **129**, 23 (1969).
[4] N. I. Abdon and M. Richter, *J. Exp. Med.* **130**, 141 (1969).
[5] L. Wofsy, J. Kimura, and P. Truffa-Bachi, *J. Immunol.* **107**, 725 (1971).
[6] J. M. Davie and W. E. Paul, *Cell. Immunol.* **1**, 404 (1970).
[7] H. Wigzell and O. Mäkelä, *J. Exp. Med.* **132**, 110 (1970).
[8] H. Wigzell, P. Golstein, E. A. J. Svedmyr, and M. Jondahl, *Transplant. Proc.* **IV**, 311 (1972).
[9] L. Wofsy, J. Kimura, and P. Truffa-Bachi, *J. Immunol.* **107**, 725 (1971).
[10] P. Truffa-Bachi and L. Wofsy, *Proc. Nat. Acad. Sci. U.S.* **66**, 685 (1970).
[11] C. Henry, J. Kimura, and L. Wofsy, *Proc. Nat. Acad. Sci. U.S.* **69**, 34 (1972).

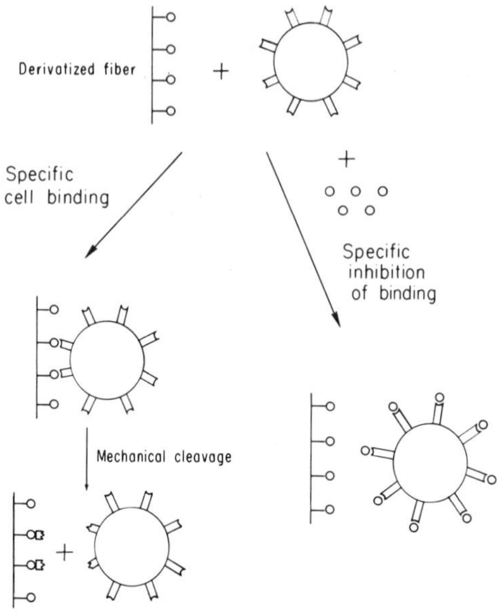

FIG. 1. General scheme of fiber fractionation.

Fiber Fractionation of Cells[12–15]

The method presented here separates cells according to their ability to bind specifically to strung fibers derivatized with molecules such as antigens, antibodies, or lectins. The particular geometry and mixing procedure were designed to avoid many of the difficulties encountered in the use of columns and at the same time to keep the separation procedure simple, efficient, and inexpensive. Fractionation can be achieved by specific binding to a unique component on the cell surface, or by differences in the binding affinity, number, and distribution of cell surface receptors of the same specificity.

The basic principles of fiber fractionation are illustrated in Fig. 1. After deciding upon the molecule or macromolecule to be used for the fractionation, it is coupled in suitable chemical form to nylon fibers that are held under tension in a frame in the desired geometrical arrangement. Dissociated cells are then shaken in a suitable medium with the fiber,

[12] G. M. Edelman, U. Rutishauser, and C. F. Millette, *Proc. Nat. Acad. Sci. U.S.* **68**, 2153 (1971).

[13] U. Rutishauser, C. F. Millette, and G. M. Edelman, *Proc. Nat. Acad. Sci. U.S.* **69**, 1596 (1972).

[14] U. Rutishauser and G. M. Edelman, *Proc. Nat. Acad. Sci. U.S.* **69**, 3774 (1972).

[15] U. Rutishauser, P. D'Eustachio, and G. M. Edelman, *Proc. Nat. Acad. Sci. U.S.* **70**, 3894 (1973).

after which unbound cells are washed away. The bound cells may be transported to other media for further characterization or released into the medium by plucking the taut fiber with a needle. This serves to shear the cells from their points of attachment. Special linkers between the fiber and ligand may also be used to remove cells by specific chemical or enzymatic cleavage.

Advantages and Disadvantages of Fiber Fractionation Techniques and Comparison with Other Methods

The fiber technique has a number of advantages over other affinity fractionation procedures. The simple spatial arrangement of the fibers permits direct observation and quantitation of the cells. The fractionated cells can be manipulated on the fibers under a variety of conditions, and the behavior of single cells can be monitored throughout an experiment.

Many cells have a natural tendency to adhere to surfaces, and a serious difficulty encountered with bead column methods is the nonspecific binding of cells.[3,12] In the case of columns, which require the cells to follow a tortuous path, this problem is aggravated by physical trapping; frequently up to 50% of the cells loaded on a column are nonspecifically retained. With the use of large undamaged agarose beads, this value can be reduced to about 5%. However, in the isolation of cells that occur at relatively low frequencies, even this amount of nonspecific binding is unsatisfactory. With fiber fractionation, the simple configuration of the solid support minimizes nonspecific binding of cells. In the presence of a competitive inhibitor of binding, less than 0.01% of the cells in the incubation mixture are bound to the fibers or the accompanying support. Adequate specificity can therefore be achieved in the isolation of even a minor subpopulation of antigen-binding cells.

Although the initial fixation of cells to columns is specific, quantitative elution of bound cells by a competitive inhibitor is often difficult to achieve. Studies with concanavalin A-derivatized beads[12] suggest that, after being specifically bound, the cell membrane interacts with the surface of the bead or fiber to form secondary adhesions that can be broken by physical shearing or when the cell is distorted by hypotonic shock. Unfortunately, osmotic elution can be achieved with erythrocytes, but not with nucleated cells. As with column methods, cells that bind to fibers are firmly attached and cannot in general be removed by incubation with a competitive inhibitor. They can, however, be rapidly and quantitatively released by plucking the taut fiber with a needle. The mechanical removal of cells also has an advantage over chemical elution in that it is not limited to cases in which a competitive inhibitor of binding is available.

A disadvantage of all solid phase methods, including fiber fractionation, is the possible perturbing effect of the ligand or surface on cell metabolism or function. This stricture also extends to the method of removing the cells from the surface. Furthermore, the accumulated experience with the method indicates that a number of variables must be controlled and studied in each case before success can be obtained. Each system has its own requirements.

Methods

Preparation of Derivatized Fiber

Selection of the Fiber. The major criteria used in selecting a fiber are its chemical reactivity, appearance and optical properties, strength, and its tendency to bind cells nonspecifically. Nylon is particularly suitable for this application in that it contains free amino and carboxyl groups, is available as a smooth transparent monofilament in a range of sizes, has exceptional strength even at a diameter of 20 μm, and does not bind cells nonspecifically after derivatization.

Monofilament is generally used in two size ranges, 125–250 μm diameter for routine fractionation and manipulation and 20–50 μm diameter for morphological observations. The larger fiber is available commercially as size 50 transparent sewing nylon (Dyno Merchandise Corp., Elmhurst, New York, or at notions stores). The smaller fibers can be obtained directly from the manufacturer (Dupont, Wilmington, Delaware). For depletion experiments, nylon meshes can be substituted for monofilament (308 gauge, square weave, Nitex; Tobler, Ernst, and Traber, Elmsford, New York).

Stringing of the Fiber. In order to derivatize the nylon, and bind, observe, and manipulate the cells, it is useful to immobilize the fiber by stringing it under tension in a supporting frame. Although a variety of designs can be used, an inexpensive and reusable frame (Fig. 2) can be cut from plastic stoppers (size 5 or 6 polyethylene or polypropylene stopper, Mallinckrodt, New York, New York). These collars fit snugly into conical 35 × 10 mm petri dishes (NUNC, Vanguard International, Red Bank, New Jersey). For routine assays, a single parallel row of 10–15 fibers is sufficient; for preparative fractionation, two rows of 24 fibers each are useful. Meshes are cut into circles of 34 mm diameter and immobilized between two Lucite rings in a 35 × 8 mm chamber. The chamber should be constructed so that the cell suspension can be added and the chamber sealed to exclude air bubbles.

Derivatization. Prior to derivatization of the fiber, it is necessary to

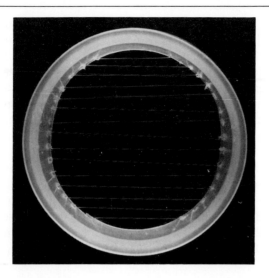

Fig. 2. Petri dish containing polyethylene frame strung with nylon monofilament. The length of the longest fiber segment is 2.5 cm.

remove surface coatings or contaminants by 10-minute extractions first with petroleum ether and then with carbon tetrachloride. The number of free amino and carboxyl groups can then be increased by partial hydrolysis of the nylon with 3 N HCl for 30 minutes at room temperature. The hydrolysis procedure is useful in the coupling of haptens or linkers directly to the fibers, but often can be eliminated in the coupling of proteins or hapten–protein conjugates. The fibers must be rinsed with water for at least 30 minutes after hydrolysis.

A great variety of procedures can be used to derivatize the fibers. A number of these can be found elsewhere in this volume. For illustrative purposes, we will present several methods for obtaining fibers derivatized with the 2,4-dinitrophenyl (Dnp) group or the protein concanavalin A (Con A).

a. Direct reaction of Dnp-sulfonic acid with fiber amino groups. Forty milligrams of Dnp-SO$_3$H and 80 mg of K$_2$CO$_3$ are dissolved in 4 ml of water and placed in the petri dish containing the strung fibers. After 1 hour at 25° the fibers are washed extensively with water.

b. Adsorption of Con A or Dnp-bovine serum albumin (Dnp-BSA). Dnp-BSA is prepared by allowing 1 g of 2,4-dinitrobenzenesulfonic acid, 2 g BSA, and 1 g of K$_2$CO$_3$ to react in 100 ml of water at 37° for 1 hour. The reaction is terminated by the addition of 1 g of glycine and dialyzing extensively against water. The hapten–protein conjugate may be lyophilized for storage. The number of haptenic groups per protein molecule should be kept low to avoid nonspecific binding in cellular fractionation. For example, 10 Dnp groups per BSA molecule will be sufficient for the case described. During a 30-minute incubation at room temperature, the nylon fiber will adsorb enough Con A or Dnp-BSA for some applications. The noncovalent binding is stable in physiological media. Because the unbound protein can be recovered intact, this procedure is particularly useful in cases where only a small amount of the protein is available.

c. Cross-linking of protein to the fiber by carbodiimide. This procedure is the one used most extensively in the applications presented here. A water-soluble carbodiimide, 1-cyclohexyl-3-(2-morpholinoethyl) carbodiimide metho-p-toluenesulfonate (Aldrich), is used to couple Con A (at 0.25–2.5 mg/ml) or Dnp-BSA (at 0.25 mg/ml) covalently to the nylon (Fig. 3). Four milliliters of these reagents in 0.15 M NaCl,

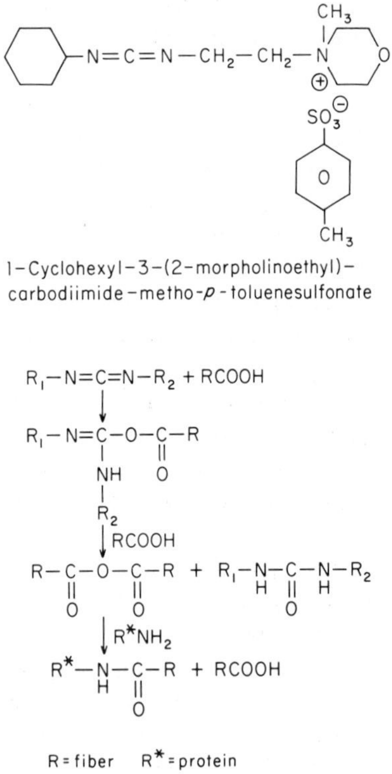

1-Cyclohexyl-3-(2-morpholinoethyl)-
carbodiimide-metho-p-toluenesulfonate

R = fiber   R* = protein

FIG. 3. Covalent coupling of protein to nylon fiber using a water-soluble carbodiimide.

pH 6–7, at a carbodiimide:protein ratio of 5:1 (w/w) is added to petri dishes containing the fibers, and the reaction mixture is gently shaken at room temperature for 30 minutes. The derivatized fibers are washed in 0.15 M NaCl, transferred to fresh petri dishes, and used the same day.

d. For reactivation of fiber carboxyl groups, thionyl chloride or phosphorus pentachloride in benzene or dioxane can be used to convert the nylon carboxyl groups to acid chlorides. The fibers can then be reacted with free amino groups on proteins or other molecules. Mild conditions (1:500 dilution of $SOCl_2$ in dioxane, 10 minutes at 25°) must be used in order to avoid extensive damage to the fiber during the activation reaction.

e. Attachment of Dnp through a chemical spacer. In some cases it may be useful to keep the binding ligand at a distance from the fiber. Kiefer[16] has reported a method for attaching Dnp groups to nylon using sequential treatment of the fiber with HCl, succinic anhydride, ethylene-diamine, and Dnp-ϵ-aminocaproic acid with a water-soluble carbodiimide. As in the case of affinity chromatography,[17] the use of spacers can affect the binding properties of the derivatized solid. For example, it has been reported[16] that binding of lymphocytes to fibers with spacers having Dnp at their termini is stable at 4°, but not above 25°. This instability may be useful for removing cells in some applications.

f. Coating fibers with derivatized gelatin.[15] The application of a thin layer of derivatized gelatin to the surface of the nylon has some unique advantages over the direct chemical methods listed above. Gelatin surfaces have been used extensively in cell culture methods and it is probable that interaction of the fiber-bound cells with gelatin is less extensive than with a nylon surface. Furthermore melting of the gelatin layer at 37° provides a rapid and gentle means of releasing bound cells into the medium. The cells may also be removed by plucking and because of the physical properties of gelatin, the shearing force necessary is milder. Dinitrophenylated gelatin (Dnp-Gel) is prepared by allowing 20 g of gelatin (Fischer Scientific) to react with 5 g of Dnp-sulfonic acid (Eastman) in 100 ml 5% K_2CO_3, for 1 hour at 37°. The Dnp-Gel is dialyzed at room temperature against water containing 0.1% Merthiolate and stored at 4° as a 5% gel. A strung frame of fibers is dipped into a 5% solution of the Dnp-Gel at 40–50° and immediately centrifuged in a clinical centrifuge (International Equipment, Needham, Massachusetts) to shake off excess gelatin. Each frame containing the fibers is placed directly in the No. 325 supporting rings of a No. 215 swinging-bucket

[16] H. Kiefer, *Eur. J. Immunol.* 3, 181 (1973).
[17] P. Cuatrecasas, *J. Biol. Chem.* 245, 3059 (1970).

TABLE I
DERIVATIZATION OF FIBERS WITH CARBODIIMIDE AND PROTEIN:
FORMATION OF NONCOVALENT AND COVALENT BONDS

Protein (mg/ml)	WSC[e] (mg/ml)	Wash[a]	Cpm of washed fiber[b]
[125I]Con A[c]			
1.25	0	PBS	18,250
1.25	0	6 M Guanidine	7,140
1.25	6.25	PBS	22,724
1.25	6.25	6 M Guanidine	18,715
[125I]Dnp-BSA[d]			
0.25	0	PBS	18,107
0.25	0	6 M Guanidine	10,260
0.25	1.25	PBS	25,472
0.25	1.25	6 M Guanidine	21,303

[a] Incubation at 25° for 4 hours.
[b] Cpm for 25 cm fiber.
[c] 2.4×10^7 cpm/mg.
[d] 3.0×10^7 cpm/mg.
[e] WSC = water-soluble carbodiimide; PBS = phosphate-buffered saline; Con A, concanavalin A; Dnp-BSA-dinitrophenylated bovine serum albumin.

head so that the fibers are horizontal and radially aligned, and centrifuged for 10 seconds at 500 rpm. After centrifugation, the coated fibers are cooled and washed with cold phosphate-buffered saline.

The Extent and Stability of Derivatizations with Protein and Carbodiimide. The derivatization with protein and carbodiimide proceeds smoothly over a wide concentration range.[12] During this reaction, a certain amount of noncovalent binding of the protein will occur, and it is valuable to test that covalent labeling has occurred. In order to determine the extent of coupling at a particular protein concentration, protein labeled with [125I] by the chloramine T method[18] is used for derivatization. In order to determine the stability of the binding, the fiber to which the radioactive protein has been attached is incubated with 6 M guanidine hydrochloride. The amount of radioactivity bound before and after treatment with guanidine is then determined (Table I). Using Con A or Dnp-BSA without the addition of carbodiimide, over 50% of the bound protein may be removed by guanidine. This noncovalent binding is, however, stable in physiological media. With the addition of carbodiimide, less than 20% of the bound protein is extracted by the guanidine wash.

[18] C. A. Williams and M. W. Chase, eds., "Methods in Immunology and Immunochemistry," Vol. 1. Academic Press, New York, 1967.

The use of carbodiimide is also useful in achieving a high degree of derivatization.

Binding of Cells to Fibers

Our experience with the fiber method indicates that a number of variables must be controlled and that experimentation in each application is required before success can be obtained. The important variables include the method of cell preparation, the temperature and mixing conditions during incubation, the composition of the separation medium, the choice of ligand, the degree of derivatization, and the number of cells added.

Preparation of Cell Suspension. In order to obtain good results with the fiber method, it is necessary to use cell suspensions that are free of aggregated material, and conditions under which the cells will not aggregate or release large amounts of DNA. Otherwise, clumps of cells will adhere firmly to the fibers and seriously reduce the quality of the fractionation. Cell populations from various sources will require different methods of preparation. In cases where enzymes or chemicals are used for cell dissociation, it is important to consider the possible effect of these reagents on the receptors by which the cell will bind to the fiber. The following section outlines a simple procedure used for cells from spleen and thymus.

Lymphoid cell suspensions are prepared by teasing the minced organs through a wire mesh into cold phosphate-buffered saline (PBS), Hank's Balanced Salt Solution (HBSS, Grand Island Biological Co., Grand Island, New York), or Minimal Essential Medium (MEM, Microbiological Associates, Bethesda, Maryland) containing 20 μg/ml DNase. The presence of DNase prevents aggregation and nonspecific binding caused by the release of DNA from damaged cells. Low speed centrifugation at 4° is used to remove aggregates and the cells are washed three times in the medium. After washing, the cell suspension is passed through fine gauge nylon mesh to remove residual aggregates. The viability of the cells treated in this manner exceeds 90% for thymocytes and 70% for spleen cells.

The Mixing Conditions, Temperature, and Medium. The specific binding of cells to the fibers involves a delicate balance between the rate and angle of collision of the fiber and cell, the dwell time of the cell at the fiber surface, and the settling of the cells due to gravity. If the cells are allowed to settle by gravity onto the fibers, excessive nonspecific binding may result. On the other hand, rapid shaking results in no binding at all. In our experience, the best results have been obtained by incubating the petri dish containing strung fibers for 1 hour with 4 ml of cell suspension on a reciprocal shaker having a 3–4 cm horizontal stroke at 70–80 oscillations per minute. The fibers must be aligned perpendicular to the direction of shaking. It is essential that the medium cover the fibers at all times during shaking.

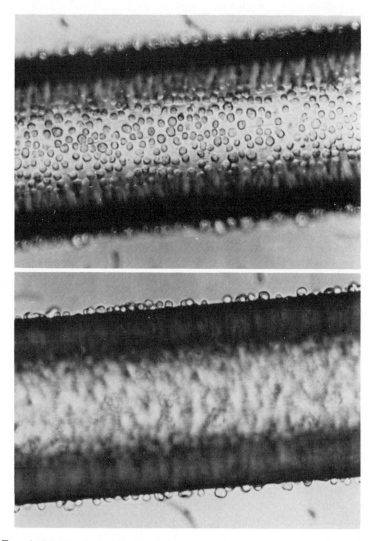

Fig. 4. Mouse thymocytes bound to a concanavalin A (Con A)-derivatized nylon fiber. *Top:* The field was focused on the face of the fiber using bright field microscopy. *Bottom:* The field was focused on the edge of the fiber using bright field microscopy. The fibers have a diameter of 250 μm.

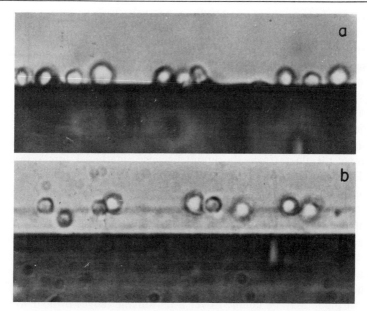

Fig. 5. Mouse spleen cells bound to (a) dinitrophenylated (Dnp) bovine serum albumin fibers, and (b) Dnp-gelatin fibers.

The binding of cells to Con A fibers (Fig. 4) or antigen fibers (Fig. 5) is not affected by the incubation temperature over the range from 4° to 25°. In preliminary experiments with transformed lymphoid cell lines, we find approximately 5-fold more binding to Con A fibers at 25° than at 4°. In many applications, however, it is desirable to keep the cells at 4° to minimize cell damage and possible effects of the fiber on the membranes and metabolism of the bound cells. Metabolic energy is not required for binding inasmuch as the presence of 50 mM NaN$_3$ does not decrease the number of cells bound. Incubation at 37° may produce nonspecific binding because of excessive death and aggregation of the cells.

Except for the addition of DNase or a specific inhibitor, the composition of the medium during the incubation at 4° does not in general affect the amount of binding. Any physiological medium can be used provided that the cells do not aggregate. Specific inhibitors are added to the cell suspension to test the specificity of the binding or to fractionate cells with respect to their affinity for the soluble inhibitor. In the binding of lymphocytes to antigen-derivatized fibers, both soluble antigen or antibodies against cell surface immunoglobulin can be used as inhibitors; with Con A fibers, α-methyl mannoside is used as an inhibitor of binding. The amount of glucose present in HBSS or MEM will decrease but not prevent binding.

Removal of Unbound Cells. To complete the isolation, it is necessary to remove all unbound cells from the fractionation apparatus. Unbound cells are removed by complete immersion of the dish in a series of larger vessels containing medium. It is also necessary to replace the petri dish in order to remove additional cells, which adhere nonspecifically to the dish bottom.

During this and all subsequent procedures care should be taken not to remove the fibers from the liquid because interfacial forces result in the release and loss of the cells. If meshes are used, this stricture does not apply, for sufficient liquid remains in the interstices to prevent interfacial changes. The cells are otherwise firmly bound, and extensive washing, mixing or manipulation is possible.

Quantitation of Bound Cells. The number of cells attached to the fibers can usually be determined by directly counting those bound to the fiber edge under the microscope. This method takes advantage of the fact that the edge of the fiber can be sharply focused (Fig. 5), and cells in the focal plane may be counted over a reasonable length of fiber to minimize error. Using ^{51}Cr-labeled cells, it has been found that the total number of bound cells is linearly proportional to the edge count. The factor to convert edge to total counts should be determined independently for a fiber of a given degree of substitution in a given experimental context. With Con A-derivatized fibers, about 10^6 lymphocytes are bound to a 48-fiber dish; with antigen-derivatized fibers the yield is about 10^5 cells per dish. Generally five different fibers are counted per dish, and the mean of these cell counts is used in calculating the total number of bound cells. The standard deviation of this mean is 5–10%. The number of cells bound to five identical dishes also has a standard deviation of the mean of 5–10%.

Removal of Cells from the Fiber. Cells can be released from the fibers by two methods: plucking the taut fibers with a needle; or, for cells bound to Dnp-Gel fibers, melting the gelatin (37°, 15 minutes). Direct visual inspection of the process of cellular removal shows that the transverse components of the mechanical vibration project the bound cells into the medium.

This shearing of the cell-fiber bond may produce a lesion in the cell surface membrane, particularly if the fiber is heavily derivatized. With the use of appropriate conditions, however, fractionated cell populations with high viabilities can be obtained routinely from either derivatized or gelatin-coated fibers. Damage to the cells is minimized by either (1) using a "soft" gelatin-coated fiber which presumably minimizes the shear required for release, or (2) avoiding shear forces altogether by "melting" the cells from a gelatin-coated fiber, and, in all cases, (3) using the appropriate medium and incubation conditions to allow repair of any lesions produced by the removal process.

The effect of these conditions on the viability of lymphocytes released from antigen-derivatized fibers is summarized in Table II. More cells survive removal from gelatin fibers than from directly derivatized fibers, especially if the cells are removed by melting the gelatin. In all cases, however, cell survival is enhanced by the presence of fetal calf serum in the medium. Furthermore, the proportion of viable plucked cells increases when the cells are incubated for 30 minutes at 37° in the presence of serum, suggesting that the cells are repairing lesions in their surface membranes. In the absence of serum, the viability of both fractionated and unfractionated cells is decreased.

With the apparatus described above, cells obtained by fiber fractionation are removed into a minimum volume of about 2.5 ml per dish. This results in relatively dilute cell suspensions (10^4 to 10^6 cells/ml). Counting the cells by hemacytometer is laborious (several chambers may have to be scanned) but satisfactory. Excellent results can be obtained by direct counting with a Coulter Counter (Coulter Electronics) or Cyto-

TABLE II
Viability of Spleen Cells Removed from Dnp Fibers[a]

Fiber	Medium	Removal	Incubation[b]	% Viable[c]
Dnp-BSA	PBS	Pluck	30 Min, 37°	21 ± 2
Dnp-BSA	PBS	Pluck	—	46 ± 2
Dnp-BSA	FCS	Pluck	—	70 ± 2
Dnp-BSA	FCS	Pluck	30 Min, 37°	81 ± 3
Dnp-Gel	PBS	Pluck	30 Min, 37°	21 ± 3
Dnp-Gel	PBS	Pluck	—	62 ± 4
Dnp-Gel	PBS	Melt	30 Min, 37°	70 ± 2
Dnp-Gel	FCS	Pluck	—	83 ± 3
Dnp-Gel	FCS	Pluck	30 Min, 37°	91 ± 2
Dnp-Gel	FCS	Melt	30 Min, 37°	97 ± 1
Unfractionated cells	PBS	—	30 Min, 37°	63 ± 2
Unfractionated cells	PBS	—	—	84 ± 2
Unfractionated cells	FCS	—	30 Min, 37°	85 ± 1

[a] Dnp = the 2,4-dinitrophenyl group; Dnp-BSA, dinitrophenylated bovine serum albumin; Dnp-Gel, dinitrophenylated gelatin; PBS, phosphate-buffered saline; FCS, PBS plus 10% fetal calf serum.

[b] Incubation was carried out in the same medium as the removal process in all cases.

[c] As determined by trypan blue exclusion. Each number is the average of 3 determinations. Standard deviations of the mean are shown.

graph (Bio/Physics Systems). Concentration of the cells by conventional centrifugation and resuspension can result in substantial losses. Although the addition of carrier particles or cells will eliminate this difficulty, it may defeat the purpose of the fractionation in some applications. In our experience, good results have been obtained by removing the strung fibers after plucking, placing the petri dish into the top of a 50-ml clinical centrifuge cup, and centrifuging at 1500 rpm for 10 minutes. Most of the medium can be removed without disturbing the sedimented cells. Inasmuch as the cells are concentrated into a monolayer, they can be used *in situ* for cytological studies or easily resuspended into a small volume.

Use of Meshes for High Yield and Depletion. Although the fibers isolate cells with high specificity, they have a relatively low capacity. For applications where high specificity and *in situ* visualization are less important, derivatized meshes can be used to deplete a population of cells of a certain subpopulation and to determine the total number of cells in that subpopulation. Cells are incubated in a 35×8 mm air-free chamber separated into two compartments by a derivatized nylon mesh. Eight milliliters of a cell suspension (1.25 to 2.5×10^7 cells/ml) is added and the chamber is placed on a horizontal rotary shaker at 200 rpm for 1 hour at 4°. The chamber is inverted every 15 minutes so that the cells filter through the mesh under unit gravity. Unlike strung fibers, the mesh can be removed from the liquid without loss of bound cells. The extent of depletion may be tested by assaying the supernatant cells for the presence of fiber binding cells using strung fibers. Although the binding of cells to meshes under unit gravity is effective for depletion and quantitation studies, the specificity (about 40–60%) of this procedure is usually inadequate for the isolation of purified cells.

In order to obtain large numbers of fractionated cells with higher specificity, a simple oscillatory machine can be used with multiple meshes[19] (Fig. 6). The meshes are oscillated at 1 cm/min through the cell suspension with a resultant increase (to 70–85%) in the binding specificity. Using 30 antigen-derivatized meshes in a 1-hour incubation period with 6×10^8 spleen cells in 100 ml, up to 2×10^7 specifically fractionated cells can be obtained. However, because of the structure of the mesh, cells cannot be released from the fibers by plucking. The bound cells are therefore recovered by rapid pipetting of medium containing serum through each mesh. The cells obtained in this way will exclude trypan blue, but their viability in terms of functional capacities has not as yet been established.

[19] U. Rutishauser and G. M. Edelman, unpublished results, 1973.

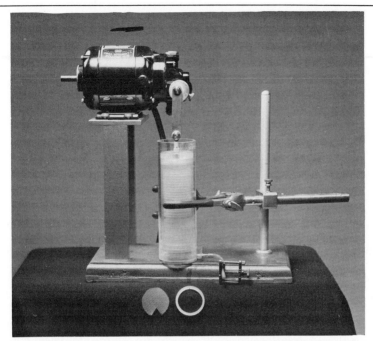

FIG. 6. Mechanical arrangement for large-scale cell fractionation using fiber meshes. One of the rings used to separate the meshes, and a mesh with a V slot to allow escape of air bubbles are shown at the base. The whole stack of 30 meshes is moved slowly as a piston fitting loosely in the cylinder containing the cell suspension.

Applications

Fiber fractionation can be used to isolate cells according to two parameters: a difference in the number or distribution of the same surface component or the presence of a unique surface component on a given cell type.

Differential Binding of Cells to Lectin-Derivatized Fibers. As a model system to demonstrate the fractionation of cells by differential-binding properties, mouse thymocytes, and erythrocytes were bound to nylon fibers that had been derivatized with different amounts of Con A (Fig. 7). The binding of cells to Con A fibers is highly specific, with over 95% inhibition of binding by 50 mM α-methyl mannoside, but not by galactose. Both thymocytes and erythrocytes have carbohydrate receptors for this lectin and will bind to heavily derivatized fibers. Erythrocytes, however, do not bind to fibers with a low Con A density. Since the thymocytes will bind to these fibers, they can be isolated. Using meshes, all thymocytes can be removed, thus separating the two cell types. This means of

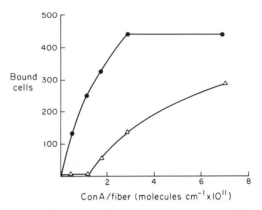

Fig. 7. Binding of erythrocytes (△—△) and thymocytes (●—●) to different fibers as a function of the number of concanavalin A (Con A) molecules per centimeter of nylon fiber; 2×10^8 cells in 4 ml of PBS were incubated with different fibers for 120 minutes with standard mixing conditions. The number of cells bound was determined by microscopic examination of a 1-mm fiber segment.

fractionation has been used to characterize villus and crypt epithelial cells from intestine.[20] Other applications might include the separation of normal and transformed cells, which are known to have different Con A-binding properties. The identification and purification of other lectins with different binding specificities[21] should also broaden the applicability of this approach.

Fractionation of Lymphoid Cell Populations. The most extensive use of the fiber method has been in the separation of lymphocytes. Several examples of this application serve to indicate how it may be used in other circumstances: (1) the isolation of cells according to the *specificity* of their receptors,[12,13] in this case cell surface immunoglobulin directed against antigens; (2) the isolation of cells according to the affinity of their receptors for a particular antigen[13]; (3) the characterization of isolated antigen-binding cells in terms of their class, the number and distribution of their immunoglobulin receptors, the presence of other cell surface receptors,[14] and their function[15]; (4) isolation of cells with antibodies specific for cell surface antigens.[19]

1. Isolation of antigen-binding cells according to their specificity for dinitrophenyl- and *p*-toluenesulfonyl group (tosyl)-derivatized fibers is summarized in Table III. More binding is obtained with cells from specifically immunized animals than from unimmunized animals, and this binding is inhibited by soluble forms of the fiber antigen. The

[20] O. K. Podolsky and M. M. Weiser, *J. Cell Biol.* **58**, 497 (1973).
[21] N. Sharon and H. Lis, *Science* **177**, 949 (1972).

TABLE III
Binding of Spleen Cells to Antigen-Derivatized Fibers

Antigen on fiber	Dnp[a]	Dnp	Tosyl[a]	Tosyl
Immunization	None	Dnp	None	Tosyl
Cells bound to fiber (per cm)	1200	4000	800	2000
% Inhibition of binding by:				
Dnp	90	95	5	10
Tosyl	1	2	75	87
Anti-Ig	85	93	73	90
High avidity cells (per cm)	<100	2800	—	—
Low avidity cells (per cm)	1200	1200	—	—

[a] Dnp = the 2,4-dinitrophenyl group; tosyl = the p-toluenesulfonyl group.

specificity of binding is demonstrated by cross inhibition studies: tosyl derivatives do not inhibit Dnp-specific binding appreciably, nor do Dnp derivatives inhibit tosyl-specific binding. Preincubation of the spleen cells with immunoglobulin isolated from a rabbit antiserum directed against mouse immunoglobulin prevents the binding of cells to antigen-substituted fibers; no inhibition is obtained with normal rabbit immunoglobulin. These data suggest that the specific attachment is mediated by specific immunoglobulin receptors on the cell surface.

The number of cells bound to a Dnp-derivatized fiber is a linear function of the cell concentration over a wide range (Fig. 8), suggesting that under the conditions used, the binding is proportional to the number of input cells, the binding of one cell is independent of another,

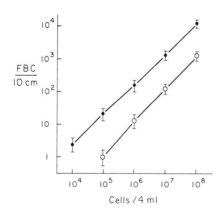

FIG. 8. Binding of spleen cells to dinitrophenylated (Dnp) bovine serum albumin fibers. Standard curves are shown for the binding of previously unfractionated spleen cells from Dnp-immune (—●—) or unimmunized (—○—) animals as a function of the cell number in the incubation mixture. Standard deviations of the mean are indicated by the bars.

and the fiber surface is not saturated with cells. As noted above, a larger proportion of the cells from immunized animals bound to the fibers than from unimmunized animals.

Further studies on specificity can be carried out by testing the ability of the fractionated cells to rebind to antigen-derivatized fibers. The efficiency with which cells removed from Dnp-fibers could rebind the same (Dnp) or a different (tosyl) antigen was determined as follows: Spleen cells were first fractionated using preparative (48-strand) dishes with Dnp-BSA or Dnp-Gel fibers. The bound cells were then removed by plucking into medium containing 10% fetal calf serum. This cell suspension was added without dilution to 12-strand dishes containing fibers derivatized with Dnp or tosyl. The specificity of the rebinding was determined by doing duplicate assays in the presence of 300 μg/ml of Dnp-BSA or tosyl-BSA. In order to determine the enrichment obtained in the initial fractionation, the binding of various amounts (cells/ml) of unfractionated cells was assayed at the same time. These experiments (Table IV) show that Dnp-fractionated cells rebind to Dnp fibers but that cells from tosyl-BSA fibers do not. Furthermore, the relatively large number of unfractionated cells required to produce an equivalent amount of binding indicated that the fractionation yields a highly enriched population of specific antigen-binding cells and the degree of enrichment can be calculated from the available parameters.

2. Fractionation of antigen-binding cells according to affinity. In a manner similar to the separation of erythrocytes and thymocytes by their differential binding to Con A fibers, a population of specific antigen-binding cells can be further fractionated with respect to their affinity for

TABLE IV
SPECIFIC REBINDING OF DNP[a] AND TOSYL-FRACTIONATED CELLS
TO DNP-DERIVATIZED FIBERS

Immunization	% Dnp-binding cells in unfractionated population	Specificity of fractionation	Observed enrichment[b]
None	1–2	Dnp	44–78×
None	1–2	Tosyl	<1×
Dnp (primary)	6–8	Dnp	12–15×
Dnp (secondary)	10–17	Dnp	6–7×
Dnp (secondary)	10–17	Tosyl	<1×

[a] Dnp = the 2,4-dinitrophenyl group; tosyl = the p-toluenesulfonyl group.

[b] Determined as the ratio of unfractionated cells to fractionated cells which give equivalent amounts of binding as defined in Fig. 9.

the same antigen. Instead of using fibers coated with different amounts of antigen, the affinity fractionation is achieved by selective inhibition using different concentrations of soluble antigen. The binding of cells to the fibers is generally irreversible. Under these conditions, inhibition by soluble antigen is the result of specific blockage of the receptors on the cell surface. Therefore, the higher the affinity or avidity of a cell, the lower the antigen concentration required for inhibition. Although the total number of fiber-binding cells obtained with immunized mice is only 2–8 times that obtained with nonimmune animals, the experiment shown in Fig. 9 reveals that the two cell populations differ greatly in their susceptibility to inhibition by soluble antigen. A clear-cut change in the number of cells of higher affinity was found after immunization (Table III). With immunized animals, over 60% of the binding is inhibited by antigen concentrations of less than 4 μg/ml; at the same inhibitor concentration the binding of cells from unimmunized animals decreases by less than 5%. These results lead to the conclusion that immunization results in a selective proliferation of those cells with a high affinity for the immunizing antigen.

3. Characterization of isolated antigen-binding cells. The ability to manipulate and visualize the cells bound to a fiber facilitates their subsequent characterization and use in functional studies. This includes the identification of T and B lymphocytes by cytotoxicity, immunofluorescence and rosette assays, the quantitation and localization of receptors with radiolabeled or fluorescent antibodies and by rosette formation, and the study of B cell function by transfer to irradiated recipients.

Mouse thymus-derived (T) and bone marrow-derived (B) cells can

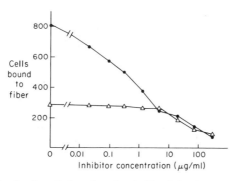

FIG. 9. Inhibition by free dinitrophenylated (Dnp) bovine serum albumin (Dnp-BSA) of spleen cell binding to Dnp$_{10}$-BSA derivatized fibers. Cell numbers represent fiber edge counts for a 2.5-cm fiber segment. Spleens from immunized mice were removed at the height of a secondary response to Dnp-BGG, and cells from several mice were pooled. ●—●, Immunized; △—△, unimmunized.

be identified by the use of specific antibodies.[22,23] B cells have much more surface immunoglobulin than T cells, and antibodies to mouse immunoglobulin (anti-Ig) can be used to detect the B cells. Antisera to the surface θ antigen of T cells may be used to detect these cells. Cytotoxicity assays can be carried out either with fiber-bound cells or with cells removed from the fibers and centrifuged to the bottom of a petri dish. The viability of cells bound to derivatized fibers exceeds 95%, even though the unfractionated cell population includes up to 35% dead cells. Although the reason for the failure to bind nonviable cells is not known, this property of fiber-fractionated cells facilitates subsequent cytotoxicity assays by eliminating the background of nonviable cells. For an *in situ* cytotoxicity assay, fiber-bound cells are incubated with a 1:10 dilution of anti-θ serum or anti-Ig serum at 37° for 45 minutes. The serum is then washed away, and the cells are incubated with a 1:10 dilution of agarose-absorbed guinea pig complement for 30 minutes at 37° with gentle shaking. When strong antiserum and complement are used, the cells bearing the target antigen are severely damaged and no longer remain bound to the fibers (Fig. 10); with less potent reagents, some dead cells may not be released. The number of T and B cells is determined by counting the number of viable bound cells before and after treatment with complement. The cytotoxic effects of anti-Ig serum and anti-θ serum on cells bound to a Dnp-BSA fiber are additive and nonoverlapping.[14] Sequential treatment with both antisera, in either order, kills and removes 97% of the fiber-fractionated cells. Thus, practically all of the bound cells are either T or B lymphocytes.

Fluorescence microscopy can also be employed to permit the detection of T and B cells.[24] In this case, care must be taken that the nylon fibers have not been treated with fluorescent additives. It is also useful to use monofilament with a small diameter (less than 50 μm) in order to minimize fluorescence by the fiber and to facilitate observation of the cells at high magnification. About 70% of the fiber-fractionated cells treated with fluorescein-labeled anti-immunoglobulin show intense fluorescent caps characteristic of B cells. With an indirect fluorescent stain for the θ antigen,[14] a significant proportion of T cells is also found among the cells isolated from the antigen-derivatized fibers, although the values are about 10% less than those determined by cytotoxicity tests. These methods can also be used to determine the distribution of receptors on the surface of fiber-bound cells.

[22] M. C. Raff, *Nature (London)* **224**, 378 (1969).
[23] T. Takahashi, L. J. Old, R. K. McIntire, and E. A. Boyse, *J. Exp. Med.* **134**, 815 (1971).
[24] M. C. Raff, *Immunology* **19**, 637 (1970).

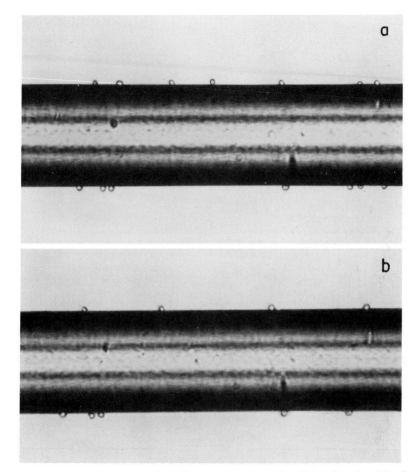

Fig. 10. Spleen cells bound to the same segment of a dinitrophenylated bovine serum albumin derivatized fiber before (a) and after (b) treatment with anti-θ serum and complement.

B cells have receptors for antigen–antibody–complement complexes.[25,26] When sheep erythrocytes are used as the antigen in these complexes, visible erythrocyte–antibody–complement rosettes, EAC rosettes, are formed as a result of their binding to B cells (Fig. 11a). T cells do not have these receptors, and cells bound to fibers can be identified as B cells by this means. Spleen cells are fractionated using Dnp-BSA derivatized fibers and incubated with 4% EAC in MEM with gentle shaking for 30

[25] C. Bianco and V. Nussensweig, *Science* 173, 154 (1971).
[26] C. Bianco, R. Patrick, and V. Nussensweig, *J. Exp. Med.* 132, 702 (1970).

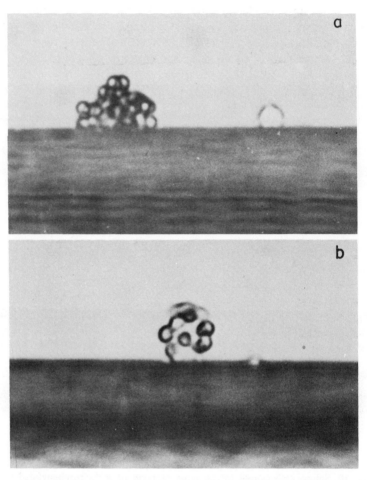

FIG. 11. (a) Section of a fiber derivatized with dinitrophenylated bovine serum albumin (Dnp-BSA) showing one cell which binds erythrocyte–antibody–complement complexes and one which does not. (b) Rosettes formed by spleen cells bound to a Dnp-BSA fiber after incubation with 2,4,6-trinitrophenyl-derivatized sheep erythrocytes.

minutes at 37°. About two-thirds of the cells bound to Dnp-BSA fibers form EAC rosettes as expected from the fact that about 60% of splenic lymphocytes are B cells.

Fiber-rosette methods can also be used to examine the distribution and specificity of antigen receptors on fiber-bound cells. To form the rosettes, spleen cells are bound to fibers derivatized with Dnp-BSA and 0.3 ml of a 33% suspension of 2,4,6-trinitrophenyl group (Tnp)-coupled sheep erythrocytes[27] is added to the dish. After a 1-hour incubation at 4° to 25° with gentle reciprocal shaking, unbound erythrocytes are washed away. Typical rosettes are illustrated in Fig. 11b. The even distribution of cells indicates that anti-Dnp Ig receptors have not all moved to the point of fiber attachment but rather are distributed over the cell surface. The specificity of the rosette formation is indicated by the fact that it is inhibited by free antigen and anti-immunoglobulin.

The relative number of immunoglobulin receptors on fractionated cells can be estimated by incubating fiber-bound cells with increasing amounts of ^{125}I-labeled rabbit anti-mouse immunoglobulin[18] for 30 minutes at room temperature, until saturation is achieved. After washing away the unbound antibody, the cells are removed from the fibers by plucking, and 10^7 carrier cells are added to facilitate further washing by centrifugation. The radioactivity of the cells is then determined using a gamma spectrometer.

The function of the antigen-specific B cells obtained by fiber fractionation can be studied in adoptive transfer experiments. Unimmunized mice that have been exposed 24 hours previously to 630–730 Rad (from a Picker cobalt-60 gamma source) are injected intravenously with 5 to 10 × 10^6 splenocytes from mice immunized with hemocyanin plus either 10^4 to 10^5 Dnp- or tosyl-fiber-binding cells, or 10^4 to 10^6 unfractionated splenocytes from mice immunized with Dnp. Control mice receive only 5 to 10 × 10^6 cells from mice immunized with hemocyanin, representing the carrier specificity. The recipient mice are then injected intraperitoneally with 200 μg of Dnp-hemocyanin adsorbed onto bentonite. All materials used in this procedure are sterile; the sterility of the fractionated cells is maintained by extensively washing the fiber-bound cells with sterile phosphate-buffered saline. Seven days after transfer, the spleen from each recipient animal is tested for antibody-secreting cells by the Jerne plaque assay,[28] modified to detect Tnp and Dnp antibodies.[27] Fractionation with Dnp-derivatized fibers yielded cells that produced a Dnp-specific response equivalent to that of at least ten times as many unfractionated cells from the same animals.[15] As expected of cells from immunized mice, enhancements were not observed in the IgM response. As a test of specificity, cells obtained by fractionation of these cell populations with tosyl-derivatized fibers did not yield Dnp plaque-forming cells above control values. There was also no detectable response if the recipient animals were not simultaneously injected with Dnp-hemocyanin. Inasmuch as the unfractionated cell populations contained 7% to 15% Dnp-binding cells, the observed

[27] M. B. Rittenberg and K. L. Pratt, *Proc. Soc. Exp. Biol. Med.* 132, 575 (1969).
[28] N. K. Jerne and A. A. Nordin, *Science* 140, 405 (1963).

enhancement in the transfer experiments approached the value expected from a population of pure Dnp antigen-binding cells of which at least the antigen-sensitive B cells are functioning normally.

4. Isolation of cells with antibodies specific for cell surface antigens. Whereas the fractionations of lymphocytes cited above involve the interaction of cell surface antibody with fiber-bound antigen, the roles of the two molecules can be reversed. The availability of antisera specific for a number of cell surface antigens (for lymphoid cells, see Takahashi et al.[23]) makes this a versatile approach. For example, fibers coated with antibodies to immunoglobulin (anti-Ig) can be used to isolate B cells. The γ-globulin fraction of the antiserum is isolated by ammonium sulfate precipitation and ion-exchange chromatography.[18] Where possible, affinity methods[18] should be used for the isolation of specific antibodies. In either case, the isolated antibody can be coupled to the fibers with carbodiimide or, if only limited quantities are available, by adsorption from solutions of 0.25 to 1.0 mg antibody/ml (see Methods). When spleen cells are incubated with anti-Ig fibers, a large number of cells are bound. All these cells will form EAC rosettes and are removed from fibers by treatment with anti-immunoglobulin serum and complement as described above.

Prospective Applications

Fiber Fractionation as a General Method for the Study of Cell Populations. The versatility of the fiber method suggests that it may be applied to a number of problems in cell biology. The geometry and manipulability of derivatized fibers allows a variety of schemes to be used in cell fractionation experiments. Successive portions of a fiber can be substituted with different binding molecules so that the isolated cells may be geometrically arranged in predetermined patterns, mechanically transported as groups, juxtaposed, and then physically removed for later analysis. A continued search for new carriers of different properties should extend the range of the method. Even with the limited experience so far gathered it is clear that besides the methods detailed above, viruses, bacteria, and eukaryotic cells may be affixed to fibers and used as "reagents" for detection of interactions with other cells as well as for the purposes of cell fractionation.

In the application of fiber fractionation techniques to cellular systems other than the immune system there are two major problems to be considered: dissociation of the cells and choice of ligand for the fiber. Obviously an important concern in cell dissociation is that the procedure employed be chosen to avoid destruction of the receptors used for cell isolation. The choice of a ligand depends upon the system and the specific purposes of the fractionation. Some obvious ones include lectins, enzymes, hormones, and antibodies directed against cell surface antigens. Some ligands, such as the lectin Con A, do not show an absolute preference for a given cell type; i.e., most cells in an animal will contain glycoprotein receptors capable of interacting with it. This lack of specificity can, in some instances, be an advantage. An example is an experiment in which it is desired to fix an already fractionated cell population on a fiber by a receptor other than one that is being studied. Obviously, it is advantageous that a competitive inhibitor (in this case, α-methyl-D-mannoside) be available. In other cases, such inhibitors may be lacking and other means must be available to test the specificity of the interaction with the fiber ligand.

Extension of the Method to Surfaces. In addition to fixation on fibers, cells may be attached to surfaces of different geometry, each with its own special advantages and disadvantages. Such arrangements open the possibility of fixing cells in various patterns according to their specificity and then examining their behavior in tissue culture. The geometrical effects in culture (topoculture) open new possibilities for studies of cell–cell interaction and cell surface–nuclear interactions.

Proteins or other ligands may be coupled to flat surfaces such as nylon films or the bottoms of plastic petri dishes. Under the appropriate conditions cells will bind to these surfaces with the same degree of specificity as to the fibers.[29] These derivatized flat surfaces have some advantages over derivatized fibers for certain kinds of experiments, particularly those requiring microscopic examinations of the bound cells. Because the optical resolution is better and because all the cells bound to the surface can be examined, the use of derivatized flat surfaces adds new applications to the general technique described here. For example, cells that bind to different ligands on surfaces may be identified by their morphology or by their ability to bind fluorescent reagents or other microscopically visible markers. The effects of various ligands on cell migration and cell growth may be continuously monitored. Because it is possible to couple more than one ligand to different areas of the same surface, one can directly compare the effects of the ligands on cell binding, migration, or growth. Derivatized flat surfaces are not very useful, however, for the recovery of specifically bound cells, for it is difficult to remove the cells quantitatively and without damage.

To manipulate the nylon film (Chieftain nylon film, tubular, autoclavable, American Hospital Supply, Evanston, Illinois) during the entire period of derivatization, incubation with cells, and examination, it is convenient to stretch the nylon over a plexiglass or stainless steel ring,

---

[29] P. Spear, J. L. Wang, and G. M. Edelman, unpublished results, 1973.

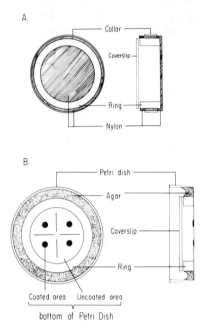

FIG. 12. (A) Nylon film stretched over a plexiglass ring and secured with a polyethylene collar, shown diagrammatically in front view and cross section. The cross-sectional view shows how the well formed by the ring and the nylon film is sealed with a coverslip after being completely filled with a cell suspension. (B) Petri dish derivatized on the encircled areas with four different proteins, shown in front view and cross section. In order to maintain the dish in a vertical orientation during incubation with a cell suspension, a well that can be sealed with a coverslip is formed by fixing a plexiglass ring to the dish.

securing it with a flexible collar slightly larger than the ring (Fig. 12A). It is necessary first to stretch the slightly opaque nylon film until it is quite transparent to obtain a material with good optical qualities. Ligands are then coupled to the film using exactly the same conditions as for the nylon fibers. The protein–carbodiimide mixture is placed in the well formed by the ring and the nylon film and incubated with the film for 0.5 hour with shaking. The derivatizing solution is then removed, and the film is washed extensively with PBS. During incubation of the film with cells the film must be oriented in a vertical position in order to allow sufficient contact between the derivatized surface and the cells while minimizing nonspecific binding of cells to the film. To accomplish this, the well of the chamber formed by the ring and film is filled to overflowing with the cell suspension. A coverslip slightly smaller than the outside diameter of the ring is dropped onto the surface of the liquid, taking care to prevent bubble formation. After excess liquid is blotted away, surface tension holds the coverslip in place, preventing the cell suspension from leaking out when the film is tilted to a vertical position. The assembled chamber is then clamped to a rotating machine (Multi-purpose rotator, Scientific Instruments, Inc.). The rotary motion keeps the cells suspended during the 1-hour incubation. After the incubation period, the coverslip can easily be removed by immersing the whole chamber in a washing solution such as PBS; the chamber is carried through several changes of this solution to get rid of unbound cells. The cells bound to the film can be examined directly or after subsequent manipulation such as staining or incubation in tissue culture medium. The number of cells per unit area can be determined by using a calibrated graticule in the eyepiece of the microscope.

Plastic petri dishes may also be derivatized with various ligands using a carbodiimide linker. Although the nature of the bond between a protein and the plastic is not known, the protein adheres very tenaciously to the plastic and does not wash off even after several days of incubation at 37°. There are advantages to using petri dishes rather than nylon films when the object of the experiment is to observe the effects of various surface-bound ligands on cell migration, growth, or cell functions rather than to fractionate cells according to their surface specificities. The plastic in the petri dishes is optically superior to nylon and is more suitable for the growth of cells. One example of the use of derivatized petri dishes comes from the work of Andersson et al.[30] They showed that when Con A is bound to a plastic petri dish, it stimulates mitosis of bone marrow-derived lymphocytes but not thymus-derived lymphocytes, whereas the reverse is true for Con A in solution.

Derivatized plastic petri dishes have also been used in a series of experiments designed to detect adventitious proteins or other ligands that happen to bind to cells of one embryonic organ but not to cells of another.[29] The problem required a method to rapidly test a large number of molecules. Therefore single petri dishes were derivatized with several ligands in discrete and well-separated areas and then incubated with dissociated cells from embryonic organs in order to determine whether the cells could bind to the derivatized areas. In order to avoid nonspecific binding of cells, it was found necessary to incubate the cells with the petri dish so that the derivatized surface was in a vertical rather than a horizontal orientation. To meet this requirement while minimizing the number of cells required, small wells or chambers were prepared from the petri dish and plexiglass rings similar to those prepared from the nylon

[30] J. Andersson, G. M. Edelman, G. Möller, and O. Sjöberg, *Eur. J. Immunol.* **2**, 233 (1972).

films and rings (Fig. 12B). The plexiglass rings (1.9 cm i.d.; 2.5 cm o.d.) were placed in 35-mm petri dishes and molten agar was then pipetted between the ring and the petri dish rim. The solidified agar held the ring in place and formed a seal between the ring and petri dish bottom. The underside of the petri dish was then marked off into quadrants, and areas to be derivatized were encircled. Small droplets of different protein–carbodiimide mixtures (1 mg protein and 5 mg carbodiimide per milliliter) were placed on the encircled areas within the well using Pasteur pipettes. The dishes were simply left at room temperature for 0.5 hour in a humidified environment. Thereafter, the droplets were aspirated and the dishes washed by immersion in a large volume of PBS. The wells were filled to overflowing with cell suspensions prepared from embryonic organs and then sealed with a coverslip as previously described. The petri dishes were oriented on a rotator so that the derivatized surfaces were vertical and incubated for 1 hour at 37° with rotary motion to keep the cells suspended. After unbound cells were washed away by immersion of the dishes in PBS, the dishes were microscopically examined for the presence or absence of bound cells in the encircled areas. By the use of a lectin that binds to many different embryonic cell types, Con A, but one whose binding can be inhibited by simple sugars, it was possible to show that the binding of cells to the derivatized areas of the petri dish was highly specific.

In many applications such as topoculture it is unnecessary to remove the cell from the surface. Fibers, meshes, and other surfaces may be moved intact with their specifically attached cells from medium to medium for biochemical and cytochemical studies and cells on two fibers may be moved into close proximity. By this means, cells in culture may be rapidly moved from reagent to reagent or environment to environment. Fluorescent markers, radiolabels and other indicators may be used to advantage, and fixation and sectioning of the entire cell-surface ensemble allows electron microscopy to be carried out. Addition of agar to the medium should permit use of the method with immunodiffusion to detect cell products.

The use of modified red cells in fiber-rosette assays suggests possibilities of extending the method for the study of cell–cell interactions. This depends upon the fact that once a cell is bound to the fiber, the presence of ligand inhibitors does not result in its removal. Thus, the specificity of cell–cell interaction can be tested in the same way that the cell–fiber interaction is tested. As shown in the formation of antigen-specific rosettes and EAC rosettes, the cell–cell interactions may be either homospecific or heterospecific with respect to the cell–fiber interaction.

Derivatized fibers can be used to ensure that a ligand *cannot* penetrate the cell but can exert its action at the surface. This opens the opportunity for study of the effect of geometry and sidedness on cell surface chemistry as well as on surface-nuclear interactions. Finally, the ability to view living cells in fixed positions under the microscope permits an opportunity to carry out biochemical experiments on single cells which, although not explored in detail, seem quite promising.

Acknowledgments

This work was supported by U.S. Public Health Service grants from the National Institutes of Health and from the National Science Foundation.

Attachment of Specific Ligands and Methods for Specific Proteins

A. COENZYMES AND COFACTORS

[16] AMP and NAD as "General Ligands"

By Klaus Mosbach

Affinity chromatography, also called biospecific affinity chromatography, uses the specificity of immobilized ligands for enzymes. In using specific ligands, each new separation problem requires different and often elaborate syntheses. This is made more complex when it is necessary to insert a spacer between ligand and matrix. It is therefore advantageous to use "general ligands" that show affinity for a broad spectrum of enzyme. We have accordingly investigated the immobilization of adenosine 5′-monophosphate (AMP) and adenine nicotinamide dinucleotide (NAD⁺). Both cofactors, when bound as analogs to suitable matrices adsorb an entire family of enzymes, which require specific elution conditions for separation. This article describes three nucleotide preparations and discusses examples of such specific elution. The broader concepts of "general ligand" and "specific elution" are treated in more detail elsewhere.[1-3]

The three nucleotide analogs, immobilized on Sepharose 4B are: (a) N^6-(6-aminohexyl)-AMP, (b) ϵ-aminohexanoyl-NAD⁺, (c) NAD⁺-N^6-[N-(6-aminohexyl) acetamide]. The nucleotide analogs for (a) and (c), consisting of nucleotide plus spacer, were synthesized first and then coupled to the matrix. For (b), NAD⁺ was coupled to a preformed Sepharose matrix substituted with ϵ-aminohexanoic acid.

Preparation of N^6-(6-Aminohexyl)-AMP-Agarose

N^6-(6-Aminohexyl)-AMP is prepared in two steps from 6-chloropurine riboside, available from Sigma Chemical Co., but may be synthesized in three steps starting from inosine.[4-6] Phosphorylation of the unprotected chloronucleoside with phosphoryl chloride with aqueous triethyl phosphate as solvent gives a product almost exclusively phosphorylated in the 5′ position. The crude 6-chloropurine nucleotide is treated in the next step with aqueous 1,6-diaminohexane yielding N^6-(6-aminohexyl)-AMP (Fig.

[1] K. Mosbach, H. Guilford, P.-O. Larsson, R. Ohlsson, and M. Scott, *Biochem. J.* **125**, 20 (1971).
[2] K. Mosbach, H. Guilford, R. Ohlsson, and M. Scott, *Biochem. J.* **127**, 625 (1972).
[3] R. Ohlsson, P. Brodelius, and K. Mosbach, *FEBS Lett.* **25**, 234 (1972).
[4] H. Bredereck and A. Martini, *Chem. Ber.* **40**, 401 (1947).
[5] J. F. Gerster, J. W. Jones, and R. K. Robins, *J. Org. Chem.* **28**, 945 (1963).
[6] G. B. Brown and V. S. Weliky, *J. Biol. Chem.* **204**, 1019 (1953).

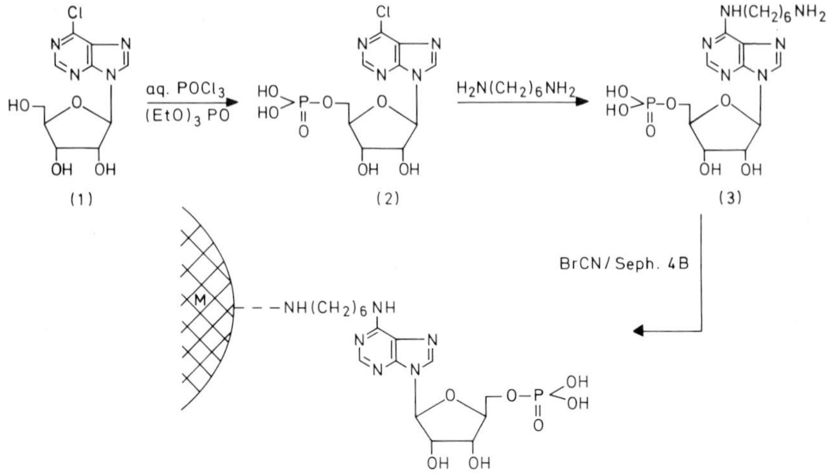

FIG. 1. Synthesis of a Sepharose-bound AMP-analog, N^6-(6-aminohexyl)-AMP (3) over 6-chloropurine riboside (1), and 6-chloropurine riboside phosphate (2).

1). The synthesis given here is an improved version[7] of the original procedure.[8]

Phosphoryl chloride is redistilled before use and triethyl phosphate is dried by filtration through a column of neutral alumina. Thin-layer chromatograms are on glass plates coated with fluorescence indicator—impregnated silica gel, 0.2 mm thick. Compounds are detected by ultraviolet light (254 nm) and by spraying with ninhydrin reagent. The following solvent systems may be used when checking the progress of the reactions: (1) methanol:ethyl acetate (1:9); R_f 6-chloropurine riboside = 0.30, 6-chloropurine riboside phosphate = 0.00; (2) ethanol:0.5 M ammonium acetate (5:2); R_f 6-chloropurine riboside phosphate = 0.60, N^6-(6-aminohexyl)-AMP = 0.30.

6-Chloropurine Riboside Phosphate

Two solutions are prepared by dissolving 6-chloropurine riboside (5.7 g, 20 mmoles) in 100 ml of triethyl phosphate and by dissolving phosphoryl chloride (5.5 ml, 60 mmoles) in a mixture of 10 ml of triethyl phosphate and 0.36 ml of water (20 mmoles). Both solutions are cooled to 0°, mixed, and stirred on a magnetic stirrer in a stoppered flask which is cooled in ice. After 10 hours about 90% of 6-chloropurine riboside has been converted. The time required for complete conversion is dependent on the amount of water added. Less water results in faster

[7] P.-O. Larsson and K. Mosbach, to be published. See also this volume [17].
[8] H. Guilford, P.-O. Larsson, and K. Mosbach, *Chemica Scripta* 2, 165 (1972).

phosphorylation, but also in an increased proportion of 2'(3')-phos-phorylated compounds.[9] The water content of triethyl phosphate will also influence the reaction time.

The brown reaction mixture is treated with 1 M LiOH to hydrolyze 5'-phosphorodichloridate (to 5'-phosphate) and excess of phosphoryl chloride. The cold solution at 0° is slowly pipetted into a magnetically stirred reaction vessel containing 100 ml of water. The vessel is kept cool on ice, and the pH is maintained at 8 by continuous addition of LiOH using an automatic titrator. After addition of the reaction mixture, the temperature is raised to 30° so as to speed up hydrolysis of 5'-phos-phorodichloridate. After 4 hours the precipitated lithium phosphates are removed by filtration, and the solution is extracted with ether (3 × 300 ml) to remove triethyl phosphate. Remaining ether is evaporated by short treatment in a flash evaporator. To remove contaminating inorganic salts that may interfere with the subsequent ion exchange chromatography step, the solution is treated with activated charcoal. Half of the reaction mixture is diluted to 500 ml, and 20 g of charcoal and 40 g of Celite are added. The pH is adjusted to 2.5 with HCl, and the mixture is stirred for 15 minutes. After filtration, the charcoal–Celite coke is washed with 1 liter of water. The nucleotide is eluted with 500 ml of a mixture of ethanol, water, and aqueous (25%) ammonia (49:49:2). The charcoal–Celite may be used again for the purification of the other half of the nucleotide solution. The combined nucleotide eluates are concentrated to 100 ml in a flash evaporator. Thin-layer chromatography generally indicates a purity of approximately 90% and a yield, based on UV measurements ($\epsilon_{264 \text{ nm}} = 8400$), of 80%.

N⁶-(6-Aminohexyl)-AMP

The solution of crude 6-chloropurine riboside phosphate (100 ml; 16 mmoles of nucleotide) is treated with a 10-fold excess of 1,6-diamino-hexane (18.6 g dissolved in 120 ml of water) at 30° for 3 hours. The reaction mixture is diluted to 1 liter and neutralized with approximately 50 g of wet Amberlite IRC 50 (H⁺) resin. The resin is removed by filtra-tion and washed with water and the solution, at a pH of 8.5, is applied to a Dowex 1X-8 column (acetate, 200–400 mesh, 5 × 45 cm).[10] The column is washed with 200 ml of water, 3 liters of 50 mM ammonium acetate at pH 7.3, 3 liters of 10 mM ammonium acetate at pH 5.5; and 3 liters of 1 mM acetic acid. These washings remove adsorbed CO_2 and nucleosides and lower the pH in the column to approximately 4.5 without eluting bound aminohexyl-AMP. A linear gradient of acetic acid is then

[9] M. Yoshikawa, T. Kato, and T. Takenishi, *Bull. Jap.* **42**, 3505 (1969).
[10] The column has sufficient capacity for batches four times as large as that given here.

applied (1 m$M \rightarrow$ 0.1 M; 6 liters). With a flow rate of 600 ml/hour and collecting 30-ml fractions, pure N^6-(6-aminohexyl)-AMP is found in fractions 65 to 140. After lyophilization of the combined fractions, 5.3 g of white powder is obtained, with an overall yield of 60% from the starting material, 6-chloropurine riboside. The pure material has a $\epsilon_{267 \, nm}^{pH \, 9.4} =$ 17,300.

Coupling of N^6-(6-Aminohexyl)-AMP to Sepharose 4B

Sepharose 4B (10 ml representing 0.4 g dry weight) is suspended in water (5 ml) and activated with a solution of cyanogen bromide (500 mg) in water (5 ml).[11] After activation for 8 minutes with pH maintained at 11 ± 0.2 with 4 M NaOH, the gel is washed with 0.1 M NaHCO$_3$ (cold) for not more than 2 minutes and is added to a cold solution of N^6-(6-aminohexyl)-AMP [200 mg (450 μmoles)] in 0.1 M NaHCO$_3$ (5 ml) (pH adjustment to about 8.9 with NaOH). The suspension is rotated gently at 4° for 24 hours. After this period, the gel is filtered and extensively washed at 4° with solutions of 0.1 M NaHCO$_3$, water, 1 mM HCl, 0.5 M NaCl, and water. After filtration on a sintered-glass filter, it is ready for use.[12] On the average, a value of 150 μmoles of nucleotide is incorporated per gram of dry gel although there is some variation (for details of this determination, see the section "Determination of Bound Nucleotide" below). The immobilized nucleotide is stored as packed gel at 4° and remains stable for several months. After repeated use as affinity chromatography material, resolution decreases but can be restored by washing the gel with 8 M urea.

Preparation of ϵ-Aminohexanoyl-NAD$^+$-Agarose

The synthesis described below is essentially identical to a previously published procedure.[13]

Sepharose 4B (200 ml, representing 8 g dry weight) is activated by the cyanogen bromide method.[11] 6-Aminohexanoic acid (20 g), dissolved in 0.1 M NaHCO$_3$ (200 ml), is added and the final pH is adjusted to 8.5 with 4 M NaOH. The gel is stirred for 15 hours at room temperature and then extensively washed with 0.1 M NaHCO$_3$, 10 mM HCl, 0.5 M NaCl, and water. The gel is removed by filtration. Eighty grams of the wet gel, corresponding to 3.9 g dry weight, is washed on a filter with a large volume of aqueous 80% (v/v) pyridine. After filtration the gel is transferred to a vessel to which are successively added NAD$^+$ (0.8 g

[11] R. Axén, J. Porath, and S. Ernback, *Nature* (*London*) **214**, 1302 (1967).
[12] The AMP analog, which has not reacted and is present in the filtrate, can be reused.
[13] P.-O. Larsson and K. Mosbach, *Biotechnol. Bioeng.* **13**, 393 (1971).

in 24 ml of water) and dicyclohexyl carbodiimide (40 g in 96 ml of pyridine). The suspension is gently agitated on a rotary shaker for 10 days at room temperature. To remove the dicyclohexylurea formed, the gel is filtered and washed successively with water, ethanol, warm (40°) butan-1-ol, ethanol, water, and aqueous 80% (v/v) pyridine; washing with methanol alone might suffice. The gel is then suspended in the first filtrate, and additional carbodiimide (10 g) is added. After one more week, the washing procedure is repeated but without the use of aqueous pyridine. In addition, the gel is washed for 30 minutes with 1 mM HCl and for at most 5 minutes with ice-cold 0.1 M NaHCO$_3$, followed by extensive washing with 0.5 M NaCl and water. All washings are carried out at room temperature. The preparations are stored as wet gels at 4° and are stable over a period of months. The amount of nucleotide bound is in the range of 50 μmoles per gram of dry polymer as determined by UV-spectra following hydrolysis of the ϵ-aminohexanoyl-NAD⁺-Sepharose preparation (details of analysis are given below under "Determination of Bound Nucleotide)."

Preparation of NAD⁺-N^6-[N-(6-Aminohexyl) acetamide]-Agarose over N^6-Carboxymethyl-NAD⁺

A general scheme of the synthesis[14] is given in Fig. 2.

N^6-Carboxymethyl-NAD⁺

NAD⁺ (2.0 g) is added to an aqueous solution of iodoacetic acid (2.0 g) neutralized with 2 M LiOH. The pH is adjusted to 6.5, and the solution with a final volume of 30 ml is kept in the dark at room temperature. The pH is checked intermittently and adjusted when necessary with 2 M LiOH. After 10 days most of the NAD⁺ has been converted to 1-carboxymethyl NAD⁺. The reaction mixture is then acidified with 3 M HCl to pH 3.5, cooled, and slowly poured into a 10-fold excess of acetone at −5°. The precipitate is collected, washed with ether, and dissolved in 50 ml of water.

To prevent hydrolysis in the subsequent rearrangement step, the nucleotide is first converted to the alkali-stable reduced form by treatment of the nucleotide solution with sodium dithionite (1 g in 100 ml of 1% NaHCO$_3$). The mixture is heated at 100° for 5 minutes and then cooled to room temperature and oxygenated for 15 minutes. To bring about the rearrangement to N^6-carboxymethyl-NADH, the pH is adjusted to 11.3 with 1 M NaOH, and the solution is heated at 75° for 1 hour.

The rearranged compound is reoxidized enzymatically to N^6-carboxy-

---

[14] M. Lindberg, P.-O. Larsson, and K. Mosbach, *Eur. J. Biochem.* **40**, 187 (1973).

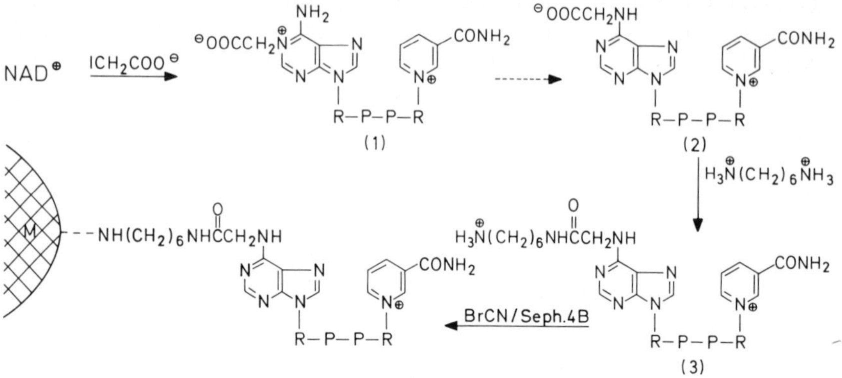

FIG. 2. Synthesis of Sepharose-bound NAD$^+$-N^6-[N-(6-aminohexyl) acetamide] (3) over 1-carboxymethyl-NAD$^+$ (1) and N^6-carboxymethyl-NAD$^+$ (2).

methyl-NAD$^+$, the form which is more stable in the acidic media used in the subsequent steps. The solution is cooled to ambient temperature, and Tris-chloride is added to give a final concentration of 0.1 M. The pH is adjusted to 7.5 with 3 M HCl, and 1 ml of acetaldehyde and approximately 1000 units of yeast alcohol dehydrogenase are added. When no further decrease in absorbance at 340 nm occurs, the oxidation is considered to be complete, and the solution is poured slowly into a 10-fold excess of acetone at $-5°$. After the preparation has stood overnight at 4°, an oily precipitate separates. The precipitate is washed with acetone and ether and dissolved in 1 liter of water followed by adjustment of pH to 7.5 with 0.1 M NaOH. The solution is applied to a Dowex 1-X8 column (Cl$^-$; 4 × 25 cm). The column is washed with 1 mM CaCl$_2$ (pH 3.0) until the effluent no longer contains UV-absorbing material and the pH has attained a value of 3. A linear CaCl$_2$ gradient is applied in which the mixing vessel contains 1 mM CaCl$_2$ (4 liters, pH 3.0) and the reservoir 50 mM CaCl$_2$ (4 liters, pH 2.0). The pooled fractions of the main peak are neutralized with Ca(OH)$_2$, concentrated in a rotary evaporator, and passed through a Sephadex G-10 column (2.5 × 90 cm). The desalted solution of N^6-carboxymethyl-NAD$^+$ is lyophilized, giving a white product, which is homogeneous in the thin-layer and electrophoresis systems tested.[14] Based on phosphate determination and UV measurements at 266 nm in Tris chloride at pH 7.5, the yield of N^6-carboxymethyl-NAD$^+$ is generally between 20 and 25%.

NAD$^+$-N^6-[N-(6-Aminohexyl) Acetamide]

N^6-Carboxymethyl-NAD$^+$ (75 mg) is dissolved in a 1 M aqueous solution of 1,6-diaminohexane dihydrochloride (5 ml). 1-Cyclohexyl-3-

(2-morpholinoethyl) carbodiimide metho-p-toluene sulfonate (300 mg) is added in portions, and the pH is maintained at 4.7 by addition of HCl. After 20 hours at 23°, thin-layer chromatography indicates almost total conversion of the N^6-carboxymethyl-NAD⁺. The reaction mixture is concentrated to 2 ml with a rotary evaporator and applied to a Sephadex G-10 column (2.5 × 90 cm). The fractions containing nucleotide are shown to be free of carbodiimide and 1,6-diaminohexane and are passed through a Dowex 1-X4 column (Cl⁻; 1 × 2 cm) to remove unreacted N^6-carboxymethyl-NAD⁺ and through a Dowex 50 W-X4 column (Na⁺; 1.5 × 1.5 cm) to remove a red impurity. In both cases elution is carried out with water and the final effluent is lyophilized. The product, NAD⁺-N^6-[N-(6-aminohexyl) acetamide], is essentially salt-free and pure by thin-layer chromatography and electrophoresis. Yields of 90% are found based on phosphate analysis and on UV data ($\epsilon_{266\ nm}^{pH\ 7.5} = 21,700$).

The coupling procedure to agarose is identical to the cyanogen bromide method given for N^6-(6-aminohexyl)-AMP except that twice the amount of BrCN per milliliter of gel is used; optimal coupling conditions have not been established for either case. The amount of NAD⁺-analog bound is approximately 45 μmoles per gram of dry gel as determined by UV spectra of the NAD⁺ analog–Sepharose preparation.

Determination of Bound Nucleotide

An important characteristic often neglected in affinity chromatography is the amount of bound ligand. This information is of value in establishing the capacity of the affinity material and also allows determination of the percentage of ligand sterically available. Three procedures suitable for the nucleotides described here have been tested.

In the first method, 0.25 g of packed gel are hydrolyzed in 0.5 M HCl (5 ml) at 100° for 15 minutes. After dilution with water to a total volume of 10 ml, the amount of nucleotide material is calculated from the absorption maxima of the adenine moieties using standard values obtained on hydrolysis of gel together with known amounts of nucleotide.[13]

Alternatively, UV spectra of the matrix-bound nucleotides are determined while the gels are kept uniformly suspended in a solution of polyethyleneglycol (Polyox WS R 301, Union Carbide). To 0.1 g of packed gel are added 2.4 ml of an aqueous 0.5% (w/v) and filtered solution of Polyox. BrCN-activated gel is used as blank. The amount of nucleotide may be calculated from ϵ_{max} since Polyox does not interfere with the nucleotide absorption.[13]

Finally, phosphate may be determined. Lyophilized nucleotide–Sepharose samples (0.5–1.25 mg) are mixed with 5 M H_2SO_4 (0.5 ml) and distilled water to a total of 1.5 ml. After heating for 3 hours at 150°, about 100 μl of H_2O_2 are added for combustion and the tubes are heated for an additional 1.5 hours. Subsequently 200 μl of 5% ammonium molybdate and 200 μl of Fiske and Subba-Row reagent[15] are added. After addition of distilled water to give 7 g, the samples are heated in a water-bath at 100° for 7 minutes and the absorbance is measured at 830 nm.[16] The amount of phosphorus in the samples may be calculated from a standard curve obtained with KH_2PO_4 in distilled water (0.07–0.24 μmole of phosphorus). Obviously, care should be taken to wash the gel free from possible contamination by phosphate buffer. The phosphate determination was tested with N^6-(6-aminohexyl)-AMP-Sepharose. The agarose component does not interfere and the linear standard curves obtained with inorganic phosphate and with free nucleotide are superimposable. In this series of analyses, nucleotide content was in the range of 0.135–0.150 μmole of AMP-analog per milligram of dry Sepharose.

Of the three procedures given, analyses based on the content of phosphorus in the samples appears to be the most reliable. Additional valuable information can be provided from complete UV spectra obtained on scanning Polyox-suspended nucleotide-Sepharose preparations.[13,14]

Specific Elution

Several elution techniques are available for affinity chromatography using a "general ligand" such as the AMP-analog and the two NAD⁺-analogs described. Examples are given here for (1) gradient elution using a cofactor[3]; (2) gradient elution using a salt solution[17]; (3) pulse elution using oxidized and reduced cofactors successively[2]; (4) elution utilizing ternary complex formation[3]; (5) elution utilizing preformed coenzyme–substrate "adducts."[18]

Other elution procedures can no doubt be devised. The following section gives a number of examples. Each type of adsorbent polymer, immobilized AMP- or NAD⁺-analogs, has been successfully applied in the

[15] C. H. Fiske and Y. SubbaRow, *J. Biol. Chem.* **66**, 375 (1925). 1-Amino-2-naphthol-4-sulfonic acid, 250 mg, is added to 100 ml of freshly prepared 15% sodium bisulfite (anhydrous), followed by addition of 1 g of sodium sulfite·7 H_2O and distilled water to give 7 ml.

[16] G. R. Bartlett, *J. Biol. Chem.* **234**, 466 (1959).

[17] C. R. Lowe, M. J. Harvey, D. B. Craven, M. A. Kerfoot, M. E. Hollows, and P. D. G. Dean, *Biochem. J.* **133**, 507 (1973). See also this volume [17].

[18] N. O. Kaplan, J. Everse, J. E. Dixon, F. G. Stolzenbach, C. Y. Lee, C. L. Lee, S. Taylor, and K. Mosbach, *Proc. Natl. Acad. Sci. U.S.* (in press).

separation and purification of nicotinamide adenine dinucleotide-dependent dehydrogenases.

It should be pointed out that in addition to their use for the isolation of specific enzymes, the general ligands applied here are also useful for a type of "group" separation. An example is provided in the removal of NAD⁺-dependent dehydrogenase impurities from a pyruvate kinase preparation.

Example 1. Gradient Elution Using a Cofactor. Elution with a weak gradient of NADH affords separation of glyceraldehyde 3-phosphate dehydrogenase from lactate dehydrogenase.[3] A mixture of 50 μg of each enzyme in 0.1 M potassium phosphate at pH 7.5, was applied to an AMP-analog-Sepharose column (100 × 5 mm, containing 2.0 g of packed gel). The enzymes were eluted with a linear gradient of 0.0–0.5 m*M* NADH in a total volume of 50 ml of the same buffer; 0.8-ml fractions were collected at a rate of 1.2 ml/hour. Glyceraldehyde-3-phosphate dehydrogenase eluted between about 0.01 and 0.1 m*M* NADH and lactate dehydrogenase between approximately 0.1 and 0.3 m*M* NADH.

The affinity of the enzymes to the immobilized AMP-analog is so high that elution by salt gradients or with high salt concentrations is not very

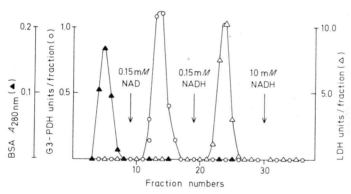

FIG. 3. Affinity chromatography of a mixture of two dehydrogenases with bovine serum albumin. The ε-aminohexanoyl-NAD⁺-Sepharose column (10.3 cm × 1.0 cm) contained about 17 μmoles of nucleotide material and was equilibrated with 0.1 M sodium phosphate at pH 7.0. A mixture containing glyceraldehyde-3-phosphate dehydrogenase (4.1 units), lactate dehydrogenase (37 units), and bovine serum albumin (0.9 mg) in the same buffer (0.2 ml) was applied and subsequent elution was carried out with 0.15 m*M* NAD⁺, 0.15 m*M* NADH, and 10 m*M* NADH, in that order. Serum albumin content was determined at A_{280}. The flowrate was 1 ml/8 minutes, and 2-ml fractions were collected. ▲, Bovine serum albumin; ○, glyceraldehyde-3-phosphate dehydrogenase activity; △, lactate dehydrogenase activity. Data from K. Mosbach, H. Guilford, R. Ohlsson, and M. Scott, *Biochem. J.* **127**, 625 (1972).

efficient whereas very weak gradients of the cofactor suffice. The resolving power of NADH gradients is also applied in the separation of the isozymes of lactate dehydrogenase discussed elsewhere in this volume.[19]

Example 2. Gradient Elution with Salt Solutions. Salt elution has been applied in a number of cases to enzymes bound to the ε-aminohexanoyl-NAD$^+$-Sepharose.[17] For example, a sample (100 µl) of the dialyzed solution of D-3-hydroxybutyrate dehydrogenase from a crude extract of *Rhodopseudomonas spheroides* was applied to a 5 mm × 20 mm column of ε-aminohexanoyl-NAD$^+$-Sepharose equilibrated with 50 mM Tris chloride at pH 8.0. Nonadsorbed proteins were washed out with 10 ml of the same buffer and the enzyme was eluted with a KCl gradient (0–0.5 M; 20 ml total volume) in 50 mM Tris chloride at pH 8.0.

The weaker affinity for the immobilized cofactor which is observed here is probably due to the attachment of the spacer to the ribose moieties of the NAD$^+$ molecule. This may allow elution with salt gradients rather than requiring the pyridine nucleotide.

Example 3. Pulse Elution with Oxidized or Reduced Cofactor. The separation of a mixture of glyceraldehyde 3-phosphate dehydrogenase and lactate dehydrogenase with weak "pulses" of 0.15 mM NAD$^+$ and 0.15 mM NADH, respectively, is shown in Fig. 3 with ε-aminohexanoyl-NAD$^+$-Sepharose. Details of the procedure are presented in the legend to the figure. The elution pattern obtained is consistent with the respective dissociation constants reported for the binary cofactor–enzyme complexes.

The same elution technique is also useful in chromatography of crude, commercially available pyruvate kinase which contains lactate dehydrogenase.[2] Crude pyruvate kinase (23 units/mg) containing lactate dehydrogenase (30 units/mg) was used with an AMP-analog-Sepharose column. The column (11 cm × 0.5 cm) was equilibrated with 0.1 M sodium phosphate at pH 7.0. After application of the crude pyruvate kinase preparation (30 µl) in buffer (0.5 ml), the column was eluted with 40 ml of the same buffer. Pyruvate kinase free from contaminating lactate dehydrogenase, appeared in the void volume. When 1 mM NADH was applied, lactate dehydrogenase was eluted. Essentially complete recovery of both enzymes was obtained.

Example 4. Elution by Ternary Complex Formation. Specific elution of one enzyme at a time can also be achieved after ternary complex formation. Thus, yeast alcohol dehydrogenase can be specifically eluted with a solution of NAD$^+$ and hydroxylamine; lactate dehydrogenase is

[19] K. Mosbach, this volume [76].

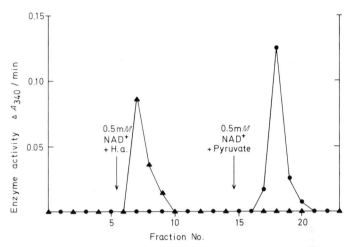

FIG. 4. Chromatography of a mixture of yeast alcohol dehydrogenase (▲—▲) and lactate dehydrogenase (●—●) using different eluent systems. A mixture of 0.1 mg of lactate dehydrogenase and 0.025 mg of alcohol dehydrogenase, dissolved in 0.5 ml of 0.1 M phosphate buffer, pH 7.5, was applied to an AMP-analog-Sepharose column (40 × 15 mm, containing 1.0 g of packed gel). Applications were made, as indicated above by the arrows, of 0.5 mM NAD⁺ plus 3 mM hydroxyl-amine (H.a.) and 0.5 mM NAD⁺ plus 5 mM pyruvate, all dissolved in the above buffer. Activities were determined for 100 μl of effluent; 2.2 ml fractions were collected at a rate of 3.3 ml/hour. Approximately 95% of the total activity of each enzyme applied was recovered. Data from R. Ohlsson, P. Brodelius, and K. Mosbach, *FEBS Lett.* **25**, 234 (1972).

eluted subsequently by a mixture of NAD⁺ and pyruvate, permitting formation of an abortive ternary complex (Fig. 4).

Purification of L-lactate dehydrogenase from a crude ox heart extract applying the above abortive ternary complex formation can also be accomplished as shown in Fig. 5. In both examples, the immobilized AMP-analog was used.

Example 5. Elution Using Dinucleotide Adducts. The preparation of homogeneous M_4-lactate dehydrogenase from crude dogfish muscle with immobilized AMP-analog, followed by elution with NAD⁺-pyruvate adduct, is summarized in the table. This procedure yields, in one step, an enzyme that can be readily crystallized and has the same specific activity as that obtained by more lengthy conventional procedures.

Evaluation of Biospecific Adsorbents and Elution
  Methods Described

Of the three nucleotide analogs described here, N^6-(6-aminohexyl)-AMP appears to have a number of advantages that make it preferable

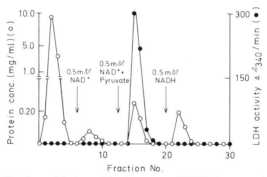

FIG. 5. Purification of lactate dehydrogenase (LDH) from ox heart by affinity chromatography. Crude extract (1.0 ml) was applied to an AMP-analog-Sepharose column (40 × 15 mm, containing 1.6 g of packed gel). The following applications were made with the various substances dissolved in 0.03 M potassium phosphate at pH 7.3 and 1 mM in cysteine: at arrow 2, 0.5 mM NAD$^+$ plus 5 mM pyruvate; and at arrow 3, 0.5 mM NADH. To ensure adequate ternary complex formation, the flow was stopped overnight for 12 hours after application of the NAD$^+$-pyruvate mixture at fraction 14, corresponding to one void volume. Total enzyme activity found per fraction of 2.3 ml and collected at a flow rate of 3.4 ml/hour is given in the figure. Data from R. Ohlsson, P. Brodelius, and K. Mosbach, *FEBS Lett.* 25, 234 (1972).

as bound ligand.[20,21] Its synthesis is relatively easily carried out and includes a spacer-arm grouping thereby eliminating the need for spacer arms on the matrix; unreacted spacer groupings on the matrix, sometimes cause nonspecific adsorption of enzymes.[22] The other adenine nucleotide substituted in the N^6-position, NAD$^+$-[N-(6-aminohexyl) acetamide], requires a lengthier synthesis, and its higher affinity for many enzymes might require more drastic elution conditions. All separations described involve the AMP-analog except for Examples 2 and 3 which use ε-aminohexanoyl-NAD$^+$. However, both ε-aminohexanoyl-NAD$^+$ and AMP-analog produce approximately equivalent elution patterns in the pulse elution experiment (Example 3).

---

[20] The N^6-(6-aminohexyl)-AMP will be commercially available from Pharmacia Fine Chemicals, Uppsala, Sweden. The synthetic procedure used is based on the one given here.

[21] Recently two analogs to N^6-(6-aminohexyl)-AMP were synthesized; these are N^6-(6-aminohexyl)-adenosine-2′,5′-diphosphate and N^6-(6-aminohexyl)-adenosine-3′,5′-diphosphate (P. Brodelius, P.-O. Larsson, and K. Mosbach, to be published). These three analogs with identical spacer substitution in the adenine moiety permit comparative binding studies as well as group separation of NAD$^+$-, NADP$^+$-, and possibly HSCoA-dependent enzymes.

[22] P. O'Carra, *in* FEBS Symposium, "Industrial Aspects of Biochemistry" (B. Spenser, ed.), p. 107. North-Holland Publ., Amsterdam, 1974. See also this volume [8].

PREPARATIVE AFFINITY CHROMATOGRAPHY OF DOGFISH M_4-LDH USING A
SEPHAROSE-BOUND AMP-ANALOG AND ELUTION WITH
REDUCED NAD⁺-PYRUVATE ADDUCT[a]

	Enzymatic activity		Protein (mg)	Specific activity (units/mg)	Purification (-fold)	Yield (%)
	(units/ ml)	(total units)				
Crude extract	178	20.470	874	23.4	1	100
AMP-analog Sepharose	640	18.560	27	696	34	91

[a] Dogfish muscle, 100 g, was minced and added to 100 ml of distilled water. After homogenization for 60 seconds in a Waring Blendor, the suspension was stirred for 30 minutes and centrifuged for 30 minutes at 29,000 g. The supernatant solution (115 ml) was applied to an AMP-analog column (6.0 × 2.5 cm) which had been previously equilibrated with 0.1 M potassium phosphate at pH 7.0. After the sample was applied, the column was washed with the phosphate buffer until the optical density of the elutant at 280 nm approached zero. The enzyme was eluted with 0.1 M potassium phosphate at pH 7.0 containing 0.12 mM of freshly prepared, reduced NAD⁺-pyruvate adduct [J. Everse, E. C. Zoll, L. Kahan, and N. O. Kaplan, *Bioorg. Chem.* **1**, 207 (1971)]. Those fractions containing enzymatic activity were pooled and concentrated in a 100-ml Diaflow cell using a PM-50 filter. Removal of the reduced pyridine nucleotide adduct by chromatography on a Sephadex G-200 column (21.0 × 2.5 cm) affords enzyme with the above specific activity. Data from N. O. Kaplan, J. Everse, J. E. Dixon, F. G. Stolzenbach, C. Y. Lee, C. L. Lee, S. Taylor, and K. Mosbach, *Proc. Natl. Acad. Sci. U.S.* (in press).

ε-Aminohexanoyl-NAD⁺ can also be applied successfully in affinity chromatography, as has been demonstrated particularly in some recent studies.[17] Its synthesis is very simple but time-consuming. However, the risk of leaving noncoupled ε-aminohexanoyl groups on the matrix is a disadvantage. In addition, the exact site and mode of attachment of the terminal carboxyl group in the ε-aminohexanoic acid-substituted Sepharose to NAD⁺ is not definitely established. The carbodiimide used as catalyst makes it probable that the ribose hydroxyls of the pyridine nucleotide are esterified, but some amide formation with the amino group in the adenine cannot be excluded. The risk of a heterogeneous preparation should therefore not be overlooked. Nevertheless, since inexpensive salt gradients are often sufficient for the elution of enzymes bound to such ligands, the ε-aminohexanoyl-NAD⁺-Sepharose is a useful alternative preparation, which, with care in interpretation, has its place in affinity chromatography.

NAD⁺-N^6-[N-(6-aminohexyl) acetamide] has only recently been synthesized and, in one case, applied in the separation of alcohol dehydrogenase from lactate dehydrogenase.[14] Further tests are needed before its value as a biospecific adsorbent can be evaluated. It may, contrary to the AMP-analog, be able to discriminate between enzymes dependent on NAD⁺ and those affected merely by AMP. All three cofactor-analogs described above have been bound to Sepharose which fulfills many of the prerequisites of a matrix suitable for affinity chromatography. This is underlined by the fact that other workers in this field have now turned from a cellulose matrix[23] to Sepharose for the binding of NAD⁴ following exactly the procedure given above for the preparation of ε-aminohexanoyl-NAD⁺-Sepharose.[17]

In this context it may be mentioned that both ε-aminohexanoyl-NAD⁺ and NAD⁺-[N-(6-aminohexyl) acetamide] are of interest because of their potential applications as active immobilized coenzymes.

[23] C. R. Lowe and P. D. G. Dean, *FEBS Lett.* **14**, 313 (1971).

[17] N^6-Immobilized 5′-AMP and NAD⁺: Preparations and Applications

By MICHAEL J. HARVEY, DAVID B. CRAVEN, CHRISTOPHER R. LOWE, and PETER D. G. DEAN

The use of group-specific adsorbents, in particular immobilized cofactors, permits the examination of a whole range of enzymes in a single affinity system. The chemical complexity of polyfunctional cofactors has, until recently, restricted the potential methods available for their immobilization. Nevertheless several preparations of chemically defined polymers are now documented.[1-8] The subject of this article is the preparation of aminoalkyl nucleotides suitable for immobilization to Sepharose.

Preparation of N^6-(6-Aminohexyl) 5′-AMP

The procedure described below involves the nucleophilic displacement of the thiol group from 6-mercaptopurine riboside 5′-monophosphate by 1,6-hexanediamine.[9] The 6-mercaptopurine riboside 5′-monophosphate

[1] C. R. Lowe and P. D. G. Dean, *FEBS Lett.* **14**, 313 (1971).

[2] P. D. G. Dean and C. R. Lowe, *Biochem. J.* **127**, 11P (1972).

[3] M. K. Weibel, H. H. Weetall, and H. J. Bright, *Biochem. Biophys. Res. Commun.* **44**, 347 (1971).

[4] J. D. Hocking and J. I. Harris, *Biochem. J.* **130**, 24P (1972).

[5] R. Barker, K. W. Olsen, J. H. Shaper, and R. L. Hill, *J. Biol. Chem.* **247**, 7135 (1972).

[6] O. Berglund and F. Eckstein, *Eur. J. Biochem.* **28**, 492 (1972).

[7] R. Lamed, Y. Levin, and M. Wilchek, *Biochim. Biophys. Acta* **304**, 231 (1973).

[8] H. Guilford, P.-O. Larsson, and K. Mosbach, *Chem. Scripta* **2**, 165 (1972).

is available commercially and as a by-product of *Brevibacterium am-moniagenes* metabolism.[10] Alternative methods for the preparation of N^6-(6-aminohexyl) 5'-AMP are cited in this volume.[11,12]

Procedure[13]

1,6-Hexanediamine, 200 mg, is added to an aqueous solution (0.6 ml) of 6-mercaptopurine riboside 5'-monophosphate (10 mg of the sodium salt) in a glass ampule (155 × 15 mm). The ampule is sealed and heated at 85–95° for 20 hours in an aluminum heating block. The reaction is surface dependent: optimal conditions are achieved by holding the ampule in a horizontal position during heating. The progress of the reaction is followed by measuring either the increased absorbance at 267 nm or the concomitant decrease at 309 nm. Quantitative conversion is usually effected in 16 hours. After cooling, the sample is diluted with 10 ml of water, and excess hexanediamine is removed by repeated lyophil-ization. The straw-colored residue is dissolved in 3 ml of water and applied at pH 10 to a column (1 × 20 cm) of Dowex 1-X2 (formate) resin, 200–400 mesh, at room temperature. An unidentified yellow impurity is eluted with 50 ml of 0.1 M ammonium formate at pH 7.0 prior to development of the column in a linear gradient of formic acid (0 to 1 M, 200 ml total volume). The substituted nucleotide is eluted at approximately 0.2 M formic acid. Following lyophilization, N^6-(6-amino-hexyl) 5'-AMP, yield approximately 80%, may be coupled to Sepharose.[14]

The 6-aminohexyl derivatives of adenine and adenosine can be prepared from 6-mercaptopurine and 6-mercaptopurine riboside, respectively, by the above procedure.[13] Furthermore, a homologous series of N^6-ω-aminoalkyl-AMP derivatives can be prepared from the corresponding α,ω-alkane diamines.[15] Using the imidazolide procedure described elsewhere in this volume,[12] the corresponding N⁶-substituted ADP and ATP can be synthesized from the N^6-ω-aminoalkyl-AMP derivatives.

Preparation of N^6-(6-Aminohexyl)-NAD⁺

Under the action of trifluoroacetic anhydride, mononucleotides will condense to the corresponding dinucleotide by a process involving anhydride exchange.[16–18]

[9] M. Saneyoshi, *Chem. Pharm. Bull.* **19**, 493 (1971).
[10] H. Tanaka and K. Nakayama, *Agr. Biol. Chem.* **36**, 464 (1972).
[11] K. Mosbach, this volume [16].
[12] R. Barker, I. P. Trayer, and R. L. Hill, this volume [56].
[13] D. B. Craven, M. J. Harvey, C. R. Lowe, and P. D. G. Dean, *Eur. J. Biochem.* **41**, 329 (1974).
[14] R. Axén, J. Porath, and S. Ernbäck, *Nature (London)* **214**, 1302 (1967).
[15] M. C. Hipwell, M. J. Harvey, and P. D. G. Dean, *FEBS Lett.* in press (1974).

Procedure[19]

A mixture of 45 mg of lyophilized NMN, 45 mg of N^6-(6-aminohexyl) 5'-AMP, and 0.5 ml of trifluoroacetic anhydride are allowed to react in a sealed test tube at room temperature. After 16 hours, volatile components of the reaction are removed under reduced pressure. The viscous light-brown residue is washed copiously with ether, and dried under reduced pressure over KOH/P_2O_5 in a vacuum desiccator. This straw-colored residue is dissolved in a minimal volume of water and applied at pH 7.0, adjusted with KOH, to a column (1.5 × 90 cm) of Sephadex G-15. The 260 nm absorbing fraction is eluted with water and the pH is adjusted to 8.0 prior to applying the sample to a column (1.2 × 15 cm) of Dowex 1-X2 (formate) resin, 200–400 mesh. After washing with 30 ml of water, the resin is developed with a linear gradient of formic acid (0 to 1 M, 300 ml total volume) collecting 2.5-ml fractions. N^6-(6-Aminohexyl)-NAD+, assayed with alcohol dehydrogenase, dinicotinamide dinucleotide and an unidentified N^6-(6-aminohexyl)adenosine nucleotide are located in the water effluent. This fraction is lyophilized, and a sample is applied at a position 10 cm from the anode on a 40-cm sheet of Whatman 3 MM paper. Electrophoresis in acetic acid:formic acid:H_2O (15:5:80, v/v), pH 1.9, at 60 V/cm for 2 hours achieves the desired separation of N^6-(6-aminohexyl)-NAD+ from the contaminating nucleotides (Table I). An additional coenzymatically active fraction, probably N^6-(6-trifluoracetyl amidohexyl)-NAD+, is eluted from the Dowex 1 column by 0.4 M formic acid. This fraction contains approximately 30% of the total coenzymatically active material applied to the column.

The yield of N^6-substituted NAD+, based on the initial N^6-(6-aminohexyl) 5'-AMP concentration, is in the region of 25%. No significant difference in the yield is apparent when dinicotinamide dinucleotide is used as the starting material. NAD+ can be used directly in this synthesis although some considerable difficulty is experienced in separating the reaction products.

Properties of N^6-Substituted 5'-AMP and NAD+

The physical and chemical properties of aminoalkyl 5'-AMP are well documented.[8,11–13] The purity of aminoalkyl-NAD+ is determined by measurement of the absorption at 338 nm of the enzymatically reduced

[16] S. M. H. Christie, D. T. Elmore, G. W. Kenner, A. R. Todd, and F. J. Weymouth, *J. Chem. Soc.* 2947 (1953).
[17] L. Shuster, N. O. Kaplan, and F. E. Stolzenbach, *J. Biol. Chem.* **215**, 195 (1955).
[18] A. M. Stein, *FEBS Lett.* **19**, 270 (1971).
[19] D. B. Craven, M. J. Harvey, and P. D. G. Dean, *FEBS Lett.* **38**, 320 (1974).

TABLE I
CHROMATOGRAPHIC AND ELECTROPHORETIC MOBILITIES OF RELEVANT NUCLEOTIDES

Nucleotide	Chromatographic R_f values[a]							Electrophoretic mobility M_L[b]
	I	II	III	IV	V	VI	VII	
NAD⁺	0.40	0.05	0.48	0.04	0.04	0.42	0.02	16
5'-AMP	0.30	0.20	0.54	0.02	0.16	ND	0.19	ND
NMN	0.82	0.13	0.30	0.04	0.16	0.57	0.18	16
Dinicotinamide dinucleotide	0.70	0.03	0.20	0.01	0.02	0.36	0.03	16
N^6-(6-Aminohexyl) 5'-AMP	0.33	0.23	0.38	0.04	0.30	0.71	0.18	100
N^6-(6-Aminohexyl)-NAD⁺	0.48	0.04	0.40	0.03	0.08	0.53	0.07	80
N^6-(6-Trifluoracetyl amidohexyl)-NAD⁺	0.19	ND	ND	ND	0.39	0.86	ND	14

[a] The solvent systems (v/v) for thin-layer chromatography on aluminum plates precoated with cellulose (0.1 mm) were: (I) $(NH_4)_2SO_4$ (70 g in 100 ml of H_2O): 0.1 M NaH_2PO_4, pH 6.0:isopropanol (79:19:2); (II) isopropanol:1% $(NH_4)_2SO_4$ (2:1); (III) pyridine:H_2O (2:1); (IV) n-butanol:pyridine:H_2O (6:4:3); (V) ethanol:1 M acetic acid, pH 3.8 with 1 M NH_4OH (5:2); (VI) ethanol:acetic acid:H_2O (174:1:173); and (VII) n-butanol:acetic acid:H_2O (5:3:2).

[b] Mobility relative to N^6-(6-aminohexyl) 5'-AMP which moved 22 cm toward the cathode; the M_L value recorded for the unidentified N^6-(6-aminohexyl)adenosine nucleotide was 122. ND, not determined.

product. Identification can also be established by the ultraviolet spectra in the presence of cyanide or sulfite.[19] The chromatographic and electrophoretic mobilities of N⁶-substituted 5'-AMP and NAD⁺ are given in Table I.

Some Parameters Relevant to Affinity Chromatography of Enzymes on Immobilized Nucleotides

The general considerations required in the preparation of adsorbents for affinity chromatography have been presented in a previous volume[20] as well as throughout this one. The parameters discussed below refer specifically to the adsorption and desorption phases of affinity chromatography and are intended to illustrate how by careful manipulation of the experimental conditions optimal separations of several enzymes may be achieved using group-specific adsorbents. The two indices used in this study to determine the effectiveness of an affinity adsorbent are (1) capacity and (2) strength of the ligand–macromolecule interaction; in this context β refers to that concentration of the eluate required to elute the macromolecule with maximal activity.

[20] P. Cuatrecasas and C. B. Anfinsen, this series, Vol. 22, p. 345.

Selection of Ligand

The interaction of dehydrogenases and kinases with nucleotides and nucleotide analogs permits a wide range of group-specific adsorbents to be used in the separation of a specific enzyme from a complex mixture of enzymes. In general, it is found that the specificity of an affinity matrix increases as the composition of the immobilized ligand approaches that of the natural cofactor of the enzyme under examination.[19,21] Thus the interaction of NAD+-dependent dehydrogenases with aminohexyl-NAD+ is tighter than that observed with aminohexyl 5'-AMP. This is particularly

TABLE II
ENZYME BINDING TO IMMOBILIZED NUCLEOTIDE COLUMNS[a] AT pH 7.5

Enzyme	Binding (β)				
	I	II	III	IV	V
Alcohol dehydrogenase (yeast)	700	400	0	415	0
Lactate dehydrogenase (pig heart)	0.35[b]	>1.0 M^c	>1.0 M^c	325	0
Lactate dehydrogenase (rabbit muscle)	0.80[b]	>1.0 M^c	>1.0 M^c	500	110
L-Malate dehydrogenase	300	65	490	45	0
Isocitrate dehydrogenase	0	0	NT	220	230
D-Glucose 6-phosphate dehydrogenase	140	0	170	125	125
D-Glyceraldehyde 3-phosphate dehydrogenase	1.0[b]	0	>1.0 M^c	>1.0 M^c	60
Glutathione reductase	75	RT	NT	120	400
Hexokinase	0	0	0	100	NT
Glycerokinase	RT[d]	122	0	NT	NT
Pyruvic kinase	100	100	110	0	NT
3-Phosphoglycerate kinase	NT[e]	70	260	NT	NT
Creatine kinase	NT	0	0	0	NT
Myokinase	50	0	380	RT	NT

[a] Columns (5 × 50 mm) were prepared containing 1 g of the following adsorbents: (I) N^6-(6-aminohexyl)-NAD+-Sepharose (2.0 μmoles/ml); (II) N^6-(6-aminohexyl) 5'-AMP-Sepharose (1.5 μmoles/ml); (III) P^1-(6-aminohexyl)-P^2-(adenosine)-pyrophosphate-Sepharose (6.0 μmoles/ml); (IV) ε-aminohexanoyl-NAD+-Sepharose (2.0 μmoles/ml); (V) ε-aminohexanoyl-NADP+-Sepharose (2.0 μmoles/ml).
[b] Elution could not be effected by 1 M KCl, enzyme eluted in NADH gradient (0–5 mM, 10 ml total volume).
[c] Elution effected by a 200 μl pulse of 5 mM NADH in 1.0 M KCl.
[d] RT, Enzyme retarded by the adsorbent and eluted after the void volume as determined by bovine serum albumin.
[e] NT, Not tested.

[21] C. R. Lowe, K. Mosbach, and P. D. G. Dean, *Biochem. Biophys. Res. Commun.* 48, 1004 (1972).

apparent with glyceraldehyde-3-phosphate dehydrogenase, which shows no affinity for the latter adsorbent (Table II). Similarly NAD$^+$-dependent dehydrogenases will bind strongly to the ill-defined aminohexanoyl-NAD$^+$ matrix and weakly, if at all, to the corresponding NADP$^+$ matrix. NADP$^+$-dependent enzymes either do not discriminate between these two matrices or bind preferentially to the NADP$^+$ matrix.[21] Thus by consideration of the data in Table II, a suitable affinity adsorbent can be selected for the purification of a specific NAD$^+$- or NADP$^+$-dependent dehydrogenase.

Immobilization of Ligand

The biological specificity of an enzyme–ligand interaction demands that the ligand must be immobilized in a manner which does not impair recognition at the enzyme active site. Thus immobilization of a ligand common to a group of enzymes in several different ways has two inherent advantages. First, the purification of an individual enzyme will be enhanced since each enzyme will not recognize the same portion of the ligand equally, and, second, some insight into the nature of each individual enzyme–ligand interaction can be obtained.

Two affinity adsorbents prepared by immobilizing 5'-AMP are illus-

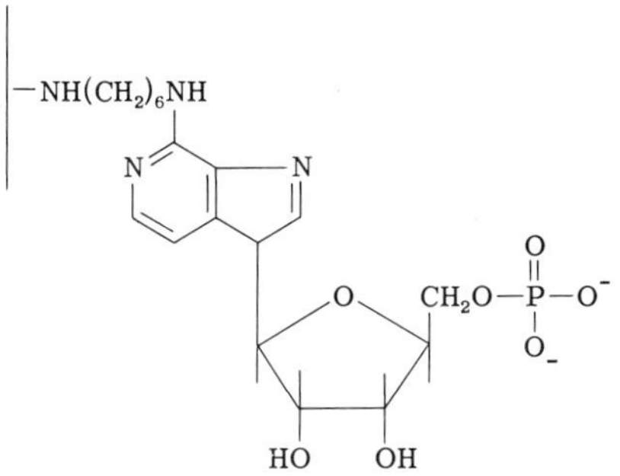

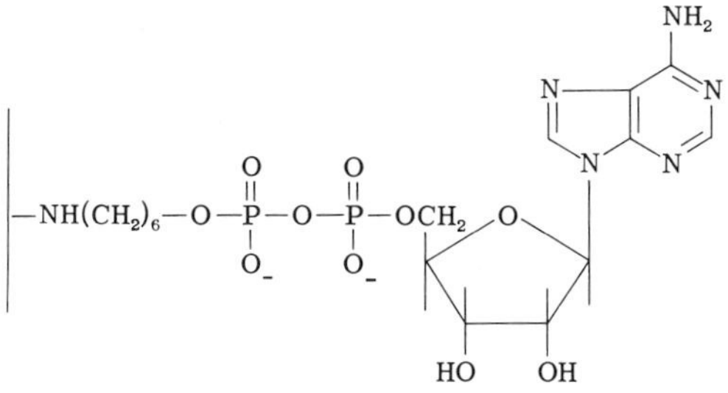

FIG. 1. Structure of immobilized 5'-AMP adsorbents. (I) N^6-(6-Aminohexyl) 5'-AMP-Sepharose; (II) P^1-(6-aminohexyl)-P^2-(adenosine)-pyrophosphate-Sepharose.

trated in Fig. 1. The immobilization procedure for the preparation of P^1-(6-aminohexyl)-P^2-(adenosine)-pyrophosphate-Sepharose leaves the adenosine moiety of the nucleotide free,[5] whereas the phosphate group is available for enzymatic interaction when 5'-AMP is linked through the N^6 of the adenine.[8,13] From the data in Table II it is seen that a free phosphate group is required for the binding of yeast alcohol dehydrogenase and glycerokinase, whereas the adenine moiety appears to be the vital portion of the nucleotide involved in the interaction of 5'-AMP with myokinase and glyceraldehyde-3-phosphate dehydrogenase. It should be noted, however, that interpretation of these data is limited owing to the different ligand concentrations of the two matrices (see below) and the involvement of an additional phosphate group in the immobilization of 5'-AMP by the procedure of Barker *et al.*[5]

Another effect of immobilization procedure is apparent when the interaction of enzymes with aminohexyl- and aminohexanoyl-NAD$^+$ is compared. With the latter polymer the ill-defined nature of the acylated-NAD$^+$ matrix apparently reduces the binding constant (β) for the NAD$^+$-dependent enzymes.

Ligand Concentration

The data in Table II imply that the binding of malate dehydrogenase to aminohexyl 5'-AMP is less than that to the P^2-immobilized ADP. However, when the concentration of immobilized ligand in the former polymer is increased to 4 μmoles per milliliter the binding (β) increases to 430 mM KCl.[22] Likewise, it would appear that the strength of the lactate dehydrogenase interaction with these two adsorbents is very similar. However, by suitably diluting the ligand concentration of these adsorbents it can be shown that the M_4 species has a preference for the N^6-substituted 5'-AMP (Table III). Two methods are available for preparing adsorbents of varying ligand concentration: (1) direct coupling of the required ligand concentration to the inert matrix and (2) dilution of an existing polymer with unsubstituted Sepharose. These methods result in two different types of ligand distribution through the gel, which, in turn, affects the apparent capacity of the adsorbent (Table III). The effect of ligand dilution with unsubstituted Sepharose on the binding characteristics of myokinase to P^1-(6-aminohexyl)-P^2-(adenosine)-pyrophosphate-Sepharose is presented in Table IV as an example of how low ligand concentration adsorbents can be utilized to facilitate elution of tightly bound enzymes.[22,23]

[22] M. J. Harvey, C. R. Lowe, D. B. Craven, and P. D. G. Dean, *Eur. J. Biochem.* **41**, 335 (1974).

[23] C. R. Lowe, M. J. Harvey, D. B. Craven, and P. D. G. Dean, *Biochem. J.* **133**, 499 (1973).

TABLE III

CAPACITY OF NUCLEOTIDE ADSORBENTS IN RELATION TO IMMOBILIZATION AND
DILUTION PROCEDURES WITH LACTATE DEHYDROGENASES[a]

Lactate dehydrogenase species:	H₄		M₄	
Method of immobilization:	N^6-(6-Aminohexyl) 5′-AMP[b]		N^6-(6-Aminohexyl) 5′-AMP[b]	P^2-ADP
Dilution method[a]:	I	II	II	II
Ligand concentration (nmoles 5′-AMP/ml)	Percent bound			
250	100	100	100	95
125	100	78	100	65
75	93	77	91	47
25	11.5	31	47	18
12.5	5.5	18	35	12
7.5	2.5	7	22	5

[a] I: ligand uniformly coupled to Sepharose; II: ligand diluted with unsubstituted Sepharose.
[b] D. B. Craven, M. J. Harvey, C. R. Lowe, and P. D. G. Dean, *Eur. J. Biochem.* **41**, 329 (1974).

TABLE IV

EFFECT OF LIGAND CONCENTRATION ON THE INTERACTION OF MYOKINASE WITH
P^1-(6-AMINOHEXYL)-P^2-(ADENOSINE)-PYROPHOSPHATE-SEPHAROSE[a]

Ligand concentration (μmoles 5′-AMP/ml)	Binding (β) (mM KCl)	Amount bound (%)	Enzyme units bound	
			Per gram of adsorbent	Per micromole of 5′-AMP
6.00	400	100	4.00	0.67
3.00	315	100	4.00	1.33
1.00	200	97.4	3.90	3.90
0.50	185	90.4	3.62	7.24
0.30	140	83.3	3.33	11.10
10.0	95	42.5	1.69	16.80
0.05	70	27.0	1.08	21.55
0.03	50	14.7	0.59	19.60

[a] The enzyme sample (100 μl), myokinase (4 U), and bovine serum albumin (1.5 mg) were applied to a column (5 × 50 mm) containing 1 g of P^1-(6-aminohexyl)-P^2-(adenosine)-pyrophosphate-Sepharose (moist weight) diluted to the appropriate ligand concentration with unsubstituted Sepharose 4B.

Column Geometry and Dynamics

The strength of the interaction (β) and the capacity of an immobilized-nucleotide adsorbent are determined by three fundamental interrelated parameters: (1) ligand concentration, (2) total amount of ligand, and (3) column geometry.[24] As the ligand concentration is decreased, the significance of column geometry increases; this effect is particularly pronounced on interacting systems of relatively low affinity. At a fixed concentration of ligand a linear relationship exists between binding (β) and the column length when the total amount of ligand is proportional to the column length (Table V). However, when the ligand concentration is varied inversely to the length, at a constant total amount of ligand, the binding (β) is inversely related to the column length. When both the ligand concentration and total amount of ligand are kept constant the binding (β) approaches a limiting value as the column length is increased. Thus, for reproducible purifications, these parameters need to be strictly controlled. The optimal column size will depend upon the enzyme under examination,[24] the adsorbent used, and the elution technique employed. The resolution and strength of interaction can also be enhanced by increasing the enzyme–ligand equilibration time, particularly under column conditions at low ligand concentrations. Under batchwise adsorption conditions the capacity of an adsorbent is related to the total immobilized ligand and also the enzyme concentration and equilibration time.[25]

Resolution of Enzyme Mixtures

A serious limitation in the use of group-specific ligands for affinity chromatography is their reduced selectivity, and thus the need exists for additional methods of resolving the complex mixtures of enzymes that can be adsorbed. Generally, for economic reasons, elution is achieved with an ionic gradient. However, in some instances desorption of an enzyme requires the application of specific elution procedures, such as the presence of the complimentary nucleotide,[21,25-27] second substrate,[26,28] or a competitive inhibitor.[25,26] In these circumstances separation from the small molecular weight eluent is required prior to kinetic studies. Other characteristics of enzymatic reactions can be utilized to increase the

[24] C. R. Lowe, M. J. Harvey, and P. D. G. Dean, *Eur. J. Biochem.* **41,** 341 (1974).
[25] C. R. Lowe, M. J. Harvey, and P. D. G. Dean, *Eur. J. Biochem.* **42,** 1 (1974).
[26] K. Mosbach, H. Guilford, R. Ohlsson, and M. Scott, *Biochem. J.* **127,** 625 (1972).
[27] M. J. Harvey, C. R. Lowe, and P. D. G. Dean, *Eur. J. Biochem.* **41,** 353 (1974).
[28] C. R. Lowe, M. J. Harvey, D. B. Craven, M. A. Kerfoot, M. E. Hollows, and P. D. G. Dean, *Biochem. J.* **133,** 507 (1973).

TABLE V

INFLUENCE OF COLUMN GEOMETRY ON THE BINDING OF ENZYMES TO N^6-(6-AMINOHEXYL) 5'-AMP-SEPHAROSE

Constants:	Ligand concentration (125 nmoles/ml) Diameter (6 mm)		Total ligand (300 nmoles) Diameter (6 mm)			Ligand concentration (1.5 μmoles/ml) Total ligand (3 μmoles)		
				β (mM KCl)			β (mM KCl)	
Variables: Column length (mm)	Total ligand (nmoles)	β (mM KCl) Lactate dehydrogenase-H₄	Ligand concentration (nmoles/ml)	Glycerokinase	Lactate dehydrogenase-H₄	Diameter (mm)	Glycerokinase	Alcohol dehydrogenase
6	20.5	110	1500	105	>1.0 M	23	100	110
14	47.7	128	750	85	>1.0 M	13.5	190	275
29	99	282				9	272	510
44	150	364	250	60	980			
62						6	330	585
88	300	672	125	45	720			
147			75	25	560			

applicability of these adsorbents to the purification of enzymes from complex mixtures.

pH. The use of pH-jump elution in affinity chromatography is well documented.[20,29-31] Providing that the enzymes under examination are stable within the experimental pH range, complex enzyme mixtures can be separated on immobilized-nucleotide adsorbents using pH gradients.[31] The elution pH values of several enzymes from aminohexyl-5'-AMP-Sepharose are documented in Table VI. Noteworthy among these results is the adsorption of glucose-6-phosphate dehydrogenase to this adsorbent (cf. Table I); and the desorption of lactate dehydrogenase-H_4, which is normally eluted only by a nucleotide pulse. Hence pH gradients may be particularly useful in effecting the desorption of tightly bound enzymes. The advantage of this technique is that it exploits a characteristic property of the enzyme–nucleotide interaction that is distinct from the overall protein charge, ionic strength, or temperature.

Temperature. The strength of the enzyme-immobilized ligand interaction is greatly reduced at increased temperatures.[27] However, prolonged exposure of such adsorbents to elevated temperatures does not alter their subsequent chromatographic properties at lower temperatures. Thus temperature gradients can be employed to resolve complex enzyme mixtures on immobilized-nucleotide adsorbents (Table VI). Also by varying the temperature of adsorption it is possible to effect a purification without

TABLE VI

RESOLUTION OF ENZYMES ON N^6-(6-AMINOHEXYL) 5'-AMP-SEPHAROSE
BY TEMPERATURE AND pH GRADIENTS

	Elution value	
Enzyme	pH	Temperature at pH 7.5
Lactate dehydrogenase (pig heart)	9.7	40°[a]
Alcohol dehydrogenase (yeast)	8.5	27°
D-Glucose-6-phosphate dehydrogenase	7.2	Void[b]
L-Malate dehydrogenase	9.8	NT[b]
Glycerokinase	8.0	20°
Hexokinase	Void	Void
Bovine serum albumin	Void	Void

[a] Elution effected by a 200-μl pulse, 5 mM NADH at 40°.

[b] NT, Not tested; void, protein not retained by adsorbent at pH 6.0 and/or 0.5°.

[29] P. Cuatrecasas, M. Wilchek, and C. B. Anfinsen, *Proc. Nat. Acad. Sci. U.S.* **61**, 636 (1968).

[30] E. Steers, P. Cuatrecasas, and H. Pollard, *J. Biol. Chem.* **246**, 196 (1971).

[31] C. R. Lowe, M. J. Harvey, and P. D. G. Dean, *Eur. J. Biochem.* **41**, 347 (1974).

any further development of the adsorbent. For example, glycerokinase and lactate dehydrogenese can be resolved by applying the enzyme mixture at 23°: this resolution can be reversed by reequilibrating the adsorbent at 4°.[27] The advantage of this procedure is that the eluted enzyme can be employed directly for kinetic studies. Temperature, being itself an elution method will obviously affect all other methods of elution, and therefore needs to be carefully controlled during all chromatographic stages.

[18] ATP- and dATP-Substituted Agaroses and the Purification of Ribonucleotide Reductases[1]

By O. BERGLUND and F. ECKSTEIN

Adenosine 5'-triphosphate and 2'-deoxyadenosine 5'-triphosphate are substrates or effectors, respectively, for a number of enzymes. To make use of the affinity of a protein for such a nucleoside 5'-triphosphate in order to purify it by affinity methods, a suitable derivative must be synthesized which allows covalent linkage to a polymer support, ATP itself is also claimed to have been coupled to Sepharose.[2] Various derivatives have been synthesized and used for the purification of enzymes. The dialdehyde obtained by oxidation of GTP[3] or ATP[4] with $NaIO_4$ has been coupled via a hydrazide to Sepharose and used for the purification of D-erythrodihydroneopterin triphosphate synthetase[3] and binding of heavy meromyosin[5] (Fig. 1).

The N^6-ε-aminocaproyl derivative of ATP as well as of adenylylimi-

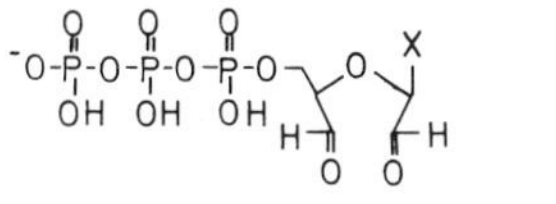

X = adenine or guanine

FIG. 1. Oxidation product of GTP and ATP.

[1] O. Berglund and F. Eckstein, *Eur. J. Biochem.* **28**, 492 (1972).

[2] E. A. Wider De Xifra, S. Mendiara, and A. M. del C. Battle, *FEBS Lett.* **27**, 275 (1972).

[3] R. J. Jackson, R. M. Wolcott, and T. Shiota, *Biochem. Biophys. Res. Commun.* **51**, 428 (1973).

[4] R. Lamed, Y. Levin, and M. Wilchek, *Biochim. Biophys. Acta* **304**, 231 (1973). See this volume [55].

[5] R. Lamed, Y. Levin, and A. Oplatka, *Biochim. Biophys. Acta* **305**, 163 (1973).

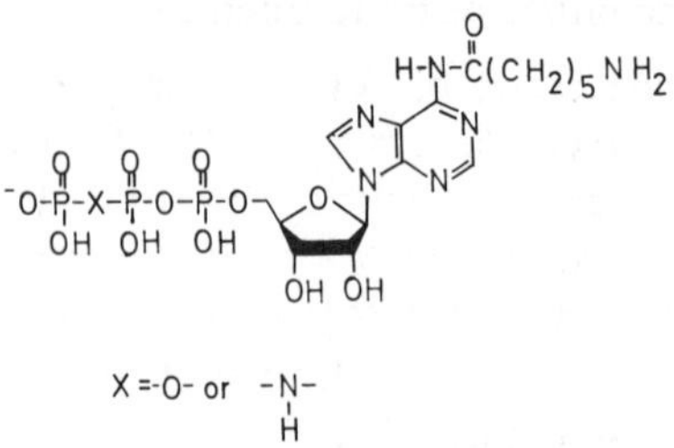

FIG. 2. N^6-ϵ-Aminocaproyl derivative of ATP.

dodiphosphate (AMPPNP) (Fig. 2) has been synthesized and coupled to Sepharose.[6]

However, this last ATP-derivative was not useful for the purification of either tRNA nucleotidyltransferase[7] or S_1-subfragment of myosin.[6] As another analog of ATP, 6-chloropurine 5'-triphosphate has been coupled to aminoethyl cellulose.[8] Esterification of the terminal phosphate group of ATP and dATP by p-aminophenol (Fig. 3), and subsequent coupling to Sepharose did not lead to a matrix suitable for the purification of tRNA nucleotidyltransferase from yeast[7,9] or, with the corresponding derivative of ADP, for the purification of polynucleotide phosphorylase from *Micrococcus luteus*.[7] However, as described below, the dATP-derivative was ideally suited for the purification of both the T4-induced ribonucleotide reductase[1,10] and the B1 subunit of *Escherichia coli* ribonucleotide reductase.[11,12]

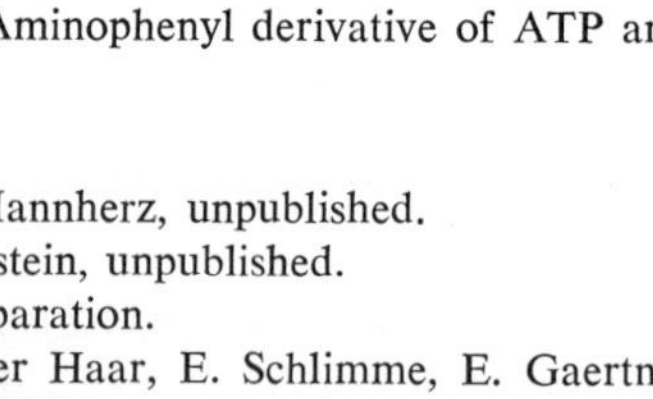

X = -H or -OH

A = adenine

FIG. 3. p-Aminophenyl derivative of ATP and dATP.

[6] F. Eckstein and H. J. Mannherz, unpublished.

[7] D. Warnock and F. Eckstein, unpublished.

[8] P. V. Sundaram, in preparation.

[9] H. Sternbach, F. von der Haar, E. Schlimme, E. Gaertner, and F. Cramer, *Eur. J. Biochem.* **22**, 166 (1971).

[10] O. Berglund, *J. Biol. Chem.* **247**, 7270 (1972).

[11] J. A. Fuchs, H. O. Karlström, H. R. Warner, and P. Reichard, *Nature (London) New Biol.* **238**, 69 (1972).

[12] L. Thelander, *J. Biol. Chem.* **248**, 4591 (1973).

Purification of T4 Ribonucleotide Reductase

The p-aminophenyl esters of ATP and dATP proved to be positive effectors for T4 ribonucleotide reductase. The K_m values in the reduction of CDP were 300 μM for the ATP derivative and 6 μM for the dATP derivative as compared to 70 μM for ATP and 0.2 μM for dATP.[13]

Columns for ATP- and dATP-Sepharose were used to purify T4 ribonucleotide reductase from a crude extract of T4-infected cells (Table I) which had been passed through a Sephadex G-25 column to remove nucleotides. The enzyme was eluted with ATP from the ATP column and with dATP from the dATP column; 10-fold purification with 25% yield, and 90-fold purification with 88% yield, respectively, were obtained.

Polyacrylamide gel electrophoresis of the purified material in the presence of dodecyl sulfate showed the presence of about 15 bands in the material eluted from the ATP-Sepharose column, but only about four protein bands eluted from the dATP-Sepharose column. When introduced at a later stage in the purification of T4 ribonucleotide reductase, i.e., after treatment of the extract with streptomycin sulfate and ammonium sulfate, chromatography on dATP-Sepharose yielded only two bands, α and β, which represent the two subunits of the reductase, the catalytically active subunit structure being $\alpha_2\beta_2$.[10]

Even from a crude extract the enzyme could be obtained in rather pure form by affinity chromatography. A densitometer tracing of a gel indicated only about 30% contamination.

T_4 ribonucleotide reductase was bound tightly to dATP-Sepharose and could not be desorbed by moderate changes in ionic strength and pH. Thus, 1 M potassium phosphate at pH 7.0, 1 M NaCl, 0.5 M MgCl$_2$ and different buffers in the range of pH 6.5 to 9.5 did not elute the

TABLE I

PURIFICATION OF T4 RIBONUCLEOTIDE REDUCTASE[a]

| | | Enzyme activity | |
Fraction	Protein (mg)	Total (units)	Specific (units/mg)
Crude extract, Sephadex G-25	85	5700	67
ATP-Sepharose eluate	2.2	1400	650
dATP-Sepharose eluate	0.8	5000	6200

[a] Taken with permission of the copyright holder from O. Berglund and F. Eckstein, *Eur. J. Biochem.* **28**, 492 (1972).

[13] O. Berglund, *J. Biol. Chem.* **247**, 7276 (1972).

enzyme. However, when various nucleotides were tried as eluents, we found that, in addition to dATP, ATP and dAMP were also effective. A comparison of elution patterns obtained with gradients of these three nucleotides is given in Fig. 4.

The concentrations of dATP (0.1 mM), ATP (3 mM), and dAMP (10 mM) required to desorb T4 ribonucleotide reductase from dATP-Sepharose were roughly proportional to the concentrations required to stimulate the enzyme (K_m for dATP = 0.2 μM, K_m for ATP = 70 μM, K_m for dAMP = 1000 μM).[13] The enzyme was not desorbed by two other effectors, dCTP (5 mM) and dTTP (5 mM), nor by its substrates, CDP (1 mM) or products, dCDP (5 mM).

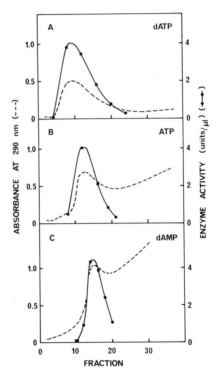

FIG. 4. Elution of T4 ribonucleotide reductase from dATP-Sepharose with different nucleotides. Enzyme (5 μg) was purified by treatment with streptomycin sulfate [O. Berglund, *J. Biol. Chem.* **247**, 7270 (1972)], precipitation with 35% saturation of ammonium sulfate, and chromatography on dATP-Sepharose. The columns were eluted with linear gradients of (A) 0–1 mM dATP, (B) 0–10 mM ATP, and (C) 0–20 mM dAMP in 0.1 M phosphate buffer pH 7.0. The total volume of the eluents was 30 ml, the flow rates were 3 ml/hour, and fractions of about 1 ml were collected and assayed for T4 ribonucleotide reductase activity. - - -, Absorbance at 290 nm; ———, enzyme activity. Taken with permission from the copyright holder from O. Berglund and F. Eckstein, *Eur. J. Biochem.* **28**, 492 (1972).

Purification of Subunit B1 of *Escherichia coli* Ribonucleotide Reductase

E. coli ribonucleotide reductase is a complex of two nonidentical subunits, proteins B1 and B2, each consisting of two very similar polypeptide chains. In contrast to T4 ribonucleotide reductase, the bacterial enzyme dissociates in dilute solution into free B1 and B2. Protein B1 contains the binding sites for the allosteric effectors, i.e., dATP and ATP.

Fuchs *et al.*[11] have chromatographed a streptomycin-treated crude extract of *E. coli* on a dATP-Sepharose column (Table II). Protein B1 bound to the column and was subsequently eluted with ATP in excellent yield. Gel electrophoresis in dodecyl sulfate of the ATP-eluate indicated that the protein was about 80% pure.

Protein B2 did not bind to dATP-Sepharose. However, as shown in Table II, protein B2 was somewhat retarded on the column presumably because of binding to protein B1.

Thelander[12] has recently introduced dATP-Sepharose chromatography in the routine large-scale purification of protein B1.

From the elution patterns which have been obtained, it is clear that only a limited number of proteins are adsorbed and specifically desorbed from each affinity column. This finding suggests that ATP- and dATP-

TABLE II
ACTIVITY OF B1 AND B2 SUBUNITS IN CRUDE AND PURIFIED FRACTIONS[a,b]

Fraction	A_{280}	B1 units	B2 units
Crude extract	63.6	1.300	2.140
Streptomycin	7.5	1.580	3.040
dATP-Sepharose			
Run-through	5.5	102	352
Buffer	—	30	2.310
Buffer + ATP	—	1.580	—
Total		1.710	2.660

[a] Data, with permission of the copyright holder, from J. A. Fuchs, H. O. Karlström, H. R. Warner, and P. Reichard, *Nature (London) New Biol.* **238**, 69 (1972).

[b] Frozen cells (3.0 g) of *Escherichia coli* CR 34 were ground with 6 g of alumina, extracted with 15 ml of 50 mM Tris chloride buffer, pH 7.0, containing 1 mM β-mercaptoethanol, and the nucleic acids were precipitated with streptomycin sulfate. Then 12 ml of these preparations was applied to columns containing 2 ml of dATP-Sepharose. After the sample had run through the column, the remaining bulk protein was eluted with five 4-ml aliquots of 0.1 M Tris chloride, pH 7.6, containing 10 mM MgCl₂ and 5 mM dithiothreitol (buffer). Protein B1 was then eluted with five 2-ml aliquots of the same buffer containing 2.5 mM ATP (buffer + ATP). Protein B1 and B2 activities were assayed with the standard assay. The results are expressed as total units/12 ml of streptomycin fraction.

Sepharose may be useful in studies of a number of proteins with the ability to bind ATP and dATP tightly. Considerably more proteins were found to bind to ATP-Sepharose than to dATP-Sepharose. This may reflect the fact that more bacterial and phage-induced proteins can bind ATP than dATP but could be due to technical reasons; i.e., the flow rate of the ATP-Sepharose column is slow, and nonspecifically adsorbed proteins are not completely washed from the column before the specific desorption with ATP is carried out.

Purification of Mammalian Ribonucleotide Reductase

Mammalian ribonucleotide reductase has not yet been isolated in pure form by conventional methods. The enzyme from rabbit bone marrow separates into two proteins (S_1 and S_2).[14] On chromatography of a crude enzyme fraction on a dATP-Sepharose column, S_1 activity was bound to the column and could be eluted with dATP. It could thus be obtained in a very pure form.[15] Results similar to these have been obtained with the reductase from Novikoff rat hepatoma.[16]

Preparation of p-Nitrophenyl Ester of Adenosine 5′-Triphosphate[17]

The sodium salt of p-nitrophenyl phosphate (740 mg, 2 mmoles) is converted to the pyridinium salt by passing a solution in 50% aqueous methanol through a column of Merck I ion exchanger (pyridinium form), and then converting to the tri-n-octylammonium salt by addition of tri-n-octylamine (0.86 ml, 2 mmoles) to a solution in methanol. The solvent is evaporated under reduced pressure and the residue is dried by repeated evaporation in pyridine and dry dimethylformamide. The residue is dissolved in dry dioxane (4 ml) and diphenylphosphorochloridate (0.6 ml, 3 mmoles) followed by tri-n-butylamine (0.96 ml, 4 mmoles) are added. The reaction mixture is allowed to stand at room temperature for 2 hours and is then evaporated under reduced pressure. Ether (40 ml) is added to the residue and, after shaking for 1–2 minutes petroleum ether (bp, 60–80°, 80 ml) is added and the mixture is allowed to stand at 4° for 1 hour. The supernatant fluid is decanted, the residue is dissolved in dry dioxane, and the solution is concentrated under reduced pressure to an oily residue, which is dissolved in 20 ml of pyridine–dimethyl-

---

[14] S. Hopper, *J. Biol. Chem.* **247**, 3336 (1972).
[15] S. Hopper, in press.
[16] E. C. Moore, personal communication.
[17] The corresponding dATP derivatives were prepared in the same way using dADP as starting material. dADP was prepared from dAMP.[18]
[18] D. E. Hoard and D. G. Ott, *J. Amer. Chem. Soc.* **87**, 1785 (1965).

formamide (1:1, v/v) (solution 1). This solution is used immediately in the coupling step described below.

Adenosine 5'-diphosphate (Na$^+$ salt, 650 mg, 1 mmole) is converted to the pyridinium salt by passage of a 50% aqueous methanol solution over a column of Merck I ion exchanger (pyridinium form). After evaporation to dryness, tri-n-octylamine (0.86 ml, 2 mmoles) and methanol (15 ml) are added to the resulting residue and the suspension is stirred for about 1 hour until solution occurs. The solvent is removed by evaporation and the residue is dried by repeated addition and evaporation of dry pyridine and dry dimethylformamide.

The activated p-nitrophenyl phosphate (solution 1) is added to the dried tri-n-octylammonium salt of ADP, and the reaction is allowed to proceed for 2 hours at room temperature. The solvent is removed by evaporation, and the residue is shaken with 20 ml of water. The aqueous extract is chromatographed on a DEAE-Sephadex A-25 column (3 × 15 cm) with a linear gradient of triethylammonium bicarbonate at pH 7.6 (0.1–0.5 M, total volume 4 liters). The elution profile will show three ultraviolet-absorbing peaks of about equal size eluting at 0.20 to 0.25 M, 0.30 to 0.35 M, and 0.40 to 0.45 M buffer, respectively. The first peak contains AMP and the p-nitrophenyl ester of ADP; the second contains ADP and the p-nitrophenyl ester of ATP. The nature of the material in third peak is not known.

The second peak is pooled, evaporated to dryness and chromatographed on six preparative silica gel plates[19] with ethanol–1 M triethylammonium bicarbonate at pH 7.6 (7:3, v/v). Two main bands are obtained, one remaining at the origin with ADP and another at $R_f = 0.80$. The latter band is removed by scraping and eluted by agitation for 1 hour with methanol–water (1:1, v/v). The silica gel is removed by centrifugation, and the supernatant liquid is evaporated to dryness (yield = 2000 A_{260} units, 0.12 mmole).

p-Aminophenyl Ester of ATP

A solution of the p-nitrophenyl ester of ATP (0.12 mmole, 2000 A_{260} units) in 5 ml of water is hydrogenated at atmospheric pressure in the presence of 50 mg of palladium on charcoal at room temperature for 3 hours. Most of the palladium is removed by filtration, and the filtrate is subjected to chromatography on a column of DEAE-Sephadex A-25 (1 × 15 cm) with a gradient of triethylammonium bicarbonate at pH 7.6 (0.1 to 0.5 M; total volume 1 liter). A single ultraviolet-absorbing peak

[19] DSC Fertigplatten, 20 × 20 cm, Kieselgel F 254 from Merck, Darmstadt. p-Nitrophenyl derivatives are detected by their ability to form yellow color after spraying of the plate with 1 M NaOH and incubation for 20 minutes at 150°.

is eluted between 0.30 and 0.35 M buffer, and the pooled fractions are evaporated to dryness. This compound has the same R_f value on thin-layer chromatography and the same mobility on electrophoresis as the nitrophenyl ester but contains less than 10% of p-nitrophenol on digestion with phosphodiesterase and phosphatase. From these observations and the coupling reaction described below, it was concluded that the compound is the p-aminophenyl ester of ATP. Yield (1140 A_{260} units, 0.067 mmole).

Coupling to Sepharose 4B

Cyanogen bromide activation of Sepharose 4B is performed[20] with 120 mg of cyanogen bromide per milliliter of settled Sepharose. Coupling is carried out for 16 hours in 0.1 M NaCHO$_3$ at pH 9.0 and at a nucleotide concentration of 0.7 to 3.0 μmole per milliliter. The amount of p-aminophenyl ATP and p-aminophenyl dATP coupled per milliliter of Sepharose was 0.3 μmole and 0.1 μmole, respectively, as measured by the release of adenosine and deoxyadenosine after incubation with phosphodiesterase and alkaline phosphatase.

The substituted Sepharose has been stored at 4° in water containing 0.02% sodium azide to prevent bacterial growth. It could be used at least five times without detectable loss in effectiveness. After use, columns are regenerated by elution with 6 M guanidine hydrochloride at 4° or 0.1% sodium dodecyl sulfate–0.1% 2-mercaptoethanol at room temperature.

Purification of T4 Ribonucleotide Reductase on ATP- and dATP-Sepharose

$E.\ coli$ B, 4.6 g (wet weight), infected with T4 am122, are ground with 9 g of alumina and extracted with 15 ml of 0.1 M potassium phosphate at pH 7.0 containing 0.5 mM dithiothreitol. Cellular debris are removed by centrifugation at 12,000 g for 20 minutes and the supernatant fluid is passed through a column (4 × 12 cm) of Sephadex G-25 which has been equilibrated with the above buffer. The eluate, 8 ml, is applied to each of two columns (1 × 1.5 cm) of ATP- and dATP-Sepharose. The ATP-Sepharose column is washed overnight with 0.1 M potassium phosphate at pH 7.0 containing 0.5 mM dithiothreitol (180 ml) until the absorbance at 280 nm of the wash fluid is 0.03. The enzyme is then eluted with 5 mM ATP in 0.1 M potassium phosphate at pH 7.0.

The dATP Sepharose column is washed with buffer (50 ml) until

[20] P. Cuatrecasas, M. Wilchek, and C. B. Anfinsen, $Proc.\ Nat.\ Acad.\ Sci.\ U.S.$ **61**, 636 (1968).

the absorbance at 280 nm is 0.01, and then eluted with 7 ml of 0.5 mM dATP in 0.1 M potassium phosphate at pH 7.0. The results of one such purification are summarized in Table I.

A procedure employing the affinity chromatography columns after an ammonium sulfate precipitation has been used successfully, and one adapted to preparative scale has been described elsewhere.[10]

Acknowledgment

We thank Dr. S. Hoppe and Dr. E. C. Moore for providing unpublished information. This work was supported by grants from the Swedish Natural Science Research Council to one of us (O.B.).

[19] Cyclic AMP: Purification of Protamine Kinase

By BENGT JERGIL and KLAUS MOSBACH

The nucleotide 3′,5′-cyclic adenosine monophosphate (cyclic AMP) has been shown to activate protein kinase obtained from various sources. The activation process involves the binding of cyclic AMP to the enzyme leading to dissociation of the inactive or partially active enzyme into a regulatory, cyclic AMP-binding component and a catalytic, fully active component.[1]

In addition to this activation of protein kinase, cyclic AMP may, as has been shown to apply for protamine kinase, be inhibitory at higher concentrations (above $10^{-4}\,M$), presumably by competing for the ATP-binding site of the enzyme.[2]

Cyclic AMP, covalently linked to Sepharose, can be used for the separation of cyclic AMP-binding and catalytic components of protein kinases.[3,4] On chromatography of these enzymes on cyclic AMP-Sepharose columns, the cyclic AMP-binding protein is adsorbed to the affinity material while the catalytic component is eluted. In addition to such separation, as first reported by Wilchek et al.,[3] purification has been achieved of the catalytic component of protamine kinase, a cyclic AMP-stimulated protein kinase from trout testis.[4] When the enzyme is applied to the ligand column at low buffer concentration, it will remain adsorbed. Subsequent

[1] M. A. Brostrom, E. M. Reimann, D. A. Walsh, and E. G. Krebs, *Advan. Enzyme Regul.* **8**, 191 (1970).

[2] B. Jergil and G. H. Dixon, *J. Biol. Chem.* **245**, 425 (1970).

[3] M. Wilchek, Y. Salomon, M. Lowe, and Z. Selinger, *Biochem. Biophys. Res. Commun.* **45**, 1177 (1971).

[4] B. Jergil, H. Guilford, and K. Mosbach, *Biochem. J.* **139** (1974), in press.

specific elution of the catalytic component can be accomplished using the inhibitor AMP as eluent whereas the cyclic AMP-binding component seems to remain bound.[4]

Preparation of 8-(6-Aminohexyl)-amino-cyclic AMP-Agarose

Cyclic AMP is brominated in position C-8 as described by Muneyama et al.[5] 8-Bromo-cyclic AMP is reacted with 1,6-diaminohexane to form 8-(6-aminohexyl)-amino-cyclic AMP,[6] which is subsequently coupled to Sepharose 4B by the cyanogen bromide method.[7] The resulting cyclic AMP-Sepharose (Fig. 1) is used for purification.

8-Bromo-cyclic AMP. Cyclic AMP (660 mg, 2 mmoles) is dissolved in 2 M NaOH (1 ml) and 12.5 ml of 1 M sodium acetate at pH 3.9. A solution of bromine (140 μl) in the above sodium acetate buffer (15 ml) is added, and the mixture is stirred at room temperature. After a few hours, 8-bromo-cyclic AMP begins to precipitate; the reaction is virtually complete after 24 hours. The precipitate is collected by filtration, washed with 15 ml of methanol, and dried to constant weight. The filtrate contains both unreacted cyclic AMP and approximately 20% of the 8-bromo derivative. To isolate the remaining 8-bromo-cyclic AMP, the filtrate is applied to a column of activated charcoal (8 g) in water. The column is washed with water before elution of 8-bromo-cyclic AMP with aqueous ethanol (1:1 v/v) saturated with benzene. The eluate is taken to dryness in a flash evaporator, washed with methanol, and dried. The total yield of 8-bromo-cyclic AMP is about 540 mg.

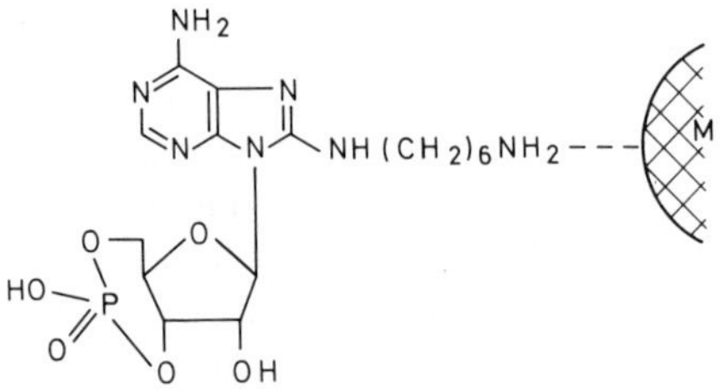

FIG. 1. Structure of Sepharose-bound 8-(6-aminohexyl)-amino-cyclic AMP. Other analogs substituted in the purine ring and found suitable for binding to Sepharose are at position C-8 substituted thioether analogs of cyclic AMP [G. I. Tesser, H.-U. Fisch, and R. Schwyzer, *FEBS Lett.* **23**, 56 (1972)] and N^6-ϵ-amino-caproyl-cyclic AMP [M. Wilchek, Y. Salomon, M. Lowe, and Z. Selinger, *Biochem. Biophys. Res. Commun.* **45**, 1177 (1971)].

[5] K. Muneyama, R. J. Bauer, D. A. Shuman, R. K. Robins, and L. N. Simon, *Biochemistry* **10**, 2390 (1971).

[6] H. Guilford, P.-O. Larsson, and K. Mosbach, *Chem. Scripta* **2**, 165 (1972).

[7] R. Axén, J. Porath, and S. Ernback, *Nature (London)* **214**, 1302 (1967).

8-(6-Aminohexyl)-amino-cyclic AMP. 8-Bromo-cyclic AMP (125 mg, 0.32 mmole) and 1,6-diaminohexane (1.1 g, 10 mmoles) are suspended in 10 ml of water and maintained at 140° for 2 hours in a thick-walled sealed tube. The mixture is applied at room temperature to a column (2 × 14 cm) of Dowex 1-X8, 200–400 mesh, in the acetate form. The column is washed with about 400 ml of water (flow rate 60 ml/hour) to remove excess diaminohexane. A linear gradient of water (500 ml) and 1 *M* acetic acid (500 ml) is applied, and fractions of 5 ml are collected. 8-(6-Aminohexyl)-amino-cyclic AMP elutes in the earlier part of the gradient, and its appearance may be monitored by UV absorbance (λ_{max} is 277 nm at pH 6.3). The product is identified by thin-layer chromatography using ethanol:0.5 *M* ammonium acetate (5:2) on glass plates coated with 0.1 mm silica gel. A ninhydrin-positive spot which is pink in UV light (254 nm) with $R_f = 0.6$ is obtained ($R_f = 0.7$ for 8-bromo-cyclic AMP). Fractions containing the 8-(6-aminohexyl)-amino-cyclic AMP are combined and lyophilized. The yield is close to 100%.

Coupling to Agarose. Sepharose is activated with cyanogen bromide according to the method of Axén *et al.*[7] Sepharose 4B (10 ml) is washed with water, and cyanogen bromide (500 mg dissolved in 7 ml of water) is added. The pH is immediately adjusted to and maintained at pH 10.5–11 for approximately 10 minutes by the addition of 4 *M* NaOH. The activated Sepharose is thoroughly washed on a Büchner funnel at 4° with cold water followed by 0.1 *M* sodium carbonate-bicarbonate at pH 9.5. The washed gel is suspended in a solution of 8-(6-aminohexyl)-cyclic AMP (50 mg) in 8 ml of the same buffer. The suspension is rotated slowly at 4° for 24 hours and filtered on a Büchner funnel. The pale-yellow gel is washed extensively with the bicarbonate buffer, with water, with 0.5 *M* sodium chloride and again with water, in that order. The moist gel may be stored at 4° for several months without essential change in properties. The extent of incorporation is approximately 45% as estimated from the amount of free nucleotide found in filtrate and washings (130 μmoles per gram of dry polymer).

Chromatography

Protamine kinase (3 mg) partially purified by chromatography on hydroxylapatite[4] was applied to a cyclic AMP-Sepharose column (0.5 × 12 cm) equilibrated with ammonium phosphate at pH 8.0, 1 m*M* in dithiothreitol. The buffer concentration was chosen so that the enzyme was retained on the column. With the cyclic AMP-Sepharose preparation described here, 0.3 *M* buffer was used. The column was washed with the same buffer and a breakthrough protein peak, but no enzymatic activity, was eluted.[8] Application of 5 m*M* AMP, an inhibitor of protamine kinase, dissolved in the buffer, released the enzyme, but very little protein. The released enzyme was not activated by addition of cyclic AMP. A 50- to 100-fold purification with 50% recovery was obtained. GMP, which is not an inhibitor of protamine kinase activity, did not elute the enzyme.

After each run the column material is washed with 8 M urea, equilibrated with an appropriate buffer and is ready for use.

When protamine kinase was applied at a higher buffer concentration, enzyme did not adsorb to the column but eluted, somewhat retarded, after the breakthrough protein peak. As with the preparation described above, the enzyme obtained at high buffer concentrations was not stimulated by addition of cyclic AMP. When the eluate from cyclic AMP-Sepharose columns was tested, neither free cyclic AMP, cyclic AMP-analog, nor cyclic AMP binding protein was detected. The eluted enzyme sedimented on centrifugation in a sucrose gradient at a slower rate than the enzyme applied to the column, but at the same rate as a sample of the applied enzyme which had been pretreated with cyclic AMP.[4] This indicates that the enzyme is dissociated on the column. Chromatography at higher buffer concentration, resulting in poor binding of the enzyme, can thus be used for the separation of the catalytic component of protamine kinase from the cyclic AMP-binding part.

Evaluation

Cyclic AMP-Sepharose can be used several times without appreciable change in its chromatography properties. Gradually, however, the retardation of protamine kinase diminishes, and the buffer concentration below which the enzyme is retained on the column decreases. It was found that different preparations of cyclic AMP-Sepharose exhibit varying chromatographic properties in that the buffer concentration below which the enzyme is retained on the column can change. The general pattern of elution of the enzyme, however, is quite consistent. On applying protamine kinase to Sepharose-bound analogs of the inhibitor, AMP, i.e., 8-(6-aminohexyl)-amino-AMP and N^6-(6-aminohexyl-AMP), binding could not be achieved.[4] In accordance with this finding is the observation that these analogs, when free in solution, are not inhibitory to the enzyme in contrast to 8-(6-aminohexyl)-amino-cyclic AMP, the compound described here as a biospecific affinity material for protamine kinase.

It should be possible to apply this cyclic AMP analog also for the purification of the catalytic component of other protein kinases.

---

[8] Assays were carried out with 10 μl of eluate for 15 minutes with protamine as substrate as described earlier[2]; partial inhibition by AMP present in the eluate was compensated for.

[20] Insolubilized Biotin for the Purification of Avidin

By ED BAYER and MEIR WILCHEK

Principle

The purification of avidin by affinity chromatography is based upon the unusually strong complex formed by this protein with polymer-bound biotin. The first specific purification of avidin was performed by McCormick, who passed an avidin solution of 10% purity over columns containing biotin esterified to cellulose.[1] However, both yield and resolution were poor. Porath[2] improved the technique by attaching biotin to Sephadex by means of a phenyl glyceryl spacer. Excellent purification was achieved, but the yield was low and the column could not be regenerated.

The best method to date was accomplished by coupling biocytin to Sepharose.[3] A 4000-fold purification was obtained in a single step, and 90% of the avidin activity was recovered directly from egg white. Two disadvantages are evident with this system. First, biocytin is not available commercially. Second, denaturing conditions were necessary in order to elute the protein from the column, although the entire activity could be restored upon regeneration (Scheme 1).

Three methods of avidin purification by affinity chromatography are described below: with biocytin-Sepharose, with biotinyl-polylysyl-Sepharose, and with a matrix consisting of 4(hydroxyazobenzyl)-2'-carboxylic acid (HABA) conjugated to polylysyl-Sepharose.

The biotinyl-polylysyl-Sepharose column is advantageous since it is easy to prepare, and the polylysine backbone is bound to agarose in numerous positions. Consequently, the possibility of the affinity ligand leaking from the matrix under severe conditions of elution is lessened.

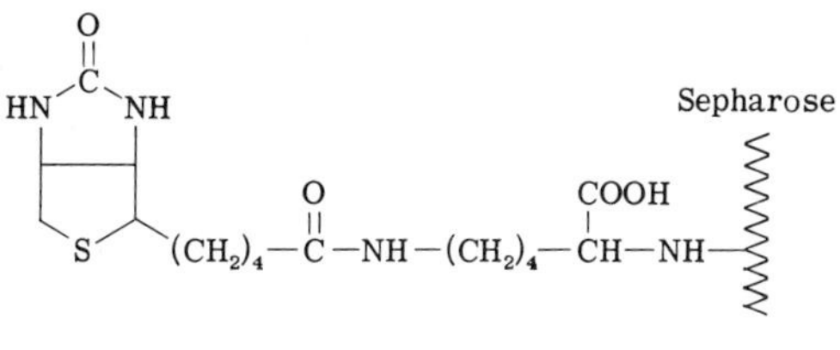

SCHEME 1. Structure of biocytin-Sepharose.

[1] D. B. McCormick, *Anal. Biochem.* **13**, 194 (1965).
[2] J. Porath, *in* "Nobel Symposium 3, Gamma Globulins" (J. Killander, ed.), p. 287. Almquist & Wiksell, Stockholm and Wiley (Interscience), New York, 1967.
[3] P. Cuatrecasas and M. Wilchek, *Biochem. Biophys. Res. Commun.* **33**, 235 (1968).

The HABA conjugated column may eventually prove to be useful. It is extremely stable and maintains its affinity for avidin even after cleansing with relatively strong acids and bases. Avidin can be eluted under much less strenuous conditions. However, this column shows preference for other acidic proteins including chymotrypsinogen, urease, and ovalbumin, but not basic proteins, such as RNase. This lack of specificity can be overcome to a certain extent by acetylation of the column with acetic anhydride, thereby blocking any free amino groups. The ultimate utility of this column is still under investigation.

Preparation of Affinity Adsorbents

Biocytin Column

Biotinyl-N-hydroxysuccinimide Ester (BNHS). Dicyclohexyl carbodiimide (DCC), 0.8 g, 4 mmoles, is added to a solution of biotin (1 g, 4 mmoles) (Sigma) and *N*-hydroxysuccinimide (0.6 g, 5.2 mmoles) in 12 ml of dimethylformamide. The suspension is stirred and the reaction allowed to proceed for 20–24 hours. The precipitate is filtered, and the filtrate is maintained at 0° overnight. Any additional precipitate is filtered, and the resulting filtrate is evaporated under reduced pressure. The residue is washed extensively with ether and recrystallized from isopropanol. Yield: 1 g (70%).

Biocytin (ε-N-Biotinyl-L-lysine). Biotinyl-*N*-hydroxysuccinimide ester (340 mg, 1 mmole) is suspended in 3 ml of dimethylformamide. α-tBoc-Lysine (400 mg) (Miles-Yeda) dissolved in 4 ml of sodium bicarbonate, is added and the suspension adjusted to pH 8. The reaction is carried out for 3 hours. The solvent is evaporated under reduced pressure and the tBoc-biocytin is precipitated by addition of 5 ml of 10% citric acid. The crystals are filtered and washed with water. A second crop of crystals can be isolated from the mother liquor. Yield: 220 mg.

The tBoc-group is removed by treatment with $4 N$ HCl in dioxane (5 ml). After 20 minutes, ether is added; the resultant crystals are filtered, washed well with ether, and dried under reduced pressure. Yield: 180 mg (45%).

Biocytin-Sepharose. Biocytin-hydrochloride is coupled to activated Sepharose as previously described.[3] About 25 mg of biocytin are used to prepare 20 ml of substituted Sepharose.

Biotin or HABA-Conjugated Polylysyl-Sepharose

The synthesis of poly-L-lysine and its subsequent coupling to Sepharose can be found elsewhere.[4,5] Poly-L-lysyl-Sepharose (3 g, containing about 60 μmoles of lysine) is suspended in 3 ml of absolute dioxane. An excess of biotin or HABA (65 mg) is added and followed by an equimolar amount of DCC (50 mg). The reaction is allowed to proceed overnight. The ligand-conjugated polylysyl-Sepharose is washed several times with large volumes of dioxane, methanol, and distilled water, respectively, and stored in about 5 ml of distilled water.

Purification of Avidin

Biocytin-Sepharose (2 ml) is added to a solution (25 ml) of 0.2 M sodium bicarbonate, pH 8.7, containing crude, dried egg white (8 g). After a few minutes the solution is centrifuged at 20,000 g for 10 minutes. The pellet is suspended in bicarbonate buffer and poured into a 1 × 5 cm glass column. The column is washed with 3 M guanidine·HCl, pH 4.5 (25 ml), and the avidin is eluted with 6 M guanidine·HCl at pH 1.5. The denatured protein is immediately reconstituted by 10-fold dilution with buffer, or by dialysis.

Commercial avidin (Sigma) has been adsorbed to HABA-polylysyl-Sepharose and could be eluted by 0.2 N acetic acid with about 70% yield. Avidin activity was determined by its capacity to bind ^{14}C-labeled biotin or by spectrophotometric titration with biotin.[6]

---

[4] M. Wilchek, this volume [5].
[5] M. Wilchek, *FEBS Lett.* 33, 70 (1973).
[6] R.-d. Wei, this series, Vol. 18A, p. 424.

[21] Coenzyme A

By ICHIRO CHIBATA, TETSUYA TOSA, and YUHSI MATUO

When a coenzyme is immobilized for affinity chromatography, the immobilized coenzyme may be used for the purification of enzymes requiring it for activity. Conversely, when an enzyme or a group of enzymes requiring a coenzyme is immobilized, the immobilized enzyme or enzymes may be useful for the purification of the coenzyme. In this article, methods for both approaches are presented and details are given for the purification of phosphate acetyltransferase and coenzyme A (CoA).

Methods of Preparation[1]

CNBr-Activated Sepharose. The preparation is essentially that of Axén *et al.*[2] Sepharose 6B (30 g packed weight) is washed with 1 liter of

---

[1] Y. Matuo, T. Tosa, and I. Chibata, *Biochim. Biophys. Acta* 338, 520 (1974).
[2] R. Axén, J. Porath, and S. Ernbäck, *Nature (London)* 214, 1302 (1967).

water and then suspended in 300 ml of water. CNBr solution (30 g per 600 ml of water) is added to the stirred suspension in a well ventilated hood, and the reaction mixture is maintained at between 10° and 20° and at pH 11–12 by adding NaOH. The reaction is complete in 8–15 minutes. The suspension is filtered on a glass filter and washed sequentially with 1 liter of cold 0.1 M NaHCO$_3$, 1 liter of cold water, and 500 ml of 0.1 M sodium acetate at pH 6.0. The resultant CNBr-activated Sepharose 6B is used immediately.

Nitroprusside Reagent. Sodium nitroprusside (1.5 g) is dissolved in 5.0 ml of 2 N H$_2$SO$_4$. To the solution are added 95 ml of methanol and then 10 ml of 28% ammonia. After removal of a white precipitate by filtration, the clear orange solution should be stored in refrigerator.

Attachment of CoA. CNBr-activated Sepharose 6B (25 g packed weight) is added to a solution of reduced CoA (46 mg) is 92 ml of 0.1 M sodium acetate at pH 6.0. After gentle stirring at 25° for 6 hours, the mixture is filtered and washed with 100 ml of 1 M NaCl. The resulting agarose-CoA derivative (immobilized CoA) is washed with 100 ml of 1% 2-mercaptoethanol, in order to provide the reduced form and washed with 0.1 M sodium acetate at pH 6.0 until washings become negative for the nitroprusside reaction. Reproducibility of the procedure is high if the initial pH remains below 7.0 and if the reaction time is 6 hours or less.

Determination of Bound CoA. The amount of CoA immobilized is calculated by the difference of the amount of CoA used and that of the unreacted CoA in the filtrate and NaCl washings described above. Reduced CoA is estimated spectrophotometrically with phosphate acetyltransferase according to the method of Michal and Bergmeyer,[3] and total CoA (reduced plus oxidized) in the presence of cysteine by the method of Stadtman et al.[4] The immobilized CoA, prepared in this manner, contains about 31 mg of CoA per 25 g (packed weight) of Sepharose 6B.

Identification and Properties

The immobilized CoA is positive in the nitroprusside reaction. As the infrared spectrum of the immobilized CoA shows the absorption band derived from R-O-CO-N at 1720 cm^{-1}, CoA is considered to be linked to Sepharose 6B in N-substituted carbamate structure, as shown in Fig. 1. The immobilized CoA does not show coenzyme activity for phosphate acetyltransferase.

[3] G. Michal and H. U. Bergmeyer, *in* "Methods of Enzymatic Analysis" (H. U. Bergmeyer, ed.), p. 512. Academic Press, New York, 1963.
[4] E. R. Stadtman, G. D. Novelli, and F. Lipmann, *J. Biol. Chem.* **191**, 365 (1951).

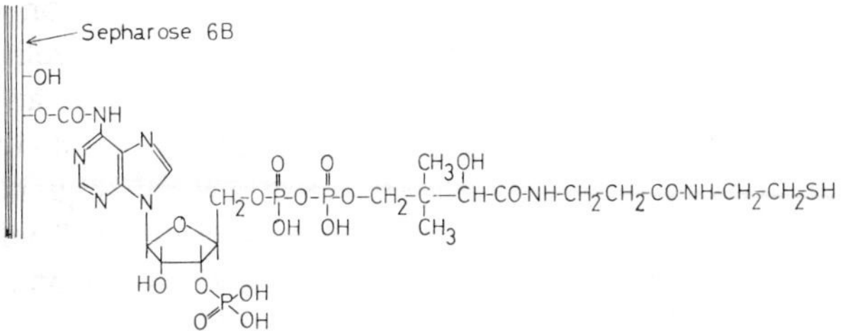

FIG. 1. Schematic structure of immobilized CoA.

Adsorption of Phosphate Acetyltransferase

Procedure. Phosphate acetyltransferase (EC 2.3.1.8) (0.5 mg) sus-pended in 3 M ammonium sulfate is dialyzed against 10 mM sodium acetate at pH 6.0 for 10 hours in an ice-bath with gentle stirring. The enzyme loses 50% of its activity during dialysis. When the dialyzed en-zyme is charged onto the immobilized CoA column, previously equilibrated with sodium acetate at an ionic strength of 0.01, the enzyme is adsorbed and may be eluted with the same buffer at an ionic strength of 0.1. Con-taminating proteins can be eluted by buffer at an ionic strength of 0.5. When used with a purified preparation of the enzyme, recovery of total protein was 98%, and that of activity 49%.

Purification of CoA

For purification of CoA, a protein with affinity for CoA may be used. Since it would be very inefficient for this purpose to prepare a highly purified enzyme, probably by affinity chromatography on immobilized CoA, we have found a relatively crude protein preparation that is easily prepared and can be used as an effective means of preferentially selecting out the coenzyme.

We have used a system easily obtainable from *Sarcina lutea*. This preparation has been shown to bind well to a column of immobilized CoA. That is, 90% of the protein in the preparation can adsorb to matrix-bound CoA, and only 10% cannot. By using the *S. lutea* preparation as a ligand and attaching it to agarose, CoA may be reversely adsorbed.

Preparation of Dialyzed Extract. S. lutea ATCC 9341 is grown with shaking at 30° for 72 hours in a medium at pH 7.0 containing 5% glu-cose, 4.4% corn steep liquor, 1.4% peptone, 1% ammonium acetate, 0.5% KH_2PO_4, 0.5% K_2HPO_4, and 0.1% $MgSO_4 \cdot 7H_2O$. The cells are harvested, washed twice with 0.9% NaCl, and lyophilized. Lyophilized cells (3 g) suspended in 100 ml of 50 mM sodium phosphate at pH 7.0 are sonicated at 20 Kc for 1 hour with cooling and are centrifuged at

20,000 g for 30 minutes. The supernatant fluid is dialyzed against 10 mM sodium acetate at pH 6.0 for 16 hours.

Preparation of Crude CoA. The process of fermentative production and preparation of crude CoA is that of Nishimura *et al.*[5] To 100 ml of the crude CoA solution is added 15 ml of 2-mercaptoethanol. The mixture is adjusted to pH 8.0 with NaOH and gently stirred for 2 hours. After adjusting to pH 3.0 with HCl, 1.2 liters of ethanol are added to remove 2-mercaptoethanol; the resulting precipitate is collected by centrifugation and dissolved in a small volume of water. Crude reduced CoA solution is prepared by repeating this ethanol treatment twice.

Preparation of Immobilized CoA-Binding Protein Column. CNBr-activated Sepharose 6B (1 g packed weight) is allowed to react with 20 mg of protein in 20 ml of the dialyzed extract from *S. lutea* at pH 8.5. After gentle shaking at 5° for 6 hours, the pH is adjusted to 6.0 with dilute acetic acid, and the mixture is shaken at 5° for 12 hours. Another 1 g (packed weight) of CNBr-activated Sepharose 6B is added, and the mixture is gently shaken at 5° for another 12 hours. The mixture is filtered and washed with 20 ml of 1 M NaCl. The amount of protein immobilized is calculated by the difference between the amount of the protein used and that of unreacted protein in the filtrate and NaCl washings. The resultant adsorbent contained about 13 mg of protein per 2 g of packed Sepharose. The immobilized CoA-binding protein is packed into a column (1 × 5 cm).

Adsorption Specificity of Immobilized CoA-Binding Protein. When the immobilized CoA-binding protein column is equilibrated with sodium acetate at pH 6.0 and an ionic strength of 0.01, reduced CoA is adsorbed and may be eluted by increasing the ionic strength with NaCl. As shown in Fig. 2, reduced CoA is eluted in two peaks. The ionic strengths of the peak fractions are 0.025 and 0.06, respectively. This result suggests the presence of at least two kinds of CoA-binding protein in the adsorbent. It is suggested that two kinds of the protein are involved, one having higher affinity than the other.

Oxidized CoA, dephospho-CoA, ATP, and ADP have also been found to be adsorbed to this adsorbent, whereas AMP was not. The ionic strength of the fractions at the respective elution peaks is 0.023 for oxidized CoA, 0.024 for ADP, 0.029 for ATP, and 0.035 for dephospho-CoA, respectively (Fig. 2).

Purification of Reduced CoA with Immobilized CoA-Binding Protein. When the adsorbent column is equilibrated with the buffer at an ionic strength of 0.04, i.e., under conditions that would utilize the adsorption

[5] N. Nishimura, T. Shibatani, T. Kakimoto, and I. Chibata, manuscript in preparation.

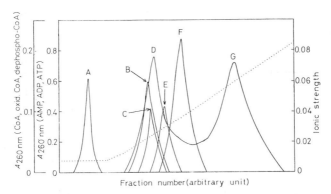

Fig. 2. Adsorption specificity of immobilized CoA-affinity protein column.[1] The column (1 × 5 cm) is equilibrated with sodium acetate buffer (pH 6.0) of ionic strength = 0.01. Chromatographic patterns of respective compounds are superposed. Compounds in eluates are determined spectrophotometrically by measuring absorbance at 260 nm. The amounts used for experiments are as follows: AMP, 58.5 μg; ADP, 218 μg; ATP, 303 μg; dephospho-CoA, 230 μg; oxidized CoA, 132 μg; reduced CoA, 415 μg. The first peak is eluted with the equilibrating buffer. Other peaks are eluted by linear gradient elution, using 50 ml of the buffer and 50 ml of 0.1 M NaCl. Fractions of 3 ml are collected. ——, Absorbance at 260 nm; -----, ionic strength. A, AMP; B, oxidized CoA; C, ADP; D, reduced CoA; E, ATP; F, dephospho-CoA; G, reduced CoA.

capacity of the "higher affinity" protein without using that of the "lower affinity" protein, reduced CoA can be selectively adsorbed without being contaminated. In one case, 336 μg of CoA was applied to the column equilibrated with sodium acetate at pH 6.0 and an ionic strength of 0.04. Most of the contaminants absorbing at 260 nm were not adsorbed. The fraction which was not adsorbed also contained 90 μg of oxidized CoA, but was free of reduced CoA. Reduced CoA was eluted by increasing the ionic strength to 0.1 with NaCl. The reduced CoA was of about 100% purity and was obtained in a 67% yield as total CoA, and 92% purity and 94% yield as reduced CoA.

The immobilized CoA-binding protein column is stable under operating conditions of pH 6–7 and at 10–25°. The adsorbing capacity is about 85 μg of reduced CoA per milliliter of adsorbent under the conditions of ionic strength equal to 0.04 and at pH of 6–7 at 10–25°. The adsorbent is stable for a month when stored at 5° in 0.1 M sodium acetate at pH 6.0 containing 1 M NaCl.

[22] Methotrexate-Agarose in the Purification of Dihydrofolate Reductase

By BERNARD T. KAUFMAN

The enzyme, tetrahydrofolate dehydrogenase (EC 1.5.1.3), which will be referred to as dihydrofolate reductase, has been purified from a variety of sources by conventional protein purification procedures. However, owing to the very small amounts of dihydrofolate reductase activity present in cells and tissues, conventional methods are frequently laborious, often incomplete, and result in relatively low yields. Thus, the availability of a purification procedure based on a strong and specific interaction between the enzyme and an inhibitor, coenzyme, or substrate covalently bound to an insoluble support material and which could be used to separate the desired enzyme from all extraneous protein would markedly facilitate the purification of such enzymes.

The potent inhibitory action of certain folate antagonists such as methotrexate (4-amino-10-methylpteroylglutamic acid) upon dihydrofolate reductases suggests that coupling of this inhibitor to an insoluble matrix should yield an exceptionally efficient affinity chromatographic system. Folate derivatives have been successfully coupled by carbodiimide-promoted reactions to the primary amino groups of polyamino acids and proteins, presumably via a peptide linkage involving the carboxyl group of the folate.[1] The obvious extension of this procedure to the attachment of folic acid or methotrexate to aminoethyl cellulose is readily accomplished as evidenced by the formation of a bright yellow cellulose derivative. However, adsorbent prepared from this material does not bind significant amounts of enzyme.

More recently, the development of affinity systems based on an insoluble agarose matrix, has been used successfully for the purification of a large number of enzymes and proteins. In most cases, effective binding of the protein under consideration to the affinity column depends on placing the ligand groups critical for the interaction at some distance from the agarose backbone. These studies suggested the following route for the preparation of a methotrexate column: (a) activation of a beaded agarose derivative (Sepharose) with cyanogen bromide, (b) formation of the ω-aminoalkyl derivative by reacting the activated Sepharose with an appropriate alkyldiamine, and (c) coupling of the carboxyl group or groups

---

[1] J. Jaton and H. Ungar-Warm, this series, Vol. 18A [189].

of methotrexate to the ω-aminoalkyl-Sepharose by means of a carbodiimide-promoted peptide synthesis.

An affinity adsorbent prepared by this approach effectively adsorbed the dihydrofolic reductase activity from a purified chicken liver extract.[2] Quantitative adsorption occurred at pH values below six and the activity could be readily eluted at pH values above 9. Although two, four, six, and twelve carbon diamines have been used to prepare the ω-aminoalkyl "spacer" derivatives of Sepharose, subsequent studies revealed no significant differences in their ability to couple with methotrexate or eventually to adsorb the enzyme from chicken liver extracts. Exactly similar results were obtained with the corresponding enzymes derived from extracts of beef liver, yeast, *Salmonella typhimurium,* and *Escherichia coli* despite the dissimilarities in the properties of the enzymes.

Attempts to further define the conditions for optimal adsorption and elution of the chicken or beef liver enzyme activity in terms of pH, ionic strength, and the presence of cofactors, such as folic acid, dihydrofolic acid, or TPNH, have not been successful either with respect to the interaction of the enzyme with the methotrexate-Sepharose column or the effect of these parameters on the degree of inhibition of the purified enzyme by methotrexate in solution. Thus, the chicken and beef liver enzymes are strongly inhibited to the same extent at all pH values from 4 to 10 regardless of ionic strength and at urea concentrations up to 5 M. Similarly, the presence of TPNH did not appear to affect the kinetics of inhibition. However, at pH values above 7.5, preincubation of the enzyme with dihydrofolate decreased the degree of methotrexate inhibition. Similar results have been reported for the enzyme from *Lactobacillus casei.*[3] On the other hand, the enzyme from L1210[3] and Ehrlich ascites cells[4] have been reported to exhibit a pH-inhibition maximum at pH 6.

Therefore, the successful adsorption and elution of the chicken and beef liver enzymes from the column may be due to a decrease in affinity resulting from the binding of the methotrexate to the Sepharose matrix. Whiteley *et al.*[3] have reported that coupling of aminopterin to soluble aminoethyl starch decreases the inhibitory effectiveness by about two orders of magnitude.

The utility of affinity chromatography is evidenced by the rapid proliferation of communications devoted to the purification of dihydrofolate reductases from a variety of sources by this method (Table I).

[2] B. T. Kaufman and J. V. Pierce, *Biochem. Biophys. Res. Commun.* 44, 608 (1971).
[3] J. M. Whiteley, R. C. Jackson, G. P. Mell, J. H. Drais, and F. M. Huennekens, *Arch. Biochem. Biophys.* 150, 15 (1972).
[4] J. R. Bertino, B. A. Booth, A. L. Bieber, A. Cashmore, and A. C. Sartorelli, *J. Biol. Chem.* 239, 479 (1964).

TABLE I

PURIFICATION OF DIHYDROFOLATE REDUCTASE BY AFFINITY METHODS

Source	Affinity system	Reference
Chicken liver	MTX-Sepharose column[a]	b,c
Rat skin	MTX-Sepharose column	d
Bacteriophage T4	N-10-Formylaminopterin Bio-Gel column	e
Lactobacillus casei	MTX-Sepharose column	c,d
L1210 Cells	MTX-Sepharose column	c,f,g
Escherichia coli	MTX-Sepharose (batch)	h
Salmonella typhimurium	MTX-Sepharose column	i
Yeast	MTX-Sepharose column	i
Hamster cells	MTX-Sepharose column	j
Calf liver	MTX-Sepharose column	i
Lactobacillus casei	Pteroyllysine-Sepharose column	k

[a] Methotrexate-aminohexyl-Sepharose.
[b] B. T. Kaufman and J. V. Pierce, Biochem. Biophys. Res. Commun. **44**, 608 (1971).
[c] J. M. Whiteley, R. C. Jackson, G. P. Mell, J. H. Drais, and F. M. Huennekens, Arch. Biochem. Biophys. **150**, 15 (1972).
[d] P. C. H. Newbold and N. G. L. Harding, Biochem. J. **124**, 1 (1971).
[e] J. S. Erickson and C. K. Mathews, Biochem. Biophys. Res. Commun. **43**, 1164 (1971).
[f] J. Gauldie and B. L. Hillcoat, Biochim. Biophys. Acta **268**, 35 (1972).
[g] P. L. Chello, A. R. Cashmore, S. A. Jacobs, and J. R. Bertino, Biochim. Biophys. Acta **268**, 30 (1972).
[h] M. Poe, N. J. Greenfield, J. M. Hirshfield, M. N. Williams, and K. Hoogsteen, Biochemistry **11**, 1023 (1972).
[i] B. T. Kaufman, unpublished results.
[j] H. Nakamura and J. W. Littlefield, J. Biol. Chem. **247**, 179 (1972).
[k] E. J. Pastore, L. T. Plante, and R. L. Kisliuk, this volume [23].

Preparation of Methotrexate-Agarose Affinity Column

Preparation of ω-Aminohexyl-Sepharose[5]

In a beaker containing a magnetic stirring bar, add 60 ml of decanted agarose (Sepharose 4B)[6] plus an equal volume of water. The electrodes

[5] Stabilized, freeze-dried CNBr-activated Sepharose 4B is commercially available (Pharmacia Fine Chemicals, Inc.) and may be used at this point in the procedure. Activated Sepharose, 15 g, is suspended in 500 ml of 0.001 N HCl and gently stirred for 10 minutes until in a uniform suspension. The swollen gel is immediately filtered on an extra-coarse sintered funnel and washed on the filter with an additional 1 liter of 0.001 N HCl. The damp gel is transferred to a 250-ml beaker, and 100 ml of 0.5 M 1,6-hexanediamine, dissolved in 0.1 M NaHCO$_3$ and containing 0.5 M NaCl, is added. The above procedure is used from this point.
[6] Sepharose 4B (Pharmacia Fine Chemicals) is thoroughly washed by elutriation in order to remove fine particles and preservatives. Settled volume is determined by allowing the gel to stand overnight in a graduate cylinder. Throughout the

of a pH meter and a thermometer are placed in the suspension. *The procedure using cyanogen bromide is performed in a well ventilated hood.* Finely divided solid cyanogen bromide,[7] 5 g, is quickly added to the stirred suspension, and the pH is immediately raised to 10.5 with 4 M NaOH. The pH is maintained at 10.5 by constant manual titration with NaOH. The temperature is maintained at about 20° by adding pieces of ice as needed. The reaction is usually complete in about 10–12 minutes as indicated by the cessation of pH change.

The reaction mixture is immediately transferred to a 500-ml coarse sintered-glass funnel equipped with a stopcock and is quickly washed under suction with approximately 1500 ml of ice cold $NaHCO_3$ (solution A).[8] The washing procedure is carried out using three 500-ml aliquots of the cold solution A, adding each aliquot as soon as the previous wash approaches the surface of the cake. The suction should be regulated in such a manner as to prevent the surface of the cake from drying.

After the last aliquot of wash is filtered from the "activated" agarose, the stopcock on the funnel is closed and the vacuum is disconnected. Immediately 100 ml of 0.5 M 1,6-hexanediamine in solution A (readjusted to pH 10) is added to the material in the funnel and gently stirred into a uniform suspension. The entire procedure of washing and adding the hexanediamine solution should take less than 90 seconds. The suspension is transferred to a beaker containing a magnetic mixing bar and gently stirred for 3 hours at room temperature. Any particles remaining in funnel may be transferred to the beaker by rinsing with small amounts of Solution A.

The ω-aminohexyl-Sepharose (AH-Sepharose) is filtered as above and washed with 500 ml of solution A. The moist cake is transferred to a beaker containing 100 ml of ethanolamine at pH 9.5. The resulting suspension is again gently stirred for either 2 hours at room temperature or overnight at 4°.

The reaction mixture is filtered, resuspended in 500 ml of solution A, filtered, and washed on the funnel with 1000 ml of H_2O. The washing procedure is continued as above, using successively 500 ml of 1 M HCl, 1000 ml of water, 500 ml of 1 M NaOH, and 2000 ml of water. The resuspended AH-Sepharose may be stored at this point.

---

procedure, particular care is taken during agitation and stirring to prevent destruction of the rather friable gel particles. This is also important during vacuum filtration.

[7] A freshly opened bottle of cyanogen bromide is used; otherwise, the resulting aminohexyl-Sepharose may exhibit a yellowish tint.

[8] Solution A consists of 0.1 M $NaHCO_3$ containing 0.5 M NaCl.

Preparation of Methotrexate-AH-Agarose[9,10]

Filter the AH-Sepharose on a coarse sintered-glass funnel and transfer the moist cake to a 100-ml beaker containing a magnetic stirring bar. Add the methotrexate solution,[11] insert the electrodes of a pH meter, and while stirring carefully adjust the pH to 6.5 with 1 N HCl. Add 1.5 g of 1-ethyl-3(3-dimethylaminopropyl) carbodiimide hydrochloride in 500-mg increments at 15-minute intervals. The pH is maintained at 6.5 by constant manual titration with 1 N HCl. After 1 hour the pH is lowered to 6.0 with 1 N HCl and the reaction is allowed to continue for a total of 3 hours.

The reaction mixture is filtered on the sintered-glass funnel and washed with 500 ml of water on the funnel. This filtrate and the following two filtrates are collected in order to obtain a measure of unreacted methotrexate.[12] The bright orange-yellow gel is washed successively with 500 ml of 1 M NaCl, 1 liter of water, 500 ml of 1 M K_2HPO_4, 1 liter of water, 500 ml of 1 M KH_2PO_4, and 2 liters of H_2O and finally resuspended in a convenient volume of 10 mM potassium phosphate at pH 5.6.

Experimental Techniques

Some features that are specific to the use of the MTX-AH-Sepharose column are briefly noted here. The affinity material for large-scale enzyme preparations is conveniently packed into a column 20 × 300 mm prepared from two coupled column units, 20 × 150 mm (Kontes Glass Co., Chromaflex extenders), with a stopcock adapter incorporating a porous glass disk at the outlet. A pad of moistened glass wool is placed over the disk to prevent clogging of the surface. The column is packed by adding the suspension of MTX-AH-Sepharose in 10 mM potassium phosphate at pH 5.6, to the column partially filled with the same buffer; the column outlet is progressively opened only after a 3-cm layer of adsorbent has settled. The material is allowed to settle to a bed height approximately 1 cm below the juncture between the two column segments. The liquid is allowed to drain to the bed surface, and a thick slurry of

[9] Methotrexate-aminohexyl-agarose is abbreviated: MTX-AH-Sepharose.
[10] Commercially available agarose derivatives possessing the appropriate functional "spacer" groups may be used at this point in the procedure: AH-Sepharose 4B (Pharmacia Fine Chemicals, Inc.) or Affinose 102 (Bio-Rad Laboratories).
[11] Methotrexate (4-amino-10-methylpteroylglutamic acid), 500 mg, 90–95% pure, is placed in a small glass tissue grinder or mortar containing 50 ml of 0.1 M NaCl. Solubilization is attained by grinding while 1 N KOH is added dropwise.
[12] The content of unreacted methotrexate in the first three filtrates is determined spectrophotometrically using the molar extinction coefficients in 0.1 N KOH, $\epsilon = 23,000$ at 257 nm and $\epsilon = 22,100$ at 302 nm. Approximately 60% of the methotrexate is coupled by the above procedure, yielding a final product containing about 10 μmoles of inhibitor per milliliter of settled gel.

Sephadex G-10 or G-25 is carefully layered over the MTX-AH-Sepharose surface to form a bed extending approximately 1 cm into the upper column segment. Thus, if precipitation occurs during the application of crude extracts, the upper column unit may be disconnected and the layer of Sephadex be removed and replaced prior to the addition of the next extract.

Experience with the MTX-AH-Sepharose[2] column suggests that despite the exhaustive washing of the gel after the coupling reaction, noncovalently bound methotrexate remains strongly adsorbed to the Sepharose. Apparently this is removed on the initial passage of enzyme containing extracts through the column as evidenced by an apparent poor recovery of enzyme activity using a newly prepared column as compared with recoveries approaching 100% on subsequent use of the same column. This problem is alleviated to some extent by pretreatment of the column with concentrated folic acid under alkaline conditions. A solution is prepared by suspending 0.5 g of folic acid in 5 ml of $1\ M$ K_2HPO_4 and adding $1\ N$ KOH dropwise until complete solution occurs. This solution is drained into the column, followed by approximately 100 ml of $1\ M$ K_2HPO_4 and 200 ml of $0.1\ M$ potassium phosphate at pH 5.6 containing $1\ M$ NaCl. The column is washed with 10 mM potassium phosphate at pH 5.6 until the absorbance of the eluate at 280 nm is essentially zero.

After elution of the sample, the affinity column may be regenerated by washing with 100 ml of $0.2\ M$ Tris base followed by approximately 1 liter of 10 mM potassium phosphate at pH 5.6. The useful life of the column appears to be limited by the eventual destruction of the beaded structure of the Sepharose resulting in a markedly diminished flow rate. Initially, removing the adsorbent from the column and defining the material by decantation may restore a reasonable flow rate. Eventually, substances in the crude extracts, particularly from bacteria and yeast, appear to destroy the Sepharose and the bound methotrexate.

Assay Method

The general method of assay for dihydrofolate reductase is that of Mathews et al.[13] However, the following modifications were found to be necessary: Tris maleate buffer is used rather than phosphate, at pH 7.5 for chicken liver and pH 6.5 for calf liver, and EDTA is added to the cuvettes at a final concentration of 20 mM. After the enzyme is eluted from the affinity column, phosphate buffers may be used with either enzyme and EDTA may be omitted from the assay mixture.

A unit of enzyme is defined as the amount that will reduce 1 μmole of

[13] C. K. Mathews, K. G. Scrimgeour, and F. M. Huennekens, this series, Vol. 6 [48].

substrate per minute at the appropriate pH and 28° based on a molar extinction coefficient of 12,300.[14]

Purification Procedure

An ideal method for the large-scale isolation of dihydrofolate reductase via affinity chromatography would involve a minimum of purification steps, particularly column procedures, prior to the affinity column. However, initial experience with the MTX-AH-Sepharose column revealed that the affinity column binds large amounts of extraneous protein if relatively crude extracts are applied. Thus two gel filtration steps are necessary prior to affinity chromatography; otherwise, almost no adsorption occurs. The following purification procedure eliminates this need for pretreatment of the crude extract by gel filtration. All operations are carried out at temperature near 0°.

Dihydrofolate Reductase from Chicken Liver

Step 1. Preparation of Crude Extract. Eight hundred grams of either fresh or frozen chicken liver are cut into small pieces and homogenized in a 1-gallon Waring Blendor (2 minutes at low speed and 2 minutes at medium speed) with 1600 ml (2 volumes) of cold 75 mM manganous chloride. The resulting homogenate is centrifuged at 10,000 rpm for 20 minutes (Sorval Model RC-2B centrifuge GSA rotor). The supernatant fluid is filtered through glass wool to remove fat particles.

Step 2. Fractionation with Zinc Hydroxide. The crude extract is placed in a 4-liter beaker surrounded with ice and equipped with an efficient stirring arrangement. Zinc sulfate (1 M) is slowly added with mixing to a final concentration of 20 mM. Stirring is continued for 5 minutes, then the pH is carefully adjusted to approximately 7.8 by the dropwise addition of concentrated NH_4OH. After an additional 5 minutes of stirring, 0.15 volume of water is added and mixed, and the extract is centrifuged at 10,000 rpm for 15 minutes as above. The large precipitate is discarded, and the supernatant fluid is transferred back to the 4-liter beaker. The pH of the extract is carefully readjusted to 5.5 by the dropwise addition of 5 N acetic acid.

Step 3. Fractionation with Ammonium Sulfate. Solid ammonium sulfate (23 g/100 ml) is slowly added and, after 15 minutes of stirring, the mixture is centrifuged at 10,000 rpm as above for 15 minutes. The precipitate is discarded and an additional 35 g of solid ammonium sulfate per 100 ml of original supernatant solution are added. After stirring for 20 minutes, the mixture is centrifuged and the precipitate is taken up in a

[14] B. Hillcoat, P. Nixon, and R. Blakley, *Anal. Biochem.* **21**, 178 (1967).

minimum volume, about 150–200 ml, of 10 mM potassium phosphate at pH 5.6.

Step 4. Affinity Chromatography. Prior to addition to the MTX-AH-Sepharose column, the preparation is centrifuged at 18,000 rpm (SS-34 Rotor, Sorval) for 20 minutes, and any precipitate is discarded. The clarified deep red solution is adsorbed into the column at a flow rate of not more than 1 ml per minute. This is immediately followed by washing with 100 ml of 50 mM potassium phosphate at pH 5.6, 50 ml of 5 M urea in 0.1 M potassium phosphate at pH 5, and 500 ml or more of 10 mM potassium phosphate at pH 5.6. At this point, the column is ready for a second application of chicken liver extract. As many as 15 applications representing 12 kg of chicken liver have been sequentially applied to the column in this manner over a period of approximately 30 days with excellent final recovery.

Step 5. Elution of Affinity Column. After the final addition to the affinity column, followed by the regular washes as previously described, the column is washed with several liters of the 10 mM potassium phosphate at pH 5.6. The elution of dihydrofolic reductase is brought about by the sequential addition of the following solutions, each carefully added the moment the surface of the column is drained free of liquid: 2 ml of 1 M K$_2$HPO$_4$, 2 ml of 0.5 M K$_2$HPO$_4$ containing 4 μmoles of dihydrofolate, and finally 250 ml of 0.1 M K$_2$HPO$_4$ containing 1 μM dihydrofolate. Fractions of 3 ml are collected with an automatic fraction collector. The active fractions, about 100 ml, are combined and immediately concentrated to approximately 8 ml by ultrafiltration (Amicon Ultrafiltration Cell, equipped with a UM-10 Diaflo membrane and operated under 30 pounds of nitrogen pressure).

Step 6. Chromatography on Hydroxyapatite. The concentrated sample is freed of dihydrofolate and equilibrated with 10 mM potassium phosphate, pH 6.8, by passage through a Sephadex G-25 column (25 × 250 mm) equilibrated with the same buffer. The combined active fractions from the gel filtration are immediately adsorbed into a hydroxyapatite column (20 × 75 mm), prepared according to Bernardi,[15] which had been washed with 500 ml of 10 mM potassium phosphate at pH 6.8. After adsorption of the enzyme, the column is washed with 500 ml of 10 mM potassium phosphate at pH 6.8. Elution of the enzyme is accomplished by a linear gradient of potassium phosphate at pH 6.8 from 10 mM to 0.3 M. The fractions exhibiting the highest specific activity are combined and immediately concentrated to approximately 12 ml by ultrafiltration. The concentrated enzyme solution is dialyzed overnight

[15] G. Bernardi, this series, Vol. 22 [29].

TABLE II
PURIFICATION OF DIHYDROFOLATE REDUCTASE FROM CHICKEN LIVER

Step	Volume (ml)	Reductase activity (μmoles/ min)	Protein (mg)	Specific activity (μmoles/ min/mg)	Re- covery (%)
1. Extraction[a]	35,000	3420	840,000	0.004	—
2. Zinc hydroxide[a]	22,000	2740	105,000	0.026	80
3. Ammonium sulfate[a]	3,500	2560	63,000	0.041	75
4,5. Affinity chromatography	120	2055	—	—	60
6A. Gel filtration	10	2050	370	5.52	—
6B. Hydroxyapatite	30	1650	195	8.50	48

[a] Represents the sum of 17 separate preparations through steps 1–3 and individually adsorbed to the affinity column as described in the text. Approximately 85–95% of the activity in step 3 was adsorbed to the affinity column.

against 2 liters of 0.15 M potassium phosphate at pH 7.8, containing 20% glycerol (v/v).

The purification procedure, based on 14 kg of chicken liver sequentially added to the MTX-AH-Column during a period of 35 days, is summarized in Table II.

Calf Liver Dihydrofolate Reductase

Step 1. Preparation of Crude Extract. Eight hundred grams of fresh or frozen calf liver are extracted by homogenization with 75 mM MnCl$_2$, treated with zinc hydroxide, fractionated with ammonium sulfate, and applied to the MTX-AH-Sepharose column by the identical procedure as previously described for the chicken liver enzyme.

Step 2. Affinity Chromatography. After the extract is adsorbed onto the affinity column, the column is washed in sequence with 100 ml of 50 mM potassium phosphate at pH 5.6, 100 ml of 0.1 M potassium phosphate at pH 5.6 containing 0.5 M NaCl, and 10 mM potassium phosphate at pH 5.6.

As is the case with the chicken liver enzyme, extracts may be sequentially added to the column. After the addition of the final extract and subsequent washing of the affinity column, the activity is eluted in the identical manner described for the chicken liver enzyme.

The combined active fractions are concentrated by ultrafiltration to about 12 ml as described and are dialyzed overnight against 0.15 M potassium phosphate at pH 7.4, containing 20% glycerol (v/v).

Step 3. Fractionation on Sephadex G-75 and Chromatography on Hydroxyapatite. The highly concentrated sample from the affinity column is applied to a Sephadex G-75 column (25 × 1000 mm) equilibrated with 10 mM potassium phosphate at pH 6.8, and eluted with the same buffer. The fractions exhibiting the maximum specific activity are combined and immediately adsorbed into a hydroxyapatite column, prepared and used in the same manner described for the chicken liver enzyme. Elution of the activity is also accomplished by the same linear gradient of phosphate buffer. The fractions exhibiting the highest specific activity are combined, concentrated by ultrafiltration to approximately 12 ml, and dialyzed against 2 liters of 0.15 M potassium phosphate at pH 7.8, containing 20% glycerol.

The purification procedure based on 8 kg of calf liver, sequentially added to the MTX-AH-Sepharose column during a period of 20 days is summarized on Table III.

TABLE III

PURIFICATION OF DIHYDROFOLIC REDUCTASE FROM CALF LIVER

Step	Volume (ml)	Reductase units (μmoles/ min)	Protein (mg)	Specific activity (units/ mg)	Re-covery (%)
1. Extraction[a]	16,500	2020	510,000	0.004	—
2. Zinc hydroxide[a]	18,000	1780	81,000	0.022	88
3. Ammonium sulfate[a]	2,000	1590	32,000	0.049	79
4,5. Affinity chromatography	95	1250	—	—	62
6. Gel filtration	78	1200	109	10.1	59
7. Hydroxyapatite	55	1070	48	22.3	53

[a] Represents the sum of 10 separate preparations through steps 1–3 and individually adsorbed to the affinity column as described in the text. Approximately 85–90% of the activity from step 3 was absorbed by the affinity column.

[23] Pteroyllysine-Agarose in the Purification of Dihydrofolate Reductase[1]

By EDWARD J. PASTORE, LAURENCE T. PLANTE, and ROY L. KISLIUK

Dihydrofolate reductase from antifolate-resistant *Lactobacillus casei* can be isolated in homogeneous form and in excellent yield by the affinity column chromatographic method described here. This method differs from

[1] Supported in part by research grants NIH CA-10914 and CA-11449 and The American Cancer Society BC-60-0.

other currently available affinity column procedures for purification of dihydrofolate reductase in using the ligand pteroyllysine,[2] a folate analog which is a substrate rather than an inhibitor, and which has only moderate affinity for the enzyme. Other affinity columns for dihydrofolate reductase use ligands which are inhibitors,[3] and most often the ligand chosen has been methotrexate,[4-9] which binds extremely tightly to the enzyme ($K_i <$ $10^{-9}\ M$).[10] The advantages of the presently described system are that enzyme elution is achieved easily under very mild conditions, thus allowing quantitative recovery of activity, and that loss of small amounts of ligand from the column during chromatography does not interfere with subsequent enzyme studies.

Assay Method

Dihydrofolate reductase is assayed by measuring the overall decrease in absorbance at 340 nm due to the oxidation of TPNH and the reduction of dihydrofolate according to Eq. (1).

$$H^+ + TPNH + \text{dihydrofolate} \rightleftarrows TPN^+ + \text{tetrahydrofolate} \qquad (1)$$

The molar extinction change at 340 nm for this reaction is approximately 12,000.[10] Assays are carried out at room temperature in 50 mM Tris chloride at pH 7.5 with 0.1 mM dihydrofolate[11,12] and 50 μM TPNH. Protein is determined by the microbiuret procedure.[13,14]

An enzyme unit is defined as that amount of enzyme which catalyzes

[2] L. T. Plante, E. J. Crawford, and M. Friedkin, *J. Biol. Chem.* **242,** 1466 (1967).
[3] J. S. Erickson and C. K. Mathews, *Biochem. Biophys. Res. Commun.* **43,** 1164 (1971).
[4] B. T. Kaufman and J. V. Pierce, *Biochem. Biophys. Res. Commun.* **44,** 608 (1971); see also Kaufman, this volume [22].
[5] M. Poe, N. J. Greenfield, J. M. Hirshfield, M. N. Williams, and K. Hoogsteen, *Biochemistry* **11,** 1023 (1972).
[6] P. C. H. Newbold and N. G. L. Harding, *Biochem. J.* **124,** 1 (1971).
[7] P. L. Chello, A. R. Cashmore, S. A. Jacobs, and J. R. Bertino, *Biochim. Biophys. Acta* **268,** 30 (1972).
[8] J. Gauldie and B. L. Hillcoat, *Biochim. Biophys. Acta* **268,** 35 (1972).
[9] J. M. Whiteley, R. C. Jackson, G. P. Mell, J. H. Drais, and F. M. Huennekens, *Arch. Biochem. Biophys.* **150,** 15 (1972).
[10] R. L. Blakley, "The Biochemistry of Folic Acid and Related Pteridines." American Elsevier, New York, 1969.
[11] R. L. Blakley, *Nature (London)* **188,** 231 (1960).
[12] S. Futterman, this series, Vol. 6 [112].
[13] A. G. Gornall, C. S. Bardawill, and M. M. David, *J. Biol. Chem.* **177,** 751 (1949).
[14] J. L. Bailey, "Techniques in Protein Chemistry," p. 341. American Elsevier, New York, 1967.

the conversion of 1.0 μmole of dihydrofolate to tetrahydrofolate per minute under the assay conditions.

Preparation of Pteroyllysine-Agarose

Procedure for N^2-Acetyl-N^{10}-trifluoroacetylpteroic Acid.[2] Pteroic acid (5 g, 15.6 mmoles) is refluxed with 150 ml of trifluoroacetic anhydride for 2 hours. The solvent is removed under reduced pressure and the residual syrup is washed three times by decantation with 200 ml of water. The amorphous solid is held under reduced pressure (1 mm Hg) to remove the last of the water, giving 5.7 g of N^{10}-trifluoroacetylpteroic acid. Acetic anhydride, 150 ml, is added, and the mixture is heated at 100° for 4 hours. The solvent is evaporated under reduced pressure and the residue is treated with 15 ml of dimethylformamide and 25 ml of water. The mixture is heated at 100° for 0.5 hour and then cooled overnight at 4°. The solid is filtered, washed with water, and dried over P_2O_5 at 1 mm Hg overnight. Approximately 4.5 g of N^2-acetyl-N^{10}-trifluoroacetylpteroic acid are obtained.

Procedure for N^α-Pteroyl-L-lysine.[2] N^2-Acetyl-N^{10}-trifluoroacetylpteroic acid (450 mg, 1 mmole) is dissolved in 8 ml of dimethylformamide containing 0.20 ml of triethylamine. The solution is cooled to 15°, and 0.2 ml of isobutyl chloroformate is added. After stirring at ambient temperature for 0.5 hour, 743 mg of N^ϵ-[*tert*-butyloxycarbonyl]-L-lysine methyl ester hydrochloride[15] and 0.45 ml of triethylamine are added. The mixture is stirred overnight at 30–35° and evaporated to dryness. The residue is triturated twice with 20 ml of water, filtered, and dried. The mixture is dissolved in 10 ml of dimethylformamide and chromatographed on a column (2.5 × 7.5 cm) of DEAE-cellulose in dimethylformamide. The fully blocked N^α-pteroyl-L-lysine is washed through with dimethylformamide, leaving the blocked pteroic acid on the column. The eluate is evaporated under reduced pressure, and the residue is washed with ether and dried, giving approximately 400 mg of fully blocked compound.

Hydrolysis (100 mg of compound in 50 ml of 0.20 N NaOH, under N_2 at 100° for 45 minutes) followed by cooling and adjustment to about pH 3.5 with 12 N HCl yields N^α-pteroyl-N^ϵ-[*tert*-butyloxycarbonyl]-L-lysine, which is separated by centrifugation and washed with water. The N^ϵ-*tert*-butyloxycarbonyl group is removed by suspending 100 mg of the compound in 25 ml of 2 N HCl. After stirring for 1 hour at between 25 and 35°, the solution, is cooled to 10–15°, adjusted to pH 3 and centrifuged. The gel is washed once by suspension with 10 ml of water,

---

[15] R. Schwyzer and W. Rittel, *Helv. Chim. Acta* 44, 159 (1961). Commercially available from Fox Chemical Co., Los Angeles, California.

centrifuged, and lyophilized to dryness. Chromatography on Whatman No. 1 paper in 0.10 M ammonium bicarbonate, ascending, shows an R_f of 0.33 (q,$R_{folic\ acid}$ = 0.50). The solid is free of pteroic acid, R_f = 0.17 (q) and N^α-pteroyl-N^ϵ-(*tert*-butyloxycarbonyl)-L-lysine, R_f = 0.60 (q).

Coupling to Sepharose. Fifty milliliters of Sepharose 4B-200 are activated with 10 g of cyanogen bromide as described[16] and coupled with 6.1 ml of ethanolamine in 50 ml of 0.1 M Na$_2$CO$_3$. The preparation is stirred overnight at 4°, then the suspension is filtered and washed with 30-ml portions of water until pH 5.5. The N-(β-hydroxyethyl)-Sepharose carbamate is activated with 10 g of cyanogen bromide and coupled with a solution of 450 mg of N^α-pteroyl-L-lysine in 50 ml of 0.1 M Na$_2$CO$_3$. The suspension, after stirring overnight at 4° (protected from light) is filtered, and the pteroyllysine-Sepharose (Fig. 1) is washed with 25-ml portions of water until pH 5.5 (the filtrate is reserved). The pteroyllysine-Sepharose is washed with 2 portions of 25 ml each of 0.1 N NaOH, followed by 25-ml portions of water until pH 5.5 is attained. The suspension is stored at 4° in 25 ml of water and protected from light. From the sodium carbonate and sodium hydroxide filtrates the recovery is about 330 mg of N^α-pteroyl-L-lysine.

Source of Enzyme

Dihydrofolate reductase is obtained from dichloromethotrexate- or methotrexate-resistant *L. casei*.[17] Cells are grown and processed exactly as described for isolation of thymidylate synthetase.[18] Dihydrofolate reductase emerges from the hydroxyapatite column in the latter portion of the 80 mM phosphate buffer, pH 6.8, wash and early fractions of the gradient used to desorb thymidylate synthetase. The pooled fractions may

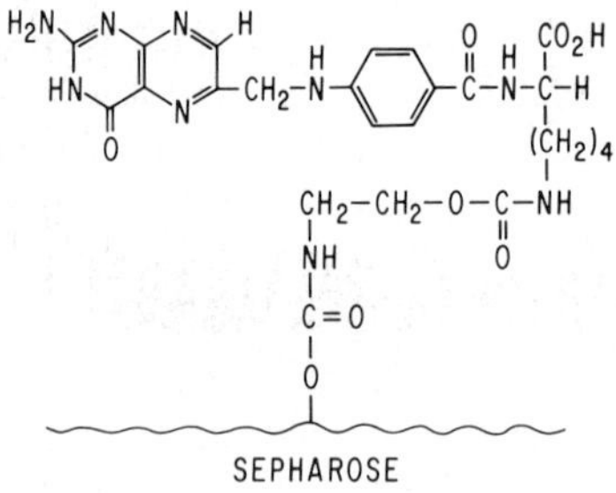

FIG. 1. N^α-Pteroyl-L-lysine-Sepharose.

[16] P. Cuatrecasas, *J. Biol. Chem.* **245**, 3095 (1970).
[17] T. C. Crusberg, R. P. Leary, and R. L. Kisliuk, *J. Biol. Chem.* **245**, 5292 (1970).
[18] R. P. Leary and R. L. Kisliuk, *Prep. Biochem.* **1**, 47 (1971).

be directly applied to the affinity column or, alternatively, may be stored indefinitely as a lyophilized powder at 5°.

Affinity Chromatography

Buffers containing folates and the affinity column should be protected from light. Adsorption and elution of enzymes may be performed at room temperature.

Adsorption and Elution of Enzyme. Pooled enzyme obtained from the hydroxyapatite column (up to 2000 units of activity and 10 g of protein) is poured through a pteroyllysine-Sepharose column (3.5 × 11 cm, ca. 100 ml) which has been equilibrated with 50 mM potassium phosphate at pH 6.7 (buffer A). Washing is continued with several column volumes of the same buffer followed by several column volumes of 50 mM potassium phosphate at pH 7.5 (buffer B). The enzyme is eluted with 200 ml of 0.2 mM dihydrofolate in buffer B and appears as a sharp peak, with 90% of the activity usually obtained in less than 100 ml. The peak from the affinity column is pooled, an aliquot is removed for activity and protein assay, and the remainder is dialyzed overnight at 5° against 20 volumes of buffer A. During chromatography, dihydrofolate present in the eluting buffer breaks down to a brightly colored product which is easily visible at a concentration of 0.2 mM. The absence of this color in the pooled enzyme indicates that there is little dihydrofolate in this fraction and that the enzyme must be displaced from the column virtually by the leading edge of the eluting buffer. Fractions following the activity peak do show the intense bright yellow color as they are being collected. Recovery of enzyme activity from the column is often 100% (see the table).

Concentration. After dialysis against 20 volumes of cold buffer A, in the dark, enzyme is concentrated to approximately 30 mg/ml in two stages using Amicon Diaflo units equipped with UM-2 membranes operating at 50 psi pressure. The initial 100 ml is reduced to approximately 10 ml in a 60-ml (43 mm in diameter) unit, and the final reduction to 3 ml is performed in a microconcentrator (25 mm in diameter). Recovery of

PURIFICATION OF DIHYDROFOLATE REDUCTASE FROM *Lactobacillus casei*

Sample	Protein (mg)	Total units	Specific activity (μmoles/ min/mg)	Purification (-fold)	Recovery (%)
Hydroxyapatite	2800	600	0.21	—	—
Pteroyllysine-Sepharose	38.3	683	17.8	85	114

the enzyme at this stage is usually 75–80%. The major loss of activity occurs during dialysis and may be due in part to penetration of the relatively small enzyme molecules through dialysis tubing. The enzyme yields a single band upon SDS-polyacrylamide gel electrophoresis and has a specific activity comparable to published values for the pure enzyme.[6,9]

Additional Elution Procedures. Enzyme may be readily eluted with 0.2 m*M* folate instead of dihydrofolate as described above. It is more convenient to use folate because it is stable in solution and gives comparable elution and recovery of enzyme. Elution may also be readily achieved with 0.5 *M* KCl in 50 m*M* potassium phosphate at pH 8.0, without either folate or dihydrofolate present; here the elution profile is somewhat broadened. In the former two procedures, i.e., with elution buffers containing either dihydrofolate or folate, the recovered enzymes have bound dihydrofolate and folate, respectively, whereas with the last procedure the enzyme is obtained free of bound folates.

Column Regeneration. Regeneration of the column after use is routinely accomplished by washing with 1 liter of cold 1 *M* NaCl in buffer B followed by 1 liter of buffer A. It has been observed that the enzyme elution peak begins to broaden after 5 to 8 runs on the same column. The performance may be completely restored by washing the pteroyllysine-Sepharose with several column volumes of cold 0.1 *N* NaOH. The base washing, carried out rapidly on a sintered-glass funnel (covered with filter paper to protect the Sepharose beads from damage by stirring under reduced pressure on the glass frit), is followed by several liters of cold water and finally 1 liter of cold buffer A. Columns regenerated in this manner have been in use for over a year without noticeable deterioration in performance. Washing and storage of the column are carried out at 5°.

Comments

Prior to undertaking synthesis of pteroyllysine-Sepharose, several attempts were made to isolate dihydrofolate reductase from antifolate-resistant *L. casei* with existing affinity column procedures. The methotrexate-Sepharose procedure of Kaufman[4] proved to be unsatisfactory because the buffers required to elute the protein from this matrix inactivated the *L. casei* enzyme. Upon a suggestion of the late Dr. B. R. Baker, we tried a Sepharose column having a ligand with only moderate affinity (a triazine, $K_i = 10^{-6}$ *M*)[19] instead of methotrexate. Despite their successful purification of rat liver dihydrofolate reductase, the triazine-Sepharose column gave low and variable recovery in our hands and was therefore unsuitable.

[19] B. R. Baker and N. J. Vermeulen, *J. Med. Chem.* **13**, 1143 (1970).

In search of a more suitable ligand for this enzyme, we applied and extended the principles initially suggested by Dr. Baker; i.e., we looked for a folate analog having only moderate affinity for the enzyme as well as an alkylamine side chain that would permit direct coupling to Sepharose with cyanogen bromide, thereby avoiding the carbodiimide procedure. The former would permit elution without large changes in pH or salt concentration, and the latter would ensure high capacity. Pteroyllysine satisfies both conditions: It has approximately the same affinity as folate for the enzyme and an alkylamine side chain. In addition, pteroyllysine is a substrate rather than an inhibitor of the enzyme. (Its dihydro form is as good a substrate as dihydrofolate itself.[2]) When coupled to Sepharose, pteroyllysine did in fact retain large amounts of the *L. casei* enzyme completely, and this activity was recovered quantitatively by elution with 0.2 mM substrate in a buffer less than 1 pH unit above that used to load the column.

The relatively moderate binding strength of the pteroyllysine-Sepharose column provides a means for obtaining the enzyme in three different forms directly without any additional steps. Elution with dihydrofolate and with folate yields enzyme with stoichiometric amounts of nondialyzable dihydrofolate and folate, respectively. An additional form of the enzyme, free of folates, may also be obtained in excellent yield by elution with 0.5 M KCl in 50 mM potassium phosphate buffer, pH 8.0.

The particular effectiveness of this column over an extended period of time raises several interesting points. For example, it is puzzling why folate itself has not been a satisfactory ligand[6,9,20] since it is very similar to pterolyllysine in structure and affinity. One possible explanation is that a substantial amount of folate coupled to Sepharose with the carbodiimide reagent is not available for binding the enzyme, e.g., linkage through the α-carboxyl group. Another possibility is that coupling of the ligand to Sepharose containing alkylamine spacer groups may be incomplete. The remaining free alkylamine side chains would act as ion exchange sites and result in nonspecific binding of protein which would contaminate the enzyme. Even more surprising is the high capacity of the pteroyllysine column. Despite equal or higher substitution on Sepharose of methotrexate,[6,9] which binds the enzyme at least a thousandfold more tightly than the pteroyllysine ligand, the activity processed by the former columns is considerably less (3.5 and 4 units/ml, respectively), than the activity processed here (20 units/ml, see Affinity Chromatography section).

The effectiveness of this column is not restricted to the enzyme from *L. casei*. Recently Blakley's laboratory[20] has obtained very similar results

[20] R. L. Blakley (1973), private communication.

with dihydrofolate reductase from beef liver; i.e., they obtained homogeneous enzyme in good yield with the pteroyllysine ligand whereas folate and the strongly binding diamines, methotrexate and aminopterin, were not satisfactory.

A final point of interest is the stability of the column despite prolonged use. Since TPNH is bound to crude dihydrofolate reductase,[21] and since pteroyllysine, like folate, is a substrate for the enzyme, some of the ligand could be converted to unstable reduced forms and column performance would deteriorate with use. The stability of the pteroyllysine-Sepharose column in our hands may reflect the fact that prior to affinity chromatography crude extract is passed through a hydroxyapatite column, thereby removing bound TPNH and avoiding reduction of the ligand. It should be kept in mind that continued processing of crude extracts without prior treatment could lead to destruction of the column.

[21] R. B. Dunlap, L. E. Gundersen, and F. M. Huennekens, *Biochem. Biophys. Res. Commun.* **42**, 772 (1971).

[24] Propyl Lipoamide Glass

By WILLIAM H. SCOUTEN

Bioselective adsorption utilizing ligands immobilized on controlled pore glass and similar inorganic matrices have recently been achieved. Although only a few examples of this technique[1] have been reported to date, this method is needed to fully utilize the potential of affinity chromatography in large-scale enzyme isolations or in modern high pressure chromatography. One of the reasons for the heretofore reluctance to use such matrices in affinity chromatography is the fear of nonspecific protein adsorption by the matrix since most of these matrices are known to adsorb proteins tightly. Such nonspecific adsorption, however, has not been observed with properly derivatized lipoamide glass.

This article is concerned with the purification of lipoamide dehydrogenase on a column of lipoamide glass.[2] This particular bioselective adsorbent is readily synthesized as shown below. Not only can this ligand-matrix function as a bioselective adsorbent, but upon reduction to

[1] See this volume [4], for a fuller discussion of the use of controlled pore glass as an affinity chromatography material.
[2] W. H. Scouten, F. Torok, and W. Gitomer, *Biochim. Biophys. Acta* **309**, 521 (1973).

$$\text{Glass}\Big)-\overset{|}{\underset{|}{Si}}-OH \quad + \quad H_5C_2-O-\overset{\overset{OC_2H_5}{|}}{\underset{\underset{OC_2H_5}{|}}{Si}}-(CH_2)_3-NH_2 \quad \longrightarrow \quad \text{Glass}\Big)-\overset{|}{\underset{|}{Si}}-O-\overset{|}{\underset{|}{Si}}-(CH_2)_3-NH_2$$

| Controlled pore glass | γ-Aminopropyl-triethoxysilane | Aminoalkyl glass |

$$\text{Glass}\Big)-\overset{|}{\underset{|}{Si}}-(CH_2)_3-NH_2 \quad + \quad \overset{O}{\overset{\|}{Cl-C}}-(CH_2)_4\underset{S\!-\!-\!-\!S}{\overset{H}{\underset{|}{C}}\overset{H_2}{\underset{|}{C}}CH_2}$$

Aminoalkyl glass   Lipoyl chloride

$$\text{Glass}\Big)-\overset{|}{\underset{|}{Si}}-(CH_2)_3-\overset{}{\underset{H}{N}}-\overset{O}{\overset{\|}{C}}-(CH_2)_3\underset{S\!-\!-\!-\!S}{\overset{H}{\underset{|}{C}}\overset{H_2}{\underset{|}{C}}CH_2}$$

N-Propyl lipoamide glass

dihydrolipoamide glass it may be utilized as an immobilized reducing agent. This use of the ligand-matrix, as a reducing agent, as well as methods for avoiding and/or eliminating nonspecific protein adsorption when using this material are also discussed.

Preparation of the Glass Bead Matrix

The proper preparation of the glass beads is one of the most critical aspects in the synthesis of lipoamide glass. The beads must not have fragile projections that will break during subsequent handling; they must be scrupulously clean, and the pore size and mesh must be chosen carefully. Very adequate beads, complete with a previously attached propyl amine group may be purchased commercially (Pierce Chemical Co.). Considerable time and effort may be saved by using these beads since they have been properly prepared and the first step in the synthesis below, aminoalkylation, has already been performed.

The commercial alkyl amine glass beads are currently available only in 80 + mesh. The author finds these difficult to use when working with narrow diameter columns (1 cm or less). Good, nonderivatized, controlled-pore glass is available from several commercial sources in a wide variety of pore sizes and mesh. Glass beads of 80–120 mesh and a pore diameter of 1200 to 1300 Å are used for the preparation of lipoamide glass.

To prepare the beads for derivatization, 10 g of beads are sonicated in 50 ml of 1 N HCl for 15 minutes using a Branson W-140 Sonifier at its highest output. This both cleans the beads and removes irregularities from the glass surface. The beads are then added to 200 ml of 1 N HCl in a 250-ml graduated cylinder and allowed to settle; the milky supernatant fluid is decanted. The beads are mixed thoroughly with another 200 ml of 1 N HCl and again allowed to settle. This process is repeated 5 or 6 times or until the supernatant liquid is clear.

The beads are collected on a Büchner funnel, washed with distilled water, and refluxed for 30 minutes in 100 ml of 1 N HNO$_3$. This reflux, *and all subsequent reflux procedures,* is carried out in a two-neck 500-ml round-bottom flask fitted with a reflux condenser in the side neck and a small overhead stirrer in the center neck. The stirrer blades are positioned well above the bottom of the flask, and the stirrer itself is attached to a variable speed motor. The rate of stirring is adjusted to the minimum needed to prevent excessive bumping. (*Never* use a magnetic stirrer since the stir bar will grind the glass beads to a fine powder.)

Silicic acid can also be used as the matrix instead of glass in the following steps. A good grade, acid-washed, commercial silicic acid (80–120 mesh) can be used without further preparation. Unisil (Clarkson Chemical Co., Williamsport, Pennsylvania) is one such preparation that has been used successfully in this laboratory.

Aminoalkylation of the Beads

Aminoalkylation is performed by a modification of the method of Weetall.[3] Ten grams of glass beads (or silicic acid) are refluxed for 16 hours in 10% α-aminopropyltriethoxysilane in dry toluene. (The reflux flask is fitted with an overhead stirrer as described.) The beads are filtered on a Büchner funnel and washed with 100 ml of 95% ethanol. The beads are refluxed for 15 minutes in 95% ethanol to complete the removal of unreacted α-aminopropyltriethoxysilane. They are filtered on a Büchner funnel and washed consecutively with 500 ml of 95% ethanol and 500 ml of distilled water. The presence of free primary amine groups in either the beads or the wash solution can be detected by heating with ninhydrin; a deep blue develops in the presence of primary amines.

Alternative methods of aminoalkylation have been reported,[4] but these have not been used in this laboratory.

[3] H. H. Weetall, *Science* **166**, 616 (1969).
[4] P. J. Robinson, P. Dunnill, and M. D. Lilly, *Biochim. Biophys. Acta* **242**, 659 (1971).

Preparation of Lipoamide Glass

Lipoyl chloride[5] is prepared by adding oxalyl chloride (2.7 ml) drop-wise to 8.0 ml of anhydrous benzene in a 50-ml round-bottom flask fitted with a calcium chloride drying tube, and chilled to 0° in an ice bath. Lipoic acid (200 mg) is added with stirring. The reaction is stirred for 15 minutes at 0°, then 100 mg of lipoic acid are added as before. After this procedure, two additional 100-mg portions of lipoic acid are added at 15-minute intervals. After the addition of the final portion of lipoic acid, the reaction is stirred for an additional 150 minutes. The reaction mixture becomes dark orange after the addition of the first 200 mg of lipoic acid and deepens to a dark orange-red when the reaction is complete. At the end of the reaction period, benzene is removed at 25° with the use of a flash evaporator. Ten milliliters of anhydrous benzene is added, and the benzene is removed as before. This is repeated twice more to eliminate the final traces of unreacted oxalyl chloride.

The orange-red oil obtained (lipoyl chloride) is dissolved in 10 ml of anhydrous dioxane, and the solution is added to 10 g of alkyl amine glass beads (either commercial alkyl amine glass beads or those prepared as described above) contained in a 20-ml screw-cap test tube. Pyridine (1 ml) is added with mixing. The tube is filled with anhydrous 1,4-dioxane and tumbled 16 hours at room temperature in a Fisher-Kendall Mixer. Dioxane is removed by flash evaporation from the reaction mixture and the resulting propyl lipoamide glass beads are washed sequentially with 500 ml of 5% HCl, 1500 ml of 10% $NaHCO_3$, 500 ml of distilled water and, finally, with 100 ml of 95% ethanol.

Blocking Unreacted Amine Groups

The lipoamide glass as prepared above may still contain free amine groups which, if left unreacted, would cause the lipoamide glass to behave as an ion exchanger rather than a bioselective adsorbent. To derivatize these unreacted amine groups, the beads are refluxed 10 minutes in acetic anhydride, filtered, and washed with 95% ethanol. A few beads are heated with ninhydrin; if a postive test indicates that unreacted primary amines persist, the above procedure is repeated.

Sulfhydryl content of the glass derivative is determined using 5,5'-dithiobis(2-nitrobenzoate).[6] Yields of from 15 to 50 μmoles of lipo-amide bound per gram of glass are usual when 1200–1300 Å pore beads

[5] A. F. Wagner, E. Walter, G. F. Boxer, M. P. Prass, F. W. Holly, and K. Folkers, *J. Amer. Chem. Soc.* **78**, 5079 (1956).
[6] G. L. Ellman, *Arch. Biochem. Biophys.* **82**, 70 (1959).

are used. The yields vary with the surface area of the original glass beads, and larger yields are obtained with 550 Å pore diameter glass or with silicic acid.

Affinity Chromatography on Propyl Lipoamide Glass

Lipoamidase, an enzyme which hydrolyzes lipoamide and its derivatives, has been reported to be present in crude lipoamide dehydrogenase containing extracts from several sources.[7] Therefore, each crude extract is heat treated (60° for 10 minutes) to destroy the heat-labile lipoamidase.

To purify pig heart lipoamide dehydrogenase, extracts are prepared from 500 g of tissue by the method of Massey[8] through the first heat step. The heat-treated extract is dialyzed for 16 hours against two changes of 1 mM potassium phosphate at pH 6.4, containing 1 mM EDTA. The dialyzed extract is brought to 10% (v/v) in acetone and clarified by centrifugation for 15 minutes at 48,000 g. The clarified extract is applied to a 2.0 × 1.2 cm lipoamide glass column. The column is washed with 5 column volumes of 1 mM potassium phosphate, pH 6.4–10% (v/v) acetone buffer. The enzyme is eluted with 50 mM potassium phosphate, pH 6.4–1 mM EDTA, followed by 1.0 M phosphate at pH 7.5. The enzyme appears in the 50 mM phosphate eluate unless nonspecific adsorption has occurred; elution of the enzyme in the 1.0 M wash implies that nonspecific adsorption has occurred. In a series of more than 20 preparations in the author's laboratory, only one preparation exhibited nonspecific adsorption and that was easily remedied as described below.

Recoveries of 75–85% of the enzyme in the 50 mM phosphate eluate are usual with the above procedure, whereas 60–80% of enzyme activity is retained by the lipoamide glass column in the absence of 10% acetone in the enzyme application buffer. The enzyme purified by this procedure is homogeneous as judged by disc gel electrophoresis.

Lipoamide dehydrogenase can be purified from other sources if heat-treated extracts are applied to the column at the pH optimum of the enzyme. For example, crude *Escherichia coli* sonicates, heat treated at 60° for 10 minutes, can be applied in 1 mM potassium phosphate at pH 8.1 and eluted in 50 mM phosphate at the same pH with results comparable to that described for the pig heart enzyme.

Nonspecific Adsorption

Properly prepared propyl lipoamide glass will not exhibit nonspecific adsorption. One reason for nonspecific adsorption is exposure of regions of nonderivatized glass, possibly due to bead breakage. The other major

---

[7] M. Koike and K. Suzuki, this series Vol. 18A [50]; see also L. J. Reed, M. Koike, M. E. Levitch, and F. R. Leach, *J. Biol. Chem.* **232**, 143 (1958).
[8] V. Massey, this series, Vol. 9 [52].

cause is the presence of free amine groups. If the glass has been treated carefully, neither of these effects will be a problem. If nonspecific adsorption does occur, i.e., the lipoamide dehydrogenase elutes with 1 M phosphate but not with 50 mM phosphate, one may be able to find the source of the difficulty by heating a few beads in ninhydrin. A deep blue color indicates the presence of free amine groups which should be derivatized by additional treatment with acetic anhydride.

If the problem is due to the presence of exposed silanol groups on broken or nonderivatized surfaces, several methods are available for desensitizing the glass surfaces. Treating the column overnight with 10% bovine serum albumin in 1 mM potassium phosphate at pH 6.4, followed by washing with 50 mM phosphate at the same pH is adequate; bovine serum albumin is not present in the final purified enzyme as judged by disc gel electrophoresis.

An alternate and more certain method is to covalently derivatize the free silanol groups. This is done in several ways, for example, by allowing the beads to stand overnight at room temperature in trimethylchlorosilane:hexamethylene disilazane:anhydrous hexane (2:1:7; v/v/v) or to stand overnight in trimethylchlorosilane:triethylamine:anhydrous hexane (4:1:10; v/v/v) or by refluxing the beads for 1 hour in 40% trimethylchlorosilane in anhydrous hexane.

Use of N-Propyl Lipoamide Glass as a Reducing Agent

In addition to its use as an affinity adsorbent, lipoamide glass can be employed, after reduction to dihydrolipoamide glass, as an immobilized reducing agent. Dihydrolipoamide tends to oxidize readily to form a stable intramolecular disulfide, in the form of a thiolane ring, while concomitantly reducing compounds of biological significance, in very much the manner that dithiothreitol reduces biological compounds. Thus the utility

of lipoamide glass appears to be comparable to that of an immobilized dithiothreitol. The major advantage of immobilized dihydrolipoamide on glass is that the reducing agent can be readily removed without danger of reoxidizing the biological compounds one is attempting to reduce. A similar immobilized dihydrolipoamide derivative has been synthesized[9] using agarose as the matrix, although it was not utilized as a bioselective adsorbent. Lipoamide glass is superior to the agarose derivative with reference to dimensional stability, resistance to microbiological degradation, and ease of storage, although the amounts of lipoamide bound to agarose is severalfold greater than that bound to glass.

Propyl lipoamide glass may be reduced very readily by several methods. Perhaps the easiest and cheapest is to allow the beads to stand overnight in 5 M β-mercaptoethanol in 0.1 M phosphate buffer. Essentially quantitative reduction of the column will occur at any pH from 6.5 to 8.5; to avoid column degradation, pH 6.5 to 7.0 is to be preferred. More rapid reduction can be performed by refluxing the beads in 5 M β-mercaptoethanol for 2 hours. A column of dihydrolipoamide glass may be stored at pH 6.5 in 5 M β-mercaptoethanol. When needed, the column is washed free of β-mercaptoethanol with an appropriate reducing buffer, and the material to be reduced is applied. We have found that merely incubating oxidized glutathione, 40 μM in 0.5 M potassium phosphate at pH 8.0, distributed throughout one column bed volume of dihydrolipoamide glass containing a 5-fold molar excess of dihydrolipoamide, will result in quantitative reduction of the glutathione in 20 minutes. More rapid reduction will occur if the column height to diameter ratio is as great as possible and the sample to be reduced is applied in a minimum volume of buffer.

Dihydrolipoamide glass can also be used batchwise, i.e., by adding a 5-fold molar excess to a solution of the material to be reduced and shaking gently. The volume of the solution should be as small as possible since the rate of reduction will depend largely upon the ratio of beads to solution. The reduction is also a function of pH and is most rapid at pH 7.5–8.0.

[9] M. Gorecki and A. Patchornik, *Biochim. Biophys. Acta* 303, 36 (1973).

[25] Pyridoxamine Phosphate

By E. BRAD THOMPSON

Enzymes containing pyridoxal phosphate as their native cofactor may, in some cases, be purified through the use of agarose columns to which pyridoxamine phosphate has been covalently bound. An enzyme for which this method has been used successfully is hepatic tyrosine aminotransferase[1] (EC 2.6.1.5). In addition, such an agarose conjugate has been used by us to enrich for polysomes synthesizing this enzyme.[2] Use of this method for polysome purification has been described in detail in another volume of this series,[3] and therefore will receive less attention here.

Several factors may affect the usefulness of agarose-bound pyridoxamine phosphate gels for enzyme purification. These include the strength of the associations between true cofactor and enzyme; bound cofactor analog and enzyme; and, in addition, enzyme stability, steric hindrance on the column, and the ability to resolve holoenzyme into apoenzyme plus cofactor. To achieve tyrosine aminotransferase binding, it was found essential to provide an inert side arm between the Sepharose and the cofactor analog, and it may be necessary to try side arms of varying length before success is obtained with any specific enzyme.

Preparation of Adsorbents

Two side arms of differing length have been useful. The methods for preparing agarose with pyridoxamine phosphate on these side arms follows. To each milliliter of wet packed Sepharose 4B, an equal volume of water is added. To the suspension at 20–25° is added, with constant gentle stirring, 250 mg of CNBr, dissolved in sufficient water to double the starting volume (HOOD). The slurry is monitored with a pH meter and maintained at pH 11 by addition of 4 N NaOH. If desired, the reaction rate may be slowed by holding the reaction mixture at lower temperatures. At room temperature the reaction is complete, as discerned by no further need for NaOH to maintain pH 11, in about 10 minutes. The activated Sepharose beads are rapidly washed with at least 20 volumes of 0.1 M NaHCO$_3$ at 1–5° on a sintered-glass funnel and are resuspended in buffer for the next addition.

To prepare the first side arm derivative, 0.3 M ethylenediamine in 0.2 M NaHCO$_3$ at pH 9.8 is allowed to react with the washed, activated beads; the ethylenediamine is added rapidly to produce a volume increase over the original CNBr reaction solution of not more than 15%. After constant stirring for 24 hours at 20–25°, the resulting ω-aminoalkyl agarose is washed with copious volumes of distilled water on a sintered-glass filter or Büchner funnel. The packed, washed, ω-aminoalkylated beads are suspended in an equal volume of water, and 2 moles of succinic

[1] J. V. Miller, Jr., P. Cuatrecasas, and E. B. Thompson, *Biochim. Biophys. Acta* **276**, 407 (1972).

[2] J. V. Miller, Jr., P. Cuatrecasas, and E. B. Thompson, *Proc. Nat. Acad. Sci. U.S.* **68**, 1014 (1971).

[3] E. B. Thompson and J. V. Miller, Jr., in this series, Vol. 40 [19].

anhydride per milliliter of the slurry are added at 1–5°. With constant gentle stirring, the pH is quickly adjusted to and maintained at 6 with 5 N NaOH. After completion of the reactions, i.e., when additional NaOH is not required, stirring is continued for 5 hours. The sodium nitrobenzene sulfurate color test[4] can be used to determine the completeness of the reaction.

The derivatized Sepharose is washed with water as before. The beads are slurried in water (1:2, volume of packed beads:water) and mixed rapidly with 5 mg of pyridoxamine phosphate per milliliter of packed beads. After the pH is adjusted to 4.7, 50 mg of 1-ethyl-3-(3-dimethyl aminopropyl) carbodiimide in about 0.15 ml water, for each milliliter of original packed beads is rapidly added to the slurry. The mixture is held at 20–25° and pH 4.7 with constant stirring for 5 hours. The final derivatized agarose beads, containing pyridoxamine phosphate on a succinyl aminoethyl side chain, are washed with 8 liters of 0.1 M NaCl over a 6-hour period to remove unbound reactants. Washing the final derivatized adsorbent with dilute glycylglycine or 1% albumin helps ensure that no exposed activated side groups remain.

Alternatively, a longer side arm can be prepared by allowing the activated Sepharose to react with 3,3'-diaminopropylamine. The procedure is exactly analogous to that for ethylenediamine, described above. The subsequent succinylation and addition of pyridoxamine phosphate are handled in an identical manner.

Example of Enzyme Purification on a Pyridoxamine Phosphate Column: Purification of Hepatic Tyrosine Aminotransferase

Liver of any suitable source is obtained as rapidly as possible after exsanguinating the animal. The wet weight of liver is estimated quickly, and an equivalent volume of 0.1 M KCl, 1 mM in EDTA, at 0° is added. The liver is minced and then homogenized in a Dounce or Potter-Elvehjem homogenizer. After centrifuging at 10,000 g for 20 minutes, the pellet is homogenized again with an equal volume of the KCl-EDTA solution and centrifuged. The supernatant fluids are combined and centrifuged at 100,000 g for 30–60 minutes. The final supernatant fluid is applied directly to the affinity column. Not all the applied activity binds to the adsorbent, even when less than saturating quantities of enzyme activity are applied. About 20–40% of the tyrosine aminotransferase activity in the extract adheres to the column. The fraction adsorbed may be increased to about 70% by prior incubation of the homogenate with 1 mM tyrosine at 37° for approximately 15 minutes followed by dialysis

---

[4] P. Cuatrecasas, *J. Biol. Chem.* **245**, 3059 (1970).

TABLE I

BINDING OF TYROSINE AMINOTRANSFERASE TO PYRIDOXAMINE PHOSPHATE–
SEPHAROSE COLUMNS[a,b]

"Side arm"	Moles of enzyme bound/mole of adsorbent cofactor
None	0
—NHCH$_2$CH$_2$NHC(=O)CH$_2$CH$_2$C(=O)NH—	2.0×10^{-5}
—NH(CH$_2$)$_3$NH(CH$_2$)$_3$NHC(=O)CH$_2$CH$_2$C(=O)NH—	2.5×10^{-4}

[a] Data were obtained from J. V. Miller, Jr., P. Cuatrecasas, and E. B. Thompson, *Biochim. Biophys. Acta* **276**, 407 (1972).

[b] Saturating quantities of enzyme in a 100,000 g supernatant solution were applied over a 12-hour period at 4° to each of the adsorbents, and the amount adsorbed was estimated by assaying the enzyme in each run-through fraction. Of the amount presumed adsorbed, 75–80% was recovered upon elution.

in the cold to remove excess tyrosine. Longer incubation of the extract with the adsorbent tends to increase the extent of enzyme binding. The extent of binding on three adsorbents differing in the length of the side arm connecting pyridoxamine to the agarose is shown in Table I.

The state of the enzyme-containing extract or solution applied to the column is important in several ways. First, it must be low in pyridoxal, which often is added to stabilize the enzyme. Second, either the enzyme must be one that readily dissociates from the cofactor, or preliminary steps must be taken to aid this dissociation. (This may be difficult or well-nigh impossible for some pyridoxal enzymes.) Third, if a tissue extract is used, it should be centrifuged or otherwise partially purified to removed lipids, membranes, and particulate matter, all of which rapidly clog the column and shorten its useful life.

After the extract is applied, the column is washed with 50 mM Tris chloride at pH 7.6, containing 25 mM KCl and 5 mM MgCl$_2$, until no further A_{280} material is observed in the washes. At this stage, the gel is washed with one column volume of the same buffer, containing 50 mM KCl.

Elution is effected with the following buffer: 50 mM Tris chloride, 25 mM KCl, 5 mM MgCl$_2$, 0.5 M NaCl, and 10 mM to 0.1 M pyridoxal phosphate. The pH of the eluting buffer, the concentration of NaCl required, and concentration of pyridoxal phosphate must be determined empirically for each new batch of adsorbent. Generally, 0.5 M NaCl and pH 9–11 are very effective in removing bound enzyme, but lower recovery may be acceptable in order to elute at a pH nearer neutrality, or in the acid range. An example of the effect of pH of eluting buffer on the recovery of enzyme is shown in Table II.

TABLE II
Elution of Bound Tyrosine Aminotransferase from
Pyridoxamine-Succinyl-Aminoethyl-Sepharose[a,b]

pH of eluting buffer	Percentage of bound enzyme eluted
4	50
5	4
6	2
7	8
8	31
9	76
10	83
11	95

[a] Equal amounts of liver extract containing tyrosine aminotransferase were applied to a series of small pyridoxamine succinyl aminoethylated Sepharose columns, and the amount bound was estimated by difference between enzyme activity in the initial extract and the initial effluent. Elution was carried out by introducing one bed volume of a buffer containing 5 mM Tris chloride, 25 mM KCl, 5 mM MgCl$_2$, 0.5 M NaCl, 10 mM pyridoxal phosphate, 30 mM β-mercaptoethanol, and 2.5 mg of albumin per milliliter. The initial pH of this buffer was varied, as shown, for each column. After collecting eluted enzyme, small aliquots for assay were quickly brought to the pH and salt concentration required for optimum enzyme activity in the assay.

[b] Data from J. V. Miller, Jr., P. Cuatrecasas, and E. B. Thompson, *Biochim. Biophys. Acta* **276,** 407 (1972).

It is suggested, in establishing elution conditions, that 0.5 M NaCl be used initially, and the pH varied. The pH may then be held constant and salt concentration varied in 0.1 M increments. Finally pyridoxal phosphate is varied in 10 mM increments between 10 and 100 mM at optimum salt and hydrogen ion concentrations. The higher concentration of cofactor partly inhibits the enzyme and therefore must be removed by dialysis or gel filtration after chromatography.

For routine purification, pH 10, 0.5 M NaCl, and 10 mM pyridoxal phosphate were satisfactory in our hands. Using this combination in the Tris–potassium–magnesium buffer, the enzyme was found to elute at the buffer front at a pH of 8.8–9.2.

To stabilize the enzyme, 30 mM β-mercaptoethanol is added to the eluting buffer, along with one of three additional stabilizers: (1) albumin, 2.5 mg/ml; (2) 0.15 M sucrose and 2 mM spermine; or (3) 10% glycerol. The albumin, although effective in preventing denaturation, complicates determination of specific activity by adding a protein contaminant. Neither spermine-sucrose nor glycerol present this difficulty; of the two, glycerol seems somewhat more effective.

With the albumin-containing eluting solution at pH 10 and with

0.5 M NaCl and 10 mM pyridoxal phosphate present, purification of 125-fold was attained.[1] It is worth emphasizing that although the apparent K_m of this enzyme for pyridoxamine phosphate is $10^{-7}\,M$ and for pyridoxal phosphate, $10^{-8}\,M$,[5] elution with an excess of cofactor alone is not adequate to remove enzyme adhering to the column. High salt and pH adjustments are required in addition.

The adsorbent can be repeatedly regenerated by washing briefly with 0.1 N NaOH or KOH, followed by copious washes of starting buffer.

Example of Polysome Purification on a Pyridoxamine
   Phosphate Column: Purification of Polysomes Capable of
   Synthesizing Tyrosine Aminotransferase

Crude ribosomes are prepared from hepatoma tissue culture cells, previously induced to their maximum synthesis of tyrosine aminotransferase by exposure for 16 hours to 1 μM dexamethasone phosphate (Merck). The cells are removed from the steroid-containing medium by centrifugation at 600 g for 5 minutes, washed twice with Tris–potassium–magnesium buffer at pH 7.6 (see below) at 4°, and then exposed to hypotonic conditions by suspending them in a volume of 5 mM MgCl$_2$ equal to half the packed cell volume, for 10–20 minutes at 4°. The swollen cell suspension is homogenized in a Potter-Elvehjem homogenizer and centrifuged at 20,000 g for 2 hours. The resulting ribosomal pellet is used for affinity chromatography either at once or after storage at or below $-90°$.

To carry out affinity chromatography, 10–15 mg of crude ribosomes are suspended in 3 ml of 50 mM Tris chloride at pH 7.6, containing 25 mM KCl, and 5 mM MgCl$_2$, and applied at 4° to a 5-ml column of a pyridoxamine phosphate–Sepharose preparation which has a succinyl aminoethyl side arm, the *shorter* side arm described above. This adsorbent is used, although its capacity was less than that with the longer side arm, because elution of enzyme from it is more effective at pH 4; elution at alkaline pH destroys the protein-synthesizing capability of the ribosomes. The sample is applied, then the column is washed with 8–10 volumes of the Tris–potassium–magnesium buffer at pH 7.6. Elution of ribosomes is achieved by applying 1–1.5 column volumes of the Tris–potassium–magnesium solution, adjusted to pH 4.0 and supplemented with 10 mM pyridoxal phosphate, 0.5 M NaCl, and 0.1% bovine serum albumin. The bright yellow fraction (due to pyridoxal), with a volume of 10–12 ml and a pH of 5.5–6, is collected and centrifuged at 100,000 g for 1–2 hours. Between 50 and 100 mg of protein are recovered in the resulting ribosomal pellet.

These ribosomes can be stored at $-90°$ or below. When made up in

[5] S. Hayashi, D. K. Granner, and G. M. Tomkins, *J. Biol. Chem.* **242**, 3998 (1967).

a protein-synthesizing system, they are active in protein synthesis. Examination of their synthesis products immunochemically and enzymatically shows them to be considerably enriched with respect to tyrosine aminotransferase synthesis.[2] Fuller details of the use of affinity chromatography for this protein-synthesizing system may be found elsewhere in this series.[3]

[26] FMN-Cellulose and Derivatives and FMN-Agarose

By Michael N. Kazarinoff, Charalampos Arsenis, and Donald B. McCormick

FMN-Cellulose and Derivatives

The following is useful for preparations of affinity materials for purifying certain FMN-dependent enzymes.[1]

FMN-cellulose is prepared by stirring in 10 g of dry pyridinium FMN prepared from the commercial sodium salt by passage through a cation exchanger, such as Amberlite IR-120 (H+), followed by neutralization with pyridine[2] and 25 g of dicyclohexyl carbodiimide to 60 g of cellulose powder (dried several hours at 50° over P_2O_5 at reduced pressure) in 400 ml of anhydrous pyridine. The suspension is kept in the dark at 30° for 5 days with shaking, and the FMN-cellulose is filtered off and washed thoroughly with cold pyridine, dilute HCl, aqueous NaCl, water, ethanol, and diethyl ether until no flavin (A_{450}) can be detected in the washes.

The FMN derivative of cellulose phosphate is made by first treating 10 g of dry riboflavin in 50 ml of anhydrous pyridine with 40 ml of $POCl_3$ added dropwise with stirring. The mixture, containing riboflavin dichlorophosphate, is kept at 50° for 3 hours and then added to a suspension of 80 g of dry cellulose phosphate in 600 ml of anhydrous pyridine. The suspension is kept at 50° for 24 hours with occasional stirring, and the FMN-cellulose phosphate is filtered off and washed as before.

The FMN derivative of DEAE-cellulose is made in the same manner as for FMN-cellulose phosphate, but more extensive washing is necessary to remove the last traces of noncovalently bound flavin.

When wet, the FMN-cellulose compounds appear brownish, but become yellow upon drying. Yields for the coupling of FMN, based on the amount of absorbance at 450 nm of filtrates obtained after hydrolysis in 1 N HCl at 100°, range from 20 to 100 mg of FMN per 100 g of

[1] C. A. Arsenis and D. B. McCormick, J. Biol. Chem. 241, 330 (1966).
[2] J. G. Moffatt and H. G. Khorana, J. Amer. Chem. Soc. 80, 3761 (1958).

RECOVERIES OF ENZYMES FROM COLUMN CHROMATOGRAMS

Enzyme	Columns and solutions of 100 ml each for linear gradient elution	Chromatographic stage	Protein (mg)	Units	Specific activity
NADPH-cytochrome c reductase[a]	1.8×40 cm FMN-cellulose phosphate; 10 mM and 1 M potassium phosphate, pH 7.5	Before	8.0	80	10
		After	0.2	85	425
Pyridoxine (pyridoxamine) 5′-phosphate oxidase[b]	1.8×20 cm FMN-DEAE cellulose; 10 mM potassium phosphate, pH 8, and the same containing 0.2 M NaCl	Before	16.5	1320	80
		After	1.1	1260	1150
Glycolate oxidase[c]	1.5×30 cm FMN-cellulose; 5 mM and 50 mM potassium phosphate, pH 8	Before	36.0	470	13
		After	0.2	720	3600

[a] Assayed by the rate of cytochrome reduction; a unit is $100 \times \Delta A_{550}$ per 5 minutes at 30°.

[b] Pyridoxal 5′-phosphate formed was converted to the phenylhydrazone measured at A_{410}; a unit is 1 nmole per hour at 37°.

[c] Measured by the rate of O_2 uptake; a unit is 1 μl per 10 minutes at 37°.

cellulose or derivative. Significant purifications of spinach glycolate apo-oxidase on FMN-cellulose, yeast NADPH-cytochrome c aporeductase on FMN-cellulose phosphate, and liver pyridoxine (pyridoxamine) 5′-phosphate apooxidase on FMN-DEAE-cellulose have been noted.[1] Typical recoveries are given in the table.

FMN-Agarose

The following procedure is an adaptation of a method used for coupling NAD[+] to commercial Sepharose.[3]

Place 1 volume of Sepharose 4B gel and 1 volume of H_2O in a beaker on a magnetic stirrer in the hood. Insert a standardized pH electrode and a thermometer. For each milliliter of gel, add 220 mg of finely divided CNBr and immediately adjust the pH to 11 by adding 2 N NaOH through a fast-running pipette. Maintain the pH between 10.5 and 11.5 by adding additional 2 N NaOH as necessary and the temperature between 18° and 22° by adding ice. After about 15 minutes base uptake ceases. A large amount of ice is added to lower the temperature to 0–4°. The suspension is poured into a chilled Büchner funnel fitted with a fritted-glass filter and washed with 20–40 ml of cold 0.1 M $NaHCO_3$ buffer (pH 10.0) per milliliter of agarose gel. One milliliter of a 50 mg/ml solution of ε-aminohexanoic acid in 0.1 N $NaHCO_3$ (pH 10.0) is then added per milliliter of the activated gel in an Erlenmeyer flask, and the mixture is stirred gently overnight in the cold room. The coupled gel is washed with 40 ml of water per milliliter of gel on a Büchner funnel and then suspended in 1 ml of 20 mM KP$_i$ (pH 7.0) in an Erlenmeyer flask covered with aluminum foil. To the suspension are added 30 μmoles of FMN and 30 μmoles of 1-ethyl-3(3-dimethylaminopropyl) carbodi-imide·HCl per milliliter of gel. The mixture is stirred for 24 hours at ambient temperature in the dark and then washed on a Büchner funnel with 50 ml of 20 mM KP$_i$ (pH 7.0) per milliliter of coupled gel.

This procedure yields 0.5 μmole of FMN coupled per milliliter of gel as indicated by a time course of the appearance of soluble 450 nm-absorbing material during hydrolysis in 1 N HCl at 100°. Since the coupling is through the polyhydroxyribityl side chain and/or the terminal phosphate, enzymes that show an absolute specificity for either of these functional groups of FMN will not bind to FMN-agarose made in the above manner. However, the FMN-agarose may be useful when interaction with the isoalloxazine ring of the flavin moiety is an important contribution to binding.

---

[3] K. Mosbach, H. Guilford, P.-O. Larson, R. Ohlsson, and M. Scott, *Biochem. J.* **125**, 208 (1971).

[27] Thiamine Pyrophosphate-Agarose and the Purification of a Thiamine-Binding Protein

By Atsuko Matsuura, Akio Iwashima, and Yoshitsugu Nose

Escherichia coli contains a thiamine-binding protein that is thought to be involved in the transport of thiamine through the cell membrane.[1,2] Because of the very small amount of the thiamine-binding protein in the periplasmic space of the cells, a more effective purification procedure than the conventional method[1] was required to isolate reasonable amounts of the highly purified protein. During an investigation into the properties of the thiamine-binding protein from *E. coli,* it has been found that the partially purified protein can bind to thiamine pyrophosphate (TPP) and thiamine monophosphate (TMP), as well as to thiamine.[3] This finding suggested that coupling of the thiamine phosphates to an insoluble matrix might yield an efficient affinity chromatographic system.

Preparation of TPP-Agarose

Fifty milliliters of packed Sepharose 6B is suspended in distilled water to achieve a total volume of 100 ml. Crushed CNBr, 6 g, is added at once to the suspension, which is vigorously mixed by magnetic stirring. The pH of the suspension is adjusted and maintained at 11–11.5 by the addition of $3 N$ NaOH for 1 hour. The suspension is filtered and washed rapidly on a funnel with 2 liters of ice-cold water and thereafter with 1 liter of ice-cold $0.1 M$ NaHCO$_3$. The activated, moist agarose cake is suspended in 100 ml of $0.1 M$ NaHCO$_3$ containing 6 g of ethylene-diamine, and the pH of the slurry is adjusted to 10 with $6 N$ HCl. The slurry is gently stirred at 4° for about 16 hours. The resulting amino-ethyl derivative of Sepharose is filtered, washed with 3 liters of ice-cold water and checked to confirm its amination by the use of 2,4,6-trinitro-benzene sulfonate.[4] The filter cake is suspended in 50 ml of ice-cold water to which is added 0.6 ml of 1-ethyl-3-(3-dimethylaminopropyl) carbodiimide. After the pH of the suspension is adjusted to 6.4 with $6 N$ HCl, 500 mg of TPP is added, and the suspension is stirred at 4° for about 16 hours. The gel is washed with about 4 liters of ice-cold water until the filtrate is negative with the thiochrome fluorescence reaction.[5]

[1] T. Nishimune and R. Hayashi, *Biochim. Biophys. Acta* **244**, 573 (1971).
[2] A. Iwashima, A. Matsuura, and Y. Nose, *J. Bacteriol.* **108**, 1419 (1971).
[3] A. Matsuura, A. Iwashima, and Y. Nose, *J. Vitaminol.* (*Kyoto*) **18**, 29 (1972).
[4] P. Cuatrecasas, *J. Biol. Chem.* **245**, 3059 (1970).
[5] A. Fujita, this series, Vol. 2, p. 622.

The coupling of TPP to the agarose can be confirmed by the thio-chrome fluorescence method as well as manometrically.[6] The amount of TPP bound to agarose is approximately 40 nmoles per milliliter of resin.

TPP-agarose is unstable, and more than 70% of bound TPP is cleaved from the agarose within 1 week at 4°. It is therefore recommended that the affinity matrix be used within a day or two after preparation and that it be washed again before use.

Among the various thiamine derivatives tested, TMP couples to amino-ethyl-Sepharose as well as TPP. On the other hand, free thiamine, thiamine sulfate ester, chloroethylthiamine, and thiamine succinic acid do not lend themselves to coupling under these conditions. TMP-agarose, however, has no significant ability to attach to the thiamine-binding protein from *E. coli*.

Purification of Thiamine-Binding Protein from *E. coli* with TPP-Agarose

A partially purified thiamine-binding protein from *E. coli,* 43 mg (29 ml) at the stage of step 3,[7] is applied to a column of TPP-agarose (1.4 × 8 cm) previously equilibrated with 50 mM potassium phosphate at pH 7.0. Virtually all of the thiamine-binding activity is adsorbed during this process. Activity of thiamine-binding protein is assayed by equilibrium dialysis[2] and the chromatography is carried out at 4°; 2-ml fractions are collected at a flow rate of 4 ml per hour. The column is first washed with 38 ml of 50 mM potassium phosphate (pH 7.0) and then successively eluted with 30 ml of 4 M urea in 50 mM potassium phosphate (pH 7.0) and 30 ml of 8 M urea in 50 mM potassium phosphate (pH 7.0). The binding of the protein to the column is sufficiently strong that elution does not occur with 4 M urea. With 8 M urea in 50 mM potassium phosphate (pH 7.0), the thiamine-binding protein is quantitatively eluted between tubes 53 and 59. The specific activity of the purified protein represents an overall purification of approximately 90-fold with a total recovery of 23%.[7] Attempts to adsorb the binding activity on TPP-agarose directly from a concentrated shock fluid from *E. coli* are also successful. With the crude protein preparation, purification is approximately 30-fold with a yield of 50–60% in a single step.

The thiamine-binding protein, purified by affinity column chromatography, exhibits a single protein band in standard polyacrylamide gel electrophoresis and has a molecular weight of 39,000 as estimated by gel filtration.[7]

---

[6] Y. Aoshima, *Seikagaku* **29**, 861 (1958).

[7] A. Matsuura, A. Iwashima, and Y. Nose, *Biochem. Biophys. Res. Commun.* **51**, 241 (1973).

[28] Vitamin B$_{12}$

By ROBERT H. ALLEN, ROBERT L. BURGER, CAROL S. MEHLMAN,
and PHILIP W. MAJERUS

Animal tissues and body fluids contain a number of vitamin B$_{12}$-binding proteins in addition to the enzymes that utilize vitamin B$_{12}$ derivatives as cofactors. Among such proteins are intrinsic factor, transcobalamin II, and the "R-type" group of proteins.[1,2] Intrinsic factor is present in gastric juice and facilitates vitamin B$_{12}$ absorption from the small intestine. Transcobalamin II is present in plasma and facilitates the cellular uptake of vitamin B$_{12}$ by a number of different tissues. The "R-type" proteins are a group of immunologically related proteins including plasma transcobalamin I and the vitamin B$_{12}$-binding proteins found in granulocytes, milk, saliva, and other tissues and body fluids; the function of the "R-type" is unknown.

The isolation of these proteins in homogeneous form using conventional methods has been difficult or impossible because of both the limited availability of certain tissues and body fluids and the fact that these proteins are present in low concentrations and require purifications of the order of 1000- to 2,000,000-fold in order to achieve homogeneity.

Affinity chromatography has had obvious appeal since these proteins have association constants for vitamin B$_{12}$ of the order of 10^{10} M^{-1} but the development of affinity adsorbents has been hindered by the fact that vitamin B$_{12}$ (Fig. 1) does not contain any primary amino, free carboxyl, phenolic, free imidazole, or sulfhydryl groups such as have been utilized in the case of other ligands to achieve covalent attachment to solid supports.[3] This article outlines methods[4,5] for preparing carboxylic acid derivatives of vitamin B$_{12}$. These derivatives are covalently linked to a substituted Sepharose for affinity chromatography of a number of different vitamin B$_{12}$ binding proteins.[5–9] Other methods for preparing vitamin B$_{12}$

[1] R. Gräsbeck, *Progr. Hematol.* **VI**, 233 (1969).
[2] C. A. Hall and A. E. Finkler, this series, Vol. 18C, p. 108.
[3] P. Cuatrecasas and C. Anfinsen, this series, Vol. 22, p. 345.
[4] R. H. Allen and P. W. Majerus, *J. Biol. Chem.* **247**, 7695 (1972).
[5] R. H. Allen and C. S. Mehlman, *J. Biol. Chem.* **248**, 3670 (1973).
[6] R. H. Allen and P. W. Majerus, *J. Biol. Chem.* **247**, 7702 (1972).
[7] R. H. Allen and P. W. Majerus, *J. Biol. Chem.* **247**, 7709 (1972).
[8] R. H. Allen and C. S. Mehlman, *J. Biol. Chem.* **248**, 3660 (1973).
[9] R. L. Burger and R. H. Allen, manuscript in preparation.

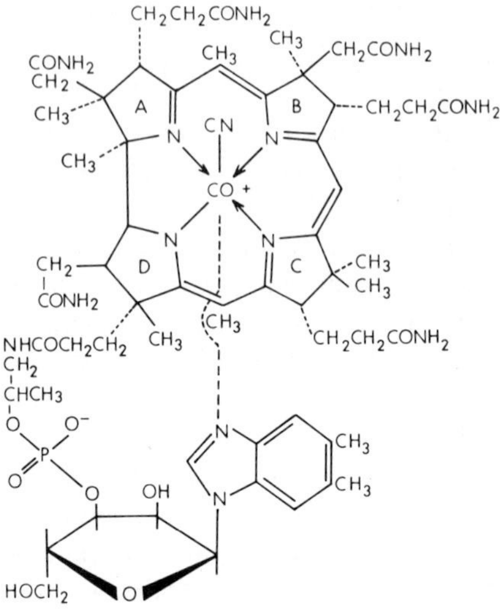

FIG. 1. Structure of vitamin B_{12}.

affinity adsorbents also have been developed recently by Olesen, Hippe, and Haber[10] and Yamada and Hogenkamp.[11]

Preparation of Affinity Adsorbents

Preparation of Monocarboxylic Acid Derivatives of Vitamin B_{12}.[4] Crystalline vitamin B_{12} (4340 μmoles containing 1 μCi of [^{57}Co]vitamin B_{12}) is dissolved in 500 ml of 0.4 N HCl. After standing in the dark at 22° for 64 hours, the deep red solution is chilled to 4° and applied to a column (4.5 × 60 cm) of BioRex AG1-X8 (acetate form, 100–200 mesh BioRad Laboratories) that has been washed with 4 liters of water immediately before use. The sample is applied at a flow rate of 200 ml per hour, and the column is eluted subsequently with 25 mM acetic acid. The eluate is collected as the red material approaches the bottom of the column. More than 95% of the applied radioactivity is present in the first 900–1000 ml of collected eluate. The eluate is lyophilized. The residue is dissolved twice in 300 ml of water followed by lyophilization each time. The final chloride-free residue is dissolved in 300 ml of water, and 55 ml of pyridine are added followed by sufficient concentrated NH₄OH (<1 ml) to adjust the pH to 9.05. The solution, 385 ml, is chilled to 4° and applied to a 2.5 × 56 cm column of QAE-Sephadex A-25 (acetate form) that has been

[10] K. Olesen, E. Hippe, and E. Haber, *Biochim. Biophys. Acta* **243**, 66 (1971).
[11] R. Yamada and H. P. C. Hogenkamp, *J. Biol. Chem.* **247**, 6266 (1972).

washed with 300 ml of 0.2 M pyridine-NH_4OH at pH 9.0. The sample is applied at a flow rate of 100 ml per hour, and 15-ml fractions are collected. After the column is eluted with 600 ml of 0.4 M pyridine, the flow rate is decreased to 50 ml per hour, and a linear acetate gradient is begun in which the mixing chamber contains 1 liter of 0.4 M pyridine and the reservoir contains 1 liter of 0.4 M pyridine–0.16 M acetic acid. A final elution is performed with 1.0 liter of 0.4 M pyridine–0.32 M acetic acid. The first major peak eluting after the start of the acetate gradient contains approximately one-third of the starting material in the form of a mixture of monocarboxylic acid derivatives of vitamin B_{12} in which an amino group is missing from one of the propionamide side chains adjacent to rings A, B, and C (Fig. 1). This material, approximately 300 ml, is lyophilized and dissolved twice in 400 ml of water followed each time by lyophilization to remove pyridine acetate. The final residue is dissolved in sufficient water such that the vitamin B_{12} derivative concentration is approximately 10–11 μmoles/ml, and the solution is stored at −20°.

COMMENT. It is important that the acid hydrolysis step and all of the lyophilization procedures be performed in the dark since at acid pH there is a light catalyzed conversion of vitamin B_{12} (cyanocobalamin), to hydroxycobalamin which is basic with a pK of approximately of 7–8. The monocarboxylic acid derivatives of hydroxycobalamin are not adsorbed to QAE-Sephadex under the conditions described above.

Preparation of Carboxylic Acid Derivatives of Vitamin B_{12} That Lack the Nucleotide Portion of the Native Vitamin.[5] Crystalline vitamin B_{12} (4340 μmoles containing 1 μCi of [^{57}Co]vitamin B_{12}) is dissolved in 500 ml of 11 N HCl and incubated in the dark at 70° for 1 hour. The solution is lyophilized in the dark and the residue is dissolved in 500 ml of water and lyophilized again. The final residue is dissolved in water to give a vitamin B_{12} derivative concentration of 17–20 μmoles/ml and stored at −20°. This preparation consists of a mixture of mono- (7.9%), di- (17.3%), tri- (32.2%), and tetra- (42.6%) carboxylic acid derivatives of vitamin B_{12} that lack the nucleotide portion of the vitamin.

Coupling of Carboxylic Acid Derivatives of Vitamin B_{12} to 3,3′-Diaminodipropylamine-Substituted Sepharose.[4,5] Thirty-four milliliters of water containing 321 μmoles of monocarboxylic acid derivatives of vitamin B_{12} are added to 30 ml of packed 3,3′-diaminodipropylamine-substituted Sepharose[3] and mixed with a magnetic stirrer at 22°. The pH is raised from between 3.0 and 4.0 to pH 5.6 with 0.2 M NaOH (about 1.0–1.5 ml) and, while stirring continues, 3 ml of 1-ethyl-3-(3-diethylaminopropyl) carbodiimide, 50 mg/ml, are added in 0.6-ml aliquots at 1-minute intervals. Gentle stirring is continued in the dark for 18 hours.

The substituted Sepharose is collected by vacuum filtration on a Büchner funnel and washed at 22° successively with 300 ml of water, 600 ml of 0.1 M glycine-NaOH at pH 10.0, 300 ml of water, and 300 ml of 0.1 M potassium phosphate at pH 7.0. The washed substituted Sepharose is deep red and contains from 0.35 to 0.70 μmoles of vitamin B_{12} derivative per milliliter of packed Sepharose, based on measurement of radioactivity. This product is referred to as vitamin B_{12}-Sepharose since the coupling of the single free carboxyl group of the monocarboxylic vitamin B_{12} derivative to the free amino group of the derivatized Sepharose presumably regenerates the amide structure of the vitamin B_{12} molecule attached to Sepharose. The amounts of each of the three monocarboxylic acid isomers that are coupled to Sepharose is not known.

The coupling reaction and subsequent washing of Sepharose substituted with vitamin B_{12} derivatives that lack the nucleotide portion of the vitamin are performed as described above except that the concentration of vitamin B_{12} derivatives in the coupling reaction, prior to the addition of carbodiimide, is 9.2 μmoles/ml. The washed Sepharose contains 0.20–0.30 μmole of vitamin B_{12} derivatives per milliliter of packed Sepharose. The amounts of the different carboxylic acid derivatives and their isomers that are coupled to Sepharose is not known.

COMMENT. Both of the reactions described above give maximal coupling at pH value between 5.0 and 6.0. Experiments have not been performed to determine how the amount of coupling varies as a function of the concentration of the specific vitamin B_{12} derivative.

Preparation of Vitamin B_{12}-Sepharose for Affinity Chromatography. Vitamin B_{12}-Sepharose has been stored at 4° in 0.1 M potassium phosphate at pH 7.0 for periods up to 3 months. Less than 5% of the covalently bound vitamin B_{12} is hydrolyzed during this time period. The rate of hydrolysis is greater if vitamin B_{12}-Sepharose is stored at acidic or basic pH.

Columns of vitamin B_{12}-Sepharose are washed immediately before use to remove the small amount of free vitamin B_{12} that has been hydrolyzed during storage. The columns are washed successively with 10 volumes of (a) 0.1 M glycine-NaOH pH 10.0; (b) 0.1 M potassium phosphate at pH 7.5; (c) 0.1 M potassium phosphate pH 7.5 containing 7.5 M guanidine·HCl; (d) 0.1 M potassium phosphate at pH 7.5; and (e) a solution appropriate to the sample that is to be applied to the column. The washing procedure is performed at a flow rate of 50–100 column volumes per hour.

The same methods are employed for storage and washing of Sepharose that contains derivatives of vitamin B_{12} that lack the nucleotide portion of the vitamin.

Affinity Chromatography

General Considerations

Column Size. Vitamin B_{12}-Sepharose has a functional capacity for vitamin B_{12} binding proteins of from 22 to 50% of its theoretical capacity.[4] In general, the optimal column is of sufficient size that it contains 10–20 times as much vitamin B_{12} as the sample to be applied contains vitamin B_{12} binding ability. Such a column will generally adsorb more than 90% of the applied vitamin B_{12} binding protein at rapid flow rates of 20–50 column volumes per hour.

Preparation of Samples. Vitamin B_{12} binding proteins can be purified directly from body fluids or cellular extracts provided that large volumes of these solutions are able to pass through columns of vitamin B_{12} Sepharose at reasonable flow rates, i.e., 20–50 column volumes per hour. Centrifugation (10,000 g for 30 minutes) followed by vacuum filtration through a layer of Celite in a Büchner funnel containing a coarse sintered-glass disk has been a suitable method of preparation in the case of human gastric juice, human granulocyte extracts, and hog gastric mucosal extracts. Centrifugation followed by vacuum filtration on a Büchner funnel containing a sheet of S & S glass wool No. 25 on top of a sheet of S & S filter paper No. 520-B has been suitable in the case of human milk and saliva. Human plasma transcobalamin II, isolated from Cohn fraction III, must be partially purified on CM-Sephadex and filtered through S & S glass wool and filter paper before its solution is capable of flowing through a column of vitamin B_{12}-Sepharose.

Washing of Vitamin B_{12}-Sepharose after Vitamin B_{12} Binding Protein Adsorption. After the sample has been applied to a column of vitamin B_{12}-Sepharose the initial wash usually consists of 5–10 column volumes of a solution that is similar to the sample in terms of pH and salt concentration. A variety of solutions of various pH and salt concentrations have been employed as additional wash steps although the necessity of including each of the individual wash steps has not been determined. A solution of 0.1 M glycine-NaOH at pH 10.0 containing 1.0 M NaCl has been found to be particularly useful in removing nonspecifically adsorbed protein from vitamin B_{12}-Sepharose; all columns are routinely washed with 50–200 column volumes of this solution. Less than 2% of adsorbed vitamin B_{12} binding protein is eluted by this washing procedure.

Elution of Vitamin B_{12} Binding Proteins from Vitamin B_{12}-Sepharose. After the washing procedure is completed, the column is equilibrated with 0.1 M potassium phosphate at pH 7.5 (5–10 column volumes) and vitamin B_{12} binding protein is eluted with the same buffer containing between 5.0 and 7.5 M guanidine·HCl. The elution of the individual vitamin B_{12}

binding proteins differs in terms of rate of release and the required guanidine-HCl concentration and is discussed in detail in the cited examples.

Renaturation of Vitamin B_{12} Binding Proteins from Guanidine·HCl. Each of the seven vitamin B_{12} binding proteins that have been studied can be renatured, i.e., vitamin B_{12} binding ability can be restored, from guanidine if a 3-fold excess of vitamin B_{12} is added to the eluted protein solution followed by dialysis for 72 hours to remove the guanidine and unbound vitamin B_{12}. Thus, after dialysis 1 mole of vitamin B_{12} is bound per mole of protein. Furthermore, the isolated human[8] and hog[5] intrinsic factors are able to facilitate vitamin B_{12} absorption from the small intestine and isolated transcobalamin II facilitates vitamin B_{12} uptake by fibroblasts.[12]

Human and hog intrinsic factors and hog nonintrinsic factor can also be completely renatured by dialysis in the absence of vitamin B_{12}, but approximately 50% of the vitamin B_{12} binding ability of the other four proteins appears to be lost during this procedure.

Re-use of Vitamin B_{12}-Sepharose. Columns of vitamin B_{12}-Sepharose can be used repeatedly after thorough washing with $0.1 M$ potassium phosphate pH 7.5 containing $7.5 M$ guanidine·HCl.

Purification of Human Vitamin B_{12} Binding Proteins

The purification of five human vitamin B_{12} binding proteins by affinity chromatography is summarized in the table.

R-Type Proteins.[6,9] Single major vitamin B_{12} binding proteins are present in granulocytes, milk, and saliva and all are "R-type" proteins that have many common properties. One of these common properties concerns the fact that these proteins are adsorbed very tightly by vitamin B_{12}-Sepharose and the affinity adsorbent can be washed with 10 column volumes of $0.1 M$ potassium phosphate at pH 7.5 containing $5.0 M$ guanidine·HCl at a flow rate of 20–40 column volumes per hour without eluting more than 2% of the adsorbed vitamin B_{12} binding protein. This is obviously a very effective washing step and is included in the washing procedure for each of these three proteins. They can be eluted with $0.1 M$ potassium phosphate pH 7.5 containing $7.5 M$ guanidine·HCl, but incubation in this solution for approximately 16 hours is required for elution of greater than 90% of the adsorbed vitamin B_{12} binding protein. This is accomplished by passing approximately two column volumes of this solution over the column after which the column is clamped for 16 hours followed by the collection of several additional column volumes of the $7.5 M$ guanidine solution.

---

[12] L. E. Rosenberg, A. Lilljequist, and R. H. Allen, *J. Clin. Invest.* **52**, 69a (1973).

PURIFICATION OF HUMAN VITAMIN B_{12} BINDING PROTEINS

Vitamin B_{12} binding protein	Starting material	Steps before affinity chromatography	Affinity chromatography on vitamin B_{12}-Sepharose							
			Applied			Recovered				
			Volume (l)	Protein (g)	B_{12} binding ability (μg)	Volume (ml)	Protein (mg)	B_{12} binding ability[f] (μg)	Purification (-fold)	Yield (%)
Granulocyte binder[a]	2.7×10^{11} granulocytes	Sonication, centrifugation, filtration	0.63	6.25	22.1	40	0.58	20.4	9860	92
Milk binder[b]	0.6 liters milk	Centrifugation, filtration	0.52	7.75	117	32	4.76	97.9	1360	84
Saliva binder[b]	4.8 liters saliva	Centrifugation, filtration	4.44	7.76	101	33	3.90	80.0	1510	79
Intrinsic factor[c]	3.3 liters gastric juice	Filtration	3.75	7.54	290	44	8.25	248	780	86
Transcobalamin II[d,e]	1400 liters plasma	Cohn fractionation, CM-Sephadex, filtration	19.7	760	447	62	24.7	353	24300	79

[a] R. H. Allen and P. W. Majerus, *J. Biol. Chem.* **247**, 7702 (1972).
[b] R. L. Burger and R. H. Allen, *J. Biol. Chem.* **249**, 1974, in press.
[c] R. H. Allen and C. S. Mehlman, *J. Biol. Chem.* **248**, 3660 (1973).
[d] R. H. Allen and P. W. Majerus, *J. Biol. Chem.* **247**, 7709 (1972).
[e] Following this step, the preparation underwent chromatography on DEAE-cellulose and Sephadex G-100. The overall purification was 2,000,000-fold with a yield of 12.8%. The final preparation contained 6.26 mg of homogeneous protein and 176 μg of vitamin B_{12}.
[f] Vitamin B_{12} content after the addition of excess vitamin B_{12} followed by exhaustive dialysis.

Intrinsic Factor.[8] Intrinsic factor is the major vitamin B_{12} binding protein present in human gastric juice although variable amounts of R-type protein that range from 0 to 40% of the total vitamin B_{12} binding ability may also be present. Most, if not all, of the R-type protein appears to be the result of contamination of gastric juice with saliva, bile, and pancreatic and intestinal juice since with careful collection of gastric juice the amount of R-type protein averages only approximately 4% of the total vitamin B_{12} binding ability.

Intrinsic factor was purified originally using gastric juice in which intrinsic factor represented 96% of the total vitamin B_{12} binding ability. Intrinsic factor is rapidly eluted with solutions of $5.0\,M$ guanidine·HCl, and this wash step could not be employed although washing with $0.1\,M$ glycine-NaOH at pH 10.0 containing $1.0\,M$ NaCl was sufficient to remove all the proteins that do not bind vitamin B_{12}. Immunological studies indicated that at least 98% of the vitamin B_{12} binding activity of the single guanidine eluate $(7.5\,M)$ was attributable to intrinsic factor.

We have recently[13] been able to obtain homogeneous preparations of intrinsic factor starting with gastric juice that contains 40% "R-type" vitamin B_{12} binding protein. This is possible since elution of vitamin B_{12}-Sepharose with 10 column volumes of $5.0\,M$ guanidine·HCl elutes intrinsic factor ($>99\%$) with only minimal release ($<1\%$) of the R-type protein. The R-type protein can be eluted subsequently with a 16-hour incubation with $7.5\,M$ guanidine·HCl.

Transcobalamin II.[7] Partially purified preparations of transcobalamin II can be purified 24,000-fold with affinity chromatography on vitamin B_{12}-Sepharose (see the table), but approximately 50% of contaminating protein remains after this purification step. Washing of vitamin B_{12}-Sepharose with solutions of varying salt concentration and pH values ranging from 4.5 to 10.0 have not resulted in elution of the contaminating protein. Because transcobalamin II is eluted with $5.0\,M$ guanidine·HCl, this wash step cannot be employed. The nature of the contaminating protein and the reason for its high affinity for vitamin B_{12}-Sepharose is unknown. Transcobalamin II can be isolated in homogeneous form after additional purification on DEAE-cellulose and Sephadex G-150.

Purification and Separation of Hog Gastric Vitamin B_{12}
    Binding Proteins[5]

Crude extracts of hog gastric mucosa contain two distinct vitamin B_{12} binding proteins that have been designated as hog intrinsic factor and hog nonintrinsic factor. The latter appears to be an R-type protein, but differs

[13] R. L. Burger, C. S. Mehlman, and R. H. Allen, unpublished observations, 1973.

from the human R-type binders in that significant amounts of it are eluted
from vitamin B_{12}-Sepharose with 5.0 M guanidine·HCl in 0.1 M potassium phosphate at pH 7.5.

The two hog gastric vitamin B_{12} binding proteins are purified by
affinity chromatography on vitamin B_{12}-Sepharose, and both proteins are
eluted together after a 1-hour incubation with 0.1 M potassium phosphate
at pH 7.5 containing 7.5 M guanidine·HCl. The relative amounts of hog
intrinsic factor (~25%) and hog nonintrinsic factor (~75%) are essentially the same as is present in the starting material. Separation of these
two proteins is then achieved by "selective" affinity chromatography as
shown in Fig. 2, using the affinity adsorbent that contains carboxylic acid
derivatives of vitamin B_{12} that lack the nucleotide portion of the vitamin.
Similar experiments performed with vitamin B_{12}-Sepharose do not result
in satisfactory separation of the two proteins.

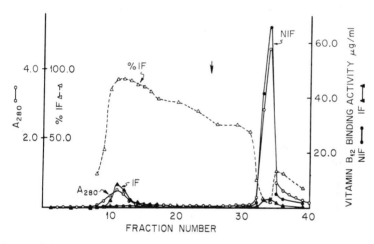

Fig. 2. Selective affinity chromatography of hog intrinsic factor and hog nonintrinsic factor. A column (2.5 × 18 cm) of substituted Sepharose containing
covalently bound derivatives of vitamin B_{12} (0.22 μmole/ml) that lack the nucleotide portion of the vitamin was prepared at room temperature and equilibrated
with 0.1 M sodium acetate, pH 5.0, containing 1.0 M guanidine·HCl. The sample
consisted of a partially purified mixture of hog intrinsic factor (520 μg of vitamin
B_{12} binding ability) and hog nonintrinsic factor (1610 μg of vitamin B_{12} binding
ability) in 11.0 ml of equilibrating buffer. The column was eluted initially with
equilibrating buffer followed, as indicated by the position of the arrow, by 0.1 M
sodium acetate, pH 5.0, containing 7.5 M guanidine·HCl. The flow rate was 50 ml
per hour, and 9.0-ml fractions were collected. ○, A_{280}; △, percentage of vitamin
B_{12} binding ability attributable to hog intrinsic factor as determined by the use of
anti-intrinsic factor antibody to block vitamin B_{12} binding; ▲, hog intrinsic factor;
●, hog nonintrinsic factor. Data from R. H. Allen and C. S. Mehlman, *J. Biol.
Chem.* **248**, 3670 (1973).

After "selective" affinity chromatography hog intrinsic factor is subjected to repeat affinity chromatography on vitamin B_{12}-Sepharose, and hog nonintrinsic factor is further purified by gel filtration followed by repeat affinity chromatography on vitamin B_{12}-Sepharose. The final preparations of both proteins appear homogeneous by gel electrophoresis and contain less than 5% contamination with each other. The overall yield for both proteins is approximately 20%.

Section III

Attachment of Specific Ligands and Methods for Specific Proteins

B. Sugars and Derivatives

[29] Monosaccharides Attached to Agarose

By ROBERT BARKER, CHAO-KUO CHIANG, IAN P. TRAYER,
and ROBERT L. HILL

Agarose derivatives of monosaccharides are easily synthesized by reaction of cyanogen bromide-activated agarose with a monosaccharide
derivative containing an amino function. Few naturally occurring monosaccharides contain an amino group, and those that have such a group
are unlikely to yield satisfactory adsorbents when coupled to agarose.
Usually it will be necessary to couple monosaccharides through a spacer
so that they extend away from the agarose matrix. Thus, the major problem in preparation of a specific agarose adsorbent is the synthesis of the
appropriate monosaccharide derivative. Four types of derivatives, which
have been synthesized and attached to agarose, have the following general
structures.

O–Glycosides   N–Acylglycosylamines   N–Acylamino sugars   Thioglycosides

The affinity adsorbents prepared by coupling these types of ligands to
agarose have been used successfully in the purification of enzymes and
monosaccharide binding proteins.[1-6] They appear to have a high degree of
specificity, are sufficiently stable to withstand repeated use, and should
find wide application in the future.

Glucose and galactose derivatives of type 4 are commercially available
and the thiogalactoside has been used to purify β-galactosidase.[6] The preparation of type 4 derivatives will not be presented in this article.

[1] R. Barker, K. W. Olsen, J. H. Shaper, and R. L. Hill, J. Biol. Chem. 247, 7135
(1972).
[2] J. A. Gordon, S. Blumberg, H. Lis, and N. Sharon, FEBS Lett. 24, 193 (1972).
[3] J. H. Shaper, R. Barker, and R. L. Hill, Anal. Biochem. 53, 564 (1973).
[4] R. Lotan, A. E. S. Gussin, H. Lis, and N. Sharon, Biochem. Biophys. Res.
Commun. 52, 656 (1973).
[5] S. Blumberg, J. Hildesheim, J. Yariv, and K. S. Wilson, Biochim. Biophys. Acta
264, 171 (1972).
[6] E. Steers, Jr., P. Cuatrecasas, and H. B. Pollard, J. Biol. Chem. 246, 196 (1971).
See also this volume [34].

Synthesis of Ligands

A. O-Glycosides (1)

The 6-amino-1-hexyl or 2-amino-1-ethyl glycosides of glucose, galactose, xylose, N-acetylglucosamine, and N-acetylgalactosamine have been synthesized by the following procedures. The synthesis of the 6-amino-1-hexyl glycoside of N-acetylglucosamine is described in detail and serves to illustrate the general procedure. The variations required in the synthesis of similar O-glycosides of other monosaccharides are given below in the section "Other O-glycosides."

6-Amino-1-hexyl 2-Acetamido-2-deoxy-β-D-glycopyranoside (10)

Scheme I shows the synthetic steps for synthesis of this ligand.

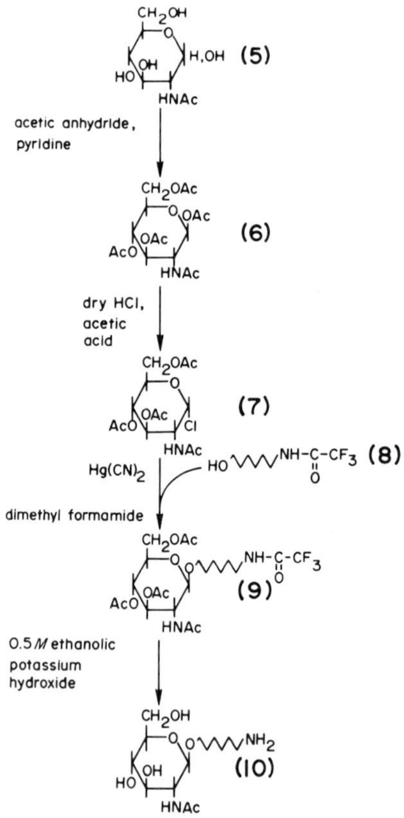

Scheme I

3,4,6,-Tri-O-acetyl-2-acetamido-2-deoxy-β-D-glucopyranosyl Chloride **(7)** *from 2-Acetamido-2-deoxy-D-glucose* **(5).** To 5 g (22.6 mmoles) of 2-acetamido-2-deoxy-D-glucose **(5)** suspended with stirring in 50 ml of dry, reagent grade pyridine[7] was added 15 ml of acetic anhydride during 5–10 minutes. Stirring was continued at room temperature until dissolution was complete (5 hours). The reaction mixture was kept at room temperature for 30 hours and 5 ml of water were added to hydrolyze excess acetic anhydride. After 2 hours the reaction mixture was concentrated at 50° at water aspirator pressure. The residue was dissolved in 100 ml of chloroform (or methylene chloride) and the solution was extracted successively with two 100-ml portions of each of the following, cold aqueous solutions; 2 N sulfuric acid, saturated sodium bicarbonate, and water. The organic layer was dried over solid anhydrous magnesium sulfate (or sodium sulfate), filtered through a sintered-glass funnel and concentrated[8] to give a clear syrup that crystallizes on standing. The yield is essentially quantitative, and further purification is not essential. The material can be recrystallized by dissolving in warm benzene (50 ml) and adding hexane or light-boiling petroleum ether to produce a slight turbidity. Refrigeration after cooling gives crystals of the pure β-anomer **(6)** 8.0 g, mp 188–189°.

Further crops can be obtained by addition of hexane to the mother liquor from the first crop.

The chloro derivative **(7)** is formed by dissolving the acetate **(6)** (3.0 g) in 40 ml of acetic anhydride, cooling the solution in an ice bath, and saturating with dry gaseous hydrogen chloride. When the solution is saturated at 4° to 5°, as judged by weighing the reaction vessel, it is tightly stoppered and kept at room temperature for 2 days. The reaction mixture is diluted with 100 ml of chloroform or methylene chloride and poured over 1 liter of chipped ice and water; the mixture is stirred vigorously. After 2 hours, during which more ice may be added, the hydrolysis of acetic anhydride is essentially complete and the organic phase is extracted with several 200-ml portions of saturated sodium bicarbonate and then with water. The organic layer is dried over either anhydrous magnesium sulfate or sodium sulfate. The drying agent is removed by filtration and solvent removed at 40°[8] to give a residue which crystallizes readily. Recrystallization can be effected by dissolution in 20 ml of chloroform followed by the addition of hexane or petroleum ether as described for the acetate above. The product **(7)** mp 120–121°, is obtained in 60–70%

---

[7] Solvents are reagent grade dried over 4 Å molecular sieves purchased from Fischer Scientific Co., Pittsburgh, Pennsylvania.

[8] Concentrations are carried out at water aspirator pressure with a rotary evaporator.

yield and is suitable for glycoside formation without further purification.

N-Trifluoroacetyl-6-amino-1-hexanol (8). To 0.1 mole (11.7 g) of 6-amino-1-hexanol[9] dissolved in 100 ml of dioxane is added 18 ml of ethyl trifluorothiol acetate,[10] in 1 ml portions during 3 hours with continuous stirring in a hood. The reaction is complete when a ninhydrin test[11] gives a faint pink color. The dioxane is removed,[8] and the residue is reconcentrated 4–5 times from dioxane. The final residual material is stored over Drierite under reduced pressure. Crystallization occurs during 24 hours; mp 54°. The yield is essentially quantitative, and the product gives the correct elemental analyses.

6-Amino-1-hexyl 2-Acetamido-2-deoxy-β-D-glucopyranoside (10). To a solution of **7** (4 g, 11.1 mmoles) in 20 ml of dry dimethyl formamide[7] are added **8** (3.0 g, 14.0 mmole) and mercuric cyanide (3.0 g, 12 mmoles). The mixture is stirred to dissolve the reactants and then stored in a vacuum desiccator over Drierite and potassium hydroxide pellets under a slight vacuum. After 2 days the solvent is removed at 50° and 0.1 mm pressure, and the residue is partitioned between methylene chloride (50 ml) and water. After several extractions with 100-ml portions of water the methylene chloride layer is dried over anhydrous magnesium sulfate, filtered, and concentrated[8] to give 3.5 g of crystalline material. Recrystallization from ethyl acetate (30 ml) by the addition of hexane gives 1.9 g of material which is recrystallized from the same solvents to give 1.6 g of pure product, *N*-trifluoroacetyl-6-amino-1-hexyl-3,4,6-tri-*O*-acetyl-2-acetamido-2-deoxy-β-D-glucopyranoside (9); mp 165.5–166.5, $[\alpha]_D^{25} = +3.0 \pm 1.0°$ (3.0 g per 100 ml of ethyl acetate).

The final product, **10**, can be obtained in a form suitable for coupling to agarose by dissolving **9** in boiling ethyl alcohol (1.0 g per 20 ml) and adding 20 ml of 1.0 *M* aqueous sodium or potassium hydroxide. Heating is continued for 30 minutes, and the volume is maintained between 30 and 40 ml by the addition of water. The solution is adjusted to pH 10 for coupling to agarose (Section D). Using radioactive 2-acetamido-2-deoxy-D-glucose it has been shown that up to 10 μmoles of ligand can be coupled per milliliter of agarose when 20 μmoles per milliliter of ligand are used in the coupling step.

Other O-Glucosides

The aminohexyl glucoside of *N*-acetylgalactosamine can be prepared readily by the procedures described above as can the 2-amino-1-ethyl

---

[9] Aldrich Chemical Co., Milwaukee, Wisconsin.
[10] Pierce Chemical Co., Rockford, Illinois.

glycosides of both monosaccharides using ethanolamine instead of 6-amino-1-hexanol. There is substantial variation, however, in the case of glycoside formation with various simple sugars by the above procedures.

In the case of the acetochloro derivatives of glucose and galactose, reaction with N-trifluoroacetylamino alcohol in the presence of $Hg(CN)_2$ is very slow, and poor yields have been obtained. To circumvent this problem, it is advisable to prepare the 6-amino-1-hexyl and 2-amino-1-ethyl glycosides of glucose, galactose, and xylose from the acetobromo derivatives of the sugars using Ag_2CO_3 in chloroform for condensation with the desired N-trifluoroacetyl amino alcohol. The acetobromo derivatives are prepared by the procedures of Fletcher.[12]

Acetobromo Sugars. To 10 mmoles of sugar acetate dissolved in 30 ml of dry 1,2-dichloroethane,[7] are added 7 ml of 30–32% hydrobromic acid in glacial acetic acid.[13] The tightly stoppered flask is kept at room temperature for 40–60 minutes. Toluene,[7] 50 ml, is added, and the mixture is concentrated.[8] The residue is dissolved in toluene and concentrated repeatedly (3–5 times). The product is suitable for use in glycosidations without purification. In some cases, crystallization occurs or can be induced by dissolution of the syrupy product in anhydrous ether followed by the addition of dry hexane to turbidity. Acetobromo sugars are not very stable and are best prepared immediately before use.

Glycoside Formation. In general, acetobromo derivatives of simple sugars give glycosides in yields of 40–60% under the following conditions. To a mixture of ethanol-free chloroform[14] (40 ml), N-trifluoroacetylamino alcohol (11 mmoles), molecular sieves 4 Å[7] (10 g), Ag_2CO_3 (15 mmoles), and a trace amount of iodine is added the acetobromo derivatives (10 mmoles). The reaction is stirred in the dark in a closed vessel at room temperature for 18 hours and then filtered over a pad of Celite, which is washed with 200 ml of chloroform. The filtrate and washings are extracted 3 times with 100 ml of water, dried over anhydrous magnesium sulfate and concentrated[8] to dryness. The residue consists of a mixture of the desired product, the isomeric orthoester, deacetylated monosaccharides, and deacetylated product as determined by thin-layer chromatog-

---

[11] Amines can be detected by spotting samples on paper and spraying with a solution containing 0.5 g ninhydrin, 20 ml glacial acetic acid, 80 ml absolute ethanol, and 1 ml s-collidine and heating for 5 minutes at 110° [A. L. Levy and D. Chung, *Anal. Chem.* **25**, 396 (1953)].

[12] H. G. Fletcher, Jr., *Methods Carbohydr. Chem.* **2**, 226 (1963).

[13] Eastman Kodak Co., Rochester, New York.

[14] Ethanol-free chloroform is prepared by passing 500 ml of reagent grade chloroform through a column packed with 200 g of chromatographic grade alumina purchased from Fischer Scientific Co., Pittsburgh, Pennsylvania.

raphy.[15] In some cases, in which the yield of glycoside is high, the desired product can be isolated by direct crystallization. To attempt this, the residue is dissolved in 30 ml of ether, and hexane is added to slight turbidity; crystallization is induced by cooling. Generally, it is advisable to purify the product by column chromatography over silica gel before crystallization is attempted.

At least 20 g of silica gel should be used for each gram of reaction mixture, and the column should have a length-to-diameter ratio of approximately 20. The column is formed from a slurry of silica gel in methylene chloride that has been well stirred to remove entrapped air. The column is packed by gravity and is not allowed to dry after being formed. The sample is dissolved in methylene chloride (3 ml/g), applied to the column and rinsed on with small portions of solvent. Development is accomplished at room temperature using methylene chloride containing 5, 10, 15, and 20% of ethyl acetate (v/v). Approximately 200 ml of each mixed solvent is used for each gram of sample applied to the column, and fractions are collected automatically. Columns are monitored by concentrating samples from fractions at 60° in a hood and examining the residues chromatographically.[15] Fractions containing the same components are pooled and concentrated.[8] In general, compounds are eluted in the fol-

R_f VALUES OF D-GLUCOPYRANOSE DERIVATIVES RELATIVE TO
β-D-GLUCOPYRANOSE PENTAACETATE[a]

Derivative	R_f relative to A
A. β-D-Glucopyranose pentaacetate	1.00
B. 2,3,4,6-Tetra-O-acetyl-D-glucopyranose	0.63
C. 2,3,4,6-Tetra-O-acetyl-β-D-glucopyranosyl bromide	1.25
D. 3,4,6-Tri-O-acetyl-1,2-O-(1-N-trifluoroacetyl 6-amino-1-hexoxy)-β-D-glucopyranose (orthoester)	0.98
E. 6-Amino-1-hexyl 2-acetamido-3,4,6-tri-O-acetyl-β-D-glucopyranoside	0.90
F. N-Trifluoroacetyl 6-aminohexan-1-ol	0.58
G. O-Acetyl-N-trifluoroacetyl 6-aminohexan-1-ol	1.38

[a] See text footnote 15.

[15] Thin-layer chromatography is performed with silica gel G as the adsorbent and ethylacetate:hexane (1:1, v/v) as developer. All compounds can be detected as brown spots by spraying with 10% (v/v) aqueous sulfuric acid and heating on a hot plate or in an oven at 110° for 15 minutes. Reducing sugars give yellow-brown spots when plates are sprayed with 1 N aqueous alkali and heated. Trifluoroacetylamino compounds give pink spots when sprayed with 1 N aqueous alkali and then with ninhydrin[10] and amines give pink spots without pretreatment with alkali.

lowing order: N-trifluoroacetylamino alcohol acetate, an unidentified monosaccharide, orthoester, glycoside, deacetylated monosaccharide, and deacetylated glycoside.

The table lists approximate R_f values for various derivatives of D-glucose relative to that of the fully acetylated compound, β-D-glucopyranose pentaacetate. Derivatives of other sugars have similar patterns of migration in this system and the R_f values in the table serve as a guide to the identification of impurities in preparations of their glycosides.

B. Acyl-N-glycosides (2)

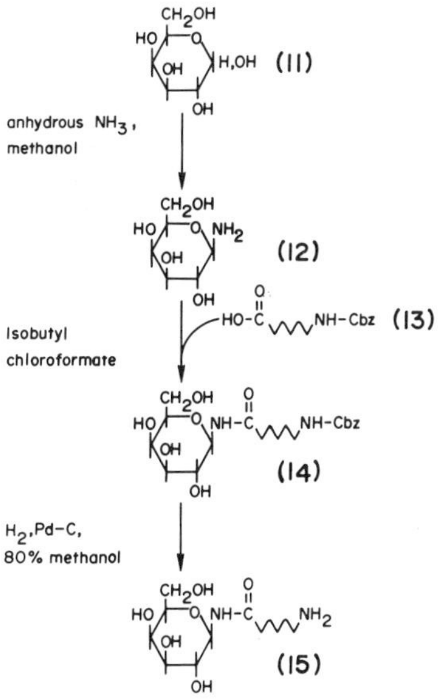

Scheme II

The series of reactions in Scheme II have been applied to the synthesis of acyl-N-glycosides of D-galactose[2] and L-fucose.[5] The synthesis of the D-galactose derivative (15) is typical.

β-D-*Galactopyranosylamine*[2,16,17] (12). D-Galactose (11) (27 g, 150 mmoles) dissolved in 200 ml of anhydrous ammonia in methanol (30% v/v) and stored in a closed vessel at room temperature. After 1 week, the precipitated α-D-galactopyranosylamine is discarded and the supernatant liquid is stored at room temperature for 3–4 days. During this period the

[16] C. A. Lobry de Bruyn and F. H. van Leent, *Recl. Trav. Chim. Pays-Bas* 14, 134 (1895).

[17] H. S. Isbell and H. L. Frush, *J. Org. Chem.* 23, 1309 (1958).

vessel is opened each day for 1–2 hours to permit the evaporation of ammonia. Crystalline β-D-galactopyranosylamine is deposited and collected by filtration, washed with absolute methanol, and stored under reduced pressure over sodium hydroxide pellets. Yield 7 g, mp 137–138°. The product gives the correct elemental analysis.

N-(ε-Aminocaproyl)-β-D-galactopyranosylamine (**15**). To *N*-Benzyloxycarbonyl-ε-aminocaprioc acid[18] (**13**, 30 mmoles) in 50 ml of dry dimethylformamide[7] at −5° is added 30 mmoles of isobutyl chloroformate[13] and 30 mmoles of triethylamine. The mixture is stirred at −5° for 20 minutes and filtered into a suspension of β-D-galactopyranosylamine (**12**, 25 mmoles) in 50 ml of dry dimethylformamide. The mixture is stirred until solution is complete (1 hour) and then stored at room temperature for 16 hours. The solvent is removed,[8] and the product is crystallized twice from ethanol. Yield 60%, mp 159–160°, the material gives the correct elemental analysis for **14**.

A solution of **14** (10 mmoles) in 80% aqueous methanol (100 ml) is hydrogenated over 10% palladium on charcoal at room temperature and atmospheric pressure. After filtration and removal of the solvent,[8] the residue is crystallized from methanol-ether; yield 75%, mp 206–208°. The material gives the correct elemental analysis for **15**.

Other Acyl N-Glycosides. Generally, glycosylamines can be prepared by the method described above[17] and are usually crystalline compounds. The reaction of glycoses with ammonia is freely reversible and, with some sugars, products other than the desired glycosylamine may be formed.[19]

As an alternative to the condensation of glycosylamine with the mixed anhydride formed from isobutyl chloroformate and *N*-benzyloxycarbonyl-ε-aminocaproic acid, the *N*-hydroxysuccinimide ester of *N*-trifluoroacetyl-ε-aminocaproic acid can be used in dry dimethylformamide solution. *N*-Trifluoroacetyl-ε-aminocaproic acid is prepared by the method of Schallenberg and Calvin[20] (as described in the following section) and its *N*-hydroxysuccinimide ester by the method of Anderson *et al.*[21]

C. *N*-Acylamino Sugars (3)

The reactions shown in Scheme III have been applied to the synthesis of an amide linkage between the amino group of glucosamine and the carboxyl group of ε-aminocaproic acid.[22] The synthesis has also been carried out successfully with galactosamine.

[18] R. Schwyzer, B. M. Iselin, and M. Feurer, *Chem. Abstr.* **56**, 4864 (1962).
[19] G. P. Ellis and J. Honeyman, *Advan. Carbohydr. Chem.* **10**, 95 (1955).
[20] E. E. Schallenberg and M. Calvin, *J. Amer. Chem. Soc.* **77**, 2779 (1955).
[21] G. W. Anderson, J. E. Zimmerman, and F. M. Callahan, *J. Amer. Chem. Soc.* **86**, 1839 (1964).

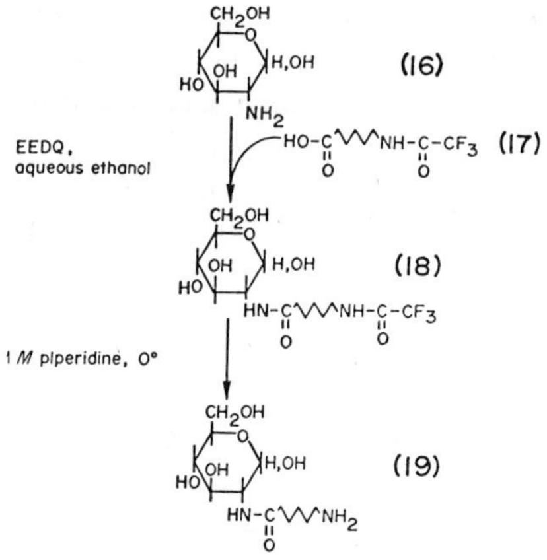

CH$_2$OH

H,OH   (16)

NH$_2$

HO-C$\sim\!\!\wedge\!\!\wedge$NH-C-CF$_3$ (17)

EEDQ,
aqueous ethanol

CH$_2$OH

H,OH   (18)

HN-C$\sim\!\!\vee\!\!\wedge$NH-C-CF$_3$

1 M piperidine, 0°

CH$_2$OH

H,OH   (19)

HN-C$\sim\!\!\vee\!\!\wedge$NH$_2$

Scheme III

N-Trifluoroacetyl-ε-aminocaproic Acid[20] (**17**). To 0.1 mole of ε-amino-caproic acid,[9] dissolved in 100 ml of 1 *N* NaOH are added 18 ml of ethyl trifluorothiol acetate[10] in 1-ml batches during 3 hours with vigorous stirring in a hood. The product is precipitated by the cautious addition of 5 ml of concentrated hydrochloric acid and separated by filtration. The residue is washed thoroughly with cold water and then recrystallized from water as needed, mp 87–88°. Additional material can be obtained from the original mother liquors by concentration. The overall yield of pure material exceeds 80%.

N-(N-Trifluoroacetyl-ε-aminocaproyl)-2-amino-2-deoxy-D-gluco-pyranose (**18**). Glucosamine hydrochloride (**16**) (2.15 g, 10 mmoles) and **17** (2.27 g, 10 mmoles) are dissolved in 400 ml of water and adjusted to pH 7.0 with 1 *N* sodium hydroxide. To this solution is added 12.36 g (50 mmoles) of *N*-ethoxycarbonyl-ethoxy-1,2-dihydroquinoline (EEDQ)[23,24] and the suspension is shaken vigorously at 37° for 16 hours. During the first 2–4 hours of the reaction, three additions of 100 ml of ethanol are made to bring the EEDQ into solution slowly so that carboxyl group activation can occur throughout the reaction.

After 16 hours, the reaction mixture is evaporated to dryness at 40–50°[8] and the residue extracted twice with 200 ml of water. The water-soluble extracts are combined, adjusted to pH 1.5 with trifluoroacetic acid, and passed over a column of 100–150 ml of Dowex 50-X8

[22] I. P. Trayer, unpublished observations, 1973.
[23] Calbiochem., Los Angeles, California.
[24] B. Belleau and G. Malek, *J. Amer. Chem. Soc.* **90**, 1651 (1968).

(H[+] form, 20–50 mesh) to remove unreacted glucosamine. The column is washed with 200 ml of water. The eluate and washings are combined and evaporated to dryness. The product contains some unreacted N-trifluoroacetyl-ε-aminocaproic acid which is removed at the next step of the synthesis.

The reaction can be monitored by high voltage paper electrophoresis at pH 6.5[25] and spraying with ninhydrin[11] or silver nitrate.[26] Based upon the disappearance of glucosamine as judged by this means, the final yield of desired product is about 60–70%.

N-(ε-Aminocaproyl)-2-amino-2-deoxy-D-glucopyranose (**19**). The above product, **18** (2.5–3 g), is dissolved in 2–3 ml of 1.0 M piperidine at room temperature, and the solution is immediately chilled in an ice bath. After standing at 0° for 2 hours, the reaction mixture is adjusted to pH 6.0 with 0.5 M acetic acid and applied directly to a column of Sephadex G-10 (140 cm × 3.0 cm) equilibrated in water. The product is located in the eluent fractions by testing aliquots for reducing sugar with ferricyanide.[27] After concentrating the appropriate fractions to dryness, the product is dissolved in 1 mM HCl and applied to a column of Dowex 50-X8 (H[+] form, 200–400 mesh, 25 cm × 0.9 cm) equilibrated to 1 mM HCl. The column is washed with about 100 ml of 1 mM HCl and a linear gradient is applied (400 ml of 1 mM HCl and 400 ml of 0.3 M lithium chloride in 1 mM HCl). The initial breakthrough peak contains ε-aminocaproic acid and neutral, silver nitrate-positive[26] material as judged by high voltage paper electrophoresis at pH 6.5.[25] The fractions giving a positive reaction for reducing sugar[27] are pooled and concentrated to dryness. This product is desalted, if required, using Sephadex G-10 as described above, but it can be used for coupling to cyanogen bromide-activated agarose without further purification.

The final product is not crystallized but can be shown to migrate as a single, silver nitrate-positive and ninhydrin-positive spot on high voltage paper electrophoresis[25] at pH 6.5, and on ascending paper chromatography in solvents 1 and 2[28] (R_f 1 = 0.65, R_f 2 = 0.56). An acid hydrolyzate (6 N HCl, 20 hours at 105°), analyzed chromatographically on an amino acid analyzer, was shown to contain only ε-aminocaproic

---

[25] High voltage paper electrophoresis is performed under Varsol with pyridine acetate buffer, pH 6.5 (pyridine:acetic acid:water, 2.5:1:225, v/v) at a potential gradient of 50 V/cm.

[26] A saturated solution of silver nitrate in water (0.1 ml) is added to 20 ml of acetone and water added until the white precipitate disappears. The dried chromatography paper is dipped in this solution and then sprayed with 0.5 M sodium hydroxide in 95% aqueous ethanol. Reducing sugars appear as dark brown spots.

[27] J. T. Park and M. J. Johnson, *J. Biol. Chem.* **181**, 149 (1949).

[28] Solvent 1: isobutyric acid:ammonium hydroxide:water (66:1:33, v/v); solvent 2: isopropanol:1:0.25 M ammonium bicarbonate (65:35, v/v).

acid and glucosamine in a molar ratio of 1:1. The structure (19) is therefore assigned to this product.

D. Preparation of Monosaccharide-Agarose Adsorbents

Sepharose 4B[29] is generally used as the agarose derivative and after treatment with cyanogen bromide reacts readily with the ligands described above. The conditions for coupling of the ligands with agarose are essentially those generally used for any primary amine,[30] and the following procedure is satisfactory.

Sepharose 4B is washed thoroughly with water and 500 g (wet weight of washed, filtered Sepharose) is suspended in 500 ml of 5 M potassium phosphate at pH 13.0 and 2–4° with efficient stirring. A freshly prepared, cold aqueous solution of cyanogen bromide (500 ml containing 100 mg/ml) is added, and the temperature is maintained at 20 ± 1° for 8 minutes. Chipped ice (400–500 ml) is added, the mixture is filtered on a 4-liter coarse sintered-glass funnel, and then washed with 8–10 liters of water at 5°. Sepharose is then added to a solution containing 2–10 mmoles of ligand in 400–500 ml of water, and the mixture is adjusted to pH 10 and stirred at 4° for 12–18 hours. The monosaccharide-agarose derivative is washed exhaustively with water by filtration and stored as a suspension in 50 mM sodium cacodylate at pH 7.5 to inhibit bacterial action. Prior to use, it is equilibrated with the appropriate buffer.

By this procedure 1–10 μmoles of ligand per milliliter of packed wet gel have been coupled to the agarose. The concentration of ligand is most easily estimated if a radioactive ligand is used. However, the filtrate and washings from the coupling reaction can be concentrated and assayed for the specific sugar by gas chromatography of the silyl derivatives.[31] The latter procedure is not recommended for galactose-containing ligands since galactose from the agarose can interfere.

E. Affinity Chromatography with Monosaccharide Derivatives of Sepharose 4B

Sepharose-monosaccharide adsorbents should have wide applicability in purification of proteins. At present, they have been used with the following enzymes or lectins.

Galactosyltransferase. The effectiveness of the N-acetylglucosamine-agarose in the purification is markedly influenced by the conditions for adsorption and elution. A significant increase in the affinity of the galactosyltransferase for N-acetylglucosaminyl-agarose was produced by the presence of UMP or UDP, which are competitive inhibitors with

[29] Pharmacia, Uppsala, Sweden.
[30] J. Porath, J. C. Janson, and T. Låås, *J. Chromatogr.* **60**, 167 (1971).
[31] C. C. Sweeley, W. W. Wells, and R. Bentley, this series, Vol. 8 [7].

SUGARS AND DERIVATIVES [29]

respect to the first-bound substrate, UDP-galactose.[1] The presence of inhibitors or substrate during adsorption may be essential for tight binding of some enzymes, such as glycosyl transferases, to monosaccharide agarose. In addition, the conditions used to elute the enzyme can significantly influence the degree of purification achieved. If N-acetyl-D-glucosamine is used to displace the enzyme, highly purified preparations were obtained, whereas elution with buffers containing borate or urea gave impure preparations.[1]

Wheat Germ Hemagglutinin. This protein has been purified by affinity chromatography on N-acetyl-D-glucosaminyl-agarose[2] and on an N-acetyl-D-glucosaminyl-agarose formed from N-(ε-aminocaproyl)-2-acetamido-2-deoxy-β-glucopyranosylamine.[32] In the former case, commercially available wheat germ lipase served as a source of agglutinin. In the latter case, the 55% ammonium sulfate fraction from an aqueous extract of defatted wheat germ was used. Both procedures gave essentially homogeneous protein of specific activity equivalent to that of material purified by other methods. Interestingly, the hemagglutinin could not be displaced from the glycosylamine column by elution with N-acetyl-D-glucosamine, but was removed with 0.1 M acetic acid.[32]

L-*Fucose Binding Proteins.* Three proteins which bind L-fucose were obtained and could be further fractionated into two components by rechromatography on the fucosaminyl agarose adsorbent, since the protein which binds least tightly could be eluted with phosphate buffer.[5]

Rat Liver Glucokinase (EC 2.7.1.2). Glucokinase has been purified directly from rat liver extracts by affinity chromatography on N-(ε-aminocaproyl)-2-amino-2-deoxy-D-glucopyranose-agarose.[33] This ligand clearly distinguishes between hexokinases which have a high affinity for glucose, and glucokinase, which have a lower affinity. Neither a commercially available yeast hexokinase nor the hexokinase found in rat liver are retarded on these columns and appear in the void volume. Glucokinase can be eluted by including either glucose, glucosamine, or N-acetylglucosamine[33] in the developing buffer, or by raising the KCl concentration to 0.2 M. Elution with potassium chloride gives a preparation of much lower specific activity. If rat liver extract is applied to a column obtained by coupling glucosamine directly to cyanogen bromide-activated agarose, the glucokinase activity is not adsorbed. With the aminocaproate "spacer" molecule, however, purifications of up to 500-fold can be obtained in a single pass of crude supernatant through the affinity column.

---

[32] R. Lotan, A. E. S. Gussin, H. Lis, and N. Sharon, *Biochem. Biophys. Res. Commun.* **52,** 656 (1973).

[33] J. M. E. Chesher, I. P. Trayer, and D. B. Walker, *Biochem. Trans.* **1,** 876 (1973).

[30] Sugars Attached to Starch

By Isamu Matsumoto and Toshiaki Osawa

Sephadex is insoluble dextran, cross-linked by epichlorhydrin,[1] which serves as a specific adsorbent for dextran binding protein, e.g., concanavalin A.[2] In an analogous manner, we have demonstrated that monosaccharides or oligosaccharides could be cross-linked to insoluble starch by epichlorhydrin under alkaline conditions, and that the resulting insoluble adsorbent is useful for the specific affinity chromatography of the corresponding carbohydrate binding protein.[3]

Procedure for the Preparation of Specific Affinity Adsorbent (Sugar–Starch Adsorbent)

Washed, finely powdered starch, 2 g, and 400 mg of sugar (L-fucose for eel serum hemagglutinin or tri-*N*-acetylchitotriose for *Cytisus sessilifolius hemagglutinin*) are suspended in 5 ml of water. To this suspension are added 7.5 ml of 5 *N* NaOH and 5 ml of epichlorhydrin under vigorous stirring at 37°. The suspension spontaneously solidifies as a gel cake. After standing for 1 hour at 37°, the resulting solid is homogenized with a Waring Blendor, and washed successively with water, with 50 m*M* glycine-HCl at pH 2.0 containing 0.5 *M* NaCl, and with 5 m*M* phosphate-buffered saline (pH 7.0). After exclusion of very fine particles by decantation, the gel is packed into a chromatography column.

Application of Sugar-Starch Adsorbent

Purification of Anti-H Hemagglutinin from Eel Serum

Since eel serum anti-H hemagglutinin are known to be inhibited by L-fucose, L-fucose-starch adsorbent is used. Fresh eel serum is directly applied to a 1.6 × 12.8 cm column of L-fucose starch, which had been equilibrated with 5 m*M* phosphate-buffered saline at pH 7.0. After washing with the same buffer, elution is carried out with 50 m*M* glycine-HCl at pH 3.0 containing 0.5 *M* sodium chloride. The hemagglutinating activity is retained by the column and is recovered in two protein peaks, fractions A and B, by elution at pH 3.0. Table I summarizes data pertaining to the purification of eel serum anti-H hemagglutinin. Fraction B,

[1] P. Flodin, "Dextran Gels and Their Applications in Gel Filtration." Pharmacia, Uppsala, Sweden, 1962.
[2] B. B. L. Agrawal and I. J. Goldstein, *Biochem. J.* **96**, 23 (1965).
[3] I. Matsumoto and T. Osawa, *Biochem. Biophys. Res. Commun.* **140**, 484 (1970).

TABLE I

DETAILS OF PURIFICATION OF EEL SERUM ANTI-H HEMAGGLUTININ

Fraction	Volume (ml)	Hemagglutinating titer for				Yield of activity (%)
		O	A	B	AB	
Crude serum	5	512	128	128	64	100
After affinity chromatography						
Fraction A	14.4	4	—[a]	—[a]	—[a]	2.4
Fraction B	9.6	160	40	40	20	60

[a] Not determined.

which represents a major active protein is homogeneous by the criteria of ultracentrifugal analysis (7.2 S) and disc gel electrophoresis at pH 8.9. The molecular weight of this fraction was estimated to be 140,000 by gel filtration on Sephadex G-200 column. Of particular note is the fact that a single-step specific purification could be effectively achieved by affinity chromatography on an L-fucose-starch column.

Purification of Anti-H Hemagglutinin from Cytisus sessilifolius

Since the most effective inhibitors for *Cytisus sessilifolius* anti-H hemagglutinin are chitin oligosaccharides, tri-N-acetylchitotriose-starch adsorbent is used.

Step 1. One hundred grams of freshly powdered *C. sessilifolius* seeds (purchased from F. W. Schumacher, Sandwich, Mass.) are suspended in 1 liter of 5 mM phosphate-buffered saline at pH 7.0 and allowed to stand overnight at 4° with continuous stirring. To the clear supernatant fluid obtained by centrifugation at 12,000 g for 20 minutes, sufficient solid $(NH_4)_2SO_4$ is added to give 0.7 saturation. The precipitate is dialyzed against distilled water until free of NH_4^+ and then lyophilized.

Step 2. In order to remove inert low molecular weight substances, which are difficult to separate by affinity chromatography, the active ammonium sulfate fraction is further purified by Sephadex G-200 gel filtration.

Step 3. Tri-N-Acetylchitotriose-Starch Column. The fractions with a high concentration of hemagglutinin are pooled and applied at 4° to a 4.5 × 20 cm column of tri-N-acetylchitotriose-starch adsorbent. Both the preparation and the column had been equilibrated with 5 mM potassium phosphate-buffered saline at 7.0. *C. sessilifolius* anti-H hemagglutinin is clearly retained by the column even after washing with the same buffer. Hemagglutinin is eluted with glycine-HCl at pH 2.0 containing 0.5 M sodium chloride.

TABLE II
DETAILS OF PURIFICATION OF *Cytisus sessilifolius* ANTI-H HEMAGGLUTININ

Fraction	Yield from 20 g of seeds (mg)	Hemagglutinating activity[a] (μg/ml)			
		O	A	B	AB
Crude extract	1139	313	625	625	1250
$(NH_4)_2SO_4$ fraction 0–0.7 saturation	600	156	313	313	625
After Sephadex G-200 gel filtration	75	39	78	78	156
After affinity chromatography	4.2	4.9	9.8	9.8	20

[a] Minimum hemagglutinating dose against human O, A, B, or AB cells.

The results of the purification of *C. sessilifolius* anti-H hemagglutinin are summarized in Table II. The minimum hemagglutinating dose of the purified hemagglutinin thus obtained (CSH) is 4.9 μg/ml against human group O erythrocytes. Purified *Cytisus sessilifolius* anti-H hemagglutinin is homogeneous by the standards of ultracentrifugal analysis (6.8 S) and disc gel electrophoresis at pH 8.9. The molecular weight has been estimated to be 110,000 by gel filtration on Sephadex G-200.

[31] Group-Specific Separation of Glycoproteins

By TORE KRISTIANSEN

The two principal methods for specific purification of glycoproteins require the use of specific antibodies for precipitation or bioaffinity chromatography, or the use of lectins for precipitation or bioaffinity chromatography. In the latter case, the method is based on the different affinities of lectins for terminal carbohydrate residues characteristic of individual glycoproteins. The first method belongs to the field of immunoprecipitation and immunosorption and will not be treated here.

Lectins[1] are defined as proteins or glycoproteins of plant (phytohemagglutinins)[2] or animal origin displaying more or less selective affinity for a carbohydrate or a group of carbohydrates.

[1] N. Sharon and H. Lis, *Science* **177**, 949 (1972); H. Lis and N. Sharon, *Annu. Rev. Biochem.* **42**, 541 (1973).

[2] J. Tobiška, "Die Phythämagglutinine," p. 191. Akademie-Verlag, Berlin, 1964. O. Prokop and G. Uhlenbruck, "Human Blood and Serum Groups," p. 72. Elsevier, Amsterdam, 1969.

The literature on purification of glycoproteins by use of lectins is already substantial and is growing rapidly. Since detailed procedures are usually found in the primary publications, the main intention of this article is to survey some of the approaches that have been successful. Lectins have been used extensively in studies of the properties of cell surfaces and in attempts to monitor changes in location and composition of surface glycoproteins associated with neoplastic transformation. The literature on this particular topic is burgeoning.[3,4]

Planning of the Purification Steps

The series of steps involved in working out a suitable procedure for purification with a lectin of a given glycoprotein or glycopeptide can be listed: (1) identification of the terminal sugar(s) in the carbohydrate moiety of the desired substance; (2) choice of a lectin with the corresponding affinity; (3) preparation or purchase of the selected lectin; (4) selection of a separation method from those using (a) direct precipitation, (b) adsorption to and desorption from a polymer of the lectin itself, and (c) adsorption to and desorption from a carrier to which lectin has been coupled covalently; (5) choice of a desorption technique from those that are nonspecific, e.g., one involving a change in pH or salt concentration, or use of a specific method, effective by reason of displacement of the adsorbed glycoprotein by competing carbohydrates.

Assuming that the terminal sugar(s) of the glycoprotein has been established, the next step is the choice of a suitable lectin. Lectins are generally not specific for only a single sugar although the degree of specificity varies widely. For instance, lectin from the seeds of *Lotus tetragonolobus* has a narrow specificity for L-fucose, whereas concanavalin A (from *Canavalia ensiformis*) has a broad specificity and will bind most glycoproteins in human serum. The material presented below classifies lectins according to their main affinities. If no other source is indicated, it is the seed of the plant which serves as source of the lectin.

Group I.   L-FUCOSE
            Lotus tetragonolobus[5–10]; weakly inhibited by L-galactose
            Ulex europaeus (gorse)[11]; contains another lectin belonging to group VII
            Ulex parviflorus[12]; weakly inhibited by other sugars

[3] M. M. Burger, *Fed. Proc., Fed. Amer. Soc. Exp. Biol.* **32**, 91 (1973).
[4] A. R. Oseroff, P. W. Robbins, and M. M. Burger, *Annu. Rev. Biochem.*, p. 647 (1973).
[5] W. T. J. Morgan and W. M. Watkins, *Brit. J. Exp. Pathol.* **34**, 94 (1953).
[6] M. Krüpe, *Hoppe-Seyler's Z. Physiol. Chem.* **299**, 277 (1955).
[7] O. Mäkelä, *Ann. Med. Exp. Biol. Fenn.* **35**, Suppl. 11 (1957).

Group II.   *N*-ACETYL-D-GLUCOSAMINE
  Group inhibited by N-acetylated chitodextrins
  Triticum vulgare (wheat germ)[1,2,13]; also inhibited by *N*-acetylneura-
  minic acid (NANA)[14]
  Solanum tuberosum (potato tuber)[15,16]; also inhibited by muramic acid

Group III.   *N*-ACETYL-D-GALACTOSAMINE
  Dolichos biflorus (horse gram)[7,17,18]
  Phaseolus lunatus (lima bean, also called *P. limensis*)[5,7,17]
  Phaseolus vulgaris (red kidney bean, black kidney bean, yellow wax
  bean; bean meal is source)[1,2]
  Vicia cracca[5,7,19]; also contains a nonspecific lectin in group VIII
  Euonymus europaeus[20]
  Helix pomatia (a snail)[21,22]

Group IV.   D-GALACTOSE
  Group also inhibited by L-arabinose, D-fucose, lactose, raffinose, and
  melibiose
  Crotalaria juncea (sunn hemp),[7,23-25] β-specific
  Ricinus communis (castor bean)[6,15,16]
  Abrus precatorius[26]
  Griffonia simplicifolia[2,27]

[8] W. M. Watkins and W. T. J. Morgan, *Vox Sang.* **7**, 129 (1962).
[9] G. F. Springer and P. Williamson, *Biochem. J.* **85**, 282 (1962).
[10] J. Yariv, A. J. Kalb, and E. Katchalski, *Nature (London)* **215**, 890 (1967).
[11] I. Matsumoto and T. Osawa, *Biochim. Biophys. Acta* **194**, 180 (1969).
[12] M. Lalaurie, B. Marty, and J.-P. Fabre, *Trav. Soc. Pharm. Montpellier* **26**, 215 (1966).
[13] Pattern of inhibition studied by A. K. Allen, A. Neuberger, and N. Sharon, *Biochem. J.* **31**, 155 (1973).
[14] P. J. Greenaway and D. LeVine, *Nature (London) New Biol.* **241**, 191 (1973).
[15] M. Tomita, T. Osawa, Y. Sakurai, and T. Ukita, *Int. J. Cancer* **6**, 283 (1970).
[16] G. I. Pardoe, G. W. G. Bird, and G. Uhlenbruck, *Z. Immunitaetsforsch. Allerg. Klin. Immunol.* **137**, 442 (1969).
[17] Z. Yosisawa and T. Miki, *Proc. Jap. Acad.* **39**, 187 (1963).
[18] M. E. Etzler and E. A. Kabat, *Biochemistry* **9**, 869 (1970).
[19] K. Aspberg, H. Holmén, and J. Porath, *Biochim. Biophys. Acta* **160**, 116 (1968).
[20] O. Mäkelä, P. Mäkelä, and M. Krüpe, *Z. Immunitaetsforsch. Exp. Ther.* **117**, 220 (1959).
[21] S. Hammarström and E. A. Kabat, *Biochemistry* **8**, 2696 (1969).
[22] S. Hammarström, this series, Vol. 28, p. 368.
[23] O. Mäkelä, *Nature (London)* **184**, 111 (1959).
[24] H. M. Bhatia and W. C. Boyd, *Transfusion* **2**, 106 (1962).
[25] B. Ersson, K. Aspberg, and J. Porath, *Biochim. Biophys. Acta* **310**, 446 (1973).
[26] M. Krüpe and A. Ensgraber, *Z. Immunitaetsforsch. Exp. Ther.* **123**, 355 (1962).
[27] V. Hořejší and J. Kocourek, *Biochim. Biophys. Acta* **297**, 346 (1973).

Group V.   N-ACETYL-D-GALACTOSAMINE AND D-GALACTOSE

These lectins are inhibited almost equally by both sugars

Sophora japonica (japanese pagoda tree)[6,20]
Glycine max (soybean),[1] α-specific
Caragana arborescens[7]
Bandaeirea simplicifolia,[23] α-specific
Bauhinia variegata, var. *candida*[7]
Momordia charantia[15]
Erythrina subrosa[24]
Coronilla varia,[6,20] α-specific
Crotalaria zanzibarica[24]
Arachis hypogea,[28,29] β-specific

Group VI.   D-GLUCOSE

Sesamum indicum[15]
Pisum sativum (garden pea); inhibited about four times better by D-mannose[27]

Group VII.   β-GLYCOSIDES AND β-N-ACETYLGLUCOSAMINIDES

Group inhibited most strongly by N,N'-diacetylchitobiose, but also by salicin (2-(hydroxymethyl)-phenyl-β-D-glucopyranoside), phenyl-β-D-glucopyranoside, and cellobiose

Ulex europaeus (gorse)[11]; contains another lectin belonging to group I
Ulex galli[7]
Ulex nanus[7]
Cytisus sessilifolius[5,7,8,23,30]; inhibited also by lactose
Laburnum alpinum[5,7,8,30]; inhibited also by lactose
Clerodendrum viscosum[31]; pulp is source

Group VIII.   Methyl-α-D-mannoside, D-Mannose, D-Glucose, N-Acetyl-D-glucosamine, L-Sorbose

Sugars listed in decreasing order of inhibition

Pisum sativum (garden pea)[6,7,32]; also inhibited by D-glucose, but only about one-fourth as efficiently[27]
Lens culinaris (common lentil)[1]
Canavalia ensiformis (jack bean; gives Con A)[1]; bean meal is source
Vicia cracca[2,7]; also contains a lectin in group III[19]
Lathyrus sativus L.[27,33,34]

[28] W. C. Boyd, D. M. Green, D. M. Fujinaga, J. S. Drabik, and E. Waszczenko-Zacharczenko, *Vox Sang.* 6, 456 (1959).

[29] G. Uhlenbruck, G. I. Pardoe, and G. W. G. Bird, Z. *Immunitaetsforsch. Allerg. Klin. Immunol.* 138, 423 (1969).

[30] T. Osawa, *Biochim. Biophys. Acta* 115, 507 (1966).

[31] G. W. G. Bird, *Nature (London)* 191, 292 (1961).

[32] M. Tichá, G. Entlicher, J. V. Koštíř, and J. Kocourek, *Experientia* 25, 17 (1969).

[33] G. Louženský, Isolation of Hemagglutinins from the Seeds of the Sweet Pea (*Lathyrus sativus* L. var. Karcagi). Thesis, Charles University, Prague, 1970.

[34] J. Kocourek and G. Louženský, *Abstr. 7th FEBS Meeting, Varna,* p. 134 (1971).

Group IX.   N-Acetylneuraminic Acid (NANA)
Limulus polyphemus (hemolymph of horseshoe crab[35])
Triticum vulgaris (wheat germ)[1,2,13,14]; also in group II

Availability of Lectins

At the time of writing only a few lectins are commercially available. Con A can be obtained from Pharmacia Fine Chemicals, Uppsala, Sweden; Sigma Chemical Co., St. Louis, Missouri; and Miles Laboratories Inc., Kankakee, Illinois. Agglutinins from wheat germ, Lotus tetragonolobus, and soybean are offered by Miles Laboratories.

Preparation of Lectins

Volume 28 of this series contains procedures for purification of the following lectins: Canavalia ensiformis, Phaseolus lunatus, Phaseolus vulgaris, Ulex europaeus, Wistaria floribunda, Lens culinaris, Dolichos biflorus, Sophora japonica, Triticum vulgaris, Lotus tetragonolobus, Glycine max, Helix pomatia.

Sharon and Lis[1] give a table of 17 references to papers describing purification of lectins by bioaffinity chromatography. In addition to those treated in Volume 28 of this series, the compilation refers to papers dealing with purification of lectins from Pisum sativum, Cytisus sessilifolius, and the N-acetyl-D-glucosamine-specific lectin from Ulex europaeus.

An additional listing is presented in Table I.

Purification of Glycoproteins

By Precipitation with Lectin

A number of reports describe the successful precipitation of glycoprotein with a lectin. These are noted in Table II.

Although convenient in many cases, purification by precipitation results in the problem of separating glycoprotein from lectin once the precipitate has been dissolved either by addition of a competing carbohydrate or by variation in salt concentration or pH. If the lectin and the desired glycoprotein have similar molecular weights, the problem is difficult. By immobilization of the lectin these difficulties are avoided.

By Direct Polymerization of Lectin

Polymerization of lectins with glutaraldehyde or L-leucine-N-carboxyanhydride has been used for preparation of adsorbents for glycoprotein. A few examples are given in Table III.

---

[35] J. J. Marchalonis and G. M. Edelman, J. Mol. Biol. 32, 453 (1968).

TABLE I

PURIFICATION OF LECTINS BY BIOAFFINITY CHROMATOGRAPHY

Lectin	Matrix	Ligand	Coupling method
Lathyrus sativus L.[a,b]	Sephadex	Direct adsorption to matrix	—
Vicia cracca, nonspecific agglutinin[c]	Sephadex	Direct adsorption to matrix	—
Phaseolus lunatus[d]	Blood group substance A	Direct polymerization	L-Leucine-N-carboxyanhydride
Canavalia ensiformis[e]	Peroxidase	Direct polymerization	Glutaraldehyde
Glycine max[e]	Erythrocyte stroma	Direct polymerization	Glutaraldehyde
Triticum vulgaris[e]	Ovomucoid	Direct polymerization	Glutaraldehyde
T. vulgaris[f]	Sepharose 4B	N-Acetyl-D-glucosamine derivative	CNBr
T. vulgaris[g]	Sepharose 4B	N-Acetyl-D-glucosamine derivative	CNBr
Ulex europaeus[h]	Polyacrylamide	Fucose	Copolymerization with D-glycosides
Pisum sativum[h]	Polyacrylamide	Mannose	Copolymerization with D-glycosides
Griffonia simplicifolia[h]	Polyacrylamide	Galactose	Copolymerization with D-glycosides

[a] G. Louženský, Isolation of Hemagglutinins from the Seeds of the Sweet Pea (*Lathyrus sativus* L. var. Karcagi). Thesis, Charles University, Prague, 1970.

[b] J. Kocourek and G. Louženský, *Abstr. 7th FEBS Meeting, Varna*, p. 134 (1971).

[c] K. Aspberg, H. Holmén, and J. Porath, *Biochim. Biophys. Acta* **160**, 116 (1968).

[d] W. Galbraith and I. J. Goldstein, *Biochemistry* **11**, 3976 (1972).

[e] S. Avrameas and B. Guilbert, *Biochimie* **53**, 603 (1971).

[f] J. H. Shaper, R. Barker, and R. L. Hill, *Anal. Biochem.* **53**, 564 (1973).

[g] R. Lotan, A. E. S. Gussin, H. Lis, and N. Sharon, *Biochem. Biophys. Res. Commun.* **52**, 656 (1973).

[h] V. Hořejší and J. Kocourek, *Biochim. Biophys. Acta* **297**, 346 (1973).

TABLE II
GLYCOPROTEINS PURIFIED BY PRECIPITATION WITH LECTINS

Material purified	Lectin
Normal IgG, myeloma IgG, several serum glycoproteins[a,b]	Concanavalin A (Con A)
Human serum glycoproteins[c]	Phaseolus vulgaris
Blood group active glycoproteins from human gastric juice[d]	Con A
Human saliva glycoproteins[e]	Phaseolus vulgaris
Antihemophilic factor (factor VIII)[f]	Con A
Blood group A active material from hog stomach[g]	Vicia cracca
Teichoic acid from Streptococcus sanguis[h]	Con A
Peptidorhamnomannan from Sporothrix schenkii[i]	Con A
Various galactomannans[j]	Ricinus communis
Onco-RNA virus[k]	Con A
Neuraminidase from influenza virus[l]	Con A

[a] M. A. Leon, *Science* **158**, 1325 (1967).

[b] S. Nakamura, S. Tominaga, A. Katsuno, and S. Murukawa, *Comp. Biochem. Physiol.* **15**, 435 (1965).

[c] B. O. Osunkoya and A. I. O. Williams, *Clin. Exp. Immunol.* **8**, 205 (1971).

[d] A. E. Clarke and M. A. Denborough, *Biochem. J.* **121**, 811 (1971).

[e] J. A. Strauchen, C. F. Moldow, and R. Silber, *J. Immunol.* **104**, 766 (1970).

[f] L. Kass, O. D. Ratnoff, and M. A. Leon, *J. Clin. Invest.* **48**, 351 (1969).

[g] T. Kristiansen and J. Porath, *Biochim. Biophys. Acta* **158**, 351 (1968).

[h] L. I. Emdur, J. G. McHugh, and T. H. Chin, *Abstr. Ann. Meet. Amer. Soc. Microbiol.* G 84 (1972).

[i] K. O. Lloyd and M. A. Biton, *J. Immunol.* **107**, 663 (1971).

[j] J. P. van Wauwe, F. G. Loontiens, and C. K. de Bruyne, *Biochim. Biophys. Acta* **313**, 99 (1973).

[k] M. L. Stewart, D. F. Summers, R. Soeiro, B. N. Fields, and J. V. Maizel, *Proc. Nat. Acad. Sci. U.S.* **70**, 1308 (1973).

[l] R. Rott, H. Becht, H.-D. Klenk, and C. Scholtissek, *Z. Naturforsch. B* **27**, 227 (1972).

Although simple and straightforward on a small scale, the technique is hampered by the amorphous and inhomogeneous character of the polymer, making it somewhat inconvenient for column operation. Furthermore, the excessive density of binding sites is unfavorable for subsequent desorption and may lead to low recovery.

By Covalent Coupling of Lectin to an Inert Carrier

This technique has the advantage of giving mechanically resistant and homogeneous adsorbents that are easy to pack for column operation. The density of binding sites can be varied at will by choosing different levels of carrier activation and carrier:lectin ratios. It should be stressed

TABLE III
GLYCOPROTEINS PURIFIED BY ADSORPTION TO AND DESORPTION
FROM POLYMERIZED LECTINS

Material purified	Lectin polymer	Polymerizer
Dextrans, yeast mannan[a]	Concanavalin A (Con A)	L-Leucine-N-carboxyanhydride
Glycogen, dextrans, IgM[b]	Con A	Glutaraldehyde
Horseradish peroxidase, glucose oxidase, human and rabbit serum glycoproteins[c]	Con A	Glutaraldehyde

[a] K. O. Lloyd, *Arch. Biochem. Biophys.* **137**, 460 (1970).
[b] E. H. Donnelly and I. J. Goldstein, *Biochem. J*. **118**, 679 (1970).
[c] S. Avrameas and B. Guilbert, *Biochimie* **53**, 603 (1971).

that the ionogenic character of the adsorbent, causing nonspecific adsorption, will inevitably increase with the degree of activation, and that all claims to the contrary are suspect. A profusion of methods for coupling of biological substances both to carbohydrate, polyacrylamide, and glass carriers have been introduced and are presented elsewhere.[36,37] One technique that has proved itself involves coupling of lectins to agarose beads by the CNBr method. Table IV lists several applications.

Availability of Coupled Lectins

Lectins coupled to agarose include Con A from Pharmacia and Miles, and wheat germ, *Lotus tetragonolobus,* and soybean from Miles.

Procedure for Cross-linking of Agarose Beads

In order to suppress the small but definite tendency to leakage from agarose, it is recommended that the beads be cross-linked before activation. The CNBr activation itself causes some cross-linking, although insufficiently. The following procedure for cross-linking with epichlorohydrin represents a slight modification from the original report.[38]

One liter of sedimented agarose beads is thoroughly washed with distilled water and mixed with 1 liter of 1 M NaOH containing 100 ml of

[36] G. Feinstein, *Naturwissenschaften* **58**, 389 (1971); G. J. H. Melrose, *Rev. Pure Appl. Chem.* **21**, 83 (1971); P. Cuatrecasas, *Advan. Enzymol. Relat. Areas Mol. Biol.* **36**, 29 (1972).
[37] J. Porath and T. Kristiansen, *in* "The Proteins" (H. Neurath and R. L. Hill, eds.), 3rd ed., Vol. I. Academic Press, New York, 1975, in press.
[38] J. Porath, J.-C. Janson, and T. Låås, *J. Chromatogr.* **60**, 167 (1971).

TABLE IV
GLYCOPROTEINS PURIFIED BY ADSORPTION TO AND DESORPTION FROM
LECTIN COUPLED TO INERT CARRIER WITH CNBr[a]

Material purified	Ligand	Carrier
Human serum components[b]	Concanavalin A (Con A)	2% Agarose beads
Human blood group substance A[c]	*Vicia cracca*	Sepharose 2B
α_1-Antitrypsin from human serum[d]	Con A	Sepharose 4B
Human chorion gonadotropin, luteinizing hormone, and follicle-stimulating hormone[e]	Con A	Sepharose 6B
Human glycogen synthetases D and I[f]	Con A	Sepharose 4B
Hepatitis B antigen[g]	Con A	Sepharose 4B
Bovine rhodopsin[h]	Con A	Sepharose 4B
Dextrans, yeast mannans[i]	Con A	Sepharose 4B
α-Glycosylated teichoic acid from *Bacillus subtilis*[j]	Con A	Sepharose 4B
Lymphocyte membrane glyco-proteins[k]	Con A	Sepharose 4B
Lymphocyte membrane glyco-proteins[l]	*Lens culinaris*	Sepharose 4B
Virus envelope glycoproteins[m]	*Lens culinaris*	Sepharose 4B
Modified rabbit interferon[n]	*Phaseolus vulgaris*	Sepharose 4B

[a] For the last entry, modified interferon, the lectin was coupled not by CNBr but by means of the N-hydroxysuccinimide ester of succinylated aminoalkyl agarose.

[b] K. Aspberg and J. Porath, *Acta Chem. Scand.* **24**, 1839 (1970).

[c] T. Kristiansen, *Biochim. Biophys. Acta* **338**, 246 (1974).

[d] I. E. Liener, O. R. Garrison, and Z. Pravda, *Biochem. Biophys. Res. Commun.* **51**, 436 (1973).

[e] M. L. Dufau, T. Tsuruhara, and K. J. Catt, *Biochim. Biophys. Acta* **278**, 281 (1972).

[f] H. Sølling and P. Wang, *Biochem. Biophys. Res. Commun.* **53**, 1234 (1973).

[g] A. R. Neurath, A. M. Prince, and A. Lippin, *J. Gen. Virol.* **19**, 391 (1973).

[h] A. Steinemann and L. Stryer, *Biochemistry* **12**, 1499 (1973).

[i] K. O. Lloyd, *Arch. Biochem. Biophys.* **137**, 460 (1970).

[j] R. J. Doyle and D. C. Birdsell, *Abstr. Ann. Meet. Amer. Soc. Microbiol.* G 82 (1972).

[k] D. Allan, J. Auger, and M. J. Crumpton, *Nature New Biol.* **236**, 23 (1972).

[l] M. J. Hayman and M. J. Crumpton, *Biochem. Biophys. Res. Commun.* **47**, 923 (1972).

[m] M. J. Hayman, J. J. Skehel, and M. J. Crumpton, *FEBS Lett.*, **29**, 185 (1973).

[n] F. Dorner, M. Scriba, and R. Weil, *Proc. Nat. Acad. Sci. U.S.* **70**, 1981 (1973).

epichlorohydrin and 5 g of sodium borohydride. The slurry is heated under gentle stirring for 2 hours at 60° and washed until neutral with *hot* distilled water. The gel is then ready for activation. This treatment causes negligible reduction in mean pore size of the gel.

Procedures for Activation of Agarose Beads with
  CNBr and Coupling

An alternative to the method described by Porath in this volume [2] involves manual or automatic titration by addition of hydroxide during the reaction. According to this procedure, a grams of CNBr are dissolved in a ml of distilled water at room temperature. It is important to check the quality of the CNBr. On standing, solid CNBr slowly decomposes and after solution will consume very little hydroxide, which can readily be tested.

The CNBr solution is mixed with $100 \times a/b$ ml of sedimented $b\%$ agarose beads previously washed with distilled water. After equilibration for a few minutes under gentle stirring with a *smooth* magnetic bar, NaOH is added at a rate that maintains pH at 11.0–11.5 for 6–10 minutes. The reaction vessel should be cooled in an ice bath, as formation of reactive iminocarbonate is favored at low temperature at the expense of side reactions giving nonreactive derivatives. With large batches, crushed ice in the reaction mixture will promote dissipation of heat. Even so, NaOH concentration should not exceed $4 N$ to avoid local overheating. The gel is then washed on a Büchner funnel with 5–6 volumes of chilled distilled water and then with $0.5 M$ $NaHCO_3$ pH 8.5. At this point there is no need to hurry before addition of the lectin solution, also in $0.5 M$ $NaHCO_3$. Within an hour near 0° there is negligible decrease in coupling capacity of the activated gel. After suspension of the washed gel in the lectin solution, the slurry is stirred gently, or preferably rocked overnight, at room temperature. Unreacted iminocarbonate is then blocked by similar treatment with a saturated solution of glycine for 24 hours. The potentially ionogenic carboxyl group of glycine is screened by the basic imino group in the iminocarbonate.[39]

Coupling of Lectins to Polyacrylamide Activated with
  Glutaraldehyde

A method introduced very recently by Avrameas and his co-workers for the preparation of immunosorbents[40] could be a very interesting alternative to the use of carbohydrate matrices activated with CNBr, because of the simplicity of the technique and the comparatively low toxicity of glutaraldehyde.[41] No experiment with lectins has been published so far, but the method deserves exploration.

[39] J. Porath and L. Fryklund, *Nature* (*London*) **226**, 1169 (1970).
[40] T. Ternynck and S. Avrameas, *FEBS Lett.* **23**, 24 (1972).
[41] Handbook of Laboratory Safety, 2nd ed., p. 774. Chem. Rubber Publ. Co., Cleveland, Ohio, 1971.

Procedure[40]

Bio-Gel P-300 is allowed to hydrate for 24 hours in distilled water and is then washed several times with the same medium. To *a* ml of hydrated gel is added 5 *a* ml of a 6% solution of glutaraldehyde in 0.1 *M* phosphate buffer at pH 7.0. The pH is adjusted if necessary before incubation overnight at 37°. The gel is washed 10 to 20 times with 5 *a* ml of distilled water each time. At this stage the activated gel can be stored at 4° for at least a week without appreciable loss of binding capacity.

Coupling of protein is performed in 0.1 *M* phosphate buffer at pH 7.4 at room temperature overnight or at 4° for 48 hours. Free, active aldehyde groups remaining are blocked by suspension of the gel in an equal volume of 0.1 *M* lysine at pH 7.4, and the solution is allowed to stand at room temperature for 18 hours. Under these conditions 1–2 mg of protein per milliliter of activated gel are coupled, depending on the number of reactive amino groups in the protein.

Desorption of Glycoproteins from Immobilized Lectins

Whenever possible, specific desorption by displacement of adsorbed glycoprotein with a competing and dialyzable carbohydrate should be used. When choosing a lectin for a given separation, it may be effective to select one that is inhibited to a greater extent by a sugar different from the one(s) in the glycoprotein, and to use this sugar to achieve efficient desorption. Thus far, there is little experience with concentration gradients of sugar for selective desorption from lectin adsorbents with multiple affinities. Preliminary results indicate that selective desorption of serum components from concanavalin A-Sepharose can be obtained with a gradient of methyl-α-D-mannoside.[42]

[42] B. Ersson and J. Porath, unpublished observations.

[32] Wheat Germ Agglutinin

By HALINA LIS, REUBEN LOTAN, and NATHAN SHARON

Affinity chromatography is being successfully used for the purification of proteins with specific binding sites, such as enzymes, antibodies, and lectins.[1-3] The latter, which have binding sites for sugars, have been purified with either commercially available carbohydrate polymers,[3] or with

[1] P. Cuatrecasas, M. Wilchek, and C. B. Anfinsen, *Proc. Nat. Acad. Sci. U.S.* **61**, 636 (1968).
[2] P. Cuatrecasas and C. B. Anfinsen, *Annu. Rev. Biochem.* **41**, 253 (1971).

adsorbents prepared by covalently binding suitable monosaccharides or glycoproteins to insoluble carriers.[3-6] In this department, affinity columns were prepared for the purification of an L-fucose binding protein from *Lotus tetragonolobus*,[7,8] of soybean agglutinin,[9,10] and of wheat germ agglutinin.[4] These columns consist of glycosylamines of L-fucose, D-galactose, and *N*-acetyl-D-glucosamine, respectively, linked to Sepharose through the seven-atom flexible chain of ε-aminocaproic acid. The glycosylamines were synthesized either by direct amination of the sugar with methanolic ammonia[11,12] or via the 1-chloride and azide.[13,14] Scheme I summarizes the steps employed in the preparation of the affinity column for the isolation of wheat germ agglutinin[4]; the detailed procedure is given below. The application of the closely related ligand, 2-acetamido-*o*-(ε-aminohexyl)-2-deoxy-β-D-glucopyranoside for the affinity chromatography of wheat germ agglutinin has recently been described.[6]

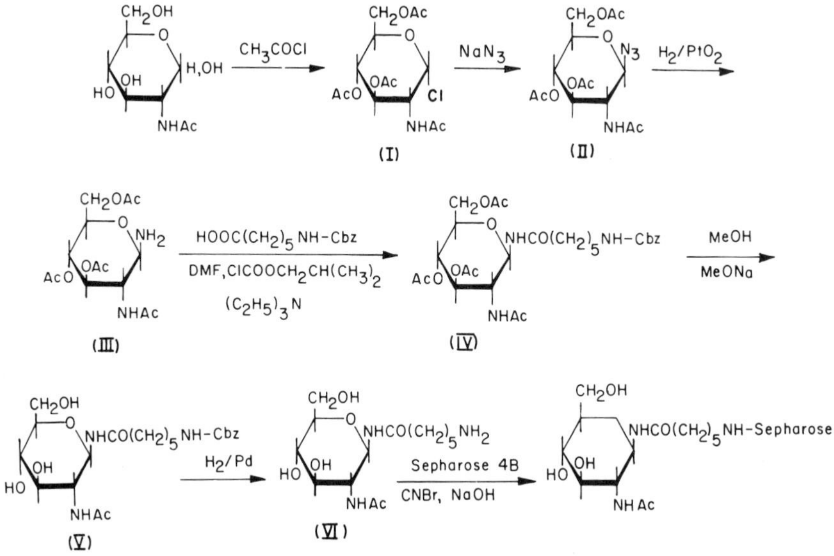

SCHEME I

[3] H. Lis and N. Sharon, *Annu. Rev. Biochem.* **42**, 541 (1973). See also Kristiansen, this volume [31] and Hořejší and Kocourek, this volume [36].

[4] R. Lotan, A. E. S. Gussin, H. Lis, and N. Sharon, *Biochem. Biophys. Res. Commun.* **52**, 656 (1973).

[5] V. Hořejší and J. Kocourek, *Biochim. Biophys. Acta* **297**, 346 (1973).

[6] J. Shaper, R. Barker, and R. L. Hill, *Anal. Biochem.* **53**, 564 (1973).

[7] S. Blumberg, J. Hildesheim, J. Yariv, and K. J. Wilson, *Biochim. Biophys. Acta* **264**, 171 (1972).

Synthesis of Affinity Column

2-Acetamido-3,4,6-tri-O-acetyl-2-deoxy-α-D-glucopyranosyl chloride (I). This compound is prepared by the method described by Horton.[13] Dried 2-acetamido-2-deoxy-D-glucopyranose (50 g) is added slowly and with stirring to 100 ml of acetyl chloride. The vessel is connected to a reflux condenser and the reaction mixture is stirred for 16 hours at room temperature. Chloroform (400 ml) is added through the condenser, and the solution is poured with vigorous stirring onto 400 g of ice and 100 ml of water. The mixture is transferred to a separatory funnel, and the organic phase is drawn off without delay into a beaker containing ice and 400 ml of a saturated solution of sodium bicarbonate, which is stirred. The organic phase is separated and treated with about 25 g of anhydrous magnesium sulfate. The drying agent is removed by filtration, washed with dry, alcohol-free chloroform and the combined filtrate is concentrated to 75 ml at reduced pressure on a rotatory evaporator at 50°. Dry ether is added to the warm solution and the mixture set aside for 12 hours at room temperature. The crystals are removed by filtration and washed with dry ether. Yield 60 g (73%); mp 130°; $[\alpha]_D^{24} + 110°$ (c, 1.04, chloroform).

2-Acetamido-3,4,6-tri-O-acetyl-2-deoxy-α-D-glucopyranosyl azide (II). This step, as well as the next, is carried out by a modification of the method of Bolton and Jeanloz.[14] A freshly prepared solution of (I) (30 g in 750 ml of chloroform) is added to a solution of sodium azide (36 g in 720 ml of distilled formamide) cooled in ice, and kept for 4.5 hours at room temperature. The reaction mixture is poured onto ice, the chloroform phase is separated, the water phase is washed twice with chloroform, and the washings are added to the chloroform phase. The combined chloroform phase is dried with sodium sulfate, the drying agent is removed by filtration and the filtrate is concentrated at reduced pressure. The residue is dissolved in 200 ml of absolute ethyl acetate, and the solution is poured into 4 liters of petroleum ether. The crystals are removed and dissolved in 200 ml of warm ethyl acetate. The solution is filtered, and 200

[8] J. Yariv, A. J. Kalb, and S. Blumberg, this series, Vol. 28, p. 356.

[9] J. A. Gordon, S. Blumberg, H. Lis, and N. Sharon, *FEBS Lett.* 24, 193 (1972).

[10] H. Lis and N. Sharon, this series, Vol. 28, p. 360.

[11] C. A. Lobry de Bruyn and F. H. van Leent, *Rec. Trav. Chim. Pays-Bas* 14, 134 (1895).

[12] H. S. Isbell and H. L. Frush, *J. Org. Chem.* 23, 1309 (1958).

[13] D. Horton, *in* "Organic Syntheses" (E. J. Corey, ed.), Vol. 46, p. 1. Wiley, New York, 1966.

[14] C. H. Bolton and R. W. Jeanloz, *J. Org. Chem.* 28, 3228 (1962).

ml of petroleum ether are added to the filtrate. The resultant crystals are removed by filtration and washed with petroleum ether. Yield 13 g (43%); mp 173°; $[\alpha]_D^{20} - 37°$ (c, 1.0, chloroform).

2-Acetamido-3,4,6-tri-O-acetyl-2-deoxy-β-D-glucopyranosylamine (III). A solution of (II) (12 g in 1 liter of ethyl acetate) is hydrogenated at room temperature and atmospheric pressure for 2 hours in the presence of 5 g of Adam's platinum oxide catalyst. The catalyst is removed by filtration and washed with ethyl acetate. The combined filtrates are shaken for 15 minutes with 3 g of charcoal, the charcoal is removed, and the clear ethyl acetate solution is concentrated under reduced pressure. The crystals thus obtained are washed under reflux for 1 hour with 600 ml of a mixture (1:1) of ethyl acetate and pentane. They are collected by filtration and washed with absolute ether. Yield 8.5 g (76%); mp 224–226° (dec); $[\alpha]_D^{20} - 14°$ (c, 0.5, chloroform).

2-Acetamido-3,4,6-tri-O-acetyl-N[N-(benzyloxycarbonyl)-ε-amino-caproyl]-2-deoxy-β-D-glucopyranosylamine (IV). Isobutyl chloroformate (1.5 ml), and triethylamine (1.7 ml) are added to a solution of N-(benzyloxycarbonyl)-ε-aminocaproic acid[15] (4 g) in N,N-dimethyl-formamide (20 ml), kept at −5°. The mixture is stirred for 20 minutes at −5° and then filtered. The filtrate is immediately added to a solution of (III) (3 g in 20 ml N,N-dimethylformamide). The mixture is stirred at room temperature for 2 hours and allowed to stand for 16 hours, after which the solvent is evaporated to dryness at reduced pressure and below 50°. The residue is crystallized twice from methanol:ether. Yield 3.68 g (62%); mp 203–205°; $[\alpha]_D^{20} + 11.9°$ (c, 1.0, N,N-dimethylformamide).

2-Acetamido-N-[N-(benzyloxycarbonyl)-ε-aminocaproyl]-2-deoxy-β-D-glucopyranosylamine (V). Freshly prepared 0.1 M methanolic sodium methoxide is added dropwise to a solution of compound (IV) (3 g, in 50 ml of absolute methanol) to give pH 8. The mixture is stirred for 5 hours at room temperature. The progress of deacetylation is monitored by thin-layer chromatography on precoated silica gel plates in acetone–methanol (9:1 v/v) [R_f of (IV) = 1.0; R_f of (V) = 0.57]. The solution is neutralized by the addition of solid CO_2, and the solvent is crystallized from methanol:ether. Yield 2.1 g (90%); mp 218–220°; $[\alpha]_D^{20} + 12.9$ (c, 1.0, N,N-dimethylformamide).

2-Acetamido-N-(ε-aminocaproyl)-2-deoxy-β-D-glucopyranosylamine (VI). A solution of compound (V) (2 g in 75 ml methanol:water, 8:2 v/v) is hydrogenated in the presence of 10% palladium-on-charcoal (50 mg) at atmospheric pressure and room temperature for 8 hours. The catalyst is removed by filtration, and the filtrate is evaporated to dryness.

[15] R. Schwyzer, B. M. Iselin, and M. Feurer, Chem. Abstr. 56, 4864 (1962).

The residue is crystallized from methanol:ether. Yield 1.2 g (80%); mp 193–195° (dec.). $[\alpha]_D^{20} + 23.3°$ (c, 1.0 water).

Binding of the Ligand (VI) to Sepharose 4B[16,17]

Cyanogen bromide (10 g) is stirred with water (100 ml) for 10 minutes; during this period most of it dissolves. To this mixture is added a slurry of Sepharose 4B (100 ml) washed with water. The pH of the reaction mixture is adjusted to 11 and maintained at this level for 6 minutes by adding 4 N NaOH. The reaction mixture is kept below 26° by the addition of crushed ice. The activated Sepharose is washed rapidly with 15 volumes of cold water on a Büchner funnel. The wet Sepharose is added quickly to a solution of (VI) (0.5 g in 40 ml of 0.5 M $NaHCO_3$), and the mixture is stirred gently for 16 hours at 4°. The resin is filtered and washed thoroughly with 0.1 M $NaHCO_3$ and with water. It is then suspended in 1 M ethanolamine, pH 8, and stirred gently for 3 hours to block unreacted active groups on the Sepharose. The amount of ligand covalently bound to the Sepharose is determined by acid hydrolysis of the conjugated resin followed by estimation of the liberated glucosamine with a Beckman-Spinco amino acid analyzer. Typically, the conjugated resin contains 2 μmoles of covalently bound ligand per milliliter of packed Sepharose.

Purification of Wheat Germ Agglutinin

Commercially available wheat germ is used as starting material for the purification of wheat germ agglutinin.[18] Wheat germ is defatted in a Soxhlet extractor with light petroleum ether (bp 40–60°) and dried in a stream of air. The defatted product (100 g) is stirred overnight with 300 ml of water in a cold room and the mixture is centrifuged at 600 g for 15 minutes; the supernatant liquid is retained. The residue is suspended in water, and the extraction procedure is repeated twice. Ammonium sulfate is added to the combined supernatant fluids to 55% of saturation at 4° (33 g per 100 ml); the precipitate is allowed to settle overnight and is collected by centrifugation at 9000 g for 20 minutes. The precipitate is dissolved in water and dialyzed extensively against water. A heavy precipitate of inactive protein is obtained, which is removed by centrifugation at 9000 g for 20 minutes. The clear supernatant liquid is applied at 4° to a column (2.8 × 17 cm) containing 100 ml of the conjugated Sepharose which has been equilibrated with saline. The column, at 4°, is eluted at the rate of 100 ml per hour, first with 600 ml of saline (fractions

[16] R. Axén, J. Porath, and S. Ernbäck, Nature (London) 214, 1302 (1967).
[17] S. Blumberg, I. Schechter, and A. Berger, Eur. J. Biochem. 15, 97 (1970).
[18] A. K. Allen, A. Neuberger, and N. Sharon, Biochem. J. 131, 155 (1973).

PURIFICATION OF WHEAT GERM AGGLUTININ (WGA) FROM WHEAT GERM

Fraction	Total protein (g)	Total activity (hemaggluti- nating units)	Specific activity (units/mg)	Yield (%)
Water extract from 100 g of defatted wheat germ	28	280,000	10	100
Ammonium sulfate precipi- tate (55%) dialyzed and centrifuged	3.5	140,000	40	50
Purified WGA after affinity chromatography	0.03	90,000	3000	32

of 15 ml) until no significant amount of material absorbing at 280 nm is detected in the effluent ($A_{280} < 0.05$), and then with 150 ml of 0.1 M acetic acid (fractions of 6 ml). The fractions eluted with acetic acid and containing UV-absorbing material are pooled (total volume 60 ml), neutralized by dropwise addition of 1 M NaOH, dialyzed extensively against water, centrifuged, and lyophilized. Yield 30 mg.

Comments

The purification procedure described above, starting directly from commercial wheat germ, involves only four steps: defatting, water extraction, ammonium sulfate fractionation, and affinity chromatography (see the table). Ammonium sulfate precipitation resulted in a 4-fold purification of wheat germ agglutinin, with about 50% yield. The bulk of the proteins present in the active ammonium sulfate fraction is not adsorbed to the column, but the hemagglutinating activity of these proteins is negligible. The agglutinin adsorbed to the column cannot be eluted with N-acetyl-D-glucosamine (0.1 M, 200 ml), but is eluted with 0.1 M acetic acid. The final product has the same activity as that described in the literature[18] and is obtained in a yield comparable to that of the conventional procedures. Its amino acid composition is very similar to that reported in the literature[6,18-20] for other preparations of wheat germ agglutinin.

Acknowledgment

We wish to thank Mr. I. Jacobson for the large-scale preparation of the ligand.

[19] D. LeVine, M. J. Kaplan, and P. J. Greenaway, *Biochem. J.* **129**, 847 (1972).
[20] Y. Nagata and M. M. Burger, *J. Biol. Chem.* **247**, 2248 (1972).

[33] α-Galactosidase

By NOAM HARPAZ and HAROLD M. FLOWERS

Conventional methods for the isolation and purification of α-galactosidases from a variety of plant and animal sources have recently been reviewed.[1] The α-galactosidase from the coffee bean is one of the few capable of hydrolyzing the nonreducing terminal α-D-galactopyranosyl residue of blood group B antigens. As part of our studies on the structure of these antigens in human erythrocytes,[2,3] we have developed a method for the rapid and complete purification of the enzyme in high yields by affinity chromatography.[4] Partial purification of the enzyme by alumina gel chromatography[5] and Sephadex gel filtration[6] have been described.

Principle. The approach here involves the synthesis of N-acylated derivatives of α-D-galactopyranosylamine conjugated with Sepharose. Similar derivatives of β-L-fucose, β-D-galactose and N-acetyl-β-D-glucosamine have recently been used for affinity chromatography of lectins.[7-9]

The α-galactosidase is specifically adsorbed onto columns of these materials, and is specifically eluted with either D-galactose or *p*-nitrophenyl-α-D-galactopyranoside. From 100 g of powdered coffee beans, 3 mg of α-galactosidase having a specific activity of 94.5 units per milligram of protein are obtained.[10]

---

[1] P. M. Dey and J. B. Pridham, *Advan. Enzymol.* **36**, 91 (1972).

[2] S. Yatziv and H. M. Flowers, *Biochem. Biophys. Res. Commun.* **45**, 514 (1971).

[3] N. Harpaz, H. M. Flowers, and N. Sharon, in preparation.

[4] N. Harpaz, H. M. Flowers, and N. Sharon, *Biochim. Biophys. Acta* **341**, 213 (1974).

[5] J. E. Courtois and F. Petek, this series, Vol. 8 [97].

[6] D. Barham, P. M. Dey, D. Griffiths, and J. B. Pridham, *Phytochemistry* **10**, 1759 (1971).

[7] S. Blumberg, J. Hildesheim, J. Yariv, and K. J. Wilson, *Biochim. Biophys. Acta* **264**, 171 (1972).

[8] J. A. Gordon, S. Blumberg, H. Lis, and N. Sharon, *FEBS Lett.* **24**, 193 (1972).

[9] R. Lotan, A. E. S. Gussin, H. Lis, and N. Sharon, *Biochem. Biophys. Res. Commun.* **52**, 656 (1973). See also Kristiansen, this volume [31].

[10] Galactosidase is assayed at 37° in a mixture containing 50 μl of enzyme, 50 μl of 5 mM *p*-nitrophenyl α-D-galactopyranoside, 50 μl of 1 mg per milliliter bovine serum albumin, and 50 μl of phosphate–citrate buffer at pH 6.0. The reaction is terminated by the addition of 0.8 ml of 0.5 M sodium carbonate. One unit of α-galactosidase is defined as the amount of enzyme that catalyzes the hydrolysis of 1 μmole of *p*-nitrophenyl α-D-galactopyranoside per minute under these conditions.

Preparation of Affinity Adsorbents

N-(N-Benzyloxycarbonyl-ε-aminocaproyl)-α-D-galactopyranosylamine. A stirred solution of 7.95 g (30 mmoles) of *N*-benzyloxycarbonyl-ε-amino-caproic acid[11] in 50 ml of dry dimethylformamide is chilled in an ice–salt bath to −5°. After the rapid addition of 3.9 ml (30 mmoles) of isobutyl chloroformate and 4.1 ml (30 mmoles) of triethylamine, stirring is continued for 20 minutes at −5°. The white precipitate which forms is removed by filtration, and the filtrate is added to a stirred suspension of 2.45 g (12.5 mmoles) of α-D-galactopyranosylamine–ammonia complex[12] ($[\alpha]_D^{23}$, 138; mp 107–109°) in 50 ml of dry dimethylformamide at room temperature. The clear solution which results after stirring for a few minutes is kept overnight. The solution is evaporated under reduced pressure below 55°, and the solid residue is dissolved in a small volume of chloroform–methanol–water (30:10:1) and applied to a column containing 800 g of silica gel (Merck, 0.05–0.2 mm, 70–325 mesh) prepared with the same solvent mixture. After elution of impurities, mainly *N*-benzyloxycarbonyl-ε-aminocaproylamide, the title compound is eluted as the second major fraction,[13] which is isolated under reduced pressure as a syrup. Yield, 91%; $[\alpha]_D^{23}$, 88.8 (*c* 2.7, methanol).

N-ε-Aminocaproyl-α-D-galactopyranosylamine. A solution of 2.72 g (6.36 mmoles) of *N*-(*N*-benzyloxycarbonyl-ε-aminocaproyl)-α-D-galacto-pyranosylamine in 100 ml of 80% aqueous methanol containing 100 mg of palladium-on-charcoal is hydrogenated at atmospheric pressure for 8 hours at room temperature. After filtration and removal of the solvent under reduced pressure, the title compound is obtained as a syrup. Yield, 99%; $[\alpha]_D^{23}$, 127.8 (*c* 2.5, methanol).

N-(N-Benzyloxycarbonyl-ε-aminocaproyl-N-ε-aminocaproyl)-α-D-galactopyranosylamine. Reaction of 318 mg (1.2 mmoles) of *N*-benzyloxycarbonyl-ε-aminocaproic acid with 292 mg (1.0 mmole) of *N*-ε-aminocaproyl-α-D-galactopyranosylamine, as described above, gives the title compound as a syrup. Yield, 95%; $[\alpha]_D^{23}$, 63.1 (*c* 3.0, methanol).

N-ε-Aminocaproyl-N-ε-aminocaproyl-α-D-galactopyranosylamine. A solution of the previous compound in 80% aqueous methanol is hydrogenated as described above, and the title compound is isolated as a syrup by removal of the solvent under reduced pressure. Yield, 95%; $[\alpha]_D^{23}$, 73.1 (*c* 1.8, methanol).

Coupling of Ligands to Sepharose. Ten milliliters of Sepharose 4B,

---

[11] R. Schwyzer, B. M. Iselin, and M. Feurer, *Chem. Abstr.* **56**, 4864 (1962).

[12] H. L. Frush and H. S. Isbell, *J. Res. Natl. Bur. Stand.* **47**, 239 (1951).

[13] Small quantities of an *O*-acylated sugar derivative are eluted just ahead of the title compound, and progress of the chromatography is followed by thin-layer chromatography on microslides.

activated with 1 g of cyanogen bromide,[14] are suspended in solutions of about 0.3 mmole of N-ε-aminocaproyl-α-D-galactopyranosylamine or of N-ε-aminocaproyl-N-ε-aminocaproyl-α-D-galactopyranosylamine in 10 ml of 0.1 M sodium borate at pH 9.5 and 4°, and stirred overnight. After thorough washing of the conjugated Sepharose with distilled water, it is stirred in 5 volumes of 0.1 M ethanolamine for 3 hours, then filtered and washed again with distilled water until the washings are neutral.

The conjugates are suspended in phosphate-citrate buffer at pH 4.5 and poured into small columns (5.5 × 1.5 cm) fitted with sintered-glass supports.

Estimation of Bound Ligand. Aliquots of the ligand solutions, before and after coupling to Sepharose, are assayed by reaction with ninhydrin to give an estimate of the total bound ligand.

Typical estimates give 13–15 μmoles of ligand bound per milliliter of Sepharose. Dilution of these conjugates with 9 volumes of unmodified Sepharose provides satisfactory adsorbents.

Purification of α-Galactosidase

Preliminary Purification. The 35–50% ammonium sulfate fraction of a saline extract[5] of 100 g of benzene-washed coffee meal is dissolved in a few milliliters of distilled water and stirred at 4° for 1 hour. The pH is brought to 4.5 by dropwise addition of 0.1 M citric acid at 4°, and the precipitate is removed by centrifugation at 30,000 g. The supernatant fluid, 30 ml, is used directly for affinity chromatography.

Affinity Chromatography. The enzyme solution at room temperature and pH 4.5, is applied to the affinity column, which is then washed with phosphate–citrate buffer at pH 4.5 until protein is no longer eluted. A solution of 50 mM p-nitrophenyl α-D-galactopyranoside in 3 ml of the same buffer is applied to the column, and washing is continued. A sharp yellow band, due to enzymatically liberated p-nitrophenolate, indicates the location of the enzyme as it descends through the column. Yellow fractions are pooled and freed from low molecular weight materials by passage through a column (35 × 2.5 cm) of Sephadex G-25 equilibrated with distilled water. The enzyme may be stored in the presence of bovine serum albumin either in solution or as a lyophilized powder.[4]

Purity of α-Galactosidase. The specific activity of purified α-galactosidase is 94.5, representing a 237-fold purification by the affinity chromatography step, and an overall purification of 8,600-fold.

Six-percent polyacrylamide gel electrophoresis of the purified enzyme

[14] R. Axén, J. Porath, and S. Ernbäck, *Nature (London)* **214**, 1302 (1967). See also Porath, this volume [2]; and Parikh et al., this volume [6].

at pH 8.6 reveals three protein-staining bands, corresponding in their migrations to three enzymatically active species in the crude saline extract; the latter are located by incubation of 1-mm polyacrylamide gel slices of the crude extract in buffered solutions of *p*-nitrophenyl α-D-galactopyranoside. The enzyme shows no activity toward the *p*-nitrophenyl glycopyranosides of β-D-galactose, α- and β-D-glucose, α-L-fucose, α-D-mannose, *N*-acetyl-β-D-glucosamine, and *N*-acetyl-β-D-galactosamine.

Inhibition of α-Galactosidase by Ligands. *N*-(*N*-benzyloxycarbonyl-ε-aminocaproyl)-α-D-galactopyranoside and *N*-ε-aminocaproyl-α-D-galactopyranoside behave in solution as competitive inhibitors of α-galactosidase, as determined by Lineweaver-Burke plots with 1.25 mM *p*-nitrophenyl α-D-galactopyranoside as substrate. $K_i = 3 \times 10^{-4} M$ at pH 4.5 and $9 \times 10^{-4} M$ at pH 6.0. Although the affinity columns are routinely operated at pH 4.5, with the introduction of substrate necessary for elution of α-galactosidase, extensive washing at pH 6.0 without substrate results in elution of two separate enzyme peaks.

Overnight incubation of the above compounds with the enzyme at various pH values does not lead to the appearance of galactose.

A different approach to the affinity chromatography of α-galactosidase has been reported very recently[15] involving synthesis of a *p*-aminophenyl derivative of the substrate melibiose, and use of a Sepharose conjugate of this compound to purify and separate serum α-galactosidases. Elution of these enzymes requires the use of a nonionic detergent.

[15] C. A. Mapes and C. C. Sweeley, *J. Biol. Chem.* **248**, 2461 (1973).

[34] β-Galactosidase

By EDWARD STEERS, JR. and PEDRO CUATRECASAS

The enzyme β-galactosidase is one of several oligosaccharide-splitting enzymes that have been purified and subsequently characterized to varying degrees both physically and mechanistically.[1-5] It is found in a wide range of biological sources in both the plant and animal kingdoms although the principal source of study has been in the bacterium *Escherichia*

[1] S. A. Kuby and H. A. Lardy, *J. Amer. Chem. Soc.* **75**, 890 (1953).
[2] M. Cohn, *Bacteriol. Rev.* **21**, 140 (1957).
[3] K. Wallenfels, M. L. Larnitz, G. Laule, H. Bender, and M. Keser, *Biochem. Z.* **331**, 459 (1959).
[4] A. S. Hu, R. G. Wolfe, and F. J. Reithel, *Arch. Biochem. Biophys.* **81**, 500 (1959).
[5] G. R. Craven, E. Steers, Jr., and C. B. Anfinsen, *J. Biol. Chem.* **240**, 2468 (1965).

coli, where it gained prominence as a result of the studies of Jacob and Monod and their collaborators in enzyme regulation.

The enzyme has been shown to catalyze hydrolytic and transfer reactions wherein a variety of substances may act as acceptor for the glycone moiety of the substrate.

Although lactose is the natural substrate of this enzyme, it has been shown that the same enzyme catalyzes the hydrolysis of *o*-nitrophenyl β-D-galactopyranoside (ONPG) resulting in a chromogen, *o*-nitrophenol, which is readily detected by its absorbance at 420 nm. Whereas the enzyme exhibits a rather stringent specificity for the glycone portion of the molecule, several different groups may serve as aglycone substituents. On this basis, a variety of substrates which successfully interact with the enzyme have been synthesized. Isopropyl-β-thiogalactoside is an excellent inducer of β-galactosidase synthesis and is resistant to hydrolysis due to the thio linkage. *p*-Aminophenyl-β-D-thiogalactoside is also capable of binding to the enzyme while being resistant to hydrolysis, and it exhibits competitive inhibition of *o*-nitrophenyl-β-D-galactoside hydrolysis by β-galactosidase of *E. coli.*

β-Galactosidase from *E. coli* has been purified by various standard procedures.[1-6] More recently, several studies have employed, with varying degrees of success, adsorption chromatography with either specific antibodies or substrate analogs. Tomino and Paigen reported purification of the enzyme with β-thiogalactose ligands attached to insoluble bovine γ-globulin.[6] Erickson and Steers purified β-galactosidase from *Aerobacter cloacae* on columns containing covalently linked specific anti-*E. coli* β-galactosidase antibodies in one case,[7] while purifying an antigenically cross-reacting protein of β-galactosidase by a combination of gel filtration and affinity chromatography using an antibody column.[8] The use of affinity chromatography with agarose derivatives of the substrate analog, *p*-aminophenyl-β-D-galactoside, for the purification of the enzyme from *E. coli*[9] and *Bacillus megaterium*[10] will be described here.

Assay Methods

Whereas both lactose and the synthetic substrate, *o*-nitrophenyl-β-D-galactopyranoside (ONPG), have been used for the detection of β-galacto-

[6] A. Tomino and K. Paigen, *in* "The Lac Operon" (D. Zipser and J. Beckwith, eds.), p. 233. Cold Spring Harbor Laboratory, Cold Spring Harbor, New York, 1970.
[7] R. P. Erickson and E. Steers, Jr., *Arch. Biochem. Biophys.* **137**, 399 (1970).
[8] R. P. Erickson and E. Steers, Jr., *J. Bacteriol.* **102**, 79 (1970).
[9] E. Steers, Jr., P. Cuatrecasas, and H. Pollard, *J. Biol. Chem.* **246**, 196 (1971).
[10] H. Pollard and E. Steers, Jr., *Arch. Biochem. Biophys.* **158**, 650 (1973).

sidase activity, the latter represents the easiest method for routine assays of enzyme activity. The free *o*-nitrophenol which is released by enzymatic hydrolysis is monitored spectrophotometrically at 420 nm. A unit of enzyme activity is defined as that amount of enzyme which hydrolyzes 10^{-9} mole of *o*-nitrophenyl-β-D-galactopyranoside in 1 minute at 25° in 0.05 M Tris·chloride at pH 7.5, containing 0.1 M NaCl, 10 mM MgCl$_2$, and 0.1 M 2-mercaptoethanol.[5]

Preparation of Inhibitor Adsorbents

Two types of insoluble supporting matrices are described for preparing the inhibitor adsorbents: synthetic polyacrylamide gel Bio-Gel P-300 (Bio-Rad Laboratories) and beaded agarose-Sepharose 4B. These two solid supports best conform to the predicted requirements necessary for an ideal inhibitor adsorbent.[11,12]

Preparation of Polyacrylamide Derivatives. *p*-Aminophenyl-β-D-thiogalactopyranoside was attached directly to the polyacrylamide backbone by procedures modified[13] from those described by Inman and Dintzis[14] for attaching compounds containing primary aliphatic amino groups. The carboxamide groups of the polyacrylamide are converted to their respective hydrazides, and these are in turn converted with nitrous acid into acyl azide derivatives. The latter react rapidly at 4° and at neutral pH with compounds having aliphatic or aromatic primary amino groups.

Five hundred milliliters of Bio-Gel P-300 are washed with and suspended in 700 ml of 0.2 N NaCl. Hydrazine is added to achieve a concentration of 3 M, and the stirred suspension is heated in a 47° water bath for 4 hours. This polyacrylamide–hydrazide derivative contains about 35 μmoles of hydrazide groups per milliliter of packed gel. The acyl azide derivative is prepared by reaction with nitrous acid. Packed hydrazide gel, 50 ml is washed, suspended in 100 ml of 0.3 N HCl, and placed in an ice bath. Ten milliliters of 1 M NaNO$_2$ are rapidly added under vigorous magnetic stirring. The acyl azide derivative is then allowed to react with a solution of 0.5 M Na$_2$CO$_3$, pH 9.4, containing 0.1 M *p*-aminophenyl-β-D-thiogalactopyranoside, 50 mM glycylglycyltyrosine, and 0.4 M ethylenediamine to prepare derivative A and the precursors of derivatives B and C, respectively (Fig. 1). Considerable shrinkage of the polyacrylamide beads occurs following the acyl azide coupling step for all acrylamide derivatives. Derivative A contains about 22 μmoles of thiogalactoside per milliliter of packed Bio-Gel.

[11] P. Cuatrecasas and C. B. Anfinsen, this series, Vol. 22, 345 (1971).
[12] P. Cuatrecasas, *Advan. Enzymol.* 36, 29 (1972).
[13] P. Cuatrecasas, *J. Biol. Chem.* 245, 3059 (1970).
[14] J. K. Inman and H. M. Dintzis, *Biochemistry* 8, 4074 (1969). See also Inman, this volume [3].

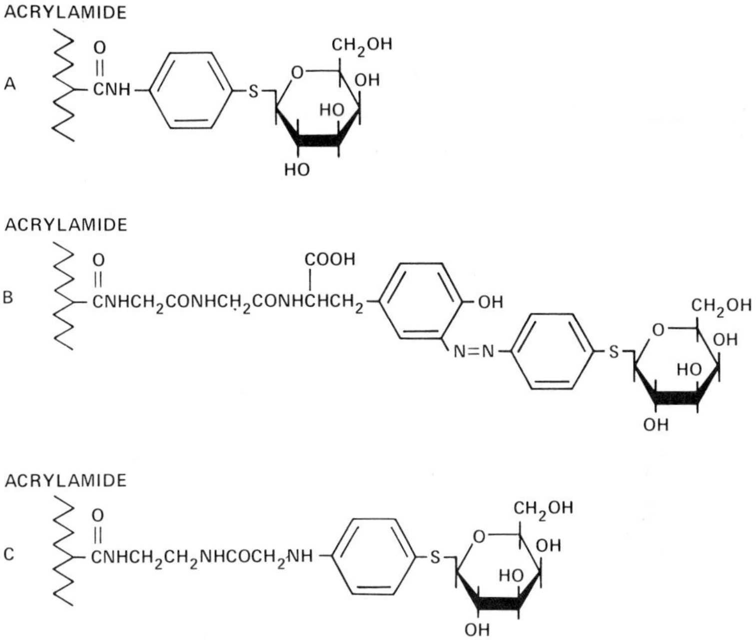

FIG. 1. Derivatives, A–C of aminoethyl polyacrylamide.

The bromoacetyl derivative of aminoethyl polyacrylamide (derivative C, Fig. 1) is prepared as follows: In 8 ml of dioxane are dissolved 1.0 mmole of bromoacetic acid and 1.2 mmoles of N-hydroxysuccinimide. To this solution 1.1 mmoles of dicyclohexyl carbodiimide are added. After 70 minutes dicyclohexylurea is removed by filtration, and the entire filtrate is added to a suspension, at 4°, containing 20 ml of packed aminoethyl polyacrylamide in a total volume of 50 ml at pH 7.5 in 0.1 M sodium phosphate. After 30 minutes the gel is washed with 2 liters of cold 0.1 N NaCl. The alkylation of p-aminophenyl-β-D-thiogalactopyranoside to this derivative is performed as described for the corresponding agarose derivative (see below). The latter derivative contains about 4 μmoles of galactoside per milliliter of packed Bio-Gel. Derivative B is prepared by azotization of the ligand to Bio-Gel–glycylglycyltyrosine (2 μmoles per milliliter of polyacrylamide) in the same manner as noted for the corresponding agarose derivative described below.

Preparation of Agarose Derivatives. p-Aminophenyl-β-D-thiogalactopyranoside is coupled to Sepharose 4B, and to derivatives of Sepharose 4B, essentially by the procedures described for derivatizations of agarose.[13]

The ligand is coupled directly to the agarose backbone after activation of the latter with cyanogen bromide (Fig. 2, derivative A). Two hundred and fifty milligrams of cyanogen bromide are used per milliliter of

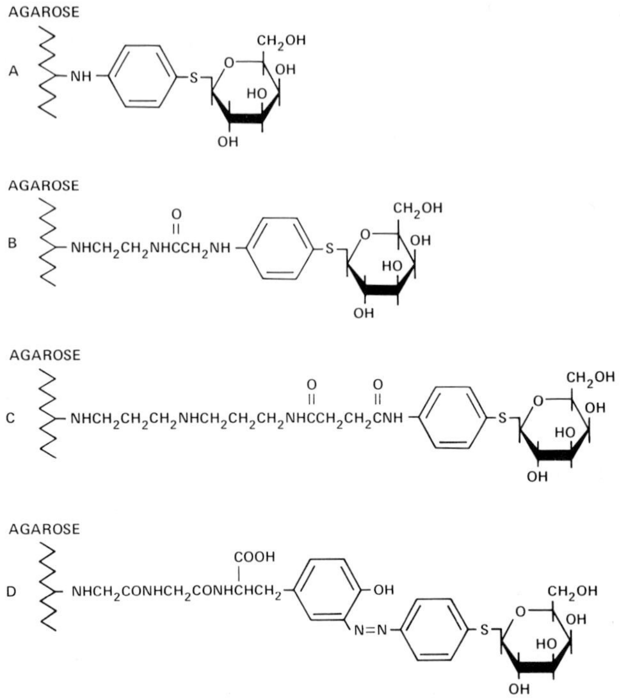

Fig. 2. Derivatives A–D of agarose.

packed Sepharose 4B, and the concentration of the thiogalactoside in the coupling step is 50 mM. The buffer used in the latter step is 0.1 M NaHCO$_3$ at pH 9.2. Judging from the amount of thiogalactoside recovered in the washes, about 6 μmoles are incorporated per milliliter of packed agarose.

Derivative B (Fig. 2) is prepared by allowing bromoacetamidoethyl-agarose to react with 5 mM p-aminophenyl-β-D-thiogalactopyranoside in 0.2 M NaHCO$_3$ at pH 8.4 for 3 days at room temperature. After thorough washing, the unreacted bromide groups are blocked by treatment with 0.2 M 2-aminoethanol for 24 hours at room temperature. The derivative has approximately 2 μmoles of ligand per milliliter of packed agarose.

Derivative C (Fig. 2) is prepared in the following way. Three hundred ml of packed Sepharose 4B is activated with 85 g of cyanogen bromide. After washing the activated agarose with 1 liter of cold 0.2 M NaHCO$_3$ at pH 9.0, 150 ml of an ice-cold 0.4 M solution of 3,3′-diaminodipropyl-amine in 0.2 M NaHCO$_3$ at pH 9.8, are rapidly added to the moist gel. The reaction is allowed to proceed for 18 hours at 4°. This ω-aminoalkyl agarose, which contains about 8 μmoles of amino groups per milliliter of packed gel, is succinylated by treatment with succinic anhydride. One milliliter of succinic anhydride is added per milliliter of packed aminoethyl-Sepharose in an equal volume of distilled water at 4°. The pH is raised to and maintained at 6.0 by titrating with 20% NaOH. When no further change in pH occurs, the suspension is left for 5 more hours at 4°. *p*-Aminophenyl-β-D-thiogalactopyranoside is then coupled at room temperature to the succinyl-agarose derivative. *p*-Aminophenyl-β-D-thiogalactopyranoside, 1.5 g, is added to a suspension containing 150 ml of packed agarose in a total volume of 250 ml of distilled water. The pH is adjusted to 4.7, and 4.5 g of 1-ethyl-3-(3-dimethylaminopropyl) carbodiimide, dissolved in 10 ml of distilled water, are added rapidly. The reaction is monitored for 2 hours with a pH meter, the pH being maintained at 4.7 with 1 *N* HCl. The suspension is stirred gently at room temperature for 3 additional hours. The final agarose adsorbent is washed with 15 liters of 0.1 *N* NaCl over an 8-hour period to remove all noncovalently bound thiogalactoside. Approximately 6 μmoles of inhibitor are covalently attached per milliliter of packed agarose. The various steps in these procedures can be followed conveniently with the color reaction produced on reaction of the agarose derivatives with 2,4,6-trinitrobenzene sulfonate in sodium borate buffer.[13]

Derivative D (Fig. 2) is prepared by coupling the diazonium derivative of *p*-aminophenyl-β-D-thiogalactopyranoside with Sepharose-glycylglycyltyrosine. The tripeptide, containing a carboxyl terminal tyrosine, is coupled to Sepharose 4B by the cyanogen bromide procedure described above. Fifteen milliliters of the packed tyrosine-containing agarose (8 μmoles of tripeptide per milliliter of packed gel, determined by amino acid analysis), are washed with ice cold, saturated sodium borate, pH 9.8, and suspended in 25 ml of the same buffer. One hundred milligram of *p*-aminophenyl-β-D-thiogalactopyranoside are dissolved in 10 ml of cold distilled water. One milliliter of 6 *N* HCl is added to the solution, which is maintained in an ice bath. The diazonium derivative is obtained by adding, over a 30-second period, 2 ml of distilled water containing 400 mg of sodium nitrite. This solution is vigorously stirred at 4° for 4 minutes, then the entire mixture is added to the tyrosine agarose suspension. The pH is raised to 9.8 with 10% NaOH. The azotization reaction is allowed to proceed for 1 hour at 4°, followed by incubation for 2 hours at room temperature. Tyrosyl azotization is essentially complete, as shown by the nearly total loss of tyrosine from acid hydrolyzates of the agarose derivative.

Purification Procedure. The procedure described here has been used successfully for the native enzyme from *Bacillus megaterium*[10] as well as from *E. coli.*[9] In addition, it has been possible to purify[9] two enzymatically inactive mutant forms of the enzyme as well as the monomer which results from dissociation of the wild-type protomer. These molecular species lack hydrolytic ability but are capable of binding to the inhibitor and can thus specifically adsorb to the affinity resin. The mutant *E. coli* K12 strain 3310 represents a point mutation that exists in the dimeric form and is presumably unable to form tetramers, whereas the mutant *E. coli* K12 strain U178, also a point mutation, exists as an inactive monomer unable to dimerize or form tetramers. Both of these inactive forms of the enzyme are capable of binding to the resin and show elution characteristics identical to the wild-type enzyme.

Cells, 100 g wet weight, are suspended in 500 ml of a buffer which consists of 50 mM Tris chloride at pH 7.5, 0.1 M NaCl, 10 mM MgCl$_2$, and 0.1 M 2-mercaptoethanol. The cell suspension is disrupted by sonication for 2 minutes at 0° with a Heat Systems Model W140D sonifier. The lysed cells are centrifuged in a Sorvall RC 2B centrifuge for 20 minutes at 48,000 g. The clear supernatant solution is fractionated by adding solid ammonium sulfate to 50% of saturation (35 g per 100 ml) followed by centrifugation at 48,000 g for 10 minutes. The precipitate is dissolved in 50 ml of the starting buffer and dialyzed against 6 liters of the same buffer at 4° for 24 hours. The dialyzed extract is layered onto the appropriate resin which is then washed with several column volumes (usually 5 volumes) of the buffer.

The various polyacrylamide affinity gels described in Fig. 1 have proved to be ineffective for the adsorption of β-galactosidase from *E. coli* and *B. megaterium,* despite the very high degree of substitution that can be achieved with these derivatives. The most likely explanation for this lack of efficacy is related to the bead shrinkage and consequent decrease in porosity which occurs upon certain types of derivatization procedures of acrylamide.[12,13] This is especially important for proteins of large molecular size, such as β-galactosidase, since such macromolecules cannot penetrate the interstices of the beads and can thus only interact with the relatively small amount of ligand which is present on the bead surface.[14]

In the case of the substituted agarose resins, variable results are obtained depending upon the conditions of elution and the particular derivative used. When the inhibitor is attached directly to the support, no detectable binding of the enzyme is observed. In this case the downward migration of the enzyme through the column is not even retarded and its chromatographic pattern is identical to that observed with unsubstituted agarose. With derivative B, wherein the ligand is separated from the resin by a short "arm" of about 10 Å consisting of a bromoacetamidoethyl side chain, significant retention of enzyme activity is not observed. The enzyme does, however, exhibit a degree of retardation since it emerges from the column slightly behind the major protein breakthrough. Although this method is not recommended for purposes of purification, it does, in principle, allow separation of the enzyme by collecting the retarded material and making successive column passes. Derivative C, wherein a rather long "arm," i.e., about 21 Å, is interposed between the ligand and the resin, is the most useful adsorbent for this enzyme. The bulk of β-galactosidase activity from *B. megaterium* and from *E. coli* binds to columns containing this adsorbent. Unlike the chromatography pattern observed with columns containing unsubstituted agarose, only a small portion, from 5 to 15%, of the applied activity emerges from the affinity column behind the major protein breakthrough while the column is washed extensively with buffer. In the case of the enzyme from *B. megaterium,* the retained and retarded enzymatically active materials exhibit patterns that are identical to their original patterns on rechromatography. If the retained material is stored in solution at 4° for 1 week before rechromatography, approximately 10% of the activity emerges in a retarded pattern. These results suggest that the retarded enzymatic material may represent enzyme which has been altered during storage and therefore does not represent a separate or unique molecular species which possesses inherently different chromatographic behavior in its native state.

Elution of the enzymatic activity which has been retained on the column may be accomplished in two ways. The first method utilizes substrate containing buffers to effectively compete with the inhibitor. In the second method the binding of protein to the ligand is disturbed by raising the pH of the eluting buffer. To elute with substrate analogs, columns are washed with Tris chloride at pH 7.5 containing either 50 m*M* lactose or 50 m*M* isopropyl β-D-galactopyranoside. Although such substrates are capable of affecting elution of the bound enzyme, the results of such methods are not entirely satisfactory. If the gel contains relatively large amounts of immobilized ligand, the enzyme can be eluted with substrate containing buffers. A serious drawback to this form of elution is the gradual nature of elution which consequently results in the recovery of the enzyme in very dilute form. The theoretical basis of these observations is discussed elsewhere.[12] It must be cautioned, however, that in certain applications—for example, those in which enzyme may be sensitive to extremes of pH—this method of elution must be utilized.

The success and utility of elution with alkaline buffers (0.1 *M* sodium borate, pH 10.0) is demonstrated by the quantitative nature of the recovery of the applied enzyme and by the appearance of the enzyme in a sharp boundary in concentrated form. After application of the "crude" extract, the column is washed extensively until the absorbance returns nearly to the base line. The elution buffer is then changed to 0.1 M sodium borate at pH 10.0, and elution is allowed to proceed until a sharp boundary of activity emerges. Recovery of the enzyme is consistently greater than 95% of the initial activity applied to the columns. Successive elution fails to recover additional enzymatic activity, and the absence of retained activity on the column can be demonstrated by washing the resin with the assay substrate, o-nitrophenyl-β-D-galactopyranoside. When a significant quantity of active enzyme is retained on the column the substrate is cleaved, resulting in a bright yellow color that is clearly visible on the column. This is a very sensitive method of detecting residual adsorbed enzyme on affinity columns. Subsequent elution of the column bed with 5 M guanidine-HCl only results in the removal of small amounts of inhibitor, which presumably had been strongly adsorbed on the gel.

Properties

Stability. The purified enzyme is stable for weeks when stored as an ammonium sulfate precipitate at 0°. The stability in buffered solutions at 0° is considerably less, with measurable losses in activity occurring in 2–3 days. The enzyme requires 2-mercaptoethanol in amounts of 50 mM to 0.1 M to present loss in activity. Solutions which are either frozen or lyophilized show a markedly increased loss of activity with the appearance of a flocculent precipitate upon thawing or dissolution.

Purity and Physical Properties. The β-galactosidase purified by affinity chromatography is homogeneous by several criteria including polyacrylamide gel electrophoresis. However, purification by this procedure does not separate the various isoenzyme forms which represent larger polymers of the basic tetramer. Separation of the latter forms of β-galactosidase on the basis or molecular weight has been accomplished by centrifugation in sucrose gradients.[15]

The enzyme purified by this procedure has a specific activity between 300,000 and 320,000 units/mg, which compares favorably with the values obtained by conventional methods.[5]

The major species of the enzyme has been shown to be a tetramer composed of identical subunits with a combined molecular weight of 540,000.[5,16,17] The minor (5%) species of higher molecular weight forms range in weight from values of 900,000 to 2,300,000 and appear to differ only in their state of aggregation.[15]

[15] S. L. Marchesi, E. Steers, Jr., and S. Shifrin, *Biochim. Biophys. Acta* 133, 454 (1967).
[16] D. Zipser, *J. Mol. Biol.* 7, 113 (1963).
[17] K. Wallenfels, M. Sund, and K. Weber, *Biochem. Z.* 338, 714 (1963).

[35] α-Lactalbumin–Agarose. A Specific Adsorbent for a Galactosyltransferase

By IAN P. TRAYER, ROBERT BARKER, and ROBERT L. HILL

The galactosyltransferase (UDP-galactose:N-acetylglucosamine galactosyltransferase), the so-called A protein of lactose synthetase[1,2] can be isolated by affinity chromatography on Sepharose–α-lactalbumin conjugates.[3] The procedures for preparing this conjugate and for operating the columns are given below.

Preparation of Agarose–α-Lactalbumin Conjugate[3]

Bovine α-lactalbumin (Pentex) may be purified by gel filtration on Sephadex G-100 in 50 mM ammonium bicarbonate as described earlier[4] or isolated from milk by published methods.[5] Sepharose-4B (500 ml of settled Sepharose, washed with deionized water on a filter flask) is suspended in water (1 liter final volume) and activated with cyanogen bromide according to the procedure of Porath and co-workers.[6,7] The suspension is adjusted to pH 11 with 4 N sodium hydroxide at room temperature and gently stirred during addition over a 2-minute period of a freshly prepared cyanogen bromide solution (500 ml, 30 mg/ml). The reaction mixture is maintained at pH 11 by addition of 4 N sodium hydroxide, and after 7–10 minutes the activated Sepharose is filtered using a coarse sintered-glass funnel with suction. It is washed immediately with ice-cold deionized water (4 liters) and then ice-cold 0.1 M sodium carbonate (10 liters). The activated Sepharose is suspended in cold 0.1 M sodium carbonate, pH 9.4 (final volume 1250 ml) and α-lactalbumin solution (100 ml, 30 mg/ml in 0.1 M sodium carbonate) added. The reaction is allowed to proceed overnight at 4° with gentle stirring whereupon coupling is found to be essentially quantitative as judged by disappearance of absorbance at 280 nm in the supernatant solution after centrifugation. The conjugate is washed by filtration successively with water (10 liters), 0.2 M sodium bicarbonate at pH 8.2 (15 liters), water

[1] U. Brodbeck and K. E. Ebner, *J. Biol. Chem.* **241**, 762 (1966).
[2] K. Brew, T. C. Vanaman, and R. L. Hill, *Proc. Nat. Acad. Sci. U.S.* **50**, 491 (1968).
[3] I. P. Trayer and R. L. Hill, *J. Biol. Chem.* **246**, 6666 (1971).
[4] F. J. Castellino and R. L. Hill, *J. Biol. Chem.* **245**, 417 (1970).
[5] W. J. Gordon and J. Ziegler, *Biochem. Prep.* **4**, 16 (1955).
[6] R. Axén, J. Porath, and S. Ernbäck, *Nature (London)* **214**, 1302 (1967).
[7] J. Porath, R. Axén, and S. Ernbäck, *Nature (London)* **215**, 1491 (1967).

(10 liters), 0.2 M ammonium acetate at pH 3.9 (15 liters), and water and then stored in water containing 0.02% sodium azide at 4°. This material was used directly in the affinity chromatography studies described below. An aliquot after acid hydrolysis was subjected to amino acid analysis and estimated to contain 90 mg of covalently bound α-lactalbumin per gram of Sepharose–α-lactalbumin conjugate (dry weight).

Affinity Chromatography of Milk Whey Galactosyltransferase on a Sepharose–α-Lactalbumin Conjugate[3]

The purification of milk whey galactosyltransferase has been performed on a Sepharose–α-lactalbumin column. Complex formation between the transferase and the Sepharose–α-lactalbumin conjugate was promoted by including glucose[3] in the developing buffer although this can also be achieved by the inclusion of N-acetylglucosamine.[8] Dissociation of the complex takes place in both instances by omitting the monosaccharide from the eluting buffer.

In a typical purification, galactosyltransferase (17.2 ml, absorbance at 280 nm = 3.0), partially purified from bovine whey by ion-exchange chromatography, is applied to a Sepharose–α-lactalbumin column (1.5 $\times$ 30 cm) and developed in 50 mM sodium cacodylate-HCl at pH 7.5 containing 5 mM MgCl$_2$, 1 mM β-mercaptoethanol, and 20 mM glucose. Fractions (5 ml) were collected at a flow rate of 20 ml per hour. After the addition of 900 ml of the above buffer, 400 ml of the same buffer from which glucose is excluded are used to elute the enzyme. Recovery of the transferase activity is about 60–65% based upon the activity of the enzyme applied. The enzyme is purified about 25- to 65-fold.[3]

It is necessary that partially purified galactosyltransferase preparations be used with the Sepharose–α-lactalbumin for adequate retention of activity by the adsorbent. Partial purification can be achieved by traditional procedures[3] or by affinity chromatography on Sepharose-UDP or N-acetylglucosamine adsorbents.[9]

[8] P. Andrews, FEBS Lett. 9, 297 (1970).
[9] R. Barker, K. W. Olsen, J. H. Shaper, and R. L. Hill, J. Biol. Chem. 53, 564 (1972).

[36] O-Glycosyl Polyacrylamide Gels: Application to Phytohemagglutinins

By VÁCLAV HOŘEJŠÍ and JAN KOCOUREK

Copolymerization of alkenyl O-glycosides with acrylamide and N,N'-methylene bisacrylamide produces neutral hydrophilic gels containing sugars linked by O-glycosidic bonds to alkyl side chains of the polymer matrix (Fig. 1). This procedure represents a versatile method for preparation of affinity adsorbents with carbohydrate affinants.[1] According to the properties and amounts of the components taken into polymerization reaction it allows: (a) preparation of polyacrylamide derivatives of sugars with specified anomeric configuration and ring form, (b) variation of the type and length of the "spacer" chain by which the affinant is linked to the polymer matrix, (c) regulation of both the sugar content

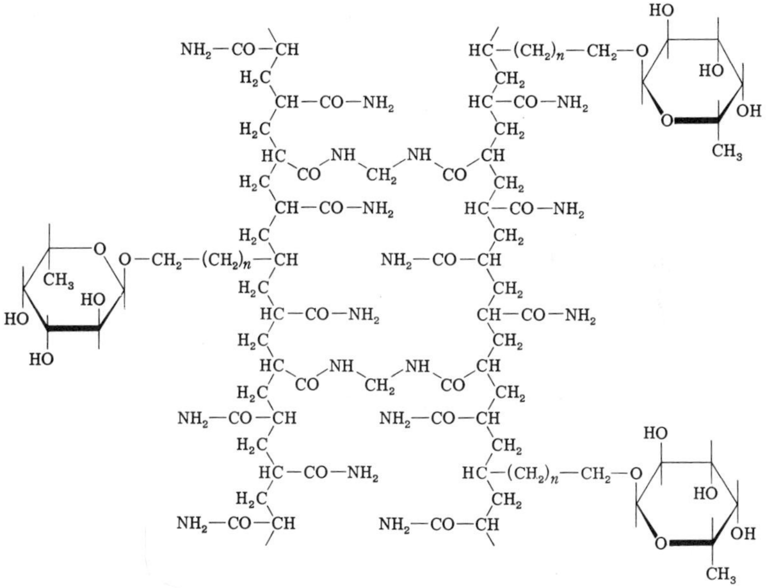

FIG. 1. Partial tentative structure of an O-α-L-fucopyranosyl derivative; when allyl α-L-fucopyranoside is used for copolymerization, $n = 0$. Reproduced from Hořejší and Kocourek[1] by permission of the copyright owner.

---

[1] The method reported here has been published by V. Hořejší and J. Kocourek, Biochim. Biophys. Acta **297**, 346 (1973).

and the pore size of the copolymer, (d) preparation of affinity adsorbents with several different carbohydrate affinants in one operation.

O-Glycosyl polyacrylamide derivatives proved very useful[1,2] in isolation of phytohemagglutinins (plant lectins), proteins with combining sites for carbohydrates.[3,4] In addition to the preparative application in affinity chromatography, these affinity adsorbents can also be used for analytical purposes in polyacrylamide gel affinity electrophoresis (this volume [13]).

It seems to be established that carbohydrate receptors serve as mediators in most of the biological activities of phytohemagglutinins, i.e., agglutination of erythrocytes, leukocytes, growth suppression of malignant cells, mitogenic activity, etc. The structural requirements of the combining sites vary widely, ranging from phytohemagglutinins strictly specific toward a single carbohydrate structure to phytohemagglutinins which combine with a number of different carbohydrates. In any case, however, the combining sites of phytohemagglutinins seem to be more easily accessible than the combining sites of antibodies and active sites of enzymes. For successful affinity adsorption they do not require long spacer chains separating the ligand from the polymer matrix. In practice allyl glycosides proved to be fully satisfactory for the preparation of reliable affinity adsorbents for phytohemagglutinins. Therefore, synthesis and utilization of allyl glycosides is presented in this article. For practical purposes it is not necessary to use highly purified glycosides for the polymerization reaction. Admixtures of β-anomers which often accompany α-anomers do not significantly influence the capacity of the affinity carrier for the particular phytohemagglutinin.

It should be emphasized that the type of affinity carrier introduced here is not usable for affinity chromatography of enzymes. However, a general applicability of the procedure for preparation of adsorbents for proteins with different combining sites seems to be only a matter of an appropriate selection and modification of the unsaturated compound taken into the polymerization reaction as a potential affinant.

Preparation of Allyl Glycosides

For the preparation of allyl glycosides almost all the standard methods of aliphatic glycoside synthesis can be adopted.[5,6] The procedures given

[2] V. Hořejší and J. Kocourek, *Biochim. Biophys. Acta* 336, 329 (1974).

[3] N. Sharon and H. Lis, *Science* 177, 949 (1972).

[4] H. Lis and N. Sharon, *Annu. Rev. Biochem.* 42, 541 (1973). See also Kristiansen, this volume [31] and Lis *et al.* this volume [32].

[5] J. Staněk, M. Černý, J. Kocourek, and J. Pacák, "The Monosaccharides," p. 257. Publishing House of the Czechoslovak Academy of Sciences, Prague, and Academic Press, New York, 1963.

[6] J. Conchie, G. A. Levvy, and C. A. Marsh, *Advan. Carbohyd. Chem.* 12, 158 (1957).

PHYSICAL CONSTANTS OF SOME ALLYL GLYCOSIDES

Glycoside	Melting point (°C)	$[\alpha]_D$ in water
Allyl α-L-fucopyranoside[a]	158–160	$-216°$
Allyl α-D-galactopyranoside[b]	144–146	$+183°$
Allyl α-D-glucopyranoside[b]	100.5–101.5	$+151°$
Allyl β-D-glucopyranoside[c]	102–103	$-40.5°$
Allyl α-D-mannopyranoside[a]	98–99	$+99°$
Allyl 2-acetamido-2-deoxy-α-D-galactopyranoside[d] (=allyl N-acetyl-α-D-galactosaminide)	198–199	$+189°$
Allyl β-lactoside[e]	171–172	$+16°$

[a] V. Hořejší and J. Kocourek, *Biochim. Biophys. Acta.* **297**, 346 (1973).
[b] E. Talley, M. D. Vale, and E. Yanovsky, *J. Amer. Chem. Soc.* **67**, 2037 (1945).
[c] E. Fischer, *Hoppe-Seyler's J. Physiol. Chem.* **108**, 3 (1919).
[d] J. Kocourek and V. Hořejší, unpublished results.
[e] V. Hořejší and J. Kocourek, *Biochim. Biophys. Acta* **336**, 338 (1974).

below are general directions valid for most of the different sugar types. However, it should be borne in mind that the synthesis of almost any individual glycoside has its optimum conditions differing from those of the synthesis of another glycoside. Up to the present the preparation of only a few allyl glycosides has been described in literature. Allyl glycosides with known physical constants are summarized in the table.

Allyl α-D- and α-L-Glycosides

In a flask equipped with mechanical stirrer and a reflux condenser with a drying tube to exclude moisture, 100 g of finely powdered monosaccharide are suspended in 200 ml of allyl alcohol which contains 3% (w/v) of dry HCl. The mixture is heated for 4 hours at 80° with stirring. During this time almost all the sugar dissolves and the solution usually turns brown. The reaction mixture is cooled and neutralized with 25% ammonia (about 12 ml is necessary) and the allyl alcohol is completely distilled off under reduced pressure at about 70°. If the residue is solid it can be crystallized directly from hot ethanol; usually two recrystallizations are necessary. If a syrup results it is mixed with about 100 g of cellulose powder (if the syrup is too thick it can be diluted with some acetone), and the wet powdery mass is then extracted thrice with 300 ml of acetone. The combined acetone extracts are evaporated under reduced pressure at about 70° to dryness. The solid or syrupy residue is dissolved in the smallest amount of hot ethanol from which it crystallizes. The yield is usually between 25 and 45% of the theoretical.

NOTE 1: Free amino sugars (2-amino-2-deoxy sugars) do not form glycosides when treated in this manner. Their N-acetyl derivatives, however, do give the appropriate glycosides in good yields. For their preparation it is advantageous to reduce the concentration of HCl to 0.5–1.0% and to shorten the reaction period to 30 minutes to 1 hour. During the reaction partial deacetylation of the *N*-acetyl amino group takes place. To obtain a readily crystallizable product it is necessary to remove the compounds with free amino groups. The product after neutralization and evaporation is dissolved in water and passed through a column of a cation exchange resin (H⁺), e.g., Dowex 50 W or a mixed bed resin such as Bio-Deminrolit. The eluate is evaporated again to dryness, and the glycoside is crystallized from ethanol.

NOTE 2: This procedure is not applicable for preparation of glycosides of oligosaccharides.

Allyl β-D- and β-L-Glycosides

Preparation of Fully O-Acetylated β-D- or β-L-Sugars

Powdered anhydrous sodium acetate, 120 g, is mixed with 150 g of anhydrous sugar and the mixture is added to 750 ml of acetic anhydride in a 2-liter round-bottom flask. It is heated on a boiling water bath under reflux condenser for 2–5 hours. Before the reaction starts it is necessary to shake the flask occasionally so as to prevent local overheating of the suspension which could trigger a violent reaction and cause caramelization of the sugar. During the given period all the sugar dissolves and only a residue of sodium acetate remains undissolved. The mixture is cooled to about 25° and poured into 10 liter of ice water. When stirred, the syrupy product usually soon solidifies. It is collected and crystallized from hot ethanol, with the use of decolorizing charcoal if necessary.

NOTE 1: Some fully O-acetylated sugars can conveniently be prepared in a quick and simple procedure using acetic anhydride and perchloric acid.[7]

Preparation of O-Acetyl α-D- or α-L-Glycosyl Bromides

To 50 g of the fully acetylated sugar, 100 g (72 ml) of the brominating agent are added; after the sugar has dissolved, the reaction mixture is maintained for 2 hours at room temperature in a glass-stoppered flask. About 150 ml of chloroform are added, and the mixture is poured into 500 ml of ice water. The chloroform layer is separated from the water layer in a separatory funnel, and the chloroform layer is extracted twice with the same volume of ice water and once with the same volume of saturated NaHCO₃ solution. The chloroform solution is dried with anhy-

[7] M. Bárczai-Martos and F. Körösy, *Nature* (*London*) **165**, 369 (1950).

drous Na_2SO_4 and, after addition of about 3 g of active charcoal, the solution is filtered and evaporated under reduced pressure at about 50–60°. The resulting syrup is dissolved in the same volume of diethyl ether, from which the O-acetylglycosyl bromide crystallizes.

NOTE 1: Preparation of the brominating agent.[8] Bromine (180 g) is added dropwise to a suspension of red phosphorus (30 g) in glacial acetic acid (300 ml) with cooling. After the reaction is completed the mixture is filtered through hardened filter paper into a dry Büchner flask by suction and kept in the cold room if not required for immediate use.

NOTE 2: The brominating agent also can be used for direct preparation of O-acetylglycosyl bromides from nonacetylated sugars.[8] Although one reaction step is saved, this procedure requires more attention and the yields are usually lower.

NOTE 3: For the preparation of O-acetylglycosyl bromides, both anomers (α- or β-) of the fully acetylated sugars may be used.

NOTE 4: Most of the O-acetylglycosyl bromides are unstable compounds, especially when not sufficiently pure and dry. It is recommended that they be used immediately for the next reaction step. If stored, it is preferable to keep them in a desiccator over CaO.

Preparation of the O-Acetylated Allyl β-D- or β-L-Glycosides

In a mixture of 100 ml of dry benzene, 30 g of Ag_2CO_3 and 50 g of Drierite are suspended. The suspension is stirred for 30 minutes and 2.5 g of iodine is added. A solution of the O-acetylglycosyl bromide (25 g of a monosaccharide derivative or 50 g of a disaccharide derivative) in 50 ml of allyl alcohol and 50 ml of benzene is added dropwise over a period of about 1 hour. The reaction mixture is stirred for an additional 4 hours, heated to boiling on a water bath, and filtered while hot. The filtrate is evaporated under reduced pressure, 100 ml of methanol is added and again evaporated, and the residue is recrystallized from ethanol.

NOTE 1: In the procedure described above, O-acetylglycosyl bromides having the bromine atom in trans position to the adjacent acetoxy group (on C-2 of aldoses such as mannose, ribose, or rhamnose derivatives) yield mixtures of O-acetylated glycosides and alkyl-1,2-orthoacetates of O-acetylated sugars. In these cases, more specific methods of synthesis must be employed.[5,6]

Deacetylation of Fully O-Acetylated Allyl Glycosides

The acetylated glycoside, 10 g, is dissolved or suspended in 50 ml of absolute methanol and 2.5 ml of 0.1 M sodium methoxide solution

[8] P. G. Scheurer and F. Smith, *J. Amer. Chem. Soc.* **76**, 3224 (1954).

in absolute methanol is added. The solution is heated to boiling for 10–60 minutes, until a drop of the reaction mixture gives no turbidity with a drop of water. The methanol is then removed under reduced pressure and the product is recrystallized from ethanol or acetone.

Preparation and Properties of O-Glycosyl Polyacrylamide Gels

Ten grams of acrylamide, 1 g of N,N'-methylene bisacrylamide, and 4.0 g of allyl glycoside are dissolved in 100 ml of water. To this solution 0.1 g of ammonium persulfate is added and the solution is heated on a boiling water bath until it polymerizes, a period of about 5 minutes. The transparent gel is homogenized in a glass homogenizer and the suspension separated from fine particles by repeated decantation with water (3 × 1 liter). The gel is then poured into a chromatographic column and washed with water until the eluate is free of nonpolymerized allyl glycoside (negative Molisch test), after which the column is ready for chromatography.

If not prepared for immediate use, the gel can be dried after washing by adding ethanol to the water suspension, decanting, and repeating the decantation with ethanol. Drying is finished in a stream of warm air. When resuspended in aqueous solutions, the dried gel readily regains its original volume.

O-Glycosyl polyacrylamide copolymers are fairly resistant to pH and temperature changes and can be regenerated for use. In general, they show all the properties of the Bio-Gel products. The carbohydrate content in the dry copolymer, prepared according to the procedures described, is usually in the range of 5–10%. This value can be changed by manipulating the amount of the components taken into the polymerization reaction.

Affinity Chromatography of Phytohemagglutinins

For the isolation of phytohemagglutinins a number of various affinity carriers have been used.[1,3] All are modified natural polymers with the exception of the O-glycosyl polyacrylamide gels introduced here. The following is a typical example of utilization of these fully synthetic adsorbents. The phytohemagglutinins of the castor bean (*Ricinus communis* L.) have combining sites for β-D-galactosyl structures[9] and can therefore be adsorbed easily onto an O-β-lactosyl carrier.

A column of O-β-lactosyl polyacrylamide gel (3 cm × 20 cm) is equilibrated by washing with 2 liters of normal saline. A solution of

[9] M. Tomita, T. Kurokawa, K. Onozaki, N. Ichiki, and T. Osawa, *Experientia* **28**, 84 (1972).

the active protein fraction in 40 ml of normal saline (see Note 1) from castor bean seeds is applied to the column and allowed to enter the gel at a rate of about 20 ml per hour. Elution of the inactive proteins is started by passing normal saline through the column at a speed of 40 ml per hour. The course of the elution is followed by UV-adsorption at 280 nm. When the elution of inactive protein is completed, a 0.2 M lactose solution in normal saline is applied to release the adsorbed phytohemagglutinins. Fractions containing the active proteins are combined (about 150 ml) and dialyzed against 3 changes of 5 liters each of deionized water; a small amount of precipitate that forms is removed by centrifugation and the supernatant liquid is lyophilized. The yield is 960 mg of protein which presents as three components on disc polyacrylamide gel electrophoresis.[10] The hemagglutinating titer of this phytohemagglutinin mixture is 16,400 (dilution limit approximately 0.06 μg/ml) whereas the titer of the active protein fraction applied to the column is 4100 (dilution limit approximately 0.25 μg/ml).

After complete desorption of the active proteins, the column is washed with 1–2 liters of normal saline to a negative Molisch test. It is then ready for use.

The capacity of this O-β-lactosyl affinity adsorbent is about 7 mg of phytohemagglutinin per milliliter of gel, corresponding to 70 mg per gram of the dry copolymer.

NOTE 1: The protein fraction with hemagglutinating activity from the seeds of *Ricinus communis* L. is obtained by normal saline extraction of the finely ground seeds (1 : 10) and precipitation with solid ammonium sulfate in the range corresponding to 40–60% saturation.[11] The yield is about 2 g from 100 g of seeds.

In the same manner as the O-β-lactosyl derivative is used for the isolation of *Ricinus communis* L. phytohemagglutinins, O-α-D-mannosyl polyacrylamide derivative can be employed for the isolation of nonspecific phytohemagglutinins from *Canavalia ensiformis* D.C., *Pisum sativum* L., *Lathyrus sativus* L., and *Lens esculenta* Moench.; O-α-L-fucosyl derivative can be utilized for the isolation of the anti-H specific hemagglutinin from *Ulex europaeus* L.; O-α-D-galactosyl derivative for the anti-B specific hemagglutinin of *Bandeirea simplicifolia* Benth. (*Griffonia simplicifolia* Baill.); and N-acetyl O-α-D-galactosaminyl derivative for the anti-A$_1$ specific hemagglutinin of *Dolichos biflorus* L.

[10] V. Hořejší and J. Kocourek, *Biochim. Biophys. Acta* **336**, 338 (1974).
[11] A. A. Green and W. L. Hughes, Methods in Enzymology, Vol. 1 [10] (1955).

[37] *araC* Protein

By Gary Wilcox and Kenneth J. Clemetson

The purification of regulatory proteins presents special problems because they are present in the cell in very low concentrations, are often unstable, and frequently are not easily assayed. The potential of employing affinity chromatography for purifying regulatory proteins was first demonstrated by Tomino and Paigen[1] when they purified *lac* repressor from crude extracts of *Escherichia coli* using *p*-aminophenyl-thiogalactoside polybovine γ-globulin as a specific adsorbent. Since in affinity chromatography the physical isolation of the protein derives from the properties of its active site, then in principle one may not require an *in vitro* assay if, for example, the appropriate genetic strains are available to use as controls.

In the bacterium *E. coli* B/r the L-arabinose operon, *araOIBAD*, consists of a cluster of three structural genes, *araBAD*, and two controlling sites, the initiator (*araI*) and the operator (*araO*).[2] Genetic and physiological studies by Englesberg and co-workers[3,4] have suggested that the expression of the L-arabinose operon is both positively and negatively controlled by the protein product of the *araC* gene. According to their model the *araC* protein exists in two functionally active conformational states, P_1 (the repressor) and P_2 (the activator). L-Arabinose induces the operon by removing P_1 from *araO* and shifting the equilibrium to P_2 which acts at *araI* to stimulate expression of the operon. The inducibility of *ara* mRNA[5] together with genetic evidence[4] suggests that both P_1 and P_2 act at the level of transcription. The ability of *araC* protein to function as either an activator or repressor of gene expression is unique among the regulatory proteins that have been studied.

Choice of a Ligand

In order to use affinity chromatography to purify *araC* protein, we had to determine which carbon atoms in the L-arabinose molecule were

[1] S. Tomino and K. Paigen, *in* "The Lactose Operon" (D. Zipser and J. Beckwith, eds.), Cold Spring Harbor Laboratory of Quantitative Biology, Cold Spring Harbor, N.Y., 1970.

[2] E. Englesberg, *in* "Metabolic Pathways" (H. Vogel, ed.), 3rd ed., Vol. V. Academic Press, New York, 1971.

[3] E. Englesberg, J. Irr, J. Power, and N. Lee, *J. Bacteriol.* **90**, 946 (1965).

[4] E. Englesberg, C. Squires, and F. Meronk, Jr., *Proc. Nat. Acad. Sci. U.S.* **62**, 1100 (1965).

[5] G. Wilcox, J. Singer, and L. Heffernan, *J. Bacteriol.* **108**, 1 (1971).

not required for recognition by the *araC* protein since the configuration about at least one carbon atom would have to change when L-arabinose was covalently bound to a solid matrix. Genetic and physiological studies had suggested that D-fucose, a compound not metabolized by *E. coli* B/r, interacts with the *araC* protein *in vivo*.[3] We screened a large number of analogs of L-arabinose and found that both the one and five carbon atoms can be substituted to produce an anti-inducer, a compound which prevents induction of the L-arabinose operon in the presence of L-arabinose. Presumably this occurs by interacting with the *araC* protein. The analog chosen must not be a substrate of any enzyme present in the crude extract. (To eliminate one such potential enzyme, a strain of *E. coli* B/r which contains a nonsense mutation in the gene coding for L-arabinose isomerase, *araA*, would be used as a source of *araC* protein.) Phenyl-β-D-fucopyranosides were found to be very good anti-inducers of the L-arabinose operon, should not be metabolized by *E. coli* B/r, and can be conveniently covalently attached to a number of solid matrices.

Sepharose appeared to be the ideal choice for use as a solid matrix.[6] It has minimal interaction with proteins, excellent flow properties, and when activated by cyanogen bromide[7] yields a product that can be coupled to unprotonated amino groups.

Preparation of Specific Adsorbent

4-Nitrophenyl-tri-*O*-acetyl-β-D-fucopyranoside is prepared by the method of Levvy and McAllan.[8]

4-Aminophenyl-tri-*O*-acetyl-β-D-fucopyranoside is prepared by hydrogenating 4-nitrophenyl-tri-*O*-acetyl-β-D-fucopyranoside (12.0 g in 200 ml of methanol) over palladium-on-charcoal (10%, 400 mg) at atmospheric pressure for 8 hours. The catalyst is removed by filtration, and the filtrate is concentrated *in vacuo* to a brown syrup (10.9 g, 98%). Thin-layer chromatography on silica gel with benzene–ether, 1:1, as eluent indicated complete conversion to the amine. The nuclear magnetic resonance (NMR) spectrum at 60 MHz showed three acetyl singlets at $\tau 8.0$, 7.93, and 7.84, a phenyl group at $\tau 3.33$ (quartet), and the methyl group of the fucose moiety at $\tau 8.80$ (doublet, spacing 3 Hz).

4-[4-(4-Nitrophenyl-)butanamido-]phenyl-tri-*O*-acetyl-β-D-fucopyranoside. 4-Nitrophenylbutyryl chloride (8.54 g) was added, with cooling, to a solution of 4-aminophenyl-tri-*O*-acetyl-β-D-fucopyranoside (10.8 g) in 80 ml of chloroform. After 40 hours at room temperature, the mixture is poured onto ice (150 ml) with stirring, and the resulting mixture is ex-

---

[6] P. Cuatrecasas, M. Wilchek, and C. Anfinsen, *Proc. Nat. Acad. Sci. U.S.* **61**, 636 (1968).

[7] R. Axén, J. Porath, and S. Ernbäck, *Nature* (*London*) **214**, 1302 (1967).

[8] G. Levvy and A. McAllan, *Biochem. J.* **80**, 433 (1961).

tracted with chloroform. The chloroform layer is washed with sodium bicarbonate solution and water, dried over anhydrous magnesium sulfate, filtered and concentrated under reduced pressure to a pale brown oil. The oil is chromatographed on a 55 × 4 cm silica gel column (Mallinckrodt SilicAR CC7, 100–200 mesh) at a flow rate of 40 ml per hour using chloroform–methanol, 5:1 as eluent. Fractions containing the fastest moving components are pooled, concentrated under reduced pressure, and rechromatographed on a 80 × 40 cm silica gel column at a flow rate of 50 ml per hour with ethyl acetate–hexane, 1:1, as eluent. Fractions containing the slower moving component are pooled and concentrated under reduced pressure to yield 9.5 g (59%) of a colorless oil. The NMR spectrum at 60 MHz showed three acetyl groups, $\tau 7.98(2)$ and 7.80, two phenyl groups at $\tau 2.14$ and $\tau 2.83$ (quartets), and the methyl group of the fucose moiety at $\tau 8.75$ (doublet, spacing 3 Hz).

4-[4-(4-Nitrophenyl-)butanamido-]phenyl-β-D-fucopyranoside. Methanol (100 ml) saturated with ammonia is added to a solution of 4-[4-(4-nitrophenyl-)butanamido-]phenyl-tri-O-acetyl-β-D-fucopyranoside (7.6 g) in 50 ml of methanol. After standing for 16 hours at 4° the solvent is removed under reduced pressure, and the residue is crystallized from methanol. Yield 5.7 g; 98%; mp 161–162°; $[\alpha]_D^{20}$ $-39°$ (c 2 in methanol). The NMR spectrum at 60 MHz showed two phenyl groups at $\tau 2.06$ and 2.33 (quartets) and the methyl group of the fucose moiety at $\tau 8.90$ (doublet, spacing 3 Hz).

4-[4-(4-Aminophenyl-)butanamido-]phenyl-β-D-fucopyranoside is prepared by hydrogenating 4-[4-(4-nitrophenyl-)butanamido-]phenyl-β-D-fucopyranoside (3.6 g, suspended in 200 ml of methanol) over palladium on charcoal (10%, 100 mg) for 2 hours at atmospheric pressure. The catalyst is removed by filtration, and the filtrate is concentrated under reduced pressure. The residue is crystallized from methanol. Yield 2.84 g; 80%; mp 202–203°; $[\alpha]_D^{20}$ $-39°$ (c 2 in methanol). The NMR spectrum at 60 MHz showed two phenyl groups at $\tau 2.80$ and 3.18 (quartets) and the methyl group of the fucose moiety at $\tau 8.95$ (doublet, spacing 3 Hz).

Sepharose 4B (Pharmacia, Uppsala, Sweden) was activated with cyanogen bromide and then allowed to react with 4-[4-(4-aminophenyl-)-butanamido-]phenyl-β-D-fucopyranoside. Sepharose 4B (50 ml) is washed by filtration, suspended in 50 ml of water and cooled to 12°. Cyanogen bromide, 10 g is added to the stirred suspension and the pH of the mixture maintained at 11 by addition of 2 N NaOH. When the reaction ceases as indicated by stabilization of the pH, the suspension is washed with water (0.5 liter) and 0.1 M sodium bicarbonate (0.5 liter), by filtration, on a sintered-glass funnel. The activated Sepharose is added to a solution of 4-[4-(4-aminophenyl-)butanamido-]phenyl-β-D-fucopyranoside (2.7 g) in a 1:2 mixture of dimethylformamide and 0.1 *M* sodium bicarbonate (75 ml), and the mixture is stirred for 16 hours at room temperature. After washing the Sepharose with 200 ml of dimethyl-formamide–0.1 *M* sodium bicarbonate, 1:2, and water (1 liter), the degree of substitution was determined by hydrolysis of samples with 6 *N* HCl (16 hours, 110°) followed by spectrophotometric estimation of aromatic amine released using *N*-1-naphthylethylenediamine as reagent.[9] Values of 6–10 μmoles of substituent per milliliter of packed Sepharose are obtained.

Assay for *araC* Protein Activity

The assay is based on the observation that *araC* protein binds specifically to *ara* DNA.[10] The binding of *araC* protein to *ara* DNA is detected using the nitrocellulose membrane filter technique developed by Riggs *et al.* for the *lac* repressor.[11] [32P]λh80dara DNA and [32P]λh80 DNA are prepared by previously published procedures[5] with the following modifications: the casamino acids (Difco) in the growth medium are treated to remove inorganic phosphate,[12] 10 μCi/ml of [32P]phosphate are added to the culture when the temperature was shifted to 42°, and only one cesium chloride equilibrium gradient centrifugation is carried out. Unlabeled DNA is prepared as above with the omission of the [32P]phosphate. Schleicher and Schuell B6 nitrocellulose membrane filters are treated in 0.4 *M* KOH for 40 minutes at room temperature as suggested by Lin and Riggs.[12] The KOH-treated filters are washed extensively with deionized water and then placed in 0.1 *M* Tris chloride, pH 7.4 at 25° for 30 minutes and stored at 4° until needed. DNA concentrations are measured spectrophotometrically at 260 nm using an extinction coefficient of 0.02 $cm^2/\mu g$. Protein concentrations are estimated by Layne's formula[13] or by the method of Lowry.[14]

In a typical assay 50 μl of an appropriate dilution of *araC* protein in BB buffer (10 m*M* Tris chloride, pH 7.4 at 25°, 5 m*M* magnesium acetate, 10 m*M* potassium chloride, 0.1 m*M* EDTA, 0.1 m*M* dithio-threitol, and 50 μg/ml of bovine serum albumin) is mixed with 1.0 ml of a 0.1 μg/ml solution of [32P]λh80dara DNA in the same buffer. After a 10-minute incubation at room temperature, two separate 0.5-ml aliquots

[9] J. Daniel, *Analyst* **86**, 640 (1961).
[10] G. Wilcox, K. Clemetson, D. Santi, and E. Englesberg, *Proc. Nat. Acad. Sci. U.S.* **68**, 2145 (1971).
[11] A. Riggs, H. Suzuki, and S. Bourgeois, *J. Mol. Biol.* **48**, 67 (1970).
[12] S. Lin and A. Riggs, *Nature (London)* **228**, 1184 (1970).
[13] E. Layne, this series, Vol. 3, p. 447.
[14] O. Lowry, N. Rosebrough, A. L. Farr, and R. Randall, *J. Biol. Chem.* **193**, 265 (1951).

are filtered. Each filter is washed with 0.5 ml of FB buffer (BB buffer without bovine serum albumin and dithiothreitol), dried under a heating lamp, and counted. The background obtained using [^{32}P]λ*h80* DNA in place of [^{32}P]λ*h80dara* DNA is subtracted from each value. Specific binding to *ara* DNA is confirmed by competition experiments using unlabeled λ*h80dara* DNA and λ*h80* DNA.[10] The quantity of protein in micrograms required to achieve one-half maximal binding to 0.1 μg of λ*h80dara* DNA in a 1.0-ml reaction mix is termed *x*. $1/x$ is the activity of *araC* protein. We have not been able to detect *araC* protein activity in crude extracts with this assay.

Growth of Cells and Preparation of Crude Extract

The source of *araC* protein is SB3102(F′*araA39C*$^+$/*araA39C*$^+$), a strain of *E. coli* B/r which contains at least two copies of the *araC* gene. The cells are grown at 37° in CHYE medium (15 g of tryptone, 10 g of yeast extract, and 5 g of NaCl per liter of deionized water) and harvested at a density of 1×10^9 cells/ml. All the following operations are performed at between 0 and 4°. Fifteen grams of cells are suspended in 30 ml of buffer A (50 m*M* Tris chloride, pH 7.0 at 20° and 1 m*M* dithiothreitol) and broken in a French press at pressures between 8000 and 12,000 psi. Cell debris is removed by centrifugation at 30,000 *g* for 30 minutes. DNase is added to the supernatant to give a final concentration of 2 μg/ml, and the solution is stirred for 1 hour.

Affinity Chromatography

The fucopyranoside-Sepharose is packed in a column 1.5 × 10 cm and washed with 10 m*M* sodium borate, pH 10 at 4°, followed by 2 *M* KCl, and then equilibrated with buffer A. The crude extract is applied to the column at a flow rate of 10 ml per hour. The column is washed with 90 ml of buffer A at a flow rate of 30 ml per hour. *araC* protein is eluted from the column with buffer B (10 m*M* sodium borate–NaOH, pH 10 at 4°, 0.1 *M* KCl, and 1 m*M* dithiothreitol). The eluent is made 10 m*M* in Tris chloride by the addition of 2 *M* Tris chloride, pH 7.0 at 20° and loaded onto the fucopyranoside-Sepharose column which had been equilibrated with buffer A. The column is developed as above and the eluent is again made 10 m*M* in Tris chloride, pH 7.0.

Upon the first passage of a crude extract through the affinity column, a considerable amount of material is bound that is released in a sharp peak when the column is eluted with buffer B. A portion of the first eluent is no longer retained on the column upon recycling. Some of the initial, presumably nonspecifically bound material, may be due to weak binding to the aromatic residues of the side chain. Extensive washing of the column is not as efficient a way to remove the nonspecific material as the recycling process. Attempts to elute the column with L-arabinose or D-fucose are not successful.

AraC protein is about 20% pure after affinity chromatography. The preparation is very unstable with a half-life of about 6 hours and only 1% is active with respect to *ara* DNA binding activity. SDS polyacrylamide gels reveal that four major protein species are present, one of which can be identified as *araC* protein.[10] The yield of *araC* protein from 15 g of cells is 2–3 μg as estimated from Lowry protein determinations and SDS polyacrylamide gels. We are unable to detect *araC* protein activity in crude extracts with our assay but, assuming 100% recovery during the purification, we have calculated that the maximum purification achieved with the fucopyranoside-Sepharose is 1000-fold.

The fucopyranoside-Sepharose becomes a yellowish brown after it has been used 3 to 4 times even though it is washed extensively with 2 M KCl and stored in 0.1% sodium azide. After the adsorbent had been used 10 to 15 times the specificity of the column began to decrease—more species of protein are released with buffer B than previously. We do not know whether the change is due to bound proteins or to a covalent modification of the adsorbent after extended contact with crude extracts.

In our initial attempts to purify *araC* protein using affinity chromatography, a specific adsorbent was prepared in which 4-aminophenyl-β-D-fucopyranoside was attached directly to Sepharose 4B. This adsorbent does not retain *araC* protein. It is necessary to place the fucopyranoside group farther away from the solid matrix by the introduction of the 4-(4-aminophenyl-)butanamido side chain in order for the specific adsorbent to be effective in purifying *araC* protein.

We have recently developed an assay for the activator form (P_2) of *araC* protein[15] by modifying the *ara* DNA-dependent protein synthesizing system described previously.[16] This assay has the advantage that it can detect *araC* protein activity in crude extracts. Using this assay we obtain a 300-fold purification of *araC* protein with the procedure described above.

Acknowledgments

We thank Professor Ellis Englesberg for advice and encouragement.

This work was supported by grants from the National Science Foundation (GB-24093) to Ellis Englesberg and from the U.S. Public Health Service (CA-10266), from the National Cancer Institute to D. V. Santi.

[15] G. Wilcox, P. Meuris, R. Bass, and E. Englesberg, *J. Biol. Chem.* **249**, 2946 (1974).
[16] G. Zubay, L. Gielow, and E. Englesberg, *Nature (London) New Biol.* **233**, 164 (1971).

Section III

Attachment of Specific Ligands and Methods for Specific Proteins

C. Amino Acids and Peptides

[38] Agarose-Anthranilate Derivatives in the Purification of Anthranilate Phosphoribosyltransferase

By STUART L. MARCUS

The anthranilate synthetase complex in *Salmonella typhimurium* is a dual-enzyme aggregate catalyzing the first two reactions in the pathway leading to the biosynthesis of the amino acid tryptophan.[1] The aggregate is composed of anthranilate synthetase (AS), the product of the *trp* A gene of the tryptophan operon[1] and anthranilate 5-phosphoribosyl-pyrophosphate phosphoribosyltransferase (anthranilate-PRtransferase), the product of the *trp* B gene of the tryptophan operon. Free anthranilate-PRtransferase or anthranilate synthetase may be obtained using strains of *S. typhimurium* carrying appropriate mutations in *trp* A or *trp* B, respectively.[1] Anthranilate synthetase catalyzes the synthesis of anthranilic acid from chorismic acid and an amino donor:

$$\text{chorismate} + \text{glutamine (or NH}_4{}^+) \xrightarrow{\text{Mg}^{2+}} \text{anthranilate} + \text{pyruvate} + \text{glutamine}$$

Glutamine can only be used as an amino donor when the synthetase is complexed to anthranilate-PRtransferase as the glutamine binding site exists on the transferase subunit.[2] Anthranilate-PRtransferase catalyzes the formation of N-(5'-phosphoribosyl) anthranilate from anthranilate and 5-phosphoribosyl 1-pyrophosphate (PRPP):

$$\text{anthranilate} + \text{PRPP} \xrightarrow{\text{Mg}^{2+}} N\text{-(5'-phosphoribosyl) anthranilate} + \text{PP}_i$$

This reaction can be catalyzed by anthranilate-PRtransferase in the free state or while complexed with anthranilate synthetase.[3]

Unaggregated anthranilate-PRtransferase is obtained from crude extracts of *S. typhimurium* strain *trp* A703 (ATCC 25567) as described previously.[4] Spectrophotofluorometric assays are performed at 24–25° as described by Henderson *et al.*[3] except that a 3-ml volume is used and 2-mercaptoethanol substituted for dithiothreitol. One unit of enzyme activity represents the disappearance of 1 nmole of anthranilate per

---

[1] R. H. Bauerle and P. Margolin, *Cold Spring Harbor Symp. Quant. Biol.* **31**, 203 (1966).

[2] H. Nagano, H. Zalkin, and E. J. Henderson, *J. Biol. Chem.* **245**, 3810 (1970).

[3] E. J. Henderson, H. Zalkin, and L. H. Hwang, *J. Biol. Chem.* **245**, 1424 (1970).

[4] S. L. Marcus and E. Balbinder, *Anal. Biochem.* **48**, 448 (1972).

minute. Protein concentration is determined by the method of Lowry *et al.*[5]

In the following studies, anthranilic acid is covalently linked to agarose through various of its functional groups. Since anthranilic acid has been shown to cause significant product inhibition of anthranilate synthetase at low concentrations,[6] the adsorption of the unaggregated form of anthranilate-PRtransferase to the various agarose-anthranilate acid derivatives and its subsequent elution is examined in an attempt to define the requirements for anthranilic acid binding by that enzyme as an aid in purification.

Preparation of Agarose-Anthranilic Acid Derivatives

Procedures are carried out at room temperature unless otherwise stated.

(a) Anthranilic acid may be coupled directly to Sepharose 2B via its free amino group (Fig. 1A). The agarose is activated after mixing with an equal volume of distilled water by the addition of finely divided cyanogen bromide,[7] using 250 mg of cyanogen bromide per milliliter of settled agarose. The pH of the mixture is brought to 10 by the addition of 5 M NaOH and maintained at that pH by manual titration with 5 M NaOH. The temperature of the reaction is maintained at approximately 20° by the occasional addition of shaved ice. After cessation of base uptake, a large amount of ice is added and the agarose washed, using a Büchner funnel with Whatman No. 1 filter paper, with 15–20 volumes of ice-cold 0.2 M sodium carbonate buffer, pH 9. To 20 ml of activated agarose is added 50 ml of a 20 mM solution of anthranilate in 0.2 M NaHCO$_3$ at pH 9.8, and the suspension is allowed to mix for 17 hours at 4°, after which the derivative is washed with 1 liter of cold distilled water. This derivative contains about 27 μmoles of anthranilate per milliliter of packed agarose.[8]

(b) Anthranilic acid may be bound by the same amino group to a derivative of agarose containing a long hydrocarbon "arm" (Fig. 1B). Aminohexamethylimino-agarose is prepared by adding 40 ml of 2 M hexamethylenediamine (Aldrich Chemical Co.), which has been titrated to pH 10 with concentrated HCl, to an equal volume of agarose activated by the cyanogen bromide procedure immediately beforehand. After stirring

[5] O. H. Lowry, N. J. Rosebrough, A. L. Farr, and R. J. Randall, *J. Biol. Chem.* 193, 256 (1951).

[6] J. C. Cordaro, H. R. Levy, and E. Balbinder, *Biochem. Biophys. Res. Commun.* 33, 183 (1968).

[7] P. Cuatrecasas and C. B. Anfinsen, this series, Vol. 22, p. 345.

[8] All agarose derivatives were analyzed for the concentrations of various ligands by Kjeldahl nitrogen determinations.

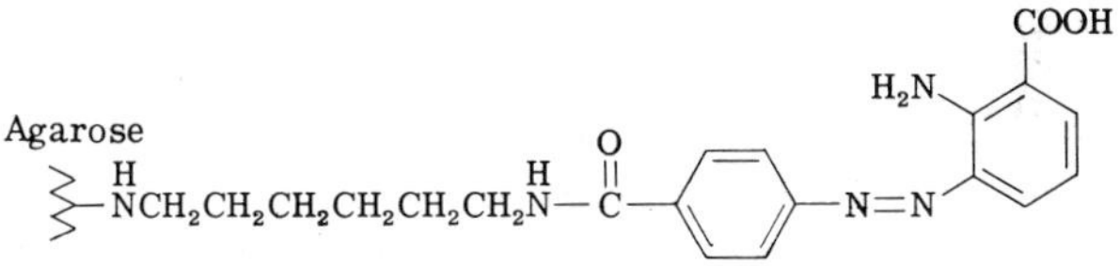

FIG. 1. Agarose-anthranilic acid derivatives prepared by attaching anthranilic acid to Sepharose 2B containing various side arms through different functional groups on the anthranilic acid molecule. Although adsorbent D represents the most probable structure for this derivative, diazotization to other portions of the benzene ring may constitute a considerable percentage of this derivative. Figure is from S. L. Marcus and E. Balbinder, *Anal. Biochem.* 48, 448 (1972).

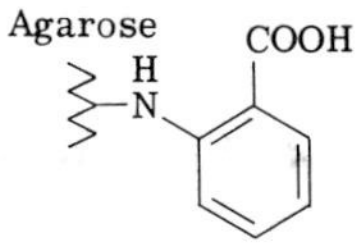

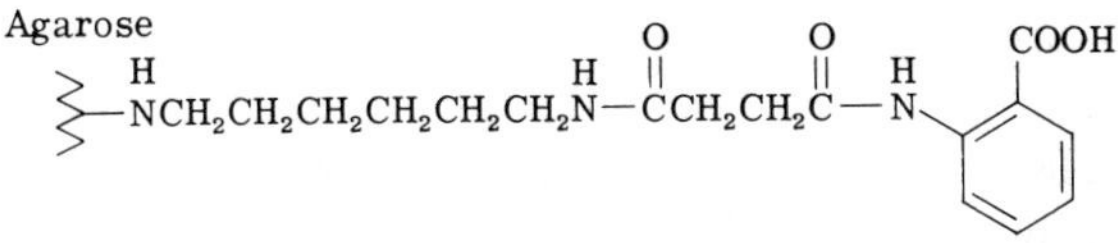

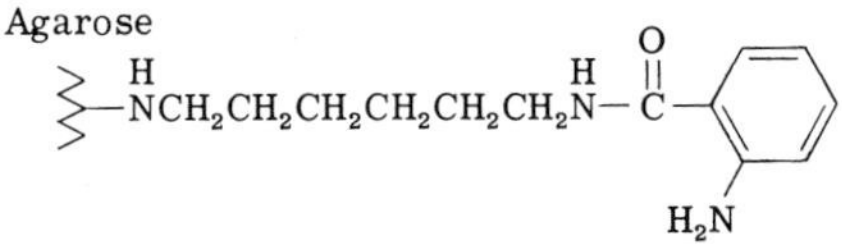

for 21 hours at 4°, the agarose derivative is washed with 1 liter of distilled water; 28 μmoles of hexamethylenediamine are bound per milliliter of packed agarose.

The free amino groups are succinylated with succinic anhydride.[7] To 10 ml of aminohexamethylimino-agarose in an equal volume of water at 4° are added 10 mmoles of succinic anhydride. The pH of the reaction is kept at 6.0 for 1 hour by manual titration with 5 M NaOH. After 5 hours at 4°, the succinylated derivative is washed with 1 liter of cold water. Complete substitution of amino groups is observed and may be followed visually using the trinitrobenzenesulfonate color reaction.[7]

Anthranilic acid is coupled to the free carboxyl group with 1-cyclohexyl-3-(2-morpholinoethyl) carbodiimide metho-p-toluenesulfonate, a water-soluble carbodiimide.[7] Anthranilic acid, 1 mmole, is dissolved in 10 ml of 40% (v/v) N,N-dimethylformamide and is added to 10 ml of the previously prepared succinylamidohexamethylimino-agarose. The pH of the suspension is adjusted to 4.9, and 1 g of cyclohexyl-3-(2-morpholinoethyl) carbodiimide metho-p-toluene sulfonate (Aldrich Chemical Co.)

is added as a suspension in 3 ml of water. The pH of the reaction is maintained at 4.9 by manual titration with 2 N HCl over a period of 2 hours and is allowed to mix for another 4 hours. The derivatized agarose is then washed with 200 ml of 40% N,N-dimethylformamide followed by 3 liters of distilled water over a period of 2 hours. The washed agarose derivative contains about 4 μmoles of anthranilate per milliliter of packed agarose.

(c) Anthranilic acid may be bound by its free carboxyl group to aminohexamethylimino-agarose (Fig. 1C). One millimole of anthranilic acid is dissolved in 30 ml of 40% N,N-dimethylformamide. To this solution is added 10 ml of settled aminohexamethylimino-agarose, and the pH is brought to 4.8 with 1 N HCl. The suspension is allowed to mix for another 17 hours, after which the derivatized agarose is washed with 2 liters of 0.1 M NaHCO$_3$ at pH 8.8 over a period of 2 hours; it is then washed with an additional 200 ml of water. About 10 μmoles of anthranilic acid are bound per milliliter of packed agarose.

(d) Anthranilic acid may also be coupled to derivatized agarose by diazotization[7] directly to the benzene ring (Fig. 1D). Aminohexamethylimino-agarose, 10 ml, is added to 30 ml of 50% N,N-dimethylformamide in 0.1 M sodium borate. p-Nitrobenzoylazide (Aldrich Chemical Co.), 470 mg, is added to the suspension, and the mixture is stirred for 1 hour. The derivatized agarose is then washed with 500 ml of 50% N,N-dimethylformamide and transferred, while moist, to a beaker containing 30 ml of 0.5 M NaHCO$_3$ and 0.1 M sodium dithionite. The beaker is placed in a water bath at 45° for 30 minutes to reduce the ligand.[7] After incubation, the gel is washed with 500 ml of cold distilled water and cooled in an ice bath for 20 minutes. The distilled water is decanted, and 15 ml of 0.5 N HCl are added together with 175 mg of sodium nitrate. The suspension is stirred for 15 minutes in an ice bath, and 40 ml of a solution of anthranilic acid (12.5 mg/ml) in saturated sodium borate is added. Stirring is continued for 17 hours at 4°, after which the gel is washed with 2 liters of cold distilled water. The color of this agarose-anthranilate derivative is a deep brownish red. The derivatized agarose contains about 8 μmoles of anthranilate per milliliter of packed agarose.

The anthranilate-agarose derivatives can be tested for their relative efficiencies at purifying anthranilate-PRtransferase after being poured into Pasteur pipettes containing glass wool as a bottom support. Because reactions condensing ligands onto hydrocarbon "arms" seldom result in the complete substitution of all the "arms" extending from the agarose matrix, columns of agarose containing the appropriate hydrocarbon "arms" but lacking anthranilic acid are used as controls in these studies. Bed volume for all columns is 1–1.2 ml. For each experiment, 2 ml of crude *S. typhi-*

murium trp A703 extract is applied to a column and washed through with 50 mM potassium phosphate, pH 7.4, containing 1 mM 2-mercapto-ethanol. All studies are carried out at 4°. Fractions of 1.7–1.8 ml volume are collected until protein is no longer detected in the wash fractions. The elution buffer contains: 50 mM triethanolamine-HCl, pH 9.5; 1 mM 2-mercaptoethanol; and 20% (v/v) glycerol. The eluate is collected in 1.7–1.8-ml fractions into tubes containing 0.1 ml of 1 M potassium phosphate at pH 7.4. All columns are equilibrated with 70–100 column volumes of 50 mM potassium phosphate, pH 7.4, containing 1 mM 2-mercaptoethanol before the addition of the enzyme extract.

Anthranilate-PRtransferase-Binding Ability of Various Agarose
Derivatives: Relative Efficiency of Purification

Agarose Derivatives Bound to the Amino Group of Anthranilic Acid. Columns prepared using unsubstituted agarose do not bind anthranilate-PRtransferase activity and the enzyme elution pattern coincides exactly with the protein elution profile. Columns prepared with agarose bound directly to anthranilic acid (Fig. 1A) do not bind large quantities of anthranilate-PRtransferase; however, it is worth noting that in column runs using this derivative, the enzyme activity peak elutes slightly after the protein peak. An example of the ability of this derivative to bind anthranilate-PRtransferase is given in the table, derivative 1. Of the 2800 units of enzyme activity applied to this column, 2000 are recovered in the wash fraction. Addition of elution buffer allows recovery of only 30% of bound anthranilate-PRtransferase activity, resulting in an increase in specific activity less than 3-fold that of the crude extract.

The insertion of a long hydrocarbon "arm" between the agarose and the anthranilic acid molecule dramatically increases the ability of the agarose-anthranilic acid derivative to bind anthranilate-PRtransferase (see the table, derivative 2). Of the 4000 units applied to a column prepared using this derivative, only 500 are recovered in the wash fractions. Upon addition of elution buffer, 63% of the bound anthranilate-PRtransferase activity is recovered; the specific activity of eluted fractions is increased 12 to 13-fold over that of the enzyme in the crude extract. When 2800 units of anthranilate-PRtransferase activity are placed on a column of an agarose derivative containing the "arm" alone (see the table, derivative 3), all the activity is recovered in the wash fractions, proving that binding to unsubstituted succinylamidohexamethylimino-agarose "arms" does not occur and that purification achieved is due to specificity of the enzyme for the anthranilic acid molecule.

TABLE

ABILITY OF VARIOUS AGAROSE DERIVATIVES TO SELECTIVELY BIND ANTHRANILATE-PRTRANSFERASE[a]

Agarose derivatives	Anthranilate-PRtransferase units			Fold purification of eluted enzyme
	Applied to column	Recovered in wash fractions	Recovered by elution	
1.	2800	2000	260	<3
2.	4000	500	2200	12

	Structure				
3.	$\approx\!\!\text{N}-(\text{CH}_2)_6-\underset{\text{H}}{\text{N}}-\overset{\text{O}}{\overset{\|}{\text{C}}}-(\text{CH}_2)_2-\overset{\text{O}}{\overset{\|}{\text{C}}}-\text{OH}$	2800	2800	—	—
4.	$\approx\!\!\text{N}-(\text{CH}_2)_6-\underset{\text{H}}{\text{N}}-\overset{\text{O}}{\overset{\|}{\text{C}}}-$ (phenyl, H_2N)	4000	3000	140	<2
5.	$\approx\!\!\text{N}-(\text{CH}_2)_6-\text{NH}_2$	2800	1800	70	<2
6.	$\approx\!\!\text{N}-(\text{CH}_2)_6-\underset{\text{H}}{\text{N}}-\overset{\text{O}}{\overset{\|}{\text{C}}}-$ (phenyl)$-\text{N}=\text{N}-$ (phenyl, H_2N, COOH)	3200	600	1300	11
7.	$\approx\!\!\text{N}-(\text{CH}_2)_6-\underset{\text{H}}{\text{N}}-\overset{\text{O}}{\overset{\|}{\text{C}}}-$ (phenyl)$-\overset{+}{\text{N}}\equiv\text{N}$	3200	1400	500	8

[a] Data from S. L. Marcus and E. Balbinder, *Anal. Biochem.* **48**, 448 (1972).

Agarose Derivative Bound to the Carboxyl Group of Anthranilic Acid. Columns prepared with anthranilic acid linked to aminohexamethyl-imino-agarose via its carboxyl group give results strikingly different from those obtained using the aforementioned agarose-anthranilate derivative. Of 4000 units of anthranilate-PRtransferase activity applied to the an-thranilate-amidohexamethylimino-agarose derivative (see the table, derivative 4), 3000 units are recovered in the wash fractions. Only 14% of the bound activity is removed upon addition of elution buffer, and the specific activity of the eluted enzyme is less than 2-fold that of the original extract. Studies with the aminohexamethylimino-agarose "arm" alone (see the table, derivative 5) show that the binding of anthranilate-PRtransferase to the agarose-anthranilate derivative of Fig. 1C is due to nonspecific charge effects. Both the aminohexamethylimino-agarose "arm" alone and the derivative produced by the attachment of anthranilic acid to that "arm" bind and release (upon addition of elution buffer) approximately the same relative percentage of anthranilate-PRtransferase activity applied to those derivatives. Since binding in these cases is nonspecific, the anthranilate-agarose derivative pictured in Fig. 1C may not be considered an "affinity matrix" for this enzyme and is therefore not useful as a means for purification.

Anthranilate-Agarose Derivative Prepared by Diazotization. The attachment of anthranilic acid via its benzene ring to an agarose-diazonium derivative results in a product (Fig. 1D) which appears to adsorb an-thranilate-PRtransferase in a specific manner (see the table, derivative 6). Of 3200 units of anthranilate-PRtransferase activity placed on a column of this derivative, only 600 units are recovered in the wash fractions. Upon addition of elution buffer, 50% of the bound activity can be recovered. The specific activity of anthranilate-PRtransferase obtained after the addition of elution buffer is increased 11-fold over that of the original extract.

Columns prepared with the agarose-diazonium "arm" alone, however, also exhibit the ability to preferentially bind anthranilate-PRtransferase activity (see the table, derivative 7). Of 3200 units of anthranilate-PRtransferase activity placed on a column of this derivative, only 1400 units are recovered in the wash fractions although anthranilic acid is not present. Only 30% of the bound anthranilate-PRtransferase activity is recovered after addition of elution buffer. The specific activity of an-thranilate-PRtransferase in the elution fractions is increased 8-fold over that of the original extract. This is the only hydrocarbon "arm" that shows true preference for anthranilate-PRtransferase activity. Since the attachment of anthranilic acid to the "arm" increases the capacity of the derivative for the enzyme and also the specificity of adsorption as noted in the relative increase in specific activity after elution, binding of PRtransferase to the diazonium "arm" may be due to a combination of charge and hydrophobic factors.

Comments

The agarose-anthranilate derivative prepared by coupling anthranilic acid via its free amino group to succinylamidohexamethylimino-agarose possesses the greatest binding capacity for anthranilate-PRtransferase and produces the best fold purification in one step of all the agarose-anthranilate derivatives tested. This derivative may be utilized as the first stage in a three-step purification procedure resulting in an anthranilate-PRtransferase preparation which is 90–95% homogeneous as determined by polyacrylamide gel electrophoresis.[9]

Binding of anthranilate-PRtransferase to agarose-anthranilate derivatives is also dependent on the nature of the functional group used to attach anthranilic acid to each "arm." Derivatives prepared by linking anthranilic acid through its free amino group or benzene ring adsorb anthranilate-PRtransferase effectively, while the derivative prepared by linking the free carboxyl group of anthranilic acid to an "arm" demonstrates a lack of preferential binding. These results suggest that the carboxyl group of anthranilic acid may be of great importance for the binding and recognition of this substrate by anthranilate-PRtransferase. The methods described here may be used as a means for providing a first indication of which functional groups on a ligand must remain free to ensure binding specificity by an enzyme. The use of derivatives containing "arms," but lacking ligand, as controls for nonspecific binding is essential for such studies, including those wholly concerned with purification.

Acknowledgment

The author wishes to express his gratitude to Dr. Elias Balbinder, in whose laboratory at Syracuse University these methods were developed.

[9] S. L. Marcus and E. Balbinder, *Biochem. Biophys. Res. Commun.* **47**, 438 (1972).

[39] Thyroxine

By JACK PENSKY and JAMES S. MARSHALL

The use of thyroxine covalently attached to agarose for the isolation of human thyroxine-binding globulin (TBG) was an early example of the application of affinity chromatography to a hormone-binding protein.[1] Batch treatment of large volumes of human serum with relatively small amounts of thyroxine-agarose, followed by washing and elution of TBG by mild changes in pH, provided a preparation highly enriched in TBG, from which the homogeneous binding protein could be isolated by two simple chromatographic steps.[1,2]

The use of 6-aminocaproyl-L-thyroxine agarose, in which thyroxine is attached to agarose by way of a six-carbon "handle," has also been reported to be effective in purifying TBG.[3]

Preparation of Thyroxine-Agarose. Thyroxine solution: L-thyroxine, 250 mg as the sodium pentahydrate salt, is suspended in 10 ml of 50 mM potassium hydroxide and 0.2 M potassium hydroxide is added dropwise with shaking until the thyroxine just dissolves. The volume is brought to 15 ml with 50 mM potassium hydroxide, and the solution is kept on ice prior to mixing with activated agarose (see below).

Activation of agarose: To a magnetically stirred aqueous suspension of Sepharose 4B (Pharmacia) (75 ml settled volume in a total volume of 150 ml) in a fume hood is added cyanogen bromide (200 mg/ml settled agarose volume) previously ground to a fine powder in a mortar. The pH of the suspension, monitored by a combination electrode, immediately falls and is brought to 11 ± 0.1 and maintained at this value by constant addition of 6 M sodium hydroxide solution. The reaction takes somewhat longer to go to completion when solid cyanogen bromide is used than when a solution of cyanogen bromide is added, but this offers the advantage that the heat produced by the reaction can be more easily controlled. Pieces of crushed ice are added as the reaction progresses to maintain the temperature of the suspension at or slightly below room temperature. After about 20 minutes the pH no longer decreases and the reaction is stopped by adding crushed ice. The chilled mixture is filtered quickly on a Büchner funnel fitted with a coarse fritted-glass disc and the activated agarose is washed on the filter several times with a total of 750 ml of ice-cold 0.1 M sodium bicarbonate. The damp cake of activated agarose is resuspended in 75 ml of cold 0.1 M sodium bicarbonate and the thyroxine solution is quickly added. The agarose-thyroxine mixture is stirred gently at 4° for 24 hours, then filtered and washed with 1 liter of distilled water and 1 liter of 0.1 M sodium bicarbonate and stored in the latter solution.

The initial filtrate of the thyroxine-agarose activation mixture appears heavily opalescent. By adding ^{125}I-labeled thyroxine in trace amount

[1] J. Pensky and J. S. Marshall, *Arch. Biochem. Biophys.* **135**, 304 (1969).
[2] J. S. Marshall, J. Pensky, and S. Williams, *Arch. Biochem. Biophys.* **156**, 456 (1973).
[3] R. A. Pages and H. J. Cahnmann, *Abstr. 4th Int. Congr. Endocrinol., Int. Congr. Ser.* No. 256, p. 241. Excerpta Med. Found., Amsterdam, 1972.

to the reaction mixture at the time of coupling the extent of attachment of thyroxine can be calculated by measuring the radioactivity of the filtrate. The amount of thyroxine bound to agarose under these reaction conditions is about 1.4 μmoles per milliliter of settled volume of agarose.

Removal of TBG from Serum by Thyroxine-Agarose. Passage of human serum through a column of thyroxine-agarose is complicated by low flow rates and/or clogging; the most efficient method of treatment is batchwise adsorption.

Pooled human serum, up to 2 liters, fresh or previously stored at −20°, is filtered through glass wool to remove all particulate matter. The filtrate is mixed with 75 ml of thyroxine-agarose, and the suspension is stirred gently with a motor-driven plastic paddle or Teflon-coated wire stirrer for 12 hours at room temperature. The thyroxine-agarose is allowed to settle overnight, and on the following day the supernatant serum is decanted and the gel is washed repeatedly with 0.1 M sodium bicarbonate on a glass Büchner funnel until 1–2 serum volumes have been used. The washed preparation is suspended in 1–2 serum volumes of 0.1 M sodium bicarbonate, stirred for 30 minutes, and filtered as before. The washed cake is resuspended in 0.1 M sodium bicarbonate and the slurry is packed into a glass chromatographic column, 2.4 × 30 cm. The column is washed with 0.1 M sodium bicarbonate until the optical density of the effluent, measured at 280 nm in a 1-cm quartz cuvette, is 0.05 or less. This may require several liters of bicarbonate, depending on completeness of prior washing. The TBG is then dissociated from the thyroxine-agarose by passing through the column 200 ml of a solution of KOH (2 mM, about pH 9.3, freshly diluted from stock 0.2 M KOH). The eluate typically emerges as two peaks absorbing strongly at 280 nm. The first major peak appears turbid to chalky and contains the eluted TBG. All fractions of this peak having an optical density of 0.8 or greater at 280 nm are pooled for further purification. The second peak is also turbid but contains little protein.[1]

Radioautography of an electrophoretogram of serum after contact with thyroxine-agarose and subsequently enriched with ^{125}I-thyroxine shows no band in the position of TBG, indicating nearly complete removal of TBG by this treatment. Similar radioautography of the 2 mM KOH eluate of thyroxine-agarose shows a strong band in the inter-alpha position characteristic of TBG.[1]

Purification of TBG. The eluate from the column contains besides TBG a small amount of albumin, γ-globulin and other nonspecifically adsorbed globulins of serum. TBG can be separated from most of these contaminants by ion exchange chromatography.

A column (1.5 × 30 cm) of DEAE-Sephadex A-50 (Pharmacia)

equilibrated with 60 mM Tris chloride, pH 8.6 is prepared and cooled to 4°. All subsequent operations are done at this temperature. The eluate containing TBG is centrifuged at 20,000 g for 15 minutes to remove non-TBG precipitated material. The supernatant solution is concentrated to between 10 and 30 ml by pressure ultrafiltration (Amicon UM-10 membrane), dialyzed to equilibrium against 60 mM Tris chloride at pH 8.6 (equilibration buffer) and mixed with 25 μl of [^{125}I]thyroxine (Tetramet, Abbott Laboratories). The radioactivity is associated almost exclusively with TBG and is thus a convenient means of locating TBG in column eluates. The dialyzed, labeled TBG is placed on the DEAE-Sephadex column and is eluted by a linear gradient of NaCl in equilibration buffer. The gradient vessels are two 500-ml cylinders of equal cross section. The limit vessel contains 500 ml of equilibration buffer, 0.2 M in NaCl, and the mixing vessel contains 500 ml of equilibration buffer. Under gravity flow, fractions of 3–5 ml are collected and assayed for [^{125}I]thyroxine in a gamma counter. The fractions containing the highest radioactivity emerge at about 0.15 M NaCl and usually contain TBG free from contaminants as judged by disc-gel electrophoresis at several gel porosities, analytical ultracentrifugation, and immunodiffusion.[4] If fractions on the ascending side of the peak are pooled, a few trace impurities appear which can be removed by concentrating the pool and subjecting it to chromatography on a 1 × 60 cm column of Sephadex G-150 equilibrated with 60 mM Tris chloride at pH 8.6.

Yield of TBG. Under the above conditions the final yield of TBG is approximately 0.3–0.4 mg per 100 ml of serum. This corresponds to an overall yield of about 25%, based on a serum content of 1.5 mg/100 ml as obtained by radioimmunoassay.[5]

Properties of TBG. Purified TBG, obtained as described above, is a glycoprotein with a molecular weight of about 63,000 and $E^{1\%}_{1\,cm} = 6.9$ and a sedimentation constant of 3.4. It binds 1 mole of thyroxine[4,6] or one mole of triiodothyronine[6] per mole of protein. The affinity constant for thyroxine was calculated to be 2.3 × 10^{10} M^{-1} at 23° by ultrafiltration.[6] The binding of thyroxine to TBG is pH-dependent; binding is maximal in the pH range 6.4–10.4, declines rapidly below pH 6.4, and is nearly absent at pH 4.2.[4] Consistent with this, fluorescence studies with purified TBG and the thyroxine competitor 1,8-anilinonaphthalene sulfonic acid demonstrated a conformational change in TBG affecting the binding site below pH 6.4 and denaturation below pH 4.0.[7]

[4] J. S. Marshall and J. Pensky, *Arch. Biochem. Biophys.* **146**, 76 (1971).
[5] R. P. Levy, J. S. Marshall, and N. L. Velayo, *J. Clin. Endocrinol.* **32**, 372 (1971).
[6] A. M. Green, J. S. Marshall, J. Pensky, and J. B. Stanbury, *Biochim. Biophys. Acta* **278**, 117 (1972).

Comment. We have observed that freshly prepared batches of thyroxine-agarose, when treated with serum and then washed and eluted under conditions known to yield TBG, often yield little or no TBG the first time they are used, but begin to function well as affinity media after washing and repeated contact with fresh serum. The reason for this behavior is unknown, but may be a reflection of more than one type of linkage of thyroxine to agarose. Initial treatment with serum could saturate some favorably situated thyroxine molecules with TBG which mild elution conditions cannot subsequently remove. An alternative explanation would invoke the combination of TBG with unremoved traces of thyroxine not covalently bound to agarose during the first exposure to serum.

Storage of Thyroxine-Agarose. It is advantageous to wash the thyroxine-agarose briefly with 1 volume of 0.2 M potassium hydroxide after elution of TBG and before storage in 0.1 M sodium bicarbonate. This treatment appears to remove or inactivate a neuraminidase which, if allowed to accumulate on the thyroxine-agarose, gives rise to partially desialylated TBG.[8]

[7] A. M. Green, J. S. Marshall, J. Pensky, and J. B. Stanbury, *Biochim. Biophys. Acta* **278**, 305 (1972).
[8] J. S. Marshall, J. Pensky, and A. M. Green, *J. Clin. Invest.* **51**, 3173 (1972).

[40] L-Tryptophan-Agarose in the Purification of an Anthranilate Synthetase Complex

By STUART L. MARCUS

The first two reactions in the pathway leading to the biosynthesis of the amino acid tryptophan in *Salmonella typhimurium* are catalyzed by the anthranilate synthetase complex,[1] a dual-enzyme aggregate consisting of anthranilate synthetase and anthranilate 5-phosphoribosylpyrophosphate phosphoribosyltransferase (anthranilate-PRtransferase). A more detailed description of the enzymes and the reactions catalyzed by them is presented elsewhere in this volume.[2] Since attempts to dissociate the native anthranilate synthetase–anthranilate–PRtransferase complex result in the loss of anthranilate synthetase activity,[3] free anthranilate synthetase or

[1] E. J. Henderson, H. Nagano, H. Zalkin, and L. H. Hwang, *J. Biol. Chem.* **245**, 1416 (1970).
[2] S. L. Marcus, this volume [38].
[3] R. H. Bauerle and P. Margolin, *Cold Spring Harbor Symp. Quant. Biol.* **31**, 203 (1966).

anthranilate-PRtransferase must be obtained from strains carrying mutations in the *trp* B or *trp* A genes, respectively. Both glutamine-dependent
(complexed) and ammonium-dependent (free) anthranilate synthetase activities are feedback-inhibited by L-tryptophan.[2,4] The existence of a
tryptophan binding site on anthranilate-PRtransferase has not been conclusively demonstrated,[5] but the complexed form is partially inhibited by
tryptophan, probably through a conformational change mediated by the
anthranilate synthetase subunit.

The anthranilate synthetase complex is obtained from crude extracts
of *S. typhimurium* strain *trp* E171 and free anthranilate synthetase and
anthranilate-PRtransferase from crude extracts of strains *trp* BEDC43 and
trp A703, respectively.[3] Cells are cultured as previously described[6] except
for *trp* E171 and *trp* BEDC43, which are grown in the presence of limiting amounts of tryptophan. The cells are harvested by centrifugation,
washed twice, and then suspended in 3 volumes of 0.1 *M* potassium phosphate buffer, pH 7.8, containing 1 m*M* 2-mercaptoethanol. Crude extracts
are immediately prepared by ultrasonic disintegration,[6] all procedures
being carried out at 4°. Complexed anthranilate synthetase activity is
assayed using the spectrophotofluorometric method of Henderson *et al.*[1]
as is anthranilate-PRtransferase activity except that in both cases a 3-ml
volume is used and 2-mercaptoethanol is substituted for dithiothreitol. Free
anthranilate synthetase activity is assayed by the method of Zalkin and
Kling.[4] Chorismic acid is prepared by the method of Gibson.[7] Protein concentrations are determined using the method of Lowry *et al.*[8] One unit of
enzyme activity is defined as the production or disappearance of 1 nmole
of anthranilic acid per minute at 24–25°.

A tryptophan-agarose matrix has been shown to be useful in the purification of chorismate mutase from *Claviceps paspari.*[9] In this chapter,
a similar matrix is described with the ability to yield partial purification
of the native anthranilate synthetase complex as well as those formed
in vitro.

Preparation of the Tryptophan-Agarose Matrix

L-Tryptophan is coupled to Sepharose 2B directly via its α-amino
group. The agarose is activated after mixing with an equal volume of dis-

[4] H. Zalkin and D. Kling, *Biochemistry* 7, 3566 (1968).
[5] E. J. Henderson, H. Zalkin, and L. H. Hwang, *J. Biol. Chem.* 245, 1424 (1970).
[6] S. L. Marcus and E. Balbinder, *Anal. Biochem.* 48, 448 (1972).
[7] R. Gibson, *in* "Biochemical Preparations" (W. E. M. Lands, ed.), Vol. 12, p.
94ff. Wiley, New York, 1968.
[8] O. H. Lowry, N. J. Rosebrough, A. L. Farr, and R. J. Randall, *J. Biol. Chem.*
193, 265 (1951).
[9] B. Sprossler and F. Lingens, *FEBS Lett.* 6, 232 (1970).

tilled water by the addition of cyanogen bromide,[10] using 100 mg of cyanogen bromide per milliliter of settled agarose. The pH of the reaction is brought to 10 by the addition of 5 N NaOH and maintained at that pH by manual titration. After cessation of base uptake, a large amount of shaved ice is added and the activated agarose washed, using a Büchner funnel with Whatman No. 1 filter paper, with 15–20 volumes of ice-cold 0.2 M sodium carbonate at pH 9. The entire washing procedure should not take longer than 2 minutes. A volume of 0.1 M L-tryptophan solution in 0.2 M sodium carbonate at pH 9 equal to that of the settled agarose is added to the agarose immediately after activation and washing. The mixture is allowed to stir for 24 hours at 4°, after which the tryptophan-agarose derivative is washed with 30–50 volumes of cold distilled water. The concentration of bound L-tryptophan may be determined by repeating the above procedure using [14C]L-tryptophan. After preparation of the [14C]L-tryptophan-agarose derivative, 1.0 ml of the packed derivative is dried, and the amount of label in various preweighed samples is determined. The concentration of bound tryptophan may also be determined through the use of Kjeldahl nitrogen determinations. The tryptophan-agarose derivative prepared by the above procedures contains approximately 2 μmoles of L-tryptophan per milliliter of packed agarose.

The tryptophan-agarose derivative is poured into Pasteur pipette columns using glass wool as a bed support. Bed volumes are 1–1.2 ml. Columns are equilibrated with 70–100 column volumes of 0.1 M potassium phosphate at pH 7.8, containing 1 mM 2-mercaptoethanol. All equilibrations and column runs are performed at 4°. Crude extracts, 1.5–2 ml, are applied to a column and washed through with the buffer used for equilibration until the 1.8–2-ml fractions contain no detectable protein. Elution buffer contains 1 mM potassium phosphate, pH 7.8, and 1 mM 2-mercaptoethanol.

Ability of Tryptophan-Agarose to Bind and Purify the Anthranilate Synthetase Complex and Its Components

Unsubstituted agarose does not bind the anthranilate synthetase complex nor either of its free components. The tryptophan-agarose derivative, however, binds large quantities of anthranilate synthetase complex activities (see the table). While 60–80% of native complexed anthranilate-PRtransferase activity bound to the column is always recovered after addition of elution buffer, only 10–25% of complexed anthranilate synthetase activity can be routinely recovered. The low yield of anthranilate synthetase activity relative to anthranilate-PRtransferase activity in column

[10] P. Cuatrecasas, *J. Biol. Chem.* **245**, 3059 (1970).

ABILITY OF TRYPTOPHAN-AGAROSE TO BIND THE ANTHRANILATE
SYNTHETASE COMPLEX AND ITS COMPONENTS[a]

Strains used	Complex components present	Enzyme units			Fold purification of eluted enzyme
		Applied to column	Recovered in wash fractions	Recovered by elution	
trp E171	Anthranilate synthetase	3000	150	320	3–4
	+ Anthranilate-PR-transferase	3200	150	2400	13–14
trp A703	Anthranilate-PR-transferase	3000	3000	—	—
trp BEDC43	Anthranilate synthetase	600	580	—	—
trp A703 + trp BEDC43	"Hybrid" anthranilate synthetase	600	50	150	ND[b]

[a] Data from S. L. Marcus and E. Balbinder, unpublished.
[b] ND, not determined.

eluates may be attributed to inactivation rather than dissociation of the complex, since the anthranilate-PRtransferase activity from tryptophan-agarose column eluates will not complex with free anthranilate synthetase to restore glutamine-dependent synthesis. Data obtained using gel filtration show that 92% of the anthranilate-PRtransferase activity in tryptophan-agarose column eluate fractions coelutes with complexed anthranilate synthetase activity at a position corresponding to that obtained using the native complex. The slight fraction of total anthranilate-PRtransferase activity eluting at a position corresponding to that of free anthranilate-PRtransferase may be due to denaturation during gel filtration. Exposure of trp E171 crude extracts to elution buffer using Sephadex G-25 desalting columns does not inactivate complexed anthranilate synthetase activity, eliminating ionic shocking as the sole factor accountable for the low yields of complexed anthranilate synthetase activity in tryptophan-agarose column eluate fractions. The specific activity of anthranilate-PRtransferase in tryptophan-agarose column eluate fractions is increased 13- to 14-fold over that of the original extract, while the specific activity of complexed anthranilate synthetase only increases 3- to 4-fold due to the partial inactivation of this enzyme. The exact mechanism by which complexed anthranilate synthetase is partially inactivated is not clearly understood.

Free anthranilate-PRtransferase (obtained using crude extracts of strain trp A703) does not bind to the tryptophan-agarose matrix (see the table). No tryptophan binding site has been conclusively shown to exist on the free anthranilate-PRtransferase enzyme.[5] Although free anthranilate synthetase (from extracts of *trp* BEDC43) possesses a tryptophan binding site,[4] it cannot be adsorbed to tryptophan in the form in which it exists on the tryptophan-agarose matrix (see the table). A "hybrid" anthranilate synthetase complex may be formed by mixing extracts of *trp* BEDC43 (containing free anthranilate synthetase) and *trp* A703 (containing free anthranilate-PRtransferase) and incubating the mixture at 2° for 1 hour.[3] Complex formation is followed by observing the appearance of glutamine-dependent anthranilate synthetase activity with time. The "hybrid" complexed anthranilate synthetase activity formed by this method binds to and elutes from the tryptophan-agarose matrix in exactly the same manner as the native anthranilate synthetase complex (see the table). This finding suggests that the formation of the quaternary complex alters the tryptophan binding site on the anthranilate synthetase subunit so that it is able to bind tryptophan in the form in which it appears on the insolubilized matrix.

Mechanism of Elution

Adsorption of the anthranilate synthetase complex to the tryptophan-agarose matrix appears to be of a functional nature and not due to non-specific charge effects, since concentrations of NH_4Cl up to 1 M fail to remove enzyme activity; activity can subsequently be eluted by the addition of 1 mM potassium phosphate buffer. The low-ionic-strength elution buffer, when used in assays of complexed anthranilate synthetase activity, does not inhibit the enzyme although it drastically decreases the ability of tryptophan to inhibit the enzyme. In the presence of 0.1 M potassium phosphate buffer, 5 μM L-tryptophan inhibits 80% of complexed anthranilate synthetase activity, whereas 100 μM L-tryptophan is required to give 70% inhibition in the presence of 1 mM phosphate under otherwise identical conditions. Full sensitivity to tryptophan inhibition may be restored by raising the potassium phosphate concentration in the reaction mixture to 0.1 M. Therefore, an abrupt decrease in ionic strength apparently lowers the affinity of anthranilate synthetase for L-tryptophan, providing a mechanism for the elution of the enzyme from the tryptophan-agarose matrix.

Comments

While the anthranilate synthetase complex from *S. typhimurium* may be partially purified using tryptophan-agarose, some inactivation of complexed anthranilate synthetase activity is observed. This inactivation appears to be due to exposure of the enzyme to elution buffer while it is complexed to tryptophan-agarose, as exposure of the enzyme to the buffer alone has no effect.

Both the native anthranilate synthetase complex and the complex formed *in vitro* bind to tryptophan-agarose, while free anthranilate synthetase does not exhibit affinity for this matrix. This finding suggests that the conformative response of the tryptophan binding site on the anthranilate synthetase subunit is altered by the formation of the quaternary enzyme complex. Such an alteration may lead to a change in the steric requirements for tryptophan binding by the enzyme. These results suggest a possible use for affinity matrices in examining ligand binding differences for two forms of an enzyme.

Acknowledgment

The author wishes to express his gratitude to Dr. Elias Balbinder, in whose laboratory at Syracuse University this method was developed.

[41] Tyrosine

By WILLIAM W.-C. CHAN and MIHO TAKAHASHI

The synthesis of 3-deoxy-D-*arabino*-heptulosonate 7-phosphate (DAHP) constitutes the first reaction in the pathway for the biosynthesis of aromatic amino acids.[1] In *Saccharomyces cerevisiae,* this reaction is catalyzed by two isozymes whose activities are feedback inhibited by tyrosine and phenylalanine respectively.[2,3] These two enzymes (to be referred to subsequently as Tyr-sensitive and Phe-sensitive DAHP synthetase) are difficult to purify by classical methods, such as ammonium sulfate fractionation or ion-exchange chromatography because of substantial loss of enzymatic activity and regulatory properties. Affinity chromatography based on the specific interaction of these enzymes with their effectors (Tyr or Phe) appears particularly attractive; the effectors are readily available, and each contains an amino group that can easily be attached to CNBr-activated agarose. Other promising features of the system are the relatively high affinity of the effectors for the enzyme (50–200 μM for 50% inhibition),[4,5] and the fact that they do not participate in the catalytic reaction.

[1] P. R. Srinivasan and D. B. Sprinson, *J. Biol. Chem.* **234**, 716 (1959).
[2] F. Lingens, W. Goebel, and H. Uesseler, *Biochem. Z.* **346**, 357 (1966).
[3] P. Meuris, *C. R. Acad. Sci.* **264**, 1197 (1967).
[4] F. Lingens, W. Goebel, and H. Uesseler, *Eur. J. Biochem.* **1**, 363 (1967).
[5] M. Takahashi and W. W.-C. Chan. *Can. J. Biochem.* **49**, 1015 (1971).

Choice of Ligand

Preliminary work using standard conditions for activating agarose[6] and for coupling ligands[7] to it indicated that L-tyrosine-Sepharose (containing 0.3–0.4 μmole of ligand per milliliter) had a slightly retarding effect on the Tyr-sensitive DAHP synthetase whereas no detectable effect was found when glycyltyrosine was the bound ligand. These results are consistent with the much lower affinity of glycyl-L-tyrosine ($K_i = \sim 1$ mM)[5] for this enzyme and suggest that the basic character of the α-amino group of tyrosine may be important in the binding process. Porath[8] has shown that this basic character may be retained in some of the products, e.g., structure (I) in Scheme I, of coupling of primary amines to Sepharose,

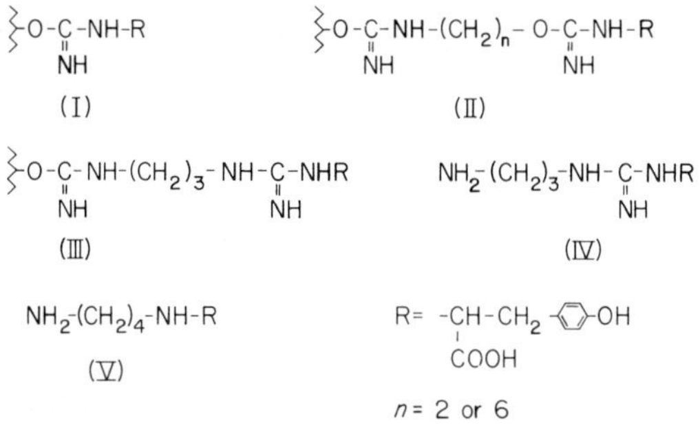

$$\xi\text{-O-C-NH-R} \qquad \xi\text{-O-C-NH-(CH}_2)_n\text{- O-C-NH-R}$$
$$\quad\ \ \overset{\|}{\text{NH}} \qquad\qquad\qquad \overset{\|}{\text{NH}} \qquad\qquad \overset{\|}{\text{NH}}$$
$$\text{(I)} \qquad\qquad\qquad\qquad\qquad \text{(II)}$$

$$\xi\text{-O-C-NH-(CH}_2)_3\text{- NH-C-NHR} \qquad \text{NH}_2\text{-(CH}_2)_3\text{-NH- C-NHR}$$
$$\quad\ \ \overset{\|}{\text{NH}} \qquad\qquad\quad \overset{\|}{\text{NH}} \qquad\qquad\qquad\qquad\qquad \overset{\|}{\text{NH}}$$
$$\text{(III)} \qquad\qquad\qquad\qquad\qquad\qquad\qquad \text{(IV)}$$

$$\text{NH}_2\text{-(CH}_2)_4\text{-NH-R} \qquad\qquad R= \text{-CH-CH}_2\ \text{-}\hspace{-2pt}\langle\ \rangle\hspace{-2pt}\text{-OH}$$
$$\text{(V)} \qquad\qquad\qquad\qquad\qquad\qquad \underset{\text{COOH}}{|}$$
$$n = 2 \text{ or } 6$$

SCHEME I

and this may explain why tyrosine-Sepharose is moderately effective in interacting with the Tyr-sensitive DAHP synthetase.

Attempts to Incorporate a Spacer

Although there are numerous methods[9,10] for the introduction of a spacer between Sepharose and the amino group of the ligand, nearly all methods result in the conversion of the amino group to an amide group and would suffer the adverse effects due to loss of the basic character of the α-amino group as in the case of glycyl-L-tyrosine. Attachment of L-tyrosine to the spacer via the α-carboxyl group was also to be avoided

[6] R. Axén, J. Porath, and S. Ernbäck, *Nature (London)* **214**, 1302 (1967).

[7] P. Cuatrecasas, M. Wilchek, and C. B. Anfinsen, *Proc. Nat. Acad. Sci. U.S.* **61**, 636 (1968).

[8] J. Porath, *Nature (London)* **218**, 834 (1968).

[9] P. Cuatrecasas, *J. Biol. Chem.* **245**, 3059 (1970).

[10] P. Cuatrecasas, *Nature (London)* **228**, 1327 (1970).

since L-tyrosine amide was shown to be a very poor inhibitor.[5] Attempts were therefore made to prepare derivatives of structure (II) (Scheme I) by coupling ethanolamine or 6-amino-1-hexanol to CNBr-activated Sepharose followed by a second CNBr treatment at pH 11 and then coupling to L-tyrosine. Although the products were effective in retarding Tyr-sensitive DAHP synthetase, they were not significantly better than L-tyrosine-Sepharose. It was not clear whether the expected structures (II) were in fact produced or whether L-tyrosine was still coupled to the hydroxyl groups of the Sepharose since the two possibilities could not be distinguished by examining the products of hydrolysis. Derivative III (Scheme I) was prepared in a similar way by coupling 1,3-propanediamine to CNBr-activated Sepharose followed by a second CNBr treatment at pH 11 and then coupling to L-tyrosine. Evidence that the expected derivative did form in this case came from amino acid analysis of the hydrolyzate of this product. A new ninhydrin-positive peak was found emerging at 99.3 minutes (Arg at 50 minutes) from the short-column of a Beckman 120C amino acid analyzer. The position of this peak is consistent with the presence of the highly basic substituted guanido-group and the hydrophobic aromatic side chain in the expected product (IV). However, the Sepharose-derivative (III) did not retard Tyr-sensitive DAHP synthetase. Compound (V) may be considered close to the ideal ligand for L-tyrosine binding proteins in general since, when it is coupled (preferably via a spacer) to CNBr-activated Sepharose through the primary amino group, it retains the ionic and hydrogen-bonding properties of both the α-amino and the α-carboxyl groups of tryosine. This compound has been synthesized recently in our laboratory[11] by condensation of commercially available (Aldrich Chemical Co.) N-(4-bromobutyl) phthalimide with excess L-tyrosine methyl ester in refluxing xylene followed by isolation of the product and hydrolysis. The Sepharose derivative of this material has yet to be tested.

Low Solubility of Tyrosine

Because of the various problems associated with the introduction of a spacer, an attempt was made to improve the procedure simply by coupling higher amounts of L-tyrosine directly to Sepharose. One difficulty was the low solubility of L-tyrosine at the usual pH (9–10) used for coupling. However, by increasing the pH to 11.0, approximately 1 mmole tyrosine could be added per milliliter of CNBr-activated Sepharose. Coupling at this pH resulted in a 10-fold increase in the amount of bound tyrosine (3.2 μmoles per milliliter of Sepharose). This product remains the best

[11] W. W.-C. Chan, T. Hussein, and G. Mazur, unpublished results.

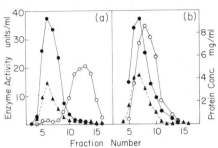

FIG. 1. Chromatography of crude yeast extract on a column of (a) tyrosine-Sepharose and (b) phenylalanine-Sepharose (details see text). ○—○, Tyr-sensitive DAHP synthetase activity; ●—●, Phe-sensitive DAHP synthetase activity, ▲----▲, protein concentration. Data presented in (a) are taken from W. W.-C. Chan and M. Takahashi, *Biochem. Biophys. Res. Commun.* **37**, 272 (1969) by permission of the copyright owners (Academic Press).

affinity chromatographic material for Tyr-sensitive DAHP synthetase[12] and gives the separation shown in Fig. 1(a). Using the dansyl group as a marker, it has been shown[5] that in this derivative the ligand is attached predominantly if not exclusively via the α-amino group, not via the phenolic group.

Phe-Sensitive DAHP Synthetase

In contrast to the Tyr-sensitive enzyme, the Phe-sensitive DAHP synthetase was not retarded by a column of either tyrosine-Sepharose (Fig. 1a) or phenylalanine-Sepharose (Fig. 1b) prepared in a similar way (containing approximately 2 μmoles of ligand per milliliter). In the latter case the lack of interaction may be due to the strict requirement for an unmodified α-amino group at the binding site since N-acetylphenylalanine is completely ineffective as an inhibitor of this enzyme. Interestingly, the phenylalanine-Sepharose derivative showed some affinity toward the *Tyr-sensitive* enzyme (Fig. 1b). This observation is consistent with the weak but significant inhibition of the Tyr-sensitive enzyme by phenylalanine.

Recommended Procedure

The preparation of suitable crude extracts of baker's yeast and the assay procedure for DAHP synthetase have been described previously.[5] Sepharose 4B (Pharmacia) (10 ml packed volume) is washed with deionized water and decanted to remove excess water. CNBr, 1 g freshly

[12] W. W.-C. Chan, and M. Takahashi, *Biochem. Biophys. Res. Commun.* **37**, 272 (1969).

dissolved in 10 ml of water, is added, and pH is adjusted to and maintained at 11.0 by adding 4 M NaOH with continuous stirring. After 8 minutes at room temperature when base-uptake has ceased, L-tyrosine (1.82 g) is added as a fine powder and pH is readjusted to 11.0 with 4 M NaOH. The mixture is stirred gently for 16 hours at 4°.

The product is washed with water, then with 0.2 M sodium phosphate (pH 6.5) and is finally equilibrated in the same buffer but containing the cofactor Co^{2+} (0.1 mM) and the protease inhibitor phenylmethylsulfonyl fluoride (0.1 mM). The sample consists of 2 ml of protamine-treated yeast extract[5] containing 9.5 mg of protein/ml and both 77 units of Tyr-sensitive DAHP synthetase and 73 units of the Phe-sensitive isozyme per milliliter. Chromatography is conducted with a column (0.6 × 15.5 cm) at 4° at a flow rate of 5 ml/hour. Fractions of 1 ml are collected, and 1 mg of bovine serum albumin is added immediately to each fraction to stabilize the enzyme.

The tyrosine-Sepharose is regenerated by washing successively with 0.1 M acetic acid and with 1 M urea solution containing 0.5% sodium dodecyl sulfate. For some preparations of tyrosine-Sepharose, the separation improves when the material is used for the second time.

[42] D-Alanine Carboxypeptidase

By MARIAN GORECKI, ATARA BAR-ELI, and AVRAHAM PATCHORNIK

D-Alanine carboxypeptidase hydrolyzes the terminal D-alanyl-D-alanine peptide bond present in the pentapeptide: UDP-MurNac-L-Ala-D-Glu-meso-Dap-D-Ala-D-Ala, which is utilized for the biosynthesis of bacterial cell wall peptidoglycan. Since the enzyme from various species is inhibited either reversibly or irreversibly by extremely low concentration of penicillin, many studies on the mechanism and mode of action of penicillin have been directed toward this enzyme.

Recently, the isolation of D-alanine carboxypeptidases from a number of organisms has been reported.[1-8] We have developed a method for the

[1] K. Izaki and J. L. Strominger, *J. Biol. Chem.* **243**, 3193 (1968).
[2] P. J. Lawrence and J. L. Strominger, *J. Biol. Chem.* **245**, 3660 (1970).
[3] G. C. Wickus and J. L. Strominger, *J. Biol. Chem.* **247**, 5307 (1972). See also Blumberg and Strominger, this volume [43].
[4] H. J. Barnett, *Biochim. Biophys. Acta* **304**, 332 (1973).
[5] J. M. Ghuysen, M. Leyh-Bouille, R. Bonaly, M. Nieto, H. R. Perkins, K. H. Schleifer, and O. Kandler, *Biochemistry* **9**, 2955 (1970).

rapid and complete purification of the enzyme from *Escherichia coli* B in high yields, by affinity chromatography. Partial purification of the enzyme by ion-exchange chromatography has been described previously.[1]

Principle. The approach we describe here involves the coupling of *p*-aminobenzylpenicillin to Sepharose. The crude extract of *E. coli* B containing D-alanine carboxypeptidase activity is specifically adsorbed to the column of this conjugate in the presence of 1% Triton X-100. After washing the column, the enzyme is eluted with buffer containing 0.5 M NaCl. From 20 g of lyophilized bacteria, 1 mg of homogeneous D-alanine carboxypeptidase is obtained.

Preparation of Affinity Column

Sepharose 4B is activated with cyanogen bromide.[9] The washed, activated Sepharose, 50 ml, is suspended in 50 ml of cold 0.1 M $NaHCO_3$ containing 0.5 g of *p*-aminobenzylpenicillin. The mixture is stirred gently at 4° for 16 hours and then washed extensively with water, with 1 M NaCl, and with an appropriate buffer. Under these conditions 1–2 μmoles of the penicillin are usually coupled per milliliter of gel as estimated by sulfur content.

Purification of D-Alanine Carboxypeptidase (D-CPase)

Preparation of Crude Extract. Lyophilized cells, 100 g, are stirred into 4 liters of water over a 10-minute period. The suspension is centrifuged at 10,000 g in a Sharples centrifuge, and the supernatant fluid, 3.5 liters, is incubated for 3 hours at 23° with 2 mg of DNase and 2 mg of RNase. The solution is chilled to 4°, and 1.4 kg of solid ammonium sulfate are added. The precipitate is collected by centrifugation with a Sharples centrifuge, redissolved in 500 ml of 50 mM Tris chloride at pH 8.2 and dialyzed at 4° against the same buffer. The insoluble material is discarded, and the solution is dried by lyophilization. The dried "crude" D-CPase is stable on storage at $-20°$ for several months.

Affinity Chromatography. The dried "crude D-CPase," 1 g, is dissolved in 100 ml of 50 mM Tris chloride at pH 8.2 containing 1 mM EDTA and 1% Triton X-100. This solution is applied to a *p*-aminobenzyl-

[6] M. Leyh-Bouille, J. M. Ghuysen, R. Bonaly, M. Nieto, H. R. Perkins, K. H. Schleifer, and O. Kandler, *Biochemistry* **9**, 2961 (1970).

[7] M. Leyh-Bouille, J. M. Ghuysen, M. Nieto, H. R. Perkins, K. H. Schleifer, and O. Kandler, *Biochemistry* **9**, 2971 (1970).

[8] M. Leyh-Bouille, M. Nakel, J. M. Frere, K. Johnson, J. M. Ghuysen, M. Nieto, and H. R. Perkins, *Biochemistry* **11**, 1290 (1972).

[9] P. Cuatrecasas, M. Wilchek, and C. B. Anfinsen, *Proc. Nat. Acad. Sci. U.S.* **61**, 636 (1968).

penicillin-Sepharose column (1.4 × 10 cm), previously equilibrated with 50 mM Tris chloride (pH 8.2)–1 mM EDTA. All the D-CPase activity is adsorbed by the column, and other proteins are removed by washing the column with the equilibrating buffer. D-CPase is eluted by raising the ionic strength of the buffer with 0.5 M NaCl.

Ammonium Sulfate Precipitation. The eluate containing the enzyme is cooled, and 48 g of ammonium sulfate are added per 100 ml of solution. The precipitate is collected by centrifugation at 15,000 g for 10 minutes, dissolved in 1 ml of 50 mM Tris chloride at pH 8.2 containing 1 mM EDTA and 0.025% Triton X-100, and is dialyzed against the same buffer without Triton X-100 for 3 hours. After this procedure, the enzyme solution is concentrated and may be stored thereafter for several weeks at −20° without detectable loss of activity. Dilute solutions of the enzyme retain their enzymatic activity at 4° for 2–3 weeks. The D-CPase preparation can be dried by lyophilization and stored dry for a prolonged period.

Yield and Purity of D-Alanine Carboxypeptidase. The purified D-alanine carboxypeptidase is obtainable in 80% yield with an overall purification was 1200-fold with respect to the crude extract. This enzyme preparation when assayed[10] on UDP-MurNAc-L-Ala-D-Glu-*meso*-Dap-Ala-D-[^{14}C]Ala released 1.6 μmole D-[^{14}C]Ala per milligram per minute at a substrate concentration of 3 mM and at 30°.

Comments

Variability of the Penicillin-Sepharose Conjugates. p-Aminobenzyl penicillin coupled to Sepharose by the method described is a very stable conjugate which retains its ability to adsorb the enzyme specifically even after storage in water at 4° for several months. However, not all freshly prepared, substituted Sepharose gels have been suitable for affinity chromatography.

Some of the conjugates adsorb the enzyme irreversibly and enzymatic activity could not be eluted by high salt concentrations. Exhaustive saturation of the gel by considerable amounts of the crude carboxypeptidase (10 times as much enzyme as used for a normal purification procedure) converted the penicillin-Sepharose column to a reversible adsorbant of the enzyme, suitable for affinity chromatography as described. The above "saturation" effect was highly specific, since no other protein, e.g., bovine serum albumin, could replace D-CPase in the above treatment. We have also noticed that prolonged storage at 4° of some of these highly adsorbing gels converted them to gels suitable for normal affinity chromatography.

[10] M. Gorecki, A. Bar-Eli, Y. Burstein, A. Patchornik, and E. B. Chain, *Biochem. J.* submitted for publication.

This change may be attributed to a decrease in the number of the highly active "adsorbing sites" on the surface of the gels, thereby allowing reversible enzyme binding to the gel-conjugate. Hydrolysis of the β-lactam ring in the penicillin nucleus is not the factor responsible for the change in the properties of the gels since prolonged treatment of the conjugate with a neutral solution of NH_2OH destroys completely their specific adsorbing properties.

[43] Covalent Affinity Chromatography of Penicillin-Binding Components from Bacterial Membranes[1]

By PETER M. BLUMBERG and JACK L. STROMINGER

The penicillin binding components of bacteria are membrane-bound proteins. They react with penicillin to form a covalent linkage, believed to be a thioester bond between a cysteine residue in the protein and the carbonyl of the β-lactam ring of the penicillin.[2] The penicilloyl-enzyme bond can be cleaved by mild treatment with neutral hydroxylamine.[2,3] Such treatment, moreover, restores enzymatic activity, at least in the case of the D-alanine carboxypeptidase of *Bacillus subtilis*.[4] The above findings suggested the feasibility of *covalent* affinity chromatography as a means of purifying penicillin binding proteins.

Preparation of Penicillin-Substituted Agarose

Succinyl-diaminodipropylamino-Sepharose 4B is prepared[5] with 250 mg of cyanogen bromide per milliliter of Sepharose in the initial coupling. The degree of substitution is monitored with the 2,4,6-trinitrobenzene sulfonate test.

Several details of the procedure should be noted. The cyanogen bromide should be finely ground with a mortar and pestle. Since base uptake slows but does not cease during the cyanogen bromide activation of the Sepharose, the activation reaction is arbitrarily halted after 12 minutes.

[1] Peter M. Blumberg was supported by a predoctoral fellowship from the National Science Foundation. This research was supported by research grants from the USPHS (AI-09152 and AM-13230) and from the NSF (GB-29747X).
[2] P. J. Lawrence and J. L. Strominger, *J. Biol. Chem.* **245**, 3653 (1970). See also Gorecki *et al.*, this volume [42].
[3] H. Suginaka, P. M. Blumberg, and J. L. Strominger, *J. Biol. Chem.* **247**, 5279 (1972).
[4] P. J. Lawrence and J. L. Strominger, *J. Biol. Chem.* **245**, 3660 (1970).
[5] P. Cuatrecasas and C. B. Anfinsin, this series, Vol. 22, p. 345.

Successful coupling of the penicillin requires complete succinylation. If the color test shows only partial substitution, the succinylation procedure is repeated.

For preparation of the penicillin derivative, 10 ml of packed succinyl-diaminodipropylamino-Sepharose is suspended in water to yield a final volume of 15 ml. Dimethylformamide (15 ml) is added, and the mixture is cooled to room temperature. The pH is adjusted to 4.8 with 1 N HCl, and 0.35 mmole of 6-aminopenicillanic acid (Sigma) is added. Next, 2.5 mmole of 1-ethyl-3(3-dimethylaminopropyl) carbodiimide dissolved in 1 ml of water is added dropwise over a 5-minute period. The pH is maintained at 4.8 for 1 hour by titration with 0.1 N HCl. The reaction mixture is gently stirred overnight at room temperature. The resin is washed on a 30-ml Büchner funnel fitted with a coarse fritted-glass disk with 500 ml of 20 mM potassium phosphate at pH 7.0 and containing 0.5 M NaCl. After washing, the resin is used immediately. Alternatively, it may be rinsed with distilled water (200 ml) and dioxane (75 ml), and then lyophilized prior to storage. The degree of substitution of the resin can be monitored with the hydroxymate assay for penicillins.[6] Routinely, several milligrams of penicillin are bound per milliliter of gel.

Isolation of Binding Components

Membranes are prepared from $B.$ $subtilis$ strain Porton[7] and stored at $-20°$ until needed. For solubilization, 5 ml of the membrane suspension (20 mg of protein per milliliter) are added to 12.5 ml of 2 N NaCl, 2.5 ml 0.5 M potassium phosphate at pH 7.0, 5 ml of 10% Nonidet P-40 (Shell Chemical Co.), and 1.75 μl of 2-mercaptoethanol. After being mixed in a vortex mixer, the solution is allowed to sit for 30 minutes at 0° and then centrifuged in the cold at 100,000 g for 1 hour to remove unsolubilized material. The supernatant solution is added to 5 ml of the 6-amino-penicillanic acid-substituted Sepharose, and the slurry is incubated with gentle mixing for 30 minutes at 25°. The mixture is poured into a wide column, e.g., a 20-ml syringe, at room temperature and rapidly washed with 150 ml of washing buffer (1.0 M NaCl, 50 mM KPO₄, pH 7.0, 0.1% Nonidet P-40, 1 mM 2-mercaptoethanol). One column volume of elution buffer (0.8 M neutral hydroxylamine, 50 mM KPO₄, pH 7.0, 1% Nonidet P-40, 1 mM 2-mercaptoethanol) is fed into the column; the flow is then halted. After 30 minutes a second column volume of elution buffer is fed into the column, after which the flow is halted for an additional 30 minutes. The second aliquot of elution buffer is finally eluted with 1

[6] G. E. Boxer and P. M. Everett, $Anal.$ $Chem.$ **21**, 670 (1949).
[7] P. M. Blumberg and J. L. Strominger, $J.$ $Biol.$ $Chem.$ **247**, 8107 (1972).

volume of washing buffer. The column eluents are pooled and dialyzed in the cold against 10 mM Tris chloride at pH 8.6, 1 mM 2-mercaptoethanol, and 0.1% Triton X-100. The protein can be concentrated by absorption to a 0.10 ml column of DEAE-cellulose (Whatman DE-52), equilibrated with dialysis buffer, followed by stepwise elution with 50 mM Tris chloride at pH 7.5, 1% Nonidet P-40, 0.5 M NaCl, 1 mM 2-mercaptoethanol. For larger batches of penicillin binding components, all quantities may be scaled up proportionately.

Recovery

There are five penicillin binding components in *B. subtilis* membranes.[7] Of these, an enzymatic activity is known only for component (V), the D-alanine carboxypeptidase. This is recovered with 50–80% yield of enzymatic activity (Table I). The yields of the components (I), (II), and (IV) relative to the carboxypeptidase are also good (Table II). Component (III), in contrast, showed only 10% recovery at best and was almost entirely lost in the DEAE-cellulose concentration procedure.

TABLE I
RECOVERY OF D-ALANINE CARBOXYPEPTIDASE DURING
ISOLATION OF PENICILLIN-BINDING COMPONENTS[a]

	Activity recovered (%)	
	Expt. 1 (50 mM Tris chloride pH 7.5)	Expt. 2 (50 mM KPO$_4$ pH 7.0)
Solubilized membranes	(100)[b]	(100)[c]
Flow-through		
(a) Control	4.8	0.9
(b) Hydroxylamine treated	20.0	2.3
1st hydroxylamine wash	33.4	54.7
2nd hydroxylamine wash	9.7	24.3
Total activity accounted for	65.1	81.3

[a] D-Alanine carboxypeptidase was assayed as described [P. M. Blumberg and J. L. Strominger, *Proc. Nat. Acad. Sci. U.S.* **68,** 2814 (1971)]. Samples eluted from the column were dialyzed against 10 mM Tris chloride (pH 7.5), 20 mM NaCl, 1 mM 2-mercaptoethanol before assay. For the assay of the flow-through fraction, sufficient 2 N hydroxylamine (pH 7.0) was added where indicated to render the solution 0.8 M in hydroxylamine. The fraction was then incubated for 30 minutes at 25° before dialysis.

[b] 2.95 × 10^7 cpm released per 30 minutes.

[c] 3.80 × 10^7 cpm released per 30 minutes.

TABLE II

YIELD OF PENICILLIN BINDING COMPONENTS ISOLATED BY
COVALENT AFFINITY CHROMATOGRAPHY[a]

Component	Proportions of isolated binding components	Proportions of components in membranes	Yield relative to carboxypeptidase (%)
(I)	14.9	19.3	77
(II)	18.1	29.1	62
(III)	—	17.1	—
(IV)	13.8	13.2	105
(V)	(100)	(100)	(100)

[a] The isolated binding components were precipitated with 80% acetone (final concentration) to remove detergent and subjected to sodium dodecyl sulfate polyacrylamide gel electrophoresis [K. Weber and M. Osborn, *J. Biol. Chem.* **244**, 4406 (1969)]. After staining of the protein with Coomassie brilliant blue, the relative amounts of protein in the bands were determined by densitometry.

High recoveries of binding components require attention to several aspects of the procedure. (1) If the phosphate buffer is replaced with 50 mM Tris chloride at pH 7.5, a substantial amount (20%) of carboxypeptidase is found in the affinity-column flow-through in an inactive form which could be reactivated by hydroxylamine. Apparently, the buffer allowed release of penicilloyl-enzyme. (2) High salt is required in the washing buffer to prevent ionic interactions between the solubilized membrane proteins and charges introduced onto the Sepharose during the derivitization. Very substantial nonspecific binding occurs in 50 mM Tris chloride at pH 7.5. For the hydroxylamine elution of the binding components, the salt concentration is dropped from 1 M NaCl to 0.8 M (the concentration of NaCl generated by neutralization of the hydroxylamine-HCl) so as to avoid the possible nonspecific release of proteins from the Sepharose due to a change in ionic strength. (3) The binding components differ in their sensitivities to release from the column by hydroxylamine. Component (I) is almost completely released by one 30-minute treatment. Component (V) requires two, and components (II) and (IV) are somewhat less readily released. (4) A side arm of succinylated ethylenediamine seemed to function as well as the longer side arm. However, 6-aminopenicillanic acid could not be coupled directly to the Sepharose with CNBr. The nature of the β-lactam antibiotic coupled to the Sepharose is also of importance. Component (V) is relatively insensitive to cephalosporin C.[7] If cephalosporin C (Eli Lilly & Co.) is coupled either directly or via a side arm to Sepharose, the affinity resin binds components (I), (II), and (IV) but only negligible quantities of (V).

TABLE III
PURIFICATION OF THE D-ALANINE CARBOXYPEPTIDASE
BY AFFINITY CHROMATOGRAPHY

	Total activity (cpm × 10^{-8})	% Yield	Total protein[a] (mg)	Specific activity (cpm/mg)
Soluble membranes	6.20[b]	(100)	986	6.3×10^5
Eluent 1 from affinity column	2.31	37.3	ND[c]	ND
Eluent 2 from affinity column	0.71	11.5	ND	ND
Eluent from DEAE-cellulose column	2.78	44.8	3.17	8.8×10^7

[a] By the ninhydrin method (C. H. Hirs, this series, Vol. 11, p. 328).

[b] 6.2×10^8 Cpm corresponds to about 10 mmoles of D-alanine released per 30 minutes at 25° under the conditions of assay.

[c] ND, not done.

Purification of *B. subtilis* Carboxypeptidase

The D-alanine carboxypeptidase is 1000-fold more resistant to the β-lactam antibiotic cephalothin[8] than are the other binding components.[7] Therefore, if cephalothin at a final concentration of 2 μg/ml is added to the solubilized membranes and incubated for 10 minutes at 25° immediately before addition of the affinity resin, the cephalothin blocks the binding to the resin of the components other than carboxypeptidase. In this manner, carboxypeptidase could be obtained at a level of purity greater than 99% in a single step. The yield of enzyme is about 50% (see Table III); up to 30 mg of enzyme could conveniently be prepared in a single batch.

[8] Cephalothin (Keflin) was generously provided by Eli Lilly and Co.

[44] Immobilized L-Aspartic Acid

By ICHIRO CHIBATA, TETSUYA TOSA, TADASHI SATO, RYUJIRO SANO, KOZO YAMAMOTO, and YUHSI MATUO

An affinity adsorbent prepared by immobilizing a ligand of high specificity may be expected to have a high adsorption specificity. On the other hand, an affinity adsorbent prepared with a ligand of lower specificity may be useful for "group-specific" adsorption. In this article we describe a method of immobilization of L-aspartic acid to agarose with interposed hexamethylenediamine as a spacer and the application of the immobilized L-aspartic acid (L-Asp) to the purification of several enzymes for which this ligand has affinity.

Method of Preparation

The method is essentially that of Tosa et al.[1] and represents a modification of the method of Kristiansen et al.[2]

Synthesis of N-(ω-aminohexyl)-L-Aspartic Acid (AHA)

D-(−)-Aspartic acid (6.7 g) and NaCl (20 g) are dissolved in 50 ml of 1 N HCl, and to the solution, 50 ml of 4% $NaNO_2$ are added dropwise with stirring at 0–5°. After further stirring for 20 hours, 50 ml of conc. HCl are added, and the reaction product is extracted 3 times with ether. By evaporation of the resultant ether layer, monochlorosuccinic acid (2.6 g) is obtained in crystalline form (mp 176°). To 50 ml of an absolute ethanol solution containing monochlorosuccinic acid (2.6 g) is slowly added hexamethylenediamine (7.5 g) in 50 ml of absolute ethanol and the mixture is refluxed with stirring at 65° for 30 hours. To the mixture are added 100 ml of water, and the solution is concentrated under reduced pressure at 50° to yield an oily product. The product is applied to a column (4 × 40 cm) of Amberlite IRA 401 (OH⁻ type). The column is thoroughly washed with water and developed with 14% NH_4OH. The eluate is evaporated to dryness under reduced pressure at 50°, and the residue is dissolved in 10 ml of water. When 200 ml of methanol are added to the solution, AHA (0.5 g) is obtained in crystalline form (mp 191°) (Fig. 1).

Preparation of Immobilized L-Asp

The activation procedure of Sepharose 6B with CNBr is essentially that of Axén et al.[3] The procedure should be carried out in a well ventilated hood. Sepharose 6B (10 g of packed weight) is washed with 500

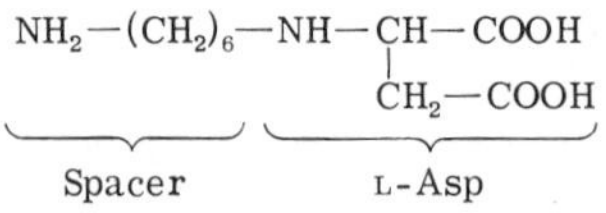

$$NH_2-(CH_2)_6-NH-CH-COOH$$
$$\underset{Spacer}{\underbrace{\qquad\qquad\qquad}}\ \underset{L-Asp}{\underbrace{\overset{\displaystyle CH_2-COOH}{\qquad\qquad}}}$$

FIG. 1. Structure of N-(ω-aminohexyl)-L-aspartic acid.

[1] T. Tosa, T. Sato, R. Sano, K. Yamamoto, Y. Matuo, and I. Chibata, Biochim. Biophys. Acta 334, 1 (1974).

[2] T. Kristiansen, M. Einerson, L. Sundberg, and J. Porath, FEBS Lett. 7, 264 (1970).

[3] R. Axén, J. Porath, and S. Ernbäck, Nature (London) 214, 1302 (1967).

ml of water and is suspended in 50 ml of water. CNBr solution (10 g per 200 ml of water) is added to the stirred suspension, and the mixture is maintained at 10–20° and at pH 11–12 by adding conc. NaOH. The reaction is complete in 8–15 minutes. The precipitate is recovered on a glass filter and washed sequentially with 500 ml of cold 0.1 M $NaHCO_3$ and 500 ml of cold water. The recovered CNBr-activated Sepharose 6B is used immediately in the next step.

AHA (0.5 g) is dissolved in 100 ml of 0.1 M $NaHCO_3$ and CNBr-activated Sepharose 6B (6 g packed weight) is added. The suspension is gently stirred at 25° for 24 hours and then washed with 200 ml of 0.1 M $NaHCO_3$ and 100 ml of 1 M NaCl.

Group-Specific Adsorption

Assay Methods

Asparaginase. Asparaginase (EC 3.5.1.1) activity is assayed according to the method of Tosa *et al.*[4] One asparaginase unit (IU) is defined as the amount of enzyme that hydrolyzes 1 μmole of substrate per minute under the standard assay conditions.

Aspartase. Aspartase (EC 4.3.1.1) activity is assayed as follows: The reaction mixture composed of 1.6 ml of sodium borate buffer (pH 8.5), 1.3 ml of 3% ammonium fumarate containing $MnCl_2$ at the concentration of 3 mM, and 0.1 ml of enzyme solution is incubated at 37° for 30 minutes. The reaction is stopped by heating for 5 minutes at 95°. The formation of L-aspartic acid is estimated by bioassay using *Leuconostoc mesenteroides* P-60 according to the method of Henderson and Snell,[5] and its activity is expressed as μmoles of L-aspartic acid produced per hour.

Aspartate β-Decarboxylase. Aspartate β-decarboxylase (EC 4.1.1.12) activity is assayed by manometric measurements of CO_2 liberated from L-aspartic acid in the presence of α-ketoglutarate (5 μmoles) according to the method of Kakimoto *et al.*[6] One decarboxylase unit is defined as the amount of enzyme forming 1 μmole of CO_2 per minute under the standard assay conditions.

L-Aspartic Acid. L-Aspartic acid in the effluent is determined by the ninhydrin colorimetric method with a Technicon Auto-Analyzer according to the method of Cadavid and Paladini.[7]

[4] T. Tosa, R. Sano, K. Yamamoto, M. Nakamura, K. Ando, and I. Chibata, *Appl. Microbiol.* **22**, 387 (1971).
[5] L. M. Henderson and E. E. Snell, *J. Biol. Chem.* **172**, 15 (1948).
[6] T. Kakimoto, J. Kato, T. Shibatani, N. Nishimura, and I. Chibata, *J. Biol. Chem.* **244**, 353 (1969).
[7] N. G. Cadavid and A. C. Paladini, *Anal. Biochem.* **9**, 170 (1964).

TABLE I
Summary of Enzymes Showing Group-Specific Adsorption
to Immobilized l-Asp Column[1]

	Nonspecific elution[a]			Specific elution[b]		
		Recovery (%)			Recovery (%)	
Enzyme	Ionic strength[c]	Protein	Activity	Ionic strength[c]	Protein	Activity
Asparaginase						
From *P. vulgaris*	0.060	92	28	0.04	94	94
From *E. coli*	0.062	82	37	0.04	94	76
Aspartate β-De-carboxylase						
Apoenzyme	0.228[d]	84	—	—	—	—
Holoenzyme	0.233	95	55	—	—	—
Aspartase	>0.275	90	95	0.175	90	99

[a] Eluted by increasing ionic strength with NaCl.
[b] Eluted by the buffer containing l-aspartic acid for asparaginase and sodium fumarate for aspartase.
[c] Ionic strength of the fraction showing the maximum enzyme activity.
[d] Ionic strength of the fraction showing the maximum protein content.

Enzyme Preparations

Apoenzyme of Aspartate β-Decarboxylase. Apoenzyme of aspartate β-decarboxylase is prepared by dialyzing the holoenzyme against 1 M sodium acetate at pH 5.0 containing 0.1 M l-aspartic acid by the method of Tate and Meister.[8]

Crude Aspartase. Aspartase is obtained from *Escherichia coli* K12 by the method of Tosa *et al.*[1]

Procedure. A group of enzymes related to the metabolism of l-aspartic acid is adsorbed onto the immobilized l-Asp column. These enzymes include both asparaginases from *E. coli* and *Proteus vulgaris,* aspartase and apo- and holoenzymes of aspartate β-decarboxylase and can be eluted by increasing the ionic strength of the eluent with NaCl (Table I). The elution profiles of respective enzymes are schematically superposed in Fig. 2. Of the several enzymes which are adsorbed, asparaginases from *E. coli* and *P. vulgaris,* as well as aspartase, can each be eluted with l-aspartic acid and sodium fumarate of lower ionic strength than is necessary for NaCl (Table I). The ionic strength and concentration of l-aspartic acid of the fraction having the maximum asparaginase activity are 0.040 and 70 mM, respectively; the values are the same for both asparaginases. The adsorbed aspartase can be eluted with 0.1 M sodium acetate at pH 7.0

[8] S. S. Tate and A. Meister, *Biochemistry* **8**, 1065 (1969).

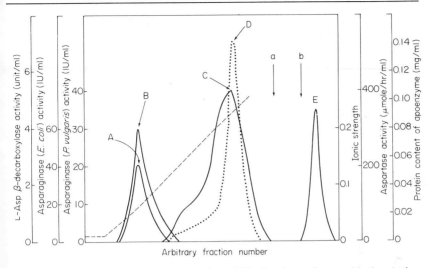

FIG. 2. Group-specific adsorption of immobilized L-Asp column with the elution profile of each enzyme superimposed. Asparaginase from *Proteus vulgaris* (3.18 mg, 800 IU), asparaginase from *Escherichia coli* (5.1 mg, 1240 IU), holoenzyme of aspartate β-decarboxylase (1.38 mg, 144.4 units), and apoenzyme of the enzyme (1.15 mg) are dissolved in sodium acetate buffer (pH 7.0, ionic strength = 0.01, 1.0 ml), and separately charged to the column. Elution is carried out with linear gradients. In the case of aspartase, the enzyme preparation is dialyzed against sodium acetate buffer (pH 7.0, ionic strength = 0.01), and dialyzed sample (15 mg, 1 ml, 1900 μmoles of L-Asp per hour) is charged to the column. Stepwise elution is carried out by using 0.1 M sodium acetate buffer containing the following concentration of NaCl: [a], 0.2 M (ionic strength = 0.275); [b], 0.3 M (ionic strength = 0.379). Fraction volume is 3.0 ml in all cases except for aspartate β-decarboxylase (2.5 ml). ——, Activity; ····, protein; – – – –, ionic strength. A, Asparaginase from *P. vulgaris;* B, asparaginase from *E. coli;* C, holoenzyme of aspartate β-decarboxylase; D, apoenzyme of aspartate β-decarboxylase; E, aspartase.

containing 0.1 M sodium fumarate (0.175 ionic strength). On the other hand, neither type of aspartate β-decarboxylase can be eluted with 0.15 M L-aspartic acid containing 0.1 M NaCl (pH 7.0, 0.113 ionic strength) or by 0.5 M L-alanine containing 0.1 M NaCl (pH 7.0, 0.103 ionic strength). The recoveries of protein and activity are summarized in Table I.

Purification of Asparaginase from *P. vulgaris*[1,9]

P. vulgaris OUT 8226 is cultured under aerobic conditions at 30° for 18 hours in 10 liters of 1% sodium fumarate and 5% corn steep liquor (pH 7.0). Approximately 210 g of cells are suspended in 500 ml of 50 mM sodium phosphate at pH 8.0 and treated with 105 mg of lysozyme;

[9] T. Tosa, R. Sano, K. Yamamoto, M. Nakamura, and I. Chibata, *Biochemistry* **11**, 217 (1972).

5 ml of toluene are added and the mixture is incubated at 30° for 24 hours with gentle stirring. After the addition of an equal volume of water, cell debris is removed. The crude extract is adjusted to pH 4.5 with 50% aqueous acetic acid and allowed to stand for 2 hours. After centrifugation the supernatant fluid is brought to 50% saturation with ammonium sulfate (38 g per 100 ml) and the precipitate is discarded. The solution is brought to 90% saturation with an additional 25.3 g of ammonium sulfate per 100 ml and the precipitate is dissolved in 10 ml of water. The enzyme is dialyzed overnight against 10 mM sodium phosphate at pH 6.8 containing 50 mM NaCl. To 42 ml of the solution sodium L-aspartate (350 mg) is added, followed by the dropwise addition of 21 ml of ethanol at 0°. The mixture is allowed to stand for 1 hour and centrifuged. An additional 21 ml of ethanol is added to the supernatant at 0°, and the resultant precipitate is dissolved in 50 ml of 10 mM sodium acetate at pH 7.0. The solution is applied to an immobilized L-Asp column (2 × 20 cm), equilibrated with 10 mM sodium acetate at pH 7.0 containing 40 mM NaCl. After the column is thoroughly washed with the equilibrating buffer, enzyme is eluted with 10 mM sodium acetate at pH 7.0 containing 70 mM L-aspartic acid. Active fractions are collected and saturated with ammonium sulfate. The resulting precipitate is dissolved in 10 mM sodium phosphate at pH 6.8 at a protein concentration of 2%, and then crystallized. A summary of this procedure is presented in Table II.

Separation of Aspartase and Fumarase

Procedure. Aspartase and fumarase are not completely separated by isoelectric focusing or by gel filtration with Sepharose 2B, indicating that they resemble each other in isoelectric point and in molecular size. The mixture of enzymes is dialyzed against 0.1 M sodium acetate at pH 7.0 containing 0.2 M NaCl (0.275 ionic strength), and the dialyzed sample (15 mg in 1 ml containing activity of 1900 μmoles of L-aspartic acid pro-

TABLE II
PURIFICATION OF ASPARAGINASE FROM *Proteus vulgaris*[1]

Fraction	Volume (ml)	Total protein (mg)	Specific activity (IU/mg)	Total activity (IU)	Yield (%)
Crude extract	1,050	19,820	1.1	21,800	100
pH 4.5	930	9,420	1.9	17,900	82
Ammonium sulfate (50–90%)	42	1,020	13.2	13,500	62
Ethanol (33–50 vol%)	50	149	70	10,460	48
Immobilized L-Asp	24	30	310	9,200	42
Crystals	1	27	300	8,080	37

duced per hour) is applied to an immobilized L-Asp column (5 × 10 cm). The column is washed with the same buffer until protein is not detected in the effluent. By this procedure, contaminating fumarase is completely removed. Aspartase is adsorbed and is specifically eluted with the same buffer supplemented to 0.1 M sodium fumarate (0.175 ionic strength). Total recovery of protein and activity are 90% and 99%, respectively.

[45] Carboxypeptidase B

By MORDECHAI SOKOLOVSKY

Since carboxypeptidase B can bind both basic and nonbasic amino acids,[1,2] several ligands may be attached; a variety of matrixes include Sepharose, 4B,[1,3,4] aminoethylcellulose,[5] and the azide of CM-Sephadex.[6] The table summarizes the various preparations and procedures described below for affinity chromatography of carboxypeptidase B. It should be noted that purification of this enzyme cannot be achieved in a single step. Affinity chromatography as well as conventional chromatography techniques[4] are required.

Preparation of Adsorbents

Via the Active Ester Derivative of Carrier.[7] Sepharose 4B, 50 ml, is activated[8,9] at pH 11 and 22° by addition of 12 g of CNBr. The reaction is allowed to proceed for 8 minutes while the pH is maintained between 10.5 and 11 by the addition of 5 N NaOH. Activation is terminated by filtration and washing of the gel rapidly with ice-cold deionized water. Finally, the gel is washed with cold 0.1 M Na$_2$CO$_3$ at pH 9.5. The activated Sepharose is suspended in 150 ml of cold 0.1 M Na$_2$CO$_3$ at pH 9.5 and mixed with 100 ml 0.1 M Na$_2$CO$_3$ containing 7 g of ε-aminocaproic acid (pH 9.5). Coupling is performed at 4° for 18 hours while the reaction mixture is gently swirled. Subsequently the gel is washed on a suction

[1] M. Sokolovsky and N. Zisapel, *Biochim. Biophys. Acta* **251**, 203 (1971).
[2] E. Winersberger, D. J. Cox, and H. Neurath, *Biochemistry* 1, 1069 (1962).
[3] M. Sokolovsky, unpublished results.
[4] G. R. Reeck, K. Walsh, and H. Neurath, *Biochemistry* 10, 4690 (1971).
[5] J. R. Uren, *Biochim. Biophys. Acta* **236**, 67 (1971).
[6] H. Akanuma, A. Kasuga, Y. Akanuma, and M. Ymasaki, *Biochem. Biophys. Res. Commun.* **45**, 27 (1971).
[7] P. Cuatrecasas and I. Parikh, *Biochemistry* **11**, 2291 (1972).
[8] R. Axén, J. Porath, and S. Ernbäck, *Nature (London)* **214**, 1302 (1967).
[9] J. Porath, R. Axén, and S. Ernbäck, *Nature (London)* **215**, 1491 (1967).

ADSORBENTS FOR AFFINITY CHROMATOGRAPHY OF CARBOXYPEPTIDASE B

	Adsorbents	Conditions		References
		Adsorption	Elution	
I	Sepharose-ε-amino-n-caproyl-D-arginine	0.8 × 7 cm column, equilibrated with 20 mM Tris pH 7.5; 3 mg enzyme	20 mM Tris–1 M NaCl, pH 7.5	a, b
II	Sepharose-ε-amino-n-caproyl-D-tryptophan	2.5 × 35 cm column, equilibrated with 20 mM Tris–50 mM NaCl pH 7.5; 80 mg enzyme	20 mM Tris–0.15 M NaCl, pH 7.5	c
III	Sepharose-D-alanyl-D-arginine	1 × 15 cm column, equilibrated with 25 mM Tris pH 7.9; 2.5 mg enzyme	20 mM Tris–0.1 M NaCl	a
IV	N-(2-Ethylcellulose)-glycyl-D-phenylalanine	1 × 15 cm column, equilibrated with 50 mM imidazole pH 6.5	50 mM imidazole–0.25 M NaCl, pH 7.5	d
V	N-(2-Ethylcellulose)-glycyl-D-arginine	As in IV	50 mM imidazole–0.25 M NaCl, pH 6.5	d
VI	CM-Sephadex-ε-amino-n-caproyl-D-arginine	1 × 11.2 cm column, equilibrated with 20 mM phosphate buffer–0.1 M pH 7.05; 5 mg enzyme	0.1 M ε-aminocaproic acid in 20 mM phosphate–0.1 M NaCl	e
VII	CM-Sephadex-ε-amino-n-caproyl-D-phenylalanine	As in VI	Ibid.	e
VIII	Polyacrylamide-ε-amino-n-caproyl-D-arginine	0.8 × 7 cm column, equilibrated with 20 mM Tris pH 7.5; 3 mg enzyme	20 mM Tris–1 M NaCl, pH 7.5	b

[a] M. Sokolovsky and N. Zisapel, *Biochim. Biophys. Acta* **251,** 203 (1971).
[b] M. Sokolovsky, unpublished results.
[c] G. R. Reeck, K. Walsh, and H. Neurath, *Biochemistry* **10,** 4690 (1971).
[d] J. R. Uren, *Biochim. Biophys. Acta* **236,** 67 (1971).
[e] H. Akanuma, A. Kasuga, Y. Akanuma, and M. Yamasaki, *Biochem. Biophys. Res. Commun.* **45,** 27 (1971).

funnel with distilled water, and 3 times with 200 ml of dioxane. The activated Sepharose is suspended in 150 ml of dioxane and gently mixed with 3 g of N-hydroxysuccinimide and 6 g of N,N'-dicyclohexylcarbodiimide for 1 hour at room temperature.[7] The reaction mixture is filtered and washed on a suction funnel with 1 liter of dioxane over a period of 5–10 minutes. This step is followed by addition of 750 ml of methanol and stirring with a glass rod for 5–10 minutes to remove all precipitated dicyclohexylurea. The preparation is washed with an additional 500 ml of dioxane and filtered. The filter cake is suspended in 300 ml of 0.1 M NaHCO₃ at pH 8.3 and coupled with 6 g of either D-arginine[3] or D-tryptophan.[3] The coupling reaction is carried out overnight at 4°, during which the reaction mixture is gently swirled. The adsorbent is then washed extensively with water and with the appropriate buffer. The average ligand content, as measured by amino acid analysis after acid hydrolysis, is 6–12 μmoles per 1 ml of resin.

By Direct Coupling of the Ligand and Carrier.[1,4] Sepharose 4B is activated with CNBr as described above. The filtered and washed activated preparation is suspended in cold 0.1 M NaHCO₃ at pH 9 to a volume about twice that of the settled volume of gel, and mixed with an equal volume of a solution of the appropriate ligand (0.2–0.3 mmole ligand per 10 ml of gel) in the same buffer. The mixture is stirred gently for 20 hours at 4°. The resin is filtered and washed thoroughly with 0.1 M NaHCO₃ and water. The amount of peptide coupled to the Sepharose is determined by amino acid analysis after acid hydrolysis. The ligand content is 7–25 μmoles per 1 ml of resin.

Ligands. The details for use of the following ligands as attached to Sepharose are summarized in the table: ϵ-Amino-n-caproyl-D-arginine for adsorbents I and VI; ϵ-amino-n-caproyl-D-tryptophan for adsorbent II; ϵ-amino-n-caproyl-D-phenylalanine for adsorbent VII; D-alanyl-D-arginine for adsorbent III.

Applications

Purification of Carboxypeptidase B.[4] A typical procedure for the purification of carboxypeptidase B is illustrated by the following example for the bovine enzyme as developed by Reeck *et al.*[4] After preliminary centrifugation, 1.6 liters active bovine pancreatic juice are applied to a soybean trypsin inhibitor–Sepharose column (5 × 35 cm) equilibrated with 20 mM Tris chloride–0.5 M NaCl at pH 7.5. Carboxypeptidase is not absorbed but is eluted just after the first 1400 ml of eluate (20 mM Tris chloride–0.5 M NaCl at pH 7.5; flow rate 600 ml per hour). This fraction is concentrated by ultrafiltration to 200 ml using an Amicon UM-10 membrane. The concentrated sample is dialyzed for 36 hours against a large excess of 20 mM Tris chloride at pH 7.5. Most of the carboxypeptidase A present in the sample precipitates during dialysis. The supernatant liquid is transferred and adsorbed onto a 2.5 × 35 cm column of Sepharose-ε-aminocaproyl-D-tryptophan previously equilibrated with 20 mM Tris chloride 50 mM NaCl at pH 7.5.

After the elution with 800 ml of the same buffer of an inactive breakthrough peak, 800 ml of 20 mM Tris chloride–0.15 M NaCl, pH 7.5, are used to elute carboxypeptidase B; the next 800 ml of eluate contain the enzyme. The solution is concentrated to 0.5 mg/ml by ultrafiltration and dialyzed overnight against 20 mM Tris chloride at pH 8.0. The preparation is then applied to a 2.5 × 40 cm column of DE-52 cellulose (Whatman) which had been equilibrated with the same buffer containing 0.1 mM ZnCl$_2$. The column is developed after application of the sample by the use of a linear gradient of from 0 to 0.1 M NaCl in a total volume of 2000 ml of 20 mM Tris chloride–0.1 mM ZnCl$_2$, pH 8.0. Inclusion of zinc in the elution buffer is necessary in order to stabilize the enzyme. This ion exchange step serves as the final one from which purified carboxypeptidase B is obtained in a overall recovery of 71% (78 mg of enzyme from 1600 ml of juice). Three forms of carboxypeptidase B have been resolved from activated bovine pancreatic juice by this procedure.[4]

Affinity Chromatography of Chemically Modified Proteins. Chemical modification of a protein may lead to heterogeneity of the resultant protein preparation; e.g., a mixture composed of modified and native enzyme may be formed or heterogeneity may be affected in the position of substitution. Affinity chromatography can be useful in such separations, as illustrated in the following two examples: (a) Bovine carboxypeptidase B that is alkylated at a glutamyl residue at the active site[10] was resolved into two components, an inactive (modified) protein and the native enzyme, by chromatography on Sepharose-L-leucyl-D-arginine. The fraction eluting in the void volume was enzymatically inactive whereas the retarded fraction was enzymatically active and essentially unmodified. (b) Porcine carboxypeptidase B that was modified with radioactive phenylglyoxal[11] has been resolved[12] into three components by affinity chromatography on Sepharose-ε-aminocaproyl-D-arginine (adsorbent I, in the table). All three fractions had radioactive label, and the last two fractions were active as esterases.

Binding Sites. Affinity chromatography technique have been applied to the study of binding of basic and nonbasic substrates to carboxypeptidase B.[6] Use of this technique revealed the presence of at least two binding subsites: one for the basic substrates and another for hydrophobic ones.

[10] T. H. Plummer, Jr., *J. Biol. Chem.* **246**, 2930 (1971).
[11] M. Werber and M. Sokolovsky, *Biochem. Biophys. Res. Commun.* **48**, 384 (1972).
[12] M. Werber, M. Moldovan, and M. Sokolovsky, submitted for publication.

[46] Chymotrypsin(s)

By G. Tomlinson, M. C. Shaw, and T. Viswanatha

Chymotrypsins are the products of the tryptic activation of the inactive precursors, the chymotrypsinogens produced by the pancreas of many species. The isolation of bovine chymotrypsinogen and its activation by trypsin were reported some forty years ago.[1-3] The initial step is the splitting of the Arg_{15}-Ile_{16} peptide bond leading to the formation of catalytically active π-chymotrypsin. Subsequent autocatalytic events give rise to δ-, κ-, and γ-chymotrypsins as intermediates in the formation of α-chymotrypsin.[4-6] The last-mentioned species, being more stable than its precursors, is the one most commonly encountered. All active forms of the enzyme are known to exhibit a marked preference in catalyzing the hydrolysis of peptide bonds on the carboxyl side of aromatic amino acid residues.

While the procedures for the isolation of chymotrypsin and chymotrypsinogen in a highly homogeneous state have been well developed,[7] the preparation of chymotrypsinogen totally devoid of chymotrypsin or of the latter free of zymogen and other enzymes normally found in the starting material is often difficult to achieve, even after repeated purification procedures. Similar considerations may apply during studies on the effects of chemical modification of the protein on the enzymatic activity, where it is often desirable to separate active from inactive species. Current interest in the homologies present among enzymes derived from different species underscores the need for a rapid, reliable technique for the isolation of these proteins. The use of affinity chromatography, a technique that exploits the biological specificity in the isolation of proteins, has proved to be extremely valuable in this connection.

Use of Agarose-ε-amino-*n*-caproyl-D-tryptophan Methyl Ester

Cuatrecasas *et al.*[8] were able to selectively adsorb chymotrypsin to Sepharose gels containing covalently bound ε-amino-*n*-caproyl-D-trypto-

[1] J. H. Northrop and M. Kunitz, *J. Gen. Physiol.* **16**, 267 (1932).
[2] M. Kunitz and J. H. Northrop, *J. Gen. Physiol.* **18**, 433 (1935).
[3] M. Kunitz and J. H. Northrop, *J. Gen. Physiol.* **19**, 991 (1936).
[4] C. F. Jacobsen, *C. R. Trav. Lab. Carlsberg* **25**, 325 (1947).
[5] H. T. Wright, J. Kraut, and P. E. Wilcox, *J. Mol. Biol.* **37**, 363 (1968).
[6] D. D. Miller, T. A. Horbett, and D. C. Teller, *Biochemistry* **10**, 4641 (1971).
[7] P. J. Keller, E. Cohen, and H. Neurath, *J. Biol. Chem.* **233**, 344 (1958).
[8] P. Cuatrecasas, M. Wilchek, and C. B. Anfinsen, *Proc. Nat. Acad. Sci. U.S.* **61**, 636 (1968).

phan methyl ester. Whereas D-tryptophan methyl ester might be considered a logical competitive inhibitor for attachment to Sepharose ($K_i \sim 10^{-4}\,M$), the above studies[8] showed that other considerations governed the relative affinity of the enzyme for the immobilized inhibitor. Thus, α-chymotrypsin, which was not retarded on unsubstituted Sepharose, was retarded but not adsorbed on Sepharose-D-tryptophan methyl ester columns. All the enzyme applied to the column could be recovered by passage of the initial buffer (pH 8.0) used in the equilibration of the column. In contrast, the enzyme was quantitatively adsorbed at pH 8.0 when applied to a column of Sepharose-ε-amino-n-caproyl-D-tryptophan methyl ester. The adsorption was selective in that chymotrypsinogen, subtilisin, DIP-trypsin,[9] and pancreatic ribonuclease were not bound. The recovery of adsorbed α-chymotrypsin could be accomplished by elution with acidic solutions. These observations show clearly that the use of a competitive inhibitor alone is insufficient to achieve efficient separation but that the distance between the binding group and the gel matrix influences the extent of selectivity that can be attained. In this case the interposing of a six-carbon "arm" between the gel surface and the binding group is apparently sufficient to minimize steric interference with the binding processes.

The use of Sepharose-ε-amino-n-caproyl-D-tryptophan methyl ester for the isolation and purification of chymotrypsin(s) has become a routine technique because of the simplicity and rapidity of the procedure. The resourcefulness of the technique was demonstrated in the purification of chymotrypsin-P, the enzyme obtained by papain activation of chymotrypsinogen. Although the ability of papain to activate chymotrypsinogen was observed as early as 1966,[10] chromatography on a column of Sepharose-ε-amino-n-caproyl-D-tryptophan methyl ester provided a single-step method for the isolation and purification of chymotrypsin-P and thus permitted its characterization.[11-13]

Use of Agarose-4-phenylbutylamine

The possibility of coupling inhibitors other than ε-amino-n-caproyl-D-tryptophan methyl ester to Sepharose was considered by Stevenson and Landman,[14] who studied the properties of Sepharose-4-phenylbutylamine

---

[9] The following abbreviations are used: DIP, diisopropylphosphoryl-; DFP, diisopropylfluorophosphate; TPCK, α-N-tosyl-L-phenylalanyl chloromethyl ketone.
[10] T. Kay, M.Sc. Thesis, University of Waterloo (1966).
[11] M. C. Shaw, T. Kay, and T. Viswanatha, *Can. J. Biochem.* **48**, 1208 (1970).
[12] M. C. Shaw and T. Viswanatha, *Can. J. Biochem.* **49**, 999 (1971).
[13] M. C. Shaw and T. Viswanatha, *Can. J. Biochem.* **50**, 846 (1972).
[14] K. J. Stevenson and A. Landman, *Can. J. Biochem.* **49**, 119 (1971).

columns. The choice of this inhibitor was based on the known binding of the phenyl moiety to chymotrypsin[15] and on the possible minimization of steric effects by virtue of the separation of the Sepharose attachment site (amino group) from the inhibitor binding site (phenyl group) by a four-carbon side chain. Although not as long an "arm" as that in ϵ-amino-n-caproyl-D-tryptophan methyl ester, it was felt that the presence of two additional methylene groups compared to D-tryptophan methyl ester would provide the minimum structural requirements needed for adsorption of the enzyme. Studies with Sepharose-4-phenylbutylamine columns were in accordance with these expectations. Thus, the affinity column was capable of adsorbing α-chymotrypsin and chymotrypsin B from solutions at pH 8.0 with recovery of these enzymes by elution at pH 3.0. The selectivity of the adsorbent was further confirmed by its inability to either adsorb or retard porcine trypsin, bovine trypsinogen, and α-chymotrypsin inhibited with TPCK[9] or DFP.[9] A slight retardation of bovine trypsin and chymotrypsinogen A was observed, whereas in the case of subtilisin, separation of two major components eluting with the pH 8.0 buffer and one minor component eluting at pH 3.0 was achieved. The Sepharose-4-phenylbutylamine columns were found to be very effective in the isolation of chymotrypsin-like proteases from crude extracts of moose pancreas.

The high concentration of 4-phenylbutylamine bound to Sepharose appeared to overcome the potential inadequacy of its being a weaker competitive inhibitor ($K_i = 2.8$ mM) relative to other inhibitors of the enzyme. These observations suggest that although inhibitors with K_i values of the order of $10^{-6} M$ and below are desirable, they are not absolutely essential for the preparation of efficient affinity columns. The use of 4-phenylbutylamine has the advantage of obviating the necessity for ϵ-amino-n-caproyl-D-tryptophan methyl ester which, despite the well-documented procedures for its synthesis,[6,8,11] could prove to be time-consuming and tedious to prepare.

Thus either of the two compounds, ϵ-amino-n-caproyl-D-tryptophan methyl ester or 4-phenylbutylamine can serve as a competitive inhibitor for attachment to activated Sepharose to produce an efficient system for the isolation of chymotrypsins.

Miscellaneous Observations

A note of caution appears essential at this juncture in the light of more recent observations. The primary consideration in the choice of a

---

[15] R. Baker, *in:* "Active-Site-Directed Irreversible Enzyme Inhibitors." Wiley, New York, 1967.

competitive inhibitor is its ability to bind specifically to the enzyme under consideration. Ideally, the inhibitor should have little or no affinity for other proteins. However, in the case of enzymes which exhibit a certain degree of overlap in their specificities, one might expect that these proteins would possess similar behavior toward a given affinity column. The isolation of stem bromelain[16] and the adsorption of lungfish procarboxypeptidase A and chymotrypsinogen[17] on Sepharose-ε-amino-n-caproyl-D-tryptophan methyl ester columns serve to illustrate this point. Likewise,[18] bovine trypsin and chymotrypsin were found to adhere to the Sepharose-ε-amino-n-caproyl-D-tryptophan column used to separate carboxypeptidases A and B.[19] This problem was overcome by the prior passage of the crude preparation through another affinity column, containing soybean trypsin inhibitor attached to Sepharose. The effectiveness of this adsorbent in the retardation of both trypsin and chymotrypsin had earlier been demonstrated by Feinstein,[20] who also showed that these two enzymes behaved similarly toward Sepharose-turkey ovomucoid. (In contrast, Sepharose-chicken ovomucoid was found to be selective for bovine, porcine, and dogfish trypsin.[21] As a further example of the multiple adsorption of proteins, the high-capacity adsorbent N-(2-ethylcellulose)-glycyl-D-phenylalanine developed by Uren[22] was shown to retard the proteolytic enzymes trypsin, chymotrypsins A and B, and carboxypeptidases A and B. However, conditions whereby separation of these proteins could be achieved were also described. Finally, the affinity column described by Ashani and Wilson[23] in the purification of acetylcholinesterase was capable of retarding α-chymotrypsin and possibly other serine esterases.

A more serious problem stems from our finding that certain preparations of both Sepharose-ε-amino-n-caproyl-D-tryptophan methyl ester and of Sepharose-4-phenylbutylamine exhibit marked variation in specificity. Whether this difference arises out of variations in the procedures used in the preparation of the gels or whether the condition manifests itself upon aging of the columns has yet to be ascertained. Nevertheless, both of the columns were found to adsorb chymotrypsinogen A, DIP-chymotrypsin and TPCK-chymotrypsin, as well as active α-chymotrypsin. The

[16] D. Bobb, *Prep. Biochem.* **2**, 347 (1972).
[17] G. R. Reeck and H. Neurath, *Biochemistry* **11**, 3947 (1972).
[18] G. R. Reeck, K. A. Walsh, and H. Neurath, *Biochemistry* **10**, 4690 (1971).
[19] G. R. Reeck, K. A. Walsh, M. A. Hermodson, and H. Neurath, *Proc. Nat. Acad. Sci. U.S.* **68**, 1226 (1971).
[20] G. Feinstein, *Biochim. Biophys. Acta* **214**, 224 (1970).
[21] N. C. Robinson, R. W. Tye, H. Neurath, and K. A. Walsh, *Biochemistry* **10**, 2743 (1971).
[22] J. R. Uren, *Biochim. Biophys. Acta* **236**, 67 (1971).
[23] Y. Ashani and I. B. Wilson, *Biochim. Biophys. Acta* **276**, 317 (1972). See also Voss *et al.*, this volume [74].

adsorbed proteins could be recovered by elution with dilute acidic solutions in the usual manner. It is pertinent to point out that these same columns consistently rejected samples of trypsin, lysozyme, ribonuclease, and subtilisin, as expected. Similar observations have been noted by Schecter.[24]

The preparation and use of Sepharose-ε-amino-n-caproyl-D-tryptophan methyl ester and Sepharose-4-phenylbutylamine is outlined below. These procedures have been found to give satisfactory results and are essentially the same as those outlined previously.[8,14,25]

Preparation of Agarose-ε-amino-n-caproyl-D-tryptophan Methyl Ester[8,25]

A given volume of Sepharose 4B that had been decanted, is mixed with an equal volume of water and 100 mg of finely divided solid cyanogen bromide is added per milliliter of packed Sepharose while the mixture is stirred. The pH is immediately raised to, and maintained at, 11.0 by titration with 2–8 M NaOH. The temperature is maintained at about 20° by the addition of pieces of ice as required. Upon completion of the reaction (8–12 minutes, as indicated by the cessation of base uptake), the Sepharose is washed on a Büchner funnel with about 20 volumes of cold 0.1 M NaHCO$_3$ at pH 9.0.

The washed Sepharose is immediately suspended in cold 0.1 M NaHCO$_3$ at pH 9.0, in a volume equal to that of the original Sepharose. The inhibitor, ε-amino-n-caproyl-D-tryptophan methyl ester, is quickly added in a solution representing 5–15% of the final volume. The amount of inhibitor used is typically about 65 μmoles per milliliter of Sepharose. The mixture is stirred gently at 4° for 24 hours, after which it is washed extensively with water and buffer until no more free inhibitor is removed. The material can be stored in 50 mM Tris chloride at pH 8.0.

The operational capacity for the adsorption of protein can be determined by adding successively small amounts of enzyme to the column until enzyme activity is detected in the eluate. The total amount of enzyme which is subsequently eluted with 0.1 M acetic acid, pH 3.0, is considered to be the operational capacity.

Preparation of Agarose-4-phenylbutylamine[14]

Cyanogen bromide activation of Sepharose is carried out as described above. To the suspension of washed, activated Sepharose in 0.1 M NaHCO$_3$ at pH 9.0 is added a 20% solution of 4-phenylbutylamine in ethanol in an amount equal to about 25% of the volume of Sepharose

[24] I. Schecter, The Weizmann Institute of Science, personal communication.
[25] P. Cuatrecasas and C. B. Anfinsen, this series, Vol. 22, p. 345.

suspension. The ethanol concentration is increased to about 35% in order to maintain the inhibitor in solution, and the coupling is allowed to proceed at 5° for 16 hours. The resultant Sepharose-4-phenylbutylamine is washed extensively with the bicarbonate buffer containing 30% ethanol and stored in 50 mM Bicine buffer, pH 8.0; 50 mM Tris chloride may be used in place of Bicine.

Illustrative Procedure: Purification of Commercial
  α-Chymotrypsin

A column of the given affinity gel is equilibrated with 50 mM Tris chloride at pH 8.0 and room temperature. The enzyme sample is applied to the column in a minimum volume of the same buffer, and the equilibrating buffer is passed at a flow rate of 30–60 ml per hour until no more material absorbing at 280 nm is eluted. Typically, the eluted material amounts to 5–10% of the initial protein applied to the column and is free of chymotryptic activity. The buffer is then changed to 0.1 M acetic acid at pH 3.0, and the remaining protein is rapidly eluted. In order to minimize autolysis, the amount of pH 8.0 buffer should be the smallest amount practically possible. Miller *et al.*[6] have reported improved recovery of activity if the enzyme is applied to the column in a small volume of 1 mM HCl, the starting buffer contains 50 mM CaCl$_2$, and the column is run at between 2 and 6°.

[47] Vertebrate Collagenases

By ARTHUR Z. EISEN, EUGENE A. BAUER, GEORGE P. STRICKLIN, and JOHN J. JEFFREY

Gallop and associates[1] were the first to recognize the high affinity of a collagenase for its substrate and to take advantage of this property in the purification of the collagenase from *Clostridium histolyticum*. The enzyme was allowed to bind to native collagen fibrils and was subsequently recovered after digestion of the substrate and release of the enzyme. Similar use of collagen fibrils for purification of vertebrate tissue collagenase is much less effective, since the limited cleavage of collagen catalyzed by these enzymes[2] makes recovery of the enzyme from the reaction mixture difficult. To overcome this problem, collagen in solution has

[1] P. M. Gallop, S. Seifter, and E. Meilman, *J. Biol. Chem.* **227**, 891 (1957).
[2] A. Z. Eisen, E. A. Bauer, and J. J. Jeffrey, *J. Invest. Dermatol.* **55**, 359 (1970).

been employed for coupling to a solid matrix (agarose) for use in affinity chromatography.[3]

Sources of Collagenase. Vertebrate collagenases, including those from human and animal sources, can be obtained from tissue culture using well-established techniques.[2] Collagenase activity is measured throughout all stages of purification by the enzymatic release of soluble [^{14}C]glycine-containing peptides from native, reconstituted guinea pig skin collagen fibrils.[4]

Affinity Chromatography. Collagen was coupled to agarose (Sepharose 4B) according to the method described for other ligands.[5] In a typical preparation, approximately 25 ml of cyanogen bromide-activated Sepharose 4B was suspended in 50 ml of 0.2 M NaHCO$_3$ at pH 9.0 and 75 mg of highly purified native guinea pig skin collagen, prepared by the method of Gross[6] in 25 ml of 0.4 M NaCl, was added immediately. The mixture was stirred gently for 18 hours at 4°, filtered, washed with water, and equilibrated with 50 mM Tris chloride at pH 7.5 containing 5 mM CaCl$_2$. The efficiency with which native guinea pig skin collagen is coupled to Sepharose was determined by measuring the hydroxyproline content of the collagen Sepharose beads at the end of a reaction. In every case, 70–95% of the collagen should be coupled to Sepharose under the above conditions, producing a slurry which contains approximately 1.8 mg collagen per milliliter of packed Sepharose.

Affinity chromatography was carried out at 4° by applying 5–10 mg of partially purified enzyme protein to a column (1.2 × 3 cm) containing collagen-Sepharose which had been equilibrated with 50 mM Tris chloride–5 mM CaCl$_2$, pH 7.5. For greater amounts of enzyme protein, larger columns were used; adequate flow rates were maintained by increasing the column diameter. The starting buffer was passed through the column until the absorbance at 280 nm returned to the base line. Elution was accomplished using the same buffer containing 1 M NaCl.

Harsher methods of elution, such as lowering the pH, have been used with other enzymes, but the lability of most human and animal collagenases at acid pH makes such an approach hazardous. Unbound enzyme protein can be subjected to rechromatography on collagen-Sepharose until no further enzyme activity remains. This is usually done after the first peak is lyophilized and reconstituted in starting buffer, but this step

[3] E. A. Bauer, J. J. Jeffrey, and A. Z. Eisen, *Biochem. Biophys. Res. Commun.* 44, 813 (1971).
[4] Y. Nagai, C. M. Lapiere, and J. Gross, *Biochemistry* 5, 3123 (1966).
[5] P. Cuatrecasas, M. Wilchek, and C. B. Anfinsen, *Proc. Nat. Acad. Sci. U.S.* 61, 636 (1968).
[6] J. Gross, *J. Exp. Med.* 107, 247 (1958).

can be performed without reducing the volume of the unbound enzyme peak.

Although collagenase can be obtained by passing crude enzyme preparations through collagen-Sepharose, a number of contaminating proteins also bind to the matrix. Thus, the use of affinity chromatography to obtain pure collagenase must be accomplished by initial partial purification of these enzymes to reduce the likelihood of nonspecific adsorption of components present in the crude mixtures. Usually, ammonium sulfate fractionation, gel filtration on Sephadex G-150, and DEAE chromatography have been employed as preliminary steps.[3] With appropriate starting solutions, the enzymatically active eluent fraction usually appears as a single band on polyacrylamide gel electrophoresis although minor contaminating protein bands may be present occasionally. In most instances these can be removed by rechromatography on freshly prepared collagen-Sepharose.

Unfortunately, the yields of homogeneous enzyme using affinity chromatography are low. Starting with 10 mg of partially purified collagenase, a single chromatographic purification may yield no more than 0.5 mg of the desired enzyme. In addition, there is a considerable loss of enzyme activity from the highly purified preparations which are rerun on collagen-Sepharose, and usually only about 2% of the material applied to the column is recovered.

In spite of some of the difficulties encountered, this approach has permitted sufficient electrophoretically pure collagenase to be obtained from human skin, rheumatoid synovium, and tadpole skin to produce functionally monospecific antisera in rabbits.[7-9]

A major limitation of this method of purification is the rather small number of rigid collagen molecules which can be spatially accommodated on an agarose bead. A yield of approximately 2 mg of collagen per packed milliliter of Sepharose appears to be about the maximum amount of this protein that can be coupled to agarose. In addition, attempts to make use of side arms to increase the extent of binding of the collagen to the matrix have been unsuccessful.

Several modifications of the method have been used in an effort to improve the yield of vertebrate collagenases. These include: (1) the use of purified alpha chains from denatured collagen, in the hope that greater molar coupling could be achieved by eliminating the restrictions imposed by the rodlike structure of native collagen. Indeed, when purified alpha$_1$

[7] E. A. Bauer, A. Z. Eisen, and J. J. Jeffrey, *J. Clin. Invest.* **50**, 2056 (1971).

[8] A. Z. Eisen, E. A. Bauer, and J. J. Jeffrey, *Proc. Nat. Acad. Sci. U.S.* **68**, 248 (1971).

[9] E. A. Bauer, A. Z. Eisen, and J. J. Jeffrey, *J. Biol. Chem.* **247**, 6679 (1972).

chains from guinea pig skin collagen were employed as a ligand, yields of 4–5 mg protein per milligram of packed Sepharose were obtained.

(2) The use of the cyanogen bromide peptide alpha$_1$-CB-7 obtained from the purified alpha$_1$ chain of rat skin collagen. This peptide was chosen as a ligand for coupling to Sepharose since it contains the site at which most vertebrate collagenases initiate their cleavage of the collagen molecule. Here too, it was reasoned that this property might enhance the specific binding of collagenase to its substrate and increase the amount of enzyme bound to the Sepharose.

Partially purified human skin collagenase, when subjected to affinity chromatography on either alpha$_1$-Sepharose or alpha$_1$-CB7-Sepharose under the same conditions as described above results in preparations purified some 3- to 4-fold with a yield of approximately 20%. However, the eluted enzyme peak still contains numerous contaminating bands when examined on polyacrylamide gel electrophoresis.

Undoubtedly the failure of alpha$_1$-Sepharose or alpha$_1$-CB-7-Sepharose to yield pure collagenase preparations is, in part, due to the fact that neither ligand possesses the native triple helical collagen structure. Thus, both are potential substrates for other, less specific proteases. Nevertheless, these ligands may be useful as an initial purification step in which yield is more important than specificity.

(3) The entrapment of collagen in a polyacrylamide matrix. This method[10] has been used for the purification of tadpole collagenase. Purified collagen was treated with Proctase[11] to remove the nonhelical (telopeptide) amino-terminal region of the collagen molecule in an effort to reduce the binding on noncollagenolytic proteases to the collagen molecules entrapped in polyacrylamide. In these studies crude tadpole collagenase preparations were pretreated with hyaluronidase and ribonuclease to improve the purity of the final enzyme preparation. Affinity chromatography was performed using 5% polyacrylamide gels containing approximately 58 mg of entrapped collagen. The collagenase was eluted with sodium acetate at pH 5.2 containing 0.2 M NaCl and 5 mM CaCl$_2$. This technique resulted in an approximate 300-fold purification of the tadpole collagenase with a yield of 69%.[10]

In our experience the use of pH 5.2 acetate buffer, rather than a neutral pH buffer, to elute human collagenases results in a loss of approximately 80% of the enzyme activity. In addition, regardless of the mode

---

[10] Y. Nagai and H. Hori, *Biochim. Biophys. Acta* **263**, 564 (1972).
[11] Proctase is an acid proteinase obtained from the broth filtrate of cultures of *Aspergillus niger* var. *macrosporus*. Presumably, the nonhelical amino terminus of the collagen molecule could be removed just as well by using pepsin.

of elution the enzyme preparations obtained contain multiple bands on polyacrylamide gel electrophoresis. Although this method may provide an excellent initial purification step, it must be followed by chromatography on collagen-Sepharose or another chromatographic procedure to obtain highly purified collagenase.

Acknowledgment

This work was supported by U.S. Public Health Service Research Grants AM 12129, AM 05611, and HD 05291.

[48] Plasminogen

By B. A. K. Chibber, D. G. Deutsch, and E. T. Mertz

Plasminogen is the zymogen of the proteolytic enzyme plasmin (EC 3.4.4.14), the enzyme responsible for the dissolution of fibrin clots in blood. In a continuing effort to resolve problems of homogeneity, yield, stability, reproducibility, solubility at physiological pH, and isolation of the native form of the zymogen, various methods have evolved over the three decades following the original euglobulin precipitation procedure for concentrating plasminogen from plasma.[1] These efforts culminated in combining gel filtration and ion-exchange chromatography techniques as the procedures of choice.[2,3] The development of affinity chromatography in recent years[4] provided an attractive alternative to conventional multi-step purification procedures as a gentle and effective technique for the purification of plasminogen.

The application of affinity methods to the purification of plasminogen was based upon the ability of this zymogen to bind various aliphatic ω-amino acids.[5-9] These amino acids inhibit the activation of the zymo-

[1] H. A. Milstone, *J. Immunol.* **42**, 109 (1941).
[2] K. C. Robbins, L. Summaria, D. Elwyn, and G. H. Barlow, *J. Biol. Chem.* **240**, 541 (1965).
[3] K. C. Robbins and L. Summaria, this series, Vol. 19, p. 184.
[4] P. Cuatrecasas and C. C. Anfinsen, this series, Vol. 22, p. 345.
[5] N. Alkjaersig, A. P. Fletcher, and S. Sherry, *J. Biol. Chem.* **234**, 832 (1959).
[6] F. B. Ablondi, J. J. Hagan, M. Phillips, and E. C. DeRenzo, *Arch. Biochem. Biophys.* **82**, 153 (1959).
[7] F. Markwardt, H. Landmann, and A. Hoffman, *Hoppe-Seyler's Z. Physiol. Chem.* **340**, 174 (1965).
[8] M. Maki and F. K. Ellery, *Thromb. Diath. Haemorrh.* **16**, 668 (1966).
[9] Y. Abiko, M. Iwamoto, and M. Tomikawa, *Biochim. Biophys. Acta* **185**, 424 (1969).

gen,[10,11] and thus possess antifibrinolytic activity. Among the better known of these inhibitors is ε-aminocaproic acid, which has found considerable clinical application as an antifibrinolytic agent.[10,11] The first applications of affinity chromatography for the purification of plasminogen[12,13] used an L-Lysine substituted agarose to obtain the zymogen from human plasma. Lysine was used as the ligand on the basis of structural similarities between the covalently linked lysine (presumably through the α-amino group), and ε-aminocaproic acid. Lysine-agarose was shown to specifically bind plasminogen from blood plasma, and the retarded zymogen was eluted with ε-aminocaproic acid. This single step purification of plasminogen provided a preparation comparable in purity and homogeneity to those obtained by conventional multistep procedures, and was the first reported preparation of a proenzyme by affinity chromatography.

Affinity Chromatography on Lysine-Agarose

The following methods are routinely employed in our laboratory for the purification of plasminogen by affinity chromatography on lysine-agarose.

Preparation of Lysine-Agarose. Agarose (Sepharose 4B) is activated with CNBr according to the following procedure.[4] One hundred milliliters of settled agarose is washed with distilled water on a Büchner funnel to remove the azide preservative, and suspended in 100 ml of distilled water; 200 ml of a 5% solution of CNBr (Mathieson, Coleman and Bell) is added to the agarose suspension and the pH is adjusted to, and maintained at, 11 by titration with 4 M NaOH. The reaction is allowed to proceed at room temperature for about 9 minutes before being terminated by washing with 1.9 liters of ice cold 0.1 M NaHCO₃ on a Büchner funnel, using a 10-μm nylon netting as the filter (Nylon Net Corporation, Memphis, Tennessee).

A 50-ml solution containing 20 g of L-lysine monohydrochloride (Sigma Chemical Co.), adjusted to pH 8.9 with 4 M NaOH, is then added at room temperature to a 150-ml suspension (0.1 M NaHCO₃, pH 8.9) containing 100 ml of activated agarose. The slurry is gently stirred for 24 hours at 5°. The removal of uncoupled lysine from the reaction mix-

[10] S. Okamoto, S. Oshiba, H. Mihaka, and U. Okamoto, *Ann. N.Y. Acad. Sci.* **146**, 414 (1968).

[11] L. Skoza, A. O. Tse, M. Semar, and A. J. Johnson, *Ann. N.Y. Acad. Sci.* **146**, 659 (1958).

[12] D. G. Deutsch and E. T. Mertz, *Fed. Proc., Fed. Amer. Soc. Exp. Biol.* **29**, 647 (1970).

[13] D. G. Deutsch and E. T. Mertz, *Science* **170**, 1095 (1970).

ture is effected by sequentially washing the suspension on a Büchner funnel with approximately 1 liter of 1 M NaCl and 3 liters of water. The lysine-agarose slurry is stored in 0.02% sodium azide and kept in the refrigerator. Under these conditions the preparation has been stable for more than two years with no loss of plasminogen binding activity. The same procedure scaled up 3 times has been employed for the preparation of 300 ml of lysine-agarose. The amount of lysine bound to agarose by this procedure is 55 μmoles per milliliter of settled agarose, as determined by amino acid analysis.

Preparation of Plasma. If freshly collected whole blood is employed, it should be collected in an anticoagulant solution and centrifuged at 4° for 15 minutes at 5000 g. A second centrifugation at higher speed (we use 27,000 g at 4°) usually removes more insoluble material. The resulting cold plasma is filtered through fluted filter paper to remove insoluble lipid materials which accumulate on the surface during the previous centrifugation. If freshly frozen plasma is available from a blood bank or other source, it should be thawed and centrifuged at a high speed for 15 minutes and filtered to remove lipid materials. Whole outdated human blood is treated exactly like freshly collected blood. In this case, the second centrifugation is necessary to remove large amounts of cellular debris.

Chromatography Procedure. A column (2.5 × 30 cm) containing 50 ml of lysine-agarose is equilibrated with 0.1 M sodium phosphate at pH 7.4. Plasma, 300–400 ml, is passed through the column at a flow rate of about 75 ml per hour, and the column is washed with 0.3 M sodium phosphate at pH 7.4 until the absorbancy at 280 nm is less than 0.05. This washing step takes at least 6 hours at a rate of about 175 ml per hour, and it is recommended that it be carried out overnight. The plasminogen is eluted from the column as a sharp peak with 0.1 M ϵ-aminocaproate at pH 7.4 (Nutritional Biochemicals Corporation), at a flow rate of about 100 ml per hour. For large-scale preparations, batchwise experiments are performed by stirring 1 liter of plasma with 50 ml of a lysine-agarose suspension for 2 hours at room temperature, or overnight at 4°. The slurry is then filtered and given a preliminary wash with approximately 2 liters of 0.3 M sodium phosphate at pH 7.4 on a Büchner funnel. The lysine-agarose is then packed on a column and treated as described above.

Separation of ϵ-Aminocaproic Acid. ϵ-Aminocaproate is removed from the plasminogen by gel filtration on Sephadex G-25 in the cold. The bed volume of the G-25 should be at least six times the volume of the applied sample. The column is equilibrated and eluted with 50 mM NH$_4$HCO$_3$ at pH 8.0. Lyophilization is used to remove the solvent as well as the

NH_4HCO_3, which sublimes. A salt-free plasminogen preparation results which may be stored at $-20°$.

Modifications and Variations

The basic procedure using lysine as an affinity absorbent for the purification of plasminogen from plasma has found wide acceptance, and has been used by a large number of workers.[14-24] The procedure has been altered and modified in different laboratories to suit various experimental conditions. Lysine has been insolubilized on polyacrylamide[25] as well as Sepharose 6B,[26] and is an equally effective ligand in otherwise identical procedures. The source of plasminogen has varied from undiluted plasma to plasminogen preparations at various stages of purification by conventional methods.

The entire procedure may be carried out in the cold room, with slight changes in the ionic strength of the buffers used. Cold room operations for the washing and elution steps eliminates the possibility of plasmin contamination due to spontaneous activation of the zymogen by plasma activators.

When the procedure is to be performed in the cold, the 0.3 M phosphate buffer used in the washing step may be replaced by a buffer of lower ionic strength. This prevents the possible loss of any retarded plasminogen.[27] Buffers that have been used are 0.01 M sodium phosphate–0.14 M NaCl (pH 7.4),[22] 0.2 M sodium phosphate,[27] and 0.1 M sodium phosphate.[15] Sodium phosphate, 0.1 M, has also been successfully employed for the washing step at room temperature.[14]

In the elution step, instead of using 0.2 M ε-aminocaproate, adjusted

[14] D. Collen, E. B. Ong, and A. J. Johnson, *Fed. Proc., Fed. Amer. Soc. Exp. Biol.* **31**, 229 (1972).

[15] L. Summaria, L. Arzadon, P. Bernabe, and K. C. Robbins, *J. Biol. Chem.* **247**, 4691 (1972).

[16] M. J. Weinstein and R. F. Doolittle, *Biochim. Biophys. Acta* **258**, 577 (1972).

[17] A. P. Kaplan and K. F. Austin, *J. Exp. Med.* **136**, 1378 (1972).

[18] K. Nagendra, N. Reddy, and G. Markus, *J. Biol. Chem.* **247**, 1683 (1972).

[19] M. Steinbuch, *Nouv. Rev. Fr. Hematol.* **11**, 650 (1971).

[20] S. V. Pizzo, M. L. Schwartz, R. L. Hill, and P. A. McKee, *J. Clin. Invest.* **51**, 2841 (1972).

[21] W. J. Brockway and F. J. Castellino, *J. Biol. Chem.* **246**, 4641 (1971).

[22] T. H. Liu and E. T. Mertz, *Can. J. Biochem.* **49**, 1055 (1971).

[23] L. A. Harker, G. Schmer, and S. J. Slichter, Third Congress on Thrombosis (Abstracts) Washington, D.C., 1972.

[24] D. Collen, E. B. Ong, and A. J. Johnson, submitted for publication.

[25] E. E. Rickli and P. A. Cuendet, *Biochim. Biophys. Acta* **250**, 447 (1971).

[26] K. Nagendra, N. Reddy, and G. Markus, *J. Biol. Chem.* **247**, 1683 (1972).

[27] L. A. Harker, personal communication.

to pH 7.4 as originally described,[13] most workers have also included a phosphate or a Tris buffer.[14,15,22,25] Lysine at 0.1 M may also be used as an eluent, and this may be desirable under certain circumstances in view of the fact that it is a relatively weak inhibitor of plasminogen activation, and studies on the zymogen have been performed in its presence.[22] However, the lysine can be removed only by extended dialysis against 1 mM HCl whereas dialysis against distilled water yields a preparation with a lysine content 10% in excess of the normal value.[22]

The authors currently use 0.1 M ε-aminocaproate for elution. A major modification of the elution step has been described[21] which entails gradient elution with this ligand. When this procedure is used, the plasminogen elutes in two distinct peaks, both containing activatable zymogen of equal purity. The procedure as originally described involved isolation of plasminogen by affinity chromatography as reported earlier.[13] This preparation was then subjected to gel filtration on Sephadex G-100, and reapplied to a lysine-agarose column. Application of a linear gradient of ε-aminocaproate to this column then eluted the plasminogen in two distinct peaks. These authors have now dispensed with both the gel filtration step and readsorption of the plasminogen preparation on lysine-agarose; they use the ε-aminocaproate gradient as the elution step in the purification.[28]

The removal of ε-aminocaproate by gel filtration on Sephadex G-25 may be replaced by dialysis.[29] However, if gel filtration is to be used, it may be useful to concentrate the plasminogen–ε-aminocaproate solution (especially with large-scale preparations) by using ammonium sulfate precipitation.[15] After removing the ligand, some workers perform a second affinity chromatography step, which seems to result in a preparation with slightly increased activity.[14]

Affinity Chromatography Using Butesin-Agarose

The attachment of lysine to agarose provided an affinity adsorbent for plasminogen by virtue of the structural similarities between agarose-bound lysine and ε-aminocaproic acid. The use of a different inhibitor of plasminogen activation as an affinity adsorbent for the zymogen was explored after the successful trials with lysine-agarose. p-Methylaminobenzoic acid[30] and its aliphatic esters are known to be competitive inhibitors of plasmin. Studies on butyl-p-aminobenzoate (Butesin) indicated that it is an inhibitor of plasminogen activation.[31] This suggested the possibility of using it as an affinity adsorbent for the purification

[28] F. J. Castellino, personal communication.
[29] E. E. Rickli and W. I. Otavski, *Biochim. Biophys. Acta* **295**, 381 (1973).
[30] F. Markwardt, H. Landmann, and P. Walsmann, *Eur. J. Biochem.* **6**, 502 (1968).
[31] R. P. Zolton, Ph.D. Thesis, Purdue Univ., Lafayette, Indiana, 1972.

of plasminogen. The attachment of Butesin to agarose is effected via the amino group by allowing it to react with CNBr-activated agarose.

Preparation of Butesin-Agarose.[31] The procedure is essentially as reported in the literature for the preparation of agarose-4-phenylbutyl-amine.[32] Agarose is activated with CNBr as described.[4] Butesin (Sigma Chemical Co.) (5 g, 26 mmoles) is dissolved in 25 ml of 95% ethanol and added with stirring to 100 ml of a suspension (0.1 M NaHCO$_3$, pH 8.9) containing 50 ml of activated agarose. The reaction mixture is stirred gently at 5° for 16 hours. Unreacted Butesin is removed from the reaction mixture by washing sequentially with 95% ethanol, 2 liters of cold water, and 1 liter of 10 mM sodium phosphate–0.14 M NaCl at pH 7.4 on a Büchner funnel. Based upon the recovery of Butesin in the washings, as determined by absorption at 260 nm, between 80 and 100 μmoles of Butesin are bound per milliliter of settled agarose. The preparation is suspended in the phosphate-saline buffer (pH 7.4) and stored in the refrigerator. It has been stable in this state for 9 months. Plasminogens from sheep plasma euglobulin and from bovine and human plasma have been successfully obtained by affinity chromatography on Butesin-agarose.[33]

A column (2 × 15 cm) containing 20 ml of Butesin-agarose is equilibrated with 10 mM phosphate–0.14 M saline buffer (pH 7.4). Plasma, 50 ml, is passed through the column at a flow rate of 60 ml per hour. The column is then washed with 10 mM sodium phosphate–0.14 M NaCl (pH 7.4) at a flow rate of about 150 ml per hour until the absorbance of the eluate at 280 nm falls below 0.05. The retarded plasminogen is eluted using the same buffer supplemented with 0.1 M lysine. After elution, the column is washed and reequilibrated with the phosphate-saline buffer (pH 7.4). Lysine may be removed from the plasminogen preparations by dialysis against 1 mM HCl for 24 hours. Alternately, 0.1 M acetic acid may be used for elution[33] although any procedure that reduces the pH below 6 may lead to altered forms of plasminogen.[34]

Preparation of Plasmin-Free Plasminogen. Affinity chromatography of plasminogen activation mixtures, containing both plasminogen and plasmin, has demonstrated that Butesin-agarose specifically binds plasminogen and has no affinity for plasmin.[33] This conclusion is based upon the observed time-dependent decrease in plasminogen content of the activation mixtures, along with increased plasmin activity monitored in the unretarded peak using the TAME (tosylarginine methyl ester) colori-

[32] K. J. Stevenson and A. Landman, *Can. J. Biochem.* **49**, 119 (1971).
[33] R. P. Zolton and E. T. Mertz, *Can. J. Biochem.* **50**, 529 (1972).
[34] N. Alkjaersig, *Biochem. J.* **93**, 171 (1964).

metric test for plasmin.[35] In view of this marked specificity for plasminogen, affinity chromatography on Butesin-agarose has been developed as a routine clinical procedure for the determination of plasminogen levels in blood.[36]

To date, preparation of plasminogen using affinity chromatography on Butesin-agarose has been restricted to our laboratories, primarily owing to the wide acceptance of the earlier procedure using lysine-agarose. However, it was recognized very early that lysine-agarose, under the conditions of the affinity procedure, contains charged carboxyl and amino groups. It is therefore possible that the lysine-agarose gel acts as an ion exchange matrix, causing retardation and elution of contaminants along with plasminogen. While plasminogen preparations using lysine-agarose are now known to be of the highest purity, this argument nevertheless was a major factor influencing the choice of Butesin, which contains no charged groups, as an affinity adsorbent for plasminogen.[31] The specificity of Butesin-agarose for binding plasminogen, particularly with the exclusion of any plasmin contamination makes this method an effective alternative to lysine-agarose. It is possible that a wider application of this procedure would lead to innovations and modifications, and produce improved methods for the isolation of plasminogen by affinity chromatography.

Comments

A comparison of yields and specific activities of plasminogen prepared by workers using affinity chromatography is shown in the table. Yields in all these preparations range between 70 and 90% (between 9 and 12 mg of plasminogen per 100 ml of plasma). This is a substantial improvement over conventional methods which have yields between 14 and 59%. It is difficult to compare specific activities of all the preparations due to a variation in assay procedures. The National Heart Institute's Committee on Thrombolytic Agents (CTA), Subcommittee for Standardization, has suggested a standard caseinolytic assay for plasminogen, with activity expressed in CTA units.[37] The assay is being adopted by most workers, and its adoption by others is recommended in the interests of standardizing results from different laboratories.

Plasminogen preparations by affinity chromatography are comparable to the purest zymogen obtained by conventional methods.[15] Both types

[35] R. J. Wulf and E. T. Mertz, *Can. J. Biochem.* **47**, 927 (1969).
[36] R. P. Zolton, E. T. Mertz, and H. T. Russell, *Clin. Chem.* **18**, 654 (1972).
[37] A. J. Johnson, D. L. Kline, and N. Alkjaersig, *Thromb. Diath. Haemorrh.* **21**, 259 (1969).

YIELD AND SPECIFIC ACTIVITY OF PLASMINOGEN
PREPARED BY AFFINITY CHROMATOGRAPHY

References	Starting fraction	Yield (%)	Yield (mg/100 ml plasma)	Specific activity (CTA units/ mg)
a, b	Outdated frozen or fresh frozen plasma	80	9	17–22
c	Fresh frozen	85–90	10–12	20–24
d	Fresh plasma	85	10	
e	Outdated human plasma	—	—	34
f	Outdated frozen plasma	75	12	17–22
g	Cohn III	70	—	20
h	Cohn III fraction	85	—	—

[a] D. G. Deutsch and E. T. Mertz, *Science* **170**, 1095 (1970).
[b] D. G. Deutsch, Ph.D. Thesis, Purdue University, Lafayette, Indiana, 1972.
[c] D. Collen, E. B. Ong, and A. J. Johnson, *Fed. Proc., Fed. Amer. Soc. Exp. Biol.* **31**, 229 (1972).
[d] E. E. Rickli and P. A. Cuendet, *Biochim. Biophys. Acta* **250**, 447 (1971).
[e] S. V. Pizzo, M. L. Schwartz, R. L. Hill, and P. A. McKee, *J. Clin. Invest.* **51**, 2841 (1972).
[f] R. P. Zolton and E. T. Mertz, *Can. J. Biochem.* **50**, 529 (1972).
[g] T. H. Liu and E. T. Mertz, *Can. J. Biochem.* **49**, 1055 (1971).
[h] W. J. Brockway and F. J. Castellino, *J. Biol. Chem.* **246**, 4641 (1971).

of preparation are homogeneous by all methods except gel electrophoresis, in which they exhibit multiple bands.[13–15,18,22,24–26,29] Studies using conventional purification procedures indicate that these bands represent the native forms of plasminogen existing in plasma.[38] These multiple forms of plasminogen have been isolated by isoelectric focusing methods, and each of them can be activated to plasmin.[15] The two types of plasminogen obtained by gradient elution with ε-aminocaproate on a lysine-agarose affinity column also exhibit multimolecular forms.[39] Preparations of the zymogen using affinity chromatography on fresh or fresh frozen plasma contain glutamic acid as the predominant N-terminal amino acid of the plasminogens.[24,25,29] On the other hand, Cohn fraction III and other commercial fractions seem to contain degraded forms of plasminogen as evidenced by the appearance of valine and lysine as the N-terminal amino acids.[15,22] In light of these results it is suggested that perhaps plasminogen be prepared from fresh or fresh frozen plasma when possible, particularly if structural studies are to be performed on the zymogen.

[38] P. Wallen and B. Winman, *Biochim. Biophys. Acta* **221**, 20 (1970).
[39] J. M. Sodetz, W. J. Brockway, and F. J. Castellino, *Biochemistry* **11**, 4451 (1972).

The plasminogen preparations using affinity chromatography are very stable. In buffer there is no appreciable loss of activity upon storage at 0° for 2 weeks.[25] The preparation is stable at room temperature for as long as 5 days.[13] Plasminogen may be stored in a lyophilized form for extended periods of time at −20°.

Acknowledgment

This work has been supported, in part, by NIH Grant HE 08401.

[49] Purification of Prekallikrein with Arginine-Agarose

By TOMOJI SUZUKI and HIDENOBU TAKAHASHI

The proteolytic enzyme kallikrein is present in mammalian plasma in an inactive form, prekallikrein. Plasma prekallikrein was partially purified from a bovine pseudoglobulin fraction by Nagasawa et al.[1] Rabbit[2] and human[3] plasma prekallikreins were isolated; the former was judged to be homogeneous by analytical polyacrylamide gel electrophoresis, but the latter was still impure. Highly purified bovine prekallikrein has now been isolated in high yield by the use of affinity techniques.[4] In contrast to plasminogen,[5] prekallikrein could not be adsorbed on lysine-Sepharose 4B, but showed high affinity for arginine-Sepharose 4B.

Purification of Bovine Plasma Prekallikrein

Assay

One unit of prekallikrein is defined as the amount of proenzyme which evolves one unit of kallikrein[6] by the action of Hageman factor. One unit of kallikrein hydrolyzes 1 μmole of TAME[7] per minute. Assay techniques include those of Takahashi et al.,[4] McConnell and Mason,[3] and Wuepper and Cochrane.[2]

Preparation of Arginine-Sepharose 4B. To 50 ml of a suspension

[1] S. Nagasawa, H. Takahashi, M. Koida, T. Suzuki, and J. G. G. Schoenmakers, *Biochem. Biophys. Res. Commun.* 32, 644 (1968).
[2] K. D. Wuepper and C. G. Cochrane, *J. Exp. Med.* 135, 1 (1972).
[3] D. J. McConnell and B. Mason, *Brit. J. Pharmacol.* 38, 490 (1970).
[4] H. Takahashi, S. Nagasawa, and T. Suzuki, *J. Biochem.* 71, 471 (1972).
[5] D. G. Deutsch and E. T. Mertz, *Science* 170, 1095 (1970).
[6] M. Yano, S. Nagasawa, and T. Suzuki, *J. Biochem.* 67, 713 (1970).
[7] Abbreviation for *N*-α-tosyl-L-arginine methyl ester.

of Sepharose 4B are added 15 g of cyanogen bromide; the pH of the suspension is maintained at 11.0 with 3.5 N NaOH. After 30 minutes, the suspension is rapidly washed with 1 liter of ice-cold 1% $NaHCO_3$ and mixed with 10 g of L-arginine monohydrochloride. The slurry is stirred for 24 hours at 4° and then washed with 1% $NaHCO_3$. The amount of arginine bound to the matrix may be analyzed with an amino acid analyzer after hydrolysis with 6 N HCl at 110° for 24 hours and was found to contain 1.9 μmoles per 1 ml of Sepharose.

Purification Procedure

The purification procedure is summarized in the table. All steps are performed in a cold room at 2–4°, and all columns are siliconized before use. The DEAE-Sephadex column methods (steps 1 and 2) are convenient for the separation of prekallikrein from Hageman factor. The dialyzed prekallikrein fraction is chromatographed on a column (2.5 × 25 cm) of CM-Sephadex C-50 (step 3), equilibrated with 20 mM sodium phosphate at pH 6.0. After washing the column with 200 ml of the equilibration buffer containing 0.1 M NaCl, proteins are eluted with a linear concentration gradient of NaCl from 0.1 M to 0.2 M. The prekallikrein fraction obtained from CM-Sephadex C-50 column chromatography is concentrated and dialyzed overnight against 2 liters of 40 mM Tris chloride at pH 8.5.

The concentrated prekallikrein fraction is applied to a column (1.5 × 6 cm) of arginine-Sepharose 4B, equilibrated with 40 mM Tris chloride buffer, pH 8.5. The column is washed with 200 ml of the equilibration buffer, then prekallikrein is eluted with 80 ml of the equilibration buffer containing 0.1 M NaCl. When a sample containing 30 mg of protein in 100 ml is applied to the column, a reproducible elution pattern is obtained. Therefore, before the affinity technique, prekallikrein fractions obtained from CM-Sephadex column chromatography are concentrated by adsorption and elution on CM-Sephadex C-50 column (2.5 × 7 cm). Lyophilization causes denaturation of prekallikrein. By the use of affinity technique with arginine-Sepharose 4B, the yield of prekallikrein is considerably increased since the highly purified proenzyme can be obtained from the subfractions of steps 2 and 3 (see footnote *b* of the table). The prekallikrein fractions are pooled and concentrated by ultrafiltration with a Diaflo UM-10 membrane to 4 ml. The concentrated material is passed through a column of Sephadex G-150.

The yield of purified prekallikrein was about 4.6 mg from 7.5 liters of plasma (see the table): a 1330-fold purification over the prekallikrein activity at step 1.

SUMMARY OF PURIFICATION OF BOVINE PREKALLIKREIN
FROM 7.5 LITERS OF PLASMA

Step	Procedures	Total protein (mg)	Total TAME[a] activity (units)	Specific activity (units/mg protein)	Yields (%)
0	Pseudoglobulin fraction	242,000	—	—	—
1	1st DEAE-Sephadex	22,600	872	0.04	100
2	2nd DEAE-Sephadex	1,820	618	0.34	71
3	CM-Sephadex	60.5	567	9.3	65
4	Affinity chromatography	8.30	289	35	33
5	Sephadex G-150	8.10	404[b]	50	46
6	2nd Sephadex G-150	4.60	240	52	28

[a] TAME = N-α-tosyl-L-arginine methyl ester.
[b] Increase of total TAME hydrolytic activity was due to the addition of prekallikrein, which was similarly purified by the affinity technique from subfractions of step 3.

Comments

1. The purified prekallikrein gave a single component on disc gel electrophoresis at pH 8.3[8] and pH 4.5,[9] on SDS-gel electrophoresis,[10] and on isoelectric focusing. Ultracentrifugal analysis of the material similarly disclosed a single component with an $s_{20,w}$ of 5.20 S at 0.68% concentration. The preparation was not contaminated with plasminogen or plasmin; it did not show any fibrinolytic activity even after preincubation of 100 μg of prekallikrein with urokinase. The preparation did not contain Hageman factor. The activities of prothrombin, thrombin, and plasma kininase were also undetectable in a 20-μg sample of this preparation. The molecular weight of bovine prekallikrein was estimated to be about 90,000 by SDS-polyacrylamide gel electrophoresis,[10] 100,000 by gel filtration on a Sephadex G-150 column,[11] and 90,000 by the Archibald method.[12] The isoelectric point determined by isoelectric focusing[13,14] was 6.98.

2. The chromatography of pseudoglobulin on a negatively charged

[8] B. J. Davis, *Ann. N.Y. Acad. Sci.* **121**, 404 (1964).
[9] R. A. Reisfeld, U. J. Lewis, and D. E. Williams, *Nature (London)* **195**, 281 (1962).
[10] A. L. Shapiro, E. Viñuela, and J. V. Maizel, *Biochem. Biophys. Res. Commun.* **28**, 815 (1967).
[11] P. Andrews, *Biochem. J.* **96**, 595 (1965).
[12] W. J. Archibald, *J. Phys. Colloid. Chem.* **51**, 1204 (1947).
[13] U. Matsuo, Y. Horiuchi, and T. Nakamura, *Kagaku No Ryoiki*, Zokan 88, 164 (1968) (in Japanese).
[14] O. Vesterberg and H. Svensson, *Acta Chem. Scand.* **20**, 820 (1966).

matrix such as CM-Sephadex C-50 or phosphocellulose, usually resulted in the spontaneous activation of prekallikrein. After the separation of Hageman factor from the prekallikrein fraction, a CM-Sephadex column can be used for further purification. Since Hageman factor was adsorbed onto arginine-Sepharose, the affinity technique must be used after the removal of Hageman factor from the prekallikrein fraction.

[50] Thermolysin and Other Neutral Metalloendopeptidases

By K. A. WALSH, Y. BURSTEIN, and M. K. PANGBURN

Thermolysin from *Bacillus thermoproteolyticus* and the neutral proteases of *Bacillus subtilis* are neutral metalloendopeptidases having similar size, metal content, amino acid composition, and substrate specificity.[1] Recent evidence[2] indicates that their amino acid sequences have regions of identity, suggesting that these enzymes have diverged from a common ancestor. Problems of self-digestion, which have been encountered in the purification of these and other proteolytic enzymes, can be minimized by rapid procedures of affinity chromatography. Since Cbz-Gly-D-Phe is a competitive inhibitor of both enzymes, affinity adsorbents were developed[3] using Gly-D-Phe covalently attached by appropriate spacers to a Sepharose matrix. The enzymes were selectively adsorbed at low pH and eluted by raising the pH.

Three effective adsorbents have been described[3] which can be used interchangeably for thermolysin and for neutral protease and differ only in the length of their spacers. Preparation of the simplest adsorbent is described below. The more elaborate syntheses of adsorbents 1 and 2 have been detailed in the primary publication.[3]

Preparation of Adsorbent 3

Briefly, CNBr-activated agarose[4] is coupled to a spacer of triethylenetetramine and the terminal amino group is allowed to react with chloroacetyl-D-phenylalanine[5] to generate Sepharose-NH-CH$_2$-CH$_2$-NH-CH$_2$-CH$_2$-NH-CH$_2$-CH$_2$-Gly-D-Phe.

[1] H. Matsubara and J. Feder, *in* "The Enzymes" (P. D. Boyer, ed.), Vol. 3, p. 721. Academic Press, New York, 1971.
[2] M. K. Pangburn, Y. Burstein, H. Neurath, and K. A. Walsh, *Fed. Proc., Fed. Amer. Soc. Exp. Biol.* **32**, 504 (1973).
[3] M. K. Pangburn, Y. Burstein, P. H. Morgan, K. A. Walsh, and H. Neurath, *Biochem. Biophys. Res. Commun.* **54**, 371 (1973).
[4] R. Axén, J. Porath, and S. Ernbäck, *Nature (London)* **214**, 1302 (1967).
[5] J. Uren, *Biochim. Biophys. Acta* **236**, 67 (1971).

Sepharose-4B (200 ml of settled gel) is suspended in an equal volume of water and activated with 40 g of CNBr at 20°. The suspension is maintained at pH 11 by the addition of 12 N NaOH. The activated Sepharose is washed rapidly with 5 volumes of cold water in a chilled fritted-glass funnel. The gel is stirred overnight with an equal volume of 2 M triethylenetetramine at pH 9.0, and excess amine is removed by repeated washing with water. The coupled Sepharose derivative is suspended in 200 ml of 1 M sodium arsenate (pH 11) and stirred with 5 g of chloroacetyl-D-phenylalanine for 20 hours at 50°. After thorough washing, the product of the reaction contains 1–5 μmoles of phenylalanine per milliliter of settled gel.

Enzymatic Assay[6]

Substrate solution containing 1 mM 3-(2-furylacryloyl)-glycyl-L-leucinamide (FAGLA) is prepared as follows. FAGLA (307 mg), obtained from Cyclo Chemical Co., is dissolved in 5 ml of dimethylformamide, and this solution is added to approximately 900 ml of buffer containing 23.8 g of HEPES[7] at pH 7.2 (adjusted with 10 N NaOH). The solution is diluted to 1.0 liter so that it contains 1 mM FAGLA, 0.1 M HEPES, and 0.5% dimethylformamide. A cuvette containing 3 ml of this solution is placed in a thermostatted spectrophotometer (25°), and hydrolysis is initiated by the addition of 25 μl of a solution containing 0.1–2 μg of enzyme. The decrease in absorbance at 345 nm ($\Delta\epsilon_m = 317$) is followed with time. The reaction rate is pseudo first order in substrate concentration and directly proportional to enzyme concentration. Second-order rate constants (k_{cat}/K_m) are 0.60, 0.15 and 0.44 sec^{-1} (mg/ml)$^{-1}$ for thermolysin, and neutral proteases A and B, respectively; extinction coefficients ($E_{280}^{1\%}$, cm^{-1}) for the three proteins are 17.6, 14.8, and 14.7.

Chromatographic Procedures

These procedures are adaptable both to small-scale purifications (1 mg of enzyme) and to large-scale preparations (5 g of dried culture filtrate containing 500 mg of enzyme) by adjusting the column size, the flow rate and the eluting pH. The capacity of an adsorbent containing 1 μmole of ligand per milliliter of settled gel is approximately 1 mg of enzyme per milliliter of gel.

---

[6] J. Feder, *Biochem. Biophys. Res. Commun.* **32**, 326 (1968).
[7] The abbreviations HEPES and MES refer to the buffers N-2-hydroxyethylpiperazine-N'-2-ethanesulfonic acid and 2-(N-morpholino)ethanesulfonic acid, respectively.

a. Thermolysin—Small-Scale Purification

One milligram of thermolysin (available from Daiwa Kasei KK, Osaka, Japan) is dissolved in 0.5 ml of a buffer containing 10 mM HEPES and 5 mM CaCl$_2$ (pH 7). The solution is then applied to a column (1.5 × 5 cm) of adsorbent previously equilibrated with the same buffer. The column is washed with buffer at about 50 ml per hour for 40 minutes, and fractions of 1–2 ml are collected. Inactive protein appears in the first 15 ml of effluent. Active enzyme is eluted at pH 9 by changing eluents to 0.1 M Tris containing 5 mM CaCl$_2$ (pH 9.0).

b. B. subtilis Neutral Protease—Large-Scale Purification

Crude enzyme (5 g of lyophilized acetone precipitate of culture filtrates prepared as described for *Bacillus cereus* neutral protease[8]) is dissolved in 50 ml of 100 mM NaCl and 5 mM calcium acetate at pH 5.0. The pH of the clear, dark brown solution is adjusted to pH 5 with acetate buffer (10% acetic acid brought to pH 4.5 with 10 N NaOH). This avoids the addition of strong acid, which inactivates neutral protease. The enzyme solution is applied to a column of adsorbent (5 × 20 cm) and washed with 3 column volumes of 100 mM NaCl, 5 mM calcium acetate (pH 5), at a flow rate of 200 ml per hour. Elution of the enzyme is accomplished with the pH 9 buffer as described above for thermolysin. To avoid autolysis at this pH, the fractions containing neutral protease should be pooled as soon as possible and adjusted to pH 5 with acetate buffer (pH 4.5) as before. If the pooled fractions contain enzyme at a concentration of 1 mg/ml or greater, the enzyme can be recovered quantitatively by the addition of 2 volumes of acetone at 4°. The precipitate obtained after centrifugation for 5 minutes at 10,000 g is dissolved in 2 mM calcium acetate at pH 5. The neutral protease is most stable when stored frozen in solution at pH 5 and −20°. The enzyme is less stable upon long storage when lyophilized.

c. Separation of Two Neutral Proteases with Different Affinities for the Glycyl-D-Phenylalanine Ligand

Some culture filtrates of *B. subtilis* NRRL B3411 contain two species of neutral protease, differing both in specific activity and in affinity for the Gly-D-Phe adsorbents (Fig. 1). Lyophilized crude enzyme, 200 mg, from *B. subtilis* is dissolved in 2 ml of the buffer described in procedure b above. The enzyme solution at pH 5 is applied to a column (0.9 × 27 cm) of adsorbent equilibrated with the same buffer and washed with

[8] J. Feder, L. Keay, L. R. Garrett, N. Cirulis, M. H. Moseley, and B. S. Wildi, *Biochim. Biophys. Acta* **251**, 74 (1971).

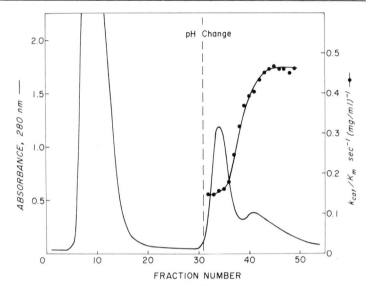

Fig. 1. Affinity chromatography of neutral protease on adsorbent 3 (1.3 μmoles of phenylalanine per milliliter of gel): 200 mg of crude enzyme, 2.6-ml aliquots were collected. Specific activities pertain to the hydrolysis of 3-(2-furylacryloyl)-glycyl-L-leucinamide. At the dashed line, the pH was changed from 5.0 to 6.25 (see text).

3 column volumes of this buffer at a flow rate of 16 ml per hour (10 minutes per fraction). Although the enzymes can be eluted together at pH 9, slower elution at pH 6.25[9] with 50 mM MES, 100 mM NaCl, and 10 mM CaCl$_2$ separates two species of enzyme differing 3-fold in their specific activity (Fig. 1). The two species are pooled separately and concentrated by ultrafiltration in an Amicon cell with a UM-20E membrane.

d. *Regeneration of Columns*

The adsorbent columns are regenerated for reuse by washing with 2 column volumes of the pH 9 buffer, then with sufficient buffer (pH 5 or pH 7) to return the effluent to the desired pH. Columns have been reused as often as 50 times without change in chromatographic behavior.

Comments

Choice of Adsorbent

Since coupling of chloroacetyl-D-phenylalanine to R-NH$_2$ generates R-Gly-D-Phe,[5] the three adsorbents have in common the terminal -Gly-D-

---

[9] The pH of each buffer is adjusted at room temperature; chromatography is done at 4°.

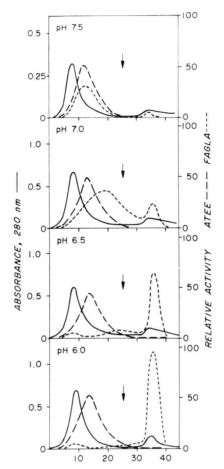

Fig. 2. The effect of pH on the adsorption of neutral protease on adsorbent 1. Crude enzyme (10 mg) was dissolved in 0.1 ml of equilibrating buffer and applied to a column (0.6 × 22 cm) containing adsorbent 1 equilibrated with 100 mM NaCl, 10 mM CaCl$_2$, containing 5 mM Tris (pH 7.5 or 7.0) or 5 mM 2-(N-morpholino)ethanesulfonic acid (pH 6.5 or 6.0). After elution for 1 hour at 25 ml/hour, each column was washed with 100 mM NaCl, 10 mM CaCl$_2$, 50 mM Tris (pH 9.0), as indicated by the arrows. Neutral protease and subtilisin are identified by catalytic activities toward 3-(2-furylacryloyl)-glycyl-L-leucinamide (FAGLA) and acetyl tyrosine ethyl ester (ATEE) [K. Morihara and H. Tsuzuki, *Arch. Biochem. Biophys.* **129**, 620 (1969)].

Phe ligand. The spacers separating Gly-D-Phe from Sepharose differ in length by various combinations of triethylenetetramine, succinic acid, and ethylenediamine. Spacers for adsorbents 1, 2, and 3 are 23, 17, and 9 atoms long, respectively. Only the latter is described here in detail since the longer spacers appear to offer no additional advantage for affinity chromatography of the proteases. A fourth adsorbent has been described with a ligand of -succinyl-D-Leu[3] attached to Sepharose by a 24-atom spacer. Although equally effective in adsorbing these enzymes, elution is accomplished at a higher pH than from the phenylalanine ligands, conditions under which the enzyme is less stable.

Sparrow and McQuade[10] have reported that a neutral proteinase from *Clostridium histolyticum* can be isolated by affinity chromatography on immobilized L-leucine substrates. However, the adsorbents lose ligand slowly and require a high pH (10) for elution of the enzyme.

pH Dependence

Maximum adsorption of neutral protease occurs at pH values corresponding to minimal K_m values.[11] The enzyme is eluted by raising the pH to values at which substrate binding is weak and the enzyme is not denatured. Between pH 5 and 6.5 neutral protease is effectively adsorbed and separated both from subtilisin and from other proteins in the culture filtrate (Fig. 2). At higher pH values neutral protease and subtilisin are ineffectively separated. The optimal pH for elution of a particular neutral protease will depend on its affinity for the ligand, the pH dependence of its affinity, and the concentration of ligand on the adsorbent gel.

Chelating Agents

Thermolysin and neutral protease are inhibited by 1,10-phenanthroline and by EDTA.[1] Either chelating agent, at a concentration of 1–5 mM, prevents adsorption of these enzymes to the adsorbent.

[10] L. G. Sparrow and A. B. McQuade, *Biochim. Biophys. Acta* **302**, 90 (1973).
[11] K. Morihara and H. Tsuzuki, *Eur. J. Biochem.* **15**, 374 (1970).

[51] Bovine Trypsin and Thrombin

By H. F. HIXSON, JR. and A. H. NISHIKAWA

Trypsin

The tendency of trypsin and trypsin-like proteases to undergo autolysis during purification and storage makes it desirable to use a simple and rapid purification procedure. The method described herein and elsewhere[1-3] is based on the affinity approach whereby these enzymes are

[1] H. F. Hixson, Jr. and A. H. Nishikawa, *Fed. Proc., Fed. Amer. Soc. Exp. Biol.* **30**, 1078 (Abstract) (1971).
[2] H. F. Hixson, Jr. and A. H. Nishikawa, *Arch. Biochem. Biophys.* **154**, 501 (1973).

first adsorbed onto polysaccharide gels containing ligands composed of synthetic inhibitors: *m*- and *p*-aminobenzamidine[4] and *p*-aminophenyl-guanidine. Jameson and Elmore have reported the use of *p'*-aminophenoxy-propoxybenzamidine in the partial but incomplete separation of α- and β-trypsins.[5] Other useful ligands might be developed by the appropriate difunctionalization of a number of the trypsin inhibitors studied by Mares-Guia and Shaw,[6] Markwardt *et al.*[7] or Marumatu and Fujii.[8] The use of a natural inhibitor for purification is considered elsewhere in this volume.[9] The specifically adsorbed trypsin is removed from the affinity matrix by elution with solutions of low pH or solutions containing benzamidine at suitable concentrations.

For successful affinity purification of trypsin, a spacer molecule is necessary to separate the ligand from the surface of the carrier gel.[2] Agarose beads, 4% or 6%, and polyacrylamide beads serve as satisfactory support matrix materials and both 6-aminohexanoic acid and mono-succinylated 1,6-diaminohexane make satisfactory spacers, whereas β-alanine is an unsatisfactory spacer. Furthermore, the concentration of the ligand (and spacer) per unit volume of gel which will be necessary to effect trypsin adsorption from solution is dependent on the binding constant of the compound used as ligand. The competitive inhibition constants (K_i) for the ligands described here are: 8 μM for *m*-amino-benzamidine[2] or *p*-aminobenzamidine,[2,6] 0.2 mM for *p*-aminophenyl-guanidine,[2] and 20 μM for benzamidine itself.[5] Our studies[2] suggest that when these compounds are covalently bonded into a carrier, K_i increases by a factor of 2–5. Taking this into account, the ligand concentration, [L], for mBz[4] or pBz[4] should be about 5 μmoles per milliliter of gel or higher and should be 30 μmoles/ml or higher for pPg.[4] However, 6% agarose appears to be more suitable for obtaining derivatives with high ligand (exceeding 50 μmole/ml) or spacer concentrations.

Ligands

m-Aminobenzamidine. *m*-Nitrobenzamidine hydrochloride (Aldrich), 25 g, and 150 ml of methanol are introduced into a 500-ml Parr glass

---

[3] A. H. Nishikawa and H. F. Hixson, Jr., U.S. Patent Number 3,746,622, filed May 12, 1971.

[4] Abbreviations: *m*-aminobenzamidine, mBz; *p*-aminobenzamidine, pBz; *p*-amino-phenylguanidine, pPg; *p*-nitrophenyl-*p'*-guanidinobenzoate hydrochloride, NPGB; 1-ethyl-3-(3'-dimethylaminopropyl) carbodiimide hydrochloride, EDC.

[5] G. W. Jameson and D. T. Elmore, *Biochem. J.* **124**, 66P (1971).

[6] M. Mares-Guia and E. Shaw, *J. Biol. Chem.* **240**, 1579 (1970).

[7] F. Markwardt, H. Landmann, and P. Walsmann, *Eur. J. Biochem.* **6**, 502 (1968).

[8] M. Marumatu and S. Fujii, *Biochim. Biophys. Acta* **242**, 203 (1971).

[9] A. Light and J. Liepnieks, this volume [52].

hydrogenation vessel. The vessel is purged of oxygen by a stream of nitrogen and stoppered. 300 mg of 10% palladium on charcoal catalyst (Engelhard Industries, Newark, New Jersey) is introduced quickly, and the vessel is restoppered. The hydrogenation vessel is then placed in a screen cage and connected to a Parr pressure reaction apparatus equipped with a mechanical shaker. The hydrogenation vessel is purged with hydrogen several times to remove oxygen and the hydrogenation of nitrobenzamidine to aminobenzamidine is conducted at an initial hydrogen pressure of 50 psi. The reaction is complete within 2 hours as shown by cessation of hydrogen uptake and disappearance of the methanol-insoluble nitro derivative. A bed of diatomaceous earth is prepared on a fine sintered-glass filter, and the catalyst is removed from the aminobenzamidine solution by filtration through the bed with suction. After filtration, 100 ml of anhydrous diethyl ether and 0.125 mole of concentrated HCl are added, and the solution is placed in a refrigerator overnight at $-15°$. The fine white needles are collected by filtration, washed with cold tetrahydrofuran, and dried in a vacuum oven (yield 20 g, mp 243°).

p-Aminophenylguanidine Hydrochloride. *p*-Nitrophenylguanidine hydrochloride is made by dissolving 83 g of *p*-nitroaniline hydrochloride (K & K Labs., the free base is available from Aldrich or Eastman) in 275 ml of *p*-dioxane with warming. Cyanamide (Eastman), 20 g, is slowly added to the stirred solution. The reaction is exothermic, and caution is advised. The orange-yellow slurry obtained is refluxed for 1 hour and then evaporated to dryness. Recrystallization of the residue from water affords fine, amber needles, mp 296°, 86% yield.

The amino derivative is obtained by catalytic hydrogenation of *p*-nitrophenylguanidine hydrochloride in an ethanolic suspension (see mBz procedure above). The straw-colored, granular crystals obtained after recrystallization from methanol and ether melted at 171°.

Procedure

General Recommendations. For preciseness in volume measurements and general ease of handling of agarose beads (they cling to wet glass), it is desirable to siliconize containers and reaction vessels. For this purpose, dimethyldichlorosilane solution in benzene (Bio-Rad Laboratories) or Siliclad (Clay-Adams) are useful. Polypropylene disposable pipettes were effective in handling small volumes of gel beads.

It is preferable to conduct the cyanogen bromide activation (always in a hood) in a jacketed reaction vessel attached to a recirculating cooling bath. The vessel should have a volume capacity at least twice that of the gel volume being treated. Magnetic stirring bars mechanically damage the agarose beads and, therefore, overhead paddle stirring, rocking or swirling are preferable.

Matrix Synthesis—mBz gel I.[4] In a well ventilated hood (!) 60 ml of settled 4% agarose beads (Sepharose 4B, Bio-Gel A-15m) are washed with distilled water to remove buffer and preservative and slurried with an equal volume of water. Finely divided cyanogen bromide (8 g) is added, and the pH is raised to 11.0, where it is maintained with 4 M NaOH and at approximately 4°. The activation reaction is complete within 15 minutes as shown by the absence of proton release. The slurry is rapidly transferred to a coarse 2-liter Büchner funnel containing 300 ml of crushed ice and washed with 1 liter of 10 mM NaHCO$_3$ at pH 8.9 with suction.

The activated, moist gel is transferred to a beaker containing 8 g of 6-aminohexanoic acid in 60 ml of 0.1 M NaHCO$_3$ at pH 8.9. The slurry is stirred slowly for 2–3 hours at room temperature and then covered with Parafilm and placed in a refrigerator at 2° overnight. The gel is washed with suction on a coarse 2-liter Büchner funnel with 2400 ml of 0.5 M NaCl used in 400-ml portions. CH-Sepharose (Pharmacia), a 4% agarose material containing 6–10 μmoles of 6-aminohexanoic acid or spacer per milliliter of swollen gel, may be substituted after it has been hydrated and washed free of preservative.

The gel is suspended in 100 ml of 0.5 M NaCl and 4 g of m-aminobenzamidine dihydrochloride are dissolved in the slurry. The slurry is placed on a Radiometer TTTI pH stat and the pH is adjusted to 4.75. Five grams of 1-ethyl-3-(3'-dimethylaminopropyl) carbodiimide hydrochloride (EDC) dissolved in 20 ml of distilled water is added, and the pH is maintained at 4.75 by the automatic addition of 1 N HCl. The surface of the slurry is swept with nitrogen, and the suspension is stirred by means of an overhead stirrer. The course of the reaction is monitored by the rate of acid addition and the end of the reaction is defined by the cessation of proton uptake. The importance of driving the coupling reactions to completion cannot be overemphasized since unreacted carboxylic acid groups will cause the affinity matrix to function as an ion-exchange resin to the detriment of protein separation. The slurry is transferred to a 2-liter coarse Büchner funnel and washed by suction with eight alternate 500-ml portions of 10 mM HCl–0.5 M NaCl and 10 mM NaOH–0.5 M NaCl. Thereafter, the gel is washed with 1 liter of distilled water until a moist cake remains. The gel product contains approximately 30 μmoles of aminohexanoic acid per milliliter of gel as determined by titration of the carboxyl group before and after ligand coupling.

mBz gel II.[4] Variations of the above procedure with 4% agarose beads yield a matrix containing 34 μmoles of ligand per milliliter of gel. Cyanogen bromide, 3 g, dissolved in 20 ml of N-methylpyrrolidinone or dimethylformamide, is used to activate 100 ml of gel. 2-Aminohexanoic acid, 2.4 g, is used to obtain the spacer coupling, and mBz is coupled to the spacer by the N-hydroxysuccinimide active ester method[9] using a 5-fold excess of the ligand. Successful matrix materials can be synthesized using pBz and pPg with the above procedures or with those of Cuatrecasas[10] for the monosuccinylated 1,6-diaminohexane spacer.

Column Chromatography. Long-term storage of affinity matrices should be in the cold. Typical capacities for the 6% agarose sorbents are 4–7 mg of trypsin per milliliter. The capacity of 4% agarose sorbents may be lower. For handling 20 mg of trypsin, a column 0.7 × 20 cm is convenient. The affinity sorbent is packed into the column with the standard buffer, 50 mM Tris/0.5 M KCl at pH 8.0. The effluent is conveniently monitored at 280 nm.

Functional testing of the column for specificity is made with a solution of 20 mg of bovine chymotrysinogen A (Calbiochem) in 2 ml of the standard buffer. With a gravity flow column it may be useful to include sucrose to increase the density of the test solution and thus sharpen the loading band. Chymotrysinogen A, which has a high isoelectric point like that of trypsin, should pass unretarded through the column. After the initial UV absorbing band has eluted from the column and the absorbance base line has been UV established, a solution of 10 mM HCl–0.5 M KCl (acid solution) is applied to the column. Any nonspecifically bound protein, if present, will be eluted at this point. Matrices which bind more than trace quantities of UV absorbing materials by this test should be rejected. Sometimes the incomplete coupling of ligand to spacer may be rectified by subjecting the gel preparation to another ligand coupling cycle.

Trypsin Purification. Bovine trypsin (Worthington Biochemical, grade TRL), 20 mg, is dissolved in the standard buffer and applied to the column. With UV monitoring of the effluent, fractions of 2 ml are collected. With the passage of nonbinding proteins through the column and establishment of the monitor base line, a solution of 10 mM benzamidine hydrochloride (Eastman) in the standard buffer is applied. The benzamidine concentration required may be higher for highly substituted matrices. The affinity bound trypsin will be eluted, and its UV absorbance will be superimposed on the background caused by the benzamidine. If a variable wavelength monitor is available, it will be convenient to observe the column effluent at 290 nm due to the high benzamidine background. After the benzamidine base line in the column effluent is stabilized, the column is eluted with the standard buffer followed by the acidic solution which should elute any nonspecifically bound protein. If no UV absorbing

[10] P. Cuatrecasas, *J. Biol. Chem.* **245**, 3059 (1970).

material is eluted at this step, this affinity sorbent can be conveniently used in subsequent experiments without the benzamidine elution step. However, unless a trypsin matrix has met all these tests, acid elution will not necessarily yield a homogeneous preparation of trypsin; only soluble benzamidine or another inhibitor will elute the specifically adsorbed trypsin.

Trypsin Assay. The active-site titration method of Chase and Shaw[11] is used. The protein content of a sample is first determined by a spectral scan from 350 nm to 250 nm. The colorimetric "burst" is measured at 410 nm in the standard buffer using 10 μl of a 10 mM p-nitrophenyl-p'-guanidinobenzoate, hydrochloride (Cyclo Chemical) in dimethylformamide. Multiplication of the burst absorbance by 6.025×10^{-5} yields the molarity of the nitrophenoxide produced and the corresponding molarity of the active trypsin present. Protein concentration is determined by multiplying the A_{280} value by the optical factor, 0.651 mg/ml/A_{280} then dividing by the molecular weight, 24,000. The fraction of active trypsin is the ratio of active-site titration molarity to the protein molarity and is typically 0.93–0.95 for affinity purified trypsin.

Thrombin

Both trypsin and thrombin recognize the same competitive inhibitors, although thrombin binds less well, i.e., K_i for thrombin is larger by a factor of 10. Thus far, only the more potent inhibitors, m- and p-aminobenzamidine, have been successful in the affinity purification of thrombin.[1-3,12] Thompson and Davie[13] have reported the use of 4-chlorobenzylamine coupled to 8% agarose gel via the 6-aminohexanoic acid spacer in a 1.5-fold purification of thrombin. Although a potent proteinaceous thrombin inhibitor (hirudin) is known,[14] its scarcity and the need for acid elution dictate the use of low molecular weight synthetic inhibitors in any practical affinity purification of thrombin.

The large dissociation constant, i.e., the poor affinity of thrombin toward the amidine ligands,[2] has required high functional group concentrations in the 6% agarose gel to effect efficient binding of enzyme. With 4% agarose gels moderate levels (6–20 μmoles per milliliter of gel) of functional groups seem to suffice. The requirements may be a consequence of the higher molecular weight of thrombin as well as secondary orders of substrate specificity which distinguish it from trypsin itself.

[11] T. Chase, Jr. and E. Shaw, this series, Vol. 19, p. 20.
[12] G. Schmer, *Hoppe-Seyler's Z. Physiol. Chem.* **353**, 810 (1972).
[13] A. Thompson and E. Davie, *Biochim. Biophys. Acta* **250**, 210 (1971).
[14] F. Markwardt, this series, Vol. 19, p. 924.

Matrix Synthesis

Two different matrix–spacer combinations can be used for thrombin purification; 4% agarose containing 6-aminohexanoic acid spacers at a substitution level of 6–10 μmoles per milliliter of swollen gel (commercially available as CH-Sepharose 4B) and 6% agarose containing monosuccinylated 1,6-diaminohexane at a substitution level of 60 μmoles per milliliter of swollen gel. Forty milliliters of agarose (Bio-Gel A-5m, 50–100 mesh; Bio-Rad Laboratories) is washed free of buffer and preservatives, slurried in an equal volume of water, and maintained at a temperature of 4° throughout the reaction course. Cyanogen bromide, 10 g, dissolved in a minimum amount of N-methylpyrrolidinone, is added, and the pH is maintained at 11.0 by adding 4 M NaOH. Upon cessation of acid evolution, the gel slurry is transferred to a 200-ml coarse Büchner funnel containing 200–300 ml of crushed ice and washed quickly under suction with 1 liter of 0.5 M NaCl and then with 1 liter of 0.1 M NaHCO₃ at pH 9. The moist gel cake is transferred to a 250-ml beaker containing 20 g of 1,6-diaminohexane in 100 ml of sodium bicarbonate at pH 9. The diaminohexane solution should be prepared in advance by titrating the amine with HCl prior to bicarbonate addition and cooling to room temperature. The gel slurry is stirred slowly for several hours at room temperature and then is covered with Parafilm and placed in a refrigerator (2°) overnight. The gel slurry is transferred to a 2-liter coarse Büchner funnel and washed with suction using 3 liters each of 10 mM HCl–0.5 M NaCl and 10 mM NaOH–0.5 M NaCl in alternate 500-ml portions, followed by 2 liters of distilled water. The gel is then slurried in an equal volume of water and adjusted to pH 6.0; 10 g of succinic anhydride are added slowly in small portions. The pH is maintained at 6.0 by the addition of 4 M NaOH. After acid evolution ceases the slurry is covered with Parafilm and placed in the refrigerator overnight. The washing described above is repeated, and the gel is tested for complete succinylation by allowing small portions of gel to react with 0.1 M fluorodinitrobenzene (FDNB) in methanol. The succinylation step is repeated until no FDNB color can be seen on the gel after ethanol and distilled water washing.

Ligand Coupling

The succinylated hexanediamine gel described above or hydrated and washed CH-Sepharose 4B may be covalently coupled to m- or p-aminobenzamidine by means of the water-soluble carbodiimide coupling reaction. Forty milliliters of gel, 6 g of m-aminobenzamidine and 5 g of the EDC[4] are used. The gel is washed with acid and base as above and then with 500 ml of 50 mM Tris–0.5 M KCl at pH 8.0, the standard buffer.

Coupling to the CH-Sepharose is accomplished similarly except that 30 ml of the gel, in an equal volume of water, is used with 2 g of *m*-aminobenzamidine and 2 g of the carbodiimide. An alternative procedure for ligand coupling is that of Cuatrecasas and Parikh[15] which utilizes the active *N*-hydroxysuccinimide ester as the coupling intermediate.

Column Chromatography

All glassware used in handling thrombin should be siliconized to prevent thrombin adsorption, and operations must be conducted at 2–4°. Column size is dictated by the amount of enzyme to be handled. For 10,000 NIH units of enzyme, a 0.7 × 20 cm column is convenient. Effluent monitoring and functional testing is performed as described for trypsin.

Thrombin Purification

Bovine topical thrombin (Parke-Davis, 10,000 NIH units) is dissolved in 2 ml of the standard buffer and applied to the affinity column. After a large band of inactive protein is eluted and the monitor base line is reestablished, the flow of liquid in the column is stopped. A second column, are of Sephadex G-25 (1.5 × 30 cm, completely filled), is placed between the affinity column and the UV monitor. The thrombin is then eluted from the affinity column with a solution of 50 mM benzamidine in the standard buffer. As liquid passes into the G-25 column, enzyme solution is separated from the soluble inhibitor, i.e., benzamidine. The enzyme will be observed in the monitor as a small peak followed eventually by the benzamidine background. This procedure is necessary since acid elution will destroy thrombin.

Assay

Thrombin esterase activity is determined by the active-site titration method of Chase and Shaw.[11] For thrombin titrations, use of the 0–0.1 absorbance slide wire is recommended. The optical factor for thrombin at 280 nm[16] is 0.51 mg ml^{-1} (absorbance unit)$^{-1}$ and the molecular weight used is 36,000.[16] Concentrated thrombin solution absorbances may be corrected for light scattering by the procedure of Baughman and Waugh,[17] but dilution is the recommended procedure. Affinity purified thrombin is greater than 90% active as measured by active site titration.

Thrombin clotting activity is determined using the procedure of Baugh-

[15] P. Cuatrecasas and I. Parikh, *Biochemistry* 11, 2291 (1972).
[16] D. J. Winzor and H. A. Scheraga, *J. Phys. Chem.* 68, 338 (1964).
[17] D. J. Baughman and D. F. Waugh, *J. Biol. Chem.* 242, 5352 (1967).

man[18] or that of Thompson and Davie.[13] Schmer[12] has reported trace quantities of plasmin in affinity-purified thrombin which could be removed by SE-Sephadex chromatography either before or after the affinity chromatography step. Such combined chromatography produces thrombin with 2000 NIH units/mg.

[18] D. J. Baughman, this series, Vol. 19, p. 145.

[52] Bovine β-Trypsin

By ALBERT LIGHT and JURIS LIEPNIEKS

Schroeder and Shaw[1] reported in 1968 that chromatography of trypsin and activated trypsinogen on columns of SE-Sephadex C-50 showed the presence of almost equal amounts of two active forms of trypsin and a small amount of inert protein. One active form is β-trypsin, an intact molecule; the second is α-trypsin, a product of autolysis with Lys 131-Ser 132 cleaved. Further slow proteolysis produces ψ-trypsin with both Lys 131-Ser 132 and Lys 176-Asp 177 hydrolyzed.[2] It is clear from these studies that commercially available trypsins are mixtures of species and that studies on the properties of trypsin should use the β-form.

High yields of β-trypsin can be obtained from bovine trypsinogen if enterokinase is used as activator since the enzyme is highly specific for the Lys 6-Ile 7 peptide bond.[3] However, autolysis of β-trypsin must be prevented; the presence of soybean trypsin inhibitor effectively inhibits β-trypsin but does not interfere with the activity of enterokinase. If the inhibitor is bound to Sepharose, affinity chromatography of the activation mixture provides a rapid procedure for the production of highly purified β-trypsin.[3a] This procedure is described here.

Assay Methods

Measurement of Trypsin Activity

Trypsin activity is measured at pH 7.9 and 25° using 10 mM N-tosyl-L-arginine methyl ester in 10 mM Tris chloride and 50 mM calcium chloride.[4]

[1] D. D. Schroeder and E. Shaw, J. Biol. Chem. 243, 2943 (1968).
[2] R. L. Smith and E. Shaw, J. Biol. Chem. 244, 4704 (1969).
[3] S. Maroux, J. Baratti, and P. Desnuelle, J. Biol. Chem. 246, 5031 (1971).
[3a] J. J. Liepnieks and A. Light, Anal. Biochem., in press.
[4] K. A. Walsh and P. E. Wilcox, this series, Vol. 19 [3].

Measurement of Enterokinase Activity

Enterokinase activity is determined by activation of 0.1 mg of trypsinogen per milliliter of 0.1 M sodium acetate, pH 5.0, and 50 mM calcium chloride. Samples of 0.5–7.0 enterokinase units are used in a 30-minute incubation at 35°. Tryptic activity is measured in 0.5 ml fractions with *N*-tosyl-L-arginine methyl ester. An enterokinase unit is defined as the amount of enzyme which produces 1 trypsin unit per milliliter of activation mixture.

Estimation of Aminopeptidase Activity

A spot test for the detection of aminopeptidase is used to examine chromatographic fractions.[5] Usually, 10-μl samples and 0.5 ml of 0.3 mM leucyl β-naphthylamide in 50 mM Tris chloride at pH 8.5, containing 5 mM calcium chloride are incubated at room temperature for 30 minutes in depressions of a porcelain spot plate. One adds 0.2 ml of naphthanil diazo blue (5 mg per milliliter of water), and the immediate appearance of a stable pink color is indicative of aminopeptidase activity. The relative intensity of color is used to judge maximum activity.

Preparation of β-Trypsin

A suitable amount of enterokinase for the activation of approximately 300 mg of trypsinogen is prepared as follows. A crude enteropeptidase preparation (3.25 g, Miles Laboratories) is extracted with 325 ml of 10 mM ammonium acetate at pH 6.0. After stirring for 1 hour at room temperature, the suspension is centrifuged and the supernatant fluid is dialyzed exhaustively against water at 4°. After lyophilization, 500 mg of a light brown powder is recovered. The powder is dissolved in 100 ml of 10 mM Tris chloride at pH 6.0, containing 50 mM sodium chloride. The insoluble material is removed by centrifugation, and the supernatant liquid is chromatographed on DEAE-cellulose (Whatman microgranular DE-52). Aminopeptidase activity is detected in the first half of the peak containing enterokinase activity. Removal of almost all the aminopeptidase contaminant is achieved by combining appropriate fractions. After dialysis against water at 4° and then lyophilization, 44 mg of a white powder is obtained containing approximately 72 units/mg.

Sepharose-bound soybean trypsin inhibitor is prepared by a slight modification of a procedure previously described.[6] Settled Sepharose 4B,

[5] J. E. Folk, J. A. Gladner, and T. Viswanatha, *Biochim. Biophys. Acta* **36**, 256 (1959).
[6] P. Cuatrecasas, M. Wilchek, and C. B. Anfinsen, *Proc. Nat. Acad. Sci. U.S.* **61**, 636 (1968).

20 ml, is treated with 3 g of cyanogen bromide at pH 11. The suspension is gently agitated for about 8 minutes with a magnetic stirrer. After washing with water and cold 0.1 M sodium bicarbonate, pH 9.5, 500 mg of soybean trypsin inhibitor are added and the suspension (40% by volume in 0.1 M sodium bicarbonate) is agitated by rotating slowly in a bottle for 18 hours at 4°. The preparation is washed with 100 ml of 0.2 M sodium formate at pH 2.2, and 0.2 M Tris chloride at pH 8.0, repeating the washing cycle four times. The second Tris buffer wash contains 15 g of glycine per liter to react with residual cyanogen bromide activated sites. The preparation of Sepharose-bound soybean trypsin inhibitor binds approximately 5 mg of trypsin per milliliter of settled gel as determined operationally by applying excess trypsin at pH 8.0 and eluting bound trypsin at pH 2.2.

Activation of bovine trypsinogen is accomplished by mixing 30 mg of zymogen with 4 mg of partially purified enterokinase (300 units) in 0.1 M Tris, pH 8.0, containing 50 mM calcium chloride. The activation mixture includes Sepharose-bound soybean trypsin inhibitor in approximately a 2-fold excess over the amount of trypsin that could be produced from the zymogen. The total volume of the mixture is 25 ml and it is incubated at 35° for 60 minutes with slow agitation obtained by bubbling nitrogen gas through the suspension. The activation mixture is transferred to a chromatography column (1.4 × 12 cm), and inert components are removed on washing the gel with 40 ml of 0.1 M sodium formate at pH 4.5, containing 50 mM calcium chloride. β-Trypsin is released from the complex with 0.1 M sodium formate at pH 2.6 containing 50 mM calcium chloride or with a pH gradient using equal volumes of the formate buffers at the two pH values. The β-trypsin fraction emerges at pH 3.5, well separated from traces of α-trypsin; α-trypsin is eluted at pH 3.8. Dialysis of the column fractions against 1 mM HCl and lyophilization results in an approximately 75% overall yield of β-trypsin.

Comments

The preparation of β-trypsin by the procedure of Schroeder and Shaw employing ion-exchange chromatography in the presence of the competitive trypsin inhibitor benzamidine required almost 7 days and could not completely separate the contaminating α-form.[1] A similar resolution was found by Robinson et al.[7] on affinity chromatography of trypsin samples on Sepharose-bound chicken ovomucoid, although the operations required less than a day. In contrast, the formation of β-trypsin from trypsinogen,

[7] N. C. Robinson, R. W. Tye, H. Neurath, and K. A. Walsh, *Biochemistry* **10**, 2743 (1971).

employing affinity chromatography on immobilized soybean trypsin inhibitor, requires 4–6 hours, and the β-trypsin fraction is completely free of α-trypsin. The yield of β-trypsin approaches the maximum possible based on the amount of trypsinogen. Furthermore, similar results were found with preparations of enterokinase that were not chromatographed on DE-52 columns if the same number of enzyme units were used.[3a] Presumably, the same procedures could be applied to trypsinogens of other species.

[53] Urokinase

By Thomas Maciag, Michael K. Weibel, and E. Kendall Pye

Urokinase, a plasminogen-activating proteolytic enzyme, was first described in 1951[1] as occurring in trace quantities in mammalian urine. Plasminogen-activating enzymes have also been shown to occur in a number of mammalian tissues[2,3] and recently have been obtained from cultures of cell lines derived from various tissues.[4,5]

A renewed interest in urokinase has now arisen because of its therapeutic potential as an effective thrombolytic agent.[6] However, large-scale clinical trials have been delayed in the United States because the recovery and purification of urokinase from human urine by classical techniques[7] has provided inadequate amounts of the highly purified enzyme. It has recently been shown that relatively high concentrations of plasminogen activators appear in the growth medium of various human tissue culture preparations,[5] and, because of the versatility and high recoveries possible with affinity chromatography, we have applied this technique to the recovery and purification of the plasminogen-activating enzyme from these sources.[8] Since there is a lack of conclusive proof that this enzyme is not urokinase and since the technique can also be used for the recovery of

[1] J. R. B. Williams, Brit. J. Exp. Pathol. 32, 530 (1951).
[2] O. K. Albrechtsen, Acta Physiol. Scand. 39, 284 (1957).
[3] O. K. Albrechtsen, Brit. J. Haematol. 3, 284 (1957).
[4] P. Kelly, personal communication, 1972.
[5] C. S. Kucinski, A. P. Fletcher, and S. Sherry, J. Clin. Invest. 47, 1238 (1968).
[6] Urokinase Pulmonary Embolism Trial Study Group, J. Amer. Med. Ass. 214, 2163 (1970).
[7] W. F. White and G. H. Barlow, this series, Vol. 19, p. 665.
[8] T. Maciag, M. K. Weibel, and E. K. Pye, in "Enzyme Engineering" (E. K. Pye and L. Wingard, Jr., eds.). Plenum, New York, 1974.

urokinase from urine, we shall refer to the enzyme as "urokinase" throughout the remainder of this article.

Considerations for the Design of Suitable Affinity Ligands

Urokinase has both proteolytic and esterolytic activity with both reactions apparently being catalyzed at the same active site.[9] A direct correlation between the proteolytic and esterolytic activities of the enzyme has been previously demonstrated.[10] Several analogs of α-amino acids have been recognized as competitive inhibitors of urokinase,[10] but their weak affinities do not make them suitable for consideration as biospecific affinity chromatography ligands. Similarities in the structures of known synthetic substrates and inhibitors of urokinase have led us to formulate the following criteria which were used in the design of more suitable competitive inhibitors for affinity chromatography ligands:

a. The α-amino group of the amino acid apparently is not necessary for binding to the active site but nevertheless must be blocked with a bulky hydrophobic group, e.g., tosyl or benzoyl.

b. It is essential that a carbonyl moiety be adjacent to the substituted alpha amine, either in the form of a carboxylic acid, ester, amide, or ketone.

c. The presence of a nucleophile (either NH_2 or OH) at least seven methylenes from the carboxylic acid end of the amino acid is required.

d. The presence of a π-orbital system adjacent to this nucleophile is not essential, but its presence improves the binding of the ligand to the enzyme.

Using these assumptions, we designed the competitive inhibitor α-benzylsulfonyl-p-aminophenylalanine (BAPA), shown in Fig. 1. The derivative has a K_i of approximately 50 μM, based on the esterase activity of urokinase. Other possible inhibitors were tested, and the most promising was α-benzoyl-L-arginine amide (BAA). Owing to the positive charge associated with BAA at neutral pH, BAPA was chosen as the

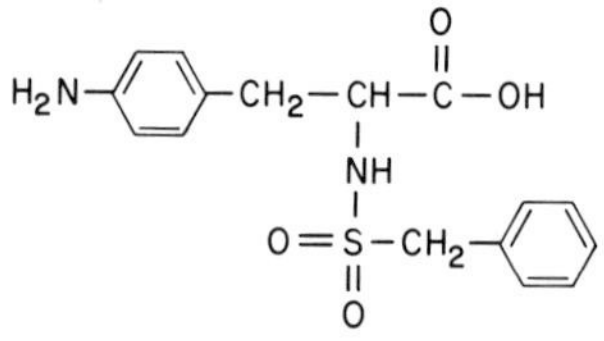

FIG. 1. α-Benzylsulfonyl-p-aminophenylalanine (BAPA).

[9] N. O. Kjeldgaard and J. Plaugh, *Biochim. Biophys. Acta* **24**, 283 (1957).
[10] L. Lorand and E. V. Condit, *Biochemistry* **4**, 265 (1969).

affinity ligand because of the neutrality associated with the aromatic amine.

Assay of Urokinase Activity (Esterolytic)

The esterolytic assay employed was based on that of Lorand and Condit.[10] Using the known extinction coefficient[10] of the p-nitrophenoxide ion at 400 nm, $E_{1\ cm} = 1.6 \times 10^4$ liters/mole/cm, and the appropriate controls for nonenzymatic ester hydrolysis, the esterolytic activity can be converted to enzyme units by defining that 1 unit of urokinase will release 1 μmole of p-nitrophenoxide ion per minute. The ratio of the conventional CTA unit, which is based upon the fibrinolytic activity of urokinase, to the esterase unit is 92.3.[10]

Synthesis of the Affinity Material

The affinity material is constructed in a series of steps. First, the affinity ligand precurser, α-benzylsulfonyl-p-nitrophenylalanine (BNPA) is synthesized. Second, Sepharose 4B is activated with cyanogen bromide and sequentially coupled to 1:6 hexanediamine, succinic anhydride and a second molecule of 1:6 hexanediamine to form a reactive pendant spacer. Next, the BNPA is covalently coupled to the gel matrix by reaction with the terminal amino group. Unreacted amino groups on the modified Sepharose 4B are blocked by acylation with acetic anhydride. Finally, the BNPA-Sepharose 4B is chemically reduced to BAPA-Sepharose 4B.

1. Synthesis of α-Benzylsulfonyl-p-nitrophenylalanine

BNPA may be synthesized as follows using the Schotten-Baumann procedure. p-Nitrophenylalanine (Aldrich), 20 mmoles, is dissolved in 1.2 equivalents of 1 N NaOH, the solution being vigorously stirred and cooled in an ice bath; 1.2 equivalents of benzylsulfonyl chloride in 1.4 equivalents of 1 N NaOH is added to the cooled p-nitrophenylalanine solution in five equal portions at 15- to 20-minute intervals. The reaction mixture is stirred at room temperature for 1 hour and then made basic to litmus paper with 1 N NaOH. The insoluble by-products are separated by vacuum filtration, and the filtrate is washed with two 20-ml portions of ether. After removal of the ether layer, the solution is acidified to Congo red with concentrated HCl, which produces a fine precipitate of the desired derivative. Recrystallization of the α-benzylsulfonyl-p-nitrophenylalanine may be performed by acidification of dilute alkaline solutions of the derivative. The melting point of the BNPA is 178°.

2. Synthesis of Sepharose 4B Coupled to BNPA

After each step in the following synthesis the modified Sepharose 4B beads are washed on a Büchner funnel with 3 liters of 0.2 M KCl followed by 3 liters of distilled water. Quantitation of functional residues on the gel matrix can be obtained with trinitrobenzenesulfonic acid,[11] isotope techniques or by titration.[12]

Activation of Sepharose 4B with Cyanogen Bromide and Coupling of Hexanediamine. A 100-ml volume of dry Sepharose 4B is washed with 3 liters of distilled water. The imino carbonate derivative of Sepharose 4B is prepared by reaction of 30 g finely ground CNBr with a 100-ml aqueous suspension of the gel, at 4° in a well ventilated hood. Immediately after addition of the CNBr, the pH of the reaction mixture is raised to pH 11.0 by addition of 10 N NaOH and sufficient ice is added to maintain the temperature below 10° during the reaction. The reaction is allowed to proceed for 15 minutes, or until stabilization of the pH occurs.[11] The reaction mixture is then rapidly filtered and the activated Sepharose washed with 1 liter of cold distilled water. The activated Sepharose is suspended in 50 ml of a previously prepared, cooled 0.2 M 1:6 hexanediamine solution adjusted to pH 10 with 5 N HCl, and the reaction is allowed to proceed at 4° for 12 hours.

Coupling of BNPA to the Modified Sepharose 4B. Distilled water, 50 ml, is added to 50 ml of the washed 6-aminohexyl-Sepharose 4B and the pH lowered to 8.0 with 1 N HCl. Succinic anhydride (5.0 g) is added to the reaction mixture at 4° and the pH maintained at pH 8.0 with 10 N NaOH. After 30 minutes the reaction mixture is filtered and washed. The matrix material is suspended in 50 ml of distilled water containing 5 ml of 1:6 hexanediamine and the pH of the mixture adjusted to pH 5.0 with concentrated HCl. Sufficient crystalline [3-(dimethylamino)-propyl] ethyl carbodiimide HCl (Eastman Kodak) is added to the suspension so that the final concentration of water-soluble carbodiimide (WSC) is 3 mM. The reaction is allowed to proceed at room temperature for 3 hours, maintaining the reaction mixture at pH 5.0 with 1 N HCl, after which the modified Sepharose is filtered and washed.

An aqueous solution, 5 ml, containing 40 mmoles of α-benzylsulfonyl-p-nitrophenylalanine is added to 25 ml of the previously washed modified Sepharose material suspended in 25 ml of distilled water. The pH is adjusted to pH 5.0 after which sufficient [3-(dimethylamino)propyl] ethyl carbodiimide HCl is added to the reaction mixture to make the final con-

[11] P. Cuatrecasas, *J. Biol. Chem.* **245**, 3059 (1970).
[12] J. K. Inman and H. M. Dintzis, *Biochemistry* 8, 4074 (1969). See also Inman, this volume [3].

centration of WSC 3 mM. The reaction is allowed to proceed at room temperature for 3 hours while maintaining the pH constant at 5.0 with 1 N NaOH. The gel is filtered and washed.

3. Blocking of Unreacted Amino Groups and Conversion of BNPA to BAPA

To 25 ml of the BNPA coupled Sepharose 4B gel in 25 ml of saturated sodium acetate is added, at 4°, 20 ml of acetic anhydride. The reaction is allowed to proceed for 30 minutes at 4°, after which the gel is filtered and washed. Suspending the gel in a 0.1 M sodium dithionite aqueous solution for 30 minutes reduces the aromatic nitro group to the arylamine. After washing, the gel is stored at 4° in 0.1 M sodium phosphate at pH 7.0, containing 1 mM sodium azide as a bactericidal agent.

Operational Characteristics of the Affinity Material for Urokinase

In each of the following experiments the tissue culture medium, containing urokinase, which was used was dialyzed against 50 mM sodium phosphate at pH 7.0 for 24 hours. The specific activity of the urokinase

TABLE I

EFFECT OF ELUTING AGENTS ON THE PURIFICATION OF UROKINASE[a]

Eluting agent[b]	Recovery[c] (%)	Specific activity (esterase units/mg)	Purification factor
8% NaCl	100	120	130
0.1 M sodium acetate, pH 4.5	0	0	0
10 mM BAA	40	1440	1560
10 mM BNPA	0	0	0
10 mM BAPA	52	978	1060
1 mM BAPA	56	1120	1220

[a] Ten milliliters of a crude urokinase preparation (0.923 esterase units/ml per milligram of protein) in 10 mM sodium phosphate at pH 7.0, was applied to the negatively charged BAPA-Sepharose 4B affinity column. The column was then washed with 25 ml of 50 mM sodium phosphate at pH 7.0, and the bound protein material was eluted with 10 ml of the eluting agent.

[b] Eluting agents were prepared in 5 mM sodium phosphate at pH 7.0. BAA, α-benzoyl-L-arginine amide; BNPA, α-benzylsulfonyl-p-nitrophenylalanine; BAPA, α-benzylsulfonyl-p-aminophenylalanine.

[c] This figure compares the esterase activity of the eluted protein fraction with the amount of esterase activity initially applied to the column. Enzyme activity was measured after dialysis of the eluted protein fraction against 1 liter of 50 mM sodium phosphate at pH 7.0 for 12 hours and the titration of this fraction to pH 8.0.

in the dialyzed crude preparation was 0.923 esterase unit per milligram of protein. Protein was estimated according to the method of Lowry et al.[13]

A 15 × 1 cm column was packed with the α-benzylsulfonyl-p-amino-phenylalanine-Sepharose 4B affinity material. This column had a void volume of 10 ml. Routinely, 5 ml of the crude urokinase preparation was diluted with 5 ml of 50 mM sodium phosphate buffer, pH 7.0, and applied to the column. The column was washed with 30 ml of 50 mM sodium phosphate at pH 7.0 and the protein fraction which was not adsorbed was free of urokinase activity. Elution of the enzyme from the column was attempted with various reagents and conditions. Table I summarizes the results. High ionic strength perturbation with 8% NaCl removed all the bound esterase activity from the column with a purification factor of only 130. Elution with competitive inhibitors of urokinase showed a dramatic increase in purification, but with only approximately 50% of the bound enzyme activity recovered. The biospecificity of the affinity column is demonstrated by the selective elution of the esterase activity from the column with the inhibitor BAPA, but not with the non-

TABLE II

EFFECT OF COLUMN MATRIX CHARGE ON UROKINASE PURIFICATION[a]

Column matrix material	Charge on column	Uro-kinase bound[b] (%)	Bound urokinase eluted with 8% NaCl (%)	Specific activity (esterase units/mg)	Purification factor
Free —NH₂ with no free —COOH groups	(+)	100	100	15.7	17
Free —COOH[c] with —NH₂ groups blocked[d]	(−)	100	100	120	130
—NH₂ groups blocked and no free —COOH groups	Neutral	57	100	102	110

[a] Ten milliliters of crude urokinase (0.923 esterase units/ml per milligram of protein) were applied to each column and washed with 25 ml of 50 mM sodium phosphate at pH 7.0. The bound urokinase was then eluted with 10 ml of 8% NaCl in 50 mM sodium phosphate at pH 7.0.

[b] The percentage of urokinase bound to the affinity column was determined by the amount of esterase activity in the breakthrough protein peak compared to the amount of esterase activity initially added to the column.

[c] Incomplete reaction with hexanediamine.

[d] Blocked by treatment with acetic anhydride.

[13] O. H. Lowry, N. J. Rosebrough, A. L. Farr, and R. J. Randall, *J. Biol. Chem.* 193, 265 (1951).

inhibitory analog BNPA. Since 8% NaCl elution removed all the bound protein from the column and the eluted urokinase was only marginally purified, it appeared possible that a considerable amount of nonspecific protein was also bound to the column. Elution with a salt gradient showed no increase in specific activity over that obtained upon elution with 8% NaCl.

In an attempt to increase the degree of purification of the enzyme eluted with 8% NaCl, we studied columns possessing the same immobilized inhibitor but with differently net charged matrices. Table II presents a summation of the results. A positively charged matrix increases the amount of nonspecific binding while a neutral matrix shows only a slight decrease in the specific activity of the 8% NaCl eluted peak compared to the negatively charged gel. An examination of the crude enzyme preparation by disc gel electrophoresis shows three major protein bands (Fig.

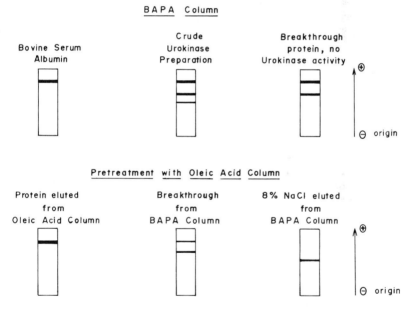

FIG. 2. Relative positions of major protein bands on polyacrylamide disc gel electrophoresis at pH 9.8. Disc gel electrophoresis was performed in standard 7% acrylamide gel preparations. Samples (0.1 ml) contained 62.5 mM Tris chloride, pH 6.8, 0.001% bromophenol blue, and 10% glycerol. A current of 4 mA per gel was applied until the tracking dye reached the bottom of the gel. Proteins were fixed in the gel with 5% trichloroacetic acid (90°, 15 minutes) and stained with 0.1% Coomassie Brilliant Blue in 50% methanol–10% acetic acid (37°, 1 hour) and destained electrophoretically with 50% methanol–10% acetic acid (40 mA per gel). Gel slices of the polyacrylamide gels were assayed for the presence of fibrinolytic activity to confirm enzyme band assignment using the fibrin plate assay [C. S. Kucinski, A. P. Fletcher, and S. Sherry, *J. Clin. Invest.* **47**, 1238 (1968)].

TABLE III

Effect of Crude Enzyme Pretreatment on Urokinase Purification

Pretreatment[a]	Activity loss (%)	Specific activity of 8% NaCl eluted protein (esterase limits/mg)	Purification factor
Change to pH 1.0 and back to pH 7.0	100	0	0
0.1% SDS	69	41.6	4.5
0.1% Tween 20	0	323	350.0
5 mM Oleic acid	0	64.5	7.0

[a] The various reagents were added to 10 ml of the crude enzyme preparation (0.923 esterase unit/ml per milligram of protein) in 50 mM sodium phosphate at pH 7.0. After incubation for 30 minutes at room temperature, the crude enzyme was applied to the negatively charged BAPA affinity column and the column washed with 25 ml of 50 mM sodium phosphate, pH 7.0. The bound protein was eluted with 10 ml of 8% NaCl in 50 mM sodium phosphate at pH 7.0.

2). The nonabsorbed material which did not contain esterase activity displays only two protein bands. We tentatively identified the third, and slower moving, protein band as containing urokinase. The major protein impurity was BSA, which is a component of the tissue culture medium. Based upon quantitated scans of the electrophoresis gels, BSA appears to be 50-fold higher in concentration than the band containing urokinase. Since BSA bears a net negative charge under our conditions for affinity chromatography,[14] the use of either a neutral or net negatively charged matrix appeared to be desirable.

To increase further the specific activity of the enzyme eluted with 8% NaCl we attempted to diminish the amount of nonspecific protein–matrix interaction with the use of detergents. Various detergents were mixed with the crude enzyme preparation and then applied to the negatively charged matrix affinity column. The results are shown in Table III. No significant increase in specific activity of the eluted enzyme was observed except in the case of Tween 20. The purification factor achieved here was 350, and no loss of enzyme activity was observed. Even with this material, polyacrylamide disc gel electrophoresis showed that BSA was still present in the peak eluted with 8% NaCl, the ratio of the BSA band to the band containing urokinase being approximately 10:1.

Peters et al.[14] recently reported success in developing an affinity material specific for serum albumin by using oleic acid or palmitic acid covalently bound to Sepharose 4B. After synthesizing such an oleic acid-

[14] T. Peters, H. Taniuchi, and C. B. Anfinsen, J. Biol. Chem. 248, 2497 (1973).

Sepharose 4B column we passed through it a crude preparation of uro-kinase and observed no binding of esterase activity to the column. The breakthrough material from this column was then loaded onto the BAPA-Sepharose 4B column (see Fig. 2). The column was washed with 50 mM sodium phosphate at pH 7.0 and the absorbed enzyme eluted with 8% NaCl in the 50 mM sodium phosphate buffer. The specific activity of this 8% NaCl eluted peak was 645 esterase units per milligram of protein. Polyacrylamide disc gel electrophoresis (Fig. 2) showed that the uro-kinase fraction resulting from the sequential use of the oleic acid-Sepharose 4B column and the BAPA affinity matrix column contained no detectable BSA.

Although the highest specific activity of urokinase is obtained upon elution from the column with soluble competitive inhibitors (Table I), the amount of enzyme recovered is only 50% of that recovered upon elution with 8% NaCl. Rechromatography of this purified material results in the same specific activity as that obtained by elution with soluble in-hibitors. For practical purposes the 8% NaCl elution is preferable to that of elution with soluble inhibitors since it provides nearly quantita-tive recovery of the enzyme with a purification factor of at least 130. Modification of the original purification procedure by addition of 0.1% Tween 20 results in purification factors of at least 350, while treatment of the crude urokinase preparation with the oleic acid-Sepharose 4B material yielded purification factors of greater than 700.

This simple technique serves as a valuble replacement for existing classical methods for the recovery and purification of urokinase. How-ever, questions concerning the ultimate purity of the enzyme obtained by this technique remain unanswered.

Acknowledgments

We wish to thank Dr. Paul Kelly of Collaborative Research Inc., for his gen-erous donation of crude tissue culture preparations of urokinase. We also wish to thank and acknowledge the contributions made by Dr. M. R. Iyengar of his time and very helpful advice.

Section III

Attachment of Specific Ligands and Methods for Specific Proteins

D. NUCLEIC ACIDS, NUCLEOTIDES, AND DERIVATIVES

[54] Nucleic Acids Attached to Solid Matrices

By ARTHUR WEISSBACH and MOHINDAR POONIAN

The attachment of DNA and RNA to insoluble matrices for subsequent purification of proteins or nucleic acids is a technique that has received wide use in molecular biology. Gilham[1] and Erhan et al.[2] have previously reported the attachment of polynucleotides to cellulose in either nonaqueous or aqueous solvents using carbodiimide reagents. Gilham has also described the binding of tRNA to aminoethyl-cellulose after periodate oxidation.[1] The binding of DNA to cellulose by adsorption has been reported by Litman[3] and well discussed by Alberts and Herrick.[4] Since these earlier reports, the binding or attachment of nucleic acids and polynucleotides to agarose, acrylic gels, and fiberglass has been accomplished. The versatility, capacity, and ease of preparation of polynucleotide matrices has established affinity chromatography as a common laboratory technique in the study of nucleic acids and enzymes.

In the following discussion, our objective is 2-fold: (a) to give detailed experimental procedures for preparing these nucleic acid–solid matrix substrates; (b) to describe some recent applications of the nucleic acid–solid matrix preparations.

Preparation of Polynucleotide Matrices

Covalent Attachment of DNA to Sephadex G-200

In principle this is an adaptation of Gilham's[1] method of condensing the 5'-phosphate end of the nucleotide with hydroxyl functions of the solid substrates. In the presence of a water-soluble carbodiimide, DNA was chemically attached to Sephadex G-200. The procedure used in the attachment is described.

Attachment of Rat Liver DNA to Sephadex.[5] Sephadex G-200, after swelling overnight in distilled water, is washed on a Büchner funnel with 6 volumes of 1 M NaCl, 10 mM Tris chloride (pH 7.4). It is further washed with 10 volumes of distilled water and finally with 2 volumes

---

[1] P. T. Gilham, this series, Vol. 21, p. 191.
[2] S. Erhan, L. G. Northrup, and F. R. Leach, *Proc. Nat. Acad. Sci. U.S.* **53**, 646 (1965).
[3] R. Litman, *J. Biol. Chem.* **243**, 6222 (1968).
[4] B. Alberts and G. Herrick, this series, Vol. 21, p. 198.
[5] D. Rickwood, *Biochim. Biophys. Acta* **269**, 47 (1972).

of ethanol before being air dried in an oven at 60°. The dried Sephadex is ground to a fine powder before use.

1-Cyclohexyl-3-(-2-morpholinethyl) carbodiimide metho-p-toluene-sulfonate, 250 mg, is dissolved in 3.75 ml of 0.04 M sodium 2-(N-morpholinoethane sulfonate (pH 6.0) containing 30 mg of sonicated DNA [Size range 6 to 7 S]. The reaction mixture is then spread evenly over the bottom of a glass dish (170 cm²) and 250 mg of dry G-200 is sprinkled evenly over the bottom of the dish.[5a] The dish containing the DNA-Sephadex mixture is placed in an oven at 42° for 8 hours and transferred to a water saturated atmosphere at 20° for 24 hours.

The DNA-Sephadex is suspended in 0.15 M NaCl, 15 mM sodium citrate, and 10 mM Tris chloride (pH 7.4). After 4 hours the swollen Sephadex is transferred to a Büchner funnel and washed with 6 volumes of distilled water, 20 volumes of 1 M NaCl–10 mM Tris chloride (pH 7.4), and finally with 6 volumes of 0.15 M NaCl–15 mM sodium citrate containing 10 mM Tris chloride (pH 7.4). It may be stored as a slurry at 0° in this medium with $CHCl_3$ as a preservative.

The amount of DNA bound to the Sephadex can be determined as follows: The gel is equilibrated with distilled water, washed twice with 10 ml of ethanol, and dried at 80°. If this residue is digested in 2 ml of 0.4 M $HClO_4$ at 90° for 20 minutes, the DNA present in the hot acid extract can be determined by the diphenylamine method.[6]

As shown by the results presented in Table I, sonication of DNA is essential to avoid nonspecific and noncovalent binding. Under optimum conditions, 38 mg of DNA were bound per gram of Sephadex G-200. In the case of Sephadex G-25, binding is very inefficient.

The DNA bound to Sephadex is susceptible to the hydrolytic action of pancreatic deoxyribonuclease. In the presence of RNA polymerase and all four nucleoside triphosphates, the Sephadex-bound DNA could direct the synthesis of RNA. However, the template capacity of the bound DNA was only one-tenth that of an equivalent amount of free DNA.

Attachment of Nucleic Acids to Agarose

Sepharose matrices, a beaded form of agarose, offer the advantage of high capacity and good flow rates. The covalent binding of nucleic acids to Sepharose was first reported by Poonian, Schlabach, and Weissbach[7] and is described here.

[5a] D. Rickwood (personal communication).
[6] W. C. Schneider, this series, Vol. 21, p. 680 (1957).
[7] M. S. Poonian, A. J. Schlabach, and A. Weissbach, *Biochemistry* 10, 420 (1971).

TABLE I
BINDING OF DNA TO SEPHADEX[a]

DNA[b]	Incubation conditions	Milligrams of DNA bound per gram of Sephadex	Binding (%)
Highly polymerized	Standard	15	50
	Omit carbodiimide	13	43
Sonicated	Standard	9	30
	Omit carbodiimide	0	0
	Dry at 20°	6	20
	75 mg of carbodiimide per reaction mix	8	27
	100 mg of Sephadex G-200 per reaction mix	10	34
	6.0 mg of DNA per reaction mix	38	32
	Replace Sephadex G-200 by Sephadex 25	0.5	1

[a] From D. Rickwood, *Biochim. Biophys. Acta* **269,** 47 (1972).
[b] DNA, highly polymerized (greater than 30 S) or sonicated (6–7 S), was incubated under the conditions shown. The DNA remaining bound to the Sephadex after washing was digested in 0.4 M HClO$_4$ at 90°, and the solubilized DNA was estimated as described in the text. All values are the mean of two separate incubations using different batches of DNA and Sephadex.

Procedure for Substrate Preparation. The method was patterned after that of Cuatrecasas[8] in which a ligand with an amino group was attached to Sepharose by forming a putative chemical link between the hydroxyl of Sepharose and nitrogen of the amino group. In preparing the nucleic acid–Sepharose the following procedure is used. A slurry of a known volume of Sepharose 4B is obtained by mixing it with an equal volume of water. The slurry container is surrounded with a jacket of iced water to maintain the temperature below 10° during the reaction. The pH of the reaction is determined by keeping the electrode of the pH meter in the reaction mixture throughout the reaction. A calculated amount of well powdered cyanogen bromide (200–500 mg/ml of Sepharose) is added in one lot to the gently stirred slurry and dropwise addition of 2–5 M sodium hydroxide is started immediately so as to maintain the pH at 11. Cessation of the decrease in pH below 11 and simultaneous disappearance of solid cyanogen bromide indicates completion of the activation reaction. The slurry of activated Sepharose is immediately cooled by adding ice to it and is then filtered and washed three times with 10 volumes of 50 mM potassium phosphate (pH 8.0) under suction without allowing the activated Sepharose to dry. The activated complex

[8] P. Cuatrecasas, *J. Biol. Chem.* **245,** 3059 (1970).

is added to the nucleic acid ligand in a volume of 50 mM potassium phosphate at pH 8.0 equal to that of packed Sepharose used in the reaction; the mixture is stirred gently at 4° for a period of 16–48 hours. The reaction slurry is packed into an appropriate column and washed with 50 mM potassium phosphate (pH 8.0) until nucleic acid no longer appears in the wash as determined by radioactivity or by $A_{260\ nm}$. To ensure complete removal of noncovalently attached material, a final wash of 0.5 M potassium phosphate at pH 8.0 is used. Quantitation of the conjugation may be based upon the difference between input and wash radioactive counts and/or A_{260}. In case of radioisotope-labeled material, direct counts were also taken on coupled Sepharose by making a slurry in appropriate buffer and counting an aliquot.

The results of coupling various DNA species to activated Sepharose are presented in Table II. With the exception of poly[d(A-T)], the other DNA species measured do not undergo a significant coupling reaction when present in a native double-stranded form. In contrast, extensive binding occurs when a double-stranded species is denatured to the single-stranded form. Thus 20% of denatured *E. coli* DNA can be attached to agarose whereas none of the original native DNA binds under these conditions. Similarly, the introduction of single-stranded ends into a double-stranded DNA permits attachment of the nucleic acid to occur. As shown in Table II, native λ DNA, a double-stranded structure containing complementary single-stranded 5' ends 12 nucleotides in length[9] does not bind to Sepharose. If 5' ends of the λ DNA are degraded 12.5% on each side (about 6000 nucleotides) with the λ exonuclease, 40% of the DNA can be attached. It is important to note that the digestion of double-stranded DNA by the λ-exonuclease to yield single-stranded ends must be carried out at high pH (9.5), where the enzyme acts in a random fashion. At lower pH values, the enzyme processively and preferentially digests the molecule to which it initially binds.[10] The dependency of binding on single strandedness is further illustrated by the results with HeLa DNA wherein the percentage of DNA which can be attached to Sepharose increases as the amount of single-stranded ends increases. However, the amount of DNA which the Sepharose can bind is apparently limited since at a high DNA:Sepharose ratio (10 A units per milliliter of Sepharose) the efficiency of attachment seems to decrease. In case of poly[d(A-T)] the pH of the solution (pH 8.0) is probably enough to "soften" the double-stranded hydrogen bonding and allow the coupling to proceed.

[9] R. Wu and A. D. Kaiser, *J. Mol. Biol.* **35**, 523 (1968).
[10] D. M. Carter and C. M. Radding, *Fed. Proc., Fed. Amer. Soc. Exp. Biol.* **29**, 896 (1970).

TABLE II

ATTACHMENT OF NUCLEIC ACIDS TO SEPHAROSE[a]

Nucleic acid	Packed Sepharose volume (ml)	CNBr (g)	Nucleic acid used		Amount coupled to Sepharose		
			A_{260}	Cpm	$A_{260/ml}$	Total A_{260}	%
[14C]Poly(dA-T)	5	1	0.91	2.2×10^5	0.136	0.68	75
3H *Escherichia coli* DNA (native)	5	1.3	29.6	1.5×10^7	0	0	0
3H *E. coli* DNA (native) (sonicated)	5	1.2	8.7	5×10^5	0	0	0
3H *E. coli* DNA (native) (coupling performed at pH 10.65)[b]	5	1.3	4.3	4×10^5	0.017	0.085	2
3H *E. coli* DNA (denatured)	5	1.3	11.05	4×10^5	0.461	2.3	20.9
3H λ DNA (native)	5	2.26	6.0	1×10^5	0	0	0
3H λ DNA (12.5% single-stranded end on each side)	10	5.0	29.8	5×10^5	1.19	11.9	40
3H HeLa DNA (native)							
2.2% single-stranded end on each side	5	1.8	5.0	5×10^5	0.043	0.215	4.3
3.8% single-stranded end on each side	10	4.95	39.0	5×10^5	0.663	6.63	17
8.3% single-stranded end on each side	5	1.5	2.0	2×10^5	0.192	0.96	24
3H HeLa DNA (denatured)	30	9.0	300	1×10^6	0.90	27	9.0
Polyadenylic acid	10	2	77.0		7.5	75	98

[a] All coupling reactions were carried out in 50 mM sodium or potassium phosphate buffer (pH 8.0) unless otherwise indicated.
[b] Sodium carbonate buffer, 50 mM was used.

With Sepharose-bound DNA it has been possible to partially purify HeLa cell DNA polymerase[7] and both bovine pancreatic and hog spleen deoxyribonuclease.[11]

Binding of RNA to Sepharose

CNBr-activated Sepharose can also be used as a matrix for the attachment of single-stranded RNA or synthetic polynucleotides. Wagner et al.[12] have coupled RNA and poly(I:C) to Sepharose and the attachment of poly(U), using this technique, has been done by Lindberg and Persson.[13] Although binding of poly(A) to Sepharose at pH 8.0 has been reported[7] (Table II), more efficient attachment (>90%) of polynucleotides to CNBr activated Sepharose is more consistently obtained at pH 6.0 under the conditions of Wagner et al.,[12] as described below.

Sepharose 4B is washed with 0.1 N NaCl and then with deionized water. For the activation reaction, 15 ml of washed Sepharose 4B are suspended in 30 ml of deionized water and cooled to about 10°. After the addition of 4.5 g of CNBr to the stirred suspension, the pH is maintained at 11 by the addition of 8 N NaOH. The temperature rises to 20° in 15 minutes. The activated matrix material is placed on a filter and washed with 600 ml of cold deionized water. It can be used promptly in the condensation reaction or after several weeks of storage in deionized water at 4°.

The binding of single-stranded RNA to Sepharose is accomplished by adding 3 volumes of a solution (3–4 mg/ml) of the ribonucleic acid in 0.2 M 4-morpholinoethanesulfonic acid buffer (pH 6.0) to 1 volume of activated Sepharose. The suspension is stirred overnight at 4°, and the product is isolated on a filter. The Sepharose-bound ribonucleic acid is packed in a column and washed continuously with 700 ml of 0.15 ml NaCl–6 mM sodium phosphate at pH 7.0. The amount of ribonucleic acid binding to activated Sepharose is established by hydrolysis of an aliquot of the product in 10 volumes of 0.25 N NaOH overnight and measurement of the ultraviolet absorption of the filtrate at pH 7.0.

Attachment of RNA to Agarose with DDC

The attachment of RNA by its 3′ terminus to agarose has been accomplished by Robberson and Davidson.[14] Their procedure used alkaline activation of agarose with CNBr followed by coupling of ε-aminocaproic

[11] J. C. Schabort, J. Chromatogr. 73, 253 (1972).
[12] A. F. Wagner, R. L. Bugianesi, and T. Y. Shen, Biochem. Biophys. Res. Commun. 45, 184 (1971).
[13] U. Lindberg and T. Persson, Eur. J. Biochem. 31, 246 (1972).
[14] D. L. Robberson and N. Davidson, Biochemistry 11, 533 (1972).

acid methyl ester to the agarose. The caprolyl ester-agarose is then converted to the corresponding hydrazide. After further blockage of resin carboxyl groups with glycinamide and water soluble carbodiimide, the treated agarose is coupled with RNA which had previously been oxidized with sodium periodate. Although rather more complicated, this procedure may have specialized use.

Entrapment of DNA in Agarose

These matrices are prepared according to the procedure of Bendich and Bolton[15] for immobilizing DNA in agar. In contrast with the covalently attached DNA-matrices discussed above, the mode of attachment here involves entrapment of single-stranded DNA molecules in the agarose matrix. DNA-agarose preparations containing 3.5 mg of calf thymus DNA per milliliter of bed volume or 0.75 mg of FD-circular DNA per milliliter of agarose have been described by Schaller et al.[16] They have shown that 6% of the soluble E. coli proteins are retained on the DNA agarose and can be eluted with increasing salt concentrations. The enzymes purified with columns of this type include E. coli DNA polymerases I and II, exonuclease III, RNA polymerase, T4 polynucleotide kinase[16] and the E. coli ribonucleases.[17]

Nucleic Acid–Cellulose Matrices

These are the most commonly used materials in nucleic acid affinity chromatography. The attachment of polynucleotides to cellulose with carbodiimide reagents has been reviewed by Gilham[1] and has been successfully used in many laboratories. DNA has also been bound to cellulose by adsorption techniques,[4] and Litman[3] has used ultraviolet light irradiation to bind DNA to cellulose. The Litman technique as described below, is very efficient and leads to high DNA ratios of cellulose.

Preparation of DNA-Cellulose.[3] Cellulose, 15 g (Solka-Floc, Brown Company, Berlin, New Hampshire), is washed by stirring for 10 minutes in 300 ml of 1 N HCl, filtered on Whatman No. 1 paper in a Büchner funnel, and rinsed thereon with water. The cellulose is then stirred with another 200 ml of 1 N HCl for 10 minutes, again filtered and washed exhaustively with water until no trace of acid remains, and finally spread out to dry in air.

DNA, 16 mg in 8 ml of 1 or 10 mM, NaCl, is mixed with 1 g of

[15] A. J. Bendich and E. T. Bolton, this series, Vol. 12B, p. 635.
[16] H. Schaller, C. Nüsslein, F. J. Bonhoeffer, C. Kurz, and I. Nietzschmann, *Eur. J. Biochem.* 26, 474 (1972).
[17] S. C. Weatherford, L. S. Weisberg, D. T. Achord, and D. Apirion, *Biochem. Biophys. Res. Commun.* 49, 1307 (1972).

acid-washed cellulose. Adequate mixing is achieved by kneading with a spatula; the paste is spread thinly over the surface of 250-ml beaker and dried with cool air for several hours. The preparation is allowed to remain at room temperature overnight and is then scraped from the sides of the beaker and suspended in absolute alcohol. A low pressure mercury lamp (Mineralight, Ultra-Violet Products, Inc., South Pasadena, California) is placed at a distance of about 10 cm from the surface of the alcohol suspension and irradiation (about 100,000 ergs/mm²), is carried out for 15 minutes with continuous slow stirring. The slurry is filtered with suction on Whatman No. 2 paper, and the filter cake is washed three times by stirring in 50 ml of 1 mM NaCl for 10 minutes. After filtration, the preparation is spread out on filter paper to dry in air. About 10% of the DNA was found to be bound to the cellulose in the absence of irradiation and 90% after irradiation.[3] Preparations may be stored at room temperature for several years.

RNA-Cellulose

This technique using ultraviolet light irradiation has been extended by Smith and co-workers[18] to the attachment of RNA to methylated cellulose as follows: 5 g of methylated cellulose are added to 8–15 mg of RNA in 20 ml of 5 mM NaCl, and the mixture is dried overnight under a stream of air. Methylation is used to block the cellulose carboxyl groups and minimizes nonspecific adsorption of nucleic acids. The dried RNA-cellulose is resuspended in 100 ml of absolute ethanol and irradiated for 30 minutes at 15 cm from a low-pressure mercury lamp (100,000 ergs/mm²) with continuous stirring. After removal of ethanol by filtration, the RNA-cellulose is washed with 4–6 liters of 5 mM NaCl. About 30% of the input RNA is bound to cellulose under these conditions.

Coupling of Nucleic Acids to Other Insoluble Matrices

The use of ultraviolet light irradiation as a "binder" has also been extended to other matrices. Sheldon et al.[19] have attached both polyuridylic and polycytidylic acid to fiberglass filters in the presence of ultraviolet light and used these filters for isolation of messenger RNA.

Preparation and Use of Poly(U) and Poly(C) Filters.[19] One hundred and fifty microliters of poly(U) solution (1 mg/ml in water) are added to a fiberglass filter (Whatman GF/C, 2.4 cm in diameter). The filter is dried at 37° and irradiated for 2.5 minutes on each side at a distance

[18] I. Smith, H. Smith, and S. Pifko, *Anal. Biochem.* 48, 27 (1972).
[19] R. Sheldon, C. Jurale, and J. Kates, *Proc. Nat. Acad. Sci. U.S.* 69, 417 (1972).

of 22 cm from a 30 W Sylvania germicidal lamp. Each filter is washed with 50 ml of distilled water to remove unbound poly(U). About 65% of the poly(U) applied is retained on the filter; ten times more poly(U) is retained if the input is increased 10-fold.

The same technique can be applied to the preparation of poly(C) filters except that the filters are equilibrated with 0.45 M NaCl–45 mM sodium citrate, and this buffer is used for the application of sample and washing of the filters. The remainder of the procedure is as described above.

Radioactive RNA to be tested for binding to poly(U) may be dissolved in any convenient amount (25 μl to 10 ml) of the binding buffer (10 mM Tris chloride of pH 7.5 containing 0.12 M NaCl). The sample is applied either without suction for very small volumes or at a suction rate of 2 ml per minute for larger volumes. The filters are washed first with water and then with buffer before the sample is applied.

The sample is applied at room temperature. Once the sample is applied, one waits a minute or two and then washes the filters sequentially with the binding buffer, with 5% TCA and with 95% ethanol. The filters are dried and counted.

If it is desired to re-use these fiberglass filters, the TCA and ethanol washes are omitted. The filters are washed instead with 0.30 M NaCl–0.03 M sodium citrate, dried and then counted in a toluene based scintillation fluid (Liquifluor). For regeneration, the filters are removed from the scintillation fluid and washed with pure toluene. After drying, they are suspended in 15 mM NaCl–1.5 mM sodium citrate (10 ml per filter) at 65° for 1 hour, and each filter is subsequently washed with 20 ml of this buffer at 65°. This procedure elutes the hybridized RNA, leaving behind the original poly(U)-fiberglass filter.[20]

Matrices of Acrylic Polymers in Agar. The coupling of periodate oxidized RNA to a linear polymer of acrylic hydrazide entrapped in agar has been reported by Petre et al.[21] and is based on a procedure first described by Knorre et al.[22] These acrylic matrices have been bound to messenger-RNA,[21] tRNA,[23] and polyuridylic acid,[24] and subsequently used for the isolation of ribosomes from *E. coli*.

[20] A. Skalka, unpublished observations.
[21] J. Petre, A. Bollen, P. Nokin, and H. Grosjean, *Biochim.* **54**, 823 (1972).
[22] D. G. Knorre, S. D. Misina, and L. C. Sandachtchiev, *Izv. Sib. Otd. Akad. Nauk SSSR Seriya Khim. Nauk* **11**, 134 (1964).
[23] O. D. Nelidova and L. L. Kiselev, *Mol. Biol.* (*USSR*) **2**, 47 (1968).
[24] L. L. Kiselev and T. A. Ardonina, *Mol. Biol.* (*USSR*) **3**, 88 (1971).

Applications of Polynucleotide Matrices

Polynucleotide-cellulose combinations provide an excellent measure of the versatility of nucleic acid matrices. The following examples show the use of polynucleotide-cellulose for the isolation of messenger RNA, the purification of an RNA dependent DNA polymerase, and as an actual substrate for the enzyme polynucleotide ligase.

Purification of Biologically Active Globin Messenger RNA by Chromatography on Oligothymidylic Acid-Cellulose[25]

A convenient technique for partial purification of large quantities of functional polyadenylic acid-rich mRNA has been used for the isolation of rabbit globin mRNA. Since many mammalian and viral mRNA's appear to be rich in polyadenylic acid, this approach should prove to be generally useful as an initial step in the isolation of such mRNA's. Oligothymidylic acid–cellulose columns are prepared according to the procedure of Gilham,[26] and RNA is chromatographed on these columns as follows:

All operations are performed at room temperature with sterile glass-ware and reagents [except for oligo(dT)-cellulose]. Crude rabbit reticulo-cyte polysomal RNA [100 A_{260} units] is dissolved in a buffer containing 10 mM Tris chloride (pH 7.5)–0.5 M KCl and applied to a 2-ml (about 0.5 g, dry weight) oligo(dT)–cellulose column equilibrated with this buffer. The nonabsorbed material is eluted by continued washing with the same buffer. The material retained by the column is then eluted sequentially with buffers of reduced ionic strength. The first elution buffer contains 10 mM Tris chloride at pH 7.5–0.1 M KCl; the second, 10 mM Tris chloride at the same pH. The material eluted by either of these buffers is precipitated by the addition of CH_3COOK [final concentration of 2%] and 2 volumes of ethanol or is frozen directly and stored in liquid nitrogen. The oligo(dT)-cellulose can be regenerated for further use by washing with 0.1 M KOH.

Over 95% of the total polysomal RNA applied to the oligo(dT)-cellulose column is eluted with the application buffer, and the rest is retained by the column. When the KCl concentration of the elution buffer was reduced from 0.5 to 0.1 M, an additional small amount of UV-absorbing material is eluted. Final elution with 10 mM Tris chloride at pH 7.5 releases an additional peak of RNA from the oligo(dT)-cellulose. This last peak exhibits the highest protein-synthesizing activity in a cell-

[25] H. Aviv and P. Leder, *Proc. Nat. Acad. Sci. U.S.* **69**, 1408 (1972).
[26] P. Gilham, *J. Amer. Chem. Soc.* **86**, 4982 (1964).

free protein-synthesizing system, and the major activity in the peak was shown to sediment in sucrose gradients at approximately 9 S.

Purification of RNA-Dependent DNA Polymerase from RNA Tumor Viruses[27]

The polymerase from Rauscher murine leukemia virus has been purified by a one-step enrichment procedure on columns of $(dT)_{12-18}$-cellulose. Advantage is taken of the property of viral polymerases to preferentially bind to $(dT)_{12-18}$ primers.[28]

Procedure. An aqueous suspension of virus is diluted with an equal volume of buffer containing 0.1 M Tris chloride (pH 7.8), 1.0 M KCl, 2 mM dithiothreitol, 2% Triton X-100, 40% (v/v) glycerol. The suspension is incubated for 30 minutes at 37° and centrifuged at 100,000 g for 1 hour at 4°. The supernatant fluid is dialyzed for 1 hour against 250 volumes of buffer containing 10 mM potassium phosphate (pH 7.1), 1 mM dithiothreitol, 60 mM KCl, 0.1% Triton X-100, 20% (v/v) glycerol, and 0.5 mM manganese acetate. The sample is applied to a $(dT)_{12-18}$–cellulose[26] column, equilibrated with the same buffer, and eluted with a linear KCl gradient in this buffer containing no $Mn.^{2+}$ The RNA-dependent DNA polymerase elutes as a sharp peak at approximately 0.25 M KCl. The columns may be washed with 1 M KCl and reused, but this should not be done more than four times.

After the $(dT)_{12-18}$–cellulose column enrichment, the enzyme can be further purified by phosphocellulose chromatography and gel filtration on Sephadex G-100.

Polynucleotide-Cellulose as a Substrate for a Polynucleotide Ligase Induced by Phage T4[29]

In this procedure polynucleotide-cellulose is used as substrate in assaying a T4-induced polynucleotide ligase. The assay is rapid and sensitive and directly measures the linkage of two polynucleotide strands. The starting material is the interrupted homopolymer pair of dI:dC schematically depicted in Fig. 1.

One strand of poly(dC) is ^{3}H-labeled and the other is covalently joined to a cellulose particle. Alkaline denaturation removes the labeled poly(dC) from the cellulose (Fig. 1b) unless the interruption has been enzymatically repaired (Fig. 1c, d). Labeled strands covalently attached

[27] B. L. Gerwin and J. B. Milstein, *Proc. Nat. Acad. Sci. U.S.* **69**, 2599 (1972).

[28] D. M. L. Livingston, E. M. Scolnick, W. P. Parks, and G. J. Todaro, *Proc. Nat. Acad. Sci. U.S.* **69**, 393 (1972).

[29] N. R. Cozzarelli, N. E. Melechen, T. M. Jovin, and A. Kornberg, *Biochem. Biophys. Res. Commun.* **28**, 578 (1967).

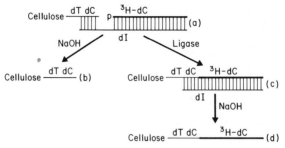

Fig. 1. Schematic representation of the assay for polynucleotide ligase as described by N. R. Cozzarelli, N. E. Melechen, T. M. Jovin, and A. Kornberg [*Biochem. Biophys. Res. Commun.* **28**, 578 (1967)]. The substrate (a) consists of dT, an oligomer of 5–10 deoxythymidylate residues esterified through its 5'-phosphoryl terminus to one of the glucose hydroxyl groups of the cellulose. The dT is in turn joined by phosphodiester linkage to dC, a polymer of 100–200 deoxycytidylate residues with a free 3'-hydroxyl group at its terminus. The dC and ^{3}H-labeled dC, a strand of about 1800 deoxycytidylate residues with free 5'-phosphoryl and 3'-hydroxyl termini, are hydrogen bonded to dI, a strand of about 1800 deoxyinosinate residues. Alkaline denaturation of (a) leads to (b), in which ^{3}H-labeled dC and dI are removed from the cellulose complex. Ligase action joins ^{3}H-labeled dC covalently to cellulose-dTdC to produce (c), alkaline denaturation of which leads to (d), in which ^{3}H-labeled dC is retained in the cellulose complex.

to cellulose are collected by filtration of the alkaline mixture and provide a measure of the number of successful strand unions.

Preparation of the Substrate. Terminal deoxynucleotidyl transferase from calf thymus[30] may be used to synthesize a poly(dC) strand attached to the oligo dT portion of oligo dT-cellulose.[26] The reaction mixture contains 20 mg of oligo dT-cellulose (about 1.5 μmoles of nucleotide residues), 200 mM potassium cacodylate at pH 7.0, 23 mM KCl, 1 mM CoCl$_2$, 1 mM β-mercaptoethanol, 1 mM dCTP, and 9 μg of purified terminal transferase in a total volume of 0.23 ml. After 4–8 nmoles of deoxycytidylate residues are attached per milligram of oligo dT-cellulose, the cellulose is sequentially washed with 50-ml portions of 10 mM EDTA, 300 mM NaCl, 50 mM NaOH, water, ethanol, and ether and dried under reduced pressure. To anneal poly(dI) to this product, 40 mg of the poly(dC)–oligo(dT)–cellulose are mixed with 1.24 ml of 1.5 mM EDTA, 0.19 mM poly(dI), and 20 mM NaOH to destroy any preexisting secondary structure. The mixture is adjusted to pH 7.8 with 20 mM sodium phosphate buffer, and made 110 mM with NaCl; it is stirred for 2 hours at 25° and for another 2 hours at 21°. The product is washed

[30] K. Kato, J. M. Goncalves, G. E. Houts, and F. J. Bollum, *J. Biol. Chem.* **242**, 2780 (1967).

3 times with 10 mM sodium phosphate at pH 7.8, containing 100 mM NaCl, and 0.5 mM EDTA. The cellulose pellet is then annealed to poly(dC) by resuspending the pellet in 0.83 ml of this buffer containing 0.11 mM ^{3}H-labeled poly(dC)[31] (7.6 × 10[7] cpm/pmole of nucleotide). The mixture is agitated for 4 hours at 21° until about 60% of the poly(dC) has been annealed. The product is washed with the sodium phosphate–NaCl–EDTA buffer and stored at 4° at a cellulose concentration of 10 mg/ml in this buffer; this is the substrate described in Fig. 1a.

The assay for polynucleotide ligase measures the covalent linking of the ^{3}H-labeled poly(dC) to the cellulose-dTdC substrate (Fig. 1).[29]

[31] M. J. Chamberlin and D. L. Patterson, *J. Mol. Biol.* **12**, 410 (1965).

[55] Immobilized Nucleotides for Affinity Chromatography

By Meir Wilchek and Raphael Lamed

Introduction and Principle

Nucleosides and nucleotide phosphates occur in many biological systems and act as enzyme substrates, as cofactors, and as coenzymes. Thus, covalently binding these compounds to insoluble matrices provides a useful tool for the study of their interactions with proteins[1,2] as well as media for affinity chromatography.[3,4]

A simple and general method for coupling nucleosides, nucleotide phosphate, and coenzymes possessing vicinal free hydroxyl groups to agarose hydrazide columns, has been described.[5,6] (Scheme I). These systems are prepared by reacting an excess of a dihydrazide of the structure

$$NH_2NH-\overset{\overset{O}{\|}}{C}(CH_2)_x\overset{\overset{O}{\|}}{C}-NHNH_2$$

to cyanogen bromide-activated polymer, in either sodium bicarbonate or acetic acid solutions. The coupling of periodate-oxidized nucleotides to the agarose hydrazide is carried out under mild conditions; the reac-

[1] M. Wilchek, Y. Salomon, M. Lowe, and Z. Zellinger, *Biochem. Biophys. Res. Commun.* **45**, 1177 (1971).
[2] R. Lamed, Y. Levin, and A. Oplatka, *Biochim. Biophys. Acta* **153**, 163 (1973).
[3] P. O. Larsson and K. Mosbach, *Biotechnol. Bioenerg.* **8**, 393 (1971).
[4] O. Berglund and F. Ekstein, *Eur. J. Biochem.* **28**, 492 (1972).
[5] R. Lamed, Y. Levin, and M. Wilchek, *Biochim. Biophys. Acta* **204**, 231 (1973).
[6] D. L. Robberson and N. Davidson, *Biochemistry* **11**, 533 (1972).

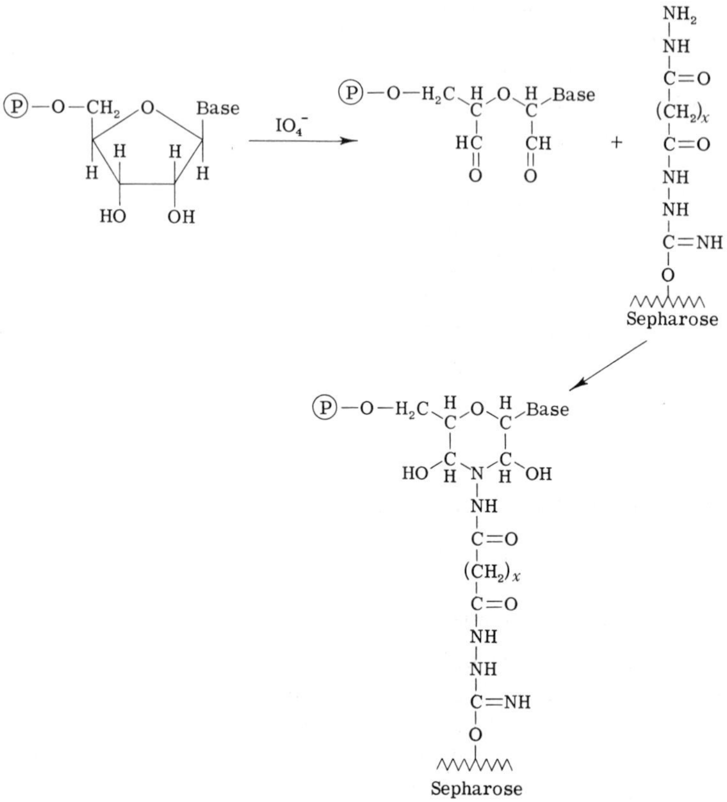

SCHEME I

tion is fast, and the binding capacity is very high.[5] The same procedure was also used for coupling to polyacrylic hydrazide. The table shows some of the nucleotides coupled by this method.

Alternatively, periodate-oxidized compounds can be allowed to react first with the dihydrazide, followed by coupling to cyanogen bromide activated agarose.

In some cases it may be of advantage to reduce the hydrazone bond formed between the oxidized nucleotides and the hydrazide with sodium borohydride. This reduction gives a different structural analog, which may be more stable.

Preparation of Selective Adsorbent[5]—Method I

Preparation of Acid Dihydrazides. One hundred milliliters of diethyl adipate, 200 ml of hydrazine hydrate (98%), and 200 ml of ethanol are

THE BINDING OF NUCLEOTIDES TO SEPHAROSE HYDRAZIDE

Nucleotide	Micromoles added per milliliter of Sepharose	Micromoles bound per milliliter of Sepharose
AMP	4.4	4
ADP	4.8	4.3
ATP	7	3.5
UMP	3.9	3.6
UTP	4.0	3.2
CMP	4.8	3.9
CTP	4.2	3.5
GMP	3.7	2.9
GTP	4.5	3.3
NAD	5.0	3.75
NADP	4.7	4.1

refluxed for 3 hours. The resulting adipic acid dihydrazide precipitates and is recrystallized twice from ethanol–water mixture (mp 169–171).

Dihydrazides of longer carbon chains are recrystallized from acetic acid or washed with ethanol.

Preparation of Agarose Hydrazide

(a) *With a Water-Soluble Dihydrazide (Adipate).* Sepharose 4B is activated by cyanogen bromide.[7,8] The activated Sepharose gel is suspended in 1 volume of a cold saturated solution of adipic acid dihydrazide (approximately 90 g/liter) in 0.1 M sodium carbonate, and the reaction is allowed to proceed overnight at 4° with stirring. The Sepharose-hydrazide gel is washed thoroughly with water and 0.2 M NaCl until samples of the washings show only minimal color reaction in the 2,4,6-trinitrobenzene sulfonate test. Columns containing 6–12 μmoles of hydrazide per milliliter of packed agarose are stored at 4°. The column is washed again before use.

(b) *With Water-Insoluble Dihydrazide (Sebacate = NH₂NH—C* $(=O)$—$(CH_2)_8$—$C(=O)$—$NHNH_2$). The cyanogen bromide-activated Sepharose gel is suspended in 2 volumes of a saturated solution of sebacic acid hydrazide in acetic acid at 20° (approximately 50 g/liter). The reaction is allowed to proceed with stirring overnight at room temperature and the product is washed thoroughly with acetic acid:water (1:1) and then with water until samples of the washings show no color on reaction with 2,4,6-trinitrobenzene sulfonide. The sebacate column

---

[7] P. Cuatrecasas, M. Wilchek, and C. B. Anfinsen, *Proc. Nat. Acad. Sci. U.S.* **61**, 636 (1968).

[8] R. Axén, J. Porath, and S. Ernbäck, *Nature (London)* **214**, 1302 (1967).

is more effective for affinity chromatography because of the longer spacer.

Preparation of Periodate Oxidized Nucleotides. The method of Gilham[9] is used. Metaperiodate solution is added to cold neutral solutions of nucleotides to a final concentration of 10 mM nucleotide and 9.5 mM periodate although more concentrated solutions may also be used. The oxidation is allowed to proceed for 1 hour in the dark at 0°.

Preparation of the Affinity Adsorbents. To 1 ml of Sepharose hydrazide in 0.1 M sodium acetate at pH 5 are added 4–5 μmoles of periodate-oxidized nucleotide in 2.5 ml of the same buffer. The suspension is stirred for 3 hours at 4°, and the product is washed with 1 M NaCl and water. The absorbance of the supernatant liquid and washings are measured to determine the extent of binding (see the table).

Preparation of Selective Adsorbent—Method II

(a) *Preparation of Nucleotide Adipic Hydrazide.* A 0.1 M neutral solution of nucleotide is oxidized by addition of solid metaperiodate (0.95 eq). The reaction mixture is stored at 0° in the dark for 1 hour and then titrated with HCl to pH 5. An equal volume of 0.1 M solution of adipic acid dihydrazide is added and the reaction is allowed to proceed at 4° for 3 hours.

(b) *Coupling of the Nucleotide Hydrazide to Sepharose.* One milliliter of cyanogen bromide-activated Sepharose is suspended in 2 ml of the nucleotide-hydrazide solution containing 0.2 M sodium carbonate at pH 8.5–9.0. The mixture is stirred at 4° for 16 hours. The resulting conjugate may still contain cyanogen bromide reactive groups, and treatment with ethanolamine or acid is recommended.

Reduction with NaBH$_4$ of (I) and (II). The Sepharose-nucleotide derivatives are suspended in 2 volumes of 0.5 M Tricine chloride at pH 8 and stirred at 4°. Three to four portions of solid NaBH$_4$ (10 mg per milliliter of Sepharose) are added at intervals of 1 hour and the resulting derivatives are washed with cold water and stored at 4°.

Conditions for Chromatography

Adsorption to and Elution of Glucose-6-phosphate Dehydrogenase from an Agarose-NADP Resin. Agarose-NADP (0.15 ml) is suspended in a total volume of 0.5 ml of 10 mM Tris chloride at pH 8.1. A suspension (10 μl) of glucose-6-phosphate dehydrogenase (Sigma, 27 μg/ml; 70 $A_{340\ nm}$ units per minute) in 2.6 M (NH$_4$)$_2$SO$_4$ is added. After stirring for 10 minutes at 25°, the suspension is centrifuged and samples from the supernatant fluid are checked for glucose-6-phosphate dehydrogenase

[9] P. T. Gilham, this series, Vol. 21, p. 191.

activity according to the method of Goldman and Lenhoff.[10] The immobilized NADP completely adsorbs the enzyme. The agarose pellet is suspended in 0.5 ml of 10 mM Tris chloride at pH 8.1 containing 4 mM NADP and is stirred for 5–10 minutes and centrifuged. The supernatant fluid is checked for dehydrogenase activity. The enzyme is readily recovered by this treatment.

Adsorption and Elution of Heavy Meromyosin (HMM) from Agarose-ATP Resin. To a suspension of 3 ml containing 1 ml of Sepharose-ATP and Ca^{2+} (2 mM) in 10 mM imidazole buffer at pH 7, 1.6 mg of heavy meromyosin are added. After 15 minutes of stirring at 0°, the suspension is centrifuged, and the ATPase activity and protein concentration are determined in the supernatant liquid. The active heavy meromyosin is almost completely adsorbed by the ATP-agarose conjugate. Mixtures containing water or Sepharose hydrazide instead of Sepharose-ATP are used as controls. To elute the HMM, concentrated solutions of ATP or ADP are added to give a final concentration of 5 mM nucleotide. After 15 minutes of stirring at 0°, the ATPase activity and the protein concentration of the supernatant are measured. Most of the adsorbed HMM is released by this treatment.

[10] R. Goldman and H. M. Lenhoff, *Biochim. Biophys. Acta* **242**, 514 (1971).

[56] Nucleoside Phosphates Attached to Agarose

By ROBERT BARKER, IAN P. TRAYER, and ROBERT L. HILL

Nucleoside mono-, di-, and triphosphates and many of their derivatives, such as the sugar nucleotides and the vitamin-containing coenzymes, are substrates, inhibitors, or cofactors for a wide variety of enzymes. Thus, nucleotide phosphates attached to agarose should have wide application as affinity adsorbents for the purification of many different enzymes.

Derivatives of nucleoside phosphates containing a primary amino group can be coupled readily to cyanogen bromide-activated agarose. The major problem in preparation of the adsorbents is the design and synthesis of the nucleotide ligands to be attached to agarose. At present, four kinds of ligand (**1, 2, 3,** and **4**) have been synthesized containing an alkyl- or aryl-amino group attached to either the phosphate, the purine base, or the ribosyl ring of the nucleotide.

Compound (**1**), 3'-(4-aminophenylphosphoryl)deoxythymidine 5'-phosphate, is a good inhibitor of staphalococcal nuclease, and reacts with agarose through the *p*-amino group to give a useful adsorbent for the

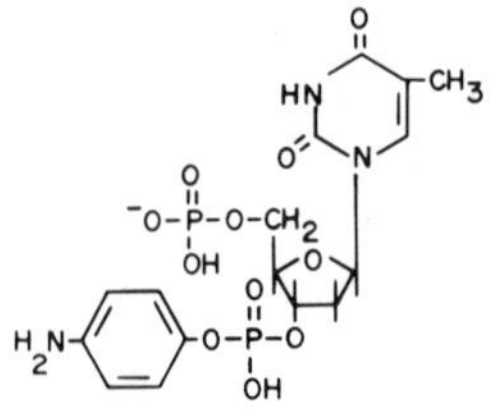

(1) 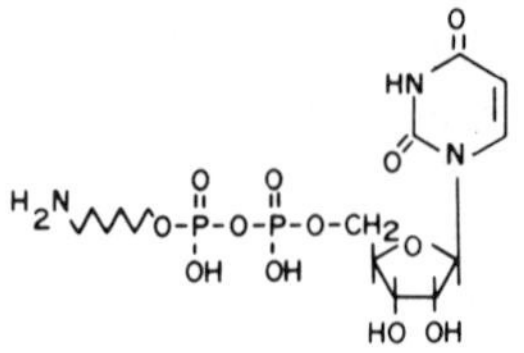   (2)

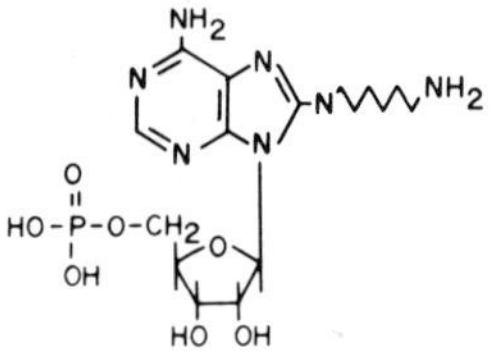

(3) 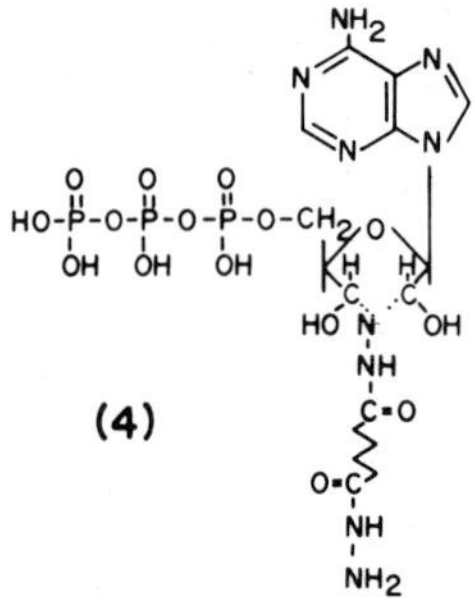   (4)

nuclease.[1] In principle the procedures used to prepare compound 1 can be applied to the synthesis of other nucleoside monophosphate ligands. Compound 2, P1-(6-amino-1-hexyl)-P2-5′-uridine pyrophosphate, couples to agarose through the primary amino group and its agarose derivative is an excellent adsorbent for certain enzymes.[2]

Compound 3, which contains a 1,6-diaminohexyl group at C-8 of the adenine ring provides a purine nucleotide with a free phosphate substituent.[3] A second ligand of this type contains a 6-aminohexyl group on the C-6 amino group rather than on the C-8 carbon of the adenine ring.[4,5] Other useful adsorbents can be prepared by reaction of periodate treated nucleoside phosphates with an acyl hydrazide derivative of agarose.[6,7] The functional ligand in these adsorbents have the general structure shown for 4. (The exact nature of the linkage between the periodate oxidized ribose moiety and the acid hydrazide has not been established.) Since attach-

[1] P. Cuatrecasas, M. Wilchek, and C. B. Anfinsen, *Biochemistry* **8**, 2277 (1969).
[2] R. Barker, K. W. Olsen, J. H. Shaper, and R. L. Hill, *J. Biol. Chem.* **247**, 7135 (1972).
[3] I. P. Trayer, unpublished observations.
[4] K. Mosbach, H. Guilford, R. Ohlsson, and M. Scott, *Biochem. J.* **127**, 625 (1972).
[5] H. Guilford, P. O. Larsson, and K. Mosbach, *Chem. Scripta* **2**, 165 (1972).
[6] R. Lamed, Y. Levin, and M. Wilchek, *Biochim. Biophys. Acta* **304**, 231 (1973); M. Wilchek and R. Lamed, this volume [55].
[7] R. Lamed, Y. Levin, and A. Oplatka, *Biochim. Biophys. Acta* **305**, 163 (1973).

ment to agarose is through the modified ribosyl moiety, both the phosphates and the pyrimidine or purine ring are unsubstituted.

The preparation and properties of agarose adsorbents containing compound 1 have been described in this series,[8] and the p-aminophenyl esters of nucleoside di- and triphosphates are described in this volume, Berglund and Eckstein [18]. This chapter is devoted to the preparation and properties of adsorbents containing ligands, similar to compounds 2, 3, and 4.

Synthesis of Ligands

A. 6-Amino-1-hexyl Nucleoside Phosphoesters

The series of reactions shown in Scheme I have been used to synthesize ligands of this type. In this synthesis any phosphate ester can be

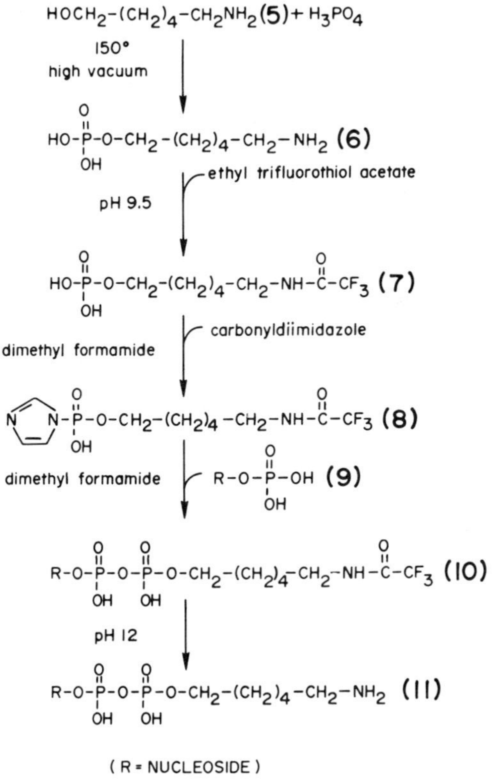

$$HOCH_2-(CH_2)_4-CH_2NH_2 \, \textbf{(5)} + H_3PO_4$$

150°
high vacuum

$$HO-\overset{\overset{O}{\|}}{\underset{OH}{P}}-O-CH_2-(CH_2)_4-CH_2-NH_2 \quad \textbf{(6)}$$

ethyl trifluorothiol acetate

pH 9.5

$$HO-\overset{\overset{O}{\|}}{\underset{OH}{P}}-O-CH_2-(CH_2)_4-CH_2-NH-\overset{\overset{O}{\|}}{C}-CF_3 \quad \textbf{(7)}$$

carbonyldiimidazole

dimethyl formamide

$$\overset{N}{\underset{N}{\diagdown}}N-\overset{\overset{O}{\|}}{\underset{OH}{P}}-O-CH_2-(CH_2)_4-CH_2-NH-\overset{\overset{O}{\|}}{C}-CF_3 \quad \textbf{(8)}$$

dimethyl formamide

$$R-O-\overset{\overset{O}{\|}}{\underset{OH}{P}}-OH \quad \textbf{(9)}$$

$$R-O-\overset{\overset{O}{\|}}{\underset{OH}{P}}-O-\overset{\overset{O}{\|}}{\underset{OH}{P}}-O-CH_2-(CH_2)_4-CH_2-NH-\overset{\overset{O}{\|}}{C}-CF_3 \quad \textbf{(10)}$$

pH 12

$$R-O-\overset{\overset{O}{\|}}{\underset{OH}{P}}-O-\overset{\overset{O}{\|}}{\underset{OH}{P}}-O-CH_2-(CH_2)_4-CH_2-NH_2 \quad \textbf{(11)}$$

(R = NUCLEOSIDE)

SCHEME I

[8] P. Cuatrecasas and C. B. Anfinsen, this series, Vol. 22 [31].

used to prepare a ligand that contains one more phosphate group than the initial phosphate ester. The following procedures describe the synthesis of the UDP-hexanolamine ester. Similar procedures have been used to synthesize the 6-amino-1-hexyl esters of GDP, ADP, ATP, dGTP,[9] and galactosyl pyrophosphate. These ligands couple readily with cyanogen bromide-treated agarose to give adsorbents with 4–6 μmoles of ligand per milliliter of agarose.

6-Amino-1-hexanol Phosphate (**6**). 6-Amino-1-hexanol (**5**)[10] (11.7 g, 0.1 mole) is mixed with 9.8 g (0.1 mole) of crystalline phosphoric acid[11] in a round-bottom flask. The flask is attached by a connecting tube with a side arm to a second flask immersed in a dry ice–ethanol bath. The apparatus is evacuated to less than 0.1 mm Hg and the reaction flask heated to 150 ± 5° in an oil bath. After heating under vacuum for 18 hours, the amber, solid reaction mixture is cooled to room temperature and dissolved in 150–200 ml of water. The solution is adjusted to pH 10.5 with 5 N lithium hydroxide, and the precipitated lithium phosphate is removed by vacuum filtration through Celite. The filtrate is adjusted to pH 3 with acetic acid and passed through a column (150 ml) of Dowex 50-X8 (pyridinium cycle; 20–50 mesh). The column is washed with 200 ml of 0.5 M acetic acid. The ninhydrin-positive eluate and washings are concentrated and then dried under reduced pressure (0.1 mm) for 2–3 hours at 100° to remove pyridinium acetate. The residue is crystallized from water (75 ml) by the addition of ethyl alcohol (75 ml). The product, 6-amino-1-hexanol phosphate (**6**) is strongly ninhydrin-positive,[12] contained a trace of inorganic phosphate,[13] and shows traces of 6-amino-1-hexanol when chromatographed in solvent 1[14] (6-amino-1-hexanol phosphate, R_f 0.54; 6-amino-1-hexanol, R_f 0.85; inorganic phosphate, R_f 0.45). In solvent 2 the product has R_f 0.35.

[9] R. L. Blakely, private communication.
[10] Aldrich Chemical Co., Milwaukee, Wisconsin.
[11] K & K Laboratories, Jamaica, New York.
[12] Amines can be detected by spotting samples on paper (10–50 μliters) and spraying with a solution containing 0.5 g of ninhydrin, 20 ml of glacial acetic acid, 80 ml of absolute ethanol, and 1 ml of *s*-collidine and heating for 5 minutes at 110° [A. L. Levy and D. Chung, *Anal. Chem.* **25**, 396 (1953)].
[13] Phosphates can be detected by spotting samples on paper (10–50 μliters) and spraying with a reagent containing 5 ml of 60% (w/w) perchloric acid, 25 ml 4% (w/v) of ammonium molybdate, 10 ml of 1 N HCl, and 60 ml of water followed by heating at 80–100° until the paper is dry and by exposure to ultraviolet light to give a blue color. Inorganic phosphate gives an immediate yellow color with this spray [R. S. Bandorsky and B. Axelrod, *J. Biol. Chem.* **193**, 405 (1951)].
[14] Ascending paper chromatography is employed on Whatman No. 1 paper. Solvent 1 is isobutyric acid–conc. ammonium hydroxide–water (66:1:33, v/v). Solvent 2 is isopropyl alcohol–1.3 M aqueous acetate buffer, pH 5 (7:3, v/v). Solvent 3 is *n*-butanol–acetic acid–water (5:2:3, v/v).

Recrystallization from water by the addition of ethyl alcohol gives material with mp 245°; yield 10.0 g, 51%. The compound has pK'_a values of 2.0, 6.3, and 11.0.

N-Trifluoroacetyl 6-Amino-1-hexanol Phosphate (7). To 2 g of **6**, dissolved in 40 ml of water at 9°, is added 2 ml of ethyl trifluorothiol acetate[15] dissolved in 10 ml of acetone. The mixture is kept below 4° in an ice bath and stirred vigorously with a magnetic stirrer to produce a fine dispersion of the ethyl trifluorothiol acetate. The reaction mixture is adjusted periodically to pH 9.5 with 5 N lithium hydroxide. The reaction is monitored intermittently by spotting 10–25 μl of the mixture on paper, spraying with ninhydrin,[12] and heating at 100° for 5 minutes. The reaction is judged to be complete after 20–30 minutes when a pink color is obtained with ninhydrin. The mixture is adjusted to pH 5 with trifluoroacetic acid, concentrated from water 4 or 5 times[16] to remove residual reagents and then dried under reduced pressure over anhydrous calcium sulfate for 24 hours. The syrupy product (yield 95% by weight) has an R_f of 0.58 in solvents 1 and 2[14] as judged with the molybdate spray for phosphate.[13] Only traces of the starting material are detected by the molybdate and ninhydrin tests.[12] The product can be stored in a desiccator for at least 2 months without decomposition. Although crystallization occurs on storage, a suitable solvent for recrystallization has not been found. The product is ninhydrin-negative except in concentrations above 5%, when a pink color is apparent. A 1 mM solution in water at pH 11 for 30 minutes is completely converted to trifluoroacetate, and the starting amine as judged by chromatography in solvents 1 and 2.[14]

N-Trifluoroacetyl-O-phosphoryl 6-amino-1-hexanol imidazolide (8). To a solution of 200 mg (0.68 mmole) of **6** in 2 ml of dry acetone or dimethylformamide[17] are added 160 mg (1.0 millimole) of 1,1′-carbonyldiimidazole.[10,18] The reagent dissolves rapidly with the evolution of a small volume of gas. The reaction is kept at room temperature under a drying tube for 6 hours, 0.015 ml (0.36 mmole) of methyl alcohol is added to hydrolyze excess diimidazole,[19] and after 20 minutes the mixture is concentrated to dryness at 45°.[16] Chromatography of the dry residue in solvents 1 and 2[14] shows for each only one product, with R_f 0.8 and 0.75,

---

[15] Pierce Chemical Co., Rockford, Illinois.

[16] Solvents are removed with a rotary evaporator at water aspirator pressure.

[17] Reagent grade organic solvents are dried over 4 Å molecular sieves (Fischer Scientific Co., Pittsburgh, Pennsylvania) for at least 24 hours and used without further purification.

[18] 1,1′-Carbonyldiimidazole (mp 113–115°) decomposes to imidazole (mp 88–90°) and carbon dioxide if not kept under strictly anhydrous conditions. Occasionally samples are partially decomposed; however, if the melting point is above 105° the reaction will proceed satisfactorily.

[19] D. E. Hoard and D. G. Ott, *J. Amer. Chem. Soc.* **87**, 1785 (1965).

respectively. The imidazolide (8) is stable in dimethylformamide solution for at least 4 weeks, but in aqueous solution is hydrolyzed about 50% in 24 hours at pH 7.5.

P^1-(6-Amino-1-hexyl)-P^2-(5'-uridine)-pyrophosphate-(2 and 11, R = 5'-Uridine). The imidazolide (8) from 3.1 g (10.5 mmoles) of compound 7 is dissolved in 4 ml of dimethylformamide,[17] and mixed with a solution containing 5 moles of uridine 5-phosphate-methyl tri-n-octylammonium salt[20] in 8 ml of dimethylformamide.[17] The mixture is stored in a desiccator over Drierite at room temperature for 24 hours and then chromatographically purified as shown in Fig. 1. The desired N-trifluoroacetyl-hexanolamine UDP is obtained in about 70% yield based on the absorbance of the starting material and the product obtained by pooling fractions 70–110. The product can be used for the preparation of the UDP-Sepharose adsorbent without further purification by removing the trifluoroacetyl group (10–20 μmoles/ml) at pH 11 for 4 hours at 25°.

N-Trifluoroacetyl-UDP-hexanolamine (10) may be characterized as follows: Fractions from the column (Fig. 1) are combined, adjusted to pH 8.0 with 1 N lithium hydroxide, and concentrated to dryness at 45° [16]; the residue is dried under reduced pressure over Drierite. Lithium chloride is removed by extraction and centrifugation of the thoroughly dried residue (5–7 times) with 50 ml of ethanol–ethyl ether (1:2, v/v). The extracted residue is ninhydrin negative,[12] but reacts strongly after incubation of pH 11–12 for a few minutes at 25°. Its molar extinction at 262 nm is 10^4 based on the molecular weight of 611 for the dilithium salt. It has R_f of 0.45 and 0.5 in solvents 1 and 2,[14] respectively. Hydrolysis with 1 N hydrochloric acid at 100° for 1 hour yields a mixture of compound 7 and UDP. Hydrolysis at pH 12 gives a product that is eluted from Dowex 1 (Cl⁻) as a single band in the same region as UMP as shown in Fig. 1, but with a strong positive ninhydrin reaction. This product after acid hydrolysis yields a mixture of compound 5 and UMP as determined by chromatography in solvents 1 and 2.[14]

Affinity Chromatography on Agarose Derivatives of Compound 2. Adsorbents containing 6-amino-1-hexyl nucleoside esters can be prepared

---

[20] Salts of nucleoside phosphates are converted to methyl tri-n-octyl ammonium salts as follows: Free acids of the nucleoside phosphates are prepared by treating solutions of either the sodium, potassium, lithium, or calcium salts with Dowex 50-X8 (H⁺ cycle; 20–50 mesh) at room temperature. The resin is removed by filtration and washed several times with water. The combined washings and filtrate are mixed with 1 molar equivalent of methyl tri-N-octylammonium hydroxide [R. Letters and A. M. Michelson, Bull. Soc. Chem. Biol. 45, 1353 (1963)] in ethanol–water solution. Ethanol is added as required to give a clear solution, and the solvents are removed under reduced pressure at 25°. The residue is dried by evaporation from dimethylformamide and stored in a desiccator over anhydrous calcium sulfate.

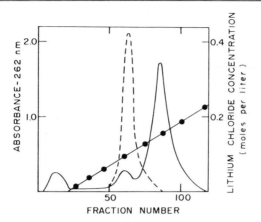

FIG. 1. The purification of trifluoroacetylhexanolamine-UDP and UDP-hexanol-amine by chromatography on Dowex 1 (Cl-cycle) [R. Barker, K. W. Olsen, J. H. Shaper, and R. L. Hill, *J. Biol. Chem.* **247**, 7135 (1972)]. Trifluoroacetylhexanol-amine-UDP, the reaction mixture from 1 mmole of UMP was applied to a column (1 × 25 cm) of Dowex 1-X2 (Cl-cycle; 200–400 mesh) and eluted with 250 ml of 50% aqueous methanol and then with a linear gradient of 500 ml of 0.01 N HCl as starting solvent and 500 ml of 10 mM HCl containing 0.4 M lithium chloride as the limit solvent. Fractions (12 ml) were collected automatically. The column was operated at 25° at a flow rate of about 60 ml per hour. The elution pattern is shown by the solid line. UDP-hexanolamine-UDP, which was prepared as described in the text, was chromatographed exactly as described above. The elution pattern is shown by the dashed line.

by reaction of the ester, e.g., compound **2**, with cyanogen bromide-treated Sepharose. Activation of the Sepharose and the coupling procedures used are essentially identical to that in this series [29, Section D]. The extent of coupling can be estimated by measuring the decrease in adsorbance during coupling or by phosphate analysis. An example of the use of UDP-hexanolamine-Sepharose for enzyme purification is the purification of a galactosyl transferase from bovine whey.[2]

B. Adenosine Phosphate Ligands Substituted in the
Adenine Ring

The synthesis of N^6-(6-aminohexyl)-adenosine 5′-phosphate is given in this volume (Mosbach [16] and Deon and Harvey [17]). This section is devoted to the 6-aminohexyl derivative of 8-aminoadenine nucleotides. The synthetic sequence used is shown in Scheme II. In principle, the monobromo-derivatives of other nucleosides[21–23] could be treated in an

[21] R. E. Holmes and R. K. Robins, *J. Amer. Chem. Soc.* **86**, 1242 (1964).

[22] A. M. Michelson, J. Dondon, and M. Grunberg-Manago, *Biochim. Biophys. Acta* **55**, 529 (1962).

[23] D. M. Frisch and D. W. Visser, *J. Amer. Chem. Soc.* **81**, 1756 (1959).

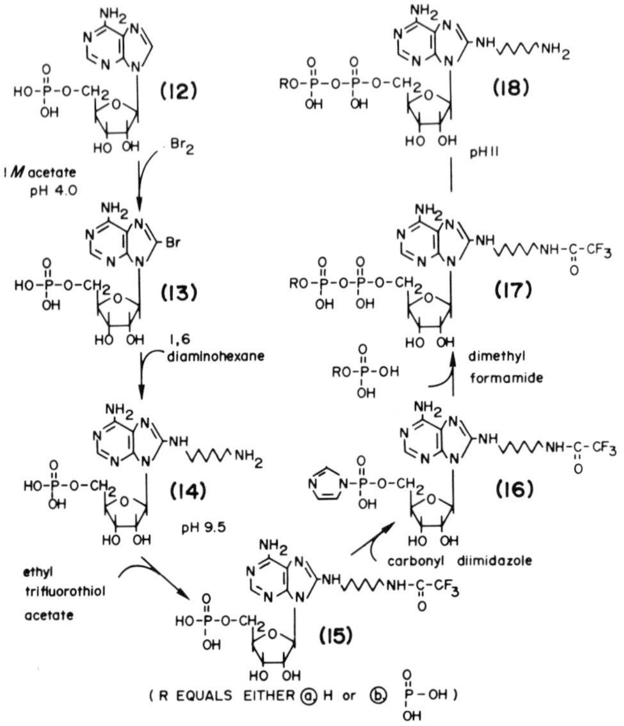

SCHEME II

analogous manner to provide a range of mononucleoside or mononucleotide derivatives suitable for affinity chromatography. Reaction of nicotinamide mononucleotide with the above imidazolide (**16**, Scheme II) should give an NAD$^+$ ligand which could be coupled to agarose through C-8 of the adenine ring. In addition, the N^6-(6-aminohexyl)-adenosine 5′-monophosphate derivative has been converted to the corresponding ADP and ATP derivatives by the imidazolide methods described below for the 8-amino derivative.

8-Bromoadenosine 5′-Monophosphate[24] (**13**). Disodium adenosine-5′-monophosphate (0.9 g, 2 mmoles) is dissolved in 1 M sodium acetate at pH 4.0 (80 ml), and bromine–water (0.15 ml bromine, 3 mmoles, dissolved in 15 ml of water with vigorous shaking) is added. After thorough mixing, the reaction mixture is stored in the dark at room temperature for 18 hours. The color of the solution is discharged by the addition of sodium metabisulfite (0.1 g) and the solvent is removed under reduced pressure.[16] The residue is concentrated several times from ethanol to give a white solid which is dissolved in 100 ml of water. The solution is added

[24] M. Ikehara and S. Uesugi, *Chem. Pharm. Bull.* **17**, 348 (1969).

to a column (20 × 1.5 cm) of Dowex 1-X8 (formate cycle, 200–400 mesh) which is washed with water (300 ml), and then eluted with 0.1 N formic acid. Two major ultraviolet absorbing peaks are obtained. The first, which contains 8 or 9% of the applied 260 nm absorbing material, appears to be AMP. The second peak ($\lambda_{max}^{H^+}$ 262.5 nm) is eluted between 1 and 2 liters and contains 90% of the material applied. Activated charcoal (1 g) is added to the fractions containing peak 2. After stirring, the charcoal is separated by filtration and the filtrate is monitored for ultraviolet absorbing material. This process is repeated until the filtrate is free of 260 nm absorbing material; 4 g of charcoal are usually used. The product is eluted by washing the charcoal several times with 250-ml volumes of 50% (v/v) aqueous ethanol containing 2% (v/v) concentrated aqueous ammonia. The combined washings (1.5–2 liters), containing 90% of the 260 nm absorbing material applied to the column, are concentrated to dryness.[16] The residue is dissolved in 20 ml of water, filtered to remove traces of charcoal, and concentrated to dryness. The product, which is the diammonium salt of **13**, can be purified by precipitation from 50% (v/v) aqueous methanol with acetone-ether,[24] but this is not necessary. The yield is 75% based on ultraviolet absorption. The product migrates as a single spot on high voltage paper electrophoresis[25] at pH 6.5 and on paper chromatography; R_f 0.31 insolvent 3.[14] Ultraviolet absorption properties: $\lambda_{max}^{pH\,1} = 262.5$ nm ($\epsilon_{262.5\,nm}^{m/1} = 16,400$); $\lambda_{max}^{pH\,11} = 264.5$ nm ($\epsilon_{264.5\,nm}^{m/1} = 15,100$).

N^8-(*6-Aminohexyl*)-*8-aminoadenosine 5'-Phosphate* (**14**). The diammonium salt at **13** (0.48 g, 1 mmole) is dissolved in water (50 ml), and 1,6-diaminohexane (3.48 g, 30 mmoles) is added. The resulting solution is sealed in six thick-wall Pyrex test tubes and heated in an oil bath at 140° for 2 hours. The cooled fractions are pooled and diluted to 100 ml with water, adjusted to pH 11.5, and applied to a column (20 × 1.5 cm) of Dowex 1-X8 (acetate cycle, 200–400 mesh). Unreacted diamine is eluted with 200 ml of water whereas the product is eluted with a linear gradient formed with 1 liter each of water and 1 M acetic acid. The major 260 nm-absorbing fractions, eluting in the first 15–20% of the gradient and containing 85% of the total 260 nm absorbing material applied to the column, are pooled, concentrated to a small volume, and lyophilized. The product migrates as a single component on chromatography in solvent 1 (R_f 0.63) and solvent 3 (R_f 0.18)[14] and is electrophoretically homogeneous at pH 6.5.[25] It gives positive tests with ninhydrin and molybdate reagents, contains 1 mole of phosphorus[26] per mole of purine base, and has

[25] High voltage paper electrophoresis is performed under varsol with pyridine acetate buffer, pH 6.5 (pyridine:acetic acid:water, 2.5:1:225, v/v) at a potential gradient of 50 V/cm.

[26] G. R. Bartlett, *J. Biol. Chem.* **234**, 466 (1959).

an ultraviolet absorption spectrum characteristic of 8-aminoalkyl derivatives of the adenine nucleus.[27] On this basis, the structure **14** was assigned to this product. Ultraviolet absorption data: $\lambda_{max}^{pH\ 1} = 277$ nm ($\epsilon_{267\ nm}^{m/1} = 17,700$) $\lambda_{max}^{pH\ 11} = 278$ ($\epsilon_{278\ nm}^{m/1} = 20,450$).

This material can be attached directly to cyanogen bromide-activated agarose or can be used for the preparation of di- and triphosphate derivatives.

N[8]-*(6-Trifluoroacetamidohexyl)-8-aminoadenosine 5'-Phosphate* (**15**). To 0.5 g, 1 mmole of **14** dissolved in water (25 ml), in an ice-bath is added 1 ml of ethyl trifluorothiol acetate.[15] The reaction mixture is vigorously stirred to produce a fine dispersion and was adjusted periodically to pH 9.5 with 5 *N* lithium hydroxide. Additional ethyl trifluorothiol acetate (3–4 ml) is added during 2 hours until an aliquot of the reaction mixture no longer gives a positive ninhydrin test.[12] The mixture is then adjusted to pH 5 with trifluoroacetic acid and concentrated to dryness several times from water.[16] The final residue is dissolved in water (10 ml), adjusted to pH 1.5 with trifluoroacetic acid, and passed over a column (6 × 0.6 cm) of Dowex 50-X8 (H+ cycle, 20–50 mesh) to remove unreacted amine. Elution with water is continued until material absorbing at 280 nm no longer appears from the column, whereupon the combined eluents are concentrated to dryness.[16] The yield is essentially quantitative. The ninhydrin-negative[12] product migrates as a single anionic spot (located by molybdate spray for phosphorus[13]) on high voltage paper electrophoresis at pH 6.5.[25] The product (**15**) may be used without further characterization in the next step.

N[8]-*(6-Trifluoroacetamidohexyl)-8-aminoadenosine 5'-Diphosphate* (**17a**). A solution of **15** (0.56 g, 1 mmole in 10 ml of water) is applied to a column (6 × 0.6 cm) of Dowex 50-X8 (pyridinium cycle, 20–50 mesh) resin, and washed with 10 ml of water. To the combined eluent and washings, tributylamine (7.5 ml, 31 mmoles) is added. The solution is concentrated under reduced pressure and the syrupy residue dried by repeated addition and evaporation[16] of anhydrous pyridine[17] (5 × 10 ml), and then of anhydrous dimethylformamide[17] (3 × 10 ml). The anhydrous tributyl ammonium salt of **15** is dissolved in 10 ml of dimethylformamide, and 1,1'-carbonyldiimidazole[10,18] (0.8 g, 5 mmoles) is added. The mixture is shaken vigorously for 15–20 minutes in a tightly stoppered container and left in a desiccator at room temperature overnight. Methyl alcohol (66 μl, 1.6 mmoles) is then added to hydrolyze excess diimidazole,[19] and, after 30 minutes, the mixture is concentrated to dryness at 40–50° under

[27] K. Muneyama, R. J. Bauer, D. A. Shuman, R. K. Robins, and L. N. Simon, *Biochemistry* **10**, 2390 (1971).

reduced pressure. The imidazolide (16) is stable for several weeks if kept in dimethylformamide under anhydrous conditions.

To the imidazolide (16) in 10 ml of dimethylformamide is added anhydrous tributyl-ammonium phosphate[28] (5 mmoles) in 10 ml of dimethylformamide and after shaking vigorously for 30 minutes, the reaction mixture is left in a desiccator at room temperature for 1 day. The precipitate of imidazolium phosphate, formed during the reaction, is removed by centrifugation, and the precipitate is extracted with three 1-ml portions of dimethylformamide. The combined extracts are added to an equal volume of methanol, and the resulting solution concentrated at 40°.[16] The residue is dissolved in 50% (v/v) aqueous ethanol (50 ml) and applied to a column (15 × 3 cm) of DEAE-cellulose. The column is developed with 50% (v/v) aqueous ethanol (300 ml), water (100 ml), and then with a linear gradient from water to 0.4 M triethyl-ammonium bicarbonate, pH 7.5,[29] over a total volume of 3.6 liters. Two major peaks absorbing at 280 nm are eluted from the column. The peak eluting at the lower salt concentration represents about 30% of the total 280 nm absorbing material applied to the column and is chromatographically and electrophoretically identical to N^8-(6-trifluoroacetamidohexyl)-8-aminoadenosine 5'-phosphate. The second peak, containing about 60%[30] of the 280 nm absorbing material applied, is pooled and evaporated to a syrup; the triethylammonium bicarbonate is removed by repeated evaporation from ethanol at 40–50°.[16] This material is homogeneous by electrophoresis at pH 6.5.[25] It contains 2 moles of total phosphorus[21] and 1 mole of acid-labile phosphorus[31] per mole of purine base.

N^8-(6-Aminohexyl)-8-aminoadenosine 5-Diphosphate (18a). The product, 17 (0.2–0.3 g), is dissolved in 30 ml of water; a few drops of methanol are added to aid dissolution if necessary, and the solution is adjusted to pH 11 with 1 N sodium hydroxide. After 3–4 hours at room temperature, the mixture is adjusted to pH 7.0 with 1 N HCl. The solution is strongly ninhydrin positive[12] and contains a single component as judged

---

[28] Prepared by passing an aqueous solution of crystalline phosphoric acid (0.49 g, 5 mmoles in 5 ml of water) over a column (6 × 0.6 cm) of Dowex 50-X8 (pyridinium cycle, 20–50 mesh) resin. To the solution of pyridinium phosphate so obtained is added tributylamine (1.2 ml, 5 mmoles) and, after mixing, the solution is concentrated.[16] The syrupy residue is concentrated from pyridine[17] (5 times) and then from 3 × 10 ml portions of dimethylformamide.[17]

[29] Triethylammonium bicarbonate solutions are prepared by bubbling CO_2 through a sintered-glass diffuser into triethylamine solutions until the pH is 7.5.

[30] The yield varies from 40% to 90% and may be a consequence of moisture in the condensation reaction, since vigorously anhydrous conditions are not stressed.

[31] Acid-labile phosphorus is the inorganic phosphorus resulting from treatment at 100°C for 30 minutes in 1 N HCl.

by electrophoresis at pH 6.5.[25] This product could be desalted on a Sephadex G-10 column (140 × 3 cm) but is generally coupled to cyanogen bromide-activated agarose without further characterization. It has the same λ_{max} and extinction coefficient as compound **15**.

N^8-(6-Aminohexyl)-8-aminoadenosine 5'-Triphosphate (**18b**). This compound can be prepared with tributyl-ammonium pyrophosphate instead of the tributyl-ammonium phosphate in the above reaction sequence. All other reaction conditions and separation procedures are identical. The triphosphate derivative is stable on removal of the trifluoroacetyl group at pH 11, and no inorganic phosphate is produced. Compound **18b** has the same ultraviolet absorption properties and molar extinction coefficient as compound **15**. It contains 3 moles of total phosphorus[26] and 2 moles of acid-labile phosphorus[31] per mole of purine and is electrophoretically homogeneous at pH 6.5.[25]

Affinity Chromatography on Agarose-Adenosine Phosphate Derivatives of Type **3**. Adsorbents are prepared by coupling ligands of type **3** with cyanogen bromide treated Sepharose under the usual coupling conditions (Section A, above). Both the N^6-(6-aminohexyl)adenosine 5'-monophosphate[3,4,32] and the N^8-(6-aminohexyl)-8-aminoadenosine 5-phosphate[3] derivatives have been used extensively for the purification of a variety of dehydrogenases. These adsorbents will bind several dehydrogenases from a mixture and allow their selective elution by gradient techniques.[4,32]

The ADP and ATP analog adsorbents have been useful for the purification of kinases, such as glucokinase and hexokinase,[3] and of the ATPase components of myosin, heavy meromyosin, and subfragment 1.[3] These adsorbents should find wide application in the purification of proteins having affinities for adenine nucleotides.

C. Coupling of Periodate Oxidized Ribonucleoside Phosphates to Agarose Hydrazides

In contrast to the other nucleoside phosphate adsorbents, these agarose derivatives are prepared by reaction of a nucleotide ligand with an agarose hydrazide rather than agarose treated with cyanogen bromide. Unless coupling with the agarose hydrazide is essentially complete, the resulting derivatives will contain not only the desired nucleoside phosphate ligand but a potentially reactive hydrazide. It is not clear that this is a serious problem, but it should be kept in mind. In contrast to this potential hazard, these adsorbents are easy to prepare and have a high content of bound ligand. The general procedure given below is essentially that described by Lamed and co-workers,[6,7] which the authors have found to be very satisfactory. It should be possible to prepare adsorbents from any

[32] R. Ohlsson, P. Brodelius, and K. Mosbach, *FEBS Lett.* **25**, 234 (1972).

ribonucleoside or ribonucleoside phosphate. At present, adsorbents of this type have been prepared from AMP, ADP, ATP, UMP, UTP, CMP, CTP, GMP, GTP, NAD⁺, and NADP⁺ and have about 3–4 μmoles of bound nucleotide per milliliter of agarose.[33]

Adipic Acid Dihydrazide. A mixture of 100 ml of diethyl adipate, 200 ml of hydrazine hydrate (98%), and 200 ml of ethanol are refluxed for 3 hours. Adipic acid dihydrazide precipitates during the reaction. After cooling to room temperature, the dihydrazide is collected by filtration, dissolved in a small amount of water, and ethanol is added until a slight turbidity appears. The dihydrazide crystallizes readily on standing. It may be recrystallized from water–ethanol (mp 169–171°).

Sepharose-Adipic Hydrazide. Sepharose 4B is activated with cyanogen bromide by any one of several methods, for example, that in Section D in this volume [29]. The washed, cyanogen bromide-activated Sepharose is mixed with an equal volume of a cold saturated solution of adipic acid dihydrazide (about 90 g/liter) in 0.1 M sodium carbonate buffer, pH 9.5. The mixture is stirred at 4° for 12–18 hours, filtered, and washed thoroughly with water and then 0.2 N NaCl.

Periodate Oxidation. A solution of nucleoside phosphate is adjusted to pH 8 with sodium hydroxide and diluted to a final concentration of 10 mM. The solution is cooled to 1–2°, and solid sodium periodate is added to give a final concentration of 10 mM. After 1 hour at 0° in the dark, the solution of oxidized nucleotide is coupled directly to the agarose-hydrazide.

Nucleotide-Hydrazide-Agarose. Periodate-oxidized nucleotide (4–5 μmoles) is dissolved in 2.5 ml of 0.1 M sodium acetate buffer, pH 5, and added to 1 ml of Sepharose-hydrazide suspended in 1 ml of the acetate buffer. The suspension is stirred at 4° for 3–4 hours. Sodium chloride (2.5 ml, 2 M) is added and the mixture stirred an additional 30 minutes. The mixture is then filtered, washed with several volumes of 0.2 M sodium chloride and then water. The amount of nucleotide bound can be estimated by measuring the absorbance of 260 nm of the supernatant after reaction, or by phosphate analysis of the final adsorbent.

Affinity Chromatography on Nucleotide-Hydrazide-Sepharose Adsorbents. A GTP-hydrazide-Sepharose adsorbent has been employed successfully in the purification of dihydraneopterin triphosphate synthetase.[34] Rabbit muscle lactate dehydrogenase can also be purified on the AMP-hydrazide-Sepharose adsorbent.[3]

---

[33] Adenosine 5'-phosphate and tRNA bind to aminoethyl cellulose after performate oxidation as described in this series (P. T. Gilham, Vol. 21 [10]; see also Weissbach and Poonian, this volume [54]).

[34] R. J. Jackson, R. M. Wolcott, and T. Shiota, *Biochem. Biophys. Res. Commun.* **51**, 428 (1973).

[57] Purification of Nucleases

By MEIR WILCHEK and MARIAN GORECKI

Principle

Nucleases from various sources can be purified rapidly and effectively by chromatography on columns containing selective adsorbents prepared by the covalent coupling of a specific nuclease inhibitor to an insoluble matrix. Nucleotide-2'(3'),5'-diphosphates are excellent inhibitors for the nucleases.[1,2] Since modification of one of the phosphates by the addition of the 4-aminophenyl group does not affect strongly the inhibitory potency of these compounds,[2,3] we have prepared 4-aminophenyl derivatives and used them for purification.[3,4] A selective adsorbent for staphylococcal nuclease was prepared by attaching 3'-(4-aminophenylphosphoryl)deoxy-thymidine 5'-phosphate to cyanogen bromide-activated Sepharose.[4] For bovine pancreated ribonuclease A, 5'-(4-aminophenylphosphoryl)uridine 2'(3')-phosphate was used (Fig. 1).[2] The latter derivative was also used for the isolation of ribonucleases from more than fifteen mammalian species.[5] 5'-(4-Aminophenylphosphoryl)guanosine 2'(3')phosphate was used to purify tobacco RNase.[6]

Nucleases are selectively adsorbed to these preparations and nearly quantitative elution is achieved by varying pH and ionic strength of buffer or by adding competitive inhibitors. The advantages of these purification procedures over the conventional methods are higher yields, and higher purity in shorter time.

A. Purification of Bovine Pancreatic Ribonuclease

Preparation of 5'-(4-Nitrophenylphosphoryl)uridine 2'(3')-Phosphate. The barium salt of uridine 2'(3')5'-diphosphate (840 mg, 1 mmole) is converted to its acidic form by passing it through a column of Dowex 50-X8 (H⁺ form). The eluate is lyophilized, suspended in 20 ml of dry

---

[1] P. Cuatrecasas, S. Fuchs, and C. B. Anfinsen, *J. Biol. Chem.* **242**, 1541 (1967).

[2] M. Wilchek and M. Gorecki, *Eur. J. Biochem.* **11**, 491 (1969).

[3] P. Cuatrecasas, M. Wilchek, and C. B. Anfinsen, *Biochemistry* **8**, 2277 (1969).

[4] P. Cuatrecasas, M. Wilchek, and C. B. Anfinsen, *Proc. Nat. Acad. Sci. U.S.* **61**, 636 (1969).

[5] R. K. Wierenga, J. D. Huizinga, W. Gaastra, G. W. Welling, and J. J. Beintema, *FEBS Lett.* **31**, 181 (1973).

[6] L. Jervis, *Biochem. J.* **127**, 290 (1972).

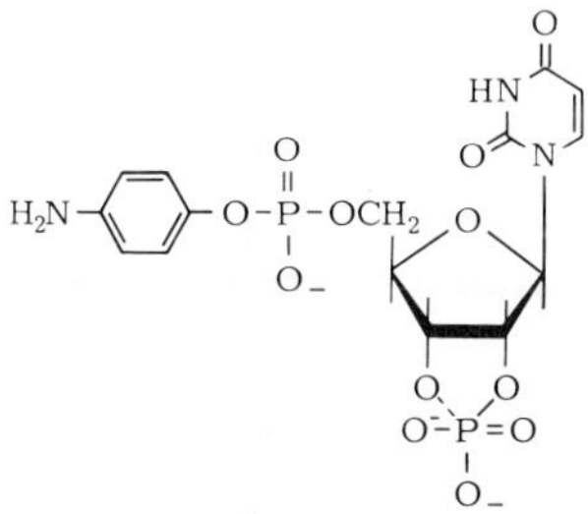

Fig. 1. 5'-(4-Aminophenylphosphoryl)uridine 2(3')-phosphate. Structure of RNase inhibitor used for attachment to Sepharose.

pyridine, and evaporated in high vacuum at room temperature. The procedure is repeated twice, and 2.8 g (20 mmoles) of *p*-nitrophenol are added; the mixture is concentrated to dryness. The residue is dissolved in 15 ml of dry pyridine, and 4.1 g (20 mmoles) of dicyclohexyl carbodiimide is added. The mixture is stirred for 3 days at room temperature. The solvent is removed under reduced pressure at 30°, and the residue is suspended in 10 ml of water and washed with ether. The aqueous solution is passed through a column of Dowex 50-X8 (H^+ form). The acid effluent is brought to pH 3.5 with sodium hydroxide and extracted with ether repeatedly until the extracts no longer form a yellow color on shaking with NH_4OH solution.

The preparation is lyophilized and the residue is dissolved in 5 ml of methanol. After removal of a small amount of insoluble material by centrifugation, 50 ml of ether are added. The white, flocculent precipitate is filtered and washed with ether.

The compound obtained is chromatographically pure on silica gel thin-layer chromatography in the solvent system, isopropanol–ammonia–water (7:1:2) (R_f 0.25). The yield is about 500 mg (85%). After incubation with snake venom phosphodiesterase, one equivalent of *p*-nitrophenol is released per equivalent of uridine as measured spectrophotometrically.

Preparation of 5'-(4-Aminophenylphosphoryl)uridine 2'(3')-Phosphate (APuP).[2] Palladium on charcoal is added to an 80% methanolic solution of 5'-(4-nitrophenylphosphoryl)uridine 2'(3')-phosphate, and reduction is carried out in a Parr instrument for 1.5 hours. The palladium is removed by filtration and the solution is concentrated to dryness. The residue is redissolved in a small volume of methanol and precipitated by addition of ether. The yield is quantitative, and the compound is homogeneous on silica gel thin-layer chromatography in isopropanol–ammonia–water (7:1:2) (R_f 0.12).

Preparation of Adsorbent. Sepharose 4B is activated by cyanogen bromide[4,7] and allowed to react with the inhibitor, APuP, in 0.1 M sodium bicarbonate at pH 9 for 16 hours at 4°. The amount of inhibitor coupled is estimated by calculating the amount of inhibitor (spectroscopically) which is not recovered in the washings.

Conditions for Adsorption of RNase to the APuP Column.[2] A sample containing 2 mg of RNase in 0.5 ml of 20 mM sodium acetate at pH 5.2 is applied to a column having dimensions of 0.5 × 5 cm, which is equilibrated with the same buffer. The column was washed until the absorbance at 280 nm of the effluent was below 0.02.

The column, when heavily loaded with APuP on aging became more efficient. Extracts from acetone powder of bovine or porcine[5] pancreas can be adsorbed by the column with 0.2 M sodium acetate at pH 5.2. The column is washed with the same buffer containing 0.2 M sodium chloride.

Elution of the RNase from the Column. The enzyme is efficiently eluted (95%) with 0.2 M acetic acid from a freshly prepared column. From an aged column, the enzyme can only be eluted at high salt concentrations and at pH 5.2, or by a linear gradient of 0.2 M NaCl–6 M NaCl in 0.2 M sodium acetate.[5] The enzyme can also be eluted with 0.1 M sodium pyrophosphate solution or with 2 M ammonium bicarbonate.

Comments

The operational capacity of the Sepharose columns for adsorption of protein was determined by two methods. In the first, an amount of enzyme in excess of the theoretical capacity was added to a small column (about 1 ml). After washing this column with buffer until negligible quantities of protein emerged in the effluent, the enzyme was rapidly removed (by acetic acid washing) and assayed. In the second method, small aliquots of purified protein were added successively to the column until significant protein or enzymatic activity emerged; the total amount added or that which was subsequently eluted, was considered the operational capacity.

The APuP column is stable between pH 3 and 11. The 4-aminophenylphosphoryl group is relatively distant from the basic structure unit recognized by the enzyme and serves as a spacer.

APuP columns were also prepared by attaching APuP to Sepharose containing a long spacer of succinylated 3,3'-diaminodipropylamine by means of a water-soluble carbodiimide 1-ethyl-3'-(3-dimethylaminopropyl) carbodiimide. These columns adsorbed RNase but also adsorbed other proteins, and incomplete elution was achieved by the eluents described above (unpublished results).

[7] R. Axén, J. Porath, and S. Ernbäck, *Nature (London)* **214**, 1302 (1967).

B. Purification of Staphylococcal Nuclease

Preparation of Thymidine 3',5'-di-(p-Nitrophenyl Phosphate) I.[3] Deoxythymidine 3',5'-diphosphate, sodium salt (Calbiochem) (586 mg, 1 mmole) is desalted by passing through a Dowex 50 (hydrogen form) column (10 × 1 cm). After lyophilization, the residue is suspended in dry pyridine and concentrated to dryness under reduced pressure at room temperature. It is resuspended in dry pyridine, 218 g (20 mmoles) of p-nitrophenol is added, and the mixture is concentrated to dryness. It is dissolved in 10 ml of dry pyridine and 4 g (20 mmoles) of dicyclohexyl carbodiimide are added. After 3 days at room temperature with stirring, the solvent is then removed under reduced pressure at 25°. The residue is suspended in 10 ml of water and washed with ether. The aqueous solution contains all the unreacted pdTp as well as about 20% of the product; 2 equivalents of NaOH are added to the ether solution, which is then extracted twice with 25 ml of water. The aqueous filtrate is extracted three times with ether and passed through a column of Dowex 50 (hydrogen form). The acid effluent is brought to pH 3.5 with NaOH and extracted with ether repeatedly until the extracts no longer form a yellow color on shaking with NH_4OH. It is then evaporated to dryness, and the residue is dissolved in 5 ml of methanol. After removal of a small amount of insoluble material by centrifugation, 50 ml of ether are added. The white, flocculent precipitate that forms is filtered and washed with ether. The compound obtained is chromatographically pure on thin-layer chromatography. Our yield was 490 mg (65%). After incubation at 100° for 4 hours in 1.0 N NaOH, 2 equivalents of p-nitrophenol are released per equivalent of pdTp as measured spectrophotometrically.

Preparation of Thymidine 5'-Phosphate 3'-(p-Nitrophenol Phosphate) *(II)*. Compound I (200 mg) is dissolved in 30 ml of 0.2 M NH_4HCO_3 (pH 8.2) containing 5 mM $MnCl_2$ and 2 mg of snake venom diesterase. The solution is left at room temperature until there is no further release of p-nitrophenol, i.e., generally overnight. After adjusting the pH 4.5 with 1 N HCl, the solution is extracted with ether until addition of NaOH no longer produces a yellow color in the ether extracts. The aqueous solution is lyophilized, dissolved in 4 ml of 50 mM ammonium acetate (pH 5.5) and passed through a 3.5 × 45 cm column of Sephadex G-15. The eluent is passed through a column (1 × 5 cm) of Dowex 50 in the hydrogen form. After lyophilization, the residue is dissolved in a small volume of water and converted into the sodium salt by bringing the pH of the solution to 3.5 with 1 N NaOH. This solution is lyophilized and the powder is dissolved in methanol and precipitated with ether. Our yield was 120 mg. The λ_{max} of the compound is 272 nm, and the E_M is 15,500.

Preparation of 3′-(4-Aminophenylphosphoryl)thymidine 5′-Phosphate (*III*). Palladium on charcoal (10%, w/w) is added to a methanolic solution of II (about 0.5 g/30 ml), and reduction is carried out at room temperature in a Parr hydrogenation apparatus for 1.5 hours at 35 psi. The palladium is removed by filtration, and the solution is concentrated to dryness. The residue is dissolved in a small volume of methanol and precipitated by the addition of ether. It is recrystallized from aqueous ethanol. The yield is quantitative. The $\lambda_{\max}$ of this compound is 268 nm, and the E_M is 8800, at pH 7.0.

Condition for Adsorption and Elution of Nuclease to the Specific Column.[4] The thymidine Sepharose column is prepared by coupling compound III to cyanogen bromide activated Sepharose. The column (0.8 × 5 cm), containing 1.5 μmoles of inhibitor, is equilibrated with 0.05 M sodium borate at pH 8.0 containing 10 mM CaCl$_2$. About 40 mg of protein containing about 8 mg of enzyme is applied in 3.2 ml of the same buffer. If a very crude enzyme solution with a protein concentration greater than 20–30 mg per milliliter is applied to the column, small amounts of nuclease appear in the first peak of protein impurities, especially if fast flow rates are used (400 ml per hour). Diluting the samples applied will prevent these leakages. After 50 ml of buffer had passed through the column, 0.1 M acetic acid is added to elute the nuclease. All the original enzyme activity is typically recovered. These columns may also be used for separation of active and inactive nuclease derivatives[8] as well as for stopping enzymatic reactions.[4]

[8] P. Cuatrecasas, M. Wilchek, and C. B. Anfinsen, *J. Biol. Chem.* **244**, 4316 (1969).

[58] Messenger RNA. Isolation with Poly(U) Agarose

By UNO LINDBERG and TORGNY PERSSON

Messenger RNA in animal cells appears to be generated by cleavage of high molecular weight precursors in the nucleus. These mRNA precursors are contained in the metabolically active pool of RNA molecules called heterogeneous nuclear RNA (HnRNA). Polysomal mRNA as well as a substantial part of the HnRNA have been shown to contain 180–200 nucleotide-long segments of poly(A) at the 3′ terminus of the polynucleotide chains. In addition to this there is a nonpolysomal pool of ribonuclear protein particles in the cytoplasm, which also contains polyadenylated RNA. At present this pool is thought to contain mRNA en route to translation. For a more detailed discussion of experimental data concerning these conclusions and for further references, the papers by Darnell et al.[1] and Jelinek et al.[2] should be consulted.

Poly(A)-containing RNA molecules can be isolated by affinity chromatography on agarose containing covalently bound polyuridylic acid.[3] The poly(A)-containing molecules are retained by the resin, apparently by their poly(A) segment, as the result of base-pairing to poly(U). Ribosomal RNA and tRNA are not retained by the column material and are thus easily removed from the poly(A)-containing RNA, which in turn can be recovered from the resin by elution with a buffer containing 90% formamide. Poly(dT)[4]- and oligo (dT)[5]-cellulose have also been used for the purpose of isolating poly(A)-containing RNA.

Procedure

Buffer Solutions

High-salt buffer I: 0.7 M NaCl, 10 mM EDTA, 25% formamide–50 mM Tris chloride at pH 7.8

High-salt buffer II: 0.5 M NaCl, 10 mM EDTA, 50% formamide–50 mM Tris chloride at pH 7.8

Elution buffer: 90% formamide, 10 mM EDTA, 0.2% sarkosyl, 10 mM potassium phosphate at pH 7.5

Coupling of Poly(U) to Agarose. Poly(U) is linked to cyanogen bromide-activated[6,7] Sepharose 4B essentially by the same procedure used for binding poly(IC) to agarose.[8] The following slightly modified procedure has been adopted. Prior to activation, 40 g of Sepharose, commercial slurry, is washed with 600 ml of deionized water and resuspended in 60 ml of ice-cold water. For activation, 9 g of cyanogen bromide dissolved in 135 ml of water is added, and the pH is raised to 11–11.5 and maintained by dropwise addition of 7 M KOH. After 15 to 20 minutes the pH remains stable and the activated Sepharose is immediately washed on a Büchner funnel with 600 ml of ice-cold 0.1 M potassium phosphate at pH 8. The activated Sepharose is resuspended in

[1] J. E. Darnell, R. Wall, M. Adesnik, and L. Philipson, "Molecular Genetics and Developmental Biology" (M. Sussman, ed.), p. 201. Prentice-Hall, Englewood Cliffs, New Jersey, 1972.

[2] W. Jelinek, M. Adesnik, M. Salditt, D. Sheiness, R. Wall, G. Molloy, L. Philipson, and J. E. Darnell, *J. Mol. Biol.* **75**, 515 (1973).

[3] U. Lindberg and T. Persson, *Eur. J. Biochem.* **31**, 246 (1972).

[4] H. Nakazoto and M. Edmonds, *J. Biol. Chem.* **247**, 3365 (1972).

[5] H. Aviv and P. Leder, *Proc. Nat. Acad. Sci. U.S.* **69**, 1408 (1972).

[6] R. Axén, J. Porath, and S. Ernbäck, *Nature (London)* **214**, 1302 (1967).

[7] J. Porath, R. Axén, and S. Ernbäck, *Nature (London)* **215**, 323 (1967).

[8] A. F. Wagner, R. L. Bugianesi, and T. Y. Shen, *Biochem. Biophys. Res. Commun.* **45**, 184 (1971).

20 ml of the 0.1 M potassium phosphate buffer and mixed with 20 ml of 0.1 M potassium phosphate at pH 8 containing 20 mg of poly-uridylic acid. The binding of poly(U) is monitored by measuring the absorbance at 260 nm of aliquots from the reaction mixture. Before reading absorbance, Sepharose beads are removed by centrifugation. Al-most all (95%) of the poly(U) is bound already after 60 minutes, but the reaction mixture is allowed to stand at 4° for another 18 hours before noncovalently linked material is removed by washing the poly(U)-Sepharose with 200 ml of elution buffer followed by 200 ml of high salt buffer I. The washed preparation has been stored at 4° for several months without deterioration, and it may be used several times if, after each experiment, it is washed immediately as indicated above.

Isolation of Poly(A)-Containing RNA. Messenger RNA from animal cell polysomes can be isolated on poly(U)-Sepharose without prior phenol extraction of the RNA. The polysomes (sample containing 10–20 OD_{260} units per milliliter) are dissociated with EDTA and sarkosyl at final concentrations of 30 mM and 1%, respectively. The sample is then diluted 5-fold with high salt buffer I and applied to the poly(U)-Sepha-rose column. When large amounts of polysomes are to be processed it is preferable to adjust the salt concentration of the sample by the addition of appropriate amounts of concentrated stock solutions of NaCl, Tris buffer, EDTA, and formamide, instead of diluting with the high-salt buffer I. After the sample has entered the resin, the column is washed with 5 column volumes of high salt buffer I. It has been found that, despite extensive washing with this buffer, 2–5% of the ribosomal RNA remains on the column. The major part of this apparently nonspecifically bound RNA can be removed in a second wash with the high-salt buffer II (containing 50% formamide). The mRNA retained on the column throughout this procedure is then eluted with 5 column volumes of the elution buffer. Up to 90% of the mRNA is recovered in the second fraction, from which it is precipitated by the addition of 25% sodium acetate (w/v) pH 6.0 to a final concentration of 2% acetate and 2 volumes of ethanol.

As much as 90–100 OD_{260} units of dissociated polysomes can be processed per milliliter of packed poly(U)-Sepharose without reaching saturation. With larger loads, however, increasing proportions of the mRNA pass through unadsorbed. It has been noted that saturation of the poly(U)-Sepharose is reached faster with mRNA of higher molecular weight than with the smaller species. Most likely this effect is due to exclusion of the largest RNA molecules from parts of the polyuridylated Sepharose matrix; i.e., more poly(U) appears to be available for binding smaller molecules than for binding larger molecules.

Similarly poly(A) containing RNA from the nonpolysomal fraction of the cytoplasma can be isolated by affinity chromatography directly after addition of EDTA, sarkosyl, and high-salt buffer I. The RNA isolated by this technique appears to be intact, as judged by comparative sucrose density gradient centrifugation before and after fractionation and by molecular weight determination in dimethyl sulfoxide–sucrose gradients and through poly(A) analysis.[3] This has been further corroborated by similar studies on adenovirus-infected cells, wherein defined species of poly(A)-containing virus-specific mRNA can be identified and followed through the purification on poly(U)-Sepharose.[9]

[9] U. Lindberg, T. Persson, and L. Philipson, *J. Virol.* **10**, 909 (1972).

[59] Carbohydrate Exchange Chromatography as a Technique for Nucleic Acid Purification

By MARVIN EDELMAN

The purification procedures for DNA in common use[1-3] are generally satisfactory for removal of proteins, lipids and RNA but are less effective for polysaccharides.[4] Part of the problem is that hydrolytic enzymes such as α-amylase cannot be usefully employed since they are normally contaminated with a variety of nucleases.[5,6] Purification by CsCl-gradient centrifugation is also ineffectual since many polysaccharides have buoyant densities similar to those of DNA.[4,7,8] To overcome these difficulties a simple and rapid procedure has been devised, based on the principles of affinity chromatography. The procedure permits purification of a wide variety of DNA's on a carbohydrate-exchange resin consisting of concanavalin A (Con A) linked to agarose.

[1] J. Marmur, *J. Mol. Biol.* **3**, 208 (1961).
[2] K. S. Kirby, *Progr. Nucl. Acid Res. Mol. Biol.* **3**, 1 (1964).
[3] See this series, Vol. 6 [100] and Vol. 12A [98].
[4] Z. M. M. Segovia, F. Sokol, I. L. Graves, and W. W. Ackermann, *Biochim. Biophys. Acta* **95**, 329 (1965).
[5] T. Ando, *Biochim. Biophys. Acta* **114**, 158 (1966).
[6] V. M. Vogt, *Eur. J. Biochem.* **33**, 192 (1973).
[7] M. Edelman, D. Swinton, J. A. Schiff, H. T. Epstein, and B. Zeldin, *Bacteriol. Rev.* **31**, 315 (1967).
[8] M. Edelman, I. M. Verma, and U. Z. Littauer, *J. Mol. Biol.* **49**, 67 (1970).

Preparation of Carbohydrate-Exchange Resin

Con A (Miles Laboratories) linked to agarose beads is available commercially.[9] Alternatively, the carbohydrate-exchange resin can be prepared from Con A[10] and Sepharose 4B by the methods of Aspberg and Porath[11] and Lloyd.[12] Suspend 50 ml of Sepharose 4B (settled beads) in 40 ml of water. Add 0.5–1.0 g of CNBr in 10 ml of water. Stir at room temperature and adjust to pH 11 with NaOH. Maintain pH between 11.0 and 11.2 for several minutes by adding more base. Filter mixture and wash successively with about two volumes each of cold water and 70 mM NaHCO$_3$.

Suspend 500–750 mg of Con A in 40 ml of 1 M NaCl and add NaHCO$_3$ to a final concentration of 0.13 M. Immediately mix with the activated Sepharose and keep overnight at 4° with occasional stirring. Drain the gel on a filter and wash out unbound protein with 200 ml of 1 M NaHCO$_3$ at pH 8.0. Then wash with 1 liter of 0.1 M sodium acetate at pH 5.0 to hydrolyze residual reactive groups on the Sepharose. Finally, wash with 10 mM sodium phosphate–0.15 M NaCl, adjusted to pH 7.0 and suspend in 100 ml of the same buffer containing 1% toluene as preservative. Maintain at 4° and do not freeze. To preserve its binding capacity the resin should be used at a pH near neutrality.

Operation of Column

Columns may be conveniently poured in shortened 5- or 10-ml pipettes with glass wool serving as a packing plug. A bed volume of 1 ml is suitable for quantitative binding of about 3 mg of polysaccharide. Before use, equilibrate the column with 20 volumes of the buffer in which the nucleic acid is to be applied, e.g., 0.15 M NaCl–15 mM sodium acetate at pH 7.3. Apply the DNA sample in a convenient volume, taking into consideration that a 2-fold dilution occurs on the column. DNA at a concentration of 1 mg/ml has been applied to a 1-ml column without serious viscosity effects developing. The carbohydrate exchange resin works efficiently at flow rates as high as 2 ml/minute. DNA is not bound by the Con A-linked agarose and is recovered with an efficiency of more than 95% in the first few fractions, whereas glycogen and starch-like

[9] Concanavalin A linked to Sepharose is available as glycosylex A from Miles Laboratories or as Con A-Sepharose from Pharmacia Fine Chemicals.
[10] Concanavalin A can be purified from Jack bean meal (General Biochemicals Co., Chagrin Falls, Ohio) by means of (NH$_4$)$_2$SO$_4$ fractionation and Sephadex adsorption as described by B. B. L. Agrawal and I. J. Goldstein, *Biochim. Biophys. Acta* 147, 262 (1967).
[11] K. Aspberg and J. Porath, *Acta Chem. Scand.* 24, 1839 (1970).
[12] K. O. Lloyd, *Arch. Biochem. Biophys.* 137, 460 (1970).

MODIFIED DNA PURIFICATION PROCEDURE[a]

	Cell lysis or breakage
	↓
(Deproteinization; lipid removal)	Chloroform, phenol or Pronase
	↓
(RNA removal)	Ribonuclease
	↓
	Alcohol precipitation
	↓
(Polysaccharide removal)	Concanavalin A-agarose chromatography
	↓
	Purified DNA

[a] General principles of DNA isolation are discussed in detail by J. Marmur [J. Mol. Biol. 3, 208 (1961)]; see also this series, Vol. 6 [100].

polysaccharides are retained by the resin and are absent from the eluted DNA sample. The bound polysaccharides may be released from the column by washing with 5–10 volumes of 5% α-methylmannoside,[13] which has a strong affinity for Con A.[14] The resin is then regenerated by removing the mannoside with 20–30 volumes of buffer. With proper care, columns may be reused many times without appreciable loss of efficiency.

Comments on the Technique

Purification of DNA. DNA extracts from mammalian cells, insect larvae, higher plants, fungal hyphae, and blue green algae have been successfully purified from endogenous polysaccharide contaminants using Con A–agarose chromatography as described above.[15] The removal of polysaccharides was monitored by analytical ultracentrifugation through gradients of CsCl. Ultraviolet absorbance and Schlieren refractive index patterns were compared[7] and polysaccharides identified on the basis of their low specific absorption, sensitivity to α-amylase and insensitivity to DNase. The DNA samples did not appear to be damaged by passage through the Con A–agarose column, as determined by CsCl band widths, thermal denaturation properties, and ability, after denaturation, to bind to cellulose nitrate filters. As a result, the general procedure for DNA purification has been modified in my laboratory to include carbohydrate-exchange chromatography (see the table).

Purification of RNA. Con A linked to agarose has also been used

[13] Methyl α-D-mannopyranoside, available from Pfanstiehl Laboratories, Waukegan, Illinois.
[14] I. J. Goldstein, C. E. Hollerman, and J. M. Merrick, *Biochim. Biophys. Acta* 97, 68 (1965).
[15] M. Edelman, unpublished data.

to separate fungal RNA from starchlike contaminants. Whole-cell RNA extracts from *Trichoderma* and *Neurospora* were obtained from columns with 95% efficiencies. The sucrose density patterns of the chromatographically purified RNA's showed that degradation had not occurred owing to contact with the resin.

Batchwise Application. The chromatographic procedure for the removal of polysaccharide contaminants from nucleic acid extracts may be modified for routine work as a batch procedure.

Add an equal volume of washed Con A-agarose to the nucleic acid sample and gently agitate for 1–2 minutes. Centrifuge the slurry at about 500 *g* for a few minutes in a clinical centrifuge and remove the upper aqueous layer which contains the polysaccharide-free nucleic acids. The pelleted carbohydrate exchange resin may be regenerated with α-methylmannoside and excess buffer as described above.

Limitations of the Column. It must be clearly understood that insoluble Con A has affinity only for those polysaccharides that conform to its specificity of binding (i.e., polysaccharides containing glucose, fructose or mannose as terminal groups[14,16]). The general usefulness of Con A-agarose for nucleic acid purification lies in the fact that many polysaccharide contaminants from a wide variety of animal and plant nucleic acid fractions are glycogen or starchlike and are bound by Con A. However, this phytohemagglutinin does not represent a "cure-all." For example, pectin, a contaminant of plant nucleic acid fractions[17] which has an unbranched chain structure consisting mainly of galacturonic acid residues, is not bound by Con A-agarose.[18] In this regard it is encouraging to note that insoluble preparations of other lectins, with specificities different from those of Con A, are becoming commercially available.[19] This should extend the usefulness of carbohydrate exchange chromatography as a technique for nucleic acid purification.

[16] I. J. Goldstein and L. L. So, *Arch. Biochem. Biophys.* **111**, 407 (1965).
[17] K. M. Jakob and I. Tal, *Biochim. Biophys. Acta* **308**, 296 (1973).
[18] K. M. Jakob and M. Edelman, unpublished data.
[19] For example: agarose–fucose binding protein; agarose–wheat germ agglutinin; agarose–soya bean agglutinin (available from Miles Laboratories, Inc.). See also Kristiansen, this volume [31].

[60] Tyrosyl-tRNA Synthetase

By HELGA BEIKIRCH, FRIEDRICH VON DER HAAR, and FRIEDRICH CRAMER

Aminoacyl-tRNA synthetases recognize their appropriate tRNA and amino acid from among at least twenty others. ATP is the third substrate of the enzymes and is needed to form the activated ester bond between tRNA and amino acid.

Tyrosyl-tRNA synthetase (EC 6.1.1.1) from baker's yeast was first isolated by van de Ven et al.[1] to 70% purity. The tyrosyl-tRNA synthetase preparation of Kijima et al.[2] was such that it showed no activity with any of the 19 other amino acids. Calendar and Berg[3] isolated tyrosyl-tRNA synthetase from *Escherichia coli* and *Bacillus subtilis,* finding molecular weights of 95,000 and 88,000, respectively.

We report here the details of the affinity method for purification of tyrosyl-tRNA synthetase from baker's yeast.[4] The product, an enzyme with a molecular weight of 42,000, is homogeneous by the criterion of sodium dodecyl sulfate gel electrophoresis and ultracentrifugation.

Affinity Chromatography of Tyrosyl-tRNA Synthetase

Tyrosine was bound to a copolymer of butanediol divinyl ether and malic acid anhydride via an amide linkage. As a spacer 1,6-diaminohexane was inserted in between (Reactions 1, 2, and 3).

$$—CO—NH—(CH_2)_6—NH_2$$
$$—COO^{\ominus} \qquad K^{\oplus} \quad \rightarrow$$
$$(I)$$

$$NH—BOC$$
$$| $$
$$—CO—NH—(CH_2)_6—NH—CO—CH—CH_2—C_6H_4—OCH_2C_6H_5$$
$$—COO^{\ominus} \qquad K^{\oplus} \qquad \qquad \rightarrow$$
$$(II)$$

$$NH_2$$
$$| $$
$$—CO—NH—(CH_2)_6—NH—CO—CH—CH_2—C_6H_4—OH$$
$$—COO^{\ominus} \qquad K^{\oplus}$$
$$(III)$$

[1] A. M. van de Ven, V. V. Koningsberger, and J. T. Overbeek, *Biochim. Biophys. Acta* **28**, 134 (1958).

[2] S. Kijima, T. Ohta, and K. Imahori, *J. Biochem.* **63**, 434 (1968).

[3] R. Calendar and P. Berg, *Biochemistry* **5**, 1681 (1966).

[4] H. Beikirch, F. von der Haar, and F. Cramer, *Eur. J. Biochem.* **26**, 182 (1972).

ISOLATION OF TYROSYL-tRNA SYNTHETASE FROM 1500 G OF YEAST

Fraction	Volume (ml)	Protein (mg)	Total activity[a] (units)	Specific activity[a] (units/ mg)	Yield (%)
Crude extract	1040	46,800	49,400	1	100
Dialyzate after ammonium sulfate fractionation	180	14,200	47,000	3.3	95
DEAE-cellulose eluate	360	13,600	46,400	3.4	94
Phosphocellulose eluate	160	256	21,300	83	43
CM-cellulose eluate	100	54	14,750	273	30
"Affinity" polymer eluate	30	7	9,900	1400	20

[a] Activity was tested in potassium cacodylate buffer with bulk tRNA.

A protein fraction containing tyrosyl-tRNA synthetase was precipitated[4] by 48% ammonium sulfate saturation from the crude extract of broken yeast cells. Nucleic acids were removed using DEAE-cellulose after which column chromatography on phosphocellulose, followed by the same procedure on CM-cellulose, yielded a protein fraction containing about 20% tyrosyl-tRNA synthetase. Starting with this material, and using the affinity polymer noted above, a homogeneous preparation of tyrosyl-tRNA synthetase was obtained (see the table). An equally purified enzyme could not be obtained by using the affinity polymer immediately after phosphocellulose chromatography.

The successful application of the affinity polymer (III) does not prove unambiguously that the retardation of tyrosyl-tRNA synthetase depends solely on the tyrosine moiety of the polymer. As mentioned above, the polymer also contains carboxyl groups, and thus might act as a cation exchanger. In the course of the purification procedure it was observed that a small amount of protein is bound more tightly to the polymer than is tyrosyl-tRNA synthetase. It is possible that these proteins also interact with tyrosine although we consider it more likely that they are linked to the carboxyl groups. No investigation with regard to the detailed mode of action of this polymer has been carried out.

Preparation of the Affinity Polymer

1,6-Diaminohexane, 30 g, is dissolved in 150 ml of dioxan and poured quickly onto 10 g of the copolymer of malic anhydride and butanediol divinyl ether (Merck, Darmstadt). The mixture is stirred overnight at 30°, suction filtered, and washed with dioxan and water until the eluate is neutral. The cake is stirred in 100 ml of 0.1 N KOH for 20 minutes, filtered, washed with water to neutrality and finally with ethanol. It is dried under reduced pressure (I).

BOC-*O*-benzyl-L-tyrosine-4-nitrophenyl ester, 16 g (33 mmoles, Fluka, Basel) is dissolved in 150 ml of dimethylformamide and, after addition of 5 ml of triethylamine, is stirred with (I) for 72 hours at 30°. The mixture is washed with 5% triethylamine in water until the polymer becomes colorless and then with ethanol; the product is dried under vacuum (II).

The product (II) is suspended in 150 ml of acetic acid, and 150 ml of 40% HBr in acetic acid are added. The mixture is stirred vigorously for 30 minutes, suction-filtered, and washed with acetic acid and water. The cake is suspended in water, titrated to pH 7.0 with 2 N KOH, filtered, washed with water and ethanol, and dried under reduced pressure. Yield 13 g (III).

Affinity Purification of Tyrosyl-tRNA Synthetase[4]

Polymer, 600 mg, is washed 3 times with buffer C (30 mM potassium phosphate pH 7.3, 0.5 mM EDTA, 1 mM dithiothreitol, and 10% (v/v) glycerol by stirring for 10 minutes and centrifuging at 20,000 g for 5 minutes. Since the structure of the polymer does not allow column chromatography with a satisfactory flow rate, the entire procedure is conducted as a batch operation. The polymer is suspended in the eluate from the CM-cellulose column purification step, stirred for 45 minutes, and centrifuged. The supernatant liquid contains neither protein nor enzymatic activity. The polymer is extracted 3 times with 10 ml each of 0.3 M KCl in buffer C; these fractions contain the purified enzyme with a specific activity of 1400 units per milligram of protein. It is completely homogeneous by the criterion of sodium dodecyl sulfate gel electrophoresis, even when large amounts of protein (20 μg) are applied. The UV characteristics are λ_{min} = 251, λ_{max} = 278 nm, $A_{278\ nm}/A_{251\ nm}$ = 2.3 and $A_{280\ nm}/A_{260\ nm}$ = 1.9. Further extraction of the polymer is carried out with 7 ml of 0.6 and 1.0 M KCl in buffer C. These later eluates show specific enzyme activities of 1000 and 310 units per milligram, respectively, and reveal impurities upon sodium dodecyl sulfate gel electrophoresis.

The solution of the homogeneous enzyme may be concentrated by dialysis against glycerol or by adsorption of the enzyme onto a phosphocellulose column (1 × 5 cm) followed by stepwise elution with 0.5 M KCl in buffer C to protein solutions of about 1.5 and 0.8 mg per milliliter, respectively. The enzyme is stable when stored in the presence of 50% (v/v) glycerol at −20° for several weeks. These solutions may be diluted or dialyzed against a buffer of choice without noticeable loss in activity.

Tyrosyl-tRNA synthetase is assayed[4-6] by measuring the incorporation

[5] R. W. Holley, I. Apgar, B. Doctor, J. Farrow, M. A. Morini, and S. H. Merill, *J. Biol. Chem.* **236**, 200 (1961).

of [^{14}C]tyrosine into tRNA bulk or tRNA^{Tyr}.[7] The enzymatic activity is calculated from the initial slope in a plot of radioactivity versus time. One activity unit is defined as 1 nmole of tyrosine fixed to tRNA per minute; the specific activity is given as nanomoles of amino acid incorporated per minute per milligram of protein. Protein is determined according to Lowry *et al.* or by using the ratio of absorbance at 280 and 260 nm.

---

[6] R. Calendar and P. Berg, *in* "Procedures in Nucleic Acid Research" (G. L. Cantoni and D. R. Davies, eds.), Vol. I, p. 384. Harper & Row, New York, 1966.
[7] D. Schneider, R. Solfert, and F. von der Haar, *Hoppe-Seyler's Z. Physiol. Chem.* **353**, 1330 (1972).

[61] Methionyl-tRNA Synthetase

By MALKA ROBERT-GERO and JEAN-PIERRE WALLER

The purification by affinity chromatography of enzymes bearing ligands with a free amino group necessary for enzyme-substrate binding, is complicated by the fact that unspecific protein retention may occur on the column owing to these charged functions. In such cases it is difficult to distinguish between retention by ion exchange and retention by affinity chromatography.

It is well established that for the enzyme–amino acid interaction of aminoacyl-tRNA synthetases, the amino function, not the carboxyl group, of the substrate must be free.[1]

The method given below describes the preparation and use of methionylhexamethylenediamine-Sepharose (VI) (see Scheme I) for the purification of methionyl-tRNA synthetase by affinity chromatography.[2] The procedure, as will be seen, can be generalized and extended to the purification of other proteins having as ligand an amino acid with a free amino group necessary for enzyme–substrate binding.

Activation Conditions of Sepharose 4B

The goal is to obtain low substrate (i.e., methionine) substitution as will be emphasized in the section dealing with the importance of the extent of substrate substitution.

Commercially available Sepharose 4B is activated as described by Axén *et al.*[3] In the case of the methionyl-tRNA synthetase from *Esche-*

---

[1] R. B. Loftfield, *Progr. Nucl. Acid. Res. Mol. Biol.* **12**, 87 (1972).
[2] M. Robert-Géro and J. P. Waller, *Eur. J. Biochem.* **31**, 315 (1972).

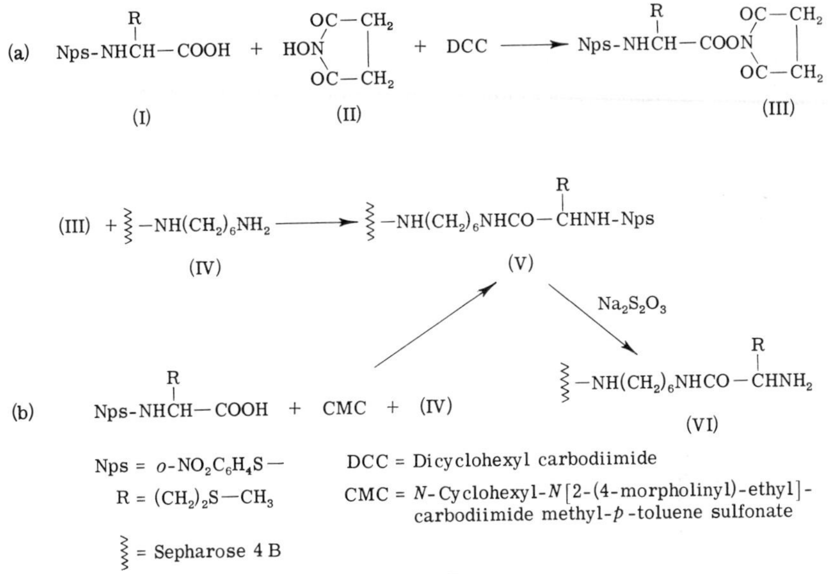

(a)   Nps-NHCH—COOH + HON(OC—CH$_2$/OC—CH$_2$) + DCC ⟶ Nps-NHCH—COON(OC—CH$_2$/OC—CH$_2$)

with R above each Nps-NHCH group.

(I)   (II)   (III)

(III) + ⅀—NH(CH$_2$)$_6$NH$_2$ ⟶ ⅀—NH(CH$_2$)$_6$NHCO—CHNH-Nps (R above)

(IV)   (V)

Na$_2$S$_2$O$_3$

(b)   Nps-NHCH—COOH + CMC + (IV)

⅀—NH(CH$_2$)$_6$NHCO—CHNH$_2$ (R above)

(VI)

Nps = o-NO$_2$C$_6$H$_4$S—   DCC = Dicyclohexyl carbodiimide

R = (CH$_2$)$_2$S—CH$_3$ CMC = N-Cyclohexyl-N[2-(4-morpholinyl)-ethyl]-
                                                  carbodiimide methyl-p-toluene sulfonate

⅀ = Sepharose 4 B

SCHEME I

richia coli preliminary experiments (see Fig. 1) showed that for maximum efficiency of the method the substrate concentration on the polymer should be 4–5 μmoles of methionine per milliliter of gel. This can be achieved with 80 mg of CNBr per milliliter of Sepharose 4B and 0.1 mmole of hexamethylenediamine per milliliter of activated gel. The CNBr is dissolved in a minimum amount of water and added at once to the Sepharose suspension. The pH of the reaction during the activation is kept constant at 10.5 by adding a solution of 4 N KOH. For details of the procedure for coupling of the "spacer molecule," see Cuatrecasas.[4]

At the end of the reaction, the amount of the diamine bound is calculated from nitrogen analysis of the substituted Sepharose (IV). Under the above conditions 0.08 mmole of hexamethylenediamine is coupled per milliliter of Sepharose 4B (80% yield), which corresponds to about 0.04 mmole of free amine per milliliter.

Choice of the Amino Protecting Group

As the carboxyl function of the amino acid is to be coupled to the hexamethylenediamine-Sepharose (IV), the amino group of the substrate must be protected. The N-blocking group of the amino acid should be easily removable without prolonged acid treatment, in order to avoid

[3] R. Axén, J. Porath, and S. Ernbäck, *Nature (London)* **214**, 1302 (1967).
[4] P. Cuatrecasas, *J. Biol. Chem.* **245**, 3059 (1970).

the irreversible shrinkage of the polysaccharide skeleton. Therefore, the use of the following amino protecting groups should be avoided: *N*-trityl, *tert*-butyloxycarbonyl, and 2-(*p*-biphenyl)isopropyloxycarbonyl (BPOC). The most convenient blocking group is the *o*-nitrophenylsulfenyl (Nps) group, which is easily removed under mild conditions as described below. Certain Nps-amino acids are commercially available as the dicyclohexylamine (DCHA) salts. However, the preparation of these Nps derivatives is simple.[5]

Importance of the Extent of Substrate Substitution

When highly substituted affinity columns, containing about 10–15 μmoles of L-methionine bound per milliliter of Sepharose are used, enzyme is strongly retained. The enzyme can be eluted from such a column by using a very large volume of buffer containing 20 m*M* substrate.[2] In spite of the high yield and a reasonable purification factor (31-fold), the method proved impracticable because of excessive dilution of the enzyme. The strong retention of the enzyme may be due to the fact that the *E. coli* methionyl-tRNA synthetase has two binding sites for methionine. It is conceivable that on a highly substituted affinity column part of the enzyme molecules are bound to two substrate molecules at the same time, rendering enzyme elution very difficult. It has been shown that enzymes with very high substrate affinity are difficult to purify by affinity chromatography. It seems that the same problem should be faced when proteins with several subunits and substrate sites are handled. In these cases it is preferable to lower the ligand concentration of the affinity column, so as to avoid such a strong retention. Dilution and mixing of highly substituted polymer with nonactivated Sepharose does not circumvent the high retention. The local concentration of the substrate is important and not its statistical distribution on the polymer.

To define the optimal amino acid concentration of the column necessary for maximal purification of any enzyme, one has to vary the amount of CNBr used for the activation, the amount of the "spacer" α-ω-diamine added remaining constant. This procedure provides polymers with different concentrations of "spacer" molecules; each of them can be coupled to the amino acid desired. The elution profile obtained after the application of an enzyme sample on such columns will define the activation conditions and the substrate concentration to be used. Figure 1 illustrates the effect of methionine concentration of the affinity column on the efficiency of the purification. When the methionine concentration of the polymer is 2 μmoles per milliliter of gel (Fig. 1a) the enzyme is not retained. A

[5] L. Zervas, D. Borovas, and E. Gazis, *J. Amer. Chem. Soc.* **85**, 3660 (1963).

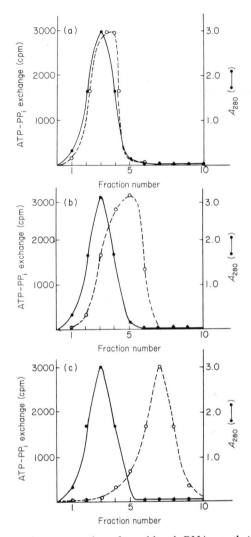

FIG. 1. Affinity chromatography of methionyl-tRNA synthetase on different columns prepared by varying the activation conditions. (a) Final methionine concentration: 2 μmoles/ml; (b) 3.5 μmoles/ml; (c) 4.5 μmoles/ml. Volume of each column: 3 ml (5 mm internal diameter). Equilibrating buffer: 0.2 M potassium phosphate, pH 7.5 containing 10 mM 2-mercaptoethanol. Load: 10 mg protein per milliliter of gel. Fraction volume: 0.8 ml. Flow rate: 3 ml/hour. All other details are given in the text. $\bigcirc$- - -$\bigcirc$, ATP-^{32}P PP$_i$ exchange (cpm); $\bullet$—$\bullet$, absorbance at 280 nm.

substitution of 3.5 μmoles of substrate per milliliter of Sepharose 4B results in a slight retardation (Fig. 1b), whereas a column with 4.5 μmoles of methionine bound per milliliter efficiently retards the enzyme and allows its complete separation from bulk protein (Fig. 1c). In this case there is no need to add the substrate in the eluting buffer and the elution procedure is also very rapid.

When interaction between the amino acid and the enzyme is too strong, a substrate analog with lower affinity for the enzyme may be used. However, in the case of *E. coli* methionyl-tRNA synthetase, an affinity column with 9 μmoles of L-ethionine bound per milliliter of Sepharose still results in strong enzyme retention, and thus its use is not recommended.

Coupling Procedures of o-Nps-L-Methionine (I) to Hexamethylenediamine-Agarose (IV)

Two procedures are described, either of which produces a 10–20% yield of compound (V) and can be used with equal facility.

Coupling via the Activated Ester. o-Nps-methionine (I) can be obtained from its dicyclohexylamine salt according to Zervas et al.[5] The purity of the compound can be controlled by thin-layer chromatography (TLC) on silica gel in chloroform–acetic acid (95:5 v/v).

Equimolar quantities of o-Nps-methionine (I), N-hydroxysuccinimide (II), and dicyclohexyl carbodiimide (DCC) are mixed on a magnetic stirrer at room temperature in the dark for 16 hours in ethyl acetate (10 ml of solvent per millimole of o-Nps-methionine). The dicyclohexylurea is removed by filtration with a Büchner funnel and thoroughly washed with ethyl acetate. The combined filtrate and washings are evaporated to dryness under reduced pressure. The yield of o-Nps-methionine succinimide ester (III) is 75–85%; it is homogeneous, migrating as a single spot on TLC in the solvent system defined above. Since decomposition of the molecule occurs during recrystallization it is advisable to perform the following step immediately after evaporation.

A suspension of hexamethylenediamine-Sepharose (IV) with a known concentration of free amino groups is decanted and suspended in an equal volume of 0.2 M sodium borate at pH 8.1 and stirred gently on a water bath at 30°. A 2-fold excess of the activated ester of the o-Nps methionine (III) is dissolved in 50% dimethylformamide (DMF) in water (100 mg of ester per milliliter of solvent) and added dropwise to the substituted Sepharose suspension (IV). After stirring for 2 hours at 30°, the suspension is filtered on a sintered-glass funnel under suction and washed with large volumes of methanol followed by water until the filtrate becomes colorless. The color of the substituted Sepharose (V) varies from yellow to orange depending on the extent of substitution. A sample is prepared for sulfur analysis by sequentially washing with methanol, DMF, water, methanol, and ether. Since there are two sulfur atoms per molecule of *o*-Nps-methionine used in the coupling, the degree of substitution of the polymer can be calculated from the sulfur analysis.

Direct Coupling of o-Nps-Methionine (I) to Hexamethylenediamine-Sepharose (IV). An alternative procedure consists in direct coupling of the free carboxyl group of the *o*-Nps-methionine (I) to the hexamethylene-diamine-Sepharose (IV) in the presence of the water-soluble carbodiimide, *N*-cyclohexyl-*N'*-2-(4-morpholinyl)ethyl carbodiimide methyl-*p*-toluene sulfonate (CMC), as follows: to a desired quantity of (IV) (with a known concentration of its free amine) suspended in 50% DMF, is added a 5-fold excess of *o*-Nps-methionine (I) in 50% DMF. The pH of the reaction is brought to 4.7 with 1 N HCl. The water-soluble carbodiimide (100 mg per milliliter of water) is then added dropwise in a 10-fold excess with respect to the concentration of the free amine of (IV). The reaction is allowed to proceed at room temperature in the dark for 24 hours. The Sepharose is filtered and thoroughly washed with water to eliminate the derivative of dicyclohexylurea, followed by repeated washing with 50% DMF and with water. A sample is prepared for sulfur analysis as described above.

Removal of the N-Blocking Group from the o-Nps-methionine-hexamethylenediamine-Agarose (V)

The resulting Sepharose (V) is suspended in an equal volume of 0.2 M sodium acetate at pH 5.1. A solution of sodium thiosulfate is added[6,7] to a final concentration of 0.4 M. The suspension is then stirred gently at 30° until the color of the gel changes from orange-yellow to faint yellow or colorless (at least 2 hours). The Sepharose (IV) is filtered over a sintered-glass funnel and washed under suction with methanol, water, aqueous DMF, and water until washings become colorless. The qualitative 2,4,6-trinitrobenzene-sulfonate test[8] shows the presence of free amino groups. A sample for sulfur analysis is prepared as described above, and the methionine concentration of the polymer is calculated.

Purification Procedure

In order to avoid unspecific retention of the enzyme on the column by ion exchange due to charged amine groups, the affinity column must

---

[6] Y. Lapidot, D. Elat, S. Rappaport, and N. de Grot, *Biochem. Biophys. Res. Commun.* 38, 559 (1970).

[7] B. Ekstrom and B. Sjöberg, *Acta Chem. Scand.* 19, 1245 (1965).

[8] J. K. Inman and H. M. Dintzis, *Biochemistry* 8, 4074 (1969). See also this volume [3].

APPLICATION OF AFFINITY CHROMATOGRAPHY AFTER DIFFERENT
STEPS OF ENZYME PURIFICATION

Material submitted to affinity chromatography			Recovery from affinity column			
Fraction	Specific activity[a] (units/ mg)	Amount (mg protein)	Protein (mg)	Activity[a] (units)	Specific activity (units/ mg)	Yield (%)
Ammonium sulfate step (30–50%)	8	5.4	0.587	44	75	100
Sephadex G-200 eluate	16	4.9	0.180	67	370	85
DEAE Sephadex pH 8.1 eluate	48	6.0	0.433	231	1000	80

[a] ATP-^{32}PPi exchange.

be equilibrated in a buffer of sufficiently high ionic strength. In the case of E. coli methionyl-tRNA-synthetase, preliminary experiments on a hexa-methylenediamine-Sepharose (IV) column showed that when 0.2 M potassium phosphate is used at pH 7.5 in the presence of 10 mM 2-mer-captoethanol, the enzyme flows through unretarded together with the bulk of the proteins. This 0.2 M buffer is used in further purification procedures to equilibrate the affinity column. Thus, the ionic strength of the equilibrat-ing buffer to be used for the purification of any other enzyme system should be the lowest at which the enzyme in question will run through with the bulk of the proteins on a hexamethylenediamine-Sepharose column.

It may occur that a one-step purification from a crude cell-free extract cannot be achieved by affinity chromatography. It then becomes necessary to determine after which step of the conventional purification procedure to introduce the affinity column to obtain the best results (see the table).

The earlier purification steps of E. coli methionyl-tRNA synthetase, ammonium sulfate precipitation, Sephadex G-200 gel filtration and DEAE-Sephadex chromatography at pH 8.1, are performed as described earlier.[9] The eluate from the last of these steps is applied on the affinity column. Figure 1c presents the elution profile. As was observed earlier, the enzyme is no longer firmly retained under these conditions, but is sufficiently retarded to give a highly purified enzyme (see the table). The yield is 80%.

The affinity column may be reused after washing with 0.5 M potassium phosphate at pH 7.5 until the $A_{280 \text{ nm}}$ reading is equal to that of the

[9] D. Cassio and J. P. Waller, FEBS Lett. 12, 309 (1971).

buffer. The column is then washed with water and reequilibrated with the starting buffer.

Conclusion

As already pointed out, this method can be generalized for the purification of any enzyme binding to an amino acid with a free amino function. The nitrophenyl-sulfenyl (Nps) protected amino acids are easy to prepare and can be readily removed under mild conditions.

The three critical points of the method are the following: (1) According to the above experiments the extent of ligand substitution of the polymer should be kept at a relatively low level to avoid strong enzyme retention. (2) The ionic strength of the buffer has to be sufficiently high as to avoid unspecific enzyme binding by ion exchange. (3) The purification step, prior to affinity chromatography, that would produce the best results must be determined.

The system should be brought to a sufficiently high state of purity so that optimum results are obtained by subsequent affinity chromatography.

[62] Phenylalanyl-tRNA Synthetase

By PETER W. SCHILLER and ALAN N. SCHECHTER

The preparation of large amounts of highly purified aminoacyl-tRNA synthetases is required for studies of the biologically specific, but still poorly understood, interactions of these enzymes with their substrates, including their cognate tRNA and amino acid. Most of the aminoacyl-tRNA synthetases known to date have been purified by means of classical isolation procedures which often are time-consuming when successful. The possibility of purification by affinity chromatography, with insolubilized ligands, of these enzymes is only now being explored.[1] So far, methods with insolubilized tRNA's have been published, which permitted partial purification in two cases.[2,3] In a more sophisticated approach which involved the use of purified tRNA[Phe], the preparation of homogeneous phenylalanyl-tRNA synthetase from yeast was achieved.[4]

This contribution describes the purification of phenylalanyl-tRNA synthetase from *Salmonella typhimurium* by affinity chromatography with the insolubilized cognate amino acid, phenylalanine. Earlier experiments with

[1] See articles [11], [60], and [61] in this volume.
[2] O. D. Nelidova and L. L. Kiselev, *Mol. Biol.* **2**, 60 (1968).
[3] S. Bartkowiak and J. Pawelkiewicz, *Biochim. Biophys. Acta* **272**, 137 (1972).
[4] P. Remy, C. Birmelé, and J. P. Ebel, *FEBS Lett.* **27**, 134 (1972).

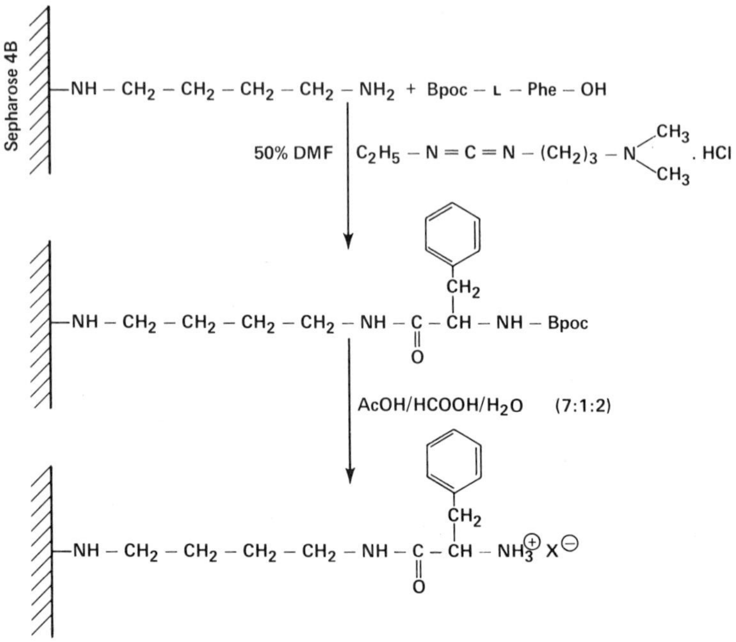

SCHEME I

insolubilized amino acids[5] and amino acid esters[6] demonstrated the necessity of coupling the ligand via an "arm" to the matrix in order to obtain satisfactory protein interaction. While the attachment of an "arm" to the α-amino group of D-tryptophan methyl ester proved successful in the purification of α-chymotrypsin,[6] such linkage was discouraged in our case by the results of an inhibitor analysis performed with phenylalanine-tRNA synthetase from *Escherichia coli.*[7] This study revealed poor affinity of α-aminoalkylated phenylalanine but still substantial binding of alkylamides of the amino acid; the inhibitor constant decreased with increasing length of the alkyl chain ($K_i \sim 1$ mmole). On the basis of these findings we chose to prepare a resin according to Scheme I. The 2-(4-biphenylyl)-2-propyloxycarbonyl protecting group (BPOC) was chosen since it is easily cleaved under mild acidic conditions[8] which do not affect the Sepharose beads.

[5] W. W.-C. Chan and M. Takahashi, *Biochem. Biophys. Res. Commun.* **37,** 272 (1969).

[6] P. Cuatrecasas, M. Wilchek, and C. B. Anfinsen, *Proc. Nat. Acad. Sci. U.S.* **61,** 636 (1968).

[7] D. V. Santi and P. V. Danenberg, *Biochemistry* **10,** 4813 (1971).

[8] P. Sieber and B. Iselin, *Helv. Chim. Acta* **51,** 622 (1968).

Preparation of the Resin

The ω-aminobutyl derivative of Sepharose 4B was prepared by the method described by Cuatrecasas.[9] The derivative (40 ml) was then combined with a solution containing 1.17 g of BPOC-L-phenylalanine DCHA salt (Fox Chemical Company, Los Angeles) in 120 ml of 50% aqueous dimethylformamide, and the pH of the suspension was adjusted to 4.8 with 1 N HCl. 1-Ethyl-3(3-dimethylaminopropyl) carbodiimide (Sigma Chemical Company, St. Louis, Missouri), 960 mg, was added to the above suspension within a 5-minute period. The pH was maintained at 4.8 with 1 N HCl during the addition and thereafter for 70 minutes. The reaction mixture was stirred gently for another 18 hours. The substituted resin was washed extensively with 4 liters of 50% aqueous dimethylformamide and subsequently with 1 liter of distilled water. The color test with sodium 2,4,6-trinitrobenzenesulfonate[10] revealed the presence of free ω-amino groups which had not undergone reaction. This finding is in agreement with earlier observations of low yields for the reaction of inhibitors with agarose derivatives.[11]

The protected Sepharose derivative, 20 ml, was suspended in 50 ml of a mixture of acetic acid–formic acid–water (7:1:2) (v/v) and stirred gently. After 1 hour the resin was filtered and subsequently resuspended in 50 ml of the same freshly prepared acid mixture for another 5 hours with gentle stirring. The resin was washed extensively until a neutral pH was reached. The amount of phenylalanine coupled to the resin was determined by amino acid analysis to be 0.25 μmole per milliliter of packed Sepharose.[11a]

Chromatography

Phenylalanine-tRNA synthetase from *S. typhimurium,* an enzyme not previously isolated, was partially purified by ammonium sulfate precipitation (30–65% fraction) and chromatography on DEAE-Sephadex A50 with a linear gradient from 50 mM potassium phosphate, pH 8.0, to 250 mM potassium phosphate, pH 6.2, in the presence of 10 mM mercaptoethanol and 0.5 mM EDTA.[12] The preparation, 30 ml, containing approxi-

---

[9] P. Cuatrecasas, *J. Biol. Chem.* **245**, 3059 (1970).

[10] J. K. Inman and H. M. Dintzis, *Biochemistry* **8**, 4074 (1969). See also this volume [3].

[11] M. Wilchek and M. Rotman, *Isr. J. Chem.* **8**, 172 (1970).

[11a] We recently found that the reaction of BPOC-L-Phenylalanine succinimide ester with the ω-aminobutyl derivative of Sepharose 4B improved the yield of the BPOC-L-Phenylalanine coupling significantly. The succinimide ester method has been used by Robert-Gero and Waller [*Eur. J. Biochem.* **31**, 315 (1972)] for the preparation of a methionine derivative of Sepharose 4B.

[12] P. W. Schiller and A. N. Schechter, unpublished experiments.

mately 150 mg of protein, was dialyzed against 50 mM potassium phosphate at pH 7.5 containing 100 mM KCl, 10 mM β-mercaptoethanol, and 0.5 mM EDTA. A column (1.2 × 15 cm) was prepared with the Sepharose derivative and equilibrated with the same initial buffer. The sample solution was applied and the column eluted with the same initial buffer. The bulk of the protein, estimated at about 90% from measurement of the absorbance at 280 nm, was removed by this procedure. On the other hand, 93% of the total applied activity was retained on the column. The column was washed extensively with 500 ml of initial buffer in order to minimize retention of protein by nonspecific (ionic) interaction. Elution of the phenylalanine-tRNA synthetase was accomplished by increasing the concentration of KCl to 0.5 M whereby about 8% of the totally applied protein containing 75% of the total initial activity were recovered in a sharp peak of about 15 ml. Disc gel electrophoresis revealed that three out of four peak fractions had phenylalanine-tRNA synthetase activity at a level of purity of at least 95% whereas the initial fraction contained several contaminants.

The undesirable ion-exchange properties which originate from the presence of nonreacted free aminobutyl groups constitute a slight drawback of the method. On the other hand, the column could be reused several times without appreciable loss in capacity for at least 3 months after its preparation.

Comment. The described method has been used successfully in the last step of the isolation of phenylalanine-tRNA synthetase from *S. typhimurium* in a highly purified state, whereby the purification obtained was 10- to 20-fold. The advantage of the procedure is its relative simplicity as compared to the use of insolubilized tRNA, which requires preliminary purification of the ligand.[4] The disadvantage of the method is the ion-exchange properties of the column material which, however, can be partially overcome by extensive washing. The method may prove most useful in the later stages of the isolation of aminoacyl-tRNA synthetases, since crude extracts are likely to contain other proteins with affinity for the insolubilized amino acid in its role as either inhibitor or allosteric effector. The usefulness of the procedure for the isolation of other aminoacyl-tRNA synthetases from this organism is presently being tested. The preparation of a phenylalanine derivative of Sepharose by different means for use in the purification of phenylalanine-tRNA synthetase from *Escherichia coli* has recently been described.[13]

---

[13] P. J. Forrester and R. L. Hancock, *Can. J. Biochem.* **51**, 231 (1973).

[63] DNase-I Inhibitor Protein. Isolation with DNase-Agarose

By UNO LINDBERG

The presence of a protein DNase I-inhibitor has been demonstrated in many different types of mammalian tissues. The inhibitor protein has been purified by conventional techniques from calf spleen[1] and thymus[2] and from rat serum,[3] and, as described here, a generally applicable method for the purification of this protein by affinity chromatography on DNase-Sepharose has been achieved.[4]

It has been found that the DNase I inhibitor is present in the cells in surprisingly large amounts. For example, this protein constitutes 5–10% of the soluble protein of thymus cells. However, nothing is yet known about its biological function or, for that matter, its relation to deoxyribonucleases in the living cell.

Procedure

Buffers

Wash buffer I: 4 M guanidine-HCl; 30% glycerol; 0.5 M sodium acetate at pH 6.5

Wash buffer II: 10 mM Tris chloride at pH 7.5; 5 mM CaCl$_2$

Extraction buffer: 10 mM Tris chloride at pH 7.5; 10 mM NaCl; 10 mM EDTA

Elution buffer I: 0.75 M guanidine-HCl; 0.5 M sodium acetate– 30% glycerol; adjusted to pH 6.5

Elution buffer II: 3 M guanidine-HCl; 1.0 M sodium acetate; 30% glycerol adjusted to pH 6.5.

Coupling of DNase I to Agarose

Activation of Sepharose with cyanogen bromide is performed as described by Kato and Anfinsen.[5] Twenty milliliters of the commercial slurry of Sepharose 4B is washed with 600 ml of deionized water and resuspended in 40 ml of water. Cyanogen bromide solution (20 ml containing 0.6 g of CNBr) is added, and the pH is brought to 11–11.5 by the addition of 2 M

[1] U. Lindberg, *Biochemistry* 6, 323 (1967).
[2] U. Lindberg and L. Skoog, *Eur. J. Biochem.* 13, 326 (1970).
[3] G. Berger and P. May, *Biochim. Biophys. Acta* 139, 148 (1967).
[4] U. Lindberg and S. Eriksson, *Eur. J. Biochem.* 18, 474 (1971).
[5] I. Kato and C. B. Anfinsen, *J. Biol. Chem.* 244, 5849 (1969).

NaOH and maintained by dropwise addition of NaOH. After 5–10 minutes, the pH remains stable and the activated Sepharose is immediately washed with ice-cold deionized water on a Büchner funnel. Bovine pancreatic DNase (Sigma)[6] 20 mg in 2 ml of 0.1 M NaHCO$_3$, is mixed with the activated Sepharose, which has been resuspended in 25 ml of 0.1 M NaHCO$_3$ at 4°. The time course of the coupling reaction is followed by the absorbance at 280 nm of aliquots of the reaction mixture after centrifugation. Within 2 hours 90% of the enzyme protein has been fixed to the gel, and essentially all after another 24 hours at 4°. The DNase-Sepharose is then washed on a Büchner funnel with 500 ml each of water, 0.1 M NaHCO$_3$, and again with water. It is packed into a column, and noncovalently bound protein is removed by passage through 50 ml of wash buffer I followed by 500 ml of wash buffer II. The DNase-Sepharose may be stored for several months at 4° without loss in effectiveness. After a purification procedure has been performed, the DNase-Sepharose is washed with wash buffers I and II as indicated above and is again ready for use.

When incubated with the DNA substrate (calf thymus DNA, Sigma) used in the standard DNase assay,[7] DNase-Sepharose conjugate causes a slow but significant degradation of the DNA. The rate of the reaction is about 1% that obtained with an equal amount of free DNase.

Purification of DNase-Inhibitor from Crude Extracts

In the preparation described here calf thymus was used as the source of the inhibitor, although the same procedure has been effective for the purification of the inhibitor from several different tissues and species including tissue-cultured human cells.

Frozen thymus tissue, 50 g, is thawed in 250 ml of extraction buffer and homogenized for 30 seconds in a high speed Turmix homogenizer after the addition of 0.1 volume of 5% NP-40.[8] The homogenate is centrifuged for 10 minutes at 12,000 g and the supernatant fluid is centrifuged again for 2 hours at 100,000 g. The supernatant liquid from this step may be stored at −20° without loss of inhibitor activity.

A sample of the crude extract (22 ml containing 450 mg of protein; a total of 870,000 inhibitor units) is applied to a column of DNase-Sepharose (2 × 1.5 cm). After the sample has entered the column, unadsorbed protein is washed out with wash buffer II. Adsorbed proteins are then eluted in two steps with elution buffer I in the first step and elution

---

[6] Before use, the DNase (DN-100, Sigma) is purified by chromatography on Sephadex G-100 in wash buffer II.
[7] U. Lindberg, *Biochim. Biophys. Acta* **82**, 237 (1964).
[8] Nonidet P-40 (NP-40) is a nonionic detergent obtained from Shell Chemical Co.

Summary of the Results of the Affinity Chromatography
of the Crude Extract of Calf Thymus[a,b]

Fraction	Activity (k units)	Activity yield (%)	Protein (mg)	Specific activity (units/ mg)	A_{260}/A_{280} (a)	A_{260}/A_{280} (b)	DNA (μg)
1. Crude extract	870	100	450	1900	1.49	—	—
2. Nonadsorbable protein	55	6	415	130	1.18	—	—
3. Lightly adsorbed protein[b]	145	17	20	7200	1.65	1.45	50
4. Firmly adsorbed protein[b]	71	8	8	8800	1.55	1.30	7

[a] From U. Lindberg and S. Eriksson, *Eur. J. Biochem.* **18**, 474 (1971).

[b] Fractions 3 and 4 were analyzed before (a) and after (b) concentration by salt precipitation (dialysis against saturated ammonium sulfate overnight). After salt precipitation, the samples were dissolved in 10 mM sodium acetate, pH 6.5, and assayed for inhibitor activity after the addition of 0.75 M guanidine-HCl–0.5 M sodium acetate–30% glycerol pH 6.5.

buffer II in the second. As much as 28 mg (6%) of the protein of the crude extract is adsorbed to the column. Of this total, 20 mg are relatively loosely bound and are removed with the first buffer whereas the remaining, more firmly attached protein consisting mainly of the DNase-inhibitor, is displaced by the second buffer. About 20% of the inhibitor activity is found in the first eluate and 10% in the second. The absorbance spectrum of the eluted fraction exhibit a maximum at 260 nm with a ratio of 1.6 between the absorbancies at 260 and 280 nm indicating the presence of nucleic acid material. It has been shown that this is due largely to RNA, but the reason for the adsorption of RNA to DNase-Sepharose has not been investigated.

That the protein in the second eluate consists mainly of the inhibitor protein has been demonstrated by polyacrylamide gel electrophoresis and by peptide mapping.[4] A summary of the affinity chromatography of the calf thymus extract on DNase-Sepharose is given in the table. It should be pointed out that the inhibitor is *not* recovered in a highly active form and that conditions have not been found as yet which increase the activity of the adsorbed and eluted inhibitor protein.

From the work on the purification of the DNase-inhibitor by conventional methods, it is known that the protein has a marked tendency to form high molecular weight aggregates with low inhibitory activity. From those studies it is also known that 0.75 M guanidine-HCl–0.5 M sodium acetate–30% glycerol is effective in dissociating such aggregates to protein monomers of high specific activity. However, these conditions do not dissociate the enzyme-inhibitor complex. It is possible therefore that the presence of inhibitor activity in the first eluate is due to release of inhibitor molecules from inhibitor aggregates bound to the DNase-Sepharose, rather than to release of active inhibitor from the DNase as such. The situation in 3 M guanidine-HCl is different. This concentration of guanidine-HCl is known to rapidly inactivate the inhibitor and would lead to dissociation of the DNase-inhibitor complex.[4]

[64] Purification of Thymidylate Synthetase with 2'-Deoxyuridylate-Agarose

By PETER V. DANENBERG and CHARLES HEIDELBERGER

The partial purification of thymidylate synthetase from a number of bacterial and animal sources has been achieved by multistep sequences of conventional procedures.[1-6] Recently Crusberg et al.[7] and Dunlap et al.[8] have reported the complete purification of thymidylate synthetase from antifolate-resistant strains of *Lactobacillus casei*. However, the purification of the mammalian enzyme has been hampered by the labile nature of this protein and low enzyme levels in cells and tissues. This disadvantage is especially evident in the case of Ehrlich ascites carcinoma cells,[6,9] and if reasonable quantities of enzyme are required for study, the usual procedures employed in purification become quite laborious. It is apparent, therefore, that a single-step procedure based on affinity chromatography should greatly facilitate the purification process as well as enhance the recovery of enzyme activity.

The present article describes the preparation of a specific adsorbent for thymidylate synthetase involving attachment of the substrate, 2'-deoxyuridylic acid (dUrd-5'-P) to agarose via an ω-aminoalkyl "arm." This mate-

[1] C. K. Mathews and S. S. Cohen, *J. Biol. Chem.* 238, 367 (1963).
[2] R. L. Blakley, *J. Biol. Chem.* 238, 2113 (1963).
[3] M. Friedkin, E. J. Crawford, E. Donovan, and E. J. Pastore, *J. Biol. Chem.* 237, 3811 (1962).
[4] D. M. Greenberg, R. Nath, and G. K. Humphreys, *J. Biol. Chem.* 236, 2271 (1961).
[5] M. Y. Lorenson, G. F. Maley, and F. Maley, *J. Biol. Chem.* 242, 3332 (1967).
[6] A. Fridland, R. J. Langenbach, and C. Heidelberger, *J. Biol. Chem.* 246, 7110 (1971).
[7] T. C. Crusberg, R. Leary, and R. L. Kisliuk, *J. Biol. Chem.* 245, 5292 (1970).
[8] R. B. Dunlap, N. G. L. Harding, and F. M. Huennekens, *Biochemistry* 10, 88 (1971).
[9] K.-U. Hartmann and C. Heidelberger, *J. Biol. Chem.* 236, 3006 (1961).

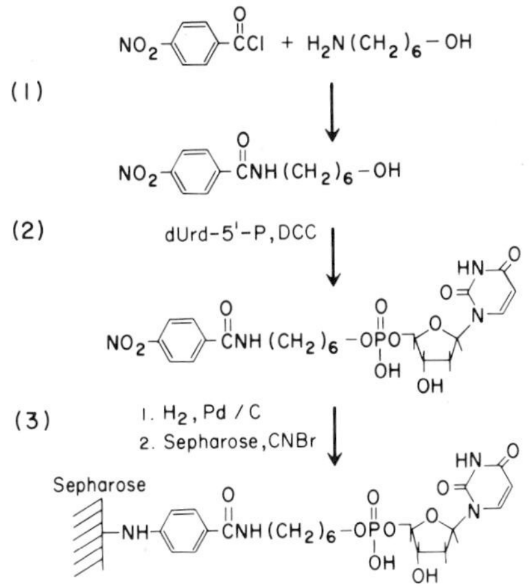

Scheme I

rial has been used successfully to purify the enzyme from a crude extract of lysed *Lactobacillus casei* cells.[10]

Preparation of the Ligand

The sequence of reactions used to prepare the affinity column is shown in Scheme I. The spacer molecule, 6-aminohexan-1-ol, is acylated with *p*-nitrobenzoyl chloride (step 1) to give 6-*p*-nitrobenzamidohexan-1-ol, thus blocking the primary amino group during the subsequent carbodiimide-mediated esterification of the 5'-phosphate group of dUrd-5'-P (step 2). The nitro group of the *p*-nitrobenzoyl moiety, after catalytic reduction, provides an aromatic amino group, which can be linked to Sepharose by cyanogen bromide (step 3) at pH values lower than those used for aliphatic amines, thereby resulting in a higher coupling efficiency.[11]

6-p-Nitrobenzamidohexan-1-ol. 6-Aminohexan-1-ol (Aldrich Chemical Co.) (4.7 g; 40 mmoles) is dissolved in 50 ml of tetrahydrofuran and treated with 7.4 g (40 mmoles) of *p*-nitrobenzoyl chloride (Aldrich Chemical Co.) and 4.0 g (40 mmoles) of triethylamine. After 1 hour at ambient temperature, the solvent is evaporated and the residue is dissolved in ethyl acetate. The solution is washed successively with 0.1 N HCl, 0.1 N NaOH, and water by agitation in a separatory funnel. The ethyl acetate layer is dried by addition of anhydrous magnesium sulfate, which is then removed by filtration. The solvent is evaporated, and the resulting solid residue is

[10] P. V. Danenberg, R. J. Langenbach, and C. Heidelberger, *Biochem. Biophys. Res. Commun.* **49**, 1029 (1972).

[11] P. Cuatrecasas, *J. Biol. Chem.* **245**, 3059 (1972).

crystallized from chloroform-benzene in the form of long needles. The yield of 6-*p*-nitrobenzamidohexan-1-ol is 8.6 g (90%), with a melting point of 81–83°.

2'-Deoxyuridine 5'-(6-p-Nitrobenzamido)hexylphosphate. The disodium salt of 2'-deoxyuridine 5'-phosphate (Sigma Chemical Co.) (0.5 g; 1.43 mmoles) is dissolved in water and converted to the pyridinium salt by passage through a Dowex 50 column in the pyridinium form. The bulk of the water containing the nucleotide is then removed by evaporation at reduced pressure at 25°, and the residue is dried by successive addition and evaporation of three 5-ml portions of dry pyridine.[12]

The syrupy residue is taken up in 50 ml of dry pyridine and treated simultaneously with 2.0 g (7.5 mmoles) of 6-*p*-nitrobenzamidohexan-1-ol and 2.0 g (10 mmoles) of dicyclohexyl carbodiimide (Aldrich Chemical Co.). The reaction is allowed to proceed for 48 hours at ambient temperature. The disappearance of dUrd-5'-P as it reacts can be monitored by paper or thin-layer chromatography. On Whatman No. 40 paper, using the solvent system *n*-butanol–acetic acid–water, 5:2:3, the product has an R_f value of 0.8, whereas the R_f of dUrd-5'-P is 0.3. When the reaction is complete, the precipitated dicyclohexylurea is removed by filtration, and the pyridine is evaporated under reduced pressure to dryness. To remove unreacted starting reagents, the residue is first washed with water, and then with chloroform. The residue should show one spot at R_f 0.8 in the above-noted paper chromatography system.

Reduction of the Nitro Group and Coupling to Agarose. The product from the preceding reaction, 2'-deoxyuridine 5'-(6-*p*-nitrobenzamido)-hexylphosphate (1.2 mmoles), is dissolved in 100 ml of methanol, and several drops of concentrated HCl are added. Hydrogenation is carried out in a Parr shaker at 40 psi in the presence of 100 mg of 10% palladium on charcoal. When hydrogen absorption ceases (usually about 1 hour), the catalyst is removed by filtration, and the methanol is evaporated. The product is dissolved in 10 ml of dimethylformamide, and an equal volume of 0.1 M NaHCO$_3$ at pH 9.0 is added. Activation of Sepharose with CNBr and coupling to the ligand is carried out at pH 9.0 in 0.1 M NaHCO$_3$.[11] After 24 hours of gentle stirring, the Sepharose is filtered and thoroughly washed with 50% aqueous dimethylformamide to remove noncovalently adhering material. This process can be monitored by checking the ultraviolet absorbance of the wash solution.

Determination of the Quantity of Ligand Attached. A small amount of the substituted Sepharose is hydrolyzed for 24 hours at 120° in 7.2 N H$_2$SO$_4$. The black material is removed by filtration, and the filtrate is

---

[12] Pyridine is dried before use by distillation from *p*-toluenesulfonyl chloride and then placed over type 4A molecular sieves (Matheson, Coleman, and Bell).

analyzed for phosphate according to the method of Meun and Smith.[13] The previously described sequence of reactions carried out in the authors' laboratory resulted in the attachment of 100 μmoles of ligand per gram of dry weight of the substituted Sepharose.

Purification of Thymidylate Synthetase from *L. casei*

The 6-*p*-aminobenzamidohexyl-dUrd-5′-P-Sepharose, prepared as described above, when used in column chromatography is capable of adsorbing thymidylate synthetase from a 30 to 70% ammonium sulfate precipitate of lysed *L. casei* cells. At buffer concentrations of up to 50 mM phosphate, pH 7.1, no enzyme activity is eluted from the column. However, binding of the enzyme is very sensitive to ionic strength: when the buffer concentration is increased to 0.1 M, the enzyme is quantitatively removed from the column in a single sharp peak with no tailing. Analysis of the peak of enzyme activity by sodium dodecyl sulfate disc gel electrophoresis[14] shows only one protein band.[10] The specific activity of the enzyme preparation is 2.4 μmoles/minute per milligram of protein, which compares favorably with the published value of 2.5 μmoles/minute per milligram of protein for the completely purified enzyme.[8] An affinity column of 1 × 10 cm has the capacity for approximately 0.8 mg of purified enzyme.

Specific elution of thymidylate synthetase by the substrate dUrd-5′-P does not occur from this column at nucleotide concentrations up to 0.1 mM, indicating that the enzyme may bind to the dUrd-5′-P column more tightly than to dUrd-5′-P itself.

Preliminary results from this laboratory indicate that thymidylate synthetase from Ehrlich ascites cells is also absorbed to the column, but purification data are not yet available.

[13] D. H. C. Meun and K. C. Smith, *Anal. Biochem.* **26**, 364 (1968).
[14] U. K. Laemmli, *Nature (London)* **227**, 680 (1970).

[65] Guanine Deaminase and Xanthine Oxidase

By HANS-ULRICH SIEBENEICK and BERNARD R. BAKER[1]

At the time when we attempted to prepare an affinity column of guanine deaminase, little was known about the characteristics and stability of the enzyme. However, extensive knowledge was available on the profile of the active site and many active-site-directed inhibitors of the enzyme

[1] Deceased, October 19, 1971.

(I)   R = H
(II) R = 3-NH$_2$
(III) R = 4-NH$_2$
(IV) R = 4-O(CH$_2$)$_2$—NH$_2$
(V)   R = 4-O(CH$_2$)$_2$—NH〜 Sepharose

had been prepared.[2] 9-Phenylguanine (I) was found to be a good reversible inhibitor of the enzyme, being complexed slightly better than the substrate.[3] The 3'- and 4'-amino derivatives of 9-phenylguanine (II and III) were also good reversible inhibitors.[4] In addition, it was known that elongation of these inhibitors by acylation at the 3'- or 4'-amino groups did not interfere with binding to the active site of the enzyme.[5] Therefore, we first selected the inhibitors (II) and (III) to be bound to Sepharose 4B as polymeric support of choice. The two compounds were directly attached to BrCN-activated Sepharose. However, both column materials failed to complex guanine deaminase. It was assumed that the complex formation was hampered because the 9-phenylguanine moiety was too close to the Sepharose backbone. When the bridge distance between the 9-phenylguanine and Sepharose was increased by insertion of an O(CH$_2$)$_2$ group into the inhibitor (IV), the resultant column material (V) readily adsorbed the crude enzyme.

Although the capacity of this column was not determined, all the guanine deaminase in 3 g of rat liver was readily retained on 1 ml of column material at 0°. The enzyme could not be eluted with 0.1 M acetic acid since it was immediately denatured by this reagent. However, elution with 0.5 mM guanine, the substrate, or 1 mM hypoxanthine, was smooth and sharp. All the enzyme appeared in the first two 2-ml fractions after addition of eluent. The guanine was converted to xanthine during this operation. Since xanthine is not an inhibitor of guanine deaminase, this conversion does not interfere with the determination of enzyme activity. However, xanthine can interfere with protein determination by the UV method[6] and, therefore, was removed from an aliquot by dialysis before assay for protein.

In separate trials, attempts were made to elute the enzyme with Tris chloride at pH 9. In this case, however, no activity could be eluted. The

[2] B. R. Baker and H.-U. Siebeneick, *J. Med. Chem.* **14**, 802 (1971) and previous papers cited therein.
[3] B. R. Baker, *J. Med. Chem.* **10**, 59–61 (1967).
[4] B. R. Baker and W. F. Wood, *J. Med. Chem.* **10**, 1101 (1967).
[5] B. R. Baker and W. F. Wood, *J. Med. Chem.* **12**, 214 (1969).
[6] O. Warburg and W. Christian, *Biochem. Z.* **310**, 384 (1942).

column treated in this way was then eluted with 0.5 mM guanine; enzyme was obtained in the first two fractions. This indicates that the enzyme is stable at pH 9.

Guanine deaminase could be recovered in a 70% yield when the non-adsorbed protein from a Sepharose column for dihydrofolate reductase[7] was passed through the Sepharose (V) column. This supports the contention that this mild and rapid affinity method could be used for isolation of several enzymes from the same extract if each column is sufficiently specific for the desired enzyme.

For reasons similar to those applicable to guanine deaminase, 9-(p-(β-aminoethoxy)phenyl)guanine (IV) is a good reversible inhibitor of xanthine oxidase.[4] Therefore, the Sepharose affinity column (V) was also tried and found to be applicable to the purification of xanthine oxidase from rat liver.[8]

In the first experiments, partially purified, commercial xanthine oxidase from bovine milk was applied to the column and readily adsorbed. Elution with 1 mM hypoxanthine, a substrate of the enzyme, gave 88% recovery of the enzyme in the first 7 ml of eluate.

Rat liver xanthine oxidase, which is obtained in the 0–45% $(NH_4)_2SO_4$ fraction of rat liver, could not be detected by spectrophotometric assay at 290 nm of the reaction hypoxanthine + $O_2 \rightarrow$ uric acid. Enzyme assay was, however, possible by the more sensitive dichlorophenolindophenol assay at 600 nm, again with hypoxanthine as substrate. In all determinations there was a large base line in the absence of hypoxanthine due to extraneous reduction of the indophenol. When the 0 to 45% $(NH_4)_2SO_4$ fraction from rat liver was passed through a column of (V), all the xanthine oxidase was readily retained while all the "activity" causing the extraneous reduction of the indophenol passed through the column. Xanthine oxidase was eluted with 1 mM hypoxanthine. The activity of the enzyme was greatly diminished after 24 hours at 3°, but was stable for at least several weeks at $-15°$. In a separate experiment, xanthine oxidase was eluted with the inhibitor 2-benzylthiohypoxanthine[9]; the latter was readily removed from the eluate by extraction with 2-octanol with no destruction of the enzyme.

Massey et al.[10] used affinity chromatography for the resolution of active and inactive xanthine oxidase from bovine milk. As affinity ligand, they used a 4-substituted pyrazolo[3,4-d]pyrimidine (VI). This compound was

[7] B. R. Baker and N. M. J. Vermeulen, *J. Med. Chem.* **13**, 1143 (1970).
[8] B. R. Baker and H. U. Siebeneick, *J. Med. Chem.* **14**, 799 (1971).
[9] B. R. Baker and J. L. Hendrickson, *J. Pharm. Sci.* **56**, 955 (1967).
[10] D. Edmonson, V. Massey, G. Palmer, L. M. Beacham III, and G. B. Elion, *J. Biol. Chem.* **247**, 1597 (1972).

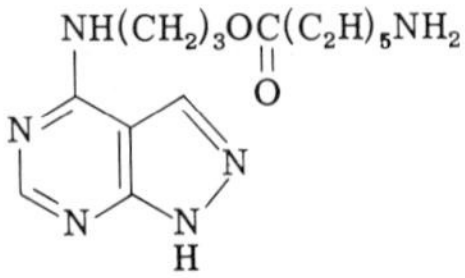

$$NH(CH_2)_3OC(C_2H)_6NH_2$$

(VI)

attached to BrCN-activated Sepharose 6B and finally converted into the 6-hydroxy derivative with xanthine oxidase. When milk xanthine oxidase was applied to this column under anaerobic conditions and allowed to equilibrate overnight, only the highly active xanthine oxidase was bound to the column, and the low-activity form appeared in the eluate. To elute the high activity xanthine oxidase, the column was first washed aerobically with cold 0.1 M PP$_i$, pH 8.5, containing 1 mM salicylate and 0.2 mM EDTA and allowed to stand in a cold room for 3–4 days to permit the slow reoxidation of reduced enzyme-bound molybdenum and finally washed out with the same buffer. The enzyme eluted in a reddish brown band and its activity was stable for several weeks in the presence of salicylate but is significantly less stable in its absence.

Reagents

The buffer was 50 mM Tris chloride at pH 7.4. Guanine, 0.5 mM, was prepared by dilution of 1 mM guanine in 0.1 M NaOH with an equal volume of buffer, resulting in a solution at pH 9. Rat liver was extracted, then fractionated into 0 to 45% and 45 to 90% saturation of $(NH_4)_2SO_4$ as previously described.[11] The protein was redissolved at 1 ml per gram of original rat liver and stored in aliquots at $-15°$; enzyme activity was stable for several months.

Enzyme Assay

Guanine deaminase was assayed spectrophotometrically with 13.3 μM guanine in buffer as previously described.[3] Xanthine oxidase could be assayed by reduction of 2.6-dichlorophenolindophenol (DCPI) as hypoxanthine was oxidized. The assay was performed in a 1-ml glass cuvette containing 0.75 ml of buffer, 50 μl of 320 μM hypoxanthine (final concentration 16 μM) in buffer, 100 μl of DCPI (0.1 mg/ml), and 100 μl of enzyme in buffer. Activity was followed by measuring the decrease in adsorbance at 600 nm in a Gilford recording spectrophotometer.

[11] B. R. Baker and G. J. Laurens, *J. Med. Chem.* **10**, 1113 (1967).

Sepharose 4B Affinity Column (V)

To a stirred mixture of 4 ml of wet Sepharose 4B in 4 ml of H_2O was added a solution of 0.4 g of BrCN in 4 ml of H_2O in one portion. The pH was maintained at 11 by addition of 4 N NaOH, reaction being complete in 8 minutes. The activated Sepharose 4B was collected in a Büchner funnel and washed with a total of 20 ml of 0.1 M $NaHCO_3$.[12,13]

The activated Sepharose 4B was added immediately to a filtered solution of 100 mg of 9-(p-β-aminoethoxyphenyl)guanine (IV)[5] in 10 ml of dimethylformamide and 15 ml of dimethylsulfoxide. The mixture is stirred at ambient temperature for about 18 hours, then filtered. The Sepharose (V) was washed with dimethylformamide and then with H_2O until absorbance at 270 nm was no longer detected. This preparation was stored at 3° as an aqueous suspension until used.

Purification of Guanine Deaminase by the Affinity Column (V)

A mixture of 1 ml of the above column material (V) and 2 ml of untreated Sepharose 4B was poured into a column 8 mm in diameter. The column was placed in an ice bath made from an inverted, bottomless polyethylene bottle and washed with cold buffer. The 45 to 90% $(NH_4)_2SO_4$ fraction from rat liver was diluted with 2 volumes of buffer; 5 ml of this solution, containing 57 mg of protein and 1.29 units[14] of guanine deaminase activity, were passed through the column. The column was washed with about 12–15 ml of ice-cold buffer until protein was no longer detectable at 280 nm. The enzyme was eluted with 0.5 mM guanine, and the eluate, in 2-ml fractions, was collected at 0°. All enzyme activity (1.13 units) appeared in the first two fractions. The two 2-ml fractions were combined, and an aliquot was dialyzed at 3° against buffer for 30 hours in a 1-ml microdialysis apparatus (Chemical Rubber Co.). Protein determination showed a total of 0.28 mg in 4 ml of eluate. The recovery of guanine deaminase activity was 87% and the purification was 200-fold; the activity was stable indefinitely at $-15°$. The column material was washed with 20 ml of buffer and was then ready for reuse.

Purification of Xanthine Oxidase by the Affinity Column V

a. Substrate Elution. The 0 to 45% $(NH_4)_2SO_4$ fraction from rat liver was diluted with 2 volumes of buffer. Of this solution, 5 ml, which contained 121 mg of protein, 16.0 units[14] of extraneous DCPI "reductase ac-

---

[12] P. Cuatrecasas, M. Wilchek, and C. B. Anfinsen, *Proc. Nat. Acad. Sci. U.S.* **61**, 636 (1968).

[13] P. Cuatrecasas, *J. Biol. Chem.* **245**, 3059 (1970).

[14] A unit of enzyme activity is arbitrarily defined as the amount of enzyme necessary to give an absorbance change of 1.0 per minute under standard assay conditions.

tivity" (base line without hypoxanthine), and 18.0 units[14] of xanthine oxidase activity, was passed through the affinity column (V) as described above for guanine deaminase. The system was washed with buffer until protein was no longer detectable; all the extraneous DCPI "reductase" was found in the washings. Xanthine oxidase was then eluted with 1 mM hypoxanthine in buffer. The activity (14.4 units, 90%) appeared in the first 5.4 ml of eluate, which contained 0.53 mg of protein (after dialysis of an aliquot). Purification was about 230-fold. The total time for adsorption and elution was about 2–3 hours.

Similarly, 13.4 units[14] of bovine milk xanthine oxidase (Sigma Chemical Co., No. X0375) could be adsorbed on the affinity column (V) and eluted with 1 mM hypoxanthine, resulting in recovery of 11.8 units (88%).

b. Inhibitor Elution. Through the affinity column (V) were passed 3 ml of diluted rat liver extract containing 15.2 units[14] of xanthine oxidase as in procedure a. The column was washed with 10 ml of buffer, then the xanthine oxidase was eluted with a 0.2 mM solution of 2-benzylthiohypoxanthine[9] prepared by dissolving 5.16 mg of the inhibitor in 5 ml of dimethylsulfoxide; the preparation was diluted to 100 ml with buffer. All the activity appeared in the first three 2-ml fractions. Each fraction was washed twice with 2 ml of water-saturated 2-octanol to remove the inhibitor. The 6 ml of eluate were then found to contain 10.3 units (68% recovery) of xanthine oxidase.

Attachment of Specific Ligands and Methods for Specific Proteins

E. OTHER SYSTEMS

[66] Covalent Chromatography by Thiol–Disulfide Interchange

By Keith Brocklehurst, Jan Carlsson, Marek P. J. Kierstan, and Eric M. Crook

Affinity Chromatography and Covalent Chromatography: General Considerations

The selective isolation and purification of enzymes and other biologically important macromolecules by "affinity chromatography" has been discussed by Cuatrecasas and Anfinsen[1] and by Porath.[2] This technique exploits the unique ability of proteins and polypeptides to bind ligands specifically and reversibly and is widely applied. Reversible binding in this context usually makes use of noncovalent bonds although covalent bonding may occur in some cases. For example, the affinity column that Blumberg et al.[3] used to purify papain may act as a virtual substrate and react with the enzyme to form an acyl enzyme or tetrahedral adduct.

A rather special type of affinity chromatography can be envisaged in which the material of the chromatography column reacts chemically with only one of the components of a mixture. After removal of the other components from the column by washing, the covalently bound component is released by reaction with a suitable reagent. The reaction that releases the component of interest should preferably leave the column in a form that can be readily "reactivated." This type of chromatography, which relies upon the scission and formation of covalent bonds, we have called "covalent chromatography."[4] In principle, it can allow the separation, not only of molecules that differ from each other in their possession or otherwise of a given functional group, but also of molecules that all possess the particular functional group. This should be possible if conditions can be found in which the reactivity toward the column of a given functional group in one type of molecule differs significantly from that in another type of molecule. One way in which this difference in reactivity might be achieved is through specific adsorptive binding forces.

In addition to our work on covalent chromatography by thiol–disulfide

---

[1] P. Cuatrecasas and C. B. Anfinsen, this series, Vol. 22, p. 345. See also Parikh *et al.*, this volume [6].

[2] J. Porath, this volume [2].

[3] S. Blumberg, I. Schechter, and A. Berger, *Eur. J. Biochem.* 15, 97 (1970).

[4] K. Brocklehurst, J. Carlsson, M. P. J. Kierstan, and E. M. Crook, *Biochem. J.* 133, 573 (1973).

interchange[4] (see below) there are two other examples of covalent chromatography of which we are aware. Ashari and Wilson[5] allowed acetylcholinesterase to react with a column containing a bound "irreversible" cholinesterase inhibitor, 2-aminoethyl-*p*-nitrophenyl methyl phosphonate and subsequently eluted the enzyme with 2-(hydroxyiminomethyl)-1-methyl pyridinium iodide. Blumberg and Strominger[6] reported the isolation of five penicillin-binding components present in *Bacillus subtilis* membranes by "covalent affinity chromatography." The penicillin binding components were covalently linked to a Sepharose column containing 6-aminopenicillanic acid and subsequently eluted with hydroxylamine which cleaves the penicilloyl-enzyme bond. Presumably, in this case, the reactive column cannot readily be regenerated.

The rest of this article deals with one particular type of covalent chromatography, covalent chromatography by thiol–disulfide interchange, and as an example of its use the preparation of fully active papain is described in detail.

Covalent Chromatography by Thiol–Disulfide Interchange

General Principle

Many proteins contain cysteine residues, and some recent work in this laboratory has been directed toward the development of a method of purification of such proteins. The method has been developed using papain as the thiol-containing protein and is currently being applied to other proteins, e.g., bromelain,[7] chymopapain,[7] ficin,[8] propapain,[9,10] creatine phosphokinase,[11] and phosphofructokinase.[11] For simplicity, a protein containing one reactive thiol per mole (ESH) is considered. This would be expected to react with a polymer containing mixed disulfide residues (P-S-S-X) by thiol–disulfide interchange (Scheme I). To facilitate isolation of ESH from a mixture, it is desirable to select P-S-S-X such that (i) the reaction of ESH producing XSH (Scheme Ia) predominates over that producing E-S-S-X (Scheme Ib); (ii) the reaction of Scheme Ia is effectively irreversible; (iii) the thiol, XSH, is readily detected by

[5] Y. Ashani and I. B. Wilson, *Biochim. Biophys. Acta* **276**, 317 (1972). See also this volume [74].
[6] P. M. Blumberg and J. L. Strominger, *Proc. Nat. Acad. Sci. U.S.* **69**, 3751 (1972). See also this volume [43].
[7] M. P. J. Kierstan, J. Carlsson, and K. Brocklehurst, unpublished work.
[8] M. Shipton and K. Brocklehurst, unpublished work.
[9] K. Brocklehurst and M. P. J. Kierstan, *Nature* (*London*) *New Biol.* **242**, 167 (1973).
[10] T. Stuchbury and K. Brocklehurst, unpublished work.
[11] J. P. G. Malthouse and K. Brocklehurst, unpublished work.

(a) P—S—S—X + ESH ⇌ P—S—S—E + XSH

(b) P—S—S—X + ESH ⇌ P—SH + E—S—S—X

(c) E—S—S—X + ESH ⇌ E—S—S—E + XSH

(d)   P—S—S—E $\xrightarrow[\text{RSH}]{\text{excess}}$ P—SH + ESH + R—S—S—R

Scheme I

physical measurement, e.g., spectrophotometry. In such a case the attachment of the thiol-containing protein to the column may be followed by monitoring the release of XSH produced in the reaction of Scheme Ia. Subsequent elution with a solution of a low molecular weight thiol (RSH) would be expected to produce protein containing 1 mole of reactive thiol per mole of protein, the thiolated column (PSH) and the disulfide derivative of the low molecular weight thiol (R-S-S-R) (Scheme Id).

If the thiol-containing protein to be purified is an enzyme containing an essential thiol group, it is possible that all the protein molecules containing thiol groups may not be catalytically active. Thus in a partially denatured preparation, for example, some molecules that contain conformationally distorted active centers may still contain reactive thiol groups. In such cases separation of catalytically active protein from catalytically inactive protein will be achieved only if conditions can be found in which the reactivities of the active and inactive thiol-containing proteins toward the mixed-disulfide column are sufficiently different. It was considered that such conditions might be found for papain (see below).

Choice of the Mixed-Disulfide Column and Its Method of Preparation

Choice of XSH. It was decided to prepare a polymer containing aliphatic thiol groups and, subsequently, to let these thiol groups react with 2,2′-dipyridyl disulfide (2-Py-S-S-2-Py) so as to produce a polymer containing alkyl-2-pyridyl disulfide residues with the release of 2-thiopyridone (Py-2-SH) (Scheme IIa). The choice of Py-2-SH as XSH of Scheme I was based on the following considerations: (1) The reactions of 2-Py-S-S-2-Py with low molecular thiols and with papain are characterized by very large equilibrium constants such that even using approximately equimolar concentrations of reactants the reactions producing Py-2-SH are essentially stoichiometric, even in solutions of low pH[12]; this suggested that reactions of ESH with P-S-S-X might be similarly effectively irreversible. (2) Py-2-SH is readily determined spectrophotometrically at

[12] K. Brocklehurst and G. Little, *Biochem. J.* **133**, 67 (1973).

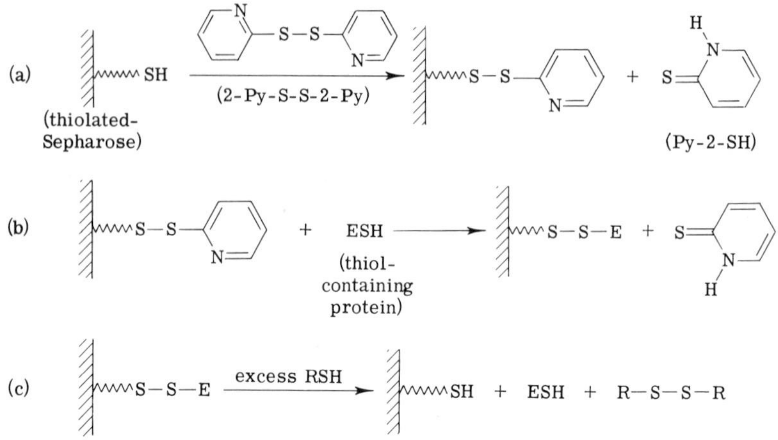

Scheme II

343 nm.[12] (3) 2-Py-S-S-2-Py increases its electrophilicity by about 4×10^3 times when it becomes protonated ($pK_{a_{II}} = 2.45$).[12,13] (4) It seemed likely that in this system the reaction of Scheme Ia would predominate over that in Scheme Ib, at least when the pyridyl moiety is protonated. (5) 2-Py-S-S-2-Py reacts specifically only with intact papain catalytic sites at pH values approximately 4.[12-14] Considerations (1)–(4) have general applicability in the choice of Py-2-SH as XSH for the preparation of any thiol-containing protein. Consideration (5) made Py-2-SH a particularly suitable choice for the preparation of fully active papain. The specific reaction of papain with 2-Py-S-S-2-Py at pH values of 4 is probably best described as an intra-complex thiol–disulfide interchange involving the nonionized thiol group of the cysteine-25–histidine-159–asparagine-175 hydrogen-bonded system of papain with 2,2'-dipyridyl disulfide hydrogen bonded at one nitrogen atom to the carboxyl group of aspartic acid-158.[15] This mechanism results in a rate of reaction of papain with 2-Py-S-S-2-Py at pH values 3.5 to 4.0 that is at least 100 times greater than that of the reaction of 2-Py-S-S-2-Py with simple thiol groups at comparable concentrations. By contrast, at pH values $\geqslant 8$, where the reactions proceed mainly by attack of the thiolate ion on unprotonated disulfide, 2-Py-S-S-2-Py reacts with L-cysteine about 10–20 times faster than with papain. If the mixed disulfide column of Scheme II possessed the high reactivity in acidic media toward the thiol group of active papain, but not toward other thiol groups, as is found with 2-Py-S-S-2-Py, it was expected that this type of chromatography would permit

[13] K. Brocklehurst and G. Little, *Biochem. J.* **128,** 471 (1972).
[14] K. Brocklehurst and G. Little, *FEBS Lett.* **9,** 113 (1970).
[15] K. Brocklehurst, *Tetrahedron,* in press (1974).

separation of active papain from other thiols including thiol-containing but inactive protein.

Choice of the Polymer–Thiol Conjugate. The two thiol-containing polymers that are available commercially, i.e., Enzacryl Polythiol (Koch-Light Laboratories, Ltd.) and Insolubilized Cysteine [Miles-Seravac (Pty.) Ltd., Maidenhead, Bucks, U.K.] were not used for two reasons. First, the thiol groups of both of these products are those of cysteine residues that are attached to the polymer chain by their N atoms. It was considered desirable to create a situation in which the thiol group may be situated at a greater distance from the polymer chain than is the case in these products, i.e., to provide a "spacer." This would be expected to decrease steric interactions of the proteins to be chromatographed with the polymer. Second, the excellent packing and flow properties and the low ion-exchange capacity of agarose were thought to be more desirable than those of either of the commercial products.

Cyanogen bromide activation constitutes a simple and convenient one-step synthesis of reactive (electrophilic) sites in polysaccharides such as Sepharose.[16-18] In this article, these sites are represented as imidocarbonates although they may exist in other forms, e.g., isourea derivatives and carbamates.[18] Reaction of these sites in the polymer with a suitable amino-thiol should result in the production of a thiol-containing polymer (see Scheme III) in which steric hindrance of the thiol groups by the polymer may not be a serious problem.

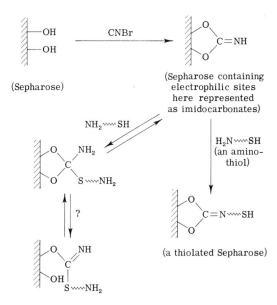

Scheme III

Our initial studies were with the readily available aminothiol, gluta-thione (γ-L-glutamyl-L-cysteinylglycine). Since it is possible that the electrophilic sites in the polymer may exist as imidocarbonates or alterna-tively that this type of structure may be an intermediate in the formation of carbamate sites, we did not protect the thiol group in glutathione for reaction with the "activated" Sepharose. If either of these situations exists, the possibility of the reactions of the thiol group with such a site pro-ceeding further than the reversible formation of a tetrahedral adduct (see Scheme III) was considered unlikely.

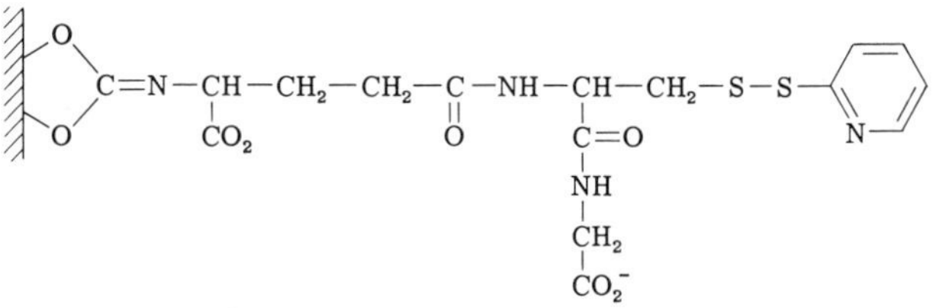

Scheme IV

Although the mixed disulfide-agarose gel (Scheme IV) so produced has proved adequate in our hands for the preparation of fully active thiol proteases, it is possible that the glutathione "arm" may be susceptible to hydrolytic cleavage. To obviate this potential instability, the glutathione moiety in the CNBr method could be replaced by a hydrocarbon moiety.

It was decided not to synthesize a thiol-containing Sepharose by the method of Cuatrecasas,[19] i.e., reaction of an ω-aminoalkyl-Sepharose with N-acetylhomocysteine thiolactone, because this route involves an extra step.

Methods

Preparation of the Agarose–Glutathione Conjugate

1. *Chemical Activation of Sepharose.* This is carried out essentially as described by Axén and Ernbäck.[18] Before use, the Sepharose is washed with 1 mM EDTA on a sintered-glass funnel. Freshly prepared CNBr solution (50 ml, 20 mg/ml) is added to 50 g of Sepharose 2B (50 g of swollen and vacuum-filtered Sepharose equivalent to 1 g dry weight equivalent to 60–70 ml bed volume). The pH of the stirred suspension

[16] R. Axén, J. Porath, and S. Ernbäck, *Nature (London)* **214**, 1302 (1967).
[17] J. Porath, R. Axén, and S. Ernbäck, *Nature (London)* **215**, 1491 (1967).
[18] R. Axén and S. Ernbäck, *Eur. J. Biochem.* **18**, 351 (1971).
[19] P. Cuatrecasas, *J. Biol. Chem.* **245**, 3059 (1970).

is adjusted to 11 by addition of 4 *M* NaOH and maintained at this value in a pH stat at 25° for 6 minutes. The activated gel is then washed rapidly ($<$15 minutes) with 500 ml of 0.1 *M* NaHCO$_3$ at 4° on a sintered-glass filter under suction. The activated gel is allowed to react with glutathione immediately after its preparation.

2. *Reaction with Glutathione.* After filtration, the activated gel is transferred to a 200-ml stoppered flask containing 1 g of glutathione in 20 ml of 0.1 *M* NaHCO$_3$ at pH 8.5. To ensure good mixing 0.1 *M* NaHCO$_3$ is added to give a total volume of 80–90 ml. The suspension is mixed by shaking for 20 hours at 22°. The gel is washed with 500 ml 0.1 *M* NaHCO$_3$ on a sintered funnel and then packed in a column (1.8 cm $\times$ 30 cm) and washed by using the following solutions at a flow rate of 10 ml per hour: (i) 0.1 *M* NaHCO$_3$ at pH 9.5, containing 1 *M* NaCl and 1 m*M* EDTA (24 hours); (ii) 0.1 *M* sodium acetate at pH 4.3, containing 1 *M* NaCl and 1 m*M* EDTA (24 hours); (iii) 1 m*M* EDTA (24 hours).

Determination of the Thiol Contents of the Agarose–Glutathione
  Gels before and after Reduction with Dithiothreitol

1. *Dry-Weight Determinations.* Samples of the unreduced and reduced gel (see below) (2 g wet weight) are suspended in deionized water (50 ml) to give suspensions of approximately 1 mg of gel per milliliter. Portions (10 ml) of the stirred stock suspensions are taken, and the gel is isolated by filtration on Millipore filters of known weight. The filters are then dried under reduced pressure over P$_2$O$_5$ to constant weight. This usually requires drying for 48 hours.

2. *Thiol Determinations.* These are made by titration of the thiol-containing gel with 2-Py-S-S-2-Py. Samples of suspension containing 5–10 mg dry weight of gel are subjected to vacuum filtration in a column (1 cm $\times$ 12 cm) fitted with a fine glass sinter, tap, and stopper. To the gel is added 3 ml of a mixture containing 0.1 *M* Tris chloride at pH 8.0, 1 m*M* EDTA, and 0.3 *M* NaCl; 1 ml of 1.5 m*M* 2-Py-S-S-2-Py is added, and the suspension is mixed by repeated inversion for 10 minutes. The mixture is subjected to vacuum filtration in the column and the concentration of 2-thiopyridone in the filtrate determined spectrophotometrically at 343 nm against an appropriate blank.[12]

Preparation of the Agarose–(Glutathione-2-pyridyl
  Disulfide) Conjugate

The Sepharose-glutathione gel, approximately 50 g wet weight, is washed by vacuum filtration with 450 ml of 0.1 *M*-Tris chloride at pH 8.0, containing 0.3 *M* NaCl and 1 m*M* EDTA. The gel is suspended in

60 ml of this buffer containing 30 mM dithiothreitol and the mixture is stirred gently for 10 minutes. The gel is separated by vacuum filtration. The treatment with dithiothreitol and subsequent vacuum filtration are repeated twice more. The fully reduced gel is washed with each of the following solutions in turn: 150 ml of 1 M NaCl, 150 ml 0.5 M NaHCO$_3$, and 150 ml 0.1 M Tris chloride at pH 8.0, each of which contain 0.3 M NaCl and 1 mM EDTA. The final wash solution is shown to be free of dithiothreitol by its lack of reaction with 2-Py-S-S-2-Py. The thiol content of a sample of gel is determined as described above. After the washing procedure, the fully reduced gel is allowed to react with 1.5 mM 2-Py-S-S-2-Py in 50 ml 0.1 M Tris chloride at pH 8.0 containing 0.3 M NaCl and 1 mM EDTA with gentle stirring for 30 minutes. The gel is then washed as described above until no 2-Py-S-S-2-Py can be detected by measuring extinction at 281 nm.

Chromatography with the Agarose–(Glutathione-2-pyridyl Disulfide) Conjugate

The gel is equilibrated as required either in 0.1 M Tris chloride at pH 8, or in 0.1 M sodium acetate at pH 4, both containing 0.3 M NaCl and 1 mM EDTA.

A batch of gel prepared as described above corresponds to a bed volume of 60–70 ml and is sufficient to prepare approximately 100 mg of papain. This amount of gel is packed into a column (1.8 cm $\times$ 30 cm) and flow rates of 30–60 ml per hour are used. To prepare 1 g of papain, it is convenient to use a bed volume of 500 ml.

Gel Regeneration

After each run, the gel is reconverted into the Sepharose-(glutathione-2-pyridyl disulfide) form by reduction with dithiothreitol and reaction with 2-Py-S-S-2-Py as described above. Only one treatment with dithiothreitol is necessary to produce the fully reduced gel.

Preparation of Fully Active Papain from Dried Papaya Latex

Crude papain is prepared from 100 g of dried papaya latex as described by Brocklehurst et al.[4]

It is dissolved in the minimum volume of either 0.1 M Tris chloride at pH 8.0, or 0.1 M sodium acetate at pH 4.0, both containing 0.3 M NaCl and 1 mM EDTA, to give a clear solution. Usually approximately 200 ml are required. The solution containing 0.1–0.2 mole of thiol per mole of protein is applied to a chromatography column (1.8 cm $\times$ 30 cm) packed with Sepharose-(glutathione-2-pyridyl disulfide) gel at room temperature. The column is eluted with the buffer used to apply the protein to the column. Elution is continued until absorbance at both 280 nm and 343 nm falls to less than 0.03. Then the gel is equilibrated with 0.1 M Tris chloride at pH 8.0, containing 0.3 M NaCl and 1 mM EDTA. Covalently bonded material is removed from the Sepharose-(glutathione-papain disulfide) column by elution with 50 mM L-cysteine in 0.1 M Tris chloride at pH 8.0, containing 0.3 M NaCl and 1 mM EDTA. Fractions are monitored for absorbance at 280 nm and 343 nm and for esterolytic activity toward α-N-benzoyl-L-arginine ethyl ester.[20] The peak-activity fractions are pooled.

At this stage, the solution may be concentrated by precipitating the protein by addition of 30 g of $(NH_4)_2SO_4$ per 100 ml of solution and dissolving the precipitate in a minimum quantity of pH 8 buffer. The protein is then separated from low-molecular-weight materials by chromatography on Sephadex G-25 in 0.1 M KCl containing 1 mM of EDTA as the eluent. To prevent the formation of papain-L-cysteine mixed disulfide during gel filtration, dithiothreitol is added immediately before application to the Sephadex G-25 column to a final concentration of 5 mM.

Preparation of Fully Active Papain from Commercial Crystallized Partially Activatable Papain

Commercial 2 × crystallized papain (200 mg) is treated with 5 mM dithiothreitol in 0.1 M Tris chloride at pH 8.0 and separated from low-molecular-weight material on Sephadex G-25. The protein (generally containing 0.4–0.6 mole of thiol per mole of protein) in 100 ml of the appropriate buffer is applied to a column (1.8 cm × 30 cm) of Sepharose-(glutathione-2-pyridyl disulfide) gel and subjected to covalent chromatography as described above.

Properties of the Gels

Thiol Content of the Agarose–Glutathione Gel

The Sepharose–glutathione conjugate, prepared as described in the Methods section, contains 1.0 ± 0.2 mg of reduced glutathione per 100 mg of dry gel (0.55 ± 0.05 μmole of thiol per milliliter of swollen gel; standard errors are based on three preparations).

After treatment of the gel with dithiothreitol as described (20 mM dithiothreitol, pH 8.0, at 22° for 30 minutes), the reduced glutathione content increases by 100%, i.e., to 2.0 mg of reduced glutathione per 100 mg of dry gel (1.1 ± 0.1 μmoles of thiol per milliliter of swollen

---

[20] K. Brocklehurst, M. P. J. Kierstan, and G. Little, *Biochem. J.* **128**, 979 (1972).

gel). No further increase in the thiol content of the gel is achieved by increasing the concentration of dithiothreitol to 100 mM or by using NaBH$_4$ as reducing agent. The fully reduced gel has been stored as a suspension in 1 mM EDTA solution for 4 weeks at 4° without a detectable decrease in its thiol content. The thiol content of the fully reduced gel is somewhat higher than that of the thiolated agarose reported by Cuatrecasas[19] (0.58 μmole of thiol per milliliter of swollen gel). The thiolated-Sepharose gel may itself have useful applications.

Capacity of the Agarose–(Glutathione-2-pyridyl Disulfide)
Column with Respect to Papain

The theoretical capacity of a Sepharose–(glutathione-2-pyridyl disulfide) column acting as a thiol–disulfide exchanger is its content of mixed disulfide residues. This is measured during its preparation by monitoring the Py-2-SH released in the reaction of the reduced Sepharose–glutathione gel with 2-Py-S-S-2-Py. The practical capacity of the column for a given thiol may be less than the theoretical capacity and will depend on the nature of the molecule containing the reactive thiol group. The results presented below demonstrate that the capacity of the column for papain is 10 ± 1% of theoretical and L-cysteine is probably essentially 100% of theoretical. The capacity for papain (substantially less than 100%) presumably arises from the inaccessibility of some of the mixed disulfide residues to the papain thiol group occasioned either by their location in hindered regions of the gel or more likely by steric shielding of some glutathione-2-pyridyl disulfide residues by glutathione-papain disulfide residues.

To determine the capacity of the column with respect to active papain it is loaded with activated, activator-free papain (0.5 mole of thiol per mole of protein) prepared from the 2 × crystallized product of BDH Chemicals Ltd. This enzyme preparation consists of 50% of papain containing 1 mole of thiol per mole of protein (that of cysteine-25) and 50% of protein that lacks a thiol group (see Brocklehurst and Kierstan[9]). A solution of this mixture of proteins is applied to the column [either in Tris chloride at pH 8.0, or in sodium acetate at pH 4.0 (Sepharose is stable in aqueous suspension in the pH range 4–9. Prolonged storage of the gel at pH 4, the low pH limit of its stability, should be avoided)] until the concentration of thiol in the eluate becomes equal to the concentration of thiol in the solution applied to the column. This occurs after 10 ± 1% of the available mixed disulfide residues have released Py-2-SH and the thiol–disulfide interchange between the column and active papain is then considered to be complete. The attached protein is then eluted from the column by 20 mM L-cysteine solution in 0.1 M

Tris chloride at pH 8.0, containing 0.3 M NaCl. In the cases in which the reaction of papain with the column is allowed to proceed at pH 4, the pH of the column is changed to 8.0 by washing with 0.1 M Tris chloride at pH 8.0, before the L-cysteine solution is introduced. Elution patterns have been presented by Brocklehurst et al.[4] Spectral scanning of the L-cysteine eluate shows that it contains both protein and Py-2-SH. When the elution of protein is complete the eluate contains protein in amount equivalent to 10 ± 1% of the theoretical capacity of the column and Py-2-SH in amount equivalent to 90% of the theoretical capacity of the column. The protein is separated from low-molecular-weight material by chromatography on Sephadex G-25. Immediately before Sephadex chromatography, dithiothreitol is added to give a concentration of 5 mM. This is necessary when aerobic solutions are used to prevent oxidation of the enzyme's thiol group in the presence of low concentrations of L-cysteine to provide the papain-L-cysteine mixed disulfide during the latter stages of the Sephadex chromatography. When this precaution is taken, the protein emerging from the Sephadex column contains 1.04 ± 0.05 moles of thiol per mole of protein and its esterolytic activity toward α-N-benzoyl-L-arginine ethyl ester is not increased by preincubation with L-cysteine or with dithiothreitol (see below). When 5 mM dithiothreitol is omitted, the thiol content of the protein is generally 0.75 ± 0.05 mole of thiol per mole of protein and its esterolytic activity is proportionately less than that of the enzyme containing 1.0 mole of thiol per mole of protein when assayed in the absence of reducing agent. When 5 mM L-cysteine is added to such assay mixtures full activity is regained.

Some Results of the Use of the Agarose–(Glutathione-2-pyridyl Disulfide) Column for Covalent Chromatography

Preparation of Fully Active Papain by Covalent Chromatography of the Commercially Available 2 × Crystallized Protein

For routine preparation of papain containing 1 mole of thiol per mole of protein from the activated activator-free 2 × crystallized product of BDH Chemicals Ltd., a procedure similar to the one described above is used except that the column is operated below the limit of its practical capacity. This is done to obtain maximum yield of active papain from a given batch of expensive commercial enzyme without having to recycle the protein. This procedure also allows the demonstration that the Sepharose–(glutathione-2-pyridyl disulfide) column is not acting to an appreciable extent as an ion exchanger in addition to acting as a thiol–disulfide exchanger. Spectrophotometric analysis of the eluate resulting from loading the column with the mixture of active and inactivatable papain under nonsaturating conditions shows that all of the thiol-containing protein becomes attached to the column and all the protein lacking a thiol group passes through. Washing the loaded column with solutions of KCl up to 1 M in concentration at pH values of 4 and 8 fail to remove protein from the column. The possibility that the column is acting as a conventional affinity column, i.e., that the glutathione-2-pyridyl disulfide moiety binds active papain without thiol–disulfide interchange is discounted because of the stoichiometric release of Py-2-SH when the thiol-containing protein binds to the column.

Preparation of Fully Active Papain from Dried Papaya Latex

Twice-crystallized papain, containing after activation approximately 0.5 mole of thiol per mole of protein, is commercially prepared by the method of Kimmel and Smith[21] (see also Arnon[22]). This method consists of (1) preparation of a crude aqueous extract, (2) removal of material insoluble at pH 9 which may contain papain inhibitors, (3) $(NH_4)_2SO_4$ fractionation, which removes another thiol-protease chymopapain (EC 3.4.4.11), (4) NaCl precipitation, (5) crystallization, and (6) recrystallization from NaCl solution.

The first three steps of this procedure perform functions that covalent chromatography may not necessarily perform, and these steps are retained. The $(NH_4)_2SO_4$ fractionation step is modified in that the second precipitation is carried out at 35% instead of 40% saturation to minimize the chymopapain content of the preparation before the covalent chromatography step. A solution of the $(NH_4)_2SO_4$-precipitated protein is applied to the Sepharose–(glutathione-2-pyridyl disulfide) column and subjected to covalent chromatography as described in the Methods section. Separation from low-molecular-weight material by chromatography on Sephadex G-25 in the presence of 5 mM dithiothreitol (see above) yields active papain containing 1.04 ± 0.06 moles of thiol per mole of protein.

Separation of Papain from L-Cysteine by Covalent Chromatography

When a mixture of fully active papain (0.1 mM) and L-cysteine (up to approximately 5 mM) is subjected to covalent chromatography at pH 4 under rigorously anaerobic conditions to prevent formation of the papain-L-cysteine mixed disulfide, all of the papain reacts with the mixed disulfide gel and the L-cysteine passes through the column without reacting

[21] J. R. Kimmel and E. L. Smith, *Biochem. Prep.* 6, 61 (1958).
[22] R. Arnon, this series, Vol. 19 [14].

with it to an appreciable extent. This result supports the view, discussed above, that the reaction of papain with the gel at pH values about 4 results mainly from reaction of the nonionized thiol group of papain in the cysteine-25–histidine-159–asparagine-175 hydrogen-bonded system with the mixed-disulfide gel protonated (or partially protonated) on the nitrogen atom of the pyridine ring. As the concentration of L-cysteine in the mixture is increased from 5 mM, an increasing proportion of L-cysteine reacts with the gel. At such concentrations there is a sufficiently large amount of the L-cysteine thiolate ion to make its rate of reaction with the protonated gel comparable with that of the nonionized thiol group of papain. By using a thiol with a higher pK_a value, e.g., 2-mercaptoethanol or with a low intrinsic reactivity, e.g., propapain,[9] it should be possible to attach papain to the gel at pH 4 in the presence of even higher concentrations of such a thiol. This would be predicted from the above discussion and from our previous observation[12,14] that it is possible to titrate papain active sites by using 2-Py-S-S-2-Py at pH values of approximately 4 in the presence of up to a 10-fold molar excess of L-cysteine and up to a 100-fold molar excess of 2-mercaptoethanol.

Properties of Papain Prepared by Covalent Chromatography

The kinetic parameters that characterize the catalysis of the hydrolysis of α-N-benzoyl-L-arginine ethyl ester by fully active papain prepared by covalent chromatography and the reactivity characteristics of its active center thiol group have been reported by Brocklehurst et al.[4]

It is well known that a free thiol group in the side chain of cysteine-25 in papain is a necessary requirement for enzymatic activity. The presence of a thiol group, however, may not be sufficient. It is possible that some papain molecules may possess a reactive thiol group, but not the active-site conformation required for catalytic activity.

The successful separation of papain from L-cysteine by covalent chromatography at pH 4 (see above) suggests that the high reactivity that the thiol group of catalytically active papain exhibits toward 2-Py-S-S-2-Py in acidic media is exhibited also toward the mixed disulfide gel and that this high reactivity is not exhibited by simple thiol groups. This is confirmed by preliminary kinetic experiments on reactions of papain and other thiols with samples of the mixed disulfide gel in stirred suspension. In view of our previous observation[14] that usually less than 10% of the papain molecules containing a thiol group do not exhibit high reactivity toward 2-Py-S-S-2-Py in acidic media, it is not surprising that papain containing 1 mole of thiol per mole of protein prepared by covalent chromatography at pH 4 has essentially the same characteristics as papain prepared by this technique at pH 8. Thus for papain, enzyme containing 1 mole of thiol per mole of protein usually contains 1 intact catalytic site per mole of protein. This may not be true for other thiol enzymes or for papain that has been subjected to certain denaturing procedures, and in these cases covalent chromatography may provide a means of separating active from inactive protein, both of which contain a thiol group.

[67] Organomercurial Agarose

By L. A. Æ. SLUYTERMAN and J. WIJDENES

General Principle

An agarose–mercurial column serves to separate thiol compounds from nonthiol compounds. The mercurial attached to the insoluble matrix binds the thiol compound:

$$\text{Matrix-HgCl} + \text{RSH} \rightleftarrows \text{matrix-HgSR} + \text{HCl}$$

whereas nonthiol compounds pass the column unchecked. Elution can be effected by a free mercurial, such as $HgCl_2$:

$$\text{Matrix-HgSR} + \text{HgCl}_2 \rightleftarrows \text{matrix-HgCl} + \text{RSHgCl}$$

The thiol compound is then obtained as its mercaptide. Elution can also be effected, especially in the case of a protein, by a second thiol compound, such as cysteine:

$$\text{Matrix-HgSR} + \text{R'SH} \rightleftarrows \text{matrix-HgSR'} + \text{RSH}$$

If necessary, the excess of the second thiol compound can be separated from the protein by dialysis. The column can be regenerated by elution of the second thiol with the aid of $HgCl_2$.

Preparation

This is a modification of the method described previously.[1] Suspend 400 ml of washed packed agarose (Sepharose 4B) in 400 ml of water and stir mechanically. Insert pH electrodes and add 100 g of solid CNBr (purchased in finely divided form from Fluka). Temperature is maintained at 20 ± 3° by addition of broken ice. The pH is maintained at pH 11 by addition of 10 N NaOH, until 80–90% of the calculated amount is consumed, a period of about 8 minutes. The suspension is filtered

[1] L. A. Æ. Sluyterman and J. Wijdenes, *Biochim. Biophys. Acta* **200**, 593 (1970).

through a Büchner funnel without filter paper onto a second Büchner funnel which does bear a filter paper; small lumps of unreacted CNBr are retained on the first funnel. The activated agarose is held on the second funnel and is quickly washed with 4 liters of cold 0.1 M sodium bicarbonate at pH 9.0.

The agarose is resuspended in 800 ml of 10 vol% dimethyl sulfoxide (DMSO) at 0°. Add 6 g of p-aminophenylmercuric acetate (Aldrich, Milwaukee) dissolved in 100 ml of DMSO. No buffer is added at this stage because salts tend to precipitate the mercurial. The suspension is stirred gently. After 20 hours at 0°, the suspension is warmed to 35° for 20 minutes and filtered. The agarose is resuspended in 400 ml of 20% DMSO at 37° for 15 minutes and filtered, a treatment which is repeated four times. The agarose is packed into a column about 4 cm in diameter and washed with 1 liter of 20% DMSO for 3 hours at room temperature. These washings serve to remove the free mercurial. The agarose is resuspended in 600 ml of 0.1 M ethylenediamine, adjusted to pH 8.0 and gently stirred. After storage overnight at room temperature, the agarose is washed, packed into a column again and treated with 2-nitro-5-mercaptobenzoic acid, as described in the next paragraph, in order to eliminate all residual reactive groups of the activated agarose and to determine the capacity of the column.

Capacity Tests

In order to assay the capacity for thiol compounds of low molecular weight, a solution of 5 mM 5,5'-dithiobis-(2-nitrobenzoic acid)[2] and 15 mM Na$_2$SO$_3$ in 0.1 M potassium phosphate at pH 8.0 is passed through the column until the effluent has the same color as the initial solution. The column is washed with the phosphate buffer until the effluent is colorless and then with 50 mM sodium acetate at pH 5.0 in order to remove all phosphate. The yellow column is eluted with 2 mM HgCl$_2$ in 50 mM sodium acetate, pH 5.0, until the column is again nearly colorless. Of the last eluate, 1 ml (total volume V_{el}) is added to 4 ml of 0.1 M cysteine hydrochloride, previously adjusted to pH 9.0, in order to liberate the free thiophenol anion. The extinction is read at 412 nm (A_{412}). The capacity of the column in micromoles per milliliter of packed material is calculated according to Eq. (1):

$$\text{Capacity (m}M) = \frac{V_{el} \times 5 \times A_{412}}{V_{col} \times 13{,}600} \tag{1}$$

in which V_{col} denotes the volume of the column and 13,600 is the molar

[2] G. L. Ellman, *Arch. Biochem. Biophys.* **82**, 70 (1959).

extinction of the thiophenol anion.[2] Capacities as high as 8 mM are found.

For testing protein binding, a column of 5–10 ml volume is used. A solution of 0.2–2% of papain, prepared according to Kimmel and Smith,[3] in standard buffer (10 vol% DMSO, 0.5% butanol as preservative, 0.1 M KCl, and 50 mM sodium acetate at pH 5.0) 1 mM in EDTA, and 10 mM in Na$_2$SO$_3$ (as activator) is passed through the column until the absorbance at 280 nm of the effluent equals the absorbance of the initial solution. The column is washed free of unbound protein with standard buffer. The active papain is eluted with standard buffer containing 0.5 mM HgCl$_2$. For the calculation of the papain capacity, a molecular weight of 23,400 and a molar absorption of 58,500 is assumed. The capacity is about 20 mg of papain per milliliter (0.85 mM).

The capacity for bovine hemoglobin, eluted with water containing 0.2 M cysteine hydrochloride and 1 mM EDTA, adjusted to pH 7.5, is about 45 mg/ml.

Application

When a dilute solution of papain is passed through the column, concentration of the retained enzyme occurs to such an extent as to make the column opaque. When this opacity extends to the bottom of the column, the operator knows that the capacity is used to its full extent.

The DMSO in the standard buffer serves to make the protein more soluble and to prevent hydrophobic interaction of the protein and the column material. In a few experiments, it was successfully replaced by acetonitrile. If the organic solvent is omitted, protein elution is retarded and the protein band is more extended.

Elution of papain can also be effected with 0.5 mM p-hydroxymercuribenzoate, 10 mM mercaptoethanol or 10 mM cysteine, although the protein band is somewhat wider in these cases. Elution does not occur with 10 mM ZnCl$_2$, CdCl$_2$, or Pb(NO$_3$)$_2$.

After constant use over several months, the column acquires a darker color. Although this hardly affects the operation of the column, the adsorbent has been replaced in order to be on the safe side.

In addition to the purification of papain, this agarose–mercurial column has been used for the purification of chymopapain,[4] Chinese gooseberry proteinase,[5] and streptococcal proteinase.[6] It has also been used for the separation of the subunits of lombricine kinase and of its tryptic thiol-

---

[3] J. R. Kimmel and E. L. Smith, *Biochem. Prep.* **6**, 61 (1958).
[4] J. C. A. Vessies, private communication.
[5] M. A. McDowall, private communication.
[6] A. A. Kortt and T. Y. Liu, *Biochemistry* **12**, 320 (1973).

containing peptides.[7] Optimal elution conditions are not the same for all proteins.

Comment

The preparation of an agarose–mercurial column with a spacer between the agarose backbone and the mercurial has been described by Cuatrecasas.[8] Its capacity was 8 μmoles of cysteine per milliliter and 40–50 mg of horse hemoglobin per milliliter. The procedures cited above indicate that, in many cases, the spacer is unnecessary, possibly owing to the high affinity between mercurial and thiol.

[7] L. A. Pradel and E. Der Terrossian, *Abstr. Commun. Meet. Fed. Eur. Biochem. Soc.* 8, (1972) No. 275.
[8] P. Cuatrecasas, *J. Biol. Chem.* **245**, 3059 (1970).

[68] Purification of Cysteine-Containing Histones

By ADOLFO RUIZ-CARRILLO

Of the major histone classes that occur in mammalian cell nuclei, only fraction F3 contains cysteine.[1-5] More recently, it has been shown that fraction F2a1 from echinoderms also contains cysteine.[6-8] The presence of one or more reactive sulfhydryl groups in certain histones has suggested a method for their separation from other histones by affinity chromatography.[9] The method to be described shows that purification of histone F3 and F2a1 (from echinoderms) is readily achieved by reversible reaction between their cysteinyl residues and an organomercurial anchored to agarose. Subsequent elution with thiols of low molecular weight, e.g., cysteine, 2-mercaptoethanol or dithiothreitol, releases the bound histone in a state of high purity and with high yields.

[1] D. M. P. Phillips, *Biochem. J.* **97**, 669 (1965).
[2] D. M. Fambrough and J. Bonner, *J. Biol. Chem.* **243**, 4434 (1968).
[3] S. Panyim, R. Chalkley, S. Spiker, and D. Oliver, *Biochim. Biophys. Acta* **214**, 216 (1970).
[4] S. Panyim, K. R. Sommer, and R. Chalkley, *Biochemistry* **10**, 3911 (1971).
[5] R. J. DeLange, J. A. Hooper, and E. L. Smith, *Proc. Nat. Acad. Sci. U.S.* **69**, 882 (1972).
[6] J. A. Subirana, *FEBS Lett.* **16**, 133 (1971).
[7] L. J. Wangh, A. Ruiz-Carrillo, and V. G. Allfrey, *Arch. Biochem. Biophys.* **150**, 44 (1972).
[8] A. Ruiz-Carrillo and J. Palau, *Develop. Biol.* **35**, 115 (1973).
[9] A. Ruiz-Carrillo and V. G. Allfrey, *Arch. Biochem. Biophys.* **154**, 185 (1973).

Preparation of Organomercurial-Agarose

The method employed for the preparation of the organomercurial derivative of Sepharose involves the coupling of sodium p-chloromercuribenzoate to aminoethyl-Sepharose by means of a water-soluble carbodiimide. Other methods for coupling organomercurial compounds by direct reaction of the cyanogen bromide "activated" Sepharose with p-aminophenylmercuric acetate have also been described.[10,11] Both methods yield organomercurial-Sepharose derivatives of comparable capacity.[9,12]

Procedure. Sepharose 4B, 250 ml, is suspended in 2 liters of cold distilled water and stirred gently for 30 minutes. The suspension is filtered through a coarse fritted-glass disk funnel, and the slurry is suspended again in cold distilled water and washed several times more until the absorbancy of the filtrate is negligible between 230 and 300 nm. The washed Sepharose is suspended in 1 volume of distilled water and transferred to a 2-liter vessel placed in a hood; the suspension is stirred. The electrode of a pH meter and a thermometer are introduced in the suspension, and 62.5 g of finely divided cyanogen bromide are added at once. The pH is maintained between 10.8 and 11.2 by constant titration with $10 M$ NaOH; ice is also added to keep the temperature near 20°. After 20–25 minutes, when the release of protons has slowed, as judged by the decreased requirement of alkali to maintain the pH constant, a large amount of ice is added and the suspension is quickly transferred to a 2-liter sintered-glass funnel and filtered under suction. The "activated" Sepharose is washed with 2.5 liters of ice-cold distilled water, a rubber stopper is introduced into the stem of the funnel to prevent liquid from flowing out, and 250 ml of an ice-cold 12% (w/v) solution of ethylenediamine, previously adjusted to pH 10 with $6 N$ HCl, are added. The washing and the addition of the ligand should take the minimum amount of time possible, because the "activated" Sepharose is unstable under those conditions. The suspension is then transferred to a flask and stirred gently for 16 hours at 4°. The substituted Sepharose is washed extensively with ice-cold distilled water until the excess of ligand is completely removed. The aminoethyl-Sepharose must give an orange color as opposed to the yellow color given by the unsubstituted Sepharose by the 2,4,6-trinitrobenzensulfonate test.[13]

To the washed aminoethyl–Sepharose slurry, 200 ml of 40% (w/v) N,N-dimethylformamide and 6.25 g of sodium p-chloromercuribenzoate

[10] L. A. Æ. Sluyterman and J. Wijdenes, *Biochim. Biophys. Acta* **200**, 593 (1970).
[11] L. A. Æ. Sluyterman, this volume [67].
[12] A. J. Barret, *Biochem. J.* **131**, 809 (1973).
[13] P. Cuatrecasas, *J. Biol. Chem.* **245**, 3059 (1970). P. Cuatrecasas and C. B. Anfinsen, this series, Vol. 22, p. 345.

are added with stirring. The pH is adjusted to 4.8 with 12 N HCl and 7.7 g of 1-ethyl-3-(dimethylaminopropyl) carbodiimide are added. The pH of the reaction is maintained at 4.8 for 1–2 hours, and the suspension is gently stirred for 16 hours at room temperature. The organomercurial–Sepharose is filtered on a glass filter and washed 10 times over a period of 48 hours at 4° with 1 liter of 0.1 M sodium bicarbonate at pH 8.8 each time. The derivative should give a yellow color by the 2,4,6-trinitrobenzensulfonate test. An orange color will indicate that amino groups remain available, and it is advisable to repeat the coupling of the sodium p-chloromercuribenzoate. The washed organomercurial–Sepharose can be stored at 4° in a tightly closed container with a few drops of chloroform as a preservative; it is stable for at least one year without noticeable alteration of its flow and capacity characteristics.

Determination of the Binding Capacity

The binding capacity is determined by reaction with [3-^{14}C]DL-cysteine (SA 40 mCi/mmole); 50 μCi are dissolved in 1 ml of 0.01 N HCl saturated with nitrogen made oxygen-free by passage through a solution of 10% (w/v) pyrogallol red in 1 M NaOH. A 100-μl aliquot is mixed with 215 μl (12.4 μmoles) of a solution of unlabeled carrier L-cysteine in 0.1 M sodium phosphate at pH 6.0 previously flushed with nitrogen. Dilutions containing increasing amounts of the amino acid are prepared in a final volume of 100 μl by dilution with the phosphate buffer. The aliquots are introduced into parallel small columns containing 0.5 ml of packed organomercurial–Sepharose washed with the buffer, and the flow is stopped for 30 minutes. The columns are then washed with 15 ml of 0.01 M sodium phosphate at pH 6.0 to remove the unbound amino acid. The capacity can be calculated by assaying aliquots of the eluent in a scintillation spectrometer and/or by removing the content of the columns and measuring the radioactivity bound to the adsorbent by combustion in an automatic Sample Oxidizer.[9] The affinity of the organomercurial–Sepharose in terms of cysteine binding capacity is 1.2–2.5 μmoles per milliliter of packed resin.

Preparation of Histones

Mammalian Tissues. Whole histones are obtained by extraction of isolated nuclei, extensively washed with isotonic saline solutions, with 0.25 N HCl.[9] Crude fraction F3 can be obtained by extraction of the washed nuclear pellet with 1.25 N HCl-ethanol (1:4, v/v) followed by precipitation from the extract by dialysis against absolute ethanol.[14]

[14] E. W. Johns, *Biochem. J.* **92**, 55 (1964).

Echinoderms. Both F2a1 and F3 histones from echinoderms contain cysteine, and therefore it is important to separate them before their purification by affinity chromatography. Fraction F2a (F2a1 + F2a2) is prepared by extraction of the saline-washed nuclear pellet with acid–ethanol mixtures and is recovered from the supernatant obtained after extensive dialysis against absolute ethanol.[8,15] Crude fraction F3 (the main contaminant being F2a2) is recovered from the precipitate that appears during dialysis of the acid–ethanol extract against absolute ethanol.[8,15]

Reduction of SH-Histones

In order to increase the interaction between the SH-histones and the organomercurial–Sepharose it is convenient to reduce any disulfide bridges that might have been formed during their extraction. The following method can be applied to the reduction of mammalian and echinoderm histones.

Procedure. One gram of either whole histone, fraction F2a, or crude fraction F3 is dissolved in 10–20 ml of a freshly prepared solution of 8 M urea in 0.1 M Tris chloride at pH 9.0 containing 1 mmole of dithiothreitol per 20 ml of solution. After warming at 40° for 1 hour in a shaker bath, the proteins are precipitated by the addition of 50% (w/v) trichloroacetic acid to a final concentration of 18% (v/v). The precipitate is washed once with a cold solution of 0.1% (v/v) HCl in acetone, three times with cold acetone, and dried under vacuum. Efficient resuspension in acetone is essential to remove the excess of dithiothreitol.

Purification of SH-Histones

The method is based on the selective retention of SH-containing histones by interaction between their reactive sulfhydryl groups and the organomercurial side chains of the modified Sepharose. Although the formation of the -S-Hg- linkage is faster at alkaline pH's, the pH of the buffers is maintained slightly acidic throughout the chromatographic procedure in order to minimize the formation of disulfide bridges. All the buffers to be used are previously saturated with oxygen-free nitrogen.

Procedure

Histone F3. Packed organomercurial–Sepharose (capacity 1.2 μmoles/ml), 40 ml, is suspended in 100 ml of 10 mM sodium phosphate at pH 6.0, and the suspension is warmed to room temperature under reduced pressure. The evacuated slurry is poured into a column of 30 cm × 1.5

[15] J. Palau, A. Ruiz-Carrillo, and J. A. Subirana, *Eur. J. Biochem.* **7**, 209 (1969).

cm and washed with 2 bed volumes of the phosphate buffer. Reduced crude histone F3 is dissolved without stirring in distilled water at a concentration of 50–100 mg/ml, and the solution (pH 4 to 5) is clarified by centrifugation if necessary. The pH is adjusted to 6.0 by the addition of 1 volume of 0.1 M sodium phosphate at pH 6.0, and the solution is introduced into the column. The flow rate is kept at 5.4 ml per hour; fractions of 7 ml are collected, and their absorbancy is measured at 275 nm. These conditions are maintained until the first peak of unretained material is eluted. The flow-through material contains the contaminating histones, mainly F2a2 and F1. The flow rate is increased to between 30 and 55 ml per hour until no more protein is eluted. The column is then washed with 10 mM sodium acetate at pH 5.0 in order to remove small amounts of proteins which are nonspecifically adsorbed to the organo-mercurial–Sepharose or to the bound F3 histone.[16] Elution of the F3 is achieved by passing 2 bed volumes of a freshly prepared 0.5 M cysteine solution at a flow rate of 5.4 ml per hour. The protein content of the fractions eluted with cysteine solutions is determined by turbidity assay, adding 250 μl of 25% (w/v) trichloroacetic acid to 50 μl of each fraction. After 70 minutes the turbidity is measured at 400 nm. It is emphasized that the flow rates during application of the sample into the column and during elution of the bound histone must be slow to ensure the maximum formation of the mercurial mercaptide. Elution of the retained histone can also be accomplished with solutions of 2-mercaptoethanol or dithiothreitol although elution with cysteine is somehow more efficient. Histone F3 obtained by this procedure shows a high degree of purity as determined both by polyacrylamide gel electrophoresis and by amino acid composition.[9] A column of the dimensions and binding capacity used is able to retain about 350 mg of calf thymus histone F3.

A similar purification is obtained when whole histone from calf thymus is applied to the organomercurial–Sepharose column. All other major histone classes pass through the column, and pure histone F3 is subsequently eluted with cysteine solutions.

Histone F2a1. Because of the unusual presence of cysteine in histone F2a1 from echinoderms[6-8] this protein when free of histone F3 can be readily purified by the method described above. However, the behavior of F2a1 is different from that of F3 during the chromatographic separation. When a solution of F2a1 + F2a2 is introduced to the organomercurial–Sepharose column the bed shrinks to about one-third of its original volume, and as a result its flow properties are affected. Nevertheless, the

---

[16] Proteins nonspecifically adsorbed can also be removed by elution with buffers containing 4 M urea. Small amounts of F3, probably not bound to the organomercurial, are also eluted during the washing.

purification achieved is comparable to that obtained for histone F3. The shrinking effect is likely to be due to cross-linking of the adsorbent by two close cysteinyl residues of the histone F2a1 molecule.[17]

Recovery

The recovery of the protein can be estimated by using a labeled SH-histone and determining the amount of radioactivity remaining on the column after elution with thiols. The recovery of the organomercurial-bound histone is essentially quantitative if the conditions for its elution are carefully observed. Under those conditions the amount of protein remaining on the column is less than 0.1% of that eluted with 0.5 M cysteine.

Regeneration of the Organomercurial–Agarose

If the adsorbent is to be used again, bound cysteine can be displaced from its mercaptide by treatment with $HgCl_2$–EDTA solutions.[12] The organomercurial–Sepharose is washed in the column with 5 bed volumes of 50 mM sodium acetate at pH 4.8, and then with 2 bed volumes of 10 mM $HgCl_2$–20 mM EDTA in 50 mM sodium acetate at pH 4.8, at a flow rate of 5.4 ml per hour. The excess of $HgCl_2$ is removed by washing with 0.2 M NaCl–1 mM EDTA in 0.1 M sodium phosphate at pH 6.0. Determination of the capacity of the resin after regeneration is suggested.

Acknowledgment

During the original research of this work the author was a Fellow of the Damon Runyon Memorial Fund for Cancer Research in the Laboratory of Dr. V. G. Allfrey at The Rockefeller University.

[17] A. Ruiz-Carrillo and L. J. Wangh, unpublished observations, 1972.

[69] Purification of an Enzyme Containing a NADP-Protected Cysteine on Organomercuri-Agarose 17β-Estradiol Dehydrogenase

By J. C. NICOLAS

Principle

17β-Estradiol dehydrogenase, among other enzymes, possesses essential cysteinyl residues[1,2] which are protected by the coenzyme against thiol reagents, such as p-hydroxymercuribenzoate (PMB), N-ethylmaleimide (NEM), or 5,5′-dithiobisnitrobenzoate (DTNB).

The incubation of an impure enzyme solution with thiol reagents in the presence of NADP results in the reaction of all cysteinyl residues that are not protected by this coenzyme. After removal of both NADP and exogenous thiol reagents, only the unmodified residues (previously protected by NADP) are able to react with an organomercuri-Sepharose derivative. These proteins are readily eluted by β-mercaptoethanol or dithiothreitol. The method provides the basis of a successful purification procedure if activity is retained.

Experimental Procedure

Preliminary purification of the 17β-estradiol dehydrogenase is carried out by automated DEAE-cellulose chromatography.[3]

The affinity matrix is prepared by the method of Cuatrecasas.[4] Packed aminoethyl Sepharose, 100 ml (capacity 5 μmoles/ml) is suspended in 150 ml of 40% dimethyl formamide. After adding 5 mmoles of p–hydroxymercuribenzoate, the pH is adjusted to 4.7–4.8 and 10 mmoles of 1-ethyl 3(3'-dimethylaminopropyl) carbodiimide are added. After reaction for 24 hours, the substituted agarose is washed with 10 liters of 0.1 M NaHCO₃. This derivative can bind 0.2 to 0.6 μmole of cysteine per milliliter of packed gel.

The dehydrogenase (100 IU; specific activity 0.05 IU/mg) is incubated at 25° with 0.5 mM NADP and 1.0 mM DTNB or NEM. After 2 hours, proteins are salted out by adding 1 volume of saturated ammonium sulfate solution, and centrifuged for 30 minutes at 15,000 g to collect the precipitate. The precipitate is dissolved in 30 mM potassium phosphate at pH 7.2 containing 20% glycerol (buffer A), and the remaining coenzyme and DTNB are removed by passage through a Sephadex G-25 column (2 × 30 cm) equilibrated with the same buffer.

The enzyme is *immediately* applied at a flow rate of 0.2 ml per minute to the organomercuri-Sepharose column (2 × 20 cm). Under these conditions, dehydrogenase activity is retained on the column, which is then washed with 200 ml of buffer A. Thereafter, the enzyme is eluted in a very sharp peak (10–20 ml) with buffer A supplemented with 1 mM β-mercaptoethanol. The resultant enzyme preparation has a specific activity of 2.5 IU/mg and represents 30- to 50-fold purification with a yield of 75% (see the table).

---

[1] J. C. Nicolas, M. Pons, A. M. Boussioux, B. Descomps, and A. Crastes de Paulet, Transactions of the 5th Meeting of the International Study Group for Steroid Hormones, Rome, 1971 (*in* "Research Steroids," Vol. V, 191 Il Pensiero Scientifico, Rome, 1972).
[2] J. C. Nicolas and J. I. Harris, *FEBS Lett.* **29**, 173 (1973).
[3] M. Pons and J. C. Nicolas, *Anal. Biochem.,* in preparation.
[4] P. Cuatrecasas, *J. Biol. Chem.* **245**, 3059 (1970).

Step	Total activity (IU)	Total protein (mg)	Specific activity (IU/mg)
DEAE-cellulose	100	2,000	0.05
Modifications of unprotected thiols	96	1,900	0.05
Ammonium sulfate	80	1,000	0.08
Organomercuri-Sepharose	75	30	2.5

After the purification steps described in the table, a contaminant of 17β-estradiol dehydrogenase is usually present as indicated by the presence of two bands on electrophoresis. However, homogeneous enzyme can readily be obtained by a second DEAE-cellulose chromatography.

Partial regeneration of the organomercuri-Sepharose can be obtained by acid cleavage of the mercaptide bond formed between the mercuri derivative and the β-mercaptoethanol of the eluent. Acid cleavage is carried out by washing the gel with 1 liter of 10 mM HCl.

Comments

Purification of estradiol dehydrogenase is incomplete since other proteins carrying thiols similarly protected by NADP would also be retained on the column. Moreover, when the cysteine residues have been allowed to react with reversible thiol reagents, exchange could occur between modified and unmodified thiol residues after removal of the protecting coenzyme. The possibility for this type of exchange can be decreased by removing the coenzyme by gel filtration and immediately applying the enzyme to the organomercuri-Sepharose column. No exchange would occur if an irreversible thiol reagent such as NEM is used, but the danger, in this case, is that of irreversible alkylation of cysteine residues which could lead to irreversible denaturation.

It should be pointed out that the presence of 20% glycerol is necessary in order to protect an essential cysteine of the 17β-estradiol dehydrogenase by NADP. Glycerol is known to protect many enzymes against inactivation and allows good recoveries of this enzyme throughout the purification process.

For purification of the 17β-estradiol dehydrogenase, the use of an organomercuri-Sepharose column was found to be equivalent to one with estradiol-Sepharose with respect to yield, extent of purification, and scale of operation. However, we were unable to use the column more than 2 or 3 times, since complete regeneration could not be obtained.

[70] The Soluble 17β-Estradiol Dehydrogenase of Human Placenta

By J. C. NICOLAS

The 17β-estradiol dehydrogenase is known to bind substrate (17β-estradiol or estrone) in the absence of its cosubstrate NADP.[1,2] We reported recently that the enzyme's active site is able to bind 17β-estradiols with bulky substituents at carbon 3 of the steroid nucleus.[3] Thus the estradiol-Sepharose described by Cuatrecasas[4] fulfills the conditions required for binding of 17β-estradiol dehydrogenase from human placenta, and we have used similar estrone derivatives for the purification of this enzyme by affinity chromatography.[5]

Experimental Procedure

Preparation of Estrone-Agarose

Two estrone agarose derivates have been synthesized by cyanogen bromide activation of Sepharose 4B (Pharmacia).[6]

Estrone-Hemisuccinate-Ethylenediamine-Agarose. This phase is synthesized according to the method described by Cuatrecasas.[4] The amount of bound steroid is determined after hydrolysis of 10 ml of the gel by 1 N NaOH, ethyl ether extraction, and gas chromatography on a 1% SE-30 column. The steroid concentration of the derivatized gel is about 20 μM.

Estrone-Aminocaproate-Agarose. One hundred milliliters of cyanogen bromide-activated Sepharose are allowed to react with 1 g of aminocaproate at pH 9.5. The resultant derivative is washed with distilled water and suspended in 200 ml of dimethylformamide. Estrone (1 g) and dicyclohexyl carbodiimide (5 g) are added to the suspension and the reaction is allowed to proceed at room temperature for 24 hours with magnetic stirring. The insoluble dicyclohexylurea and the excess of steroid are removed by washing the gel successively with 500 ml of dimethylformamide and 500 ml of methanol-HCl (9:1). The steroid concentration of the gel is greater than 0.1 mM.

[1] J. Jarabak and G. H. Sack, *Biochemistry* **8**, 2203 (1969).
[2] G. Betz, *J. Biol. Chem.* **246**, 2063 (1971).
[3] B. Descomps and A. Crastes de Paulet, *Bull. Soc. Chim. Biol.* **51**, 1591 (1969).
[4] P. Cuatrecasas, *J. Biol. Chem.* **245**, 3059 (1970).
[5] J. C. Nicolas, M. Pons, B. Descomps, and A. Crastes de Paulet, *FEBS Lett.* **23**, 175 (1972).
[6] J. Porath, R. Axén, and S. Ernbäck, *Nature (London)* **214**, 1302 (1967).

Chromatography Procedures

Preliminary purification is carried out by automated DEAE-cellulose chromatography.[7] The enzyme, eluted in 80 mM potassium phosphate at pH 7.2 containing 20% glycerol (buffer A), has a specific activity of 0.05 IU/mg. This enzyme solution is further purified by affinity chromatography on either of the estrone-Sepharose derivates.

Use of a Column with a High Concentration of Bonded Estrone (>0.1 *mM*). One liter of enzyme solution (4 g of proteins, 200 IU) is applied to a 2 × 10 cm column of estrone–Sepharose. The column is washed with 100 column volumes of buffer A. Under these conditions, the enzyme is strongly adsorbed by the gel; more than 90% of the initial activity is retained. The enzyme can be eluted by adding a 0.3 mM estrone hemisuccinate to the buffer, but elution is slow and a yield of 50% is attained. A better system of elution involves the addition of 3 M urea and 0.1 mM estradiol to buffer A. In the presence of estradiol–urea, the enzyme is eluted in a very sharp peak at a specific activity of 1–1.2 IU/mg. The protein is readily reactivated: urea is removed by chromatography on a DEAE-cellulose column (2 × 15 cm) and the purified enzyme (4 IU/mg) is eluted with a gradient of salt from 0.03 to 0.1 M potassium phosphate at pH 7.2.

After 5 or 6 runs on the same column of estrone-Sepharose, the amount of enzyme adsorbed decreases, owing in part to the hydrolysis of the phenol ester bond.

Use of a Column with a Low Concentration of Bonded Estrone (*10* μ*M to 0.1 mM Estrone*). The enzyme solution (3 g protein, 150 IU) is applied to a 2 × 10 cm column of estrone-Sepharose containing between 0.4 and 4 μmoles of bound estrone. Application is in buffer B, which consists of buffer A supplemented to 1 M in ammonium sulfate. The 17β-estradiol dehydrogenase is quite soluble in this medium, and 150 IU of enzyme are retained on the column. The gel is washed with 600 ml of buffer B, and the enzyme is eluted by lowering the ammonium sulfate concentration. The enzyme elutes in 250 ml and is concentrated by ammonium sulfate precipitation (2 M salt). The precipitate is dissolved in buffer A. The specific activity of this solution is about 1 IU/mg; homogeneous enzyme is obtained by subsequent chromatography on DEAE-cellulose as described above.

Comments

When applied in the usual buffers, 17β-estradiol dehydrogenase is strongly retained on columns containing the higher estrone concentration conjugated to Sepharose. Addition of estrone or estradiol to the buffer, did

---

[7] M. Pons and J. C. Nicolas, *Anal. Biochem.* in preparation.

not result in elution of the enzyme because efficient competition for the enzyme requires a higher concentration than the low solubility of estrogen allows. The enzyme can be eluted either by 0.3 mM estrone hemisuccinate or by urea plus estradiol; the weak competition of estrone hemisuccinate results in the elution of dilute enzyme (0.1 mg/ml) whereas urea and estradiol produce a higher enzyme concentration in the eluate (1 mg/ml). Although estradiol is added to protect the enzyme against irreversible inactivation by urea, a simultaneous role of this molecule in the elution process cannot be excluded.

The 17β-estradiol dehydrogenase is retained on low-estrone concentration Sepharose only in the presence of ammonium sulfate, and elution is easily achieved by decreasing the ammonium salt concentration. This technique does, in fact, represent affinity chromatography; the estrone molecule bonded to Sepharose is essential. Even when applied in 1 M ammonium sulfate, estradiol 17β-dehydrogenase is not retained on columns packed with Sepharose itself or with ethylenediamine–Sepharose or aminocaproate–Sepharose. The ammonium sulfate concentration necessary to achieve retention of the enzyme is related to the steroid concentration of the solid phase.

This procedure results in a 20- to 25-fold purification and homogeneous enzyme is obtained after subsequent DEAE-cellulose chromatography. The estrone–Sepharose containing a low concentration of conjugated estrone is particularly useful since elution is easily obtained. This material can be used more than 20 times with reproducible results; one column has allowed the purification of more than 400 mg of enzyme from 30 kg of placenta.

[71] Steroid-Transforming Enzymes

By Ann M. Benson, Anthony J. Suruda, Evelyn R. Barrack, and Paul Talalay

Affinity chromatography on adsorbents containing steroidal substrate analogs covalently linked to agarose[1,2] promises to be an extremely valuable technique both for the purification of steroid-transforming enzymes, and for studies on their mechanism. When *Pseudomonas testosteroni* is grown on media containing certain steroids, crude extracts of lyophilized or sonically disrupted cells are a rich source of various induced steroid-transforming enzymes.[3,4] The isolation from these mixtures of discrete

[1] P. Cuatrecasas and C. B. Anfinsen, this series, Vol. 22 [31] (1971).
[2] P. Cuatrecasas, *Advan. Enzymol.* 36, 29 (1972).

catalytic proteins, in purified form and uncontaminated by each other, has been a challenging task. The availability of such purified enzymes would greatly augment their suitability for the microanalysis and determination of steric configuration of steroids.[5]

Development of a procedure[6,7] for large-scale affinity chromatography of Δ⁵-3-ketosteroid isomerase (EC 5.3.3.1) from partially purified extracts of *P. testosteroni* and insertion of this step into the purification procedure[8] of the isomerase has enhanced both the yield and the ease of purification of this enzyme. The use of affinity chromatography techniques has also provided information on the effects of a number of organic solvents and inorganic salts on the ability of the isomerase to bind steroids.

Extracts of *P. testosteroni* also contain 3α- (EC 1.1.1.50), 3β-, and 17β-hydroxysteroid dehydrogenases (EC 1.1.1.51).[3,9] Affinity chromatography of these enzymes,[7] while still in an early stage of development, has completely separated the isomerase from the hydroxysteroid dehydrogenases and has provided the 3β- and 17β-dehydrogenases completely free of contaminating 3α-enzyme, as well as achieving partial resolution of the 3β- and 17β-hydroxysteroid dehydrogenases. Furthermore, affinity techniques have indicated that the β-enzymes do not bind strongly to steroids in the absence of NAD⁺, thus suggesting that these oxidations may involve an obligatory ordered addition in which the binding of NAD⁺ precedes that of the steroidal substrate.

Preparation of Affinity Adsorbents

The two types of adsorbents employed in these procedures are: 17β-estradiol-3-*O*-succinyldiaminodipropylamino-agarose (estradiol-agarose) and 19-nortestosterone-17-*O*-succinyldiaminodipropylamino-agarose (nortestosterone-agarose). The structures are shown in Fig. 1.

[³H]17β-Estradiol-3-*O*-hemisuccinate is synthesized from [³H]17β-estradiol and succinic anhydride[10] and coupled to diaminodipropylamino-Sepharose 4B as described by Cuatrecasas.[11] The product is washed as described below.

19-Nortestosterone-17-*O*-hemisuccinate is prepared by an adaptation

---

[3] P. Talalay, this series, Vol. 5 [69] (1962).

[4] F. S. Kawahara, this series, Vol. 5 [70a] (1962).

[5] P. Talalay, *Methods Biochem. Anal.* 8, 119 (1960).

[6] A. M. Benson, A. J. Suruda, R. Jarabak, and P. Talalay, 162nd National Meeting Amer. Chem. Soc., Washington, D.C. (1971), Division of Biological Chemistry, Abstract 119.

[7] A. M. Benson, A. J. Suruda, E. Barrack, R. Shaw, and P. Talalay, unpublished observations, 1970–1973.

[8] R. Jarabak, M. Colvin, S. H. Moolgavkar, and P. Talalay, this series, Vol. 15 [31] (1969).

[9] J. Boyer, D. N. Baron, and P. Talalay, *Biochemistry* 4, 1825 (1965).

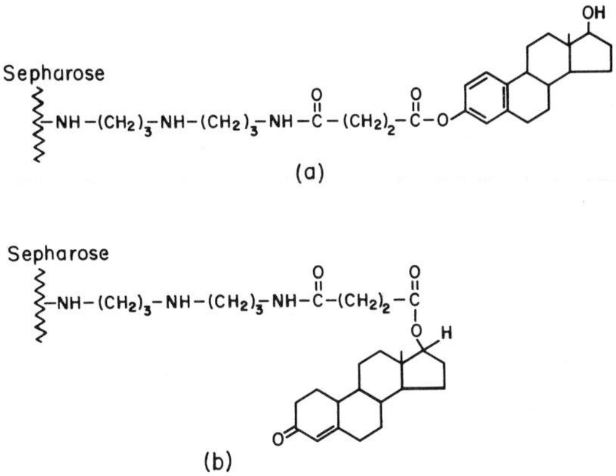

FIG. 1. Structures of steroidal affinity adsorbents: (a) 17β-estradiol-3-O-succinyldiaminodipropylamino-agarose (estradiol-agarose), and (b), 19-nortestosterone-17-O-succinyldiaminodipropylamino-agarose (nortestosterone-agarose).

of the procedure of Giannini and Fedi.[12] One gram of 19-nortestosterone (10 μCi of [4-¹⁴C]19-nortestosterone[13]) and 3 grams of succinic anhydride (recrystallized from acetic anhydride) are dissolved in 30 ml of pyridine (dried over BaO) and heated to 100°, with stirring, in an oil bath under nitrogen. When the reaction is complete (after approximately 6 hours), as shown by thin-layer chromatography on silica gel (GF) in chloroform: methanol (9:1), the mixture is cooled and slowly poured, with stirring, into 75 ml of cold 6 M HCl; the resulting precipitate is allowed to harden, collected, washed with 200 ml of cold 0.1 M HCl, and dissolved in 50 ml of 0.2 M Na_3PO_4. The solution is then filtered, decolorized briefly with charcoal, and acidified. The crystals that form at 0° are collected and recrystallized from acetone/petroleum ether. The ester may be purified further by chromatography in chloroform:methanol (9:1) on a 2.5 × 80 cm column of dry silica gel.

Nortestosterone-agarose is prepared by the coupling procedure described by Cuatrecasas.[11] Diaminodipropylamino-Sepharose 4B (100 ml) is mixed with 100 ml of dimethylformamide and the pH is adjusted to 4.9 with HCl. Radioactive 19-nortestosterone-17-O-hemisuccinate (750 mg) in 20 ml of ethanol is added, followed by 1.5 g of 1-ethyl-3-(3-dimethylaminopropyl) carbodiimide. The mixture is stirred for 36 hours at room temperature, and then washed as described below.

---

[10] V. Sica, I. Parikh, E. Nola, G. A. Puca, and P. Cuatrecasas, *J. Biol. Chem.* **248**, 6543 (1973); see also T. O. Yellin, *J. Lipid Res.* **13**, 554 (1972).
[11] P. Cuatrecasas, *J. Biol. Chem.* **245**, 3059 (1970).
[12] M. Giannini and M. Fedi, *Boll. Chim. Farm.* **99**, 24 (1960); *Chem. Abstr.* **54**, 11084 (1960).
[13] [4-¹⁴C]19-Nortestosterone was obtained from Amersham/Searle.

The choice of solvent for the coupling reaction can be critical. An aqueous solution containing 50% dimethylformamide and 10% ethanol is satisfactory with respect to both the solubility of the steroidal derivatives and the structural stability of the agarose beads. The use of 50% ethanol has yielded products with poor flow properties, and the use of 50% aqueous dimethylformamide may lead to the appearance of the steroid as an oily layer on the surface of the reaction mixture. The latter effect may be avoided by equilibration of the agarose with the solvent before addition of the steroidal compound to be coupled.

The extent of coupling is determined by measurement of the radioactivity of the adsorbent after thorough washing to remove unbound steroid.[11] The adsorbents are washed briefly on a Büchner funnel with 500 ml of 50% aqueous dimethylformamide, packed into a column, and the washing continued until radioactivity is no longer detectable in the effluent. About 5 liters of solvent are required.

The batches of nortestosterone-agarose and estradiol-agarose used in the following affinity chromatography procedures contained, per milliliter of packed adsorbent, 1.4 μmoles of bound 19-nortestosterone and 4.9 μmoles of bound 17β-estradiol, respectively.

Affinity Chromatography of Δ^5-3-Ketosteroid Isomerase

Both 17β-estradiol ($K_i = 10 \ \mu M$) and 19-nortestosterone ($K_i = 5.2 \ \mu M$) are competitive inhibitors of Δ^5-3-ketosteroid isomerase[14,15] and both of the affinity adsorbents derived from these inhibitors are capable of binding this enzyme. All chromatographic procedures were carried out at 4°, unless otherwise specified. The pH values of all Tris buffers refer to 25°.

Preparative Scale Chromatography on Nortestosterone Agarose

Large-scale affinity chromatography procedures have been developed by which Δ^5-3-ketosteroid isomerase, partially purified (7–8% pure[16]) by previously published procedures through the DEAE-cellulose chromatog-

---

[11] F. S. Kawahara, S.-F. Wang, and P. Talalay, *J. Biol. Chem.* **237**, 1500 (1962).

[15] S.-F. Wang, F. S. Kawahara, and P. Talalay, *J. Biol. Chem.* **238**, 576 (1963).

[16] Assays for Δ^5-3-ketosteroid isomerase (this series, Vol. 5 [70a] 1962) are carried out at 25° in systems of 3.0 ml containing: 100 μmoles of potassium phosphate buffer (pH 7.0), 0.175 μmole of Δ^5-androstene-3,17-dione in 0.05 ml of methanol, and an appropriate amount of enzyme to initiate the reaction. The absorbance at 248 nm is measured against a blank containing all components except the steroid. One unit of enzyme activity is defined as that amount which under these conditions causes the isomerization of 1 μmole of Δ^5-androstene-3,17-dione per minute to Δ^4-androstene-3,17-dione ($a_m = 16,300 \ M^{-1} \ cm^{-1}$).

raphy step,[17] is obtained in 70–80% purity with yields approaching 100% by means of a single chromatography on nortestosterone-agarose. In a typical purification, combined fractions (180 ml) from DEAE-cellulose chromatography, containing 74 mg of the isomerase (specific activity, 3.53 × 10³ units/mg; purity 7%) and a total of 1060 mg of protein in approximately 70 mM Tris phosphate at pH 7.0, were applied to a column of nortestosterone agarose (21 × 236 mm; bed volume, 82 ml) which was equilibrated and developed with 0.1 M Tris phosphate at pH 7.0. Fractions of about 50 ml were collected. A large amount of unadsorbed protein was detected in the effluent by its absorbance at 280 nm, which reached a maximum of 1.6 at 250 ml of effluent and then declined sharply to less than 0.1 at 400 ml of effluent. At this stage a yellow material which had initially been adsorbed as a clearly visible band near the top of the column began to appear in the effluent. An additional 200 ml of 0.1 M Tris phosphate (pH 7.0) were required to completely elute the yellow band. After passage through the column of 100 ml of 0.2 M Tris phosphate at pH 7.0, elution with 350 ml of 1 M sodium acetate at pH 5.5 in 20% glycerol gave quantitative recovery of the Δ^5-3-ketosteroid isomerase. The major peak fractions (135 ml) contained 71 mg of the isomerase (96% recovery) with a specific activity of 38.2 × 10³ units/mg (76% pure). By use of moderate hydrostatic pressure it is possible to achieve flow rates of 150 ml per hour during application of sample and elution with Tris buffers, and of 75 ml per hour during elution with acetate-glycerol buffer. Chromatography may thus be completed in about 8–10 hours. After dialysis and concentration, the eluted isomerase can be easily obtained in a homogeneous form by crystallization from ammonium sulfate solution.[8] The specific activity of the crystalline isomerase is identical to that found previously for the pure enzyme.[8]

This procedure is now a routine step in the large-scale preparation of crystalline Δ^5-3-ketosteroid isomerase. The affinity chromatography step

---

[17] The initial steps of a typical large-scale purification (this series, Vol. 15 [31] 1969) of this enzyme involve extraction of lyophilized cells of *Pseudomonas testosteroni* with a mixture of equal volumes of 20 mM Tris hydrochloride (pH 7.0) and 95% ethanol, followed by centrifugation (specific activity, 200 units/mg; 0.32% pure) and precipitation of the enzyme in the presence of 5 mM MgCl$_2$ and 80% ethanol. The precipitated enzyme is dialyzed and applied to a DEAE-cellulose column in 1 mM Tris phosphate (pH 7.0) and eluted with a linear gradient to 0.4 M Tris phosphate of pH 7.0. (Specific activity varies from 4000 to 7000 units/mg protein; 6–11% purity.)

At subsequent stages of purification, calculation of the protein concentration from the ultraviolet absorbance is based on an absorbance in a 1-cm light path of 0.336 at 280 nm for a solution of 1.0 mg of the pure isomerase per milliliter (A. M. Benson, A. J. Suruda, and P. Talalay, unpublished results).

has certain distinct advantages: (i) It eliminates the tedious calcium phosphate gel chromatography procedure[8] and the time-consuming preparation of this adsorbent. (ii) It can raise the specific activity of lower activity fractions than those here illustrated (e.g., DEAE-cellulose chromatography side fractions) although it is not suitable for direct purification of crude initial extracts (see below). (iii) The nortestosterone agarose is highly stable on storage at 4° for at least two years, and can be repeatedly regenerated after use by washing with 6 M guanidine hydrochloride[11] and reequilibrating with initial buffer. (iv) The adsorbent gives highly reproducible chromatographic patterns, even after repeated uses.

Chromatography of Partially Purified Δ^5-3-Ketosteroid Isomerase on Estradiol Agarose

Estradiol-agarose is also effective in selectively adsorbing Δ^5-3-ketosteroid isomerase. A partially purified extract of *P. testosteroni,* containing 5.9 mg of the isomerase (specific activity, 3.4 × 10³ units/mg; 7% pure) was applied in 63 ml of 50 mM sodium phosphate of pH 7.0 to a 9 × 250 mm column of estradiol agarose. A large amount of material absorbing at 280 nm appeared in the effluent during the sample application and subsequent washing with 32 ml of 50 mM sodium phosphate at pH 7.0. The isomerase was eluted in 70% yield by 70 ml of 1.75 M sodium citrate at pH 5.5 in 25% glycerol. The specific activity of the enzyme in the eluate (16.4 × 10³ units/mg; 32% purity) represents a 4.8-fold purification.

Chromatography of Crude Δ^5-3-Ketosteroid Isomerase on Nortestosterone Agarose

The isomerase at an earlier stage of purification (dialyzed magnesium precipitate[17]) was chromatographed on a 9 × 30 mm column of nortestosterone-agarose. The sample, containing 6.0 × 10³ units of quite impure isomerase (344 units/mg; 0.7% pure), was applied in 5.0 ml of 50 mM potassium phosphate at pH 7.0. Most of the material absorbing at 280 nm was not adsorbed and could be removed by passing 40 ml of 50 mM potassium phosphate, pH 7.0, through the column. The yellow band which was previously noted was not eluted by this buffer but was eluted by 1 M sodium bicarbonate at pH 8.8 in 20% glycerol. After the column had been washed with starting buffer to remove bicarbonate, the isomerase was eluted in 1 M sodium acetate at pH 5.5 in 20% glycerol. The yield of the isomerase was quantitative and a 25-fold increase in purity was achieved. However, the final specific activity was only 8.7 × 10³ units/mg (17% pure).

Elution of Δ⁵*-3-Ketosteroid Isomerase by Alcohols*

In two of the procedures described above, acetate buffers containing glycerol were used to elute the isomerase from the affinity columns. Glycerol appears to promote elution of the isomerase, since in the presence of glycerol, lower concentrations of acetate are effective eluents. Alternatively, this enzyme may be eluted by including methanol in the column eluent. In the latter procedure, the partially purified enzyme is applied to a column of nortestosterone agarose in 0.1 M potassium phosphate, pH 7.0, at 4°. The column is washed with the same buffer, also at 4°, until the effluent no longer absorbs at 280 nm. If washing is continued with the same buffer containing 10% methanol, the enzyme is eluted in a very broad band. However, this difficulty may be overcome by raising the temperature to 25°. In a typical experiment, partially purified isomerase (27.8 × 10³ units; specific activity, 5.9 × 10³ units/mg; 11.6% pure) in 28 ml of 0.1 M potassium phosphate of pH 7.0 was applied at 4° to a 9 × 80 mm column of nortestosterone agarose. After unretained material had been removed by washing at 4° with 33 ml of 0.1 M potassium phosphate of pH 7.0, the column was developed at 25° with the same buffer containing 10% methanol. Δ⁵-3-Ketosteroid isomerase was eluted in a rather broad band beginning at about 80 ml (total effluent volume), reaching a maximum concentration of 630 units/ml at 105 ml, and then declining gradually throughout the next 100 ml of effluent. Although the low absorbance of the eluate at 280 nm makes calculation of the protein concentration uncertain, the specific activity in the main portion of the isomerase peak (85–122 ml of effluent) is estimated at approximately 18 × 10³ units/mg (35% pure). The total recovery of the enzymatic activity was 91%. The slow rate of elution is apparently not due to slow dissociation of the enzyme–inhibitor complex since turning off the flow of eluent for 16 hours did not result in any increase in the rate of release of the isomerase when elution was resumed.

Although the elution with buffers containing methanol has not been perfected, it would seem that this technique is capable of further development.

General Observations on Affinity Chromatography of
Δ⁵*-3-Ketosteroid Isomerase*

The capacity of these adsorbents to bind the isomerase and the final specific activity attained in these affinity chromatography procedures are greatly dependent upon the extent of prior purification of the enzyme. Attempts to obtain isomerase of high purity directly from crude bacterial extracts have invariably failed. However, application of these procedures at a stage wherein the enzyme is already 7–8% pure has been highly successful.

The effect of acetate in promoting the elution of the isomerase is quite specific. Although inclusion of glycerol (20–25% by volume) in the sodium acetate buffers results in a decrease in the volume required, complete elution of the isomerase from these affinity adsorbents also occurs in 2 M to 4 M sodium acetate at pH 5.5 without glycerol. Elution is not accomplished by sodium citrate of the same pH and of equivalent or even higher ionic strength, nor is the isomerase eluted by 1 M sodium bicarbonate at pH 8.8 in 20% glycerol. Methanol is known to be an inhibitor of the isomerase.[18] Its effectiveness in eluting the isomerase from affinity columns suggests that it interferes with steroid binding. Ethanol has a similar effect, and its presence greatly decreases the binding of isomerase by the affinity adsorbents.

Affinity Chromatography of Hydroxysteroid Dehydrogenases

Nicolas et al.[19] have described the purification of the 17β-hydroxysteroid dehydrogenase of human placenta on estrone-3-hemisuccinate coupled to various amino-Sepharose derivatives. Procedures are currently being developed[7] for affinity chromatography of the 3α-, 3β-, and 17β-hydroxysteroid dehydrogenases of P. testosteroni.[3,9] Chromatography on nortestosterone-agarose of a dialyzed ammonium sulfate (20–70% saturated) precipitate of a streptomycin-treated supernatant fraction of a cell sonicate is shown in Fig. 2. The 9.5-ml sample (32.3 mg of protein[20]) contained 3α-, 3β-, and 17β-hydroxysteroid dehydrogenases in amounts oxidizing 34.1 μmoles of androsterone, 6.12 μmoles of epiandrosterone, and 7.04 μmoles of 17β-estradiol per minute,[21] and 2.8×10^3 units of Δ^5-3-ketosteroid isomerase. The buffer used for equilibration of the column, application of the enzyme sample, and the initial stage of elution was 0.1 M sodium pyrophosphate, pH 8.9, containing 1 mM EDTA, 7.2

[18] J. B. Jones and D. C. Wigfield, Can. J. Biochem. 46, 1459 (1968).

[19] J. C. Nicolas, M. Pons, B. Descomps, and A. Crastes de Paulet, FEBS Lett. 23, 175 (1972). Also see Nicolas, this volume [69] and [70].

[20] O. Warburg and W. Christian, Biochem. Z. 287, 291 (1936); this series, Vol. 3 [73] (1957).

[21] The determination of hydroxysteroid dehydrogenase activity (this series, Vol. 5 [69] 1962) is based on spectrophotometric measurement of the rate of reduction of NAD+. Assays are carried out at 25° in systems of 3.0 ml containing: 100 μmoles of sodium pyrophosphate buffer (pH 8.9); 0.5 μmole of NAD+; 30 μg of androsterone (3α-enzyme), epiandrosterone (3β-enzyme), or 17β-estradiol (17β-enzyme), each in 0.02 ml of purified dioxane; and an appropriate quantity of enzyme to initiate the reaction. The absorbance at 340 nm is measured against a blank containing all components except steroid.

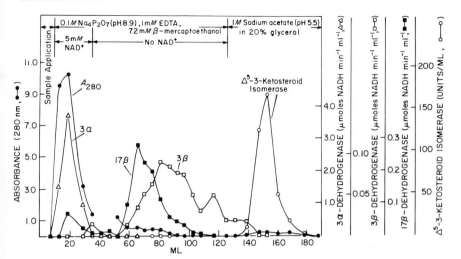

FIG. 2. Chromatography on a 10 × 100 mm column of nortestosterone-agarose of a partially purified *Pseudomonas testosteroni* cell sonicate containing 3α-, 3β-, and 17β-hydroxysteroid dehydrogenases, Δ^5-3-ketosteroid isomerase, and 32.3 mg of protein. The 9.5-ml sample was applied in 0.1 M sodium pyrophosphate of pH 8.9 (1 mM in EDTA, 7.2 mM in 2-mercaptoethanol), containing NAD+ (5 mM), and washed initially with 25 ml of the same buffer. All the 3α-hydroxysteroid dehydrogenase, 11% of the 17β-enzyme, 2% of the 3β-enzyme, and 91% of the material absorbing at 280 nm were unretained by the adsorbent and appeared in the first 35 ml of effluent. The 3β- and 17β-enzymes were then eluted in 90 ml of the same buffer, from which NAD+ was omitted. Finally, 65 ml of 1 M sodium acetate of pH 5.5 in 20% glycerol was passed through the column to elute the Δ^5-3-ketosteroid isomerase.

mM 2-mercaptoethanol, and 5 mM NAD+. The column (10 × 100 mm) was equilibrated with this buffer, loaded with the enzyme solution, and developed at a flow rate of about 25 ml per hour. The initial peak of unretained material accounted for 91% of the absorbance at 280 nm present in the enzyme preparation applied to the column. All of the 3α-hydroxysteroid dehydrogenase activity, 11% of the 17β-activity, and only 2% of the 3β-activity, were found in this fraction. Subsequent passage through the column of the same buffer, from which NAD+ was omitted, resulted in elution of 89% of the 17β-hydroxysteroid dehydrogenase and 65% of the 3β-hydroxysteroid dehydrogenase. Finally, Δ^5-3-ketosteroid isomerase could be eluted with 1 M sodium acetate, pH 5.5, containing 20% glycerol, and in this case was obtained in 91% yield.

While this affinity chromatography procedure is still in a rather early stage of development, it has provided a means of obtaining in a single step the β-hydroxysteroid dehydrogenases completely free of 3α-hydroxysteroid dehydrogenase activity. This separation is difficult to achieve by other available methods, and important in the use of these enzymes for analytical purposes.[5] In addition, it has been shown (Fig. 2) that the hydroxysteroid dehydrogenases possess no intrinsic Δ^5-3-ketosteroid isomerase activity, a point not previously established. Furthermore, the elution pattern indicates the partial resolution of 3β- and 17β-hydroxysteroid dehydrogenase activities, and clearly establishes that these reactions are promoted by distinct enzymes. Earlier methods of purification had failed to give any indication that these activities were separable.

A most interesting finding is the failure of any of the hydroxysteroid dehydrogenases to be retained by the nortestosterone-agarose in the absence of NAD^+. One interpretation of this observation is that the β-hydroxysteroid dehydrogenases do not bind steroids in the absence of NAD^+, and consequently their mechanism probably involves compulsory ordered addition of NAD^+ followed by the steroid. This technique, possibly in conjunction with NAD-affinity columns could provide information on the detailed mechanisms of the hydroxysteroid dehydrogenase reactions.

Acknowledgment

These studies were supported by Research Grant AM07422 from the U.S. Public Health Service and by a Henry Strong Denison Fellowship of The Johns Hopkins University School of Medicine (A.J.S.).

[72] Chromatography of Detergent-Dispersed Mitochondrial ATPase on an Inhibitor Protein Column[1]

By PATRIK SWANLJUNG and LUCIANO FRIGERI

The basis of the method is to bind ATPase in detergent-dispersed mitochondrial membranes to an ATPase inhibitor protein[2] covalently coupled by the CNBr-method to Sepharose 4B.[3,4] The enzyme obtained with this technique is cold-stable and potentially sensitive to oligomycin.

[1] A major part of this work was done at the Department of Biochemistry, University of Stockholm, Sweden.
[2] M. E. Pullman and G. C. Monroy, *J. Biol. Chem.* 238, 3762 (1963).
[3] P. Swanljung and L. Frigeri, *Biochim. Biophys. Acta* 283, 391 (1972).
[4] P. Swanljung, L. Frigeri, K. Ohlson, and L. Ernster, *Biochim. Biophys. Acta* 305, 519 (1973).

Assay Methods

The ATPase is assayed in the presence of phosphoenolpyruvate and pyruvate kinase to provide a system regenerating ATP from the product inhibitor, ADP.[5] Pyruvate formed in the regenerating reaction may be measured enzymatically using lactate dehydrogenase. This permits continuous spectrophotometric monitoring of the ATPase. Alternatively, the enzyme is incubated with the regenerating system only, and the inorganic phosphate which is liberated is determined. Suitable assay conditions using both procedures have been described.[4]

Determination of Protein

Protein is determined with a biuret method[6] in the case of submitochondrial particles or with the method of Lowry et al.[7] in the case of purified enzyme where no or low concentrations of Triton X-100 are present. With more than 0.3% Triton X-100, the Lowry method is difficult to use. In this case, the filter paper staining method of Bramhall et al.[8] has been found to be satisfactory. The xylene brilliant cyanine dye can be replaced by 0.05% amido black in 7% acetic acid, followed by washing in 7% acetic acid.

Preparation of Adsorbent

The ATPase inhibitor protein is prepared by the procedure of Horstman and Racker,[9] except that the final heat treatment step is omitted. Alternatively the method of Nelson et al.[10] can be used. With either procedure the inhibitor protein is dissolved after the last ethanol precipitation in 0.1 M $NaHCO_3$–0.5 M NaCl, to a protein concentration of 0.6 mg/ml.

One gram of CNBr-Sepharose 4B (Pharmacia Fine Chemicals, Uppsala) is swollen and washed 15 minutes on a glass filter with 200 ml 1 mM HCl. The washed gel is mixed with 5 ml of ATPase inhibitor solution in a stoppered test tube, which is then rotated end-over-end for 2 hours at room temperature. The gel is then washed with 100 ml of 0.1 M $NaHCO_3$–0.5 M NaCl and is mixed with 1 M ethanolamine at pH 8 and

[5] M. E. Pullman, H. S. Penefsky, A. Datta, and E. Racker, *J. Biol Chem.* **235**, 3322 (1960).
[6] E. E. Jacobs, M. Jacob, D. R. Sanadi, and L. B. Bradley, *J. Biol. Chem.* **223**, 147 (1956).
[7] O. H. Lowry, N. J. Rosebrough, A. L. Farr, and R. J. Randall, *J. Biol. Chem.* **193**, 265 (1951).
[8] S. Bramhall, N. Noack, M. Wu, and J. R. Loewenberg, *Anal. Biochem.* **31**, 146 (1969).
[9] L. L. Horstman and E. Racker, *J. Biol. Chem.* **245**, 1336 (1970).
[10] N. Nelson, H. Nelson, and E. Racker, *J. Biol. Chem.* **247**, 7657 (1972).

rotated 2 hours as before. Finally, the gel is washed with 200 ml 0.1 M sodium acetate–1 M NaCl at pH 4, followed by 200 ml 0.1 M sodium borate–1 M NaCl at pH 8. The washing step is repeated twice. The gel, having a volume of approximately 3 ml, is packed into a column.

Purification

Reagents

Several different buffer solutions are used in the purification procedure, for which the preparation is outlined below.

> Solution A: 40 ml of 0.5 M sucrose, 7.5 ml of 0.2 M Tris + TES, pH 6.6, 20 μl of 3 M MgSO$_4$, 0.2 ml of 0.3 M ATP, and 10 ml of 10% Triton X-100 are diluted to 100 ml with water. Prepare immediately before use
>
> Solution B: 40 ml of 0.5 M sucrose, 22.5 ml of 0.2 M Tris + TES, pH 8.8, 10 ml of 3 M KCl, and 10 ml of 10% Triton X-100 are diluted to 100 ml with water
>
> Solution C: 40 ml of 0.5 M sucrose, 7.5 ml of 0.2 M Tris + TES, pH 6.6, and 10 ml of 10% bovine serum albumin are diluted to 100 ml with water. Prepare before use
>
> Solution D: 40 ml of 0.5 M sucrose and 7.5 ml of 0.2 M Tris + TES, pH 6.6, are diluted to 100 ml with water
>
> Solution E: As Solution B, but omitting Triton X-100

Preparation of ATPase in Triton X-100

Nonphosphorylating submitochondrial particles (EDTA-particles) are prepared from beef heart mitochondria as described by Lee and Ernster.[11]

The adsorbent column is equilibrated by passing approximately 30 ml of Solution A through it at a flow rate of 15 ml per hour; 120 mg of EDTA particles are sedimented at 104,000 g for 30 minutes at 0° and are resuspended in 6 ml of Solution A. The suspension is centrifuged as above, and about 5 ml of the yellowish supernatant is collected. To this extract, 1 μl of 3 M MgSO$_4$ and 10 μl of 0.3 M ATP are added. The extract is applied to the column, followed by 5 ml of Solution A, at a flow rate of between 6 and 8 ml per hour. The flow rate is then raised to 12–14 ml per hour; 20 ml of Solution A is passed through the column at this rate. Finally the flow rate is decreased to 1–2 ml per hour and the purified ATPase is eluted from the column with Solution B. The

---

[11] C.-P. Lee and L. Ernster, this series, Vol. 10, p. 543.

TABLE I

PURIFICATION OF MITOCHONDRIAL ATPASE BY AFFINITY CHROMATOGRAPHY

Step	Volume (ml)	Protein (mg)	Total ATPase (μmoles/ min)	Specific activity (μmoles/ min/mg protein)	Yield (%)
Submitochondrial particles	6	120	250	2.1	100
Triton extract	5	18.5	37	2.0	15
Affinity chromatography	6	1.2	15.0	12.5	6

column is operated at room temperature. When eluted, the enzyme is immediately transferred to an ice bath.

A summary of one preparation is given in Table I.

Preparation of ATPase Free of Triton X-100

The procedure described above is followed until the point at which the flow rate is raised to 12–14 ml per hour. Then 10 ml of Solution C is passed through the column at this rate, followed by 10 ml of Solution D. Finally the flow rate is decreased to 1–2 ml per hour, and the purified enzyme is eluted from the column with Solution E. The ATPase prepared in this way requires added phospholipids or Triton X-100 for full activity.[4]

Reuse of Adsorbent

After use, the column is washed with 30 ml of 0.1 M acetate buffer, 1 M NaCl, pH 4, followed by 30 ml of 0.1 M sodium borate–1 M NaCl at pH 8, at a flow rate of 15 ml per hour. The column is equilibrated with Solution A and is then ready for another purification. If the adsorbent is to be stored for a longer period between uses, it is best kept at 4°, in 0.1 M sodium borate–1 M NaCl at pH 8, containing 1% (w/v) sodium benzyl penicillin and 1% (w/v) streptomycin sulfate. Under these conditions the adsorbent is active for several months after preparation.

Comments on Operating Conditions

Choice of Detergent

The detergent used must fulfill the following criteria: (i) it should effectively disperse the mitochondrial membrane, (ii) it must not inactivate the ATPase, and (iii) it must not inhibit the interaction between the ATPase and the ATPase inhibitor protein. Some anionic and nonionic detergents have been tested against these criteria, for which the results

TABLE II
Use of Detergents for Affinity Chromatography
of Mitochondrial ATPase

Detergent	Dispersion of membrane	ATPase activity	Interaction of enzyme with inhibitor
Sodium dodecyl sulfate	Very effective	Lost	−
Potassium deoxycholate	Effective	Inhibited	−
Potassium cholate	Fair	Slightly inhibited	−
Triton X-100	Fair	Unaffected	+
Triton X-114	Fair	Unaffected	+
Tween 20	Low	Unaffected	+
Tween 80	Low	Unaffected	+

are summarized in Table II. It can be seen that among the detergents tested, only the Tritons satisfactorily fulfill all criteria.

Effect of Load and Flow Rate

A large part of the ATPase is not adsorbed in its passage through the column. Even with small sample loads, the yield of ATPase attached to the column is below 50%. With larger sample loads the yields are still smaller, but the absolute amounts of enzyme attached are increased.[3] A possible explanation is that only some ATPase molecules in the starting material, perhaps those free of endogenous ATPase inhibitor,[9] can become firmly bound to the adsorbent.

The yield is not greatly improved by decreasing the flow rate during sample application. On the other hand, a slow flow rate is essential during desorption of the ATPase from the column in the final elution step. A flow rate of 1–2 ml, as described in the procedures above, will give a concentration of purified enzyme in the order of 0.2 mg of protein per milliliter. If desired, this can be increased by further lowering the flow rate. Raising the pH or the ionic strength during the final elution step does not appreciably speed up the desorption rate.

Evaluation of Purification Procedures

Affinity chromatography of detergent-dispersed mitochondrial membranes, using an ATPase inhibitor protein attached to a Sepharose column as in the purification procedures described, gives an ATPase virtually free of cytochromes. In contrast to soluble mitochondrial ATPase,[5] this enzyme is not sensitive to storage at 0°; if Triton is removed, the enzyme is activated by phospholipids and inhibited by oligomycin.[4] Purification in terms of specific activity is 3- to 10-fold. However, in view of the much greater instability of the purified ATPase as compared with ATPase in native membranes, the real purification obtained by affinity chromatography is likely to be higher.

No binding is observed between ATPase in detergent-dispersed membranes and a CNBr-Sepharose column prepared without ATPase inhibitor, using the described conditions. It therefore seems probable that the attachment of ATPase to the adsorbent is a biospecific phenomenon. The dependence of ATP, Mg^{2+} for the interaction to occur[3] in analogy to the binding of free inhibitor to mitochondrial membranes,[9] would appear to support this conclusion.

The most difficult aspect of the procedure is the isolation of the ATPase inhibitor protein in sufficient amounts. Once the inhibitor-Sepharose column is prepared, the method is very convenient for repeated purification of small amounts of ATPase. The whole purification can be carried out in half a day. This is especially valuable as the instability of the enzyme makes the use of fresh, daily preparations desirable.

[73] Acetylcholinesterase

By Yadin Dudai and Israel Silman

Acetylcholinesterase (EC 3.1.1.7) (AChE) is widely distributed in excitable membranes of nerve and muscle. Its molecular properties are of interest because of its involvement in nervous transmission.[1] Most of the work on the molecular structure of AChE has utilized enzyme purified from the electric organ of the electric eel, *Electrophorus electricus*. This organ, because of its highly specialized function, contains relatively large amounts of AChE. However, even in the electric organ there are only about 50–70 mg of enzyme per kilogram of wet tissue, whereas in other tissues the amounts of enzyme seem to be much lower if the same specific activity is assumed for AChE from different sources.[2] Obviously, the application of affinity chromatography to the purification of AChE is desirable.

AChE was first purified from electric organ tissue by conventional techniques as a soluble globular protein with a sedimentation coefficient

[1] D. Nachmansohn, *Science* 168, 1059 (1970).
[2] L. T. Potter, *in* "Handbook of Neurochemistry" (A. Lajtha, ed.), Vol. IV, p. 263. Plenum Press, New York, 1970.

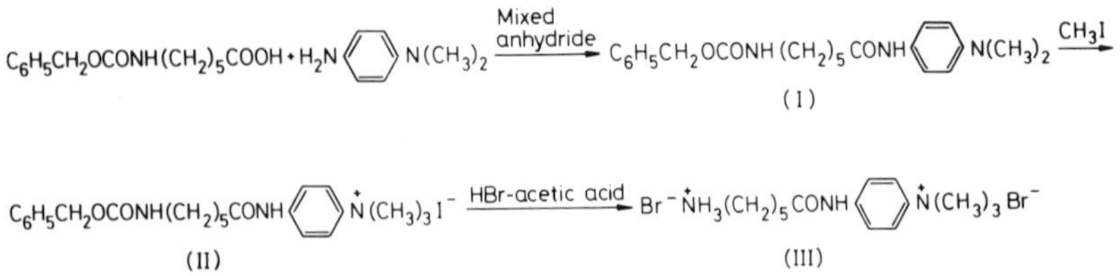

FIG. 1. Synthesis of [N-(ε-aminocaproyl)-p-aminophenyl]trimethylammonium bromide hydrobromide(III), the AChE inhibitor attached to Sepharose.

of about 11 S,[3,4] and subsequently such a purified preparation was crystallized.[5] The yields of purified enzyme obtained were rather low (less than 10% of the initial activity). This 11 S AChE was subsequently shown to be a product of autolysis of the native forms of the enzyme present in fresh electric organ tissue[6] (see below). The autolysis apparently occurred during prolonged storage of electric organ tissue under toluene which was the first step in the purification procedures mentioned above, and had initially been intended for the removal of mucins.[7]

The use of affinity chromatography in the purification of AChE was introduced by Kalderon et al.[8] An AChE-Sepharose conjugate was prepared by synthesis of the inhibitor [N-(ε-aminocaproyl)-p-aminophenyl)]-trimethylammonium bromide hydrobromide (Fig. 1), which was covalently linked to Sepharose activated with cyanogen bromide. This compound was chosen since the phenyltrimethylammonium moiety is a good competitive inhibitor of AChE[9] and the ε-aminocaproyl group can serve as a "spacer." Procedures were subsequently worked out for purification of electric organ AChE, which involved either prolonged autolysis or controlled tryptic digestion, to convert the "native" forms of AChE (see below) to 11 S AChE.[6] These procedures all yielded a homogeneous product comparable in specific activity and several other properties to the 11 S AChE previously isolated by conventional procedures.[3,4] One of these affinity chromatography procedures will be given in detail in the Experimental section. Subsequently other groups have employed affinity

[3] L. T. Kremzner and I. B. Wilson, J. Biol. Chem. 238, 1714 (1963).
[4] W. Leuzinger and A. L. Baker, Proc. Nat. Acad. Sci. U.S. 57, 446 (1967).
[5] W. Leuzinger, A. L. Baker, and E. Cauvin, Proc. Nat. Acad. Sci. U.S. 59, 620 (1968).
[6] Y. Dudai, I. Silman, N. Kalderon, and S. Blumberg, Biochim. Biophys. Acta 268, 138 (1972).
[7] M. A. Rothenberg and D. Nachmansohn, J. Biol. Chem. 168, 223 (1947).
[8] N. Kalderon, I. Silman, S. Blumberg, and Y. Dudai, Biochim. Biophys. Acta 207, 560 (1970).
[9] I. B. Wilson and J. Alexander, J. Biol. Chem. 237, 1323 (1962).

columns containing either the phenyltrimethylammonium ligand[10,11] or the N-methylpyridinium ligand[12] linked to Sepharose for purification of eel AChE either from toluene-treated electric organ tissue or from partially purified commercial preparations, in both cases the enzyme presumably being in the 11 S form. In addition, affinity chromatography procedures have been described, using phenyltrimethylammonium-Sepharose conjugates, for purification of AChE from bovine erythrocytes,[10] mouse neuroblastoma cells,[13] and bovine brain.[14]

AChE is present in salt extracts of fresh or frozen electric organ tissue as three main components, which can be distinguished by their sedimentation coefficients (about 18, 14, and 8 S) on sucrose gradient centrifugation at high ionic strength.[15] As was mentioned above, the 11 S AChE previously purified is a degradation product, formed from the "native" forms either by autolysis or tryptic treatment of the electric tissue.[6,16] Attempts on our part to utilize the phenyltrimethylammonium-Sepharose conjugates for purification of the "native" forms of AChE encountered difficulties since the 14 and 18 S forms, which comprise about 90% of the total AChE activity in the homogenate, aggregate at low ionic strength.[6,15] As a result they are poorly adsorbed on the resin, and appear, when adsorbed, to be associated with other components. At high ionic strength, for example in $1 M$ NaCl, the 14 S and 18 S forms are not aggregated. However, like the 11 S form, they are not adsorbed to the resin at high ionic strength. This is probably because of the combined effect of decreased affinity of AChE for the quaternary ligand with increasing ionic strength,[6,17] together with decreased nonspecific electrostatic interaction between enzyme and resin.[6] Our observation that the N-methylacridinium ion is a powerful AChE inhibitor, with a K_i of about 0.3 μM even in $1 M$ NaCl, prompted us to synthesize an N-methylacridinium derivative with an elongated "spacer" chain (Fig. 2), which could be covalently linked to Sepharose so as to yield an affinity column which would adsorb AChE even at high ionic strength.[18] The N-methylacridinium-Sepharose conjugate was successfully employed for purification of the native forms

[10] J. D. Berman and M. Young, *Proc. Nat. Acad. Sci. U.S.* **68**, 395 (1971).
[11] T. L. Rosenberry, H. W. Chang, and Y. T. Chen, *J. Biol. Chem.* **247**, 1555 (1972).
[12] G. Mooser, H. Schulman, and D. S. Sigman, *Biochemistry* **11**, 1595 (1972).
[13] R. Siman-Tov and L. Sachs, *Eur. J. Biochem.* **30**, 123 (1972).
[14] S. L. Chan, D. Y. Shirachi, H. N. Bhargava, E. Gardner, and A. J. Trevor, *J. Neurochem.* **19**, 2747 (1972).
[15] J. Massoulié and F. Rieger, *Eur. J. Biochem.* **11**, 441 (1969).
[16] J. Massoulié, F. Rieger, and S. Tsuji, *Eur. J. Biochem.* **14**, 430 (1970).
[17] J.-P. Changeux, *Mol. Pharmacol.* **2**, 369 (1966).
[18] Y. Dudai, I. Silman, M. Shinitzky, and S. Blumberg, *Proc. Nat. Acad. Sci. U.S.* **69**, 2400 (1972).

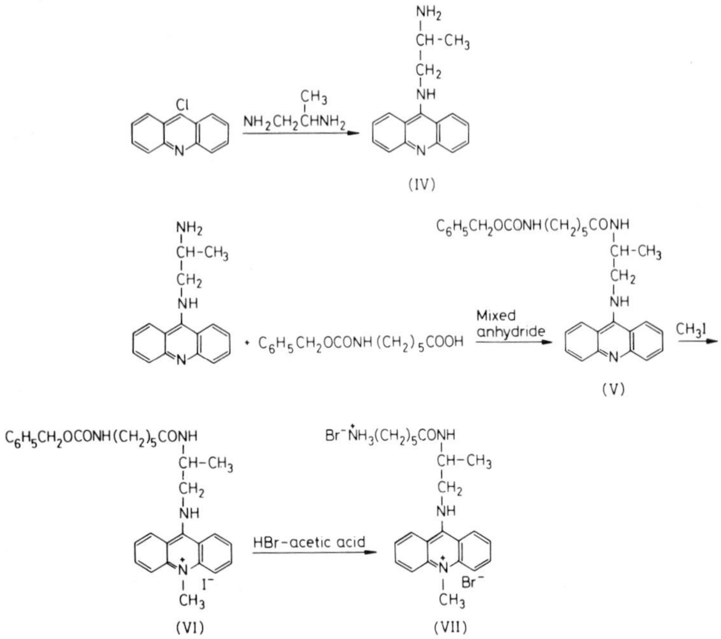

Fig. 2. Synthesis of 1-methyl-9-[N^β-(ϵ-aminocaproyl)-β-aminopropylamino]-acridinium bromide hydrobromide(VII), the AChE inhibitor attached to Sepharose.

of AChE, which were directly adsorbed on such an affinity column from a 1 M NaCl extract of fresh electric organ tissue, and subsequently eluted with an AChE inhibitor. The purified 14 S + 18 S AChE so obtained could subsequently be separated into the 14 S and 18 S forms by preparative sucrose gradient centrifugation.[18,19] The affinity chromatography procedure was the first method to be described for purification of the native forms of AChE, and is given below.

Finally, mention should be made of the procedure of Ashani and Wilson[20] for the purification of electric eel AChE by covalent affinity chromatography. In their procedure, a Sepharose-organophosphate conjugate is prepared which can react covalently with the active site serine of the enzyme, just like a soluble organophosphate enzyme inhibitor. This procedure is covered elsewhere in this volume.[21]

The methods of assaying and characterizing acetylcholinesterase and

[19] Y. Dudai, M. Herzberg, and I. Silman, Proc. Nat. Acad. Sci. U.S. 70, 2473 (1973).
[20] Y. Ashani and I. B. Wilson, Biochim. Biophys. Acta 276, 317 (1972).
[21] H. F. Voss, Y. Ashani, and I. B. Wilson, this volume [74].

other purification methods, have recently been reviewed by Silman and Dudai.[22]

General

Electric eels (*Electrophorus electricus*) are purchased from Paramount Aquarium (Ardsley, New York).

Unless otherwise specified the following two buffers are employed: buffer 1, 10 mM phosphate–0.1 M NaCl (pH 7.0); buffer 2, 10 mM phosphate–1.0 M NaCl (pH 7.0).

Assay Methods

Enzymatic activity is determined by the pH-stat method using acetylcholine as substrate, with assay conditions similar to those of Kremzner and Wilson.[23] The assay is performed at pH 7.0 and 25° under nitrogen, and the reaction mixture, in a final volume of 10 ml, contains 2.5 mM acetylcholine, 0.1 M NaCl, 20 mM MgCl$_2$, and 0.01% gelatin.

Specific activity of AChE solutions is expressed as units/A_{280} where 1 unit of enzyme is the amount hydrolyzing 1 μmole of acetylcholine per minute under the standard assay conditions specified above, and A_{280} is the absorbance of the enzyme solution at 280 nm.

Purification of 11 S Acetylcholinesterase (AChE)

Preparation of the Acetylcholinesterase Inhibitor
[N-(ϵ-Aminocaproyl)-p-aminophenyl]trimethylammonium
Bromide Hydrobromide (Fig. 1)

Synthesis of N-(N-Benzyloxycarbonyl-ϵ-aminocaproyl)-N',N'-dimethyl-p-phenylenediamine (*I*). *N*-Benzyloxycarbonyl-ϵ-aminocaproic acid (60 mmoles, 15.9 g) is dissolved in 200 ml of ethyl acetate, cooled in a salt–ice bath and stirred vigorously. Triethylamine (60 mmoles, 6.1 g) is added and then isobutylchloroformate (60 mmoles, 8.2 g). After 20 minutes, *p*-dimethylaminoaniline (60 mmoles, 8.2 g) dissolved in 50 ml of cold ethyl acetate is slowly added, with continued stirring, and the reaction mixture is left for 5 hours at room temperature with stirring. A large part of the product precipitates directly; this material is filtered, washed with water, ethyl acetate and light petroleum (b.p. 40–60°), and dried by suction. Yield of crude material 14.5 g (63%).

Additional crude material can be obtained by extracting the ethyl acetate supernatant twice with equal volumes of water, drying the ethyl

[22] I. Silman and Y. Dudai, *in* "Research Methods in Neurochemistry" (N. Marks and R. Rodnight, eds.), Vol. III. Plenum Press, New York (in press).

[23] L. T. Kremzner and I. B. Wilson, *Biochemistry* 3, 1902 (1964).

acetate solution with anhydrous Na_2SO_4, evaporating to dryness under reduced pressure, and triturating the product with light petroleum. The crude material can be purified by treatment with activated charcoal in hot methanol and crystallization from the same solvent. The total yield of recrystallized product is 13.1 g (57%), mp 110–111°.

Synthesis of [N-(N-Benzyloxycarbonyl-ε-aminocaproyl)-p-aminophenyl]trimethylammonium Iodide (II). Compound I (11.5 g, 30 mmoles) is suspended in 50 ml methanol. Methyl iodide (7.14 g, 150 mmoles) is added. The reaction mixture is refluxed for 3 hours, then taken to dryness under reduced pressure, triturated with 250 ml absolute diethyl ether, and the product (II) is crystallized from 100 ml of ethanol. Yield 10.8 g (68%), mp 143–145°.

Synthesis of [N-(ε-Aminocaproyl-p-aminophenyl]trimethylammonium Bromide Hydrobromide (ε-Aminocaproyl-PTA, III). Compound (II) (5.25 g, 10 mmoles) is dissolved in 10 ml of glacial acetic acid, and 20 ml of anhydrous HBr in glacial acetic acid are added. After 40 minutes at 25°, the product is precipitated with 400 ml of dry diethyl ether, triturated with successive batches of diethyl ether until it solidifies, and left overnight under reduced pressure over dry NaOH pellets. Compound (III) (ε-aminocaproyl-PTA) is crystallized from 200 ml of ethanol and then recrystallized from the same solvent. Yield 3.15 g (74%), mp 191–194°.

ε-Aminocaproyl-PTA has an absorption maximum at 245 nm in 0.1 M sodium borate, pH 9.3, with $\epsilon = 15,500$.

Preparation and Characterization of ε-Aminocaproyl-PTA-Agarose Conjugates

Sepharose was activated with cyanogen bromide and allowed to react with ε-aminocaproyl-PTA.[24] The detailed experimental procedure was that described by Blumberg et al.,[25] starting from 70 ml of Sepharose 2B slurry. Solutions containing 15–250 μmoles of ε-aminocaproyl-PTA in 30 ml of 0.5 M $NaHCO_3$ are added to the activated Sepharose. The amount of the reagent coupled to the resin is estimated by spectrophotometric determination of the unreacted inhibitor in the supernatant liquids and washings. Resins containing 0.1–2.5 μmoles of ε-aminocaproyl-PTA per milliliter of resin are thus obtained; they are stored at 4° until used. Before use, the resin is washed extensively with the appropriate buffer to remove traces of soluble ligand released on prolonged storage.

[24] R. Axén, J. Porath, and S. Ernbäck, *Nature (London)* **214**, 1302 (1967).
[25] S. Blumberg, I. Schechter, and A. Berger, *Eur. J. Biochem.* **15**, 97 (1970).

Purification of 11 S AChE

Fresh or thawed frozen tissue from the main organ of *Electrophorus electricus* is freed of superficial connective tissue and cut into 0.5–1.0 cm cubes. One hundred grams of tissue is homogenized with 250 ml of buffer 1 at 4° for 4 minutes using a Sorvall Omnimixer at full speed; the homogenization vessel is immersed in ice. The homogenate is treated with a solution of 35 mg of trypsin (Worthington) in 5 ml of water and incubated at 25° for 16 hours with stirring. Soybean trypsin inhibitor (Worthington; 35 mg in 5 ml of water) is then added. At this stage one can verify, by sucrose gradient centrifugation, that the AChE has been converted to the 11 S form.[6] After 30 minutes of additional incubation, the homogenate is centrifuged for 50 minutes at 105,000 *g*, and the pellet is discarded. The supernatant is dialyzed for 5 hours against buffer 1. After dialysis the supernatant fluid containing about 60,000 units of enzyme is applied at a rate of 30–40 ml per hour to an affinity column with a total bed volume of 12 ml (ε-aminocaproyl-PTA content, 1 μmole per milliliter of resin), previously equilibrated with the same buffer. After application of the protein solution, the column is washed with 5 column volumes of buffer 1. AChE is then eluted with 4–5 column volumes of a solution of 10 m*M* decamethonium bromide (K and K) in buffer 1; fractions of 2 ml are collected. The column is then washed with 4–5 column volumes of buffer 2. Finally the column is washed with 50 column volumes of buffer 1 before reutilization.

The fractions eluted with decamethonium are assayed for enzymatic activity, and those with the highest specific activity (1500–3500 units/A_{280}) are pooled and dialyzed for 72 hours at 4° against buffer 1, with several changes of the buffer. The total activity obtained is 22,000–32,000 units. The enzyme is then rechromatographed exactly as above, using 1 ml of resin for every 5000 units of AChE applied. On elution with decamethonium, a peak is obtained with a uniform specific activity of 4700–5100 units/A_{280}. The total amount of AChE activity obtained is 11,000–15,000 units, corresponding to 1.3–1.8 mg of AChE, assuming an extinction coefficient of $\epsilon_{280}^{1\%} = 17.6$.[6]

The purified AChE has a sedimentation coefficient of about 11 S on sucrose gradient centrifugation,[6] appears homogeneous on analytical ultracentrifugation[19] and exhibits only one band on polyacrylamide gel electrophoresis.[6]

The purification procedure given above is essentially that described for the enzyme referred to as AChE-A by Dudai *et al.*[6] Two somewhat different procedures, utilizing the same affinity column and yielding two preparations, AChE-B and AChE-C, both with properties very similar to AChE-A, are described in the same paper. These two procedures both utilize toluene-treated tissues, and may involve more extensive autolytic degradation of the enzyme (see Dudai et al.[19]). For this reason, the method given above is to be preferred, since it is better controlled, even though the procedure used for obtaining AChE-B may often yield a more homogeneous product in the first stage of affinity chromatography (see Dudai et al.[6]). On rechromatography, AChE-A, AChE-B, and AChE-C are all obtained with the same high specific activity.

Purification of 14 S + 18 S Acetylcholinesterase

Preparation of the AChE Inhibitor 1-Methyl-9-[Nβ-(ε-aminocapyroyl)-β-aminopropylamino]acridinium Bromide Hydrobromide (MAC) (Fig. 2)

Synthesis of 9-(β-Aminopropylamino)acridine (IV). 9-Chloroacridine (10 g) is mixed with phenol (60 g) and heated to 110°. The melted mixture is cooled to 70°, and 1,2-diaminopropane (50 ml) is added. The reaction mixture is then maintained at 125° for 1 hour, cooled, and neutralized with 800 ml of 0.75 M NaOH. The product precipitates as an oil which solidifies on trituration overnight in the alkaline medium. The solid is filtered, washed first with 0.2 N NaOH and then with water, and dried under reduced pressure. The yield of crude material is 9.9 g (84%). Recrystallization from benzene yields 7.2 g (61%), mp 131–133°.

There are two possible isomeric products in the reaction of 1,2-diaminopropane with 9-chloroacridine, since both the amino group on C_1 and that on C_2 can participate in the reaction. In the scheme shown in Fig. 2, the reaction has been assumed to proceed exclusively with the amino group on C_1, but in fact a mixture of the two possible products may have been obtained and employed in the subsequent synthetic steps.

Synthesis of 1-Methyl-9-[Nβ-(N-benzyloxycarbonyl-ε-aminocaproyl)-β-aminopropylamino]acridinium Iodide (VI). N-Benzyloxycarbonyl-ε-aminocaproic acid (7.9 g, 30 mmoles) is dissolved in dry ethyl acetate (150 ml), cooled in a salt–ice bath to below −10°, and stirred vigorously. Trimethylamine (3.05 g, 30 mmoles) is added, followed by isobutylchloroformate (4.1 g, 30 mmoles). After 20 minutes at −10°, the reaction mixture is filtered by suction and the precipitated triethylamine hydrochloride is washed with 20 ml of dry ethyl acetate which is then combined with the filtrate. The reaction mixture is returned to the ice–salt bath, and treated with a solution of (IV) (6.3 g, 25 mmoles) dissolved in dry dimethylformamide (100 ml) previously cooled to −10°. After 20 minutes at −10°, the reaction mixture is left at room temperature overnight. Thin-layer chromatography on silica gel using glacial acetic acid shows that the coupling reaction has proceeded to completion to yield (V). The reaction mixture is then evaporated to dryness under reduced pressure and the product dissolved in 50 ml of absolute methanol. Methyl iodide, 8 ml, is added and the solution is refluxed for 4 hours. It is left overnight at room temperature, at which stage thin-layer chromatography on silica gel in ethyl acetate shows that the quaternization reaction is complete. The solution is evaporated to dryness under reduced pressure, extracted three times with 100-ml portions of ethyl acetate, and the product, compound (VI), crystallized from 100 ml of 2-propanol. The product is filtered by suction, washed with ice-cold 2-propanol followed by ether, and dried under reduced pressure. Yield 7.0 g (44%), mp 156–157°.

Synthesis of 1-Methyl-9-[N$^\beta$-(ϵ-aminocaproyl)-β-aminopropylamino]-acridinium Bromide Hydrobromide (MAC, VII). Compound (VI) (4.1 g) is dissolved in anhydrous glacial acetic acid (40 ml, analytical grade). Anhydrous HBr in glacial acetic acid (80 ml) is then added. After 30 minutes at room temperature the product is precipitated with dry diethyl ether and triturated with successive batches of diethyl ether until it solidifies. The solid is filtered by suction, washed with diethyl ether and left overnight under reduced pressure over dry NaOH pellets. The crude product is crystallized from absolute ethanol (50 ml). Yield 2.75 g (80%), melting with decomposition above 240°.

MAC has absorption maxima in the visible region at 393 nm, 410 nm, and 431 nm in 0.1 M sodium bicarbonate at pH 8.3, with ϵ = 7,880, 12,050, and 10,150, respectively.

Preparation and Characterization of N-Methylacridinium-Agarose Conjugates

The procedure is exactly the same as described for preparation of ϵ-aminocaproyl-PTA-Sepharose conjugates. Resins containing 0.5–1.5 μmoles of the N-methylacridinium ligand are thereby obtained. Here too, extensive washing of the resin is needed to remove traces of soluble ligand.

Purification of 18 S and 14 S AChE

Fresh tissue from the main organ of *Electrophorus electricus* is freed of superficial connective tissue and cut into 0.5–1.0-cm cubes; 275 g of tissue is homogenized in 700 ml of buffer 2. The homogenate, containing 100,000–140,000 units of enzyme activity, is centrifuged at 10,000 g for 30 minutes, and the pellet is discarded. The supernatant fluid is centrifuged at 100,000 g for 60 minutes and the pellet is also discarded. The supernatant liquid, containing between 80,000 and 120,000 units of enzyme activity, is applied to a column containing 20 ml of the MAC–

Sepharose conjugate. Application is performed at 4° and at a flow rate of 40 ml per hour. After application, the resin is washed with 7 column volumes of buffer 2; 10–20% of the total enzyme applied can be detected in the effluent and washings. When 20 mM decamethonium bromide in the same buffer is applied to the column, a peak of protein and enzymatic activity is eluted with a specific activity of 1000–3300 units/A_{280}. About 50% of the enzyme applied can be eluted.

The fractions eluted with decamethonium are pooled and dialyzed for 72 hours at 4° against buffer 2, with several changes of the buffer. The enzyme is then chromatographed in the same manner with 1 ml of resin for every 4000–6000 units of enzyme. On elution with decamethonium, a peak is obtained of specific activity 3800–4000 units/A_{280}. The total amount of AChE activity obtained is 11,000–26,000 units, corresponding to 1.6–3.9 mg of AChE, assuming an extinction coefficient $\epsilon_{280}^{1\%} = 17.6$.[6]

It is advisable not to use the same resin which was used for initial chromatography for the second chromatographic step since difficulties are encountered in washing the resin free of all adsorbed contaminations. However, the resin can be reutilised for the first chromatography step.

On acrylamide gel electrophoresis, in the presence of sodium dodecyl sulfate and β-mercaptoethanol, the purified enzyme display a major band with a molecular weight of about 80,000. The sedimentation pattern of the AChE preparation can be analyzed during purification by performing sucrose gradients, as previously described.[18]

Separation of Purified 18 S AChE and 14 S AChE

The purified 14 S + 18 S AChE can be separated into its two components i.e., purified 14 S AChE and purified 18 S AChE, as follows: Purified 14 S + 18 S AChE (10–20 ml; A_{280} = 0.3–0.8) is concentrated by pressure dialysis at 4°, until A_{280} = 2.0–3.0. A 1.8-ml aliquot of the concentrated enzyme containing 13,000–20,000 enzyme units is layered on a 5–20% linear sucrose gradient in buffer 2, of total volume 28 ml, with a 5-ml cushion of 60% sucrose in the same buffer at the bottom. Centrifugation is performed in an SW 27 rotor with an L2 65-B Beckman preparative ultracentrifuge at 27,000 rpm at 4° for 20 hours. The 18 S and 14 S peaks are collected and dialyzed against buffer 2. The properties of the 14 S AChE and 18 S AChE obtained by this method have been described by Dudai et al.[19]

[74] A Covalent Affinity Technique for the Purification of All Forms of Acetylcholinesterase

By HOUSTON F. VOSS, Y. ASHANI, and IRWIN B. WILSON

The purification of the enzyme acetylcholinesterase (EC 3.1.1.7) has attracted much attention since it was first recognized to be involved in nervous function. The study of acetylcholinesterase was facilitated by Nachmansohn's introduction of the electric organ of *Electrophorus electrophorus* as a source of the enzyme. Rothenberg and Nachmansohn using ammonium sulfate fractionation obtained a preparation that is now known to have been about 70% pure.[1] This enzyme preparation had a very high molecular weight that was estimated at about 3×10^6. Much later Kremzner and Wilson[2] described a chromatographic technique for purifying this enzyme, using benzylated DEAE-Sephadex gel in an attempt at affinity chromatography. Although effective in purifying the enzyme this gel did not function as an affinity support. This enzyme preparation was nearly pure, had a low molecular weight of about 2.4×10^5, and behaved like a globular protein. It had a sedimentation coefficient of about 11 S. Shortly thereafter, Leuzinger and Baker,[3] using the same chromatographic method succeeded in crystallizing the enzyme. It is now known that the globular 11 S enzyme is not the native enzyme.[4-7] Freshly homogenized electric organ contains three nonglobular forms with sedimentation coefficients of about 9 S, 14 S, and 18 S; these, on standing, are converted to the 11 S form.[7]

The development of affinity techniques, especially the side-arm extension by Cuatrecasas,[8] supplied the impetus for many laboratories to rethink their approaches to the purification of biologically active proteins. The nature of the active site of acetylcholinesterase[9] suggests two possible approaches that can be taken in designing an affinity column. The first approach makes use of the properties of the anionic subsite as shown in

[1] M. A. Rothenberg and D. Nachmansohn, *J. Biol. Chem.* **168**, 223 (1947).
[2] L. T. Kremzner and I. B. Wilson, *J. Biol. Chem.* **238**, 1714 (1963).
[3] W. Leuzinger and A. L. Baker, *Proc. Nat. Acad. Sci. U.S.* **57**, 446 (1967).
[4] J. Massoulié, F. Rieger, and S. Bon, *C. R. Acad. Sci. Ser. D* **270**, 1837 (1970).
[5] J. Massoulié, F. Rieger, and S. Tsuji, *Eur. J. Biochem.* **14**, 430 (1970).
[6] J. Massoulié, F. Rieger, and S. Bon, *Eur. J. Biochem.* **21**, 542 (1971).
[7] F. Rieger, S. Bon, J. Massoulié, and J. Cartaud, *Eur. J. Biochem.* **34**, 539 (1973).
[8] P. Cuatrecasas, *Nature (London)* **228**, 1327 (1970).
[9] H. C. Froede and I. B. Wilson, Acetylcholinesterase, *in* "The Enzymes" (P. D. Boyer, ed.), 3rd ed., Vol. 5, p. 87 (1971). Academic Press, New York.

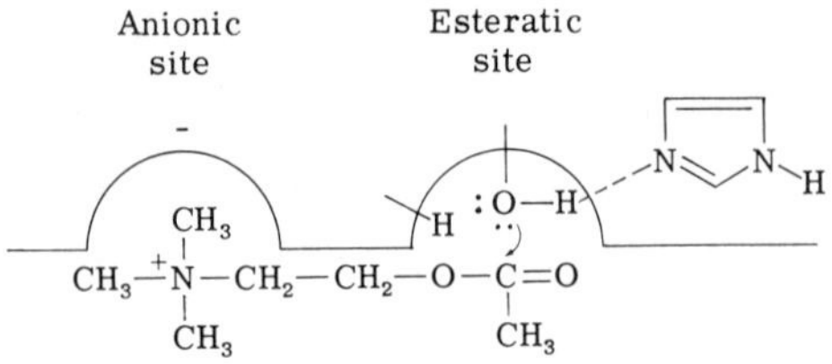

Anionic   Esteratic
site   site

Fig. 1. The enzyme–substrate complex and one of the more commonly suggested mechanisms whereby the hydroxyl group of serine functions as a nucleophile to displace choline and form an acetyl enzyme. The imidazole of histidine serves as a general base, and a proton is shown at the esteratic site to envisage the possibility that a proton may be involved in the catalysis, perhaps by transfer to make choline (rather than the choline dipolar ion) the leaving group. From "The Enzymes" (P. D. Boyer, ed.), Vol. 5, p. 93, Academic Press, New York, 1971.

Fig. 1. This approach, based upon reversible inhibitors, is reviewed by Dudai and Silman in this volume [73].

A second approach to affinity chromatography of acetylcholinesterase involves the other region of the active site, the esteratic subsite.[10,11]

The type of affinity chromatography for acetylcholinesterase that we discuss here is based upon ligands which covalently react with the enzyme at the esteratic subsite. The ligands are based upon the well known reaction of serine esterases with organophosphate and organophosphonate esters that contain a good leaving group.[12] The reaction here illustrated with O,O-diethylphosphofluoridate;

$$\underset{EtO}{\overset{O}{\underset{\diagdown}{\overset{\|}{P}}}}-F \; + \; E \;\; \underset{k_r = 10}{\overset{k_i = 2 \times 10^5}{\rightleftharpoons}} \;\; \underset{EtO}{\overset{O}{\underset{\diagdown}{\overset{\|}{P}}}}-E \; + \; F^- \qquad (1)$$

is essentially reversible, but the equilibrium lies far to the right (as indicated by the values of the second-order rate constants given in liters mole^{-1} min^{-1} for acetylcholinesterase from electric eel),[12-14] and the reaction is usually unidirectional. The enzymatic nucleophile is the same serine hydroxyl group that is normally acetylated during the course of the

[10] Y. Ashani and I. B. Wilson, *Biochim. Biophys. Acta* **276**, 317 (1971).
[11] H. F. Voss and Y. Ashani, *Fed. Proc., Fed. Amer. Soc. Exp. Biol.* **32**, 553 (Abstract) (1973).
[12] For a discussion of this and other relevant points, see G. B. Koelle, "Cholinesterases and Anticholinesterase Agents," Vol. 15 of "Handbuch der Experimentellen Pharmakologie" (O. Eichler and A. Farah, eds.). Springer-Verlag, Berlin, 1963.
[13] I. B. Wilson and R. A. Rio, *J. Molecular Pharmacol.* **1**, 60 (1965).
[14] E. Heilbronn, *Acta Chem. Scand.* **18**, 2410 (1964).

hydrolysis of acetylcholine. The phosphoryl enzyme is relatively stable and hydrolyzes only very slowly in aqueous solution. However, potent nucleophiles, especially site-directed ones such as N-methyl pyridinium-2-aldoxime (2PAM, PAM2), can react rapidly with this inhibited enzyme to give free enzyme, as shown in Eq. (2).[15,16]

$$(2)$$

Rationale of a Covalent Affinity Column for Acetylcholinesterase

A common way of affixing the ligand to the gel (in our case agarose) is to incorporate an amine function in the ligand structure and join it to the side-arm extension via an amide linkage. The ligand we have selected is 2-aminoethyl p-nitrophenyl methylphosphonate (I).

This ligand is a very weak inhibitor at best, but it becomes a very potent one when acetylated ($k_i = 6 \times 10^5$) so that when it is incorporated into the insoluble support via an amide linkage it becomes a rapid reactant toward acetylcholinesterase. p-Nitrophenol is the leaving group. An outline of the process we envisaged is given in Fig. 2. The affinity support will trap acetylcholinesterase and very likely other serinesterases. Probably not all serinesterases will be trapped, because it is reasonable to suppose that some may not react rapidly with this particular ligand. Trapping acetylcholinesterase would be of little value if we did not have some

[15] I. B. Wilson and S. Ginsburg, *Biochim. Biophys. Acta* **18**, 168 (1955).
[16] A. F. Childs, D. R. Davies, A. L. Green, and J. P. Rutland, *Brit. J. Pharmacol.* **10**, 462 (1955).

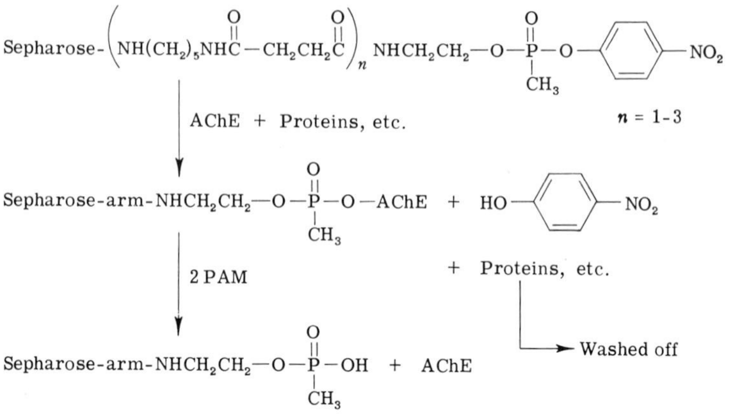

$$\text{Sepharose-}\left(\text{NH(CH}_2)_5\text{NH}\overset{\overset{\displaystyle O}{\|}}{C}-\text{CH}_2\text{CH}_2\overset{\overset{\displaystyle O}{\|}}{C}\right)_n \text{NHCH}_2\text{CH}_2-\text{O}-\overset{\overset{\displaystyle O}{\|}}{\underset{\underset{\displaystyle CH_3}{|}}{P}}-\text{O}-\bigcirc-\text{NO}_2$$

ACfE + Proteins, etc. $n = 1\text{-}3$

$$\text{Sepharose-arm-NHCH}_2\text{CH}_2-\text{O}-\overset{\overset{\displaystyle O}{\|}}{\underset{\underset{\displaystyle CH_3}{|}}{P}}-\text{O}-\text{AChE} \;+\; \text{HO}-\bigcirc-\text{NO}_2$$

2 PAM   + Proteins, etc.

$$\text{Sepharose-arm-NHCH}_2\text{CH}_2-\text{O}-\overset{\overset{\displaystyle O}{\|}}{\underset{\underset{\displaystyle CH_3}{|}}{P}}-\text{OH} \;+\; \text{AChE}$$

Washed off

Fig. 2. Proposed scheme for action of covalent affinity support. In step 1 the column would be treated with a mixture of acetylcholinesterase (AChE) and other proteins. The AChE would be covalently bound, along with any other serine esterases, and nonreacted proteins could be washed from the column. In step 2 the column would be treated with a solution of 2PAM, selectively releasing the AChE.

means of effecting its release. We have already indicated in Eq. 1 that the basic reaction is reversible. Thus fluoride can react with the inhibited enzyme to yield active enzyme by acting as a nucleophile toward phosphorus; the enzyme is the leaving group in this reactivation process. Thus it would appear reasonable to suppose that the bound enzyme could be released from the gel with high concentrations of fluoride ion. However other nucleophiles were more attractive as releasing agents because they are more effective in reactivating the enzyme. In particular N-methyl pyridinium-2-aldoxime, Eq. (2), appeared most promising because it is rather specific as a reactivator for acetylcholinesterase. Thus it was thought that specificity could be achieved in releasing the enzyme from the column.

Actual Mechanism of the Column

Although we have obtained a successful column with the insoluble support described above,[10,11] the column does not function in the manner indicated for the release of covalently bound enzyme. However, it does appear to trap enzyme in the manner envisaged. Thus at a flow rate that allows 3 minutes of contact between the solution and the gel with a ligand concentration of 20 μM almost all the enzyme is trapped. This rate of reaction corresponds to an effective second-order rate constant greater than 10^5 M^{-1} min^{-1}. If the enzyme is completely inhibited with an organo-

Capacity

The capacity of this covalent affinity column is remarkably high. We found that about half the ligands can react with chymotrypsin. Thus 0.1 μmole of chymotrypsin, or 2.5 mg of protein per milliliter of gel, was bound to a gel containing 0.2 mM bound ligand. The fact that half of the bound ligand reacted with chymotrypsin suggested that perhaps only one of the enantiomeric forms of the ligand can react. This is not the case; the difference in reaction rate of the two optical isomers with chymotrypsin is relatively small. Possibly about half the ligands are inaccessible to chymotrypsin. If this kind of explanation is correct, accessibility might be lower for larger molecules, and other enzymes might distinguish more drastically between the two enantiomeric forms of the ligand.

We do not know the full capacity of the column for eel acetylcholinesterase, but it is quite high, at least 1.0 mg/ml, when the bound ligand is 0.5 mM.

Methods of Achieving Specificity

If acetylcholinesterase or chymotrypsin is completely inhibited with an organophosphorous inhibitor prior to applying it to the column, none of the enzyme is retained by the column. This result is anticipated from the envisaged mode of binding of the column. We would also expect that, if a specific reversible competitive inhibitor were added to the enzyme solution prior to its application to the column, there would be little retention of the enzyme, yet other serinesterases would still be bound. We have found this to be the case with acetylcholinesterase using 3 hydroxyphenyl trimethylammonium ion as a specific reversible inhibitor. The enzyme was then separated from the reversible inhibitor by gel filtration and reapplied to a column containing fresh gel for further purification. Alternatively if the crude enzyme solution is first treated with an organophosphorous inhibitor such as O,O-diethylphosphofluoridate, all proteins which can react with the column probably will be inhibited. It would then be possible to selectively reactivate the acetylcholinesterase with low concentrations of 2PAM before applying the solution to the column. We have not yet tried this procedure. Other methods of achieving specificity readily come to mind.

Preparation of the Column

Synthesis of the Ligand

The procedure for the synthesis of the ligand is given below:

(a) $(CH_3O)_2\overset{\overset{O}{\|}}{P}-CH_3$ $\xrightarrow{PCl_5}$ $CH_3-\overset{\overset{O}{\|}}{P}Cl_2$

(II)

(b) $CH_3\overset{\overset{O}{\|}}{P}Cl_2$ + $HO-\langle\text{arene}\rangle-NO_2$ $\longrightarrow$ $CH_3-\overset{\overset{O}{\|}}{\underset{\underset{Cl}{|}}{P}}-O-\langle\text{arene}\rangle-NO_2$

(III)

(c) $CH_3-\overset{\overset{O}{\|}}{\underset{\underset{Cl}{|}}{P}}-O-\langle\text{arene}\rangle-NO_2$ + $H_2NCH_2CH_2OH\cdot HCl$ $\longrightarrow$ $CH_3-\overset{\overset{O}{\|}}{\underset{\underset{O-CH_2CH_2NH_2\cdot HCl}{|}}{P}}-O-\langle\text{arene}\rangle-NO_2$

(I)

(a) Fifty grams of $CH_3P(O)(OCH_3)_2$ (Aldrich) are gradually added to 200 g PCl_5 in a 500-ml round-bottom flask equipped with a condenser. When the exothermic reaction subsides ($\approx$30 minutes), the mixture is refluxed gently for 60 minutes at 100–103°; a clear solution is attained after 10–15 minutes of refluxing. The completion of the reaction may be observed by the disappearance of the methoxy signal in the nuclear magnetic resonance (NMR) spectra (doublet centered at 3.82 ppm relative to TMS, at 60 Mc). Unreacted methoxy groups are easily converted completely to the corresponding $CH_3P(O)Cl_2$ by adding 10–15 g of fresh PCl_5. $POCl_3$ is removed by vacuum distillation (b_{10}, 25–28°) and the main product, methyl phosphonodichloride (II), is collected at b_{10}, 53–55°.[17] Yield 30 g. Compound (II) usually solidifies in the receiver at room temperature (mp 33°). NMR (60 Mc) shows one doublet centered at 2.54 ppm (J_{CH_3-P}, 19 Hz).

(b) Compound (II) (10 g) and 12 g of p-nitrophenol are heated for 15 hours at 120°, according to the instructions given by deRoos.[18] The unreacted dichloro compound (II) is removed by distillation (b_{10}, 53°). The final product, p-nitrophenyl methyl phosphonochloridate (II) is collected at $b_{0.1}$, 135–137° and solidifies in the receiver (mp 72–76°). NMR spectra agrees with the expected structure (1 eq of p-nitrophenol was incorporated).

(c) Compound (II) (4.0 g) and 1 g of dry 2-aminoethanol hydrochloride are stirred for 1 hour at 100°. The homogeneous mixture is cooled to room temperature, triturated in absolute ethanol and filtered. The insoluble fraction has a melting point of 120° and has been identified as bis(p-nitrophenyl) methylphosphonate. The product, 2-aminoethyl

[17] H. Tolkmith, *Ann. N.Y. Acad. Sci.* **79**, 189 (1959).
[18] A. M. deRoos, *Rec. Trav. Chim.* **78**, 145 (1959).

p-nitrophenyl methylphosphonate·HCl (III) is precipitated by adding anhydrous ether to the filtrate. The precipitate is recrystallized from ethanol–ether (anhydrous). The first crop yields 500 mg of (III) with a melting point of 148–151°. Treatment with 0.1 N NaOH releases 46.3% p-nitrophenol (theor. 47.0%).

Attachment of the Arm to the Matrix

The side-arm extension is attached to Sepharose 4B (Pharmacia) in two steps.

(a) 1,5-Diaminopentane is first attached to the matrix by the CNBr reaction.[19] Sepharose 4B (100 ml of packed gel) is washed with 2 liters of sodium carbonate at pH 9.0 (μ = 0.1), and is followed by 4 liters of distilled water. Cyanogen bromide (5.0 g) is added in one step to the stirred washed gel in a total volume of 300 ml of distilled water. The pH is maintained at 11.0 by additions of 25% NaOH. The temperature during the activation step is kept below 25°. When the pH stabilizes, the gel is washed at 4° with 2 liters of sodium carbonate at pH 10.0 (μ = 0.1) followed by 2 liters of 0.2 N NaCl and finally by 2 liters of distilled water. The activated gel is immediately taken up in a total volume of 250 ml of distilled water, and the pH is adjusted to 10.0. 1,5-Diaminopentane (5.0 g) (Aldrich) in 15 ml of water (pH adjusted to 10.0 with conc. HCl) is added and allowed to shake overnight at 4°. The gel is then washed with 2 liters of 0.2 N NaCl and distilled water until the filtrate contains no trinitrobenzene sulfonate (TNBS)-reacting amine. The presence of attached amine is determined by the reaction of TNBS with the washed gel.[20]

(b) Succinylation of the attached amino groups is carried out at pH 6.0 (4°). Succinic anhydride is added to the stirred gel in small batches of 200–300 mg and allowed to react until the pH stabilizes. Small samples of the mixture are then washed and tested for the presence of amino groups with TNBS. The stepwise addition is continued until all the amino groups are succinylated. We found that this stepwise succinylation served to decrease succinylation of the hydroxyl groups of the Sepharose matrix, thus decreasing the rate of release of soluble inhibitor from the column. The succinylated gel is washed with 0.1 N NaOH for 60 minutes to remove any succinyl ester linkages which have formed.[21] This treatment further reduces the rate of release of soluble inhibitor from the column.

The succinylated gel is thoroughly washed as before, and the ligand is either coupled at this stage (1× extension) or the side arms are fur-

[19] R. Axén, J. Porath, and S. Ernbäck, Nature (London) 214, 1302 (1967).
[20] H. Filippusson and W. E. Hornby, Biochem. J. 120, 215 (1970).
[21] P. Cuatrecasas and I. Parikh, Biochemistry 11, 2291 (1972).

ther extended by the addition of 0.5 g of 1,5-diaminopentane and 500 mg of N-ethyl-N'-dimethylaminopropyl carbodiimide (EDAC) (Bio-Rad) to 10 ml of packed gel. The mixture is allowed to react for 18 hours at 4°, and the presence of attached amino groups is determined with TNBS as mentioned previously after washing. Succinylation is carried out exactly as before and the resulting preparation (2× extension) could be either further extended or the ligand could be attached at this stage.

Attachment of the Ligand

The ligand is attached to the extended gel at pH 5.0, 4° with the soluble carbodiimide EDAC. Typically, 10 mg of the ligand and 100 mg of EDAC are added to a stirred solution of 10 ml of packed succinylated gel in a total volume of 20 ml. The mixture is shaken overnight at 4° and then washed extensively with 0.2 N NaCl and distilled water. Once the ligand has been attached, the gel is stored at 4° in sodium acetate at pH 5.0 to minimize loss of the labile phosphonate ester. In all operations only the mildest stirring or shaking conditions are used to avoid breaking the agarose beads.

Use of the Column

These columns are very simple to use because of the nature of the reaction. Enzyme solutions can be applied to the column in volumes as small as 1 μl or as large as several hundred milliliters. Enzyme solutions can be loaded at any ionic strength up to 2.0 which enables one to trap and purify the nonglobular native molecular forms of acetylcholinesterase found in *Electroplax*.[4-7] These forms aggregate at low ionic strength and have proved to be extremely difficult to purify by reversible affinity chromatography although a reversible column developed by Silman[22] seems to have overcome this difficulty. After enzyme has been trapped on the column, the column should be washed with buffer having an ionic strength of at least 0.5 to remove absorbed protein. Washing with very large volumes of buffer is possible because of the covalent nature of the binding of the enzyme to the gel.

In one case we passed 100 ml of a crude 100,000 g supernatant of electric eel electroplax (homogenized in 10% NH_4SO_4, 50 mM potassium phosphate at pH 7.0–1.0 M NaCl) containing ≈1.5 mg of acetylcholinesterase through a 1-ml column, which was 0.5 mM in bound ligand at a flow rate of 0.75 ml min⁻¹. Approximately 85% of the acetylcholinesterase was trapped. Within experimental error, all the protein came

[22] Y. Dudai, I. Silman, M. Shinitzky, and S. Blumberg, *Proc. Nat. Acad. Sci. U.S.* **69**, 2400 (1972).

through, i.e., very little protein was trapped. The column was then washed with 50 ml of 50 mM potassium phosphate at pH 7.0 containing 1 M NaCl; less than 2% of the trapped activity was eluted.

Five milliliters of 10 mM 2PAM (pH 8.0, 50 mM potassium phosphate, 1 M NaCl) were passed through the column, and the flow was stopped with 1 ml of solution above the gel. The column was stored at room temperature, and after 4 days 2 ml of the 2PAM solution were used to displace the fluid in the gel. The 2 ml of displaced solution contained 25% of the trapped enzyme. (This case represented the slowest release we have ever observed.) The enzyme solution was separated from 2PAM with a 50 ml Sephadex G-50 column. As expected the enzyme came out at the void volume. The specific activity of the initial solution was 1 unit (1 mmole of acetylthiocholine, 0.5 mM, hydrolyzed per hour in 0.1 M NaCl, 20 mM MgCl$_2$, 20 mM phosphate pH 7.0, 25°; Ellman method[23]) per milligram of protein ($E_{280}^{1\%} = 10$). The released acetylcholinesterase had a specific activity of 310 units/mg ($E_{280}^{1\%} = 18$).[2] Sucrose gradients showed that both the initial homogenate and the released enzyme contained no 11 S enzyme and had the usual distribution of molecular forms, i.e., 9 S, 14 S, 18 S, reported by Massoulié.[5-7] Since no attempt was made to obtain specificity in this run, this purified preparation may contain other esterases which were trapped and released in an inhibited state along with acetylcholinesterase. However, this example shows the tremendous capacity of this column for trapping and purifying the native molecular forms of this enzyme.

Problems

As previously mentioned, the column does not function as initially planned. Two major problems were encountered in the development of this procedure. The rate of release of a potent soluble cholinesterase inhibitor from the gel was decreased by techniques used during the succinylation stage of the arm extension. Second, the failure of 2PAM (or F⁻, NH$_2$OH, TMB$_4$) to release trapped acetylcholinesterase from the gel is difficult to explain since enzyme inhibited by the soluble inhibitor is readily reactivated. The fortuitous slow release of the protein in an inhibited state which can be reactivated by 2PAM has enabled us to use this gel for the purification of the enzyme.

We have tried various side arms to prevent the release of the soluble inhibitor from the column and to bring about the expected release of the trapped protein with 2PAM. Our efforts in this direction have been un-

---

[23] G. L. Ellman, K. D. Courtney, V. Andres, Jr., and R. M. Featherstone, *Biochem. Pharmacol.* **7**, 83 (1961).

successful. The release of soluble inhibitor can be greatly decreased by using side arms that contain a carboxyl group as well as an amino group and thus do not require succinylation. We have tried 6-aminocaproic acid, 8-aminocaprylic acid, and bis(6-aminocaproic acid) as side arms. Columns prepared with these arms do release soluble inhibitor at much slower rates than any column which we have prepared using a succinylation step. However, columns prepared with 6-aminocaproic acid do not trap the enzyme. Enzyme was trapped by columns prepared with the two other side arms but was not released with 2PAM and was not released spontaneously. It is quite remarkable that the change from 6-aminocaproic acid to 8-aminocaprylic acid should have such a dramatic effect on the ability of the gel to trap enzyme.

The slow release of enzyme, especially the native forms, from the column prompted us to investigate means of improving the rate of release. We found that treatment of the aminopentylagarose (the product just before the succinylation step) with 10 mM HCl for 60 minutes doubled the rate of release of trapped enzyme and also improved the trapping of enzyme. The release of soluble inhibitor was also increased but this was not troublesome.

Our current thoughts on the reactivation problem is that some association of the protein with either the side arm or the carbohydrate matrix is occurring; an association which is probably responsible for steric masking of the organophosphorus–serine ester preventing nucleophilic attack by even the smallest of nucleophiles. We are currently looking to the synthesis of alternate ligands which will have a different stereochemistry at the active site. We are also exploring the possibility of using polypeptides, as suggested by Cuatrecasas et al.,[24–25] to extend the ligand much further from the polysaccharide matrix.

Finally, we should point out that this method of covalent affinity chromatography is not a chromatographic process but is convenient to use in the same manner as a chromatographic process.

Acknowledgment

This work was supported by National Institutes of Health grants Nos. NS09197 and NS07156.

[24] V. Sica, E. Nova, G. A. Puca, I. Parikh, and P. Cuatrecasas, Fed. Proc., Fed. Amer. Soc. Exp. Biol. 32, 454 (Abstract) (1973).
[25] I. Parikh and P. Cuatrecasas, Fed. Proc., Fed. Amer. Soc. Exp. Biol. 32, 559 (Abstract) (1973).

[75] Purification of Proteins Involved in Ca(II)-Dependent Protein–Protein Interactions (Coagulant Protein of Russell's Viper Venom)

By BARBARA C. FURIE and BRUCE FURIE

The application of affinity chromatography to the purification of enzymes that catalyze reactions involving protein substrates has been limited until recently to proteases for which specific inhibitors or ligands are known. In the absence of the availability of highly specific inhibitors or synthetic substrates, an alternative chromatographic method is proposed for the purification of enzymes involved in Ca(II)-dependent protein–protein interactions. This general method employs the lanthanide ions (Nd(III), Gd(III), Pr(III), Tb(III), etc.), whose ionic radii closely resemble that of Ca(II), as substitutes for Ca(II) in Ca(II)-dependent protein complex formation and as inhibitors of Ca(II)-dependent enzyme catalysis.

The interaction of trivalent lanthanide ions with the metal-binding sites of Ca(II)-binding proteins has been demonstrated for albumin,[1,2] trypsinogen,[3] amylase,[4-6] thermolysin,[7] staphylococcal nuclease,[8] and bovine Factor X.[9] Although the Ca(II) requirement of some enzymatic activities may be replaced by lanthanide ions,[5] the lanthanide ions do not promote staphylococcal nuclease activity.[8] It appears that, although the lanthanides competitively inhibit nuclease activity by displacement of Ca(II) from the Ca(II)-binding site of the enzyme, the lanthanides do facilitate the formation of a ternary complex composed of enzyme:metal: substrate. With the nuclease system as a model it appeared feasible to purify by affinity chromatography enzymes that participate in Ca(II)-dependent enzyme–protein interactions with lanthanide ions. By covalently

[1] E. R. Birnbaum, J. E. Gomez, and D. W. Darnall, *J. Amer. Chem. Soc.* **92**, 5287 (1970).
[2] J. Reuben, *Biochemistry* **10**, 2834 (1971).
[3] D. W. Darnall and E. R. Birnbaum, *J. Biol. Chem.* **245**, 6484 (1970).
[4] A. Levitzki and J. Reuben, *Biochemistry* **12**, 41 (1973).
[5] G. E. Smolka, E. R. Birnbaum, and D. W. Darnall, *Biochemistry* **10**, 4556 (1971).
[6] D. W. Darnall and E. R. Birnbaum, *Biochemistry* **12**, 3489 (1973).
[7] P. M. Colman, L. H. Weaver, and B. W. Matthews, *Biochem. Biophys. Res. Commun.* **46**, 1999 (1972).
[8] B. Furie, A. Eastlake, A. N. Schechter, and C. B. Anfinsen, *J. Biol. Chem.* **248**, 5821 (1973).
[9] B. C. Furie and B. Furie, *Fed. Proc., Fed. Amer. Soc. Exp. Biol.* **33**, abstr. 2066 (1974).

coupling one of the protein pair to agarose, the other protein might be purified by exploiting the high specificity of the interaction between the two proteins and the nonproductive binding of the proteins in the presence of lanthanide ions. As a model to test this general hypothesis, the coagulant protein of Russell's viper venom has been purified by affinity chromatography using bovine factor X.

Factor X is a circulating plasma zymogen which participates in an intermediate step of the blood coagulation cascade. In addition to physiological modes of activation, factor X may be activated by a coagulant protein in Russell's viper venom. The reaction of coagulant protein on factor X is proteolytic, has an absolute requirement for Ca(II), and proceeds at physiological pH. Lanthanide ions are competitive inhibitors of the Ca(II)-mediated factor X activation by the coagulant protein; a K_i of 1 to 4 μM has been estimated.[9] The single-step purification of the coagulant protein from crude Russell's viper venom described below is based on the strategy that lanthanide ions facilitate binding of coagulant protein to factor X, but do not promote catalysis, i.e., activation of factor X.

Bovine factor X, prepared by conventional chromatographic techniques,[10] was coupled to agarose by the methods of Cuatrecasas et al.[11] Activated agarose was prepared with 300 mg of cyanogen bromide and 2 ml of washed Sepharose 4B. Three milliliters of factor X (2.5 mg/ml) in 20 mM potassium phosphate–0.13 M NaCl at pH 6.8, were added to the activated Sepharose suspension, and the mixture was stirred for 18 hours at 4°. The Sepharose–factor X conjugate was washed sequentially in a sintered-glass funnel with 10 ml each of 0.1 M ammonium acetate at pH 7.1; 1 M NaCl; 6 M guanidine·HCl; 25 mM imidazole–0.15 M NaCl–10 mM–NdCl$_3$ at pH 6.8; and 25 mM imidazole–0.15 M NaCl–10 mM EDTA adjusted to pH 6.8.

A 0.7 × 3-cm column of Sepharose–factor X was equilibrated at 4° with 25 mM imidazole–0.5 M NaCl–10 mM NdCl$_3$ at pH 6.8. Crude Russell's viper venom, 20 mg, dissolved in 1 ml of 25 mM imidazole–0.5 M NaCl at pH 6.8, was dialyzed for 2 hours at 4° against 25 mM imidazole, 0.5 M NaCl, pH 6.8. The sample was then made 10 mM with respect to NdCl$_3$ and incubated at 4° for 2 hours. The precipitate which formed was removed by centrifugation and the supernatant fluid, containing essentially all of the original coagulant protein activity, was charged onto the column. Most of the crude venom protein did not adhere to the Sepharose–factor X column; the bound protein, about 5–10% of the total recovered protein, was eluted with 10 mM EDTA–25 mM imidazole–0.5 M NaCl, at

[10] M. P. Esnouf, P. H. Lloyd, and J. Jesty, Biochem. J. 131, 781 (1973).
[11] P. Cuatrecasas, M. Wilchek, and C. B. Anfinsen, Proc. Nat. Acad. Sci. U.S. 61, 636 (1968).

pH 6.8. This protein fraction showed an average 10-fold increase (range = 8- to 12-fold increase) in specific activity of the coagulant protein and appeared to be electrophoretically homogeneous. Of the 75% of the original coagulant activity recovered, 40% of the original coagulant protein activity was associated with the bound protein fraction. Application of excess quantities of crude venom to the column maximized the amount and the specific activity of the coagulant protein associated with the bound protein fraction. None of the crude venom protein was bound to the Sepharose–factor X column in the absence of metal ions. The Sepharose–factor X column may be recycled into starting buffer and used repeatedly.

Comments

Systematic study of the optimal conditions for affinity chromatography has suggested the importance of certain parameters. Chromatography at 23° yielded material with lower specific activity than when chromatography was performed at 4°. Nonspecific protein binding could be minimized by the use of buffers containing 0.5 M NaCl; the occasional removal of non-covalently bound protein from the Sepharose–factor X column with 6 M guanidine after multiple exposures of the column to crude protein preparations similarly reduced nonspecific protein binding. It is unclear which of these findings will be generally true for other enzyme systems; in each case optimal conditions for affinity chromatography should be determined.

The application of this method to the purification of enzymes involved in Ca(II)-dependent protein–protein interactions requires that (1) lanthanide ions competitively inhibit catalysis during enzyme–protein substrate interaction, (2) one protein of the protein pair must be available in purified or at least partially purified form for coupling to Sepharose, (3) the association constant of the protein complex must be sufficiently high at a pH of 7.0 or lower to permit tight binding (the lanthanide ions form insoluble hydroxides at alkaline pH), (4) the protein to be purified must not be precipitated by the lanthanide ion concentrations employed. This approach may prove to be useful for the purification of the components of Ca(II)-requiring ternary complexes or the isolation and identification of substrates of Ca(II)-dependent enzymes.

[76] Separation of Isozymes

By KLAUS MOSBACH

The separation of enzymes that occur in multimolecular forms into their iso(en)zymes has until recently been carried out by methods based primarily on differences in the charge on the enzyme molecules.[1,2] It is also possible to use specific ligands, i.e., affinity chromatography; separation of isozymes by this approach has recently been accomplished. A detailed procedure is given below for the separation of L-lactate dehydrogenase into its five isozymes.[3]

Affinity chromatography has also been used recently for the separation of carbonic anhydrase into its isozymes B and C with sulfanilamide coupled to Sepharose.[4] The stronger affinity of halide ions for enzyme form B, relative to the bound inhibitor sulfanilamide, allows form B to be eluted with sodium iodide. Complete separation was obtained on eluting successively with 0.1 M NaI and 10 mM KCNO.

The ligand used for the separation of LDH isozymes is N^6-(6-amino-hexyl)-AMP. The preparation of this material and its binding to Sepharose 4B, is described in a previous article.[5] On application of a mixture of proteins to the column, the five isozymes of lactate dehydrogenase remain bound to the ligand, as expected from the K_i values of about 0.1 mM obtained for free N^6-(6-aminohexyl)-AMP. Elution of the lactate dehydrogenases from the AMP-column is subsequently effected with a gradient of NADH at a low concentration. This elutes the five isozymes in the order H_4, H_3M, H_2M_2, HM_3, M_4 (H = heart type, M = muscle type) in accordance with reported dissociation constants for the binary complexes: H_4–NADH = 0.39 μM[6] and M_4–NADH = 1.4–2.0 μM.[7] It is noteworthy that an affinity chromatography separation can be obtained when the dissociation constants between the two extreme forms, H_4 and M_4, differ only by a factor of 4. However, in addition, in each case there are differences not only in the binding strength of the isozyme-coenzyme complex but probably also between that of isozyme and AMP-analog. The dual

[1] J. H. Wilkinson, "Isoenzymes." Spon, London, 1965.
[2] C. L. Markert, *Ann. N.Y. Acad. Sci.* **151**(1), 14 (1968).
[3] P. Brodelius and K. Mosbach, *FEBS Lett.* **35**, 223 (1973).
[4] S. O. Falkbring, P. O. Göthe, P. O. Nyman, L. Sundberg, and J. Porath, *FEBS Lett.* **24**, 229 (1972).
[5] K. Mosbach, this volume [16].
[6] S. R. Anderson and G. Weber, *Biochemistry* **4**, 1948 (1965).
[7] R. A. Stinson and J. J. Holbrook, *Biochem. J.* **131**, 719 (1973).

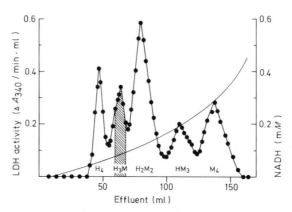

FIG. 1. Elution of lactate dehydrogenase (LDH) isozymes with a concave gradient of NADH. Protein, 0.2 mg, in 0.2 ml of 0.1 M sodium phosphate buffer, pH 7.0, 1 mM β-mercaptoethanol, and 1 M NaCl was applied to an AMP-analog–Sepharose column (140 × 6 mm, containing 2.5 g of wet gel) equilibrated with 0.1 M sodium phosphate buffer, pH 7.5. The column was washed with 10 ml of the latter buffer, then the isozymes were eluted with a concave gradient of 0.0 to 0.5 mM NADH in the same buffer, containing 1 mM β-mercaptoethanol. Fractions of 1 ml were collected at a rate of 3.4 ml per hour. The hatched area indicates the pooled fractions that were rechromatographed.

Rechromatography of the H_3M isozyme peak (not shown here) was likewise carried out with a linear gradient of NADH. Dialyzed enzyme solution, 4.0 ml, was applied to the above AMP-analog–Sepharose column equilibrated with 0.1 M sodium phosphate at pH 7.5. The isozyme was eluted with a linear gradient of 0.0–0.2 mM NADH in the same buffer, containing 1 mM β-mercaptoethanol. Data from P. Brodelius and K. Mosbach, FEBS Lett. 35, 223 (1973).

operation of these affinity properties ensures high specificity in gradient elution. Five peaks were obtained, corresponding to the five isozymes found by polyacrylamide gel electrophoresis (Fig. 1). Both H_4 and M_4 appear to be homogeneous, but the remaining isozymes are somewhat contaminated with those from the preceding peak. Rechromatography of the pooled fractions of the H_3M peak yields homogeneous H_3M, as shown by gel electrophoresis. The results obtained illustrate the great sensitivity inherent in the affinity chromatography technique. It should be possible to apply such isozyme separation in clinical diagnosis. One advantage in this context is that isozyme mixtures containing inactive isozyme will be easily resolved even if the latter are only modified in their active sites since these will not adsorb to the column material and therefore will not interfere with the elution pattern. In addition to cofactor–gradient elution, other elution procedures presented elsewhere[5] may be useful in isozyme separations.

Procedure

Affinity Material. The AMP-analog, N^6-(6-aminohexyl)-AMP, was bound to Sepharose 4B as described.[5] It contained about 150 μmoles of nucleotide per gram of dry polymer.

L-*Lactate Dehydrogenase Isozyme Mixture.* A mixture of beef heart lactate dehydrogenase (35 μg, mainly H_4 and H_3M), beef muscle enzyme (40 μg, mainly M_4) and hybridized enzyme (300 μg, containing also H_2M_2 and H_3M) is prepared. The hybridized enzyme is made following the "quick-freeze"–"slow-thaw" method[8] by dissolving heart lactate dehydrogenase (0.45 mg) and muscle lactate dehydrogenase (0.55 mg) together in 0.4 ml of 0.1 M sodium phosphate at pH 7.0 containing 1 mM β-mercaptoethanol. The preparation is dialyzed against the same buffer (500 ml) for 5 hours and is diluted with the same buffer, containing 2 M NaCl, to give a protein concentration of 1 mg/ml. This solution is rapidly frozen with a mixture of acetone and dry ice. It is then allowed to thaw slowly at room temperature to yield about 90% of the initial enzyme activity.

Specific Elution. A concave gradient of NADH is applied using two parallel-sided containers with a cross section area ratio of 2 (mixing chamber):1 (reservoir containing NADH). This concave gradient is used so as to shorten the total time of elution although roughly the same pattern is obtainable with a linear gradient.

The fractions containing H_3M are rechromatographed after dialysis for 5 hours against 0.1 M sodium phosphate at pH 7.5, containing 1 mM β-mercaptoethanol, to remove nucleotide. Although it is to be expected that even after dialysis the enzyme will be present as binary complex, the enzyme will bind to the AMP-analog–Sepharose because of the relatively high concentration of the analog in the column.

Evaluation

The reproducibility of the procedure described, as indicated by the elution pattern obtained, is good. For instance, the isozyme H_3M was eluted in three successive runs at the same NADH molarity, 95, 98, and 96 μM, respectively. The recovery of enzyme activity is about 60%. The loss in activity is probably due to partial denaturation as a result of the dilute solution. The total amount of NADH used for the 0.0–0.4 mM gradient is 20 mg. A recent successful separation of the isozymes of liver alcohol dehydrogenase by this technique[9] suggests that the procedure described will have wider applicability.

[8] O. P. Chilson, L. A. Costello, and N. O. Kaplan, *Biochemistry* 4, 271 (1965).
[9] L. Andersson, H. Jörnvall, Å. Åkeson, and K. Mosbach, to be published.

[77] Lactate Dehydrogenase: Specific Ligand Approach[1]

By PÁDRAIG O'CARRA and STANDISH BARRY

Rationale and General Comments

One approach to affinity chromatography of NAD-dependent dehydrogenases involves the use of immobilized analogs of the common or general substrate, NAD. This approach has been termed "general-ligand affinity chromatography."[2] An alternative approach involves the use of immobilized analogs of the specific substrates (e.g., lactate or pyruvate in the case of lactate dehydrogenase). This latter approach requires the development of a different immobilized ligand for each dehydrogenase, but it is potentially much more specific than the general-ligand approach.[3] Both approaches have been applied in affinity chromatography of lactate dehydrogenase,[2-7] but in our experience the specific-ligand approach, using an immobilized pyruvate analog, has yielded the most clear-cut and efficient purifications.[3]

In the search for an effective immobilized ligand, attention was concentrated on analogs of pyruvate rather than of lactate, since kinetic studies[8] indicate that pyruvate has considerably higher affinity for the enzyme. As explained below, the kinetic mechanism of the enzyme requires the addition of NADH to the irrigating buffer during affinity chromatography. Under these conditions catalytically susceptible immobilized analogs (e.g., β-mercaptopyruvate, attached through the β-thiol group) underwent enzymatic reduction when the enzyme was applied, and such analogs proved to be unsatisfactory for affinity chromatography.[9] Oxamate is a structural analog of pyruvate (cf. Fig. 1) which is not catalytically susceptible, but retains the binding characteristics of pyruvate for lactate dehydrogenase.[10,11] This analog, immobilized through the amide grouping

[1] Based on research supported by grants from the Medical Research Council and National Science Council of Ireland.
[2] K. Mosbach, H. Guilford, R. Ohlsson, and M. Scott, *Biochem. J.* **127**, 625 (1972).
[3] P. O'Carra and S. Barry, *FEBS Lett.* **21**, 281 (1972).
[4] R. Ohlsson, P. Brodelius, and K. Mosbach, *FEBS Lett.* **25**, 234 (1972).
[5] K. Mosbach, this volume [76].
[6] S. Barry and P. O'Carra, *Biochem. J.* **135**, 595 (1973).
[7] C. R. Lowe and P. D. G. Dean, *FEBS Lett.* **14**, 313 (1971). See also Dean and Harvey, this volume [17].
[8] A. D. Winer and G. W. Schwert, *J. Biol. Chem.* **231**, 1065 (1958).
[9] S. Barry and P. O'Carra, unpublished work (1973).
[10] W. B. Novoa, A. D. Winer, A. J. Glaid, and G. W. Schwert, *J. Biol. Chem.* **234**, 1143 (1959).

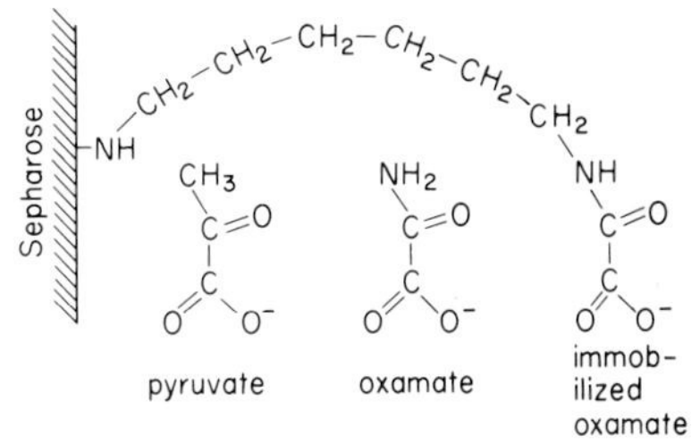

FIG. 1. Structural analogy of pyruvate, oxamate, and the immobilized Sepharose-linked oxamate derivative. From P. O'Carra and S. Barry, *FEBS Lett.* **21**, 281 (1972).

as shown in Fig. 1, proved to be a very effective affinity chromatographic ligand. The amide nitrogen group corresponds to the methyl group of pyruvate and was chosen as the point of attachment because available specificity studies[12] indicated that pyruvate analogs in which the methyl group is substituted remain effective as substrates for lactate dehydrogenase (e.g., α-ketobutyrate and other α-keto acids).

The affinity chromatographic results confirmed the already considerable evidence for a compulsory order of ligand-binding to L-lactate dehydrogenase, with NADH binding first, followed by the pyruvate analog. To achieve biospecific adsorption of the enzyme on the immobilized oxamate gels, NADH must be added to the applied samples and to the irrigating buffer. When the NADH is subsequently omitted from the irrigant, the lactate dehydrogenase is eluted spontaneously.

This behavior renders the method doubly specific, since the adsorption selects the enzyme on the basis of both oxamate specificity *and* NADH specificity. Some of the adsorbed protein remaining after discontinuation of NADH may be eluted by addition of pyruvate to the irrigant. This fraction has not been identified.

The lactate dehydrogenase isoenzymes have similar kinetic properties as regards the formation of the ternary complex with NADH and oxamate or pyruvate. In line with this, no significant distinction between the isoenzymic forms was observable during chromatography on the immobilized oxamate gels with NADH as the "promoter ligand." However, the H-type isoenzyme also binds pyruvate in the presence of NAD+, although much more weakly than in the presence of NADH whereas the M-type does not, or does so to a much lesser degree. The resulting ternary lactate dehydrogenase–NAD+–pyruvate complex is termed the abortive complex,

---

[11] G. W. Schwert, *in* "Pyridine Nucleotide-Dependent Dehydrogenases" (H. Sund, ed.), pp. 135–143. Springer-Verlag, Berlin, Heidelberg, New York, 1970.
[12] A. Meister, *J. Biol. Chem.* **184**, 117 (1950).

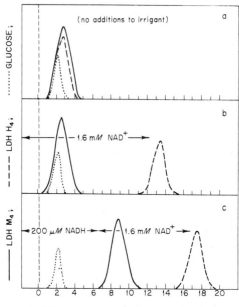

FIG. 2. Separation of the M_4 and H_4 isoenzymic forms of porcine lactate dehydrogenase by chromatography on immobilized oxamate. The irrigant throughout was 0.3 M KCl in 50 mM sodium phosphate at pH 6.8, with the additions indicated on the elution profiles. ———, M_4 isoenzyme; – – –, H_4 isoenzyme; ····, glucose, added to applied samples and monitored in effluent to mark the "straight-through" or "break-through" position [P. O'Carra and S. Barry, *FEBS Lett.* **21**, 281 (1972)]. Data from P. O'Carra, S. Barry, and E. Corcoran, unpublished.

and the difference between the H and M forms with regard to the formation of this complex has been rationalized by Kaplan and co-workers as a key functional property, reflecting distinct biological roles for the two isoenzymic forms.[13,14] The chromatographic behavior of the isoenzymes on the oxamate gel in the presence of NAD+ reflects this kinetic distinction and allows a separation of the M_4 and H_4 types—the M_4 type does not bind at all whereas the H_4 type binds weakly.[15] As explained below, the weak "abortive complex affinity" may be reinforced to allow a clear separation of the two main isoenzymic types as illustrated in Fig. 2b.

Figure 2c illustrates the tactic adopted in the one-step specific isolation of lactate dehydrogenase-M_4. The sample is applied and irrigated in NADH-containing medium so that the total lactate dehydrogenase is

[13] N. O. Kaplan, J. Everse, and J. Admiraal, *Ann. N.Y. Acad. Sci.* **151**, 400 (1968).
[14] J. Everse and N. O. Kaplan, *Advan. Enzymol.* **37**, 61 (1973).
[15] P. O'Carra, S. Barry, and E. Corcoran, *FEBS Lett.,* in press.

strongly adsorbed and separated from "nonspecific protein," most of which is washed through. The NADH in the irrigant is then replaced by NAD+ and the M_4 isoenzyme alone dissociates from the gel and is spontaneously eluted. The H_4 type as well as the hybrid forms containing H-type subunits remain adsorbed and may be subsequently eluted simply by discontinuing the NAD+ in the irrigant. Another affinity chromatographic method allowing the separation of lactate dehydrogenase isoenzymes is described by Mosbach elsewhere in this volume.[5]

To allow full operation of the biospecific element in affinity chromatography, careful control of nonspecific adsorption effects is essential.[16,17] In the present instance, biospecific adsorption of lactate dehydrogenase is obscured by nonspecific adsorption of the enzyme and of many other, unrelated proteins on the affinity gel unless the irrigating buffer is maintained at a high ionic strength. Addition of 0.5 M NaCl or KCl to the buffer seems to largely eliminate such nonspecific binding of lactate dehydrogenase so that the binding depends entirely on bioaffinity.[18] Under such conditions, the bioaffinity in the continuing presence of NADH is strong enough to allow the columns to be loaded with samples many times the volume of the column bed and then washed with several column-volumes of irrigant without loss of any significant proportion of the lactate dehydrogenase in the effluent.

However, the "abortive-complex affinity" of the H-type isoenzyme is very much weaker, and practical separations based on this phenomenon are greatly improved by reinforcement of the bioaffinity. The affinity is stronger at lower temperatures so that the isoenzyme separation should be carried out at 0° to 4°. Further reinforcement is achieved by taking advantage of the compound-affinity effect.[6,16,17] The ionic strength is judiciously balanced so that the bioaffinity is synergistically reinforced by nonbiospecific interactions which are themselves insufficiently strong to cause significant adsorption on their own. With most of the gels used in the present work, a very effective compound-affinity was achieved by decreasing the NaCl or KCl concentration in the irrigant to about 0.3 M.[18] At this ionic strength, the H_4 isoenzyme was bound much more strongly than at the higher salt concentration although no detectable affinity was

---

[16] P. O'Carra, in FEBS Symposium on "Industrial Aspects of Biochemistry." (B. Spenser, ed.), p. 107. North-Holland Publ., Amsterdam (1974).

[17] P. O'Carra, S. Barry, and T. Griffin, this volume [8].

[18] With some highly substituted gel preparations, the degree of nonbiospecific adsorption has considerably exceeded that encountered in the studies on which this article was based. To achieve satisfactory results in such cases, it has been necessary to increase the KCl concentrations in the irrigant, sometimes to as high as double those described in this article.

evident in the absence of NAD$^+$; the isoenzyme was eluted from the gels immediately after NAD$^+$ was discontinued in the irrigant. The M$_4$ isoenzyme remained unadsorbed both in the presence and absence of NAD$^+$. The results illustrated in Fig. 2 were obtained at this salt concentration and reflect compound affinity rather than pure bioaffinity.

The high degree of specificity of this affinity chromatographic method has allowed the one-step isolation of lactate dehydrogenase from crude extracts of tissues such as placenta,[3] skeletal muscle, heart muscle, and kidney,[19] in almost quantitative yields. The method has not as yet been applied on a large scale, but has been mainly used for the small-scale isolation of human lactate dehydrogenase. Placenta is by far the most readily available human source of this enzyme, but conventional purification methods had failed to produce the highly purified lactate dehydrogenase preparation from extracts of this tissue.[20] The affinity chromatographic method was effective in completing the purification of the isoenzymic forms after initial separation by ion-exchange chromatography.[21] It also allowed the isolation of the mixed isoenzymes or of the M$_4$ isoenzyme directly from ammonium sulfate fractions of placental extracts or from the unfractionated crude extracts. Direct isolation from crude placental extracts has the advantage of minimizing problems of instability with the M$_4$ isoenzyme from this source,[20] but there are indications that such crude extracts may contain factor(s) that gradually destroy the effectiveness of the affinity gel preparations.

This affinity system has also been found useful as a negative step in the purification of an unusual and unstable form of pyruvate kinase found in human placenta.[22] Conventional separation methods failed to free this enzyme from contaminating lactate dehydrogenase, but the separation was readily achieved by passage of the preparations through the oxamate gel in the presence of NADH.

Preparation and Stability of the Affinity Gel

The immobilized oxamate derivative illustrated in Fig. 1 is prepared by first coupling diaminohexane to agarose and then condensing the terminal amino group with an oxalate molecule via an amide bond. This condensation is achieved by activating the oxalate with a water-soluble carbodiimide. The "activated oxalate" is very unstable, decomposing apparently to CO and CO_2, and, owing to this side reaction, the preparative

---

[19] D. Nugent and P. O'Carra, (1973), unpublished work.
[20] C. M. Spellman and P. F. Fottrell, personal communications.
[21] C. M. Spellman and P. F. Fottrell, *FEBS Lett.* **21**, 186 (1972).
[22] C. M. Spellman and P. F. Fottrell, *Commun. FEBS Spec. Meet., Dublin* (1973), Abstract No. 91.

procedure as originally described[3] gives somewhat erratic results. More consistent results are obtained by carrying out the reaction at pH 7.5 rather than pH 4.7 and by increasing the concentrations of oxalate and carbodiimide used. Even with these modifications, complete substitution is not always achieved. The degree of substitution may be checked by subjecting a little of the gel product to the trinitrobenzene sulfonate test for free amino groups.[23,24] If the gel gives a considerable positive reaction, i.e., an orange color, the coupling with oxalate should be repeated.

In the modified preparative procedure, all operations are performed at 15–20° unless otherwise specified. Sepharose 4B (Pharmacia) is activated with 200 mg of CNBr per milliliter of packed gel and coupled with 230 mg of diaminohexane per milliliter of gel.[23] The resulting aminohexyl-Sepharose gel is washed with water, and to each milliliter of gel is added 224 mg of potassium oxalate dissolved in 1 ml of water, followed by 148 mg of 1-ethyl-3-(3-dimethylaminopropyl) carbodiimide, dissolved in 0.3 ml of water immediately before use and added dropwise over a period of about 5 minutes. Over this period, and for the following hour, the mixture is stirred gently and the pH is maintained at 7.5 with 0.5 N HCl. The pH tends to drift upward owing to decomposition of a proportion of the activated oxalate, which also gives rise to some bubbling. The mixture is stirred gently for a further 20 hours at room temperature without further adjustment of the pH. The gel is finally washed on a sintered glass funnel with about 100 ml of 0.1 M NaCl per milliliter of gel. The gel is further washed with distilled water and stored at 2–4°.

Unused gel preparations showed no deterioration after 6 months' storage at 4°. The stability did not seem to be significantly affected by repeated use of the gels for the chromatography of partly purified lactate dehydrogenase preparations or of crude extracts from some tissues, e.g., heart muscle. However, considerable deterioration of chromatographic effectiveness was observed on several occasions after exposure of the gels to crude extracts from some other tissues, notably fresh placenta. Such deterioration was not invariably observed, and when it occurred it was a slow process that did not seriously affect the immediate experiment. Effectiveness was not restored by washing the gels with 8 M urea, indicating that the deterioration is unlikely to be attributable to masking effects of nonspecifically adsorbed protein. It seems possible that this erratic damage may be caused by degradation of the immobilized ligand by unidentified enzyme(s) in these particular extracts.[16]

[23] P. Cuatrecasas, *J. Biol. Chem.* **245**, 3059 (1970). See also Inman, this volume [3].
[24] P. Cuatrecasas and C. B. Anfinsen, this series, Vol. 22 [31].

Chromatographic Procedures

Columns of the affinity gel (7 mm internal diameter, 4 ml bed volume) have been used for most of the isolations hitherto performed, but operations can be scaled-up without difficulty. The gel presents no particular packing problems. When the 4-ml columns are operated under a 25-cm hydrostatic head of pressure, a flow rate of about 1 column-volume per 10 minutes is obtained with no detectable slowing after weeks of operation; 0.5 M KCl (or NaCl) in 50 mM potassium phosphate at pH 6.8 is regarded as the standard irrigant.[18] In preparation for chromatography, columns are equilibrated to temperature and irrigated with at least 5 column-volumes of this irrigant followed by at least 1 column-volume of the standard irrigant containing 200 μM NADH. Samples for chromatography are applied in the same NADH-containing medium.

Where it is desired to chromatograph a tissue extract as rapidly and directly as possible, the tissue is homogenized in an equal volume of the standard irrigant and the homogenate is clarified by centrifugation at 35,000 g for 30 minutes. The pH of the supernatant liquid is adjusted to 6.8, the KCl concentration is adjusted to approximately 0.5 M, and NADH is added to give a final concentration of about 300 μM. To minimize possible destruction of the NADH by the extracts, this last addition should be made just before the sample is applied to the column. We find it generally preferable to subject the clarified homogenates to an ammonium sulfate fraction step[25] before affinity chromatography, both to concentrate the lactate dehydrogenase and to ensure removal of microsomal and other particulate material. The packed ammonium sulfate precipitate is redissolved in a minimal volume of the standard irrigant, the pH is adjusted to 6.8, and any undissolved material is removed by centrifugation. NADH is added just before application of the sample to the column. Lyophilized material, acetone powders, etc., are similarly prepared for application to the affinity columns.

The applied samples are allowed to sink into the column bed and are washed in with a little of the standard irrigant containing 200 μM NADH. Very viscous samples containing a high concentration of protein may be difficult to apply to the columns, and such samples are better diluted (with the NADH-containing irrigant) to reduce the viscosity to a manageable level. Samples of over 10 column-volumes may be applied without significant loss of lactate dehydrogenase in the effluent. After the last of the applied sample has been washed into the column, irrigation is continued with the NADH-containing irrigant for at least 3 column-volumes. If the NADH is then omitted from the irrigant and irrigation is continued with

[25] F. Stolzenbach, this series, Vol. 9 [53].

phosphate before being applied to the column, no acetylcholinesterase is trapped.

There is also a steady slow release of a potent inhibitor from the gel. This inhibitor has a rate constant for inhibition of eel acetylcholinesterase of 10^6 M^{-1} min^{-1} and is released from the gel, depending strongly on the particular gel preparation, at a rate of about $10^{-2}\%$ of the gel ligand per minute. Enzyme inhibited by this soluble inhibitor is readily reactivated by 2PAM. The inhibitor released from the column presumably contains some galactose and anhydrogalactose residues, because gel filtration and sucrose gradient sedimentation indicate a broad molecular weight range much higher than that of the ligand and arm alone.

A solution of 2PAM releases all p-nitrophenol from the column, but little enzyme. Thus the anticipated release of enzyme does not occur. However, a column washed with 2PAM, on standing in the presence of 2PAM, releases active acetylcholinesterase at the rate of about 5–20% per day at room temperature; in the absence of 2PAM inactive enzyme is released at a comparable rate. The inactive enzyme is readily reactivated by 2PAM.

It appears that all trapped protein is released in this manner. Chymotrypsin is released at the rate of about 25–50% per day. It is possible that acetylcholinesterase (eel) is released more slowly because it has four active sites per molecule (11 S) whereas chymotrypsin has only one. The slower rate of release of acetylcholinesterase can be explained by assuming reaction of the ligand with more than one site per molecule of acetylcholinesterase.

The scheme for the release of enzyme is indicated in Fig. 3.

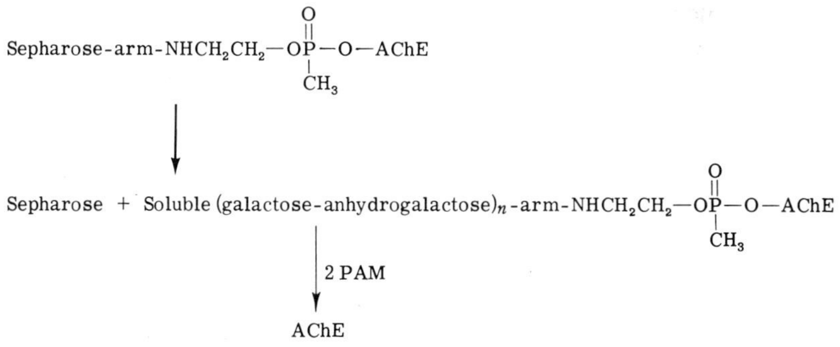

$$Sepharose\text{-}arm\text{-}NHCH_2CH_2\text{-}O\underset{\underset{CH_3}{|}}{\overset{\overset{O}{\|}}{P}}\text{-}O\text{-}AChE$$

$$\downarrow$$

$$Sepharose + Soluble\ (galactose\text{-}anhydrogalactose)_n\text{-}arm\text{-}NHCH_2CH_2\text{-}O\underset{\underset{CH_3}{|}}{\overset{\overset{O}{\|}}{P}}\text{-}O\text{-}AChE$$

$$\downarrow\ 2\ PAM$$

$$AChE$$

FIG. 3. Actual release mechanism. Step 2 of Fig. 2, release of trapped acetylcholinesterase (AChE) by 2PAM, is inoperative. As shown here the inhibited AChE is released spontaneously from the column and can then be reactivated by 2PAM. The presence of sugar residues on the inhibitor is implied from investigations into the nature of the soluble inhibitor released from these columns (see text).

the standard irrigant alone, the total lactate dehydrogenase is sharply eluted from the column, beginning to emerge 1 column-volume later. This chromatographic isolation of total lactate dehydrogenase has been successfully performed at temperatures ranging from 2° to 20°, but where it is desired to isolate the M_4 isoenzyme separately, the affinity chromatography is best carried out at 2–4°. The preparation and application of the sample, and the initial irrigation of the column with NADH-containing irrigant are performed in the same way as described above, but at the point where the NADH was previously omitted, it is now replaced by 1.6 mM NAD⁺. At the same time the KCl concentration is decreased to a predetermined level which allows some reinforcement of the NAD⁺-dependent binding of the H-type isoenzyme without allowing nonspecific adsorption of the M_4 isoenzyme. The precise concentration of KCl necessary to achieve this balance varies with the degree of substitution of the affinity gel; 0.3 M KCl in 20 mM potassium phosphate at pH 6.8 has been found satisfactory with most of the gels used in our work, but these were of a relatively low substitution level. The behaviour of each gel should be checked experimentally in advance and the KCl concentration be adjusted if necessary.[18] The change to NAD⁺-containing irrigant results in the elution of the M_4 isoenzyme 1 column-volume later. The other isoenzymes remain adsorbed, but the hybrid forms usually begin to leak off the column slowly. The hybrids together with the H_4 isoenzyme are eluted sharply 1 column-volume after the NAD⁺-containing irrigant is discontinued and replaced by the standard irrigant.

Note Added in Proof

Since this article was written, an adaptation of the immobilized oxamate method allowing the separation of the testicular isoenzyme X from the other lactate dehydrogenase isoenzymes has been described by H. Spielmann, R. P. Erickson, and C. J. Epstein [*FEBS Lett.* **35**, 19 (1973)]. These authors also note difficulties in the preparation of the affinity gel and introduce modifications in the synthetic procedures analogous to those described here.

[78] Phosphodiesterase from Snake Venom[1]

By F. ECKSTEIN and A. M. FRISCHAUF

Phosphodiesterases have been isolated from a variety of snake venoms[2,3] as well as other sources.[4–6] In many cases these enzymes have not been thoroughly characterized because of difficulty in obtaining

[1] A. M. Frischauf and F. Eckstein, *Eur. J. Biochem.* **32**, 479 (1973). The table is taken from this reference with permission of the copyright owner.

homogeneous preparations. Since we had made the observation that dithymidine 3′,5′-thiophosphate was resistant to action by snake venom and to spleen phosphodiesterases,[7] we have attempted to couple a thiophosphate diester to agarose with the intent of using this system for the purification of phosphodiesterases. Although initial experiments with 5′-deoxy-5′-aminothymidine 3′-O-thiophosphate 5′-O-thymidine ester (K_i = 10 μM) coupled to Sepharose looked promising for such an approach, we sought a more readily accessible compound as a ligand. One such compound, O-(4-nitrophenyl) O′-phenyl-thiophosphate proved to be resistant to phosphodiesterases and was a competitive inhibitor with K_i = 10 μM for snake venom diesterases. Its synthesis is simple, and it can readily be reduced to the amino derivative necessary for coupling to agarose by known procedures.[8,9]

However, attempts at obtaining a homogeneous enzyme from crude *Bothrops atrox* venom by affinity chromatography as the sole purification step failed. Purification factors of only 30 could be obtained. As explanation for this limited effect, we suggest the occurrence of nonspecific binding of other basic proteins in the venom to the anionic substituted agarose.

By subjecting the crude enzyme preparation to prior chromatography on phosphocellulose, the subsequent use of an affinity column allowed an overall purification factor of about 200. Since the specific activity of the phosphodiesterases from different batches of venom was not constant, overall purification factors varied. The final specific activity, however, was always the same. That the purification on the affinity column was really due to a specific interaction between enzyme and anionic ligand was indicated by the failure of unsubstituted agarose, as well as of phosphocellulose, to retain the enzyme in the presence of 0.1 M NaCl.

[2] M. Laskowski, *in* "Procedures in Nucleic Acid Research" (G. L. Cantoni and D. R. Davies, eds.), Vol. 1, pp. 154–187. Harper & Row, New York, 1966.

[3] M. Laskowski, Sr., *in* "The Enzymes" (P. D. Boyer, ed.), 3rd ed., Vol. IV, pp. 313–328. Academic Press, New York, 1971.

[4] Reviewed *in* "The Enzymes" (P. D. Boyer, ed.), 3rd ed., Vol. IV. Academic Press, New York, 1971.

[5] C. Harvey, K. C. Olson, and R. Wright, *Biochemistry* 9, 921 (1970).

[6] B. Lerch and G. Wolf, *Biochim. Biophys. Acta* 258, 206 (1972).

[7] F. Eckstein, *J. Amer. Chem. Soc.* 92, 4718 (1970).

[8] P. Cuatrecasas, *J. Biol. Chem.* 245, 3059 (1970).

[9] P. Cuatrecasas, M. Wilchek, and C. B. Anfinsen, *Proc. Nat. Acad. Sci. U.S.* 61, 636 (1968).

We also attempted the purification of the diesterase from *Crotalus adamanteus* venom by the same procedure. Affinity chromatography alone yielded a purification of about 5-fold as compared to 30-fold with *Bothrops atrox* venom. The combination of phosphocellulose and affinity chromatography resulted in a purification of about 100 although the enzyme was not homogeneous.

Methods

Assays of phosphodiesterase activity using either thymidine 5'-(4-nitrophenyl)phosphate or bis(4-nitrophenyl)phosphate as substrate were performed as described.[1] Protein concentrations were estimated according to Warburg.[10] In crude *Bothrops atrox* venom, the 280:260 ratio was 1.4 to 1.5, and in the pure enzyme, 1.75. It was assumed that 1 A_{280} unit corresponds to 1 mg of protein.

Synthesis of Substituted Sepharose

Synthesis of O-(4-Nitrophenyl) O'-phenyl-thiophosphate. A solution of 1.69 g of phenol (18 mmoles) in 15 ml of anhydrous dioxane and 1.3 ml of pyridine is added dropwise to a stirred solution of 4.6 g of O-(4-nitrophenyl)-thiophosphoryl chloride[11] (18 mmoles) in 30 ml of anhydrous dioxane at 35°. After stirring overnight the precipitate is filtered and a solution of 5.8 ml of pyridine (73 mmoles) in 15 ml of water is added to the filtrate with cooling by ice. The solution is taken to dryness, and the residue is dissolved in $ChCl_3$–MeOH (95:5, v/v).

The solution is chromatographed on a silica gel column (40 × 6.5 cm) by elution first with $CHCl_3$–MeOH (95:5, v/v). On appearance of nitrophenol in the eluate, the eluent is changed to chloroform–methanol (80:20, v/v). The yield is approximately 30%. Fractions are analyzed by thin-layer chromatography on SiO_2 plates with acetone–benzene–water (8:2:1, v/v).

The compound has the same mobility as dithymidine 3',5'-phosphate on electrophoresis at pH 7.4. Spraying the electropherogram with 1 N NaOH and heating in a drying oven results in development of a yellow color coinciding with an ultraviolet-active spot. A sulfur reagent[12] also yields a positive reaction. The following physical constants were found: λ_{max} (water) = 290 nm (ϵ = 18,500); R_f on SiO_2 plates (benzene–acetone–water, 8:2:1, v/v) 0.76.

Synthesis of O-(4-Aminophenyl) O'-phenyl-thiophosphate. O-(4-Ni-

---

[10] O. Warburg and W. Christian, *Biochem. Z.* **310**, 384 (1942).
[11] M. C. Tolkmith, *J. Org. Chem.* **23**, 1685 (1958).
[12] One gram of NaN_3 and 1 g of water-soluble starch dissolved in 100 ml of water; iodine is added until a deep color remains.

trophenyl) O'-phenyl-thiophosphate (11,500 A_{290} units, 0.6 mmole) is dissolved in 45 ml of water. Palladium:charcoal catalyst (180 mg) is added, and the system is hydrogenated under atmospheric pressure for 4 hours. The catalyst is removed by filtration. The filtrate is stored overnight, twice decolorized by addition of charcoal, shaken, filtered, taken to dryness, and dissolved in 10 ml of 0.1 M $NaHCO_3$ at pH 9.0.

The electrophoretic mobility at pH 7.4 was the same as that of the nitrocompound; at pH 3.5 it was 25% of that of the nitrocompound. R_f on SiO_2 plates (benzene–acetone–water, 8:2:1, v/v) was 0.51. The product gave one positive spot with ninhydrin coincident with that observed with the sulfur spray. The product decomposed on standing and was therefore used directly for coupling to activated Sepharose.

Coupling to Sepharose. Sepharose was activated with CNBr essentially as previously described.[8,9] CNBr, 1.5 g in 5 ml of water is added to 10 ml of washed Sepharose 4B in 10 ml of water. The pH is brought to 11 with 2 N NaOH and maintained between 10.5 and 11 by addition of 2 N NaOH. To keep the temperature below 20°, small pieces of ice are added. After about 15 minutes, an equal volume of ice is added and the suspension is filtered. The activated Sepharose is washed with 300 ml of cold 0.1 M $NaHCO_3$ at pH 9.0. The inhibitor, 200 μmoles, is dissolved in 10 ml 0.1 M $NaHCO_3$ at pH 9.0 and the washed activated Sepharose added to it. The Sepharose is stirred for approximately 24 hours at 4°, then filtered and washed with 200 ml of water. To determine the amount of inhibitor coupled to Sepharose, a phosphate analysis is carried out before and after coupling. The extent of coupling is usually in the range of 5 to 20 μmoles/ml of Sepharose. The yield of the reaction is between 30 and 40%. Before use for enzyme purification, the column is washed with 500 ml of 0.1 M ethanolamine at pH 9.0 in order to inactivate remaining reactive groups on the polymer.

Purification

Phosphocellulose Chromatography of Phosphodiesterase from Bothrops atrox Venom

One gram of *Bothrops atrox* venom (Sigma) is dissolved in 50 mM Tris chloride at pH 8.2 and applied to a column (3.5 × 40 cm) of phosphocellulose P-11 equilibrated with the same buffer at 4°. The column is washed with this buffer until the absorbance at 254 nm begins to fall. A linear gradient of 50 mM Tris chloride (pH 8.2) and 50 mM Tris chloride (pH 8.2)–0.5 M NaCl (800 ml each) is applied. Fractions of 6 ml are collected. Most of the diesterase activity appears before the bulk of the protein. Fractions 158–190 are pooled, diluted with half the volume of 50 mM Tris base, and brought to pH 9.15 (4°C) with 2 N NaOH.

Affinity Chromatography

This solution, approximately 300 ml, is applied to a column (2.5 × 9 cm) of O-(4-aminophenyl) O'-phenylthiophosphate coupled to Sepharose (45 ml) and equilibrated with 50 mM Tris chloride at pH 9.15. Essentially all of the diesterase activity is adsorbed. The column is washed with 1 volume of the same buffer made 0.1 M in NaCl and then eluted with a linear gradient consisting of 120 each of 0.1 M NaCl–50 mM Tris chloride at pH 9.15, 1.0 M NaCl–50 mM Tris chloride at the same pH. Fractions of 3.2 ml are collected, and fractions 104 through 120 are pooled. They contain about 70–80% of the applied activity at an average purification of 10-fold. Purification is usually better in cases in which the purification factor of the prior phosphocellulose step is smaller than 20. The total purification factor for both steps is generally about 200.

Concentration

The eluate from the inhibitor-Sepharose column, about 50 ml, is concentrated by centrifugation in 4 tubes for 12 hours at 4° and 60,000 rpm in a Ti-65 rotor (L2-65-B Beckman centrifuge) with 0.2 ml of glycerol per tube as a "cushion." Approximately 9 ml of the solution is removed with a pipette. More than 90% of the activity is found in the remaining 0.5 ml. The protein concentration of this solution is between 0.5 and 2 mg/ml. The enzyme is stable at 4° for several weeks. In addition to concentration of the protein solution, centrifugation produces additional purification. The results of a typical purification are summarized in the table.

After purification as described above, the protein contains phosphatase activity as measured by dephosphorylation of [³H]AMP. In order to re-

PURIFICATION OF PHOSPHODIESTERASE FROM *Bothrops atrox*[a]

Fraction	Total protein (mg)	Total activity (U)	Specific activity (U/mg)	Yield (%)
Crude venom	1000	960	0.9	100
Phosphocellulose	43.5	840	19.2	87
Affinity chromatography	2.5[b]	570	229	59
Concentration	1.4	515	350	53

[a] Activity was measured with thymidine 5'-(4-nitrophenyl)phosphate as substrate.
[b] This value is only approximate because of the high dilution encountered.

move it, an additional step, chromatography on hydroxyapatite,[1] is necessary.

Regeneration of Affinity Columns

The substituted Sepharose used here is regenerated by washing with 4 *M* NaCl and equilibrating with 50 m*M* Tris chloride at pH 9.1 (4°C). The substituted Sepharose may be used at least up to five times without loss of binding capacity.

[79] Ganglioside-Agarose and Cholera Toxin

By INDU PARIKH and PEDRO CUATRECASAS

Cholera toxin, an exotoxin from *Vibrio cholerae,* is the protein responsible for the gastrointestinal manifestations of clinical cholera.[1] This toxin is present in low concentrations in the culture medium of *Vibrio cholerae* (up to 1 mg per liter of medium). It consists of at least two subunits.[2,3] The toxin binds very strongly and specifically to certain brain gangliosides,[4] and there are indications[5] that membrane-localized monosialogangliosides (G_{M1}) are the natural receptors with which cholera toxin specifically interacts to stimulate the activity of adenylate cyclase in mammalian tissues.

Various insoluble agarose derivatives containing covalently linked gangliosides have been prepared and compared for their abilities to extract and purify cholera toxin by affinity chromatography (Table I).[2] Since cholera toxin also binds to certain glycoproteins, such as fetuin and thyroglobulin,[4] agarose derivatives containing these glycoproteins have also been prepared and tested. The ganglioside-agarose adsorbents may also be useful in the purification of various other toxic proteins, such as that derived from *Escherichia coli,* and botulinum and tetanus toxin, which are known to complex very tightly with free gangliosides.

Preparation of Specific Adsorbents

Ganglioside-agarose derivatives are prepared either by coupling the ganglioside carboxylic group to an aminoalkyl-agarose derivative or by

[1] R. A. Finkelstein, *CRC Critical Reviews in Microbiol.* 2, 553 (1973).
[2] P. Cuatrecasas, I. Parikh, and M. D. Hollenberg, *Biochemistry* 12, 4253 (1973).
[3] R. A. Finkelstein, M. K. La Rue, and J. J. Lo Spalluto, *J. Infect. Immunity* 6, 934 (1972).
[4] P. Cuatrecasas, *Biochemistry* 12, 3547 (1973).
[5] P. Cuatrecasas, *Biochemistry* 12, 3558 (1973).

TABLE I
AFFINITY ADSORBENTS FOR CHOLERA TOXIN

A. Ganglioside-diaminodipropylamino-agarose
B. Ganglioside-poly-L-Lysine-agarose
C. Ganglioside-poly(L-lysyl-DL-alanine)-agarose
D. Ganglioside-albumin-agarose
E. Fetuin-agarose
F. Thyroglobulin-agarose

reaction of functionalized (aldehydic) gangliosides to aminoalkyl- or hydrazido-agarose. The coupling yields are relatively low owing to problems of solubility, micelle formation, and possible steric hindrance in the ganglioside molecule.

Ganglioside-agarose adsorbents containing polyfunctional spacer arms proved to be superior to those containing the conventional diaminodipropylamino-spacer arm. Poly-L-lysine or poly(L-lysyl-DL-alanine) is coupled to CNBr-activated agarose by previously described methods.[6,7] The derivatives used contain about 1.2 mg of the copolymer per milliliter of agarose and about 1.4 mg of the homopolymer per milliliter of agarose. Albumin has also been used as a macromolecular spacer. Albumin is coupled to CNBr-activated agarose in the absence (native) or presence (denatured) of 10 M urea, as described earlier.[6,7] These derivatives contain 2–3 mg of albumin per milliliter of gel. 3,3'-Diaminodipropylamine, fetuin, and thyroglobulin are similarly coupled to agarose; these derivatives contain about 10 μmoles, 6 mg, and 8 mg, respectively, of the ligand and the proteins per milliliter of agarose.

Method A: Activation of the Ganglioside Carboxyl Groups

The single carboxylic group of the monosialoganglioside can be coupled by a variety of condensing agents to alkylamino-agarose, such as diaminodipropylaminoagarose or to agarose containing polyfunctional spacer arms, such as poly-L-lysine, poly(L-lysyl-DL-alanine), native albumin, or denatured albumin. The ganglioside carboxylic group is activated prior to coupling by generating *in situ* its active ester, mixed anhydride, or activated complex with a carbodiimide reagent.

Carbodiimide. Twenty-five milliliters of the derivatized agarose (containing amino groups) are washed and suspended in 50 ml of 50% (v/v) aqueous dioxane. Bovine brain ganglioside (50 mg, type III, Sigma) is added, and the suspension is shaken gently for 15 minutes at room tem-

---

[6] V. Sica, I. Parikh, E. Nola, G. A. Puca, and P. Cuatrecasas, *J. Biol. Chem.* **248**, 6543 (1973).
[7] I. Parikh, S. March, and P. Cuatrecasas, this volume [6].

perature. Two portions of 100 mg each of 1-ethyl-3-(3-dimethylamino-propyl) carbodiimide are added 6 hours apart, and the suspension is shaken gently on a mechanical shaker. The gel is washed (under gravity) on a coarse-disc sintered-glass funnel with 500 ml of water, 500 ml of 90% (v/v) aqueous methanol, 250 ml of 6 M guanidine·HCl, and 500 ml of water. The substitution of ganglioside, as judged by the recovery of unreacted ganglioside, is between 0.2 and 0.5 mg per milliliter of gel. Coupling can also be performed in an organic solvent by allowing 20 mg of ganglioside and 5 mg of dicyclohexylcarbodiimide to react in 10 ml of dioxane for 30 minutes at 15°. This is added to 20 ml of albumin-agarose suspended in dioxane in a total volume of 40 ml. After reacting for 15 hours at room temperature the gel is washed with 500 ml of dioxane, 500 ml of 90% (v/v) methanol, 250 ml of 6 M guanidine·HCl, and 500 ml of water.

N-Hydroxysuccinimide Ester. Twenty milligrams of ganglioside are allowed to react with 2.5 mg of N-hydroxysuccinimide and 2.5 mg of dicyclohexyl carbodiimide for 30 minutes at 15° in 10 ml of dioxane. The solution is added to 20 ml of albumin-agarose suspended in dioxane (total volume, 40 ml). After shaking for 15 hours at room temperature, the gel is washed as described above for the carbodiimide reaction.

Mixed Anhydride. One hundred microliters of 0.1 M N-methylmorpholine in tetrahydrofurane are added to a solution of anhydrous tetrahydrofurane containing 20 mg of ganglioside. After stirring the solution for 10 minutes at 0°, 100 μl of 0.1 M isobutylchloroformate[8] in tetrahydrofurane are added and the reaction is allowed to continue for 20 minutes at 0°. The reaction mixture is added to 20 ml of the amino agarose derivative suspended in dioxane (total volume, 40 ml). After reacting for 15 hours at room temperature the gel is washed as described above.

Although all of these methods result in effective adsorbents with substitution of ganglioside varying between 0.2 and 0.5 mg/ml, the best results are consistently obtained with derivatives prepared with the water-soluble carbodiimide and with the mixed anhydride.

Method B: Preparation of Aldehyde Derivatives of Gangliosides

Mild periodate oxidation of ganglioside generates an aldehydic function which readily reacts with the hydrazide groups of a hydrazido-agarose derivative. The relatively unstable Schiff's bases are subsequently reduced with NaBH$_4$, resulting in stable bonds between ganglioside and agarose. The hydrazido derivatives of agarose are prepared as described earlier.[7]

Bovine brain ganglioside (type III, Sigma), 22 mg (10 μmoles), is

[8] R. A. Finkelstein and J. J. Lo Spalluto, *J. Infect. Dis.* 121, 563 (1970).

dissolved in 100 ml of water, and 120 μl (12 μmoles) of 0.1 M NaIO$_4$ are added. After stirring the mixture for 30 minutes at room temperature, it is added to 20 ml (packed gel) of poly(L-lysyl-DL-alanyl-hydrazido)-agarose.[7] The agarose suspension is shaken gently on a mechanical shaker for 4–6 hour and 2 ml of a 5 M aqueous solution of NaBH$_4$ are added in small portions over a 6- to 10-hour period. The ganglioside-agarose is then washed extensively as described. The substitution of ganglioside varies between 0.2 and 0.5 mg per milliliter of packed gel.

General Properties of Ganglioside Affinity Adsorbents

Whereas ^{125}I-labeled cholera toxin does not bind to unsubstituted agarose, 80–100% of radioactive toxin does adsorb to columns containing one of the various ganglioside-agarose adsorbents described in these studies (Fig. 1). Affinity columns containing fetuin-agarose or thyroglobulin-agarose remove only 50–60% of the biologically active toxin from the samples. The specificity of the adsorptive process on the affinity columns is illustrated by demonstrating that incubation of the ^{125}I-labeled cholera toxin with gangliosides before chromatography effectively prevents the subsequent adsorption of radioactivity to the column (Fig. 1,D).

The binding of cholera toxin to the affinity columns is so strong that to achieve elution it is necessary to use buffers that are likely to unfold and possibly denature the protein. The denaturation of cholera toxin by acid, urea, and guanidine·HCl is reversible if the protein concentration is 0.1 mg/ml or greater. In the experiments described in Fig. 1, elution of the adsorbed ^{125}I-labeled toxin is achieved with 7 M guanidine·HCl. The strength with which the toxin binds to the ganglioside-agarose adsorbents is evident from the inability to achieve elution with 0.1 M acetic acid or with 3 M guanidine·HCl containing 1 M NaCl. Even 4 M guanidine·HCl yields only one-third of the bound toxin; 0.1 N HCl elutes about 90% of the toxin. However, since in the presence of these denaturants the toxin is dissociated into small subunits that pass through dialysis membranes extremely rapidly, dialysis should always be performed after diluting or neutralizing the denaturant. The concentration of guanidine·HCl or of urea should be decreased to 1 M. Sephadex G-15 should not be used to remove the denaturants since one of the toxin subunits adsorbs very strongly to this gel and such procedures will result in substantial losses of material.

Chromatography on ganglioside-agarose columns can release significant quantities of free ganglioside. This material can interfere with assay of toxin as well as interfere with biospecific adsorption of the toxin. Leakage of free gangliosides is much less marked in those adsorbents containing macromolecular spacers (Table II). These adsorbents are also much

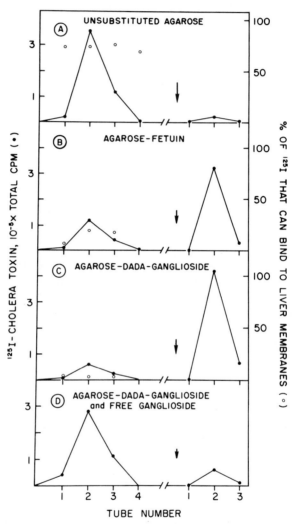

FIG. 1. Affinity chromatography of ^{125}I-labeled toxin from *Vibrio cholerae*. Pasteur pipettes containing 0.5 ml of the indicated gel are washed for 4 hours at 24° with 50% (v/v) methanol followed by 40 ml of 7 *M* guanidine·HCl and 40 ml of Krebs-Ringer-bicarbonate buffer containing 0.1% albumin. The samples (0.5 ml) applied on the columns contained 5.1×10^5 cpm of ^{125}I-labeled cholera toxin (6.7 µCi/µg) in Krebs-Ringer-bicarbonate buffer, 0.1% albumin. The columns are washed at 24° with 20 ml of the same buffer (note break in axis) before elution with 7 *M* guanidine·HCl (arrows). Radioactivity in each fraction (0.5 to 1 ml) is determined. Column D is exactly like that described in C except that the sample is incubated with 100 µg per milliliter of gangliosides for 40 minutes at 24° before application to the column. The derivative described in C and D is that prepared with a water-soluble carbodiimide.

TABLE II
CHROMATOGRAPHY OF ^{125}I-LABELED CHOLERA TOXIN
ON VARIOUS AFFINITY ADSORBENTS[a]

Agarose derivative[b]	Percent of ^{125}I-Labeled toxin		Ganglioside in elution (% inhibition)
	In breakthrough	In elution	
Unsubstituted agarose	93	3	—
Diaminodipropylamino-agarose	91	7	—
A-DADA-gang			
Undiluted	21[c]	74	48
1:10	27[c]	70	20
1:50	48	41	0
A-PLL-gang			
Undiluted	18[c]	76	10
1:10	21[c]	75	5
1:50	40	46	0
1:200	68	25	0
A-PLL-Ala-gang			
Undiluted	17[c]	80	5
1:10	19[c]	78	0
1:50	20	76	0
1:200	52	42	0
A-Nat Alb-gang			
Undiluted	18[c]	78	10
1:10	20[c]	81	0
1:50	21[c]	75	0
1:200	38	52	0
1:600	68	30	0
A-Denat Alb-gang			
Undiluted	24[c]	72	21
1:10	29[c]	69	0
1:50	48	47	0

[a] Columns (Pasteur pipettes) containing 1 ml of the specified gel are washed with 50% (v/v) methanol, 7 M guanidine · HCl, and Krebs-Ringer-bicarbonate buffer. ^{125}I-Labeled cholera toxin (2.1 × 10^5 cpm, 12 μCi/μg), in 0.5 ml of Krebs-Ringer–bicarbonate buffer, 0.1% albumin, is applied to each column. The columns are washed and eluted with 5 M guanidine · HCl in 50 mM Tris chloride, pH 7.4, 0.1% albumin. The adsorbents are diluted serially with unsubstituted agarose before use. The eluted samples are examined for the presence of free gangliosides by incubating (24°, 40 minutes) aliquots (20 μl) with 0.1 ml of Krebs-Ringer–bicarbonate buffer, 0.1% albumin, containing 5 × 10^5 cpm of fresh ^{125}I-labeled cholera toxin. The percent inhibition of the binding of this iodoprotein to liver membranes is a measure of the free gangliosides present in the eluates. None of the breakthrough samples contained free gangliosides. Data from P. Cuatrecasas, I. Parikh, and M. D. Hollenberg, *Biochemistry* 12, 4253 (1973).

[b] The ganglioside-agarose derivatives (Table I, A–D) are abbreviated as follows: (A) A-DADA-gang. (B) A-PLL-gang. (C) A-PLL-Ala-gang. (D) A-nat Alb-gang or A-denat Alb-gang.

[c] Less than 5% of the radioactivity in these samples can bind to liver membranes. In contrast, approximately 80% of the total radioactivity present in the samples which are applied to the column can bind to liver membranes.

more effective in extracting the toxin from samples. Gels containing these polymeric spacers are very effective even after they are diluted substantially with unsubstituted agarose.

Assay Methods

Tracer amounts of ^{125}I-labeled toxin[4] are added to the toxin preparations. The effluent and eluent fractions from each affinity column are assayed for protein, free ganglioside, and ability of the ^{125}I-labeled toxin to bind to various cell membranes.[2] Each fraction is also tested for its lipolytic activity.

The presence of free gangliosides in the column effluents is determined by measuring the ability of the effluent sample to block the binding of ^{125}I-labeled toxin to liver membranes.[2]

Affinity Chromatography of Purified Cholera Toxin

Cholera toxin purified by conventional methods can be purified to homogeneity (1.8-fold) by affinity chromatography (Table III). Unless stated otherwise, all the chromatographic steps are performed at room temperature. Agarose-albumin-ganglioside (0.5 mg ganglioside per milliliter of gel) is diluted 1:5 (v/v) with unsubstituted agarose. The diluted adsorbent (1.5 ml) is packed in a 5-ml disposable glass pipette and washed with 2 liters of 50% aqueous methanol over a 2-day period followed by 100 ml of 7 M guanidine·HCl. The column is equilibrated with Krebs-Ringer-bicarbonate buffer at pH 7.4. Partially purified toxin (2.4 mg), dissolved in 1.4 ml of the same buffer and containing a tracer quantity of ^{125}I-labeled toxin (14 μCi/μg), is applied to the column at 4° over a 30-minute period. The column is washed with 15–20 ml of the above buffer at a flow rate of 3 ml per hour. The column removes about 70% of the protein and more than 95% of the original lipolytic activity. The protein in the breakthrough fractions is virtually without lipolytic activity and does not bind to liver membranes. Toxin is eluted from the affinity column with 50 mM Tris chloride at pH 7.4 containing 5 M guanidine· HCl. The eluted samples are immediately diluted 5-fold with 50 mM Tris chloride at pH 7.4, and dialyzed for 16 hours at 4°. About 90% of the lipolytic activity applied to the affinity column is recovered in the eluate (Table III).

Affinity Chromatography of Crude Cholera Toxin

Columns containing ganglioside-agarose derivatives can extract more than 99% of the lipolytic activity and virtually all of the tracer ^{125}I-labeled toxin present in large quantities (40 g of lyophilized material) of crude culture medium from *Vibrio cholerae* (Table IV).[2] Small quantities of

TABLE III
AFFINITY CHROMATOGRAPHY OF PURIFIED CHOLERA TOXIN ON GANGLIOSIDE-AGAROSE[a]

	Total protein (mg)	Cholera toxin		Specific activity	
		Radioactivity (cpm)	Activity[b] (units)	Cpm/mg protein	Units/mg protein
Applied on column	2.40 (100%)	8.8×10^5 (100%)	2.4 (100%)	3.7×10^5	1.0
Breakthrough	0.76 (32%)	1.2×10^5 (14%)	0.1 (4%)	1.6×10^5	0.1
Elution	1.19 (50%)	6.0×10^5 (68%)	2.1 (88%)	5.0×10^5 $(1.4)^c$	1.8 $(1.8)^c$

[a] The data summarize the experiment described in the text. The purification is more marked when expressed according to protein content than when expressed by radioactivity. The apparent discrepancy is explained by the fact that the ^{125}I-labeled cholera toxin was prepared from partially purified (by gel filtration) toxin. About 85% of the iodoprotein which was added to the sample as a tracer adsorbs to affinity columns in the absence of unlabeled toxin, and the breakthrough radioactivity does not bind to cell membranes.
Data from P. Cuatrecasas, I. Parikh, and M. D. Hollenberg, Biochemistry 12, 4253 (1973).
[b] Lipolytic activity is described in arbitrary units which are defined on the basis of comparison with the starting cholera toxin.
[c] Purification achieved by these procedures.

TABLE IV

SUMMARY OF AFFINITY CHROMATOGRAPHY OF CRUDE FILTRATE CHOLERA TOXIN FROM *Vibrio cholerae* ON A COLUMN OF A-NAT ALB-GANG[a]

	Volume (ml)	Protein (mg)	Cholera toxin			Yield	
			From tracer ^{125}I-toxin		From biological activity[b] (mg/ml)	From ^{125}I-toxin (%)	From activity (%)
			Cpm/mg protein	Total cpm			
Applied on column	50	210	5.2×10^4	11×10^6	0.07	100	100
Breakthrough	60	186	1.2×10^4	2.3×10^6 [c]	0.001	21[c]	0.8
Washes	500	—	—	0.4×10^6	0	4	0
Elution	9.6	1.3	4.7×10^6	6.1×10^6	0	55[d]	0

[a] The data are obtained from the experiment described in the text. The eluted and dialyzed sample is lyophilized, redissolved in 0.8 ml of water, and chromatographed (to remove the guanidine) on a Bio-Gel P-2 column (1.7×45 cm) equilibrated with 10 mM sodium phosphate buffer, pH 7.4. The radioactive material, present exclusively in the void volume, was pooled, lyophilized, and redissolved in 0.4 ml for studies of biological activity. Data from P. Cuatrecasas, I. Parikh, and M. D. Hollenberg, *Biochemistry* **12**, 4253 (1973).

[b] Determined by lipolysis, comparing with standard curves performed with native cholera toxin.

[c] About 20–25% of the ^{125}I-labeled protein in the original preparation was initially denatured or represents contaminating protein since it does not adsorb to any affinity column and does not bind to cell membranes. Therefore there is virtually quantitative adsorption of active toxin to the column in this experiment.

[d] This represents at least 70% of the active toxin which adsorbed to the column.

protein and at least 70% of the ^{125}I-labeled toxin can be recovered upon elution with 50 mM Tris-chloride at pH 7.4 containing 5 M guanidine-hydrochloride. On the basis of radioactivity, the toxin is purified more than 90-fold by these procedures.

Unfortunately, the material eluted from this column is virtually devoid of biological activity. The lack of activity in this material is not explained by the presence of gangliosides, residual guanidine, or biologically inactive choleragenoid. The loss of activity may result from dissociation of the toxin into subunits.[2] This process is essentially irreversible when the concentration of cholera toxin is very low, and substantial losses of these subunits occur during the dialysis procedure used to remove guanidine.

[80] Morphine and Related Drugs

By ERIC J. SIMON[1]

In our laboratory, efforts are being made to isolate and purify opiate receptors from the central nervous system of animals and man and to purify antibodies to morphine. For both purposes affinity methods are being utilized. We describe here methods for covalently attaching morphine and its analogs to agarose and glass beads. We also summarize briefly some results obtained with these beads.

Preparation of Morphine-Agarose Beads

It is best to couple a drug to its support via a functional group known not to be required for pharmacological activity. The hydroxyl group in position 6 of morphine is such a group.[2] The specific introduction of a succinyl group into the 6 position by esterification is readily achieved by refluxing morphine with excess succinic anhydride in a solvent such as benzene.[3] After decantation of benzene the residue is dissolved in water. Unreacted morphine is removed by filtration at pH 9 where succinyl-morphine remains in solution. The pH is then adjusted to 5 where 6-succinylmorphine crystallizes at 4° overnight. The yield is 60–70% of theoretical.

[1] Career Scientist of the Health Research Council of the City of New York. This work was supported by grant DA-00017 from the National Institute of Mental Health.
[2] E. J. Simon, W. P. Dole, and J. M. Hiller, *Proc. Nat. Acad. Sci. U.S.* **69**, 1835 (1972).
[3] Pyridine was used in earlier experiments, but yields and purity of product were better in benzene.

The 6-succinyl morphine is coupled via the free carboxyl group to agarose beads containing side arms ending in free amino groups. Sepharose 4B with hexylamine side arms (available from Pharmacia as AH-Sepharose 4B) is used as a matrix. The coupling is done by a procedure similar to that described by Cuatrecasas[4] for coupling succinyl estradiol to agarose, except that the reaction is carried out in aqueous medium. 6-Succinylmorphine (50 mg), or 6-succinyl-7,8-[³H]dihydromorphine to facilitate determination of bound residues, is dissolved in 20 ml of water and added to 12 ml of AH-Sepharose 4B beads. The pH is adjusted to 5 and 250 mg of 1-ethyl-3-(3-dimethylaminopropyl) carbodiimide are added over a 5 minute period. The pH is monitored frequently and readjusted to 5 with 0.1 N HCl. The reaction is allowed to proceed at room temperature for 20 hours. The beads are washed in a column successively with 1 liter of 0.1 N NaCl–0.01 N Tris chloride at pH 7, 100 ml of 1 N NaCl, and again with 1 liter of the NaCl-Tris buffer. The yield of bound morphine is 0.3–0.5 μmole per milliliter of settled beads. Conjugated beads are stored with sodium azide (0.025%) and washed extensively immediately before use to remove the azide and small amounts of released morphine.

Preparation of Morphine Derivatives Covalently Attached to Glass Beads

The use of glass beads with side arms which terminate in aromatic amine groups allows the coupling by diazotization of a variety of morphine derivatives. The drugs can be used directly without previous derivatization. The method used is essentially that described for immobilizing catecholamines.[5]

Glass beads containing alkylamine side chains are obtained from Pierce Chemical Co. (GAO-3940). Arylamino groups are added by the introduction of nitrobenzene groups and subsequent reduction with sodium dithionite as described in a brochure provided by Pierce. The resulting arylamine glass is diazotized and coupled to a drug as follows: 10 ml of 2 N HCl are added to 1 g of glass beads in an ice bath followed by 250 mg of sodium nitrite. The nitrous oxide gas evolved is evacuated by placing the reaction mixture in a vacuum desiccator attached to a water pump for 30 minutes. The solution is decanted and the beads are washed on a coarse sintered-glass filter with large quantities of cold water, followed by cold 1% sulfamic acid and again by cold water.

---

[4] P. Cuatrecasas, *J. Biol. Chem.* **245**, 3059 (1970).
[5] J. C. Venter, J. E. Dixon, P. R. Maroko, and N. O. Kaplan, *Proc. Nat. Acad. Sci. U.S.* **69**, 1141 (1972).

The diazotized beads are coupled to a morphine derivative by adding 5 ml of a solution of the drug (10 mg/ml) to the beads and incubating at 0° to 4° for 60 minutes; by this time the beads will assume a deep yellow to red color. The color frequently intensifies upon overnight incubation in the refrigerator. The beads are washed on a coarse glass filter with 1 liter of HCl (pH = 1) and 1 liter of water. Under these conditions, approximately 50 μmoles of drug are bound per gram of beads. The preparation is stored at 4° in 0.1 N HCl as a slurry. The procedure has been used to attach morphine, levorphanol, dextrorphan, and etorphine to the glass beads.

Use of Morphine-Agarose for Purification of Antibodies

Antibodies to morphine were obtained by injection of rabbits with 6-succinylmorphine-BSA[6] together with complete Freund's adjuvant. Similar antibodies have been prepared by Wainer et al.[7] Spector and Parker[8] had previously prepared antibodies to morphine by use of the immunogen 3-carboxymethylmorphine-BSA. Morphine antibodies have been shown to be specifically removed from antisera by morphine-Sepharose beads[2] and such beads have been used for the purification of the antibodies as described here.[9]

The γ-globulin fraction is isolated from rabbit antiserum by ammonium sulfate precipitation.[10] The titer of the whole antiserum and of the γ-globulin fraction is determined by the hemagglutination assay developed by Adler and Liu[11] with formalinized sheep red blood cells. The γ-globulin is equilibrated with phosphate-buffered physiological saline (PBS, pH 6.9, 0.1 M phosphate). A column (8 mm in diameter, 8 cm long) of morphine-Sepharose is washed with 2 liters of PBS. The wash is checked for absence of free morphine by a hemagglutination inhibition assay.[12] The PBS equilibrated γ-globulin, 3–4 ml, is layered on top of the column at room temperature, and the column is washed with 3 hold-up volumes of PBS, thereby eluting the bulk of unadsorbed protein. The protein peak, as monitored at 280 nm, is pooled and equilibrated against PBS at pH

---

[6] The immunogen injected was 6-succinyl-7,8-[³H]dihydromorphine-BSA, but for simplicity the antibodies will be referred to as morphine-antibodies.
[7] B. H. Wainer, F. W. Fitch, R. M. Rothberg, and J. Fried, *Science* 176, 1143 (1972).
[8] S. Spector and C. W. Parker, *Science* 168, 1348 (1970).
[9] M. C. Walker and E. J. Simon, unpublished results.
[10] D. H. Campbell, J. S. Garvey, N. E. Cremer, and D. H. Sussdorf, "Methods in Immunology," 2nd ed., pp. 189–191. Benjamin, New York, 1970.
[11] F. L. Adler and C.-T. Liu, *J. Immunol.* 106, 684 (1971).
[12] F. L. Adler, C.-T. Liu, and D. H. Catlin, *Clin. Immunol. Immunopathol.* 1, 53 (1972).

7.6 and is used in the hemagglutination assay. The column is washed further with saline to remove the PBS and traces of protein. Elution of antibody is achieved with glycine·HCl buffer, 0.55 M at pH 2.6, containing 0.85% NaCl. The protein peak is monitored at 280 nm, pooled, quickly neutralized to pH 7 with dilute NaOH, and dialyzed against PBS at pH 7.6 overnight. A second, smaller antibody peak is eluted with 0.1 M acetate at pH 3. Regeneration of the column is achieved by passing 1 M salt solution through it, followed by washing with PBS.

All three protein peaks are concentrated about 6-fold by ultrafiltration using Amicon centriflo membrane cones and centrifugal force. Determination of antibody titers by hemagglutination and protein concentrations by the method of Lowry[13] establish that our early experiments give a recovery of about 40–50% of antibodies and approximately a 20-fold purification.

Use of Opiate-Glass Beads for Removal of Membrane-Bound Opiate Binding Sites

The existence of stereospecific binding sites for opiates in rat brain has been demonstrated in our laboratory[14] by the use of the potent narcotic analgesic, etorphine, labeled with tritium of high specific activity. Similar results were obtained by Pert and Snyder with the tritiated antagonist, naloxone[15] and by Terenius with tritiated dihydromorphine.[16] Each of the laboratories used a modification of a method for measuring stereospecific binding first proposed by Goldstein et al.[17] To date, none of these laboratories has succeeded in solubilizing these opiate-binding components.

We have therefore prepared a membrane fraction and demonstrated specific removal of membrane-bound binding sites by glass-bead-bound opiates.[18]

A mitochondrial-synaptosomal (P_2) fraction is prepared from rat brain homogenate in 0.32 M sucrose by the procedure of Whittaker.[19] This fraction is treated with hypo-osmotic medium (50 mM Tris chloride at

[13] O. H. Lowry, N. J. Rosebrough, A. L. Farr, and R. J. Randall, *J. Biol. Chem.* **193**, 265 (1951).

[14] E. J. Simon, J. M. Hiller, and I. Edelman, *Proc. Nat. Acad. Sci. U.S.* **70**, 1947 (1973).

[15] C. B. Pert and S. H. Snyder, *Science* **179**, 1011 (1973).

[16] L. Terenius, *Acta Pharmacol. Toxicol.* **32**, 317 (1973).

[17] A. Goldstein, L. I. Lowney, and B. K. Pal, *Proc. Nat. Acad. Sci. U.S.* **68**, 1974 (1971).

[18] N. Clendeninn and E. J. Simon, unpublished results.

[19] E. G. Gray and V. P. Whittaker, *J. Anat.* **96**, 79 (1962).

pH 7.4) to break up particulates. Of the resulting membrane suspension (1:30 dilution w/v, based on original brain wet weight), 50 ml is incubated for 5 minutes with 10^{-6} M dextrorphan to minimize nonspecific binding. Aliquots (9 ml) are incubated as follows: (1) no glass beads, (2) control alkylamino beads, (3) dextrorphan beads, (4) levorphanol beads, (5) etorphine beads. The tubes are incubated for 5 minutes at 37° on a gently rotating wheel. They are then placed in ice, and the supernatant above the beads is removed as soon as the beads settle. An assay for the stereospecific binding of [^{3}H]etorphine is carried out on the supernatant liquid as previously described.[14] In several experiments the levorphanol and etorphine beads consistently remove 60–70% of the binding capacity from the membrane suspension whereas no decrease of binding is seen in the presence of uncoupled beads or beads coupled to dextrorphan, the inactive enantiomorph of levorphanol. The amount of protein removed is very small. The experiments suggest the specific removal of membrane fragments that contain opiate binding components. Attempts to elute the purified membranes from the glass beads are in progress.

[81] Tubulin

By H. SCHMITT and U. Z. LITTAUER

Microtubular proteins[1-4] can be broadly classified into three groups: those associated with motile systems (cilia, flagella), those that assume a structural role in cell shaping, and those involved in the spindle apparatus. At least the first two groups share the property of binding 1 mole of colchicine[5,6] and 1–2 moles of GTP per 6 S protein monomer (molecular weight 120,000 containing two subunits). Mitotic and cytoplasmic microtubules are reversibly disrupted by colchicine whereas flagellar microtubules, once assembled, are insensitive to the alkaloid. The vinca alkaloids, vinblastine and vincristine, also bind to microtubular proteins,

[1] R. E. Stephens, Microtubules, *in* "Subunits in Biological Systems" (S. N. Timasheff and G. D. Fasman, ed.), Part A, Dekker, New York, 1971.
[2] M. L. Shelanski and H. Feit, *in* "The Structure and Function of Nervous Tissue" (G. H. Bourne, ed.), Vol. VI. Academic Press, New York, 1972.
[3] J. B. Olmsted and G. G. Borisy, *Annu. Rev. Biochem.* **42**, 507 (1973).
[4] L. Wilson and J. Bryan, *Advan. Cell. Mol. Biol.* 3 (1973) in press.
[5] R. J. Owellen, A. H. Owens, and D. W. Donigian, *Biochem. Biophys. Res. Commun.* **47**, 685 (1972).
[6] G. G. Borisy and E. W. Taylor, *J. Cell. Biol.* **34**, 525 (1967).

but at a different site than colchicine, and precipitate them in the presence of magnesium ions.[7]

The broad involvement of microtubules in mitosis, cell shaping, membrane phenomena and axoplasmic transport, calls for a rapid isolation method for this protein. This has been achieved by affinity chromatography.

For this procedure, one could make use of either of the two alkaloids that bind to tubulin: colchicine or vinblastine. The former was chosen because of its less complex chemical structure and greater commercial availability.

Purification of Tubulin

Our work has centered on mouse brain, from which tubulin can be conveniently purified by the method of Eipper,[8] or that of Weisenberg.[9] Homogenization of the tissue is best performed in 0.25 M sucrose, 10 mM magnesium chloride, 10 mM potassium phosphate at pH 7.4, and 0.5 mM GTP. The presence of GTP during purification is essential, to avoid aggregation and loss of colchicine binding activity (half-life = 11 hours). Alternatively, a high concentration of sucrose (0.8 M) or glycerol (4 M) may be used.[10] The purified protein is stored in liquid air and thawed only once. Another convenient source of soluble microtubular protein is the sea urchin egg during cleavage.[11] Preparation of microtubules from the mitotic apparatus has also been described.[12]

Preparation of the Sepharose-Bound Colchicine Derivative

We have used agarose-bound colchicine in which the active groups were separated from the agarose backbone by a benzamidoethyl spacer. This derivative was synthesized according to the Scheme I.

Step A. Coupling of Ethylenediamine to Agarose.[13] Washed Sepharose-4B is mixed with an equal volume of water and activated by reacting it with finely ground cyanogen bromide (300 mg per milliliter of gel). The reaction is carried out with stirring at 20°C, and the pH is maintained above 11 by the addition of 8 M NaOH. The washed, activated resin is immediately mixed with 1 volume of a solution of ethylenediamine in

[7] L. Wilson, J. Bryan, A. Ruby, and D. Mazia, *Proc. Nat. Acad. Sci. U.S.* **66**, 807 (1970).

[8] B. A. Eipper, *Proc. Nat. Acad. Sci. U.S.* **69**, 2283 (1972).

[9] R. C. Weisenberg, G. G. Borisy, and E. W. Taylor, *Biochemistry* **7**, 4466 (1968).

[10] F. Solomon, D. Monard, and M. Rentsch, *J. Mol. Biol.* **78**, 569 (1973).

[11] R. A. Raff, H. V. Color, S. E. Selvig, and P. R. Gross, *Nature (London)* **235**, 211 (1972).

[12] G. G. Borisy and E. W. Taylor, *J. Cell Biol.* **34**, 535 (1967).

[13] P. Cuatrecasas, *J. Biol. Chem.* **245**, 3059 (1970).

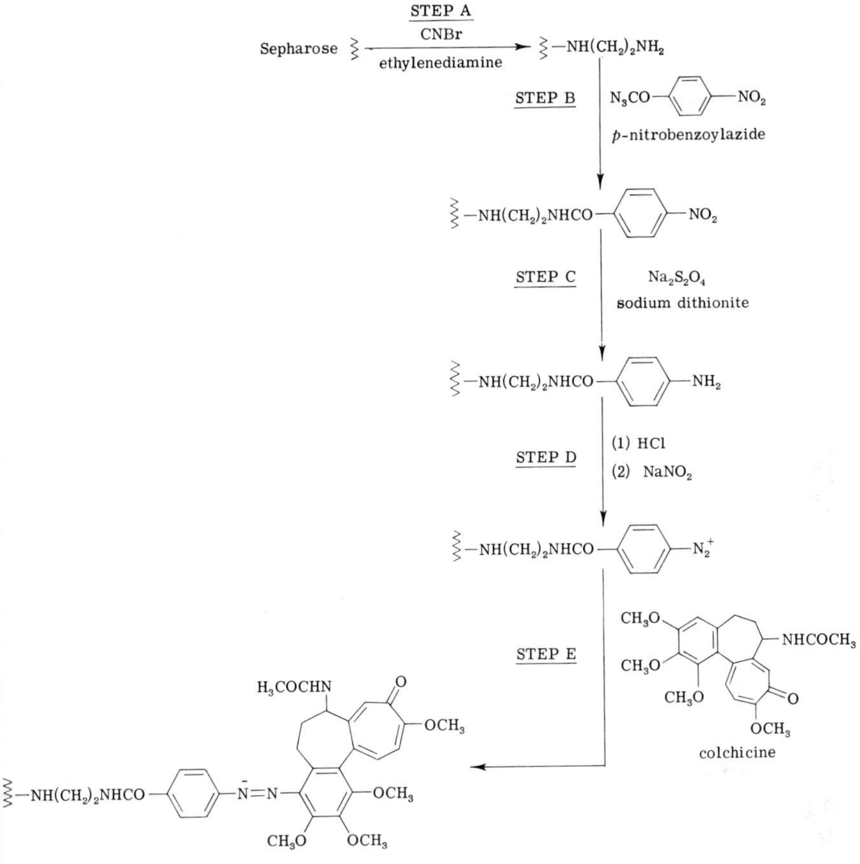

SCHEME I

water (2 mmoles per milliliter of agarose gel), previously titrated to pH 10 with HCl. The reaction is allowed to proceed for 16 hours at 4°. The gel is then washed with a large excess of water.

Step B. Preparation of p-Nitrobenzamidoethyl Agarose.[13] The aminoethyl-Sepharose is treated for 6 hours at room temperature with 2 volumes of a prefiltered solution containing 0.2 M sodium borate at pH 9.3, 40% dimethylformamide (v/v), and 70 mM p-nitrobenzoylazide. The pH is checked and eventually readjusted to 9.3 with NaOH. The completeness of the substitution can be established by the loss of color after allowing an aliquot to react with sodium trinitrobenzene sulfonate.[13] The p-nitrobenzamidoethyl-Sepharose is washed extensively with 50% dimethylformamide and water.

Step C. Reduction of the p-Nitro Group.[13] The p-nitrobenzamidoethyl-

Sepharose is reduced with 0.1 M sodium dithionite in 0.5 M NaHCO$_3$ at pH 8.5. The reaction is carried out for 4 hours at room temperature.

Step D. Diazotization of the p-Aminobenzamidoethyl-Agarose.[13] The washed Sepharose derivative is suspended in 3 volumes of ice-cold 0.5 M HCl and stirred for 10 minutes at 0°. A solution of 1 M NaNO$_2$ is then added to give a final concentration of 0.1 M, and the mixture is further incubated for 10 minutes at 0°. The resulting diazonium salt is very unstable, and further operations must be performed quickly and in the cold.

Step E. Coupling of the Colchicine. The resin is washed twice with cold water on a sintered glass funnel and added to 1 volume of a colchicine solution (10 μmoles per gram of resin). The pH is adjusted to 10 with NaOH, and the reaction is allowed to proceed overnight in the cold and in the dark. After adjustment of the pH, the resin turns immediately from a white to a red-brown color. One gram of packed Sepharose gel binds about 6 μmoles of colchicine. This can be assayed either by measuring the decrease in colchicine absorbance at 350 nm, or by using [^{3}H]colchicine and estimating the loss of radioactivity in the supernatant fluid (Sepharose quenches so that measurement of the total radioactivity bound to the gel is not an accurate estimate of the real binding value.)

Another alternative is to measure the colchicine quantitatively released from the gel by the reduction of the azo bond with 0.1 M sodium dithionite in 0.2 M sodium borate.

Binding of Tubulin to Agarose-Bound Colchicine

High speed mouse brain supernatant is diluted to 2 mg of protein per milliliter with 10 mM phosphate-Mg buffer (pH 7.4) containing 0.5 mM GTP. To 1 ml of the protein solution, 200 mg of Sepharose-bound colchicine are added, and the suspension is incubated with occasional stirring at 37° in the dark. Under these conditions, over 90% of the colchicine-binding protein is removed from the solution. The reaction is complete in 1 hour, whereas at 0° only 80% binding is obtained. The suspension is packed into a column and washed with the phosphate-Mg buffer.

Electrophoretic analysis, on SDS-polyacrylamide gels, of the mouse brain 105,000 g supernatant fluid, before and after incubation with Sepharose-colchicine, clearly shows specific adsorption of tubulin by the column. The bound protein can be eluted from the column with 1% SDS. Electrophoretic analysis of this fraction shows that it is composed of at least 90% pure tubulin. The amount of bound tubulin can be quantitatively measured by releasing the colchicine from the column on reduction of the azo bond with 0.1 M sodium dithionite in 0.2 M sodium borate. About 60% of the bound tubulin can be eluted from the column with 1 M NaCl in 10 mM phosphate-magnesium buffer at pH 7.4 containing 40% (v/v) ethyleneglycol.

Recently, Hinman *et al.*[14] have published a similar procedure coupling directly diacetylcolchicine with Sepharose without spacer groups. This method yields a seemingly less stable complex between colchicine and tubulin that enables partial elution of the protein with 0.1 M NaCl.

[14] N. D. Hinman, J. L. Morgan, N. W. Seeds, and J. R. Cann, *Biochem. Biophys. Res. Commun.* **52**, 752 (1973).

Section IV
Purification of Synthetic Macromolecules

[82] Functional Purification of Synthetic Proteins and Protein Fragments

By IRWIN M. CHAIKEN

Principle

Purification of *synthetic* proteins and protein fragments by conventional separatory methods such as ion exchange chromatography, gel fractionation, or crystallization has often proved inadequate. This problem exists for solution syntheses, owing largely to the insolubility of large intermediate fragments, but especially for solid phase synthesis, because of the presence of unwanted failure sequences that may differ from the desired polypeptide by only subtle amino acid sequence variations. Inasmuch as purified synthetic proteins are of profound potential importance in the elucidation and engineering of conformational and functional features of proteins by planned chemical mutation of important residues, efforts have been made to develop efficient and specific purification techniques for such products. These techniques, categorized generally as functional purifications, are based on a property, or set of properties, exhibited by the correct sequence of the peptide or protein in question but not by at least the large majority of contaminant sequences.

Approach

In practice, functional purification includes manipulations that vary, depending on sequence-specific properties of the particular species being isolated. Generally, these manipulations are applied in two stages (Fig. 1). The first consists of procedures, usually performed in solution, which are designed to transform either the desired product or the unwanted contaminants so as to make the desired product uniquely recognizable. Among the functional properties applicable at this stage of the purification process are: (a) binding of the desired synthetic product to a ligand or complementary polypeptide; (b) proteolytic refractivity of folded polypeptides or polypeptides in complex with either a ligand or a complementary polypeptide; and (c) correct disulfide bond formation. Such properties require a high degree of specificity, such as is typically found in biological systems. The second stage of the purification involves chromatographic or other fractionation procedures designed to separate the now-distinct correct synthetic product from the impurities. Since such separatory steps usually have been developed for the native polypeptide, they are specific for the corresponding synthetic product of similar characteristic behavior. Thus,

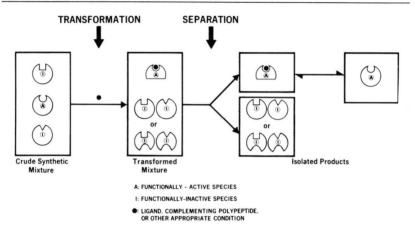

A: FUNCTIONALLY - ACTIVE SPECIES

I: FUNCTIONALLY-INACTIVE SPECIES

●: LIGAND, COMPLEMENTING POLYPEPTIDE, OR OTHER APPROPRIATE CONDITION

FIG. 1. Schematic format for functional purification.

protein-specific properties may be exploited in both the *transformation* and *separation* steps. Such a superimposition of several conditions for isolation, either simultaneously or in sequence, can lead to a quite stringent resolving condition.

A specific and widely applied variety of functional purification is affinity chromatography,[1] which involves separation on a solid support to which specific ligand is bound. As described elsewhere in this volume, this procedure has been important for the purification of native proteins and protein fragments, for which contaminants vary widely in ligand-binding properties. On the other hand, the approach is of less use for synthetic products, for which impurities can exhibit a continuous gradation of binding affinity. This results in the inability to separate a desired product from impurities which binds to a ligand more weakly than normal although sufficiently to be retained along with desired product. Similarly, binding products which are otherwise nonproductive also are not eliminated. Because of these weaknesses, affinity chromatographic separations of synthetic polypeptides have been most useful as applied for analysis of binding properties of crude synthetic products and for partial purification.

Example 1. Semisynthetic Staphylococcal Nuclease-T′

Approach. Staphylococcal nuclease-T′[2,3] is the enzymatically active noncovalent complex of the two tryptic fragments of staphylococcal nuclease, nuclease-T-(6–48)[4] and nuclease-T-(49–149).[5] Synthetic-(6–47),

---

[1] P. Cuatrecasas and C. B. Anfinsen, in this series, Vol. 22, p. 345.

[2] H. Taniuchi and C. B. Anfinsen, *J. Biol. Chem.* **243**, 4778 (1968).

[3] The designation prime (′) is used for a reconstituted fragment complex, formed by addition of separated component polypeptide fragments and subsequent removal of excess fragments which remain unbound.

corresponding to nuclease residues 6 through 47, has been obtained by the solid phase procedure[6] and found to generate, in the crude form, about 2–4% nuclease-T′ enzymatic activity when added to nuclease-T-(49–149) in 1:1 molar amounts. As a means of purifying the crude synthetic-(6–47) peptide, the preparation of semisynthetic nuclease-T′ has been developed.[7] In this procedure, complementation of synthetic-(6–47) and native nuclease-T-(49–149) is allowed to occur in solution. Additionally, the incubation is carried out in the presence of the active-site ligands, Ca^{2+} (a metal essential for nuclease and nuclease-T′ enzymatic activity), and the competitive inhibitor, deoxythymidine-3′,5′-diphosphate (pdTp). Both nuclease-T′ formation and ligand binding are specific for the correct (6–47) sequence, with even subtle changes in many of the residues leading to greatly altered properties.[8,9] The resultant mixture of peptides and ligands is treated with trypsin. Under the conditions used, all nuclease-T′-like complexes which form and bind ligands effectively survive the protease treatment; all noncomplexed fragments, as well as species of complex which do not bind ligands, are destroyed by trypsin. The complex surviving this multistepped transformation procedure is then isolated by phosphocellulose chromatography under conditions specifically designed for the separation of native nuclease-T′[2,7].

Sample Procedure.[7] Crude synthetic-(6–47), 200 mg, prepared essentially as originally described by Ontjes and Anfinsen,[6] is incubated at room temperature with 20 mg of nuclease-T-(49–149) in 20 ml 0.4% NH_4HCO_3, pH 8, containing 0.1 mM pdTp and 10 mM $CaCl_2$. After standing for 5–10 minutes, 3 mg of trypsin (DFP-treated) are added and incubation continued for 1 hour. Finally, 9 mg of soybean trypsin inhibitor are added and the solution is mixed, allowed to stand for 10 minutes, and lyophilized. After a similar preparation starting with an additional 200 mg of synthetic-(6–47), the pooled lyophilized digestion mixture is dissolved in a few milliliters of 0.3 M ammonium acetate, pH 6, and applied to a phosphocellulose column (1 × 20 cm) equilibrated with the pH 6 buffer. Chromatography is carried out by elution first with the pH 6 ammonium

---

[4] Native and synthetic fragments are named according to the prototype, trivial name-(x-y). Here, the trivial name denotes the origin, and x and y denote the NH_2- and COOH-terminal amino acid residues, respectively.

[5] Nuclease-T-(49–149) is usually obtained as an equimolar mixture of fragments, containing either residues 49 through 149 or 50 through 149 (Ref. 2). Both of these fragments have been isolated and shown to have similar potential enzymatic activity when added to nuclease-T-(6–48). The nuclease-T-(49–149) referred to here is such a mixture; the notation used is for simplicity.

[6] D. A. Ontjes and C. B. Anfinsen, *Proc. Nat. Acad. Sci. U.S.* **64**, 428 (1969).

[7] I. M. Chaiken, *J. Biol. Chem.* **246**, 2948 (1971).

[8] I. M. Chaiken and C. B. Anfinsen, *J. Biol. Chem.* **246**, 2285 (1971).

[9] I. M. Chaiken, *J. Biol. Chem.* **247**, 1999 (1972).

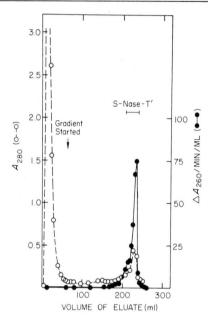

FIG. 2. Separation of semisynthetic staphylococcal nuclease-T' by phosphocellulose chromatography [I. M. Chaiken, *J. Biol. Chem.* **246**, 2948 (1971)]. S-Nase-T' denotes the peak of semisynthetic complex. Enzymatic activity, given in units of ΔA_{260}/minute/ml, is against 50 μg of heat-denatured salmon sperm DNA per milliliter in 50 mM Tris chloride, pH 8.8, 25°. Absorbances at 280 nm are direct readings.

acetate buffer and thereafter with a gradient from 0.3 M ammonium acetate at pH 6 (200 ml), to 1.0 M ammonium acetate buffer at pH 8 (200 ml). As shown in Fig. 2, about 15 mg of the semisynthetic complex is isolated in a discrete, well resolved peak. Based on the enzymatic activity exhibited (see below), the amount of complex obtained represents a yield of 0.75% versus the initial amount of crude synthetic-(6–47) used.

Results. As obtained from phosphocellulose, semisynthetic nuclease-T' has the phosphocellulose chromatographic mobility and amino acid composition expected for a nuclease-T' derivative. The semisynthetic complex exhibits at least 90% of the enzymatic activity of native nuclease-T' against DNA, RNA, and deoxythymidine-3'-phosphate-5'-*p*-nitrophenylphosphate. The K_m against this last substrate and the binding affinity to pdTp[10] also agree well with the values obtained for native nuclease-T'. Furthermore, the native and semisynthetic derivatives are structurally very similar, if not identical, as judged by fluorescence emission, sensitivity of activity to temperature, and chromatography on pdTp-Sepharose. By all tests applied,

[10] I. M. Chaiken and G. R. Sánchez, *J. Biol. Chem.* **247**, 6743 (1972).

therefore, the semisynthetic nuclease-T' obtained is enzymatically, chemically, and conformationally similar to native nuclease-T'.[7,10]

Remarks. The experiment leading to the procedure for the preparation of semisynthetic nuclease-T' was an effort to purify synthetic-(6–47) on an affinity column of [nuclease-T-(49–149)]-Sepharose.[6] Such chromatography did lead to significant purification of crude synthetic peptide, to about 30% enzymatic activity. A higher degree of purification apparently was prevented by the existence of synthetic peptide which bound, at least weakly, to the complementary native fragment but was either nonproductive or only partially productive enzymatically. It is the elimination of these latter classes of contaminant which appear to be the key to the much higher purity of synthetic-(6–47) obtained as part of semisynthetic nuclease-T'.

The ability to prepare semisynthetic complex provides a means to obtain chemically mutated nuclease-T' derivatives with sequence changes which are not attainable by other chemical procedures. Thus, in one case, an active site glutamic acid has been replaced by an aspartyl residue, with the resultant [aspartic acid⁴³]semisynthetic nuclease-T' species obtained and shown to be structurally like native nuclease-T'; it is capable of binding active site ligands but inactive enzymatically.[10] Other active site semisynthetic analogues have also been prepared more recently,[11] and there is also the possibility to make conformationally perturbed analogs as well.[9]

Example 2. Semisynthetic Bovine Pancreatic Ribonuclease-S'

Approach. Ribonuclease-S' is, like nuclease-T', an enzymatically active, noncovalent protein complex, which, in this case, contains two ribonuclease-A fragments, ribonuclease-S-(1–20) and ribonuclease-S-(21–124).[12] The ribonuclease peptide containing residues 1 through 15 can replace ribonuclease-S-(1–20) in this fully active complex.[13] Again, existence of the noncovalent complex provides the opportunity to prepare synthetic derivatives of the (1–20) or (1–15) sequence and to study the effects of synthetic alteration in the reconstituted complex with ribonuclease-S-(21–124). As with the synthetic-(6–47) fragment of nuclease, the solid phase synthesized (1–15) fragment is only partially active when obtained after removal of deblocking groups and Sephadex fractionation.[14,15] However, purification has been possible through the preparation

[11] G. R. Sánchez, I. M. Chaiken, and C. B. Anfinsen, J. Biol. Chem. 248, 3653 (1973).
[12] F. M. Richards and P. J. Vithayathil, J. Biol. Chem. 234, 1459 (1959).
[13] J. T. Potts, Jr., D. M. Young, and C. B. Anfinsen, J. Biol. Chem. 238, 2593 (1963).
[14] I. Kato and C. B. Anfinsen, J. Biol. Chem. 244, 5849 (1969).
[15] I. M. Chaiken, M. H. Freedman, J. R. Lyerla, Jr., and J. S. Cohen, J. Biol. Chem. 248, 884 (1973).

of semisynthetic ribonuclease-S',[15] a tactic used originally for fragment (1–20) synthesized by solution techniques.[16]

Preparation of highly purified semisynthetic ribonuclease-S' is, in principle, similar to that used for nuclease-T'. Crude synthetic-(1–15) is allowed to form a complex with native ribonuclease-S-(21–124) under conditions appropriate for native S' formation. In this case, the correctly formed semisynthetic ribonuclease-S' complex can be isolated by direct chromatography of the fragment mixture on sulfoethyl-Sephadex. Elution is accomplished with phosphate buffers. Since phosphate is a competitive inhibitor of ribonuclease-S', with a K_i = 6.5 mM at pH 6.5,[17] the phosphate buffers used probably serve to stabilize the semisynthetic complex present. Uncomplexed synthetic peptide, uncomplexed ribonuclease-S-(21–124), and semisynthetic complex are all distinguished in the ion exchange procedure.

Sample Procedure.[18] Crude solid-phase derived synthetic-(1–15) (40 mg of about 35% potential enzymatic activity) is mixed with 80 mg of ribonuclease-S-(21–124) in 4 ml of 0.1 M sodium phosphate buffer, pH 6.5, at 7° for 30 minutes. The mixture is fractionated on a sulfoethyl-Sephadex C-25 column, 2 × 150 cm, by elution first with 0.1 M sodium phosphate at pH 6.5, and thereafter with 0.2 M sodium phosphate at the same pH (Fig. 3). In this separation, uncomplexed synthetic material elutes with 0.1 M buffer; excess ribonuclease-S-(21–124), when it exists, elutes with 0.2 M buffer after the normal S' complex, although the relative positions of complex and (21–124) fragment can vary in cases where (1–15) species of analog sequence are involved.[19]

Results. Semisynthetic ribonuclease-S' obtained as described above is essentially 100% as active enzymatically as fully native ribonuclease-S'. In addition, the K_m for cyclic cytidine-2',3'-monophosphate is similar for native and semisynthetic complexes.[15]

Remarks. The procedure for preparing semisynthetic ribonuclease-S' includes polypeptide incubation for complex formation (transformation) and subsequent chromatography (separation). In striking contrast to the procedure used for semisynthetic nuclease-T', no steps are required in the transformation process to destroy fragments which form nonproductive or weak complexes. In part, this suggests that most of the synthetic-(1–15) which binds strongly to ribonuclease-S-(21–124) must form fully productive complex. If contaminant synthetic peptide exists which forms a significant amount of nonproductive or weakly productive complexes, these

[16] K. Hofmann, M. J. Smithers, and F. M. Finn, *J. Amer. Chem. Soc.* 88, 4107 (1966).

[17] D. G. Anderson, G. C. Hammes, and F. G. Walz, Jr., *Biochemistry* 7, 1637 (1968).

[18] I. M. Chaiken, unpublished results.

[19] B. M. Dunn and I. M. Chaiken, unpublished results.

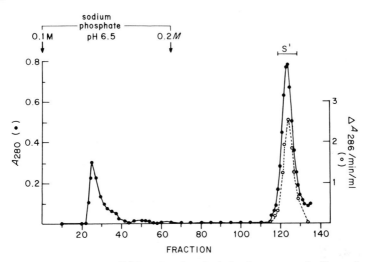

FIG. 3. Separation of [His^{12}-^{13}C$_{ImC2}$]semisynthetic bovine pancreatic ribonuclease-S′ (complex with, at residue position 12, histidine containing 95% carbon-13 at the imidazole C-2 position) by sulfoethyl-Sephadex chromatography (I. M. Chaiken, unpublished results). Enzymatic activity, given in units of ΔA_{286}/minute/ml, is against 0.25 mM cyclic cytidine-3′,5′-monophosphate in 50 mM Tris chloride, pH 7.13, containing 55 mM NaCl, at 25°. Here, S′ denotes the resolved complex. Absorbances at 280 nm are direct readings.

complexes must be chromatographically distinct from normal semisynthetic complex. Here again, it is the apparent existence of nonproductive complex-forming peptides which has interfered with purification of solid phase derived synthetic-(1–15) by affinity chromatography on [ribonuclease-S-(21–124)]-Sepharose.[14]

In addition to studies of the intact semisynthetic ribonuclease-S′ complex, this material acts as an intermediate for the further isolation of purified synthetic-(1–15), by fractionation of the complex on Sephadex G-75 in 50% acetic acid.[19] The (1–15) peptide so obtained can be used to study both kinetic and equilibrium properties of interaction with the complementing ribonuclease-S-(21–124) fragment.

Example 3. Synthetic Ribonuclease-A

Gutte and Merrifield[20] have synthesized the complete sequence of bovine pancreatic ribonuclease-A by the solid phase procedure. The crude synthetic product obtained after removal of side chain protecting groups required extensive further fractionation. In this case, purification was based on the application of several solution properties of the correct se-

[20] B. Gutte and R. B. Merrifield, *J. Biol. Chem.* **246**, 1922 (1971).

quence, including, most notably, proper disulfide bond formation upon reoxidation of sulfhydryl groups and resistance of the proper disulfide-containing folded polypeptide to tryptic destruction. In the former instance, products of incorrect sequence formed incorrect disulfide pairs, leading either to aggregates or to incorrectly folded molecules. In the latter case, contaminant polypeptides which assumed improperly folded forms not resembling native ribonuclease-A were destroyed. For both cases, resultant peptide of correct structure was isolated by specific chromatographic procedures. The combined use of these function-based procedures was critical to the ability to obtain a synthetic ribonuclease-A with close to 80% of the specific enzymatic activity of the native enzyme.

Limitations

In the application of functional purification to synthetic proteins, it is implicit that a binding or other specific property be available that can be exploited efficiently. When such properties are not available or cannot be applied conveniently, this approach to purification becomes ineffective. In addition, even when functional purification of a particular species is possible, it may not be applicable to a synthetic analog of that species if the intended alteration in sequence results in too drastic a change in the functional property. Finally, even when functional purity is achieved, chemical purity is not necessarily a result.

Conclusion

Functional isolation methods provide a mechanism wherein synthetic proteins and protein fragments can be obtained. The approach can be applied for a variety of systems, including, in addition to the types described here, hormones, protease inhibitors, and pseudosubstrates of enzymes. Where the purification procedures can be applied, they are generally efficient and allow synthetic products to be prepared that are of a quality suitable for detailed physicochemical and functional characterization to a degree expected for native proteins and protein derivatives. Thus, functional purification allows the protein chemist to use synthesis as a tool for predetermined chemical mutation, which, especially in conjunction with a knowledge of three-dimensional structures determined to atomic resolution, can permit the structural detail of proteins to be manipulated with great latitude and at a quite subtle level.

[83] Purification of Native and Synthetic Lysozyme with
Tri-(*N*-acetylglucosamine)-Agarose

By DENNIS A. CORNELIUS, WILLIAM H. BROWN,
ANDREW F. SHRAKE, and JOHN A. RUPLEY

Affinity chromatography of lysozyme has been carried out using a
variety of adsorbents (descriptions of work not cited here can be found
in reference 1): dispersed chitin,[2-7] carboxymethylated chitin,[8,9] deami-
nated chitin,[10,11] and chitin-coated cellulose.[12,13] We describe here the preparation and use of an affinity adsorbent based on a low molecular weight
substrate analog of lysozyme. The procedure for preparation of the adsorbent is general and could be used for coupling other saccharides to
agarose.

The ligand is the $\beta(1 \to 4)$-linked trisaccharide of *N*-acetylglucosa-
mine (GlcNAc) joined to an arm containing an amino group and suitable
for CNBr coupling to Sepharose. This compound (III, Fig. 1) was prepared by reductive-hydrogenolysis of the phenylhydrazone of (GlcNAc)$_4$
as described in Fig. 1 and in the following section. The use of a trisaccharide derivative was dictated by the fact that smaller ligands bind less
strongly, and large oligomers no more strongly, than the trimer.[1]

Preparation of the Affinity Adsorbent

Phenylhydrazone of (*GlcNAc*)$_4$. A solution of 1 g (1.2 mmoles) of the
tetrasaccharide of *N*-acetylglucosamine (I) and 2 ml of phenylhydrazine

[1] T. Imoto, L. N. Johnson, A. C. T. North, D. C. Phillips, and J. A. Rupley, *in* "The
Enzymes" (P. D. Boyer, ed.), Vol. 7, p. 665. Academic Press, New York, 1972.
[2] H. Nozu, *Osaku Daigaku Igaku Zasshi* **12**, 1531 (1960).
[3] I. A. Cherkasov, N. A. Kravchenko, and E. D. Kaverzneva, *Dokl. Akad. Nauk
SSSR* **170**, 213 (1966).
[4] I. A. Cherkasov and N. A. Kravchenko, *Mol. Biol.* **1**, 381 (1967).
[5] I. A. Cherkasov and N. A. Kravchenko, *Mol. Biol.* **1**, 326 (1967).
[6] I. A. Cherkasov, N. A. Kravchenko, and E. D. Kaverzneva, *Mol. Biol.* **1**, 41
(1967).
[7] I. A. Cherkasov and N. A. Kravchenko, *Biokhimiya* **33**, 761 (1968).
[8] T. Imoto, K. Hayashi, and M. Funatsu, *J. Biochem. (Tokyo)* **64**, 387 (1968).
[9] T. Imoto, Y. Doi, K. Hayashi, and M. Funatsu, *J. Biochem. (Tokyo)* **65**, 667
(1969).
[10] *Chem. Eng. News,* September 4, 1972, p. 16 (P. P. McCurdy, ed.); F. Katz, *J. Biol.
Chem.* in press (1974).
[11] I. A. Cherkasov and N. A. Kravchenko, *Biokhimiya* **34**, 1089 (1968).
[12] T. Imoto and K. Yagishita, *Agr. Biol. Chem.* **37**(3), 465 (1973).
[13] T. Imoto and K. Yagishita, *Agr. Biol. Chem.* **37**(5), 1191 (1973).

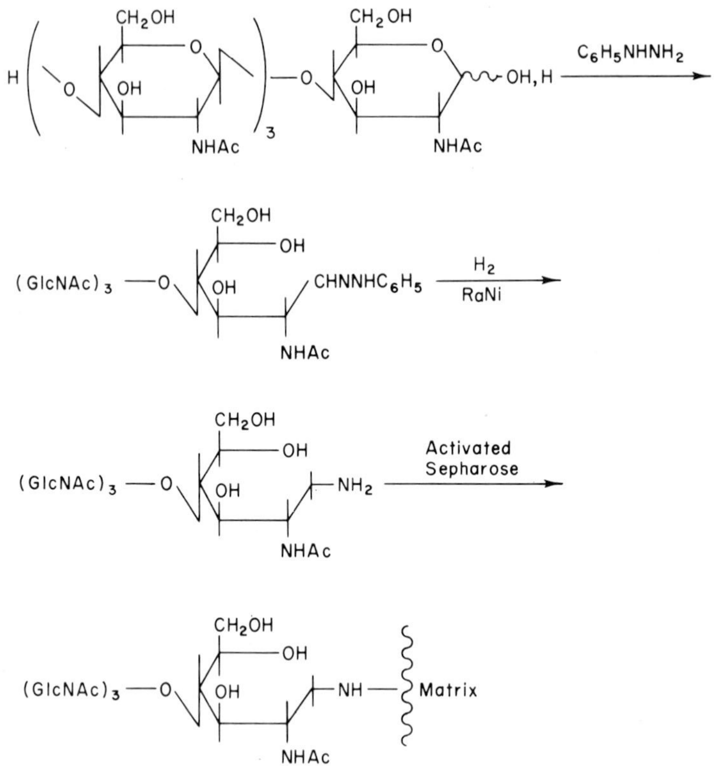

Fig. 1. Synthesis of tri-(N-acetylglucosamine)-Sepharose affinity resin.

in 3.5 ml of water and 4.2 ml of ethanol is refluxed gently for 15 days. The solution is cooled to room temperature, water (30 ml) is added, and the solution is extracted with five 25-ml portions of ether. The aqueous layer is concentrated to dryness under reduced pressure, and the resulting solid material is collected, washed with ether, and finally dried under reduced pressure to provide 890 mg (80%) of yellow solid. This material (II) is used without further purification.

Reductive Hydrogenolysis of (II). A mixture containing 880 mg of the crude phenylhydrazone (II), 40 ml of water, and ca. 0.5 teaspoon of freshly prepared Raney nickel W-2 catalyst is shaken for 24 hours on a Parr apparatus under hydrogen at 35–40 psi. The reaction mixture is filtered through a Celite pad to remove catalyst. The filtrate is concentrated to ca. 10 ml on a rotary evaporator, then charged onto a 1 × 30 cm column of Dowex 50-W X2 (H⁺ cycle). The column is first eluted with 150 ml of water. The entire eluate is collected and lyophilized, giving

210 mg of unreacted tetramer. The column is next eluted with 200 ml of 2 M NH₄OH. The entire eluate is again collected and lyophilized to yield 430 mg (54%) of the amine (III) as a tan solid.

Coupling of (III) to the Matrix.[14] The solution obtained from the careful addition of 10 g of cyanogen bromide to 80 ml of water is added with magnetic stirring to 60 g of well washed and packed wet Sepharose 4B. The pH of the solution is quickly raised to 11.0 by the dropwise addition of 4 N NaOH. The reaction temperature is maintained as close as possible to 20° by the addition of ice to the thermostatting bath, and the pH of the reaction is kept near 11 by manual titration. Reaction is complete in ca. 15 minutes as indicated by the cessation of base uptake. The reaction mixture is poured quickly over ca. 200 ml of coarsely ground ice, filtered to partial dryness, washed with 500 ml of ice-cold 0.1 M sodium borate at pH 10, and then quickly transferred to a 250-ml flask. Fifty milliliters of the ice-cold borate buffer and 100 mg of the amine (III) are added to the activated Sepharose with external cooling. The reaction mixture is stirred at room temperature for 24 hours, after which the adsorbent is filtered and successively washed with 200 ml of water and 200 ml of 0.1 N acetic acid. The adsorbent is stored under refrigeration in 0.1 N acetic acid containing a few drops of chloroform to retard bacterial growth.

More highly substituted Sepharose adsorbents are obtained by increasing the amount of amine (from 100 mg to 1.0 or 2.5 g) while keeping other reaction parameters the same.

The extent of coupling of the ligand may be determined by acid hydrolysis of the adsorbent (3 M HCl, 14 hours at 110°) and analysis for glucosamine with an amino acid analyzer. The amount of ligand bound was directly proportional to the concentration in the reaction mixture and was 2% of the ligand added.

Preparation of Columns. A slurry containing an appropriate quantity of resin in the storage solution is degassed and freed of chloroform under an aspirator. Columns packed with degassed adsorbent are successively washed with several column volumes of starting buffer, at least 5 volumes of 0.1 M acetic acid containing 1 M NaCl and 0.1 mM (GlcNAc)₃, and finally with 3 column volumes of 6 M guanidine hydrochloride (Ultrapure, Mann). These washings are repeated, then the column is washed with several column volumes of starting buffer to ensure complete removal of all guanidine hydrochloride.

Columns are reconditioned after each experiment by washing with

[14] P. Cuatrecasas and C. Anfinsen, this series, Vol. 22, p. 345.

6 M guanidine hydrochloride before equilibration with starting buffer. For storage, columns are equilibrated with 6 M guanidine hydrochloride and kept at 5°. There was no apparent change in operation of a column during a period of more than one year.

Methods of Use

Native Lysozyme. The table gives the recoveries of native lysozyme from an affinity column using various buffers for elution. The composition of the buffer in which the protein was applied to the column did not affect retention or elution behavior: retention was quantitative with 0.1 M ammonium acetate, 0.1 M acetic acid, 0.1 M sodium phosphate, or 1.0 M

ELUTION OF LYSOZYME FROM AFFINITY ADSORBENT[a]

Expt. No.	Second buffer	Percent of protein sample recovered
1	1.0 M NaCl, 0.1 mM (GlcNAc)$_3$, 0.1 M HOAc, pH 2.5	100
2	0.3 M NaCl, 0.1 mM (GlcNAc)$_3$, 0.1 M HOAc, pH 2.5	95
3	0.1 M NaCl, 0.1 mM (GlcNAc)$_3$, 0.1 M HOAc, pH 2.5	Broad protein peak; recovery not determined
4	1.0 M NaCl, 0.1 mM (GlcNAc)$_3$, neutral pH	80
5	1.0 M NaCl, 0.1 M HOAc, pH 2.5 (no inhibitor added)	Broad protein peak; recovery not determined
6	1 mM (GlcNAc)$_3$, 0.1 M HOAc, pH 2.5 (no salt added)	<10
7	1.0 M NH$_4$OAc, 1 mM (GlcNAc)$_3$, 2.5 M HOAc, pH 4	100
8	1.0 M NH$_4$OAc, pH 7, 1 mM (GlcNAc)$_3$	80
9	0.1 M NH$_4$OAc, pH 7, 1 mM (GlcNAc)$_3$	65
10	1.0 M NH$_4$Cl, 0.1 mM (GlcNAc)$_3$, 0.1 M HOAc, pH 2.5	85
11[b]	0.1 M Na phosphate, pH 5.7, 0.1 mM (GlcNAc)$_3$	80

[a] Conditions: 0.1 mg of lysozyme was applied in 0.1 ml of the starting buffer to a 0.35 × 22 cm column (2 ml volume). Elution was at a flow rate of approximately 5 ml per hour and was first with 3 column volumes of starting buffer and then with the second buffer as indicated. Protein concentration was determined by measurement of absorbance of the effluent at 280 or 254 nm. The starting buffer was 0.1 M acetic acid, pH 2.5, or 0.1 M NH$_4$OAc, pH 7, except for the experiment indicated by superscript b.

[b] Starting buffer was 0.1 M sodium phosphate of pH 5.7.

NaCl. The column quantitatively adsorbed lysozyme from dilute solution (0.1 mg of lysozyme from 10 ml of 0.8 M Tris chloride at pH 8).

The elution behavior of the protein depended on the buffer used (see the table) and can be summarized as follows: (a) High salt concentration, low pH, and inhibitor [0.1 mM or greater of (GlcNAc)$_3$] are required for quantitative elution (experiments 1 and 2). (b) The protein was leached from the column by high salt in acetic acid [no (GlcNAc)$_3$], but not by inhibitor in acetic acid with no salt (experiments 5 and 6, respectively). (c) Useful extents of recovery of protein ($\geq 80\%$) were obtained using other buffers. The use of volatile buffers can be of sufficient advantage to compensate for the lower recovery of sample from the column.

The experiments described in the table and the preceding paragraph were carried out in order to determine optimum elution conditions for handling small amounts of lysozyme such as are manipulated during purification of synthetic protein. If the amount of protein used is greater, recovery is higher; under the conditions of experiment 9 of the table, recovery at 5 times greater load was 90%. The adsorbent used in the experiments of the table was the lightly substituted sample, the preparation of which was described above. Recoveries were comparable from adsorbents that were more highly substituted with the inhibitor ligand.

The elution of native lysozyme using a gradient in (GlcNAc)$_3$ following a wash with starting buffer is shown in Fig. 2. For each of the three runs, the amounts of native enzyme applied to the column, the column conditions, and the compositions of starting and finishing buffers were approximately the same; the volumes of the solutions used to establish the gradient were increased successively, however, in order to vary the rate of change of (GlcNAc)$_3$ concentration. The recoveries of protein were approximately 90%. The concentration of (GlcNAc)$_3$ at which lysozyme eluted was approximately 5×10^{-5} M for all three gradients of Fig. 2. Elution at this concentration is expected from the association constant for the binding of (GlcNAc)$_3$ to lysozyme ($K_{assoc} = 10^5$ M^{-1}).[1]

Oxindole-62 Lysozyme. Oxindole-62 lysozyme has been identified and characterized as the principal oxidation product of the reaction of N-bromosuccinimide with native lysozyme.[15] Because other oxidation products and native enzyme are also present in the reaction mixture, it is necessary to purify the oxindole-62 protein. The oxindole-62 lysozyme was separated from native lysozyme using a 1 × 15 cm affinity column

---

[15] K. Hayashi, T. Imoto, G. Funatsu, and M. Funatsu, *J. Biochem. (Tokyo)* **58**, 227 (1965).

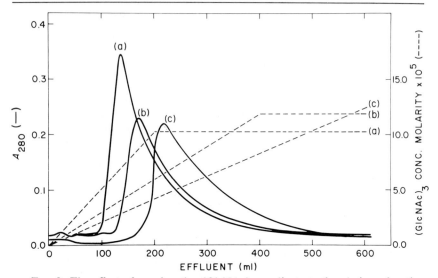

FIG. 2. The effect of varying the (GlcNAc)₃ gradient on the elution of native lysozyme from the affinity column. Fifteen milligrams of native protein in 1 ml of starting buffer (0.1 M NH₄OAc, pH 7.0) was applied to a 1.7 × 26 cm column. The gradients were begun after washing the column with ca. 5 column volumes (5 × 35 ml). The finishing buffer was 0.1 mM (GlcNAc)₃ in starting buffer. The gradients were formed using: 100 ml of starting buffer vs. 100 ml of finishing buffer (a); 200 ml vs. 200 ml (b); and 300 ml vs. 300 ml (c). The flow rate was 23 ml per hour.

equilibrated with 0.1 M ammonium acetate. About 10 mg of this binary mixture in the same buffer were applied to the column; the flow rate was 24 ml per hour, and the fraction volume was 2.4 ml. Oxindole-62 lysozyme was retained only slightly; a solution of 0.1 mM trimer in starting buffer was used to elute native lysozyme. The purity and homogeneity of the oxindole-62 lysozyme component was demonstrated by the constant ratios of absorbance at 280 nm to absorbance at 250 nm and by the constant values of the specific lytic 'activity for the peak fractions.

Synthetic Lysozyme. Polypeptide material with the enzymatic activity of lysozyme has been prepared using the Merrifield procedures by groups at La Jolla and Tucson.[16,17] Affinity adsorbents have been used to purify the crude products from these syntheses. In one set of experiments 10 mg of crude product (synthesis 4, described by Barstow and co-workers[18] con-

[16] J. J. Sharp, A. B. Robinson, and M. D. Kamen, *Fed. Proc., Fed. Amer. Soc. Exp. Biol.* **30**, 1273 (1971).
[17] L. E. Barstow, V. J. Hruby, A. B. Robinson, J. A. Rupley, J. J. Sharp, and T. Shimoda, *Fed. Proc., Fed. Amer. Soc. Exp. Biol.* **30**, 1274 (1971).

taining activity corresponding to 70 mg of native lysozyme, was separated on a 0.2 × 15 cm affinity column, giving a single enzymatically active component that was eluted by $(GlcNAc)_3$. This material had 20% the specific activity of native lysozyme, based on the rate of cell lysis and on protein concentration determined by absorbance. This specific activity is 30-fold above that of the crude product. The material from the affinity column was heterogeneous, as judged by a chromatographic separation using Bio-Rex 70 which provided an active component with greater than 70% of the specific activity of native lysozyme.

We note, however, that affinity chromatography has not been successful with all preparations of the synthetic material. Polypeptide samples obtained from the same batch of resin peptide but using different cleavage conditions behaved differently on the affinity adsorbent. Native lysozyme mixed with a sample of inactive synthetic protein did not bind to the adsorbent in 0.1 M acetic acid, although it did in 0.1 M ammonium acetate. Apparently, impurities in the synthetic mixture affect elution behavior.

---

[18] L. E. Barstow, D. A. Cornelius, V. J. Hruby, T. Shimoda, J. A. Rupley, J. J. Sharp, A. B. Robinson, and M. D. Kamen, *in* "Chemistry and Biology of Peptides" (J. Meienhofer, ed.), p. 331. Ann Arbor Science Publishers, Ann Arbor, Michigan, 1972.

[84] Rapid Separation of Synthetic Oligonucleotide Mixtures on Aromatic Cellulose Derivatives

By M. FRIDKIN, P. J. CASHION, K. L. AGARWAL, E. JAY, and H. G. KHORANA

The chemical synthesis of oligonucleotides usually involves the coupling of a protected segment, containing a free 5'-phosphate group, with the 3'-hydroxyl end group of a growing oligonucleotide chain, whose 5'-hydroxyl end generally carries a trityl derivative.[1] One of the major progress-limiting factors in such syntheses is the difficulty associated with separation of the desired product from the multiple components of a reaction mixture. Presently used fractionation procedures involve mainly anion-exchange chromatography and might be tedious, time-consuming, and often unsuccessful.

---

[1] K. L. Agarwal, A. Yamazaki, P. J. Cashion, and H. G. Khorana, *Angew. Chem. Int. Ed. Engl.* **11**, 451 (1972).

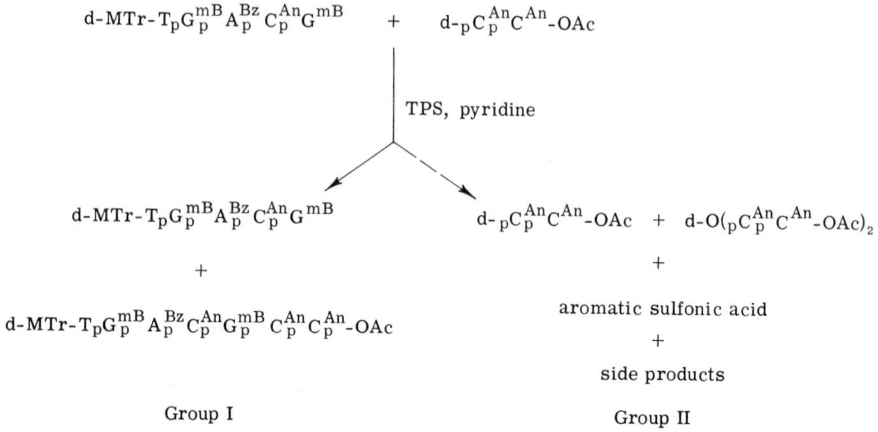

$$d\text{-MTr-}T_pG_p^{mB}A_p^{Bz}C_p^{An}G^{mB} \quad + \quad d\text{-}_pC_p^{An}C^{An}\text{-OAc}$$

TPS, pyridine

$$d\text{-MTr-}T_pG_p^{mB}A_p^{Bz}C_p^{An}G^{mB}$$

$+$

$$d\text{-MTr-}T_pG_p^{mB}A_p^{Bz}C_p^{An}G^{mB}C_p^{An}C_p^{An}\text{-OAc}$$

$$d\text{-}_pC_p^{An}C^{An}\text{-OAc} \quad + \quad d\text{-O}(_pC_p^{An}C^{An}\text{-OAc})_2$$

$+$

aromatic sulfonic acid

$+$

side products

Group I   Group II

SCHEME I. A typical reaction mixture and products.

In general the components in a synthetic mixture (Scheme I)[2] may be divided into two subgroupings: those terminated by a trityl group (type I, see Scheme I) and those that lack this aromatic functionality (type II, see Scheme I). Rapid and simple separation of the two groups from each other can be accomplished by taking advantage of the more lipophilic nature of the trityl-terminated oligomers. These derivatives are retained specifically on trityl and naphthylcarbamoyl derivatives of cellulose (see Fig. 1).[3] Further fractionation of group I may then be easily and efficiently achieved by anion-exchange chromatography.

Materials and Methods

Cellulose powder (CF-11, Whatman Chemical Co.) is dried before use for 10 hours at 80–100° under reduced pressure. Trityl chloride

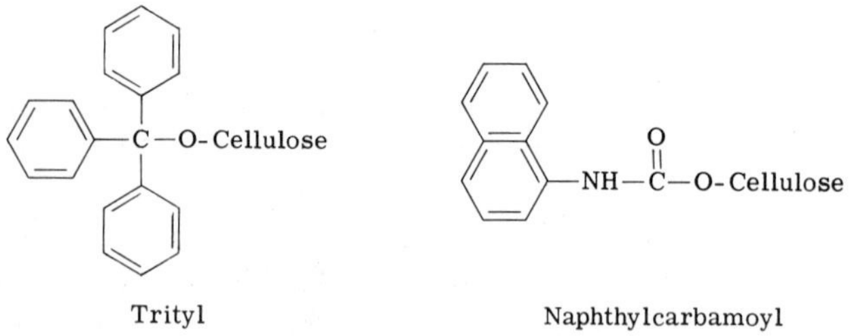

Trityl   Naphthylcarbamoyl

FIG. 1. Aromatic cellulose derivatives prepared.

[2] Abbreviations used are: TPS, 2,4,6-triisopropylbenzenesulfonyl chloride; TEAB, triethylammonium bicarbonate pH 7.5–8.0; Bz, benzoyl; An, anisoyl; iB, isobutyryl: mB, 2-methylbutyryl; MTr, monomethoxytrityl; TPM, p-(triphenylmethyl)aniline.
[3] P. J. Cashion, M. Fridkin, K. L. Agarwal, E. Jay, and H. G. Khorana, Biochemistry 12, 1985 (1973).

(Fisher Chemical Co.), crystallized before use from anhydrous n-pentane in the presence of $SOCl_2$. 1-Naphthyl isocyanate is purified by fractional distillation. Reagent grade pyridine is purified and dried as described elsewhere.[4]

Preparation of Trityl Cellulose

Anhydrous pyridine (180 ml) is added to a mixture of 15 g of dried cellulose (93 mmoles of glycosyl residue) and 38 g of recrystallized trityl chloride (136 mmoles) in a round-bottom flask fitted with a reflux condenser. The resultant suspension is stirred slowly and refluxed gently for 4 hours, whereupon 100 ml of 10% aqueous ethanol are added to terminate the reaction. The reaction mixture is cooled and filtered, and the resinous product is packed into a column and washed thoroughly with ethanol to remove pyridine and trityl alcohol. Washing is continued until the effluent is free from ultraviolet absorption at 260 nm and at 430 nm after acidification with 60% perchloric acid. After air-drying, approximately 19 g of a white-brownish powder are obtained.

For determination of the trityl content, a dried sample of about 10 mg is suspended in 60% perchloric acid and the absorption of the released, soluble, trityl-cation is measured at 430 nm ($E_{430} = 35,200$). In general, about 0.2 trityl group is linked per glucosyl residue of cellulose.

Preparation of Naphthylcarbamoyl Cellulose

A suspension of dried cellulose, 60 g, in 300 ml of anhydrous pyridine is stirred mechanically while 29.5 g of 1-naphthyl isocyanate (0.17 mole) are added. The reaction mixture is heated at 70° for 16 hours with stirring and exclusion of moisture. To the brown reaction mixture is added a mixture of $CHCl_3$–MeOH (600 ml; 1:2 v/v), and the insoluble product is filtered and washed with the same solvent until the washings are free from ultraviolet absorption at 280 nm. The resin is then washed with ethanol and dried under reduced pressure to yield about 85 g. Derivatization, as calculated on the basis of the increase in weight of cellulose, was found to be 15% of the total hydroxyl groups.

Chromatographic Procedures

Columns are packed and separations are usually performed at 4°. A suspension of the cellulose derivative in 25–40% ethanol, containing 50–100 mM triethylammonium bicarbonate at pH 7.5 is poured into a column of desired size and allowed to pack at a fast rate (about 5–10 ml per minute). The synthetic oligonucleotide mixture (concentration about 1 mM) in the same solvent is applied to the column. Washing is

[4] K. L. Agarwal, A. Yamazaki, and H. G. Khorana, *J. Amer. Chem. Soc.* 93, 2754 (1971).

TABLE

FRACTIONATION OF OLIGONUCLEOTIDE MIXTURES ON AROMATIC CELLULOSE DERIVATIVES

| Cellulose derivative | Oligonucleotide mixture | | % Ethanol-TEAB(M) for elution of oligonucleotides | |
	Reactants	Product	Non-trityl compounds	Trityl compounds
Trityl	d-MTr-T + d-pC^{An}-OAc	d-MTr-T$_p$C^{An}	40–0.10	90–0.10
	d-MTr-T$_p$C$_p$A^{An}G$_p$^{iB}A^{Bz} + d-pC$_p$A^{An}C^{An}-OAc	d-MTr-T$_p$C$_p$A^{An}G$_p$^{iB}A^{Bz}$_p$C$_p$A^{An}C^{An}	45–0.10	75–0.50
	d-MTr-T$_p$G^{mB}$_p$A^{Bz}$_p$C^{An}$_p$G^{mB} + d-pC^{An}$_p$C^{An}-OAc	d-MTr-T$_p$G^{mB}$_p$A^{Bz}$_p$C^{An}$_p$G^{mB}$_p$C^{An}$_p$C^{An}	40–0.10	75–0.50
	d-MTr-T$_p$C^{An}$_p$G^{iB}$_p$A^{Bz}$_p$C^{An}$_p$C^{An}$_p$T$_p$G^{iB} + d-pC^{An}$_p$A^{Bz}$_p$G^{iB}$_p$A^{Bz}-OAc	d-MTr-T$_p$C^{An}$_p$G^{iB}$_p$A^{Bz}$_p$C^{An}$_p$C^{An}$_p$T$_p$G^{iB}$_p$C^{An}$_p$A^{Bz}$_p$G^{iB}$_p$A^{Bz}	45–0.10	75–0.50
	d-TPM-T$_p$T + d-$_p$T-OAc	d-TPM-T$_p$T-OAc	25–0.10	70–0.10
Naphthyl-carbamoyl	d-MTr-G^{mB}$_p$G^{mB}$_p$T + d-$_p$C^{mB}$_p$G^{mB}-OAc	d-MTr-G^{mB}$_p$G^{mB}$_p$T$_p$G^{mB}$_p$G^{mB}	25–0.05	40–0.50
	d-MTr-T$_p$C^{mB}$_p$A^{Bz}$_p$C^{mB}$_p$G^{mB} + d-pC^{An}$_p$C^{An}$_p$C^{An}-OAc	d-MTr-T$_p$C^{mB}$_p$A^{Bz}$_p$C^{An}$_p$G^{mB}$_p$C^{An}$_p$C^{An}	40–0.10	75–0.10
	d-TPM-$_p$T + d-$_p$T-OAc	d-TPM-T$_p$T-OAc	25–0.10	70–0.10

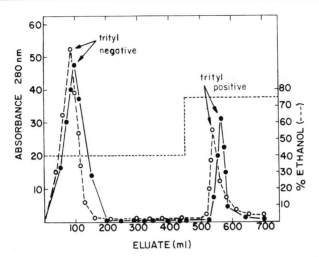

ELUATE (ml)

Fig. 2. Comparison of the fractionation of a reaction mixture, obtained on coupling of d-MTr-T$_p$G$^{mB}_p$A$^{Bz}_p$C$^{An}_p$G^{mB} with d-pC^{An}pC^{An}OAc. The reaction mixture was diluted with 200 ml of 40% ethanol–0.10 M TEAB and applied in equal portions to both tritylcellulose and naphthylcarbamoyl cellulose columns (each 1.4 × 10 cm, equilibrated with 40% ethanol–0.10 M TEAB). After application the columns were washed with 1 liter of 40% ethanol–0.10 M TEAB followed by 1 liter of 75% ethanol–0.50 M TEAB for the trityl cellulose column (—○—), and by 1 liter of 75% ethanol–0.10 M TEAB for the naphthylcarbamoyl cellulose column (—●—).

continued with the same solvent until the removal of the nontrityl components are completed. The trityl-containing components are then eluted by increasing the ethanol and salt concentrations. Flow rates are markedly increased by increasing the ethanol concentration of the eluent.

Conditions for the fractionation of several synthetic mixtures of oligonucleotides, on trityl and naphthylcarbamoyl cellulose, are listed in the table. A typical separation pattern is shown in Fig. 2. The experimental findings show that both aromatic derivatives of cellulose are, in general, satisfactory for the separation of the trityl from nontrityl products. However, it is clear that trityl-containing compounds are absorbed to trityl cellulose more strongly than to naphthylcarbamoyl cellulose and, consequently, higher salt concentration are required for elutions larger segments from trityl cellulose. Elution from naphthylcarbamoyl cellulose usually requires low salt.

Both cellulose-derivatives show considerable affinity for other aromatic protecting groups used in oligonucleotide synthesis including TPM-phosphoramidates[4] and the naphthylcarbamoyl[5] group.

[5] K. L. Agarwal and H. G. Khorana, *J. Amer. Chem. Soc.* **94**, 3578 (1972).

Section V

Purification of Receptors

[85] Insulin Receptors

By PEDRO CUATRECASAS and INDU PARIKH

Structures capable of specifically binding insulin, and having properties consistent with those expected for specific receptor interactions, have been identified in many cell types and in isolated membrane preparations (reviewed in references cited in footnotes 1–3). The receptors for this as well as other peptide hormones appear to be localized exclusively to the surface membranes of cells.

The ability to extract and solubilize the binding proteins in intact form from fat- and liver-cell membranes with nonionic detergents, and the availability of a rapid, quantitative assay for the soluble receptors, provide the necessary basis for studies designed to purify these membrane proteins.[4,5] The extraordinarily small quantity of insulin receptors in mammalian-cell membranes and the absolute requirement for the presence of detergents to maintain their solubility pose special problems in purification studies. For example, the difficulties associated with purification of this receptor can in part be appreciated by considering that despite the large size (about 300,000) of the macromolecule, it represents only about $2 \times 10^{-4}\%$ of the protein of the liver homogenate and about $4 \times 10^{-3}\%$ of the membrane protein. A total of about 0.2 mg of receptor protein would be expected to be present in the liver homogenate obtained from 40 rats, which contains about 90 g of protein. Thus, it is estimated that to achieve complete purity would require a 400,000-fold purification of the crude liver homogenate. The formidable nature of this purification can be contrasted with the purification of another membrane-localized receptor, that for acetylcholine, which requires only 300- to 1000-fold purification in order to attain a homogeneous protein.[6,7]

The ability of insoluble insulin-agarose derivatives to simulate the biological effects of native insulin in a variety of cells[3] and the direct interaction of fat cells with insulin agarose beads[8] suggest that such in-

[1] P. Cuatrecasas, *Fed. Proc., Fed. Amer. Soc. Exp. Biol.* **32**, 1838 (1973).
[2] B. Desbuquois and P. Cuatrecasas, *Annu. Rev. Med.* **24**, 233 (1973).
[3] P. Cuatrecasas, *Annu. Rev. Biochem.* **43** (in press) 1974.
[4] P. Cuatrecasas, *Proc. Nat. Acad. Sci. U.S.* **69**, 318 (1972).
[5] P. Cuatrecasas, *J. Biol. Chem.* **247**, 1980 (1972).
[6] E. Karlsson, E. Heilbronn, and L. Widlund, *FEBS Lett.* **28**, 107 (1972).
[7] R. W. Olsen, J. C. Meunier, and J.-P. Changeux, *FEBS Lett.* **28**, 96 (1972).
[8] D. D. Soderman, J. Germershausen, and H. M. Katzen, *Proc. Nat. Acad. Sci. U.S.* **70**, 792 (1973).

soluble hormone derivatives may interact effectively with membrane structures and are, thus, potentially useful for purification of the receptors by affinity chromatography. The selective adsorption of liver-membrane fragments that contain glucagon receptors to glucagon-agarose columns[9] also points toward the promise of these procedures in receptor purification.

It has been shown that conventional procedures can be used for purification of the solubilized insulin receptors, and that the receptor proteins can be adsorbed to, and eluted from, affinity columns with an overall purification that probably exceeds 2×10^5-fold.[10]

Special Procedures

Assays for the Binding of Insulin to Membranes and to
  Solubilized Receptors

The binding of [^{125}I]insulin (1200–1800 mCi/μmole) to liver membranes is described elsewhere.[11] The assay of soluble insulin–receptor complexes is based on the differential precipitation of the complex by polyethylene glycol in the presence of carrier γ-globulin.[4,5]

Fractionation of Tissue and Preparation of Membranes

Liver. The method used is based on conventional differential centrifugation procedures designed to yield large quantities of a microsomal preparation. The livers (370 g total) of 40 male Sprague-Dawley rats (150 g; 2 months old) is suspended in 2 liters of ice-cold 0.25 M sucrose and homogenized with a Polytron PT35ST for 3 minutes at a setting of 3.5. The homogenate is centrifuged at 600 g for 10 minutes, and the supernatant is centrifuged at 12,000 g for 30 minutes. This supernatant fluid, after it is adjusted with NaCl and MgSO$_4$ to achieve concentrations of 0.1 M and 0.2 mM, respectively, is centrifuged at 40,000 g for 40 minutes. The pellet is suspended in 1200 ml of 50 mM Tris chloride at pH 7.4 by homogenizing (Polytron) for 12 seconds, and the suspension is centrifuged at 40,000 g for 40 minutes. This step is repeated, and the pellet is suspended in a total volume of 200 ml with the Tris chloride buffer. Specific binding of insulin is determined after incubation of diluted samples with [^{125}I]insulin (7 $\times$ 10^5 cpm/ml) for 20 minutes at room temperature. No significant insulin-binding activity can be detected in the

[9] F. Krug, B. Desbuquois, and P. Cuatrecasas, *Nature* (*London*) *New Biol.* **234**, 268 (1971).
[10] P. Cuatrecasas, *Proc. Nat. Acad. Sci. U.S.* **69**, 1277 (1972).
[11] P. Cuatrecasas, *Proc. Nat. Acad. Sci. U.S.* **68**, 1264 (1971); *J. Biol. Chem.* **246**, 7265 (1971); *J. Biol. Chem.* **246**, 6522 (1971).

TABLE I
Specific Insulin-Binding Activity and
Fractionation of Liver Homogenates[a]

Sample	Volume (ml)	Protein (g)	Insulin-binding activity (cpm/mg)
Crude homogenate	2000	88 [100][b]	1,800 [100]
600 g Supernatant	1650	50 [62]	3,900 [122]
12,000 g Supernatant	1400	28 [32]	5,750 [101]
40,000 g Pellet	200	4.6 [5]	37,000 [108]

[a] Details of this experiment are described in the text. Data from P. Cuatrecasas, *Proc. Nat. Acad. Sci. U.S.* **69**, 318 (1972).
[b] In square brackets are percentages of total.

first 40,000 g supernatant with the polyethylene glycol assay. The results of this study are depicted in Table I. The insulin receptor of liver-cell homogenates is purified about 20-fold during preparation of the "membranes." However, these steps are not designed to achieve purity but rather to obtain in high yield, and by rapid and simple procedures, large amounts of material with insulin-binding activity ("membranes") that could then be used as starting material for subsequent purification procedures. The binding activity of these membranes is essentially unchanged after storage at $-20°$ for 3 months. It is notable that during preparation of the membranes the yield of binding activity is very high.

Human Placenta. Microsomal preparations are obtained by procedures identical to those described above from liver. These preparations are especially promising since the specific activity of insulin binding in these microsomes is some 20- to 30-times greater than those from liver. The procedures which are used to solubilize the receptors from these microsomes are, however, different from those used for liver microsomes.

Fat Cells. A crude, total particulate preparation of cell homogenates is used to achieve high yields of material by rapid procedures. Fat cells, suspended in Krebs-Ringer-bicarbonate-0.1% albumin, are homogenized (Polytron, 60 seconds) and centrifuged at 23,000 g for 30 minutes. The pellet is suspended in the buffer used for detergent solubilization.

Solubilization of Receptors with Detergents[4,5]

A membrane suspension (1–30 mg of protein per milliliter) in 50 mM Tris chloride at pH 7.4, is adjusted with Triton X-100 to a final concentration of 1 or 2% (v/v). The higher concentration of Triton X-100 is used with suspensions containing more than 15 mg of protein per milliliter, and the lower Triton X-100 concentration, 1%, is used for the less concentrated membrane suspensions; under these conditions more than 80% of the total binding activity is solubilized. The membrane suspension is incubated with Triton X-100 for 20–30 minutes at room temperature. The suspension is centrifuged for 70–90 minutes at 44,000 rpm in a Spinco 65 rotor at 2°. The supernatant fluid is used directly or, in certain cases, is dialyzed for 16 hours against 0.1 M sodium phosphate at pH 7.4, containing 0.1% Triton X-100. Dialysis under these conditions does not lead to loss of insulin-binding activity although a fine inactive precipitate frequently forms which can be removed by centrifugation at 30,000 rpm for 30 minutes.

With these extraction procedures, the insulin-binding activity which disappears from the membrane pellet is recovered quantitatively in the soluble fraction. Red blood cell membranes subjected to similar procedures produce solubilized proteins with insignificant capacity for specific binding of [^{125}I]insulin. The concentration of Triton X-100 in the supernatant solution can be decreased by dilution to 0.05% (v/v) without resulting in immediate precipitation of the receptor. With Triton X-100 concentrations of 0.05% or less, however, precipitation of the receptor occurs gradually during storage over several days. Dialysis of the Triton X-100 extracts against sodium phosphate (pH 7.4) or Krebs-Ringer-bicarbonate buffers which contain no detergent results in gradual but virtually complete precipitation of the insulin-binding proteins. This facile transition from the soluble to the particulate state, with no accompanying loss in insulin-binding activity, is also observed with Triton X-100 extracts of delipidated or phospholipase-digested membranes.

Another nonionic detergent, Lubrol-PX, is nearly as effective as Triton X-100 in solubilizing the insulin receptor from liver or fat-cell membranes. The presence of this detergent in the incubation medium, however, interferes more seriously with formation of the insulin-receptor complex, as determined by the polyethylene glycol binding assay. Attempts to obtain soluble insulin-binding materials by extraction of membranes with sodium dodecyl sulfate, dimethylsulfoxide, dimethylformamide, hexafluoroisopropanol, and pyridine are not successful. About 10% of the binding activity of liver-cell membranes is extracted by vigorous agitation in distilled water. Solubilization of the insulin receptor appears to be nearly quantitatively reversible, since dialysis of the extracts against detergent-free buffers at neutral pH results in the formation of precipitates that contain virtually all the insulin-binding activity originally present in the solution.

The insulin-binding macromolecules from human placental microsomes are not readily solubilized with Triton X-100. In this tissue Lubrol-PX is a much more effective reagent.

Preparation of Affinity Adsorbents

Conventional Agarose Derivatives

Some of the more useful insulin-agarose derivatives are depicted in Fig. 1. Derivatives A and B are prepared by attaching insulin by the N-terminal residue of the B chain (Blphe) or by the single lysyl residue (B29) directly to the polymer backbone, by the cyanogen bromide procedure.[12] C, one of the best derivatives, is prepared by the use of active carboxyl esters of agarose. The N-hydroxysuccinimide ester[13] of 3,3'-diaminodipropylaminosuccinyl agarose is reacted at 4° with porcine insulin (5 mg/ml) in 0.1 M sodium phosphate at pH 6.4 containing 6 M urea. After 1 hour, glycine is added and the reaction is continued at room temperature for 3 hours. Such derivatives contain 0.3–0.5 mg of insulin per milliliter of agarose. Derivative D is similarly prepared except that the buffer used for coupling is 0.1 M NaHCO₃ at pH 9.2. Derivative E is prepared by reacting the N-hydroxysuccinimide ester of diaminodipropyl-

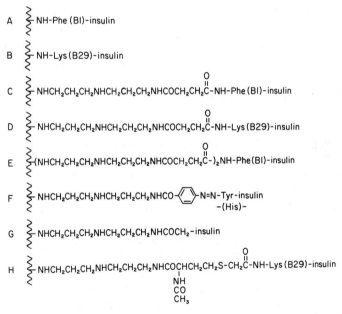

FIG. 1. Various insulin-agarose derivatives for affinity chromatography of soluble insulin receptors of liver membranes.

[12] P. Cuatrecasas, *Proc. Nat. Acad. Sci. U.S.* **63**, 450 (1969).
[13] P. Cuatrecasas and I. Parikh, *Biochemistry* **11**, 2291 (1972); See also this volume [79].

aminosuccinyl-agarose with a large excess of 3,3′-diaminodipropylamine, succinylating with succinic anhydride,[13,14] forming the corresponding N-hydroxysuccinimide ester, and coupling with insulin as described for D. Derivatives described by F are prepared by procedures similar to those used for preparation of azoglucagon-agarose.[9] At pH values near 5.5, the predominant azo linkage is with histidyl residues, whereas at pH values near 7.5–8, a mixture of azohistidyl and azotyrosyl linkages is present. Derivative G, obtained by reacting the bromoacetyl agarose derivative[15] with insulin in 0.1 M NaHCO$_3$ buffer (pH 9.0) for 2 days at 24°, contains a mixture of linkage forms (amino, histidyl, and tyrosyl). Derivative H is obtained by reacting a sulfhydryl agarose[15] with an insulin derivative having both N-terminal residues blocked with acetyl groups.[12]

Macromolecular Agarose Derivatives

Poly-(L-*lysyl-*DL-*alanyl-p-aminobezamido*)-*agarose.* (i) PREPARATION OF AGAROSE. Phenolic and histidyl groups of ligands can readily be coupled in good yields to macromolecular spacer arms[14] attached to agarose beads and which have been substituted with a diazotizable aromatic amine. Poly-(L-lysyl-DL-alanine)-agarose[14] (1.2 mg of copolymer per milliliter of gel) is suspended in 1 volume of 0.2 M NaHCO$_3$ at pH 8.5 and treated with three volumes of 50 mM p-nitrobenzoylazide (Eastman) in dioxane for 4 hours at 20°.[15] The substituted agarose is washed on a coarse-disc sintered-glass funnel (without suction) with 10–20 volumes of dioxane. The p-nitrobenzamido agarose derivative is reduced to the corresponding p-aminobenzamido derivative by reaction with an equal volume of 0.2 M sodium dithionite (Baker) in 0.5 M NaHCO$_3$ at pH 8.5 for 60–90 minutes at 20°. The poly-(L-lysyl-DL-alanyl-p-aminobenzamido)-agarose is extensively washed with 10–20 volumes each of water and dioxane.

ii. COUPLING OF INSULIN. A suspension of 7 ml of p-aminobenzamido-agarose derivative in 7 ml of 0.5 N HCl is magnetically stirred in an ice bath. An ice-cold solution (2 ml) of 1 M NaNO$_2$ is added to the agarose suspension over a 2-minute period. After stirring for an additional 5–7 minutes in an ice bath, the diazotized agarose is filtered under suction and washed rapidly with 50 ml of cold 0.05 N HCl. The agarose cake is quickly added to a solution of 70 mg of insulin (24 U/mg; Eli Lilly) dissolved in 7 ml of 0.5 M NaHCO$_3$ at pH 9.0 containing 3.5 × 10^6 cpm of [125I]insulin (1 Ci/μmole). The coupling reaction is performed for 6 hours at 4°, and for an additional 10 hours at room temperature. The

[14] I. Parikh, S. March, and P. Cuatrecasas, this volume [86].
[15] P. Cuatrecasas, *J. Biol. Chem.* **245**, 3059 (1970).

insulin-agarose is washed on a sintered-glass funnel with 500 ml of 0.2 M NaHCO$_3$. The derivatized agarose is incubated with 10 ml of 0.5 M histidine at pH 8.5 for 2 hours at room temperature to mask the un-reacted diazonium groups on the agarose matrix. The insulin-agarose is packed in a 10-ml column and washed with 3–4 liters of 5 M urea at pH 8.5 over a 4- to 5-day period at 4°. The substitution of insulin, as calculated from the radioactive content of the adsorbent, is between 4.5 and 5.0 mg per milliliter of gel.

Poly- (L-*lysyl*-DL-*alanyl-N-carboxymethyl-hydrazido*) -*agarose.*   (i) PREPARATION OF AGAROSE. The free α-NH$_2$ groups of the poly-(L-lysyl-DL-alanine)-agarose[14] react readily with ethyl bromoacetate in the presence of a base. The ethyl ester functions upon mild hydrozinolysis yield the hydrazide derivative of agarose. This derivative can be used very conveniently by the azide method for coupling peptides and proteins containing free NH$_2$ groups. Ten milliliters of poly-(L-lysyl-DL-alanine)-agarose, washed previously with 20–30 ml of saturated sodium borate at pH 9.3 is suspended in 50 ml of the same buffer and 1.5 ml of ethyl bromoacetate (Baker). The suspension, placed in a tightly capped poly-ethylene bottle, is gently shaken on a mechanical shaker for 8–10 hours at room temperature. The agarose derivative is filtered (in a well ven-tilated hood), and washed with 100 ml each of dioxane and water, and incubated in 50 ml of 3 M aqueous hydrazine for 4–6 hours. The hydra-zide derivative of agarose is then extensively washed to remove the un-reacted hydrazine. The hydrazido-agarose containing the macromolecular spacer arm can be coupled to peptides or proteins containing free amino groups.[14] The corresponding derivatives containing albumin instead of the copolymer are prepared by similar methods.

ii. COUPLING OF INSULIN. The hydrazide groups of the agarose deriva-tive are converted to azide groups in order to couple peptides and pro-teins by known methods. One milliliter of ice-cold 1 M NaNO$_2$ is added dropwise over a 1-minute period to an ice-cold suspension of 5 ml hydrazido-agarose in 7 ml of 0.3 N HCl. The suspension is magnetically stirred for an additional 2 minutes at 4°, and is rapidly filtered with suction through an ice-cold coarse-disc sintered-glass funnel. The agarose azide is washed rapidly with 30 ml of ice-cold 0.05 N HCl. The outlet of the funnel is covered with parafilm, and a solution of 150 mg of insulin (24 U/mg; Eli Lilly) in 5 ml of 0.5 M NaHCO$_3$ at pH 9.0 (con-taining 7.5 × 10^6 cpm of [^{125}I]insulin; 1 Ci/μmole) is added to the agarose azide while stirring with a glass rod. The final pH of the reaction mixture is 8.5. Since the azide function is not very stable in aqueous medium, the procedures for washing of the agarose azide and the mixing with insulin should be performed within 2 minutes or less. The insulin-agarose suspension is gently shaken for 4 hours at 4° and for 10 hours at 20°. After washing with 500 ml of 0.5 M NaHCO$_3$ at pH 8.5, the insulin-agarose is incubated while on the sintered-glass funnel with 1 M glycine, pH 8.5, for 2 hours. The insulin-agarose is packed in a 10-ml column and washed continuously with 3–4 liters of 5 M urea, pH 8.5, over a 4–5-day period at 4°. The substitution of insulin, judged by its ^{125}I-content, is between 3 and 3.5 mg per milliliter of packed gel.

With lower concentrations of insulin (6 mg/ml), substitution of the order of 0.4 mg of insulin per milliliter of gel is attained. Glucagon has also been coupled by the same procedures, with 3 mg of glucagon per milliliter of hydrazido-agarose. The substitution of glucagon is about 0.3 mg per milliliter of gel.

The exact degree of substitution of insulin or of glucagon by this method is subject to variation, in part due to the relative instability of the intermediate agarose-azide derivatives. High degrees of substitution depend on the rapidity and temperature of the washing procedures, and on the time between washing of the gel and coupling of the hormone.

Diazonium and Hydrazido Derivatives of Albumin-agarose. Insulin is coupled to *p*-aminobenzamido-albumin-agarose, prepared from albumin-agarose, by the same procedures described above for the polyamino acid derivative. Insulin is also coupled to albumin-agarose via the azide method by the same procedures described for poly-(L-lysyl-DL-alanyl-*N*-carboxymethyl-hydrazido)-agarose.

N-Hydroxysuccinimide Derivative of Poly-(L-lysyl-DL-alanyl)-agarose. Ten milliliters of the active *N*-hydroxysuccinimide ester of the polyamino acid polymer,[14] are mixed at 4° with 10 ml of 0.1 M NaHCO$_3$, pH 8, containing 40 mg of insulin and 8 × 10^6 cpm of ^{125}I-labeled insulin. After 3 hours, 0.2 M glycine is added and the reaction is allowed to continue for 4 hours at room temperature. The insulin content is 0.24 mg of insulin per milliliter of gel.

Coupling of Insulin to Poly-(L-lysyl-DL-alanine)-agarose with Glutaraldehyde. The polyamino acid agarose derivative is treated for 16 hours at 4° in 0.1 M NaHCO$_3$ at pH 8.8, containing 40 mg of insulin, 8 × 10^6 cpm of ^{125}I-labeled insulin, and 1% glutaraldehyde. After thorough washing (5 days), the insulin content is determined as about 2.5 mg/ml. Attempts to reduce the Schiff bases formed with the agarose by treating with NaBH$_4$ result in facile reductive cleavage of the insulin disulfide bonds. In separate experiments it was determined that it is not possible with NaBH$_4$ to selectively reduce and stabilize the glutaraldehyde-linked bonds without destroying the biological activity of the hormone. Because of the relatively unstable nature of the bonds in the derivatives, the resultant leakage of insulin is relatively high.

Purification of Receptors

Ammonium Sulfate Fractionation

Although ammonium sulfate fractionation of the solubilized insulin receptor (Table II) can be performed satisfactorily despite the presence of detergents in the buffers, the receptor molecules coprecipitate with the bulk of the protein in the sample and the final purification that is achieved is poor. Nearly 90% of the starting activity can be recovered during these procedures.

Ion Exchange Chromatography

Substantial purification of the insulin receptor can be achieved by chromatography on DEAE-cellulose with buffers that contain Triton X-100. The insulin-binding protein adsorbs to this ion-exchange resin at pH 6.3; lower pH values cannot be used because of loss of activity. During dialysis against buffer at pH 6.3 a large amount of protein precipitates with only minimal loss of insulin-binding activity so that a 3-fold purification can be achieved by this simple procedure. The receptor proteins are purified nearly 20-fold overall.

Procedure. Liver membranes are extracted with 1% (v/v) Triton X-100 and the high-speed (50,000 rpm, 70 minutes) supernatant is

TABLE II

PURIFICATION OF INSULIN RECEPTOR OF LIVER-CELL MEMBRANES
BY AMMONIUM SULFATE FRACTIONATION[a]

Fraction	Protein (mg/ml)	Insulin-binding activity			Purification (-fold)[b]
		Protein (cpm/mg)	Total cpm	Recovery (%)[b]	
Membrane	9.4	38,000	—	—	—
Triton extract	4.8	88,000	2.1×10^6	100	0
(NH$_4$)$_2$SO$_4$ fractions					
0–20%	1.8	167,000	3×10^5	14	1.9
20–40%	5.2	260,000	1.4×10^6	67	3
40–60%	3.0	67,000	2×10^5	9	—
40–60% supernatant	—	5,000	—	—	—

[a] Liver membranes are extracted with 1% (v/v) Triton X-100 and centrifuged for 60 minutes at 44,000 rpm (130,000 g). The supernatant is dialyzed overnight against 0.1 M sodium phosphate buffer (pH 7.4)–0.1% Triton. The ammonium sulfate (at 4°) pellet is dissolved in about 1 ml of the above phosphate buffer and dialyzed overnight before assaying for specific [^{125}I]insulin binding. Data from P. Cuatrecasas, *Proc. Nat. Acad. Sci. U.S.* **69**, 318 (1972).

[b] Relative to Triton extract of membranes.

dialyzed (at 4°) for 18 hours against 50 mM sodium acetate, pH 6.3–0.2% Triton. The large precipitate that formed during dialysis is removed by centrifugation; about 15% of the initial insulin-binding activity is in this pellet. Of the supernatant liquid (5 mg of protein per milliliter), 12 ml are charged onto a column (0.5 × 5 cm) of DEAE-cellulose at 4°, which has been equilibrated with the dialysis buffer described above. The column is washed further with about 40 ml of the same buffer, and elution is achieved with a linear two-component gradient prepared from 25 ml of 0.1 M ammonium acetate, pH 6.3–0.2% Triton, and 25 ml of 1.0 M ammonium acetate (pH 6.3)–0.2% Triton at a flow rate of about 8 ml/hour with fractions of 1.8 ml. [^{125}I]Insulin binding is determined by the polyethylene glycol assay. Under these conditions the sample applied to the column binds 1.5 × 10^5 cpm of [^{125}I]insulin per milligram of protein, and the peak elution fraction (tube 49) can bind 2.5 × 10^6 cpm per milligram of protein, indicating purification of about 17-fold. Recovery of insulin-binding activity (fractions 45–55) is about 65%.

Affinity Chromatography

Washing of Insulin-Containing Adsorbents. As described elsewhere,[15] exhaustive and meticulous washing of the insulin-substituted gels to remove all noncovalently adsorbed insulin is an essential and crucial step in the proper use of these derivatives.[10,12] Because of the unusually high propensity for adsorption of insulin to a variety of surfaces, extraordinarily thorough washing procedures must be used, especially with certain of the agarose derivatives. The adsorbents are washed at room temperature, while packed in a column, with a continuous flow of buffers containing 5 M guanidine·HCl or 6 M urea for 3–14 days, followed by 0.1 M NaHCO$_3$ at pH 8.5 for 3–4 days. The washing is checked by the release of radioactivity, by the ability to compete with [^{125}I]insulin for binding to liver membranes, or by assays of biological activity. The length of time required for washing depends greatly on the quantity of adsorbent, on the degree of substitution, on the nature of the specific gel, and on whether the specific adsorbent has been diluted with unsubstituted agarose. This last, as described for the purification of estradiol adsorbents,[14,16] greatly facilitates washing and diminishes the potential problems that can result from leakage. Washed and stored derivatives must be washed thoroughly again before use. Even after the gels have been so thoroughly washed, as described above, the final small affinity column should be washed continuously for 2–3 days (24°) before application of the sample.

[16] V. Sica, I. Parikh, E. Nola, G. A. Puca, and P. Cuatrecasas, *J. Biol. Chem.* **248,** 6543 (1973); see also this volume [86].

The flow rate should not be stopped during or just preceding application of the sample. Performance of affinity chromatography at 4° will further decrease the probability that significant quantities of the free hormone will be released during the experiment.

General Comparisons of Adsorbents. Affinity chromatography experiments with the adsorbents depicted in Fig. 1 reveal that those derivatives in which a spacer or "arm" separates the insulin molecule from the matrix backbone are the most effective in extraction of the insulin receptor. The only exception is derivative F, which is ineffective in either the predominantly azohistidyl form or in the form containing a mixture of azohistidyl and azotyrosyl bonds. Derivatives C, D, E, and H (Fig. 1) retain a greater proportion of the chromatographed receptor than derivative G, and derivative H appears to be less stable during storage than derivatives C, D, and E. No clear differences can be discerned in the apparent effectiveness of these last three derivatives. Similarly, essentially the same results are obtained when the same substitutions are performed on agarose beads of various porosity. Various analogous insulin derivatives prepared with porous glass beads instead of agarose are ineffective in chromatography.[10] These derivatives demonstrate continual leakage of insulin in the buffers used for equilibration of the columns, presumably a reflection of the lability of the basic silane linkage bonds.

The macromolecular agarose derivative containing insulin coupled by azo linkage is relatively effective despite substantial leakage of radioactivity, i.e., presumably insulin. Although the leakage in this case is a serious complication, it does not prohibit the use of such columns since the insulin which is released has relatively low biological activity. The glutaraldehyde-linked derivative cannot be used because of the prohibitively high rates of release of active insulin. The best macromolecular adsorbents are those prepared from N-hydroxysuccinimide esters and from hydrazido derivatives of agarose.

Specific Adsorption of Receptors to the Gel. Affinity chromatography of the insulin receptor illustrates that although no loss or retardation of insulin-binding activity is detectable with columns containing unsubstituted agarose, considerable adsorption occurs to columns containing the selective adsorbent. It is frequently advantageous to dilute the adsorbent 1:5 to 1:20 with unsubstituted agarose before use, especially when the macromolecular arm derivatives are used. Depending on the conditions of the experiment, between 20 and 90% of the binding activity can be extracted with such columns. However, extraction of activity by passage of the sample through the column is not by itself proof that the adsorbent is selectively binding the protein. The binding protein may, for example, be degraded during chromatography or it may bind free ligand (insulin) that is desorbed from the gel or slowly cleaved during the experiment. Therefore, effluents are tested for the presence of native insulin by examination of their ability to displace the binding of [^{125}I]insulin to the unchromatographed receptor. However, the failure to find displacement does not exclude the presence of a significant amount of insulin–receptor complex. The protein–ligand complex that may emerge in the effluent, because of the very high affinity of the complex, would not be detected by the usual assay procedures. In view of these considerations, it is clear that proof of adsorption to the gel requires that the protein be *recovered* after appropriate elution procedures.

The specificity of the process of adsorption of the insulin-binding macromolecules to the column can be demonstrated directly by experiments in which native insulin is added to the sample before chromatography. The breakthrough protein, after passage through a Sephadex G-100 column to separate free native insulin, is examined for binding of [^{125}I]-insulin under conditions (40°, 90 min) that promote exchange of free and bound insulin. The same binding activity is present in the samples that are chromatographed on insulin and on unsubstituted agarose columns. Furthermore, adsorption of the insulin-binding protein to the affinity column can be effectively prevented by first digesting the insulin-agarose with Pronase, or by reducing and alkylating the insulin adsorbent before use.

A specific example is illustrated in Fig. 2. Liver cell membranes are homogenized (Polytron), extracted with 2% Triton X-100 by shaking at room temperature for 40 minutes, and centrifuged at 50,000 rpm for 70 minutes. The supernatant liquid is dialyzed for 16 hours at 4° against Krebs-Ringer-bicarbonate buffer (pH 7.4) containing 0.1% Triton X-100, and the precipitate that forms during dialysis is removed by centrifugation for 30 minutes at 35,000 rpm. Of the supernatant, 2 (A,B) or 12 (C) ml are slowly chromatographed at 24° over affinity columns (V_t 1.3 ml, in Pasteur pipettes) that have been washed for 20 hours with 0.1 M NaHCO$_3$ at pH 8.4 and followed by equilibration for 2 hours with Krebs-Ringer-bicarbonate buffer containing 0.1% Triton X-100. The columns are washed thoroughly (note break in abscissa) before elution.

Elution. The insulin-binding macromolecules can be eluted by buffers containing urea at a relatively low pH. Urea dissociates the insulin–receptor complex, and this effect is completely reversed by reducing the urea concentration by dilution or dialysis. Urea concentrations higher than 4.5 M result in much lower recoveries of insulin-binding activity, consistent with the irreversible denaturation which is observed. Elution cannot be achieved by simply using buffers of low pH, since inactivation of binding activity occurs below pH 5.5. Under the conditions described below,

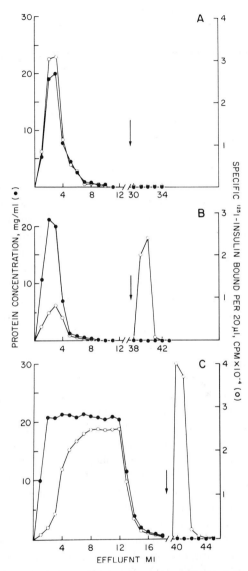

FIG. 2. Affinity chromatography of detergent-solubilized insulin receptors of liver-cell membranes on affinity columns containing unsubstituted agarose (A) and derivative C (Fig. 1), diaminodipropylaminosuccinyl-*N*-phenylalanyl-insulin-agarose (B,C). See text for details of experiment. Data from P. Cuatrecasas, *Proc. Nat. Acad. Sci. U.S.* **69**, 318 (1972).

between 50 and 80% of the total activity that is presumably adsorbed to the column can be consistently recovered with virtually all the derivatives tested.

The specific method illustrated in Fig. 2 involves elution (arrow) with 50 mM sodium acetate at pH 6.0 containing 4.5 M urea and 0.1% Triton X-100. After application of this buffer to the column, the flow is stopped for 15 minutes before resumption of chromatography. Fractions, 1 ml, are collected for determination of protein, and specific binding of [^{125}I]insulin is determined with the polyethylene glycol assay; the effluent samples (20 μl) are added directly to the assay mixture (0.2 ml). Frequently it is desirable to immediately decrease the concentration of urea in the eluted fractions by adding an equal (or 2-fold) volume of a urea-free buffer of neutral pH. The conditions used to elute the receptor proteins sometimes lead to the concomitant release of sufficient free insulin from the gel. Since this will tend to result in an underestimation of the total binding activity eluted, it may be necessary to (a) perform binding assays under conditions (37°, 120 minutes, high concentrations of [^{125}I]-insulin) which promote an exchange reaction so as to determine whether a substantial increase in binding activity results, and (b) chromatograph samples of the eluted material on Sephadex G-50 (fine), equilibrated with 50 mM NaHCO$_3$ at (pH 8.0)–0.1% albumin, to determine whether insulin binding (exchange reaction) increases after removal of the excess free insulin by this procedure.

Because of the extraordinarily small quantities of receptor protein which are being handled during small-scale, pilot experiments, it is wise in all preliminary studies to include albumin (0.1%) in the eluting buffer to prevent nonspecific and adsorptive losses of the receptor proteins. In experiments wherein the degree of purity is to be estimated, aliquots are removed and added to albumin-containing buffers (for binding assays). The remaining material is analyzed for protein. Since most of the affinity columns which are effective result in extremely low yields of protein, accurate estimations depend on acid hydrolysis followed by amino acid analysis of the pooled samples.

Purification. With the use of these affinity adsorbents it is possible to purify the insulin-binding macromolecules from the Triton X-100 extracts by about 8000-fold (Table III). However, it has not been possible to obtain large quantities of the binding protein by these procedures. As demonstrated in Fig. 2C, the effective "capacity" of these columns is rather low. Chromatography of large quantities of extract does not lead to correspondingly greater adsorption of binding activity. The low capacity does not result from an insufficient amount of insulin on the adsorbent, since only a very small fraction of the total amount present would be

TABLE III
Purification of Soluble Insulin Receptor of Liver Membranes on an Insulin-Agarose Affinity Column[a]

Sample	Volume (ml)	Protein concentration (mg/ml)[b]	Insulin-binding activity			
			Specific activity (cpm/mg of protein)	Total activity (cpm)	% Recovery of activity	Purification (-fold)
Sample applied to column (Triton extract)	3.0	26.2	36×10^3	2.8×10^6	—	—
Column effluent, breakthrough	4.5	16.1	7.9×10^3	5.8×10^5	20	—
Elution	1.8	3×10^{-3}	2.9×10^{8} [d]	1.6×10^6	57[c]	8000[d]

[a] The high-speed supernatant of a Triton X-100 extract of liver membranes is chromatographed and eluted on an affinity column [V, 1.7 ml of derivative E (Fig. 1)] as described in Fig. 2. Similar results can be obtained with Triton extracts of fat-cell membranes. Data from P. Cuatrecasas, *Proc. Nat. Acad. Sci. U.S.* **69**, 318 (1972).

[b] Determined by amino acid analysis, on the basis of leucine content and by use of crystalline bovine albumin as standard. This was the only way that permitted detection of protein in the elution samples.

[c] Represents 70% recovery of the activity that was presumably adsorbed to the column.

[d] Because of the difficulty in determination of such low concentrations of protein in the effluent, these figures must be considered provisional.

expected to participate in the specific interaction with the binding protein that is adsorbed. It is probable that the low capacity is related to interfering reactions that result from the application of very large volumes of crude, concentrated protein solutions through these small columns. Although proportionately larger quantities of the binding protein can be purified with larger columns, the quantities of protein obtained and the size of the columns required make such procedures impractical. Considerably higher capacities and better results are obtained if the receptor proteins are partially purified by salt fractionation and ion exchange chromatography before application to affinity columns.

The affinity columns appear to have a much greater capacity for the insulin-binding protein of Triton X-100 extracts of fat-cell membranes than for proteins prepared by the same method from liver-cell membranes. The use of placental microsomal membranes promises to be helpful in attempts to achieve large-scale purification since this tissue is available in large quantities and since the specific activity of insulin binding is so much greater than that of liver microsomes.

The need for the continued presence of detergents to maintain the insulin-binding proteins of liver- or fat-cell membranes in a soluble state does not appear to seriously handicap the application of conventional procedures or of affinity chromatography for purification of this macromolecule. The efficacy of the various procedures when used in combination in the purification is given in Table IV. Although a protein of approximately theoretical purity is attained, the difficulties encountered in large-scale purification of these proteins have not yet been resolved.

The Use of Plant Lectin Adsorbents

Since solubilized insulin receptors appear to be glycoproteins which are capable of interacting with certain plant lectins, adsorbents containing such lectins may be used to purify insulin receptor structures.[17,18]

Preparation of Adsorbents

Agarose derivatives of the lectins are prepared by the CNBr procedure or by reaction with activated N-hydroxysuccinimide esters of diaminodipropylaminosuccinyl-agarose.[14] Forty milliliters of Sepharose 4B are activated with 6 g of CNBr and allowed to react with 60 ml of ice-cold 0.1 M sodium phosphate at pH 7.4 containing 500 mg of concanavalin A and 0.1 M α-methyl-D-mannopyranoside. After 16 hours at 4°, 2 g of glycine are added, and the incubation is continued for 8 hours at room

[17] P. Cuatrecasas and G. P. E. Tell, *Proc. Nat. Acad. Sci. U.S.* **70**, 485 (1973).
[18] P. Cuatrecasas, *J. Biol. Chem.* **248**, 3528 (1973).

TABLE IV

RESUME OF PROCEDURES USED IN THE PURIFICATION OF THE
INSULIN RECEPTOR OF LIVER-CELL MEMBRANES[a]

Preparation or procedure	Insulin-binding activity (pmoles/mg of protein)	Purification
Crude liver homogenate	0.008	0
Liver membranes	0.15	20[b]
Triton extract of membranes	0.26	1.7[c]
(NH$_4$)$_2$SO$_4$, fraction 20–40%	0.75	3[d]
DEAE-cellulose chromatography	14	60[d,e]
Affinity chromatography[f]	About 2000	About 8000[d] About 250,000[b]

[a] The Triton extract was used for (NH$_4$)$_2$SO$_4$ fractionation, DEAE-cellulose chromatography, and affinity chromatography. Data from P. Cuatrecasas, *Proc. Nat. Acad. Sci. U.S.* **69**, 318 (1972).

[b] Compared to crude liver homogenate.

[c] Compared to liver membranes.

[d] Compared to Triton extract of liver membranes.

[e] Dialysis of Triton extract results in a 3-fold purification; DEAE-chromatography results in a further purification of about 20-fold.

[f] These are tentative figures because of the difficulty in accurately determining the small amounts of protein obtained by these procedures.

temperature. This adsorbent contains 5.5 mg of protein per milliliter of gel. Wheat germ agglutinin (1.4 mg/ml) is similarly coupled to activated agarose in the presence of 0.1 M N-acetyl-D-glucosamine; 1.1 mg of protein is coupled per milliliter of gel. ^{125}I-Labeled plant lectins may be used during the coupling procedures. By phase-contrast microscopy, the lectin-coupled beads become heavily coated with erythrocytes when these are mixed at 24° for 1–2 hours. The cells can be rapidly desorbed with 50 mM N-acetyl-D-glucosamine.

Affinity Chromatography

Columns (1 ml, Pasteur pipettes) containing wheat germ agglutinin-agarose and concanavalin A-agarose are equilibrated at room temperature for 3 hours with Krebs-Ringer-bicarbonate buffer containing 0.1% (v/v) Triton X-100. Two milliliters of liver membrane extract (14 mg of protein per milliliter) are applied to each column, and the columns are washed with 10 ml of the buffer. The adsorbed proteins are eluted with 0.1 M sodium bicarbonate at pH 8.4 containing 0.1% Triton X-100, 0.1% (w/v) albumin, and either 0.3 M α-methyl-D-mannopyranoside (for concanavalin A column) or 0.3 M N-acetyl-D-glucosamine (for wheat germ agglutinin column); the column flow is stopped for 3 hours after addition of the eluting buffer. About 60–80% of the insulin-binding activity of the sample is extracted by the columns. Wheat germ agglutinin appears to be more effective than concanavalin A. If the specific sugar is added to the crude sample before chromatography, the receptor does not adsorb to the column. The protein content of the pooled material of the void volume is the same as that of the control column, indicating that only a very small fraction of the total protein is actually adsorbed. The binding protein is purified about 3000-fold, and the recovery of binding activity has been greater than 90%.

Concanavalin A and wheat germ agglutinin bind quite well to the insulin-binding protein in the presence of 0.2% Triton X-100, which is used in the chromatography buffers. Fortunately, most of the lectin-binding glycoproteins of the membrane bind very weakly in the presence of this detergent. Thus, very small quantities of protein adsorb to these affinity columns, and it is possible to achieve substantial purification (3000-fold) of the insulin-binding protein by these procedures. Some advantages of these adsorbents, compared to insulin-agarose derivatives, include the ease of elution, the high capacity for the binding protein, and the avoidance of possible contamination by insulin.

[86] Estrogen Receptors

By INDU PARIKH, VINCENZO SICA, ERNESTO NOLA, GIOVANNI A. PUCA, and PEDRO CUATRECASAS

Estrogen receptor proteins exist in at least three different molecular forms (8.6 S, 5.3 S, and 4.5 S) and are present in the cytoplasm of target tissues in very low amounts. Attempts to purify these relatively labile proteins by conventional methods have resulted only in modest purification and recovery has been poor.[1]

The reversible nature of the estradiol–receptor interaction together with its stereochemical specificity and high affinity, i.e., a dissociation constant about $10^{-9} M$ at $4°$, suggests that the receptor proteins may be ideally suited for purification by affinity chromatography. Although it has been reported that columns containing estradiol linked to benzyl cellu-

[1] G. A. Puca, E. Nola, V. Sica, and F. Bresciani, in "Advances in the Biosciences" (G. Raspe, ed.), Vol. VII, p. 97. Pergamon, New York, 1971.

lose[2] and to polyvinyl-(N-phenylenemaleimide)[3] can remove estradiol-binding activity from solutions of uterine cytosol, it was not possible to subsequently elute specific estrogen-binding proteins from these columns.

Principal Problems Encountered in the Purification of Estradiol Receptors by Affinity Chromatography

Scarcity of Receptor Proteins in Biological Materials

The extremely small quantities and the relatively labile nature of the estrogen binding proteins from different target tissues presents special practical difficulties in the handling and isolation of these receptors. It has been estimated that complete purification of the receptor proteins of calf uterine preparations would require 20,000- to 100,000-fold purification.[4,5]

Strong Adsorption of Estradiol to Polymers

Virtually all polymers used as solid supports in affinity chromatography exhibit marked adsorptive properties for free estradiol.[5,6] Desorption of free estradiol during chromatography of samples containing estrogen receptors can result in "inactivation" or *apparent* removal of the estrogen-binding activity from the sample.[5-7] Although free estradiol may not be present when the derivatized gel is washed with simple buffers, application of protein-containing solutions, e.g., albumin or uterine cytosol, alter the partitioning properties of estradiol between the gel and the aqueous medium and can thus markedly enhance the release of adsorbed hormone from adsorbents washed only with simple buffers. This serious problem can generally be avoided by exhaustive washing of the adsorbent with organic solvents.[5]

Stability of Estradiol–Gel Bonds

The ester and azo bonds used in coupling the ligand to agarose or to other polymers are relatively unstable and may slowly release the ligand during chromatography.[5] The lability of the bonds may depend on the pH and ionic strength of the buffer used, and the presence of reducing

[2] P. W. Junglbut, I. Hatzel, E. R. De Sobre, and E. V. Jensen, *Colloq. Ges. Physiol. Chem.* 18, 58 (1967).

[3] B. Vonderhaar and G. C. Mueller, *Biochim. Biophys. Acta* 176, 626 (1969).

[4] V. Sica, E. Nola, I. Parikh, G. A. Puca, and P. Cuatrecasas, *Nature (London) New Biol.* 244, 36 (1973).

[5] V. Sica, I. Parikh, E. Nola, G. A. Puca, and P. Cuatrecasas, *J. Biol. Chem.* 248, 6543 (1973).

[6] P. Cuatrecasas and C. B. Anfinsen, *Annu. Rev. Biochem.* 41, 259 (1971).

[7] J. H. Ludens, J. R. DeVries, and D. D. Fanestil, *J. Biol. Chem.* 247, 7533 (1972).

substances. Similarly, nucleophiles in the tissue extract containing the estrogen receptors may catalyze the hydrolysis or cleavage of such bonds.[5]

The release of some free estradiol from the gel during chromatography of receptor-containing samples does not necessarily mean that the adsorbent will be ineffective in selectively extracting the receptor from the sample. If the affinity of the free hormone released from the gel is not very different from that of the gel-bound hormone, the small amount of free hormone will not compete effectively with the much greater amount of immobilized hormone, and the receptor will thus preferentially bind to the solid support. If, however, the free hormone has a much greater affinity, e.g., by a factor of 1000, a very small fraction, 0.1%, of the total hormone will interfere with selective adsorption if it is present in free form. For this reason the release of adsorbed hormone (see above) is generally a more serious problem, and instability of the cyanogen bromide bonds formed are less likely to lead to the release of hormone in a form that effectively competes with the matrix-bound material.

Since the release of adsorbed estradiol during chromatography of the sample may lead to the erroneous conclusion that the column is removing the receptor from the sample, it is essential that steps be taken to examine the chromatographed samples for free estradiol as well as for the presence of estradiol–receptor complexes. Estradiol-agarose adsorbents containing macromolecular spacer arms, due to their multipoint attachments to the matrix, increase the chemical stability of the agarose bound estrogen and provide other advantages in affinity chromatography.[5,8]

Selection of Estradiol Derivative for Attachment to Gel

The estradiol derivatives which have proved[5] to be most useful in the purification of receptors are those in which the hormone is attached to agarose through position 17 of the estradiol molecule (Fig. 1). Although such estradiol derivatives demonstrate lower affinity than native estradiol for the receptor, they retain sufficiently high affinity to be useful in affinity chromatography (Table I). The affinity of 17-β-estradiol-17-hemisuccinate, the estradiol derivative which has proved to be the most useful, is only 300 times lower than that of 17-β-estradiol. Immobilization of these derivatives on certain of the agarose gels does not cause a further, serious decrease in the affinity for the receptor.[5] For example, attachment of 17-β-estradiol-17-hemisuccinate to diaminodipropylamine agarose (Fig. 1, A) results in a 13-fold fall in affinity (K_i of about $10^{-5} M$), whereas substitution on albumin-agarose only leads to a 2-fold fall in affinity (K_i of about $10^{-6} M$).[5]

---

[8] I. Parikh, S. March, and P. Cuatrecasas. This volume [6]; see also Wilchek and Miron [5].

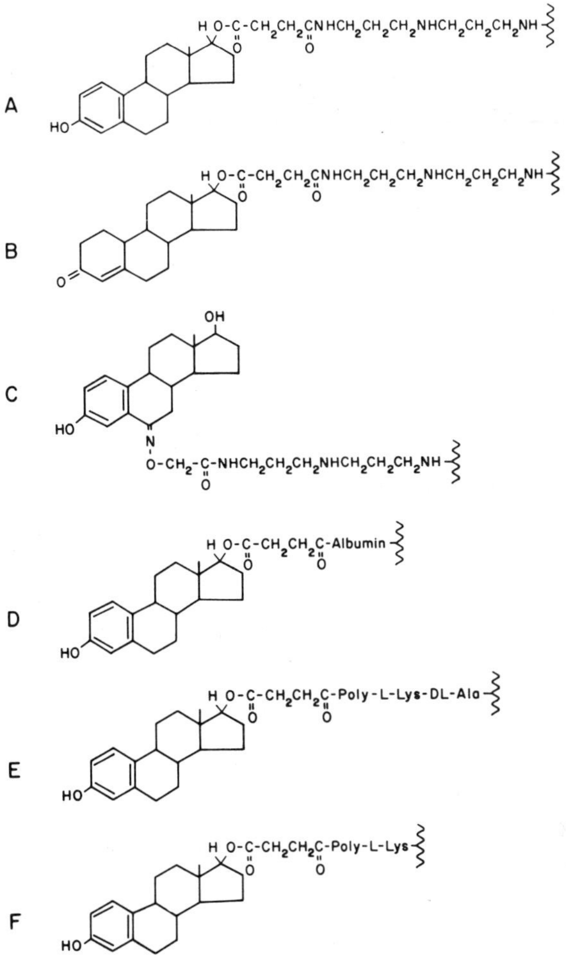

FIG. 1. Major agarose adsorbents used in the purification of estradiol receptors. (A) 17-β-Estradiol-17-hemisuccinyldiaminodipropylamino-agarose (A-DADA-Succ-E); (B) 19-nortestosterone-17-O-succinyldiaminodipropylamino-agarose (A-DADA-Nor-T); (C) 17-β-estradiol-6-(O-carboxymethyl) oxime-diaminodipropylamino-agarose (A-DADA-Carboxy-E); (D) 17-β-estradiol-17-hemisuccinyl-albumin-agarose (A-Alb-Succ-E); (E) 17-β-estradiol-17-hemisuccinyl-poly-L-(lysyl-DL-alanine)-agarose (A-PLL-Ala-Succ-E); (F) 17-β-estradiol-17-hemisuccinyl-poly-L-lysine-agarose (A-PLL-Succ-E). The copolymer used in these derivatives is a multichain copolymer of L-lysine ("backbone") and DL-alanine (side chains) in a 1:15 ratio (MW 37,500). The agarose matrix is indicated by the wiggled line in the figures. Data from V. Sica, I. Parikh, E. Nola, G. A. Puca, and P. Cuatrecasas, *J. Biol. Chem.* **248**, 6543 (1973).

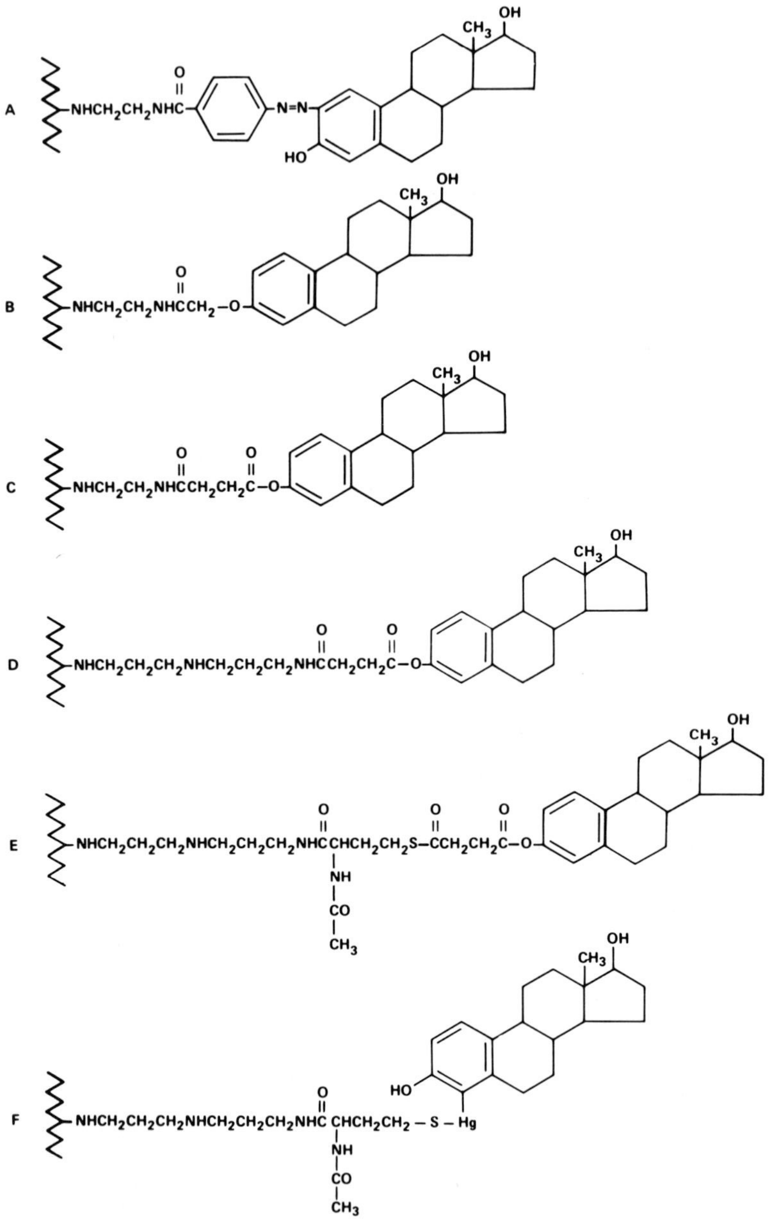

FIG. 2. Estradiol derivatives of agarose which proved ineffective as adsorbents for the purification of uterine receptors for estradiol. 17-β-Estradiol is allowed to react with a diazonium (A) or a bromoacetyl (B) derivative of agarose. 17-β-Estradiol-3-O-hemisuccinate is coupled with carbodiimide reagents to agarose deriva-

TABLE I

RECEPTOR AFFINITY OF SOME HORMONES USED IN AFFINITY ADSORBENTS[a]

Derivative	Dissociation constant
17-β-Estradiol	$2.4 \times 10^{-9}\ M$
19-Nortestosterone-17-hemisuccinate	$1.4 \times 10^{-5}\ M$
6-Carboxymethyl oxime-17-β-estradiol	$1.3 \times 10^{-4}\ M$
17-β-Estradiol-17-hemisuccinate	$7.0 \times 10^{-7}\ M$

[a] Samples containing 15 μg of partially purified receptor protein were incubated for 5 hours at 4° in 1 ml of 10 mM Tris chloride buffer, pH 7.5, containing 0.2 M KCl, 1 mM EDTA, 0.5 mM dithiothreitol, [³H]estradiol in concentrations varying from $5 \times 10^{-10}\ M$ to $5 \times 10^{-8}\ M$, and the estradiol derivative to be tested. Estradiol-binding activity was determined as described in the text. The relative affinities of the different hormones were calculated from double reciprocal plots using data obtained with 1.2 μg of 19-nortestosterone-17-hemisuccinate, 11 μg of 6-carboxy-methyl oxime-17-estradiol, and 60 ng of 17β-estradiol-17-hemisuccinate. It is estimated that coupling of these hormones to agarose results in about 300-fold reduction in its affinity toward the receptor. Data from V. Sica, I. Parikh, E. Nola, G. A. Puca, and P. Cuatrecasas, *J. Biol. Chem.* **248**, 6543 (1973).

The specific estrogen derivative used has important implications for the subsequent steps of receptor elution from the gel. Because the estradiol derivatives which are used have substantially lower affinity than 17-β-estradiol for the receptor (Table I), it is not necessary that the amount of free estradiol which is added to the eluting medium be in great excess compared to the gel-bound ligand. Since the "excess" of free estradiol required for effective competitive exchange is related to both the concentration and the affinity of the particular gel derivative used, the specific ratio used in most work is adapted to reflect the particular conditions of the experiment. It is thus possible to use very low concentrations of estradiol of very high specific activity in the exchange reaction used in elution. This permits the use of the same radioactive estradiol for the subsequent assay of the binding activity present in the eluted sample, and it avoids contamination with large amounts of free, native estradiol.

Most gel derivatives which contain estradiol linked to agarose by bonding to the A-ring of estradiol (Fig. 2) possess very low affinity for the receptor.[5] This is indicated by the total inability of the very stable 3-O-ether derivatives (Fig. 2, B) to bind estradiol receptors. This low affinity, coupled with the chemical instability of some of the other A-ring deriva-

tives containing ethylenediamine (C), diaminodipropylamine (D), or *n*-propyl-*N*-acetylhomocysteine (E). 4-Mercuri-17-β-estradiol is coupled to a sulfhydryl derivative of estradiol (F). Data from V. Sica, I. Parikh, E. Nola, G. A. Puca, and P. Cuatrecasas, *J. Biol. Chem.* **248**, 6543 (1973).

tives (which release free estradiol) probably explains the failure of these adsorbents.[5] The specific problems arising from the use of these derivatives are described below.

Selection of Polymer and the Use of Macromolecular Spacers

Unfortunately, polyacrylamide beads cannot tolerate the organic solvents used for properly washing the derivatives. Glass beads may potentially be very useful because they tolerate organic solvents, possess rather rigid and stable structures, and can be obtained in highly porous form. However, the basic silane bonding used as the initial glass–ligand linkage does not appear to be sufficiently stable for the kinds of experiments required in receptor purification studies.

The most useful estradiol adsorbents appear to be those that are prepared with "macromolecular" arms interposed between the agarose backbone and the hormone molecule.[5] These derivatives have been used with the anticipation that the macromolecular polymers would attach to the agarose backbone at several points, thus greatly increasing the stability of subsequently attached ligands.[8]

The macromolecular estradiol-agarose adsorbents demonstrate strong affinity for the estrogen receptor even when they are substantially diluted with unsubstituted agarose. As described in Table II, small columns containing 100-fold diluted samples of four different macromolecular estradiol derivatives of agarose can still extract a significant portion of the estradiol receptor present in 10-ml samples of uterine cytosol. By the criteria described in Table II, the most effective adsorbent appears to be A-Alb-Succ-E (Fig. 1, D). In separate experiments it is found that the use of this adsorbent, though being highly effective in extracting the receptor proteins from the sample, results in the recovery of only about 25% of the adsorbed receptor while 90% of the receptor can be eluted from A-PLL-Ala-Succ-E (Fig. 1, E). Furthermore, with the latter the receptor can be purified 10,000- to 100,000-fold whereas only a 4000 fold purification is achieved with the albumin derivatives. The derivatives that contain albumin are thus more difficult to handle in the elution steps, presumably because the estradiol added for the exchange reaction has a certain affinity for the albumin molecules which are present in excess compared to the receptor molecules. Therefore, the best overall derivative appears to be that containing the branched polyamino acid "arm."

Elution of the Receptor Protein

Irreversible denaturation of the receptors occurs when tightly bound receptor on the adsorbent is eluted by buffers which are sufficiently extreme with regard to the presence of guanidine·HCl or urea or to changes

TABLE II
CAPACITY FOR RECEPTOR BINDING OF MACROMOLECULAR ESTROGEN
ADSORBENTS DILUTED WITH UNSUBSTITUTED AGAROSE[a]

Adsorbent	Dilution	Receptor protein adsorbed (%)
A-Alb-Succ-E	1:1	80
	1:20	60
	1:100	45
A-dent-Alb-Succ-E	1:1	37
	1:20	20
	1:100	18
A-PLL-Ala-Succ-E	1:1	40
	1:20	27
	1:100	20
A-PLL-Succ-E	1:1	40
	1:20	22
	1:100	18

[a] One-milliliter columns (in Pasteur pipettes) containing the selective adsorbent diluted with unsubstituted agarose as indicated are washed with 400 ml of 90% aqueous methanol and equilibrated with 20 ml of 10 mM Tris chloride at pH 7.5 containing 0.2 M KCl, 1 mM EDTA, and 0.5 mM dithiothreitol. Ten milliliters of a calf uterine supernatant are applied to the columns at 4° with a flow rate of 20 ml per hour. The effluent samples (5 ml) are examined for the presence of estradiol binding activity. None of the effluent samples (5 ml) contained free estradiol as determined by the ether extraction procedure. Each of the adsorbents tested contained the same degree of estradiol substitution. Data from V. Sica, I. Parikh, E. Nola, G. A. Puca, and P. Cuatrecasas, *J. Biol. Chem.* **248**, 6543 (1973).

in pH, to cause the necessary dissociation. This problem can be overcome[5] by combining the use of native estradiol as a specific competing ligand with the selection of experimental conditions that enhance the rate of dissociation of the gel bound estradiol–receptor complex, thereby facilitating exchange of the intact receptor from an insoluble to a soluble ligand-bound state.

Assay Methods

Assay for Estradiol Binding Activity

The standard assay is based on the facile separation of [³H]estradiol–receptor complexes from free estradiol by gel filtration on Sephadex G-25 chromatography as described elsewhere.[5,9] It is known[9] that the specific binding of estradiol to receptor proteins is destroyed sharply by exposure to a temperature of 65° whereas nonspecific binding is actually enhanced by this treatment. All estrogen-binding proteins eluted from affinity columns are therefore examined for thermolability properties. The sample, in 0.1 to 0.5 ml of water, is heated for 5 minutes at 65°. The sample is quickly cooled to 4°, and the estradiol-binding activity is determined by gel filtration.

Detection of Free Estradiol in the Column Effluents[4]

The concentration of free estradiol present in the column buffers may be sufficiently high to interfere with the proper interaction of the receptor proteins with the gel bound estradiol and yet be too low to detect by conventional methods. The method of detection of free hormone consists of ether extraction of estradiol present in the affinity column effluents and assaying this extract for its ability to inhibit the binding of [3H]estradiol to fresh uterine receptor preparation. Two uterine cytosol samples (5 ml each) taken before and after the passage through the affinity column are extracted twice each with 10 ml of ether. The combined ether extracts are dried and redissolved in 0.2 ml of 50% (v/v) aqueous dimethylformamide. Fresh uterine cytosol samples (0.2 ml) are incubated with 20 μl of the ether extract (or with 20 μl of dimethylformamide in the control samples) and [3H]estradiol (3×10^5 dpm) for 7 minutes at 37°. The binding of [3H]estradiol is determined by the gel filtration method.[5,9] The presence of free estradiol in the effluents from the affinity column is determined in terms of inhibition of binding of [3H]estradiol to the receptor protein present in the 0.2 ml of uterine sample. In this test, 10% inhibition of binding signifies the presence of a significant quantity of free estradiol in the original 5 ml sample. It is necessary in such cases to repeat the washing procedures for the adsorbent with organic solvents. If, on the other hand, the inhibition is less than 10%, the presence of possible estradiol–receptor complexes in the affinity column effluent is determined as described below.

Detection of Estradiol–Receptor Complexes in the Column Effluents[4]

The test described above does not reliably detect the presence of estradiol–receptor complexes which can result from the leakage (bleeding) of small quantities of estradiol from the gel during chromatography. The procedure used depends on an enhanced rate of exchange at higher temperatures between excess of [3H]estradiol and the unlabeled hormone

[9] G. A. Puca, E. Nola, V. Sica, and F. Bresciani, Biochemistry 10, 3769 (1971).

present in the estradiol–receptor complex. A sample of an affinity column effluent or of fresh uterine cytosol (0.2 ml) is incubated with 0.8 ml of 10 mM Tris chloride at pH 7.5, containing 10 mM KCl, 1 mM EDTA, and 3×10^5 dpm of [³H]estradiol for 7 minutes at 37° followed by an additional 60 minutes at 4°. The control sample is incubated in an identical manner except that the incubation at 37° is omitted. Both samples are subjected to Sephadex G-25 gel filtration to determine the estradiol binding activity. The difference in estradiol binding activity between the two samples is an accurate measure of the presence of estradiol–receptor complexes in the column effluents provided that no significant free estradiol is present in the effluent samples. It has been determined that estradiol–receptor complexes, which have been incubated with radioactive estradiol, will exchange about 70–90% of the bound estradiol under the conditions used in this heat exchange test. The apparent removal by affinity columns of estradiol receptor proteins from chromatographed samples is always substantiated with this test before assuming that biospecific adsorption had indeed occurred.

Preparation of Selective Adsorbents[5]

17-β-Estradiol-17-hemisuccinyldiaminodipropylamino-agarose
(Fig. 1, A)

The preparation of the ω-aminoalkyl agarose has been described.[8,10] The succinylated estradiol derivative is coupled to the diaminodipropyl-amino-agarose with a water-soluble carbodiimide.[10] Ten milliliters of diaminodipropylamino-agarose is washed with 2–3 volumes of 70% aqueous dioxane and suspended in 10 ml of 70% dioxane. A solution of 10 mg (27 μmoles) 17-β-estradiol-17-hemisuccinate (Steraloids, Inc., Pawling, New York) in 1.0 ml dioxane is added to the agarose suspension. One hundred milligram of 1-ethyl-3-(3-dimethylaminopropyl) carbodiimide (Pierce Chem. Co.) are added and the suspension is gently shaken in a tightly capped polyethylene bottle on a mechanical shaker for 6 hours at room temperature. Another portion, 100 mg, of the carbodiimide reagent is added and after an additional 10 hours the substituted agarose is filtered on a coarse-disc sintered-glass funnel and washed with 500 ml of dioxane. The estradiol-agarose is packed in a glass column (diameter 1.0 cm) and washed further with 5 liters of 80% aqueous methanol for 3–4 days followed by 200–500 ml water. The substitution of 17-β-estradiol varies between 0.5 and 1.0 μmole per milliliter of packed agarose.

[10] P. Cuatrecasas, *J. Biol. Chem.* **245**, 3059 (1970).

19-Nortestosterone-17-O-succinyldiaminodipropylamino-agarose
(Fig. 1, B)

This derivative is prepared by coupling the activated N-hydroxysuccinimide ester of the succinylated derivative of 19-nortestosterone to an ω-aminoalkyl derivative of agarose. Five hundred milligrams (1.33 mmoles) of [^{14}C]19-nortestosterone-17-O-succinate[11] (10 μCi/mmole) are dissolved in 30 ml of CH_2Cl_2; 310 mg (1.5 mmoles) of dicyclohexylcarbodiimide and 185 mg (1.6 mmoles) of N-hydroxysuccinimide are added. After stirring for 4 hours at room temperature, the precipitated urea is removed by filtration and the filtrate is evaporated to dryness. The residue is dissolved in 30 ml of dioxane and added to 30 ml of diaminodipropylamino-agarose,[10] which has previously been washed with 200 ml of dioxane by filtration on a Büchner funnel. The coupling reaction is performed at 24° for 24 hours. The agarose is filtered and washed with 2 liters of dioxane and 2 liters of 50% (v/v) aqueous methanol. The ligand substitution is 5 μmoles per milliliter of packed gel. A similar adsorbent was also prepared using the water-soluble carbodiimide, 1-ethyl-3-(3-dimethylaminopropyl) carbodiimide, in 50% (v/v) aqueous dimethylformamide. This derivative contained 1.5 μmoles of 19-nortestosterone per milliliter of packed gel.

17-β-Estradiol-6-(O-carboxymethyl) Oxime-diaminodipropylamino-agarose (Fig. 1, C)

17-$β$-Estradiol-6-(O-carboxymethyl) oxime is synthesized according to the method described by Dean *et al.*[12] An activated form of this compound is coupled to an ω-aminoalkyl agarose derivative. A suspension containing 120 mg of the oxime (0.33 mmole) in 20 ml of dioxane is stirred in a water bath at 12° to 15° with a magnetic stirrer. N-Methylmorpholine (0.4 mmole) is added, followed by 0.33 mmole of isobutyl chloroformate. After stirring for 25 minutes, the mixture is added to 25 ml of diaminodipropylamino-agarose which had previously been washed with 200 ml of dioxane. The coupling reaction is carried out for 2 hours at 12° to 15°, followed by 15 hours at room temperature. The substituted agarose is washed with 2 liters of dioxane (2 days) followed by 2 liters of a 1:1 mixture of dioxane and methanol (2 days) and finally with 500 ml of water (30 minutes). The substitution of 17-$β$-estradiol-6-(O-carboxymethyl) oxime is about 3 μmoles per milliliter of packed gel.

[11] A. M. Benson, A. J. Suruda, E. R. Barrack, and P. Talalay, this volume [71].
[12] P. D. G. Dean, D. Exley, and M. W. Johnson, *Steroids* **18**, 593 (1971).

Macromolecular Spacer Arm Derivatives of Agarose[5,8]

Poly-(L-lysyl-DL-alanine) (Miles Laboratories), is a branched-chain copolymer of L-lysine (backbone) and DL-alanine (side chains). The copolymer contains a Lys to Ala ratio of 1:15 and an average molecular weight of 37,500. The copolymer (150 mg) is coupled to cyanogen bromide-activated agarose (100 ml) by a method described elsewhere in this volume.[8] The substitution of poly-(L-lysyl-DL-alanine) on the agarose is 1.2 mg per milliliter of packed gel as judged by the recovery of the polymer in the washing solutions. Albumin is similarly coupled to agarose in the presence of a high concentration of urea to promote the coupling of this protein in its unfolded state, thus increasing the likelihood of its multipoint attachment to agarose.[8] Native albumin is coupled to agarose as described for denatured albumin except that the albumin is dissolved in 0.2 M NaHCO$_3$ at pH 9, in the absence of urea.

17-β-Estradiol-17-hemisuccinyl-poly-(L-lysyl-DL-alanine)-agarose (Fig. 1, E)

The succinylated estradiol derivative is coupled[5] to the polyamino acid agarose derivative with a water-soluble carbodiimide by procedures similar to those described above for A-DADA-E. Ten milliliters of poly-(L-lysyl-DL-alanine)-agarose and 10 mg (27 μmoles) 17-β-estradiol-17-hemisuccinate are coupled in the presence of 1-ethyl-3-(3-dimethylaminopropyl) carbodiimide. The substituted agarose is filtered and washed as described above for A-DADA-E derivative. The substitution of 17-β-estradiol is 2.5 μmoles per milliliter of packed gel.

17-β-Estradiol-17-hemisuccinyl-albumin-agarose (Fig. 1, D)

The coupling reaction and washing procedures are carried out using albumin-agarose (or denatured albumin-agarose) as described for the preparation of the A-DADA-E derivative. The substitution of the ligand is 2.6 μmoles per milliliter of packed gel.

Other Estradiol-agarose Derivatives[5,10]

A number of estradiol adsorbents are prepared by chemical procedures that result in the modification of the A-ring of estradiol (Fig. 2). The basic procedures used to prepare most of these derivatives have been described. Derivatives of estradiol attached to agarose by azo linkage at positions 2 and 4 of the A-ring (Fig. 2, A) are prepared by adding 50 ml of dimethylformamide containing 7 mM [^{3}H]estradiol (2 μCi/μmole) to a suspension of diazotized agarose which had been prepared by reacting

(7 minutes at 4°) 20 ml of *p*-aminophenylethylenediamino-agarose in 36 ml of 0.5 *N* HCl with 0.1 *M* sodium nitrite. The pH of the suspension is raised to 9.5, and the reaction is continued for 6 hours at 4°. The agarose derivative is washed at room temperature with 50% (v/v) aqueous dimethylformamide for 5 days. The derivative contained 0.6 μmole of estradiol per milliliter of gel.

Estradiol is coupled to agarose by alkylation of the phenolic hydroxyl group (Fig. 2, B) by reacting 7 m*M* [³H]estradiol (2 μCi/μmole) with bromoacetyl ethylenediamino-agarose in 50% (v/v) dimethylformamide titrated to pH 9.9 with triethylamine. After reacting for 3 days at 24° aminoethanol (0.2 *M*) is added, and the incubation is continued for 24 hours to block the unreacted alkylating groups. The derivative, washed as described above, contains 0.2 μmoles of estradiol per milliliter of gel.

17-β-Estradiol-3-*O*-hemisuccinate is prepared by allowing 4 g of [³H]17-β-estradiol (0.21 μCi/μmole) to react (24 hours, 40°, in the dark) with 16 g of succinic anhydride in dimethylformamide–pyridine (1:1, v/v). The material is evaporated to dryness, and 50 ml of water (4°) are added to hydrolyze the excess succinic anhydride over a period of 16 hours at 4°; the succinic acid is removed by repeated extractions with water. The 3-*O*-succinylestradiol derivative is coupled to ethylenediamino-agarose (Fig. 1, C), to diaminodipropylamino-agarose (Fig. 1, D), and sulfhydryl-agarose[10] derivative (Fig. 1, E) with a water-soluble carbodiimide as described earlier.[10] These derivatives contain between 0.1 and 1.4 μmoles of [³H]estradiol per milliliter of gel.

[³H]4-Mercuri-17-β-estradiol (0.1 μCi/μmole) is prepared as described by Chin and Warren.[13] This compound is coupled to a sulfhydryl-agarose (Fig. 1, F) derivative which is prepared by reacting *N*-acetylhomocysteine thiolactone[10] with diaminodipropylamino-agarose. The sulfhydryl-agarose was incubated with 50 m*M* dithiothreitol in 0.1 *M* sodium phosphate buffer, pH 7, for 2 hours at 24° and washed with distilled water immediately before use. The agarose (containing 1.7 μmoles of sulfhydryl per milliliter of gel) is reacted at room temperature in 60% (v/v) aqueous dimethylformamide–0.05 *M* sodium borate at pH 8.2, with a 3-fold excess (relative to SH group) of the mercuriestradiol derivative; the reaction is complete in 15 minutes at 24°, as determined by the disappearance of sulfhydryl groups.[14] The derivative contains about 1.5 μmoles of [³H]estradiol per milliliter of packed gel. Treatment of this adsorbent with 5 m*M* dithiothreitol (1 hour at room temperature, 0.1 *M* sodium phosphate at pH 7) results in the release of virtually all the bound estradiol into the medium.

[13] C. C. Chin and J. C. Warren, *J. Biol. Chem.* **263**, 5056 (1968).
[14] G. L. Ellman, *Arch. Biochem. Biophys.* **82**, 70 (1959).

Determination of Degree of Ligand Substitution[5]

The degree of substitution of estrogen derivatives attached to the matrix through the 17-hemisuccinyl function is determined by measuring the estradiol which is released from the gel into the medium following hydrolytic cleavage. The released estrogen is measured by a modification of the competitive binding method described by Korenman.[15] The estrogen-agarose derivative (0.5 ml) is incubated with 0.5 ml of 0.5 N NaOH at room temperature for 6 hours. After the addition of 0.5 ml of 0.5 N HCl the free estrogen is extracted with four 5-ml portions of ether. The combined ether extracts are evaporated and redissolved in 1 ml of 50% (v/v) aqueous dimethylformamide. After appropriate dilution (50- to 200-fold) with water, samples of 5, 10, 20, and 50 μl are incubated in the presence of 400 pg of [^{3}H]estradiol (48 Ci/mmole) with 0.2 ml of freshly prepared uterine supernatant at 4° for 5 hours. The samples are chromatographed on standardized columns containing Sephadex G-25, as described in the estradiol-binding assay procedure. The content of estradiol in other derivatives (Fig. 2) was determined by direct measurements of radioactivity. The ester bonds in these derivatives can also be cleaved (and released into the medium for counting) by 1 M neutral hydroxyl-amine, and the diazonium derivative can be cleaved by reduction with sodium dithionite.[10]

Washing of Estradiol Adsorbents

The importance and difficulty of washing affinity adsorbents sufficiently well to remove the free, noncovalently bound steroid hormone has been emphasized.[4-7] In most cases the use of large volumes of organic solvents over a prolonged period of time is the most practical and effective way of washing. Sometimes the adsorbent must in addition be washed with solutions of proteins which display affinity for estrogens. 17-β-Estradiol-17-hemisuccinyldiaminodipropylamino-agarose (Fig. 1, A), for example, must be washed with large volumes of the uterine cytosol protein. In such cases the adsorbent is washed thoroughly with solutions containing urea or guanidine·HCl to remove the adsorbed protein prior to its use in purification experiments. The most generally effective solvent for removing the estradiol which is not covalently bound is 80–90% aqueous methanol (1 liter per each milliliter of packed adsorbent) over 12 to 24 hours at room temperature. The macromolecular agarose derivatives which are diluted with unsubstituted agarose before use are washed adequately with 50–100 ml of 90% methanol per 1 ml of packed gel over 6–8 hours. The column is then equilibrated with 10 mM Tris chloride

[15] S. G. Korenman, *Endocrinology* 87, 1119 (1970).

at pH 7.5, containing 0.2 M KCl, 1 mM EDTA, and 0.5 mM dithiothreitol. The thoroughness of the washing procedure is tested by chromatographing a sample of uterine protein solution (5 ml per milliliter of adsorbent). The eluate is examined for the presence of free hormone by the heat exchange and ether extraction methods described above. If there is more than 10% inhibition, the entire washing procedure is repeated.

Elution of Adsorbed Receptors[4,5]

The elution procedure is performed in a batchwise manner, with the gel suspended in the form of a slurry. Dilution of the gel decreases the concentration of the gel-bound ligand and facilitates the competition-exchange reaction upon addition of the free estradiol. In some cases dilution is necessary to permit the addition of estradiol in concentrations which are compatible with solubility and yet sufficiently high to substantially exceed the concentration of gel-bound ligand. In the general eluting procedure, the adsorbent to be eluted is removed from the column and suspended in a volume of 10 mM Tris chloride at pH 7.5, containing 0.2 M KCl, 1 mM EDTA, and 0.5 mM dithiothreitol. A solution of [³H]estradiol (4–8 Ci/mmole; 1 mCi/25 ml) in benzene–methanol (9:1) is placed in an Erlenmeyer flask and evaporated to dryness. The agarose slurry is added to the flask, and the suspension is incubated at 30° for 15 minutes. The slurry is filtered and washed with an equal volume of buffer heated to 30°. The recovery of [³H]estradiol in the filtrate varies from 70 to 80%. After cooling the filtrates to 4°, they are incubated for 1 hour at the same temperature. The bulk of the free estradiol is extracted[16] by the addition of 0.25% (w/v) dextran charcoal[5] (1.2 ml per milliliter of filtrate) and incubation for 5 minutes at 4°. The protein-bound radioactivity in the supernatant liquid (2 minutes, 5000 g) is determined directly by gel filtration on Sephadex G-25. The use of estradiol of high specific activity in the eluting process thus permits direct determination of the estradiol-binding proteins present in the eluate. Quantitative elution of the adsorbed receptor proteins is specifically dependent on the presence of an excess of free estradiol in the eluting medium. The specificity of the estradiol-binding protein eluted is always confirmed by demonstrating inactivation of estradiol-binding by heating at 65° for 5 minutes as described in an earlier section.

Preparation of Calf Uterus Extract for Affinity Chromatography

Calf uteri weighing more than 20 g are homogenized at 4° in 3–4 volumes of 10 mM Tris chloride, pH 7.5, containing 1 mM EDTA and 1 mM dithiothreitol. The homogenate is centrifuged at 2° for 1 hour at 100,000 g.

---

[16] G. A. Puca, E. Nola, V. Sica, and F. Bresciani, *Biochemistry* 11, 4157 (1972).

Purification of Estrogen Receptor

The procedures described here are for the 4.5 S receptor protein, a form that is stable at low ionic strength and which has a low tendency for aggregation. This form is easier to handle than the 8 S and 5 S estrogen receptors. For higher recovery and better purification, the receptor protein is partially purified by ammonium sulfate precipitation at 4° prior to affinity chromatography step.

Step 1. Pretreatment of Cytosol. Solid KCl is added to the 100,000 *g* supernatant to achieve a final concentration of 0.4 *M*. This is followed by dropwise addition of 0.2 *M* CaCl₂ solution until a final concentration of 4 m*M* is achieved. The pH of the solution is adjusted to 7.5 to 8, if necessary, by 0.1 *M* NaOH. The solution is then incubated for 45 minutes at 4° to transform the estrogen receptor to 4.5 S form.[5]

Step 2. Ammonium Sulfate Precipitation. Finely powdered (NH₄)₂SO₄ is slowly added (0.144 g/ml; 25% saturation) with continuous magnetic stirring. After standing for an additional 60 minutes, the suspension is centrifuged for 20 minutes at 18,000 *g* at 2°. The pellet is dissolved in a volume of 10 m*M* Tris chloride at pH 7.5 containing 0.2 *M* KCl, 1 m*M* EDTA, and 2 m*M* dithiothreitol equivalent to one-tenth of the original supernatant volume. After centrifugation for 30 minutes at 250,000 *g*, the supernatant liquid is used for the next step.

Step 3. Affinity Chromatography (Table III). Agarose adsorbents containing macromolecules (e.g., albumin or branched-chain copolymers

TABLE III

PURIFICATION OF ESTRADIOL BINDING PROTEIN BY AFFINITY CHROMATOGRAPHY
ON 17β-ESTRADIOL-17-HEMISUCCINYL-POLY(L-LYSYL-DL-ALANYL)-AGAROSE
COLUMN (0.9 × 8.7 cm)[a]

Step	Volume (ml)	Protein (mg)	Specific binding[b]		Total recovery (%)
			Dpm/mg	Total	
Extract	440	6248	1.8×10^4	11.0×10^7	100
(NH₄)₂SO₄	48	662	13.0×10^4	8.6×10^7	78
Effluent	48[d]	662	2.7×10^4	1.8×10^7	16
Eluate	26	0.26[c]	2.3×10^8	5.9×10^7	53.6

[a] The experiment is performed as described in the text. Data from V. Sica, I. Parikh, E. Nola, G. A. Puca, P. Cuatrecasas, *J. Biol. Chem.* **248**, 6543 (1973).

[b] Specific binding is based on the thermolability (65°, 5 minutes) of the estradiol-binding activity of the receptor [G. A. Puca, E. Nola, V. Sica, and F. Bresciani, *Biochemistry* **10**, 3769 (1971)].

[c] Amino acid analysis of an acid hydrolyzate of 1 ml of this eluate revealed undetectable protein, which under the conditions used indicates a protein concentration of less than 10 μg/ml.

[d] Flow rate is 6 ml per hour at 4°.

of Lys and Ala) to which 17-β-estradiol-17-hemisuccinate is covalently attached have proved most useful in the purification of estrogen receptors. A well washed 17-β-estradiol-17-hemisuccinyl-poly-(L-lysyl-DL-alanine)-agarose (estradiol substitution 2.5 μmoles per milliliter of packed gel) is diluted 1:20 with unsubstituted agarose. Five milliliters of this diluted adsorbent is packed in a glass column ($d = 0.9$ cm) and equilibrated with 50 ml of 10 mM Tris chloride at pH 7.5, containing 0.2 M KCl, 1 mM EDTA, and 2 mM dithiothreitol.

The product of Step 2, about 50 ml, is applied to the column at a flow rate of 6 ml per hour at 4°, and 5-ml fractions are collected. The affinity column adsorbs between 70 and 80% of the estradiol binding activity applied to the column. The column is washed with 200 ml of 10 mM Tris chloride at pH 7.5, containing 0.2 M KCl, 1 mM EDTA and 1 mM dithiothreitol and 100 ml of the above buffer containing 1.0 M KCl (instead of 0.2 M). For elution of the estradiol binding protein the gel is removed from the column and suspended in 5 ml of 10 mM Tris chloride at pH 7.5 containing 0.2 M KCl, 1 mM EDTA, 0.5 mM dithiothreitol and 0.4 mCi of [³H]estradiol (4 Ci/mmole). The suspension is incubated for 15 minutes at 30° and filtered through a Büchner funnel. The resin is washed with 5 ml of the same buffer which has been heated to 30°. The combined filtrate is kept in an ice bath for 1 hour and extracted with dextran-coated charcoal. The suspension is centrifuged at 5000 g for 5 minutes. The dextran-coated charcoal removes about 95% of the free estradiol. The supernatant is tested for binding (Sephadex G-25) before and after heating for 5 minutes at 65°. The binding activity is completely destroyed by heating at 65°. The breakthrough protein from the affinity column does not contain free hormone or hormone-receptor complexes. Table III summarizes the results of this experiment. Between 11,000- and 13,000-fold purification of estrogen receptor is obtained by this method. In large-scale experiments in which it has been possible to quantitate accurately (by amino acid analysis) the amount of eluted protein, purification in excess of 100,000-fold has been achieved. Affinity chromatographic procedures similar to that described above, using 17-β-estradiol-17-hemisuccinyl-albumin-agarose (Fig. 1,D) affords about 5,000-fold purification, whereas 17-β-estradiol-17-hemisuccinyldiaminodipropylamino-agarose (Fig. 1,A) gives only about 400-fold purification.[5]

Estradiol Adsorbents That Were Ineffective for
    Purification (Fig. 2)

Although a variety of derivatives containing estradiol linked to agarose by the A-ring of the steroid have been tested repeatedly and under a variety of conditions, only the salient features of these experiments will be described because the ultimate practical value of these adsorbents is limited. Agarose derivatives containing estradiol linked by azo bonds (Fig. 2, A) can be washed free of adsorbed estradiol after many days of continued washing with organic solvents. However, the azo bond is too unstable under the conditions required for the chromatography of uterine samples. Although it is possible to demonstrate the disappearance of estradiol-binding activity in chromatographed samples, in all cases this results from the release into solution of the column-bound estradiol with the formation of very stable, soluble estradiol–receptor complexes. Perhaps substances in the crude tissue extracts catalyze the reductive cleavage of the azo bonds.

Agarose derivatives containing estradiol which is coupled to the support through ether linkage at the 3-hydroxyl of the A ring (Fig. 2, B) can be washed satisfactorily. However, it is not possible to demonstrate significant adsorption of estradiol receptors to these derivatives. Similar derivatives prepared with much longer hydrocarbon extensions (e.g., diaminodipropylamine) are similarly ineffective. The covalently bound estradiol in these derivatives is very stable and free estradiol is not detected in the chromatographed tissue extracts.

Agarose derivatives in which estradiol is linked by ester linkage to the phenolic hydroxyl group of the A ring (Fig. 2, C, D, E) are presumably not sufficiently stable since they nearly always result in the release of free estradiol when incubated with crude protein solutions or with solutions containing high concentrations of albumin. The ineffectiveness of these derivatives may also be explained by their poor affinity for the receptor. Despite the presence of traces of estradiol in the column buffer (about 10^{-8} M), the very large excess of gel-bound estradiol (about 1 mM) should be sufficient to compete effectively and thus to adsorb receptor molecules if its affinity is not seriously impaired. It is pertinent, for example, that A-DADA-Carboxy-E derivative (Fig. 1, C) is effective in extracting receptors from uterine supernatants despite leakage of some estradiol into the medium.

The agarose adsorbent containing 4-mercuriestradiol bound through thio ether linkage (Fig. 2, F), is washed relatively easily and demonstrates no significant leakage of estradiol when chromatographed with the customary buffers. The application of albumin-containing buffers or of uterine supernatants, however, results in the release of very large quantities of estradiol. This effect is similar to the release observed by exposure to buffers containing dithiothreitol.

Many of the derivatives described in Fig. 2 have been prepared with acrylamide and glass beads by procedures very similar to those described elsewhere.[5,10] Without exception the acrylamide derivatives could not be washed free of adsorbed estradiol by practical procedures. These derivatives display the serious disadvantage of not tolerating exposure to the organic solvents, dioxane and methanol, which are useful for washing the agarose adsorbents.

Estradiol-glass adsorbents similar to those of Fig. 2 are similarly difficult to wash, despite the ability of these materials to tolerate organic solvents. Even after washing such derivatives for many days at room temperature with absolute methanol and with 100% dioxane, exposure of the beads to aqueous buffers and to tissue extracts results in the release of very substantial quantities of free estradiol into the medium. The basic silane linkage which is used as the first step in coupling all of the ligands to glass may be too unstable to permit the procedures necessary for receptor purification.

[87] A Membrane Receptor Protein for Asialoglycoproteins

By William E. Pricer, Jr., Roger L. Hudgin, Gilbert Ashwell, Richard J. Stockert, and Anatol G. Morell

Specific receptor sites have been identified on hepatic plasma membranes which appear to be involved in the catabolism of plasma glycoproteins by virtue of their ability to bind selectively to asialoglycoproteins, i.e., glycoproteins from which the terminal sialic acid residues have been removed.[1,2] These receptors have been solubilized from particulate liver preparations by extraction with the nonionic detergent Triton X-100[3] and further purified by affinity chromatography on a column of Sepharose to which asialo-orosomucoid has been covalently attached.

Binding of the receptor protein to the immobilized ligand occurs at a neutral or slightly alkaline pH and has an absolute requirement for calcium; elution is accomplished either by reducing the pH to between 5.6 and 6.4 or by addition of the calcium chelator EDTA. Excess Triton X-100 can be readily removed, prior to elution of the receptor by washing the column with a neutral, calcium-containing, aqueous buffer.

[1] W. E. Pricer, Jr. and G. Ashwell, *J. Biol. Chem.* **246**, 4825 (1971).
[2] A. G. Morell, G. Gregoriadis, I. H. Scheinberg, J. Hickman, and G. Ashwell, *J. Biol. Chem.* **246**, 1461 (1971).
[3] A. G. Morrell and I. H. Scheinberg, *Biochem. Biophys. Res. Commun.* **48**, 808 (1972).

Preparation of the Affinity Resin

Sepharose 4B (Pharmacia) is washed several times with 10 volumes of distilled water and the fines removed by decantation. To 50 ml of settled gel is added an equal volume of water and 50 ml of a 10% solution of cyanogen bromide (Eastman-Kodak). The pH of the reaction mixture, which falls as the reaction proceeds, is promptly brought to 11 and maintained there by the continuous addition of 4 N NaOH until no further drop in pH is observed. The reaction is usually completed in 8–10 minutes at room temperature.

The activated gel is collected on a Büchner funnel and washed rapidly with 1 liter of cold 0.1 N NaHCO$_3$ at pH 9.1. The residue, suspended in the same buffer, is brought to a final volume of 100 ml, and 5 ml of a 2% solution of orosomucoid[4] is added. The coupling reaction is allowed to continue overnight at room temperature with gentle mixing on a rotating wheel.

On the following day, the suspension is poured into a chromatographic column and washed with several bed volumes of distilled water until the A_{280} is less than 0.05. The filtrate and washings are pooled, concentrated, and the unreacted glycoprotein is recovered, after dialysis, by lyophilization.

To remove the terminal sialic acid residues from the Sepharose-bound orosomucoid, the gel is suspended in an equal volume (50 ml) of 0.1 M sodium acetate at pH 5.6 and incubated at 37° for 6 hours in the presence of 20 μg of neuraminidase.[5] From the amount of sialic acid liberated, as determined by the Warren assay,[6] it was estimated that between 0.3 and 0.5 mg of orosomucoid is bound per milliliter of settled gel. The same figure is obtained when the extent of the reaction is monitored by the inclusion of trace amounts of radioactive orosomucoid in the coupling mixture or by simply determining the amount of unreacted orosomucoid recovered. When not in use, the gel may be stored at 4° in the presence of 0.01% sodium azide. The gels have been used repeatedly over a period of months without apparent loss of activity.

Solubilization of the Receptor Protein

The solubilization of the receptor protein from a crude rabbit liver particulate fraction has been described.[3] An alternative procedure, employing a rabbit liver acetone powder, has been found to be a convenient and

---

[4] L. Van Lenten and G. Ashwell, *J. Biol. Chem.* **246**, 1889 (1972).
[5] *Clostridium perfringens* neuraminidase was obtained from Worthington Biochemical Corp. and was estimated to contain 0.58 unit per milligram of protein.
[6] L. Warren, *J. Biol. Chem.* **234**, 1971 (1959).

stable starting material. The dry powder is stirred for 30 minutes at 4° with 40 volumes of a buffer containing 0.1 M sodium acetate at pH 6.0 10 mM EDTA and 0.2 M NaCl, centrifuged, and the supernatant fluid discarded. This step is repeated with the same buffer and then with cold distilled water. The residual pellet is resuspended by homogenization in 40 volumes of an extracting buffer consisting of 10 mM Tris chloride at pH 7.8, 0.4 M KCl, and 1% Triton X-100. After stirring for 30 minutes at 4°, the crude, slightly turbid extract is recovered by centrifugation and made 50 mM in $CaCl_2$. The recovery of receptor protein ranges from 90 to 100% of that present in the acetone powder.

Binding Assay

A binding assay has been reported whereby the complex of receptor protein and radioactive asialo-orosomucoid is precipitated with cold acetone, and the pellet is freed from contaminating radioactivity by washing with aqueous buffer prior to counting.[3] An alternative procedure involving selective precipitation of the complex with ammonium sulfate is described here.

The assay is carried out in disposable glass tubes, 10×75 mm. The incubation mixture contained, in a final volume of 0.50 ml, the following components: Tris chloride at pH 7.8, 25 μmoles; NaCl, 500 μmoles; $CaCl_2$, 20 μmoles; Triton X-100, 0.1% (w/v); bovine serum albumin, 0.6% (w/v); ^{125}I-asialo-orosomucoid, 1 μg (0.5–0.8 μCi/μg[4]); purified binding protein, 1–4 μg (or equivalent amount of crude extract). The complete mixture is incubated at 25° for 30 minutes and then chilled in ice. The labeled complex is precipitated by the addition of 0.5 ml of a cold, saturated solution of ammonium sulfate, adjusted to pH 7.8 with solid Tris. After 10 minutes at 4°, the suspension is filtered onto Whatman GF/C discs (2.4 cm) under reduced pressure and the incubation tube is washed 3 times with 1.0 ml of 0.4 saturated ammonium sulfate, pH 7.8, containing 10 mM $CaCl_2$ and 0.1% bovine serum albumin. The filter is then washed with an additional 7 ml of the same ammonium sulfate solution. The disc is removed and counted in a Packard Autogamma spectrometer (60% efficiency). Appropriate blanks are prepared by incubating the receptor without added ligand, precipitating as above, and adding 1 μg of ^{125}I-asialo-orosomucoid immediately before filtration. The values so obtained, approximately 0.2% of the total radioactivity added, are subtracted from each sample. Under these conditions, the assay is linear and proportional to protein concentration.

Chromatographic Purification

After equilibration of the affinity column with a loading buffer consisting of 10 mM Tris chloride at pH 7.8, 50 mM $CaCl_2$, 1.25 M NaCl, and 0.5% Triton X-100, the crude extract is added in a ratio of 20 volumes of extract to 1 volume of resin. The column is then washed with several additional bed volumes of this buffer prior to elution of the receptor protein by a solution consisting of 20 mM ammonium acetate at pH 6.0, 1.25 M NaCl and 0.5% Triton X-100. The major portion of the binding activity is recovered in the second bed volume of eluting buffer and the protein content of the eluate is determined by a minor modification of the Lowry procedure.[7] Upon readjustment of the pH to 7.8 and addition of CaCl$_2$ to 50 mM, this material is adsorbed onto a second, smaller affinity column containing 1 ml of resin for each 3 mg of protein present. Triton X-100 is removed by washing the bound receptor on the column with a solution containing 10 mM Tris chloride at pH 7.8, 50 mM CaCl$_2$ and 1.25 M NaCl. When the bulk of the detergent has been removed, as determined by monitoring the absorbance at 280 nm, the receptor protein is eluted in 3–4 bed volumes of a buffer containing 20 mM ammonium acetate at pH 6.4, and 0.5 M NaCl. Those fractions containing the maximum activity, as determined by the binding assay, are pooled and concentrated by dialysis against Aquacide II (Calbiochem). The overall recovery of binding activity varies from 55 to 75%.

Comments

From 10 g of acetone powder, approximately 2 to 3 mg of purified receptor protein is obtained. This material has a binding capacity of 50–100 ng of asialo-orosomucoid per microgram of protein and represents a purification of 160- to 320-fold over that present in isolated plasma membranes.[1] The aqueous solution is stable for periods in excess of 2 months when stored at 4° but is promptly destroyed by freezing.

Upon gel filtration in Sepharose 6B, the curves for binding activity and protein content are superimposable; both emerge together in the void volume and trail throughout the first third of the column. When subjected to disc gel electrophoresis, the binding protein fails to penetrate more than 1–2 mm into the gel. It is of some interest to note that, in the presence of calcium, the receptor protein binds to the unsubstituted Sepharose and can be eluted with EDTA. The binding capacity of the native Sepharose, however, appears to be small compared to that of the affinity column prepared by the covalent attachment of asialo-orosomucoid.

Acknowledgment

Recent work in the laboratory of two of us (RJS and AGM) has been partially supported by Grant AM 01059 from the National Institute of Arthritis, Metabolism, and Digestive Diseases, Bethesda, Maryland 20014.

[7] O. H. Lowry, N. J. Rosebrough, A. L. Farr, and R. J. Randall, *J. Biol. Chem.* **193**, 265 (1951).

[88] Thyrotropin Receptors and Antibody

By RAMON L. TATE, ROGER J. WINAND, and LEONARD D. KOHN

Thyrotropic hormone (TSH) can be coupled to beaded agarose derivatives which have been activated by cyanogen bromide treatment. The agarose coupled TSH specifically binds both antisera prepared to TSH and solubilized TSH receptors from thyroid tissue plasma membranes; it may be used for their purification.

Preparation of Agarose–Thyrotropic Hormone

Sepharose 4B is activated by the cyanogen bromide method.[1] The activated beads are rapidly filter-washed with 1–2 liters of cold 0.1 M potassium phosphate at pH 6.5. A Büchner-type of funnel with a coarse sintered-glass disc is used; the activated Sepharose is not allowed to dry out, i.e., it is to remain as a thick slurry or moist cake. The washed, activated Sepharose is suspended in cold 0.1 M potassium phosphate at pH 6.5 to yield a final volume of 30 ml containing approximately 50% beads.

A cold (2°) solution of 30 mg of TSH[2] in 3 ml is prepared in 0.1 M potassium phosphate at pH 6.5. One may wish to add [³H]TSH[5] (1 ×

---

[1] See this volume, Porath [2] and Parikh et al. [6].

[2] Bovine TSH was purified as described.[3] An additional step, chromatography on Sephadex G-100,[4] was added to the procedure in order to eliminate residual contamination of the TSH by its subunits or by the β subunit of luteinizing hormone.

[3] R. J. Winand and L. D. Kohn, J. Biol. Chem. 245, 967 (1970).

[4] T.-H. Liao and J. G. Pierce, J. Biol. Chem. 245, 3275 (1970).

[5] [³H]TSH was prepared by means of our modification[3,6–8] of the technique developed by Morell et al.[7] To obtain high specific activity [³H]TSH, tritium-labeling procedures were applied to purified rather than to crude TSH preparations. The [³H]-TSH so prepared was rechromatographed twice on Sephadex G-100[3,4,6,8] to eliminate residual nonprotein-bound tritium, polymerization products, or subunits created by the reductive procedure. As described before,[3,6,8] each batch of [³H]TSH was shown to have the same chromatographic behavior on DEAE- and CM-cellulose as did nonlabeled TSH, the same molecular weight, amino acid and carbohydrate composition, and the same disc gel electrophoretic behavior.

[6] L. D. Kohn and R. J. Winand, J. Biol. Chem. 246, 6570 (1971).

[7] A. G. Morell, C. J. A. Van den Hamer, I. H. Scheinberg, and G. Ashwell, J. Biol. Chem. 241, 3745 (1966).

[8] S. M. Amir, T. F. Carraway, Jr., L. D. Kohn, and R. J. Winand, J. Biol. Chem. 248, 4092 (1973).

10^7 cpm or 0.1–0.5 mg) to the preparation in order to subsequently evaluate the extent of coupling and the efficacy of washing.

The cyanogen bromide activated agarose suspension, 30 ml, is added to the TSH solution and stirred for 12–20 hours to effect coupling. This step is performed at between 0° and 4°. The coupled gel is charged onto a small column and washed with 1–2 column volumes of 1 M ethanolamine·HCl at pH 8.0. It is then washed sequentially with 200 ml of each of the following solutions: 0.1 M sodium acetate at pH 4 containing 1 M sodium chloride; 0.1 M sodium borate at pH 8.5 containing 1 M sodium chloride; and 0.1 M sodium chloride. All washing procedures are performed at 0–4°.

Comments

Tritiated TSH is an optional inclusion in the coupling procedure and is used to quantitate both the extent of coupling and the efficacy of the washing procedure for removal of uncoupled hormone. [^{125}I]TSH can replace the tritiated derivative as a "marker," but its shorter half-life restricts its value when TSH-agarose preparations are used after prolonged periods of storage. As described, the procedure results in an 80% to 90% coupling of the hormone to agarose.

Several modifications in the procedure are possible without significant changes in the effectiveness of the TSH-agarose preparations for purification purposes. For example, a commercially available preparation of activated Sepharose 4B (Pharmacia) may be used. Prior to use, 7.5 g of this activated gel are diluted with 10 ml of cold 1 mM HCl and washed with an additional 500 ml of cold 1 mM HCl over the course of 20 minutes. A Büchner type of funnel with a coarse sintered-glass disc is used; the beads are not allowed to dry at any time. The beads are washed thoroughly with cold distilled water and then with 250 ml of cold 0.1 M potassium phosphate at pH 6.5. The final wet slurry or moist cake is suspended in the same phosphate buffer to yield a final volume of 30 ml which is added to the TSH solution as described above.

In another modification 0.1 M NaCHO$_3$ at pH 8 to 9 can, in all steps, replace the 0.1 M potassium phosphate at pH 6.5. The concentration of the TSH solution used in the coupling step can be varied from 5 to 12.5 mg/ml, and the coupling procedure can be scaled up or down by keeping the approximate proportions of 10 ml of activated agarose suspension (about 50% beads) to between 5 and 12.5 mg TSH. Slow rotation in a stoppered tube can replace mixing with a magnetic stirrer for the coupling step, and batchwise centrifugation at 500 g can be used to replace the initial column exposure to ethanolamine.

The TSH coupled agarose may be stored for at least 4 months at 0–4° without adverse effects for purification. Sodium azide or a few drops of toluene may be used to inhibit bacterial contamination during storage and the washing procedure is repeated, with the exception of exposure to ethanolamine hydrochloride, immediately prior to use.

Purification Procedure

TSH-agarose has been used to purify antisera to TSH, TSH receptors from cultured thyroid cells treated with trypsin, and TSH receptors from thyroid plasma membranes solubilized by lithium diiodosalicylate.

For purification of antisera, 10 ml of a TSH-agarose suspension, approximately 50% beads, are used to form a column with dimensions 0.7 × 9 cm. The column is washed with 0.1 M NaCl and 5 ml of whole serum are applied. The serum is allowed to slowly percolate into the beads and to remain exposed to the beads for 30 minutes at room temperature or overnight at 0° to 4°. Subsequently, the column is washed sequentially with 0.1 M NaCl and with 0.1 M sodium borate at pH 8.5 containing 1 M sodium chloride until the absorbance of the effluent at 280 nm has returned to the base line. Elution of the TSH antibody takes place with 0.1 M glycine·HCl at pH 2.2 to 2.5, or with 0.1 M sodium citrate at pH 2.2. The eluted peak is titrated to pH 7.0 with 1 M NaOH, lyophilized, made up to between 2 and 5 ml with water, and dialyzed against saline for 24 hours. Any precipitate which forms is removed by centrifugation. Recovery of antibody activity is between 58% and 74% of that applied; the protein behaves electrophoretically as pure γ-globulin.

In the case of TSH receptors from cultured thyroid cells, the sample applied to the column is the supernatant fluid from cells exposed to 0.25% trypsin in calcium- and magnesium-free phosphate-buffered saline (Dulbecco) and partially purified by passage over Sephadex G-50. In the case of TSH receptors from bovine thyroid plasma membranes, the sample is a receptor preparation partially delipidated by exposure to 70% ethanol–5% acetic acid and solubilized in 0.1 M lithium diiodosalicylate; after centrifugation at 30,000 g, the receptor solution is dialyzed and concentrated against water with hollow fiber beakers (Dow) which have a 5000 molecular weight exclusion limit. In each case, sample volumes of 1–2 ml are applied to the columns. Elution procedures are entirely analogous to those described above for the TSH antisera. Recoveries are monitored by assays of binding activity.[8] The recovery of trypsin-released receptor activity is 40–70% of that applied to the column. The recovery of lithium diiodosalicylate solubilized activity is 15–50% of that applied. All binding activity is adsorbed from the former preparations; 75% of the binding activity is adsorbed from the latter preparations.

The trypsin-released receptor is a 10,000 to 30,000 molecular weight glycoprotein containing at least 10% sialic acid. The lithium diiodosalicylate solubilized receptor activity is heterogeneous in molecular weight with species estimated at MW 280,000, 140,000, 70,000–80,000, and 10,000–30,000 by sucrose density gradient centrifugation. The last component is analogous to the trypsin-derived component, i.e., it is a glycoprotein containing at least 10% sialic acid which is insensitive to trypsin digestion.

[89] Adrenergic Receptors and Binding Proteins

By DONALD S. O'HARA and ROBERT J. LEFKOWITZ

The biological effects of catecholamines are initiated by their binding to specific adrenergic receptor binding sites.[1] These receptors trigger subsequent events that lead to such diverse responses as smooth muscle contraction or relaxation, increased force of cardiac muscle contraction, increased rate of glycogenolysis, and increased rate of lipolysis. A number of actions of catecholamines are brought about by stimulation of the enzyme adenylate cyclase.[2] The receptors which mediate catecholamine stimulation of this enzyme generally have the characteristics of so-called "β-adrenergic receptors." As with other hormone receptors, these appear to be proteins localized in membranes of responsive tissues.[3-7]

Recently methodology has been developed for preparation and study of adrenergic binding proteins from membranes of several tissues.[8] These binding proteins, which may function as part of the adrenergic receptor complex, can be solubilized[7,9] and purified by affinity chromatography.[9]

[1] R. P. Ahlquist, *Amer. J. Physiol.* **153**, 586 (1948).

[2] G. A. Robison, R. W. Butcher, and E. W. Sutherland, "Cyclic AMP," p. 152. Academic Press, New York, 1971.

[3] R. J. Lefkowitz and E. Haber, *Proc. Nat. Acad. Sci. U.S.* **68**, 1773 (1971).

[4] R. J. Lefkowitz, G. Sharp, and E. Haber, *J. Biol. Chem.* **248**, 342 (1973).

[5] R. J. Lefkowitz, D. O'Hara, and J. Warshaw, *Nature (London) New Biol.* **244**, 79 (1973).

[6] M. Schramm, H. Feinstein, E. Naim, M. Long, and M. Lasser, *Proc. Nat. Acad. Sci. U.S.* **69**, 523 (1972).

[7] J. Bilezekian and G. Aurbach, *J. Biol. Chem.* **248**, 5577 (1973).

[8] R. J. Lefkowitz, *in* "Frontiers in Catecholamine Research" (E. Usdin and S. Snyder, eds.), p. 361. Pergamon, Oxford, 1974.

[9] R. J. Lefkowitz, D. O'Hara, and E. Haber, *Proc. Nat. Acad. Sci. U.S.* **69**, 2828 (1972).

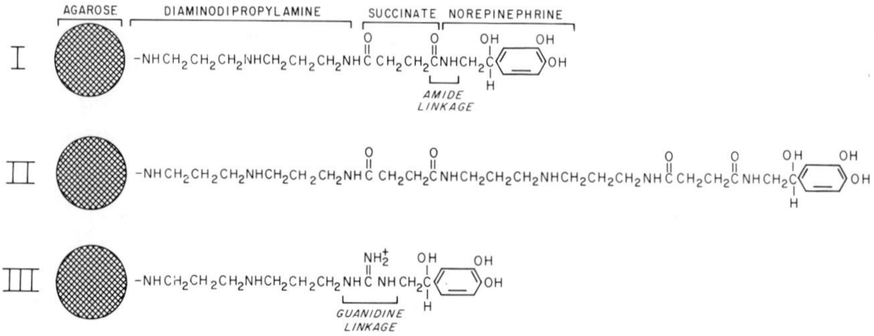

FIG. 1. Agarose derivatives I, II, and III.

Preparation of Agarose Derivatives

Catecholamines are covalently bound to agarose modified by the attachment of hydrocarbon side chains of 10–30 Å length by an adaptation of techniques which have been previously described.[10-13] Ligand content is generally in the range of 3–8 μmoles per milliliter of packed resin. Figure 1 illustrates the three derivatives whose preparation is given below.

Derivative I

a. Activation. Twenty-five grams of 4% agarose beads (Pharmacia or Bio-Rad) are washed on a coarse sintered-glass funnel with 2 liters of water. The moist agarose cake is transferred with rapid stirring to 150 ml of freshly prepared 100 mg/ml cyanogen bromide in a fume hood, and the reaction mixture is maintained at pH 10.8–11.2 for 10 minutes by addition of 5 M KOH. Ice is added as required to keep the temperature between 15° and 20°. The agarose is then washed on a funnel within 2 minutes with 2 liters of cold 0.5 M NaHCO$_3$–NaOH, pH 10, and briefly sucked free of excess fluid.

b. Coupling of Amine. The activated agarose filter cake is immediately added to 2.5 g of diaminodipropylamine (DPA)[14] in 50 ml of 0.5 M NaHCO$_3$ at pH 10 and stirred gently in the cold overnight. The resin is then washed successively with 1 liter each of 5 mM NaOH, water, 5 mM HCl, and water.

c. Succinylation. The material from step b is suspended in 50 ml of

[10] R. Axén, J. Porath, and S. Ernbäck, *Nature (London)* **214**, 1302 (1967).

[11] J. Porath, R. Axén, and S. Ernbäck, *Nature (London)* **215**, 1491 (1967).

[12] P. Cuatrecasas, *J. Biol. Chem.* **245**, 3059 (1970).

[13] P. Cuatrecasas and C. Anfinsen, this series, Vol. 22 [31].

[14] Abbreviations used: DPA, *N,N'*-diaminodipropylamine; CDI, 1-cyclohexyl-3-(2-morpholinoethyl) carbodiimide metho-*p*-toluene sulfonate; PCMB, *p*-chloromercuribenzoate.

water at room temperature. Succinic anhydride, 8 g, is added in 0.5-g aliquots over a 45-minute period to the stirred suspension, and the pH is maintained between 7 and 8 with $5 N$ NaOH. After standing in the cold overnight, the agarose is washed as in step b.

d. Coupling of Norepinephrine. The succinyl derivative of step c is suspended in 35 ml of water at room temperature. To this is added 5 ml of a solution containing 400 mg of L-norepinephrine·HCl and 5 μCi of [³H]norepinephrine, and the pH is adjusted to 4.8 with 0.1 N NaOH. (The [³H]norepinephrine, obtained as an ethanol–acetic acid solution, is lyophilized and redissolved in 0.1 N HCl before use to avoid coupling by the acetate.) The coupling of norepinephrine is initiated by adding, over a 30-minute period, 2-ml aliquots of a stirred slurry of 8 g of 1-cyclohexyl-3-(2-morpholinoethyl) carbodiimide metho-*p*-toluene sulfonate (CDI) in 15 ml of H_2O. The pH is maintained between 4.8 and 5.5 with 0.1 N HCl during the addition of CDI and for another 30 minutes thereafter. After stirring gently at room temperature overnight, the adsorbent is washed with 3 liters of 5 mM HCl.

Derivative II

The washed agarose from step c for derivative I is suspended in 50 ml of water at room temperature, 1 g of DPA is added, and the pH is adjusted to 4.8 with $6 N$ HCl. Coupling of the amine is carried out by addition of CDI as described in step d. The resin is washed with 3 liters of 5 mM HCl and 1 liter of water, and the derivatization is completed as described in steps c and d above.

Derivative III

The free amino group on the substituted agarose from step b is converted to a reactive cyanamide[10,15] by activation with BrCN as described in step a. The cyanamide derivative is washed rapidly with cold 0.5 M KH_2PO_4–KOH at pH 7.5 and briefly subjected to suction to remove excess buffer. The agarose is then immediately transferred to a solution of 2 mmoles of L-norepinephrine·HCl and 5 μCi [³H]norepinephrine in 50 ml of the same phosphate buffer. Coupling is allowed to continue overnight at 2–4°. The norepinephrine solution is made by dissolving the tracer and 411 mg of L-norepinephrine·HCl in 5 ml of water and adding this to the phosphate buffer just before the addition of the activated agarose. Since cyanamides react with secondary as well as primary amines, derivatives containing epinephrine or propranolol can be made in the same way.

[15] "Rodd's Chemistry of Carbon Compounds" (S. Coffey, ed.), Vol. I, Part C, pp. 344 and 375. Elsevier, Amsterdam, 1965.

Determination of Ligand Content

Approximately 1 g of washed, substituted agarose is washed again with 50 ml of water followed by 50 ml of p-dioxane. The moist resin is suspended in 10 ml of dioxane and centrifuged for 1 minute at 1000 g. The supernatant fluid is completely removed by aspiration, and an aliquot of the packed agarose is weighed in a counting vial; 10 ml of a 2:1 toluene:Triton X-100 (Rohm and Haas) based fluor is added for counting in a scintillation spectrometer. The miscibility of dioxane with organic solvents allows thorough penetration of the fluor into the beads with negligible quenching for quantities of resin less than 0.5 g.

Storage

Conjugates may be stored indefinitely without loss of activity in 10 mM EDTA at pH 4.5 and 2–4°.

Operation of Column

Affinity chromatography is carried out at room temperature with adsorbent beds 1 × 3 cm which are washed with 50–100 ml of 0.15 M potassium phosphate or 0.3 M Tris chloride (pH 7.5) to remove the EDTA used for storage.

Buffers

The ionic strength of buffers used to equilibrate, load, or flush the columns are above 0.2 to minimize ion-exchange effects and nonspecific adsorption. At an ionic strength of 0.01, as much as 50% of the applied protein is nonspecifically retarded.

The presence of the detergent Lubrol-PX (ICI America) in the sample prevents the retention of receptor protein and is eliminated by prior passage of the preparation over Amberlite XAD-2 resin (Rohm and Haas) equilibrated with the same buffer used for the affinity column. Sodium deoxycholate, however, at concentrations of 0.25 mg/ml or less does not diminish adsorption.

Loading

Sample loads consisting of 5 mg or less of solubilized[9] microsomal protein at concentrations no greater than 1 mg/ml are allowed to enter the column at a rate of 15 ml per hour.

Flushing

Unretarded material is flushed from the column with 30 ml of the same buffer used for equilibration.

Elution

Receptor protein is recovered after flushing by elution at a rate of 15 ml per hour with 20 ml of 0.1 M epinephrine–0.15 M NaH$_2$PO$_4$ at pH 3.8. Epinephrine is then removed by dialyzing four times against 200 volumes of 50 mM sodium phosphate at pH 7.5 in the cold over a 2-day period. Saturated solutions of epinephrine at pH 7.5 elute only 40% of the resin-bound activity. However, the greater effectiveness of 0.1 M epinephrine at pH 3.8 is not due solely to the lower pH, since 1 mM HCl does not elute binding activity. Other solutions which do not elute receptor protein include 0.5 mM PCMB,[14] 50 mM EDTA, and 2 M ammonium acetate at pH 7.5.

Occasionally a light purple color develops on the affinity resin at neutral pH because of oxidation of the bound catecholamine. The colored material, which is eluted by 0.1 M epinephrine in an oxidation-reduction reaction, does not perceptibly lower the capacity of the column or the binding properties of the eluted protein.

After removal of epinephrine, eluted samples may be applied to a second affinity column with quantitative adsorption and elution of binding activity. Resin beds are discarded after one use.

Performance Variables

With derivative II, at ligand concentrations of 3–8 μmoles/ml, 85 ± 12% of the loaded norepinephrine-binding activity is retained by the column after flushing.[9] Of this activity, 120 ± 30% is recovered in the epinephrine eluate with a purification over the load of 260-fold. Since no significant difference is found between the performance of derivatives I and II, the length of the hydrocarbon extension is probably not important beyond about 13 Å. At ligand concentrations of 14 μmoles/ml, adsorption is quantitative but only 2–5% of the bound receptor can be eluted.

Derivative III generally adsorbs a greater fraction of the loaded β-adrenergic binding activity than the other two but results in less complete elution. The net effect is that approximately the same total amount of loaded activity is recovered during elution from all three types of resin. The difference in the characteristics of derivative III may reflect a greater affinity for receptor protein due to the positive charge[10] on the guanidine linkage between the cyanamide and the catecholamine in contrast to the uncharged amide linkage formed in derivatives I and II (Fig. 1).

Other Derivatives

Venter *et al.*[16] have described a physiologically active derivative prepared by coupling the catechol nucleus of epinephrine, norepinephrine or isoproterenol to a diazonium salt covalently bound to glass beads. Although not designed for this purpose, these glass preparations should serve as effective resins for affinity chromatography. A resin identical to derivative I has been shown by Johnson et al.[17] to induce fat cell lipolysis.

In addition to receptors, a variety of other biological processes involve the binding of catecholamines. These include metabolizing enzymes such as catechol-O-methyltransferase[18] and various neuronal uptake mechanisms.[19] The agarose derivatives described here may be useful also for purifying the macromolecules involved in these other processes.

Note Added in Proof

M. S. Yong[20] has observed that catecholamines bound to agarose or glass are unstable at neutral and alkaline pH. This limitation must be considered when testing derivatives for biological activity but does not diminish their usefulness for affinity chromatography in short-term experiments.

[16] J. Venter, J. Dixon, P. Maroko, and N. Kaplan, Proc. Nat. Acad. Sci. U.S. 69, 1141 (1972).
[17] C. Johnson, M. Blecher, and N. Giorgio, Jr., Biochem. Biophys. Res. Commun. 46, 1035 (1972).
[18] J. Axelrod and R. Tomchick, J. Biol. Chem. 233, 702 (1958).
[19] A. von Euler, Circ. Res. 20, Suppl. 3, 5 (1967).
[20] M. S. Yong, Science 182, 157 (1973).

Section VI

Immunological Approaches

[90] Immunoadsorbents

By JOHN B. ROBBINS and RACHEL SCHNEERSON

Hardly a journal devoted to immunology or related to the study of purified proteins appears today without an article indicating the utilization of immunoadsorbents. In addition to a review article in this series,[1] several treatises have dealt with related subjects.[2-6] In addition, there are several articles in this volume with detailed methods of preparation for protein, polysaccharide, and organic radical derivatives of various insoluble matrices. Accordingly, emphasis will be directed toward new developments and uses of immunoadsorbents, evaluation of their effectiveness, and suggestions for further applications.

Protein Immunoadsorbents

Procedures for the purification of antiprotein antibodies from serum, secretions, or culture fluids based upon differences in their physicochemical properties due to their antigenic specificities are limited because of their structural heterogeneity. Affinity chromatography, which utilizes the specific and reversible interaction of antibodies with their antigen, has permitted the isolation and characterization of antibodies from a wide variety of sources. Isolation of myeloma proteins with antibody activity secreted from human and murine lymphoid malignancies, antibodies from immunized and diseased individuals, and minute quantities of antibodies synthesized *in vitro* may be accomplished with immunoadsorbents.

Immunoadsorbents may also be used as antibodies. If not extensively derivatized, antibodies retain their reactivity with their specific antigen. Preparation of antibody immunoadsorbents has facilitated the isolation of proteins, polypeptide chain components of proteins, or their degradation products.

Based upon their properties of high flow rate and capacity, ease of preparation, and relatively low nonspecific interaction, agarose deriva-

[1] P. Cuatrecasas and C. B. Anfinson, this series, Vol. 22, p. 345.
[2] P. Cuatrecasas, *Advan. Enzymol.* 36, 29 (1972).
[3] I. Silman and E. Katchalski, *Annu. Rev. Biochem.* 35, 873 (1966).
[4] R. Goldman, L. Goldstein, and E. Katchalski, *in* "Biochemical Aspects of Reactions on Solid Supports" (G. R. Stark, ed.). Academic Press, New York, 1971.
[5] M. Wilchek and D. Givol, *in* "Peptides" (H. Nesvadba, ed.), p. 203. North-Holland Publ., Amsterdam, 1973.
[6] Pharmacia Publication, May 1972-1. Appelbergs, Sweden.

tives have been preferred as antigen and antibody immunoadsorbents.[7-10] The preparation of representative agarose derivatives is described elsewhere in this volume. However, two problems observed with these and other insoluble matrices deserve mention. The first, is their reactivity with proteins and cells. The basis for this nonspecific reactivity has not been identified, but the activity is probably due to the amphoteric nature of the bound moieties and/or weak bonds formed with the insoluble backbone. This nonspecific reactivity is especially significant when the ratio of concentration of adsorbed moiety to immunoadsorbent is low. The second problem is the solubilization of the antigen or antibody from the immunoadsorbent by the eluting procedure. Although immunoadsorbents may be extensively washed with the eluting solvent before adsorption without detectable release of the bound moiety, solubilization of the ligand may occur when the antigen–antibody bond is dissociated during purification.

Immunoadsorbents have been especially effective for the purification of antihapten antibodies. In this instance, attachment of the monovalent ligand to "spacers" increases the efficiency of antibody binding.

There is continued use of other insoluble matrices for the preparation of insoluble antigens and antibodies. Cellulose-based ligands seem to have less nonspecific interaction with some proteins and cells than the derivatized agarose products. Accordingly, the preparation of some of these protein immunoadsorbents is described.

Example 1. Bromoacetyl Cellulose Derivatives (BAC)[11,12]

Acetone and anhydrous dioxane-washed cellulose (Whatman), is dried over P_2O_5 to constant weight. Bromoacetic acid, 100 g, dissolved in 30 ml of dioxane is added to 10 g of cellulose and stirred at room temperature for 20 hours in a tightly stoppered flask. Bromoacetyl bromide, 75 ml, is added with vigorous stirring for 20 hours at room temperature. The HBr

[7] R. Axén, J. Porath, and S. Ernbäck, *Nature (London)* **214**, 1302 (1967).
[8] J. Porath, R. Axén, and S. Ernbäck, *Nature (London)* **215**, 1491 (1967).
[9] P. Cuatrecasas, M. Wilchek, and C. B. Anfinsen, *Proc. Nat. Acad. Sci. U.S.* **61**, 636 (1968).
[10] J. Porath and L. Sundberg, *Protides, Biol. Fluid Colloq.* **18**, 401 (1970).
[11] A. T. Jagendorf, A. Patchornik, and M. Sela, *Biochim. Biophys. Acta* **78**, 516 (1963).
[12] J. B. Robbins, J. Haimovich, and M. Sela, *Immunochemistry* **4**, 11 (1967).

generated by the reaction is removed with a NaOH trap. The resultant solution is slowly poured into a vigorously stirred ice–water mixture (6 liters) and the precipitated cellulose washed successively with 0.1 M $NaHCO_3$ and water and stored at 4° in the moist state. It has been found that covalent binding of the protein to the BAC occurred in two steps. The first represents physical adsorption of the protein to the BAC. For each protein, a different pH is optimal for maximal physical adsorption to BAC. To determine the optimal pH, 10 mg of moist BAC is added to 10 ml of antigen solutions (1.0 mg/ml) prepared in 0.15 M citrate phosphate buffers ranging from pH 2 to 7. The suspensions are homogenized for 5 minutes (low speed Sorvall omnimixer) at room temperature, and the unbound protein is measured. When the optimum pH has been determined, 300–500 mg of the antigen is dissolved in the appropriate buffer and added to 1 g (moist weight) of the bromoacetylcellulose. The suspension is agitated for 24 hours at room temperature in the presence of Dow Corning Anti-Foam. The reaction mixture is centrifuged and suspended in 30 ml of 0.1 M $NaHCO_3$ at pH 8.9 and slowly tumbled for 24 hours. The antigen cellulose is extensively washed with 0.1 M $NaHCO_3$ in 50 mM 2-aminoethanol at pH 8.9 to block unreacted bromine. The sample is washed in 8 M urea to remove noncovalently bound antigen, washed with saline, and stored at 4° in saline–0.01% sodium azide at pH 7.4.

There is a large variation in the extent of attachment and antigenicity of several BAC protein antigens. Egg-white lysozyme binds well to the cellulose derivative (0.185 μg/mg BAC) but adsorbs low amounts of antibody compared to its reactivity in solution. In contrast, ribonuclease is comparatively poorly bound but retains most of its antigenic reactivity. A high yield of antibody is obtained by elution of the adsorbed anti-protein antibodies with acid or with monovalent ligands for antihapten antibodies. When the antibodies are equilibrated with buffered saline by dialysis, a small amount of precipitate forms, which is due to small amounts of the antigen present in the eluate. This loss can be reduced by passage of the eluted antibody through a 0.2 μm membrane (Millipore) before equilibration with neutral buffer; this procedure might be considered for other immunoadsorbents.

In addition to its use as an immunoadsorbent for purification of antibodies, BAC antibody derivatives are useful as adsorbents for the isolation and measurement of antigens including other immunoglobulins.[13]

---

[13] S. H. Polmar, T. H. Waldmann, and W. D. Terry, *J. Immunol.* **110**, 1253 (1973).

Example 2. Protein Immunoadsorbents Prepared with Ethyl Chloroformate[14]

This method of insolubilizing proteins is frequently used for preparation of monospecific antiimmunoglobulin antisera, and for purification of serum proteins present in low concentration. Of all immunoglobulins, IgG is the most easily prepared free of other serum proteins. Since, in many cases, whole immunoglobulins are used as immunogens because they are more effective than class-specific heavy chains, the resultant antisera may be reactive with all immunoglobulins because of shared antigenic sites. Because IgG is frequently used to remove these shared antigenic determinants, its insolubilization is cited as a model.

IgG, 150 mg, is dissolved in 0.1 M potassium phosphate at pH 7.2, and 0.1 ml of ethyl chloroformate is added with gentle stirring. A faint precipitate forms within 5 minutes, and a turbid suspension results within 1 hour. The precipitate is removed by centrifugation and washed with water, homogenized with a ground-glass piston, and washed with glycine· HCl at pH 2.4 until protein is no longer detected in the supernatant liquid. Appropriate volumes of antiserum (an estimate can be made by assuming that the adsorbent retains about 10% of its antigenic activity) are mixed without foaming for 2 hours at room temperature and then centrifuged. If the adsorbed antiprotein antibodies are to be recovered, the washed adsorbent is suspended in 0.1 M glycine·HCl at pH 2.2 or 5.0 M KI or 3.0 M KCNS. The elution should be carried out at room temperature with gentle stirring for 2 hours with a volume of eluate equal to or greater than that of the serum used for adsorption. The adsorbent is centrifuged, and the eluate is passed through a 0.4 μm Millipore filter and equilibrated with neutral buffer. The yields of antibody are high (30–60%) and active. There are several advantages to this method of insolubilization which make it a convenient tool: the relative ease of the reaction and the stability of the product with respect to time. However, several proteins, e.g., lysozyme and hemoglobin, are soluble to eluting buffers.

The antigenic or the antibody activity of the insolubilized product is, in many cases, more closely related to the pH at which cross-linking is carried out rather than to the concentration of ethyl chloroformate used. In general, polymerization in the pH range from 4 to 5 results in an efficient antibody adsorbent. Since the reaction of proteins with ethyl chloroformate is between unprotonated amino and carboxyl groups, the loss of antibody activity at a pH greater than 7 suggests that excessive

[14] S. Avrameas and T. Ternyck, *J. Biol. Chem.* **242**, 1651 (1967).

participation of lysyl residues reduces antigenic and antibody activity of IgG. A similar pH dependence of antibody activity has been observed with derivatives activated by cyanogen bromide.

Example 3. Antibody Immunoadsorbent for Purification of Proteins

Macromolecules such as serum proteins may have properties sufficiently similar that physicochemical techniques may be inadequate for purification. For example, proteins A and B may have identical physicochemical properties at the pH and ionic strength necessary for maintaining the biological activity of one component. However, one component can be purified at other than ideal conditions to prepare monospecific antiserum for an antibody immunoadsorbent. Purification of serum α_1-antitrypsin (A1AT) with physicochemical methods is complicated by the fact that in the pH range 5–9, necessary for the preservation of biological activity, the enzyme's net charge and gel filtration characteristics are almost identical to serum albumin.[15,16] However, serum albumins can be purified, free of A1AT, and monospecific antialbumin produced. An equal volume of an IgG solution of goat antimouse albumin (20 mg/ml) in 0.1 M sodium citrate at pH 6.0 is added to cyanogen bromide-activated Sepharose[17]; approximately 80–90% of the globulin protein is bound. Based upon measurement of the precipitating antibody content of the IgG solution (2 mg of antibody protein per milliliter) the antibody-Sepharose retains 90% of the albumin-binding capacity of the IgG solution. A solution containing only A1AT and albumin, previously partially resolved from the other serum proteins by anion exchange chromatography, is applied to the immunoadsorbent. Although the eluate shows a 40% loss of A1AT, all detectable albumin is retained, and the resultant A1AT preparation has high specific activity. This method leads to a single component by polyacrylamide electrophoresis and induction of a single precipitating antibody component when injected into animals.

Example 4. Purification of a Peptide Component of a Complex Protein

A unique polypeptide, designated J chain, is associated with polymeric immunoglobulins.[18] This peptide component comprises about 5% of the

[15] R. L. Myerowitz, A. Chrambach, D. Rodbard, and J. B. Robbins, *Anal. Biochem.* 48, 394 (1972).

[16] R. L. Myerowitz, Z. T. Handzel, and J. B. Robbins, *Clin. Chim. Acta* 39, 307 (1972).

[17] P. Cuatrecasas, *J. Biol. Chem.* 245, 3059 (1970).

[18] S. Morrison and M. E. Koshland, *Proc. Nat. Acad. Sci. U.S.* 69, 124 (1972).

weight of polymeric immunoglobulins and has properties similar to those of the L (light) polypeptide chains possessed by all immunoglobulins, and the secretory peptide of secretory IgA. Isolation of the J chain from polymeric immunoglobulins, such as serum IgM or colostral (secretory) IgA, has been facilitated by the use of anti-L-chain antibody-Sepharose columns. The J and L chains are separated from the heavy polypeptide chains of two polymeric immunoglobulins, serum IgM and colostral IgA, by reduction and alkylation of their interchain disulfides followed by chromatography of the reaction mixture on Sephadex G-100. L chains are prepared by the same technique from serum IgG, a nonpolymeric immunoglobulin, i.e., without J chains. The serum IgG L chains are used to prepare antiserum and an L-chain Sepharose derivative. With these reagents, purified anti-L-chain antibodies are prepared by passing the serum IgG L-chain antiserum through the L-chain immunoadsorbent and eluting the purified anti-L-chain antibodies with 0.1 M glycine·HCl at pH 2.5. The purified antibodies are coupled to cyanogen bromide-activated Sepharose 4B. Passage of an L and J chain mixture, obtained by reduction and alkylation of the polymeric immunoglobulins, through the L-chain antibody immunoadsorbent, permits separation of the J chains from the L chains.

Example 5. Immunoadsorbents for Measurement of Proteins

Agarose or bromacetyl cellulose (BAC) coupled to antibody have been used in radioimmunoassays to measure IgE.[13,19-21] This immunoglobulin is present in minute quantities in normal human serum and in secretions. Comparison between the Sepharose and the BAC-bound IgE radioimmunoassays with the recently developed double radioimmunoassay showed good agreement among the three methods when IgE levels greater than 200 ng are measured. However, falsely high levels were obtained with the immunoadsorbents when IgE was measured at the low levels present in certain immunodeficient, normal adult, or young patients. Similar findings were obtained when IgE was measured in urine with the Sepharose-anti IgE. The IgE quantitation in the 3 radioimmunoassays depends on inhibition of the binding of purified labeled IgE with specific anti-IgE by the IgE in the tested sample. At low levels, there is an unidentified inhibition of binding not due to IgE when immunoadsorbent

[19] S. G. O. Johansson, H. Bennich, and L. Wide, *Immunology* **14**, 265 (1968).
[20] A. S. Levin, M. O. Pipkin, and H. Fudenberg, *Vox Sang.* **18**, 459 (1970).
[21] M. Bazaral, M. Orgel, and R. N. Hamburger, *J. Immunol.* **107**, 794 (1971).

techniques are used. The discrepancy might be due to the high anionic charge of the immunoadsorbent acting as an exchange resin binding non-IgE molecules, to steric inhibition by other molecules adhering to the other immunoglobulins on the resin (non-anti-IgE), or to a specific inhibitor in the serum or secretion as found in certain malignancies.

Example 6. Measurement of Antibodies at Low Concentration

Solid matrix immunoadsorbents have been extremely useful in permitting the measurement of low levels of antibodies. Allergen-linked immunoadsorbents have been prepared for the measurement of reaginic (skin fixing, histamine releasing) antibodies. These antibodies, presumed to be the mediators of immediate hypersensitivity reactions, are IgE immunoglobulins and exert a profound biological effect despite their low concentration in serum and external secretions.[22-24] Traditionally, the Prausnitz-Küstner test, an *in vivo* assay, was used to measure antibodies of the IgE class. This method is not ideal because it is not quantitative and utilizes human subjects as indicators. Furthermore, assay of many samples is time consuming.

The allergen, crude short ragweed extract (Center Laboratories, Port Washington) or purified protein E (AgE), is attached to microcrystalline cellulose (Avicel, Brinkman). Cellulose, 500 mg, are suspended in 20 ml of water, and 1–2 g of cyanogen bromide dissolved in 2 ml of dimethylformamide are added. The pH was maintained at 11 with 4 N NaOH at 4° for 30 minutes. The activated cellulose is reacted with 50 mg of partially purified ragweed extract or with 25 mg of purified AgE. After thorough washing to remove noncovalently bound allergen, the cellulose is suspended at a concentration of 1 mg/ml in 0.1 M potassium phosphate at pH 7.4, containing 0.2% bovine serum albumin, 1% Tween, and 0.1% sodium azide. Allergen-immunoadsorbent, 0.5 mg, is mixed with 0.05 ml of the test serum and 0.2 ml of the diluent in a 10 × 4.4-ml tube and gently rotated overnight at room temperature. The allergen-cellulose is centrifuged and washed, and ^{125}I-labeled antihuman IgE is added and mixed thoroughly. The nonreacted radiolabeled antibody is removed by repeated centrifugation. Bound radioactivity is recorded as a measure of IgE antiallergen antibody. A fundamental defect in this assay system is the absence of a purified human IgE antiallergen antibody to permit an

[22] L. Wide, H. Bennich, and S. G. O. Johansson, *Lancet* 2, 1105 (1967).
[23] J. W. Yuniger and G. J. Gleich, *J. Clin. Invest.* 52, 1268 (1973).
[24] G. J. Gleich, R. T. Larson, R. T. Jones, and H. Baer, *J. Allergy Clin. Immunol.* 1974, in press.

absolute reference standard to be studied by several laboratories. Accordingly, the method is difficult to evaluate with precision. However, controls (normal, nonallergic or fetal calf serum) yield a maximum of 12% binding and accuracy controls had a standard deviation of about ±10%. This assay is capable of measuring extremely low levels of antibody (2–5 ng/ml) and produces important information regarding the effect of therapy upon the patient's reaginic antibody which closely parallels more tedious *in vitro* or *in vivo* bioassays.

Extension of this method of antibody analysis has taken two directions. The first is to use purified, radiolabeled antibodies directed to all of the immunoglobulin classes and subclasses. Thus, it is possible to measure the class and subclass composition of antibodies present in serum or secretions to an antigen-matrix with the method outlined above. Recently, investigators have succeeded in immobilizing haptens and a bacterial protein, cholera toxin, to disposable polystyrene tubes (11 × 70 mm, Nunc, Roskilde, Denmark). After removal of the nonbound antigen by washing, the tubes are incubated with 1 ml of the antiserum diluted in PBS containing 0.05% Tween. The nonbound material is removed by washing with PBS-Tween. An anti-immunoglobulin-enzyme conjugate (alkaline phosphatase) is then allowed to react with antigen-bound antibodies of various classes. The tubes are thoroughly washed, and 1 ml of nitrophenyl phosphate (1 mg/ml) is added as a substrate. The enzyme substrate reaction is stopped after an appropriate time, the solution is made alkaline, and the yellow product is measured at 400 nm. With appropriate controls and purified class specific antibodies, it is possible to determine the concentration and class composition of antihapten and antiprotein antibodies.[25,26]

To study *in vitro* antibody biosynthesis, sensitive detection techniques are needed.[27-29] Minute quantities of antiviral antibodies may be measured with immunoadsorbents. Sephadex G-25 is coupled to rabbit γ-globulin (RGG) through diazo bonds and then saturated with donkey anti-RGG. The attached donkey anti-RGG remains able to react with rabbit γ-globulin and has been used to measure minute quantities of rabbit antiphage (φX174) antibodies. Serum dilutions in 0.05 ml are incubated with an equal volume of phage suspension (2 × 10⁹ PFU/ml) at 37° for 1 hour,

[25] E. Engvall and P. Perlmann, *J. Immunol.* **109**, 129 (1972).
[26] J. Holmgren and A. M. Svennerholm, *Infection and Immunity* **7**, 759 (1973).
[27] A. E. Gurvich, O. B. Kuslova, and G. E. Ozlov, *Lab. Delo* **12**, 732 (1967).
[28] A. E. Gurvich and G. I. Drizlikh, *Nature (London)* **203**, 648 (1964).
[29] V. T. Skvortsov, *Immunochemistry* **9**, 366 (1972).

and at 4° overnight. Aliquots of 0.01 ml of the mixture are then incubated at 37° for 2 hours with 20 mg (dry weight) of the anti-RGG adsorbent. The immunoadsorbent is washed 12 times with Tris chloride at pH 7.4, and the washings are tested for plaque formation. The reproducibility of the technique is within 10%. The lowest level of detected antibodies is 5×10^{-11} mg/ml; the sensitivity of this method is 3 orders higher than that arrived at by phage neutralization.

Example 7. Antibody Immunoadsorbents for Isolation of "Active Site" Peptides

A very important application of immunoadsorbents has been to isolate peptides from antibodies and other ligand binding proteins that comprise or are near the "active" site of these macromolecules. This has been accomplished by extensive proteolysis under controlled conditions.[30] Another approach for characterization of peptides from the antibody-active site is the use of site-affinity reagents.[31]

Structural studies of immunoglobulins have been aided by analyses of myeloma proteins. These proteins are highly homogeneous and have been shown to be analogs of immunoglobulins; accordingly, analysis of their amino acid sequence and tertiary structure is easier than analysis of purified antibodies. Of importance is identification of the amino acid residues of the polypeptide chains that form the binding site for the antigen. Since many myeloma proteins have specificity for the hapten DNP, affinity labeling has been accomplished with the use of DNP derivatives.[6,32,33] The DNP portion of the derivative is bound owing to the antibody specificity of the myeloma protein, and the covalent attachment is provided by a variety of active groups attached to the hapten. Pepsin digests of the Fab fragment of myeloma protein are passed over a anti-DNP antibody Sepharose adsorbent. Those peptides at the antibody-combining site that reacts with the DNP affinity label, adsorb to the immunoadsorbent, and are eluted with DNP-glycine. This method offers the possibility of efficiently isolating site-affinity labeled peptides from other binding proteins, provided the monovalent ligand can be used as a hapten or will bind to the affinity label.[6,33]

Another approach that utilizes antibody immunoadsorbents for study of

---

[30] D. Inbar, J. Hochman, and D. Givol, *Proc. Nat. Acad. Sci. U.S.* **69**, 2659 (1972).
[31] S. J. Singer, *Advan. Protein Chem.* **22**, 1 (1967).
[32] M. Wilchek, V. Bocchini, M. Becker, and D. Givol, *Biochemistry* **10**, 2828 (1971).
[33] M. Wilchek, this volume [14].

the site that bind ligands is to extensively degrade proteins by proteolytic enzymes.[30,33] Under specific conditions, the antibody combining activity of the small peptides is retained after extensive proteolytic digestion of the Fab fragment of anti-DNP antibody fragment and can be adsorbed to a DNP lysine Sepharose column. Thus, immunoadsorbents can be used as antigens or antibodies to isolate critical peptides to permit characterization of active sites.

Polysaccharide Immunoadsorbents

Antipolysaccharide antibodies are of interest to immunologists for several reasons. When injected as purified polysaccharides, these antigens elicit low levels and prolonged synthesis of antibodies that have a restricted structural homogeneity. When whole bacteria are used as the antigen, remarkably high levels of homogeneous antibodies are elicited in a small percentage of laboratory animals. Accordingly, several different approaches have been proposed for the preparation of polysaccharide immunoadsorbents for the purification of antipolysaccharide antibodies.

Example 1. Purification of Antidextran Antibodies Using
Insolubilized Dextran (Sephadex)

A novel approach for the purification of antipolysaccharide antibodies was advanced by Schlossman and Kabat.[34] These investigators used Sephadex G-75 for isolation of antidextran antibodies from humans immunized with purified dextrans extracted from *Leuconostoc mesenteroides*. Human sera were applied to Sephadex G-75, and the antidextran antibodies were found to be specifically retained by the insolubilized polysaccharide. Elution of various subpopulations of the antidextran antibodies was accomplished with an isomaltose series of oligosaccharides. It could be shown that antibodies of different specificity are eluted with oligosaccharide derivatives ranging from 2 to 6 residues. The availability of the insoluble dextran, its susceptibility to hydrolysis with dextranase,[35] and the wide prevalence of purified dextrans and bacteria with dextranlike antigens for immunization stimulate us to suggest that this immunoadsorbent might be utilized for studies of immune cells synthesizing antidextran antibodies.

Example 2. Poly L-Leucine Derivatives of Blood Group Substances

The *N*-carboxy-L-leucine anhydride (Leuch's anhydride) (Pilot Chemicals Inc., Watertown, Massachusetts) is reactive with nonprotonated

[34] S. F. Schlossman and E. A. Kabat, *J. Exp. Med.* **116**, 535 (1962).
[35] S. F. Schlossman and L. F. Hudson, *J. Immunol.* **110**, 313 (1973).

amine groups at neutral pH. Kaplan and Kabat[36] prepared insoluble deriv-atives of purified blood substances as originally described by Tsuyuki *et al.*[37] Purified blood group substance is dissolved in 7 mM sodium bi-carbonate, pH 8.3, and equilibrated at 5°. An equal weight of solid crystalline *N*-carboxy-L-leucine anhydride is added to the blood group substance with constant stirring, and the reaction is allowed to continue for 48 hours. The insoluble residue is centrifuged and washed with PBS repeatedly. The blood group substance immunoadsorbent may be used as prepared or packed into a column when mixed with cellulose or poly-acrylamide beads (Bio-Gel P-5). Elution of the absorbed antibodies is accomplished with freshly prepared, cold sodium acetate at pH 3.6 (10% yield) or with *N*-acetyl-D-galactosamine for blood group A antibodies and D-galactose for blood group B antibodies (approximately 70% yield). This method may be used to prepare insoluble substances that have their active moiety distant from available amino groups.

Another method of preparing insoluble blood group substances was advocated by Porath and his colleagues on the basis of their work with cyanogen bromide activation.[7,8,38] It was reported that highly stable im-munoadsorbents resulted from the coupling of blood group substances to Sepharose 2B and aminoethyl cellulose. Since the blood group substance was a complex mucopolysaccharide isolated from ovarian cyst fluid, the authors proposed several methods for preparation of an insoluble antigen. The first was to partially deacetylate the blood group substance with *p*-toluenesulfonic acid, thus, exposing amino groups that were not involved in the antigenic site of the molecule. This "amino" derivative could then be coupled to cyanogen bromide-activated Sepharose.[37] Alternatively, the blood group substance could be activated directly with cyanogen bromide and then allowed to react with an amino derivative of cellulose (amino-ethyl) or agarose (lysine-Sepharose 2B). These methods have been used for another carbohydrate, and recent procedures advocated for complex polysaccharides are more fully described in Example 3.

Example 3. Polysaccharide-Agarose Immunoadsorbents

A successful application of affinity chromatography has been to isolate streptococcal group-specific polysaccharide antibodies in rabbit hyperim-mune sera.[39] Antisera had been prepared by hyperimmunization with streptococcal vaccines, as the result of which a small percentage of rabbits

[36] M. A. Kaplan and E. A. Kabat, *J. Exp. Med.* **123**, 1061 (1963).

[37] H. Tsuzuki, H. Van Kley, and M. A. Stahmann, *J. Amer. Chem. Soc.* **78**, 764 (1956).

[38] T. Kristiansen, L. Sundberg, and J. Porath, *Biochim. Biophys. Acta* **184**, 93 (1969).

[39] K. Eichmann and J. Greenblatt, *J. Exp. Med.* **133**, 424 (1971).

reacted by synthesizing high concentrations of "monocloncal" group C anticarbohydrate antibodies (highly homogeneous). The immunoadsorbents were prepared by partial deacetylation of the streptococcal polysaccharide, which resulted in free amino groups. The aminopolysaccharide derivative was bound directly to a cyanogen bromide-activated or to a lysine-Sepharose 4B derivative.

For the direct binding to Sepharose, 100 mg of the streptococcal group C carbohydrate is partially deacetylated in 100 ml of 20% p-toluenesulfonic acid for 6 hours at 50°. The resultant preparation is dialyzed against water and lyophilized. The availability of free amino groups is verified by the ninhydrin reaction. The lyophilized and partially deacetylated carbohydrate is dissolved in 50 ml of 0.5 M sodium bicarbonate at pH 8.5. The polysaccharide solution is mixed with 100 ml of cyanogen bromide-activated Sepharose and allowed to react by tumbling gently in the cold for 24 hours.[40] For coupling to a lysine derivative of Sepharose 4B, 100 ml of cyanogen bromide-activated Sepharose is allowed to react with 4 g of lysine (free base) in 100 ml of 0.5 M sodium bicarbonate at pH 8.5. After 24 hours of gentle tumbling in the cold, the gel is exhaustively washed with phosphate-buffered saline at pH 7.4; 100 mg of the group C streptococcal carbohydrate is activated by the addition of 100 mg of cyanogen bromide for 6 minutes at pH 11 and maintained at that pH by the continuous addition of 2 N NaOH. The pH of the activated polysaccharide is brought to 8.5 with 1 N HCl and added to 100 ml of the lysine-Sepharose derivative, which is then tumbled in the cold. The gels are packed into columns and exhaustively washed with phosphate-buffered saline.

Antisera are charged onto the columns and are followed by saline until the absorbance of the effluent falls below 0.08 at 280 nm. The absorbed antibodies are eluted by application of a linear gradient of 10 mM sodium acetate-buffered saline, pH 5.0, to 0.5 M acetic acid, 1 M NaCl, pH 2.5 resulting in a sigmoidal decrease of pH and a linear increase in salt concentration. The eluted antibodies emerge as sharp peaks that are electrophoretically homogeneous. In general, antibodies with the slowest electrophoretic mobility are eluted at the highest pH and lowest salt concentration. Antibodies with the most rapid electrophoretic mobility require the highest hydrogen ion and salt concentration for elution from the immunoadsorbent. It was concluded that since the hydrogen ion and salt concentration have a reciprocal relationship to their electrophoretic mo-

---

[40] It has been the experience of the authors and others that slow rotation of the slurry, i.e., about 140 rpm over a 20-cm radius, results in a more effective agarose derivative than that obtained after vigorous stirring with a magnetic stirring bar.

bility, elution conditions are a useful measure of the "affinity" of the anti-carbohydrate antibodies to the antigen column.

Example 4. Haemophilus influenzae Type b Capsular Polysaccharide Immunoadsorbent

Sepharose 4B is activated at 20° with 250 mg of cyanogen bromide per milliliter of packed Sepharose at pH 11 for 30 minutes. The gel is washed exhaustively with $0.1 M$ sodium bicarbonate and then mixed with 1,6-diaminohexane (10 mg per milliliter of packed gel) and tumbled in the cold overnight. The gel is packed in a column and washed with $0.1 M$ sodium bicarbonate followed by phosphate-buffered saline at pH 7.4. *H. influenzae* type b polysaccharide is dissolved in water (10 mg/ml) and mixed with a freshly prepared solution of cyanogen bromide, 50 mg/ml at a ratio of 0.25 mg of cyanogen bromide per milligram of polysaccharide. The pH of the reaction mixture is maintained at 11 by the continuous addition of $0.2 N$ NaOH.[41,42] After this activation step, 0.5 mg of the polysaccharide per ml of Sepharose is suspended in an equal volume of $0.1 M$ sodium bicarbonate. The mixture is tumbled overnight, packed into a column and washed at 4° with phosphate-buffered saline at pH 7.4 until orcinol positive material is absent from the effluent, i.e., approximately 20 column volumes. More than 90% of the *H. influenzae* type b poly-saccharide is bound by the adsorbent, which is capable of reacting with approximately 1.0 mg of antibody per milliliter of agarose. Elution of the adsorbed antibodies from hyperimmune rabbit serum or "natural" antibodies from human sera was accomplished with $0.1 M$ acetic acid or $3.0 M$ potassium isothiocyanate. In some cases, a significant amount of polysaccharide is released from the Sepharose, but only when the eluting solvent was applied to the immunoabsorbent after the antibodies were adsorbed. When applied singly or after normal serum, there is no detectable loss of polysaccharide from the Sepharose. This release of antigen with the eluting solvent reduces the yield of purified antibody and has been postulated to result from hydrolysis of *O*-glycosidic bonds, possibly facilitated by the bound antibodies.[43]

Alternatives to the 1,6-diaminohexane "spacer" have been suggested

[41] E. C. Gotschlich, M. Rey, R. Triau, and K. J. Sparks, *J. Clin. Invest.* **55**, 89 (1972).

[42] J. B. Robbins, J. C. Parke, Jr., R. Schneerson, and J. K. Whisnant, *Pediat. Res.* **7**, 103 (1973).

[43] S. Chipowsky, Y. C. Lee, and S. Roseman, *Proc. Nat. Acad. Sci. U.S.* **70**, 2309 (1973).

which could avoid the antigen release noted when the absorbed antibody is eluted. A multipoint attachment of the amino derivative was achieved with polylysine, which offers a more stable derivative than the mono-attached aminohexyl Sepharose.[44]

Some polysaccharides are not attacked by cyanogen bromide. For the meningococcal group B polysaccharide, activation of the sialic acid polymer has been attained with cyanuric chloride (2-amino-4,6-dichloro-*s*-triazine).[45-47]

Immunoadsorbents for Cells

Immunocompetent cells are a heterogeneous population with varied morphology and function. Preparation of homogeneous cell populations has been considered an important objective because both antibody biosynthesis and cell-mediated immunity require several different interacting cells. Isolation of homogeneous cell populations would permit more precise characterization of their activity and of the cell products involved in this multicell interaction. Immunologically competent mammalian cells have two origins at birth: thymus and bone marrow. Thymus-derived cells are distinguished by unique cell surface antigens and by reactivity with certain lectins. In contrast, bone marrow cells are concerned primarily with antibody biosynthesis and are characterized by readily detectable surface immunoglobulins that are antigen reactive.[48,49] These different properties have been utilized to separate cells with restricted function from the heterogeneous populations encountered in the adult lymphoid tissues. Immunoadsorbents are useful in separating lymphoid cells with a wide variety of immune capabilities, utilizing the reactivity of specific immunoglobulin surface receptors as antigens or their activities as specific antibodies.

Originally, specific binding of cells to immunoadsorbents permitted depletion of cell populations of specific functions. Later, it was possible to recover the absorbed cells, thereby permitting the study of enriched cell populations with respect to function or other characteristics associating with binding to the immunoadsorbent. Thus, it has been possible to prepare cell populations that are depleted or enriched with T (thymus-

[44] M. Wilchek, *FEBS Lett.* 33, 70 (1973); see also this volume [5].
[45] G. Kay and M. D. Lilly, *Biochim. Biophys. Acta* 198, 276 (1970).
[46] R. J. Fielder, C. T. Bishop, S. F. Grappel, and F. Blank, *J. Immunol.* 105, 265 (1970).
[47] E. C. Gotschlich, personal communication.
[48] R. A. Good, *J. Amer. Med. Ass.* 214, 1389 (1970).
[49] M. C. Raff, *Transplant. Rev.* 6, 52 (1971).

derived) or B (bone marrow-derived) cells of wide or narrow specificities. The techniques are in their infant stage but have been especially important in identifying the function of cells and may provide a method of cell replacement in genetically determined or acquired deficiency states.[50]

Most techniques that use immunoadsorbents for fractionation of cells have several drawbacks. The first is that there is a nonspecific interaction of lymphoid cells with the insoluble matrix. Most investigators have tried to avoid the resultant loss of cells by saturation of the immunoadsorbent with normal serum. The second drawback is that release of the specifically bound cells by inhibitors of binding has not been effective with macro-molecular antigens such as proteins. Cited below are representative procedures for isolation of immune cells with immunoadsorbents with emphasis on their limitations.

Example 1. Cell Separation Using Columns of Antigen-Coated Particles

Columns of glass and plastic beads coated with antigenic proteins have been used for depleting cell populations of specific antibody-producing and specific memory cells. Glass beads of 200 μm average diameter (Super-brite, 100-1005, 3M Co., St. Paul, Minnesota) or polymethylmetaacrylic plastic beads of 250 μm average diameter (Degalan, V26, Degussa Wolf-gang Ag, Hanau am Main, Germany) have been coated with human or bovine serum albumin, or with hen ovalbumin.[51] The optimal concentration of protein for coating (5 mg/ml) was determined using ^{125}I-labeled albumin. Before packing into the column the beads were allowed to remain in 10% normal mouse or rabbit serum at room temperature for at least 1 hour.[52] The treated beads were poured into 1.5 × 100 cm glass tubes. Fractionation was carried out at 4°. A minimum of 3 × 10⁸ cells were applied to the column, followed by cooled Eagle's Medium in Earle's (EME) at about 3 ml per minute. The slower the flow rate, the more non-specific was the retention of cells. The selective elimination of cells was about 50–100%. No difference was noted between the glass and the plastic beads with respect to the elimination of cells although a higher density of albumin molecules could be attached to the plastic beads. Attempts to mechanically separate the cells from the beads yielded a high percentage of nonviable cells and a loss of immune reactivity.

[50] T. L. Takahashi, L. J. Old, R. K. McIntire, and E. A. Boyse, *J. Exp. Med.* **134**, 815 (1971).

[51] H. Wigzell and B. Anderson, *J. Exp. Med.* **129**, 23 (1969).

[52] The authors explain that this procedure substantially increases the total amount of cells recoverable from the column.

Example 2. Antigen-Agarose Immunoadsorbents for Cell Fractionation

The specificity of antigen-binding receptors of lymphoid cells, taken from guinea pigs immunized with 2,4-dintrophenyl homologous albumin (DNP-GPA), has been examined by removing cells reactive with the carrier protein (homologous albumin) or with the hapten (2,4-dinitrophenyl derivatives of heterologous albumins and lysine). The effectiveness of the immunoadsorbent was assayed by the decreased antigen-stimulated DNA synthesis of the unretarded cells as compared to the reactivity of the original inoculum. This technique permitted study of the carrier specificity as well as the hapten effect in cell-mediated responses compared to antibody-producing cells.[53]

The larger Sepharose 2B beads are selected by decanting quickly the supernatant liquid in a large volume of water. Conjugation is carried out at 4°. The beads are suspended in a large volume of water and stirred with an equal volume of water saturated with cyanogen bromide. The pH of the reaction mixture is kept at 11 with 5 N NaOH for 3–5 minutes. The beads are filtered through a coarse glass filter and washed in a large volume of chilled 0.1 M sodium bicarbonate at pH 8.5. They are suspended in twice their packed volume in this buffer, and the antigen, 1 mg per milliliter of packed gel, is added. After stirring for 16–20 hours, the beads are washed extensively on a coarse fritted-glass filter with phosphate-buffered saline at pH 7.2, followed by 50% acetic acid, phosphate-buffered saline, fetal bovine serum, and phosphate-buffered saline. The beads are heated at 70° for 2–6 hours and again washed as above with sterile solutions. They are stored at 4° in tissue culture medium. The degree of conjugation of the most used conjugate, DPN-GPA, is about 2% of the dry weight of the conjugate as measured by acid hydrolysis and amino acid analysis; 3 to 8 × 10⁶ lymphoid cells are added to 2 ml of Sepharose beads (1 ml of packed gel), and the volume is brought to 5 ml with tissue culture medium containing 10% serum. The suspension is mixed in plastic culture dishes on a rotary shaker at room temperature for 1 hour. Unabsorbed cells are recovered by passing the suspension through a thin layer of glass wool in a 5-ml syringe. The recovery of cells is between 20 and 50%. The loss is unrelated to the antigen—an unrelated antigen retained a similar amount of cells—but seems to be due to a nonimmune interaction between the lymphoid cells and the agarose.

The reactivity of the DNP antigens was illustrated by the observation that they were three times more active than comparable concentrations of the soluble antigens in inducing *in vitro* DNA synthesis of immune lymphoid cells.

[53] J. M. Davie and W. E. Paul, *Cell. Immunol.* 1, 404 (1970).

This technique permits depletion of specific antihapten antibody-producing cells as well as cells reactive with the carrier and the carrier hapten complex. The degree of depletion, as measured by DNA stimulation of the eluted cells, ranges between 50 and 70%.

Example 3. Preparation of Relatively Purified Antihapten Specific Lymphocytes

Immunoadsorbent columns can be used for the preparation of specific antihapten lymphocytes and for enrichment of the active antibody-producing population. A method is described in which large polyacrylamide beads with azophenyl-β-lactoside (Lac) hapten groups are used.[54,55] Bio-Gel P-6 beads, 16–25 mesh (Bio-Rad, Richmond, California) are coupled to histamine and covalently linked to Lac.[54] Columns of 2 ml of beads in sodium phosphate at pH 7.2 are prepared in 5-ml plastic pipettes. Mice spleen cells, 10^8, (5×10^6/ml) are passed through the column at room temperature and at a flow rate of 30 drops per minute. After elution with cold buffer at 5°, buffer prepared in 1.5 μM Lac-hapten is allowed to penetrate the column for 15 minutes at room temperature followed by 15 minutes at 5°. Fractions are collected and assayed for their antibody synthesizing capability by enumeration of plaque-forming cells (PFC's). In these experiments, approximately 30% of PFC's applied to the column are recovered. The specific activity of the hapten-eluted cells is about 500 times that of the original cell population.

Example 4. Immunoadsorbent That is Enzymatically Degradable

Bone marrow, spleen, or lymph nodes from adults contain heterogeneous mixtures of immunopoietic cells. "B" cells, i.e., immunoglobulin-producing cells, may be identified by readily detectable surface immunoglobulins when studied immediately after removal from the animal. A method for obtaining *enriched* populations of B cells with surface immunoglobulins was devised using a rabbit antimouse IgG-Fab Sephadex G-200 column.[35]

Adult spleen cells, 10^7, suspended in Eagle's Minimal Essential Medium with 5% fetal calf serum and 5 mM EDTA, are passed through 8 ml of rabbit antimouse Fab G-200 Sephadex suspended in a 20-ml syringe with a porous filter (0.81 mm Vyon, Porrair, Ltd., Norfolk, England). A

[54] J. K. Inman and H. M. Dintzis, *Biochemistry* 8, 4070 (1971).
[55] P. Truffa-Bachi and L. Wofsy, *Proc. Nat. Acad. Sci. U.S.* 66, 685 (1970); L. Wofsy, J. Kimura, and P. Truffa-Bachi, *J. Immunol.* 107, 725 (1971).

40–50% depletion of surface immunoglobulin-bearing lymphocytes is observed. Dextranase (Worthington) is added to the Sephadex G-200 (1.0 mg per 8 ml), and the syringe is capped, and incubated for 1 hour at 23°. The cells are removed from the enzymatically solubilized immunoadsorbent by repeated centrifugation. The eluted cells are viable as measured by trypan blue dye exclusion (greater than 90%) and display a high degree of homogenicity as measured by surface immunoglobulin immunofluorescence and lack of reactivity with anti-T cell serum.

This method of solubilizing the immunoadsorbent has been verified using human lymphoid cells and should permit isolation of viable cells with other surface components, such as the C3 component of complement, antigen–immunoglobulin complexes, and thymic antigens.[56] It has been reported that recoveries of the total amount of the eluted cells achieved with this technique are as high as 100% of the theoretical yield.[35]

Example 5. Fiber Fractionation of Lymphoid Cells

Cells specific for an antigen have been removed from the general population of immunized animals and recovered for further studies by attachment to antigen-derivatized nylon fibers.[57–59] The method is covered in detail in this volume [15].

Example 6. Cellular Immunoadsorbents

Separation of specific cell populations from lymphoid tissues can be achieved by utilizing their immunological properties of interacting with cells. T cells react with cellular antigens as shown by binding and/or by cytolysis of target cells. Accordingly, this interaction may be used to separate this type of lymphoid cell from B cells in a heterogeneous population such as adult lymph nodes. Since some cells, e.g., those containing histocompatibility antigens such as fibroblasts, are glass adherent, T-cells that specifically reacted with these cellular antigens could be isolated.[60,61]

The fibroblast monolayers are prepared from trypsinized, 16-day-old

[56] D. B. Thomas and B. Philips, *J. Exp. Med.* **138**, 64 (1973).

[57] G. M. Edelman, U. Rutishauser and C. F. Millette, *Proc. Nat. Acad. Sci. U.S.* **68**, 2153 (1971).

[58] U. Rutishauser, C. F. Millette, and G. M. Edelman, *Proc. Nat. Acad. Sci. U.S.* **69**, 1596 (1972).

[59] U. Rutishauser and G. M. Edelman, *Proc. Nat. Acad. Sci. U.S.* **69**, 3774 (1972).

[60] P. Lonai, H. Wekerle, and M. Feldman, *Nature (London) New Biol.* **235**, 235 (1972).

[61] H. Wekerle, P. Lonai, and M. Feldman, *Proc. Nat. Acad. Sci. U.S.* **69**, 1629 (1972).

mouse embyros in Waymouth medium enriched with 5% calf serum in plastic petri dishes (Falcon Plastics) and are irradiated with 2000 R. Rat lymph node cells, suspended in Eagle's Medium containing 15% horse serum, are plated on the mouse fibroblast monolayers for 0.5 to 2 hours. The medium containing nonadherent cells is removed after 5 minutes of rotary shaking. The adherent cells are incubated for 5 days, at which time the reactive lymphoid cells undergo blast transformation and separate from the adherent mouse fibroblasts. The released cells are transferred onto another layer of fibroblasts previously labeled with $Na_2^{51}CrO_4$, and the degree of cytolysis caused by the sensitized lymphocytes after 18–40 hours of incubation is measured by ^{51}Cr release. Spontaneous release of ^{51}Cr is between 8 and 24%. The adherence of the lymphocytes to the fibroblasts is strain specific. The first separation removes 96% of the cells capable of inducing lysis. Repeating the separation procedure three times with the supernatant cells removed almost all sensitized cells reactive with mouse histocompatibility antigens as measured by this system. Since no antibodies to the histocompatibility antigens were detected in the supernatant fluid, it may be concluded that the separated cells are solely T cells.

Another example of cell separation achieved by specific interaction with other cells is illustrated by the experiments of van Boxel et al.[62] Mouse B lymphocytes, bearing a receptor for antigen–antibody–complement complexes (CRL) were isolated by complex formation with sheep red blood cells (SRBC) previously coated with rabbit anti SRBC antibody and complement (antigen–antibody–complement macromolecule). This cellular interaction involves the rather symmetrical arrangement of several of the mouse B lymphocytes bearing this specific receptor around the coated sheep erythrocyte, forming a rosette.

Washed 4×10^8 mouse BALB/c spleen cells are layered on a 23–33% bovine serum albumin gradient (Pathocyte, Pentex Inc., Kankakee, Illinois), and the cells that remained in the gradient are harvested. This preliminary step removes erythrocytes, nonviable cells, and macrophages. Rosettes are formed by incubation of 2×10^8 washed cells in 10 mM EDTA with the sensitized sheep erythrocytes. One milliliter of the incubation mixture containing 8×10^7 cells is relayered on the same albumin gradient and centrifuged at 10,000 g for 30 minutes in an SW 50L rotor (Beckman). The pellets from these gradients contain the rosettes and can be dissociated by incubation with fresh plasma. The eluted cells demon-

[62] J. A. van Boxel, W. E. Paul, M. M. Frank, and I. Green, J. Immunol. 110, 1027 (1973).

strate highly specific cytolytic activity as compared to the whole and depleted cell populations. Heterogeneity with respect to cytolytic activity is observed in CRL preparations from the spleen and lymph nodes.

Example 7. Charge Fractionation of Antibody-Producing Cells on Glass Beads

Although not strictly an immunoadsorbent technique, depletion of a thymus-derived precursor population of antibody-producing cells has been achieved by filtration of spleen, bone marrow, and thymus cells through native (negatively charged) or polylysine-coated (positively charged) glass beads. This technique of depletion of a specific antibody-producing cell population is important because it is apparently due to an electrostatic and nonimmune interaction of the thymus cells with the glass beads, and permits studies of differentiation of isolated cells without prior contact with antigen. Glass beads (Superbrite, 100-5005, Minnesota Mining and Mfg. Co., St. Paul, Minnesota) were prepared according to Wigzell and Anderson.[51,63-65] A positive charge was conferred by incubation of the uncoated beads in a solution of 5 mg of poly-L-lysine hydrochloride (average degree of polymerization 95) per milliliter of phosphate-buffered saline for 1 hour at 45° and then for 18 hours at 4°. Columns were prepared with glass pipettes, 22.5 cm long and 0.86 cm wide, supported by fine-mesh wire to retain the beads. Thymus, spleen, and bone marrow cell suspensions containing 1 to 2 × 10⁸ cells per milliliter in a volume of 0.75–1.5 ml were passed through the columns. When the filtered cells were compared to nonfiltered cells by injection into irradiated syngeneic mice, a specific depletion of thymic precursor cells was observed. Thymus cells filtered through native glass beads (negatively charged) gave a low level of restoration of antihapten antibody when the ligand was attached to a positively charged carrier. Conversely, depletion of antihapten antibodies occurred when thymus cells taken from an animal immunized with a negatively charged carrier-hapten antigen were passed through a poly-lysine-coated glass bead column. These findings indicate that thymus-derived cells may interact with immunogens on an nonimmune basis.

[63] H. Wigzell and B. Anderson, *Annu. Rev. Microbiol.* **25**, 291 (1971).
[64] M. Sela, G. M. Shearer, and Y. Karniely, *Proc. Nat. Acad. Sci. U.S.* **67**, 1288 (1970).
[65] Y. Karniely, E. Mozes, G. M. Shearer, and M. Sela, *J. Exp. Med.* **137**, 183 (1973).

[91] Immunoaffinity Chromatography of Proteins

By DAVID M. LIVINGSTON

With an ever increasing need for the rapid purification of proteins, immunoaffinity chromatography has become a valuable technique. Because of the extreme specificity and potentially high affinity of antibodies for antigens, this method can be highly efficient in the isolation of any protein for which the investigator has specific, high-affinity antibody.

In principle, protein antigen X, contained in a mixture of many other proteins, will bind selectively to insolubilized anti-X immunoglobulin (IgG). The other proteins for which the antibody has no affinity can be washed away and discarded. Thereafter, protein X is eluted from the adsorbent under conditions that disrupt the immune complex. The procedure has become widely applicable, having been developed for the purification of a variety of enzymes and other proteins.[1]

To establish such a procedure for the purification of hypothetical protein, X, the primary required reagent is specific anti-X antibody, covalently bound to an insoluble matrix. In addition, one must establish chromatographic conditions that permit only *specific* binding of protein X to a column of the immunoadsorbent. Thus, a similar column containing covalently linked nonimmune IgG should bind little or no protein X. Moreover, under ideal conditions, X will be the only macromolecule binding to the specific immunoadsorbent. Because of strong antigen binding, elution conditions which, of necessity, may not be as gentle as in ionic adsorption chromatography, must be both sufficiently drastic to remove the protein completely and sufficiently gentle to permit its isolation in biologically active form.

Preparation of Antibody

Immunization

Several small and medium-sized animal hosts are commonly used, including rabbits, guinea pigs, hamsters, and rats. When larger volumes of serum, and thus, antibody, are required, goats are frequently employed. Immunization procedures are, however, reasonably constant for all these species. A useful protocol will be described for a given protein antigen, assuming a rabbit host, although many variations of immunization pro-

---

[1] P. Cuatrecasas and C. B. Anfinsen, *Annu. Rev. Biochem.* **40**, 259 (1971).

cedures exist.[2] New Zealand white rabbits, 4–6 weeks of age, usually serve as good sources of specific antibody after immunization. In order to obtain highly specific antibody, the protein antigen should consist of a homogeneous preparation.

Day 1. Bleed each of two animals of 5–10 ml of blood from the marginal ear vein, let the serum clot at 4°, and allow several hours for the clot to retract. Centrifuge the sample at room temperature, collect the serum, and label it "prebleed control serum."

Day 2. Emulsify the antigen, dissolved in standard aqueous buffer at neutral or near neutral pH, in an equal volume of complete Freund's adjuvant (CFA) at room temperature. Emulsification should be complete. It is most thoroughly accomplished either in an electrical homogenizer or by repeated mixing and passage through a No. 20–25 gauge needle. The material is ready to inject when it is a thick and pasty emulsion.

Inject into each site 1 μg to 1 mg of protein emulsified in 1–4 ml of CFA in a number of equal aliquots: intradermally, into each of the four footpads; subcutaneously, into the region of each scapula; and, subcutaneously, along the back of each of at least two animals. Six to 12 equal immunization sites, and hence antigen aliquots, are reasonable. A plastic disposable syringe with a No. 25 needle is an effective injection device. The skin in the area to be injected should be shaved, and if the animal can be thoroughly restrained, e.g., by tying with rope on a board, there is no necessity for general anesthesia. Great care should be taken not to inject oneself with CFA since at least one constituent of this preparation, mycobacterial protein, will lead to an intense inflammatory reaction at the site of injection. Should this occur, the wound must be thoroughly cleaned and medical attention sought immediately.

Day 30. The animals are bled from the marginal ear vein of a total of approximately 5–10 ml. The serum is collected and tested for antibody activity. Whether detected or not, the animals should be given "booster" doses of antigen. If specific antibody is detected, the booster dose can be from 10 to 50% of that given initially, again in the emulsified form (1:1 with CFA). The emulsification and injection procedure are the same as for day 2. If antibody activity is not detected, a full dose, equivalent to that given initially, is administered by the same technique.

Days 37–40. On one day in this period, each animal is again bled of 5–10 ml, and the serum is collected. If the antibody titer is sufficient at this time, each animal can be bled again in 2 days of up to 20–25 ml. The sera from the two bleedings are pooled. If antibody activity is absent, the animal should again be boosted as on day 30 and another bleed under-

[2] E. Kabat, "Structural Concepts in Immunology and Immunochemistry." Holt, New York, 1968.

taken 7–10 days later. This cycle of boosting and bleeding may be repeated until the animal's serum is found to contain antibody and no further increase in antibody titer appears 7–10 days after another booster dose.

Animals may be bled of 12–20 ml twice or even three times at well-spaced intervals each month. When the level of antibody drops 5- to 10-fold, the animals can be boosted, i.e., receive another dose of antigen, according to the procedure noted above, and bled 7–10 days later. A successful booster immunization may be followed by as much as a 3- to 10-fold increase in antibody and a substantial increase in binding affinity of antibody for antigen.

Purification of IgG

The antibody activity detected after initial immunization, followed by one or more booster injections, will be largely associated with IgG molecules,[3] and it is desirable to isolate this serum fraction. For all the animal species mentioned in this section, purified IgG can be prepared as follows: Serum is rendered 50% saturated in ammonium sulfate (31.3 g per 100 ml), and the precipitate is collected and dissolved in 10 mM potassium phosphate at pH 6.8, the volume being equivalent to that of the original serum sample prior to salt addition. This solution is once again made 50% saturated in ammonium sulfate, and the resulting protein precipitate is again dissolved as before. The solution is dialyzed against three changes of 100 volumes each of 10 mM potassium phosphate at pH 6.8. The sample is applied to a column of DEAE-cellulose (DE-52, Whatman) previously equilibrated in the same buffer. Between 10 and 15 ml of packed DEAE-cellulose are used per milliliter of serum to be fractionated. After sample application, the column is washed with the equilibration buffer, and IgG, which does not bind to the adsorbent under these conditions, washes through the column and is collected. The absolute amount of IgG obtained can be calculated by measuring the 280 nm absorbance of the pooled sample. A 1 mg/ml solution of homogeneous IgG in a light path of 1 cm has a 280 nm absorbance of 1.46 at neutral pH. The IgG isolated by the above method is highly purified at this stage. It should be concentrated or diluted to between 3 and 5 mg/ml and can be stored at −70° in 10 mM potassium phosphate at 6.8–8.0 with insignificant losses in activity over many months.

Immunoadsorbent Preparation

In most of the individual procedures to be considered, purified IgG has been coupled to Sepharose by the cyanogen bromide procedure. A

---

[3] G. J. V. Nossal and G. L. Ada, "Antigens, Lymphoid Cells, and the Immune Response." Academic Press, New York, 1971.

particularly effective CNBr-Sepharose coupling procedure has been reported by Wilchek et al.[4]

Sepharose 4B is washed with 40 volumes of distilled water on a sintered glass funnel. It is then suspended in water at a concentration of 333 mg/ml; 3 N NaOH is added to the suspension until the pH is 11.2. It is then stirred gently on a magnetic mixer. Solid cyanogen bromide (33 mg/ml) is added, and the solution is constantly monitored on a pH meter. The pH falls rapidly and must be rapidly brought back to 10.8–11.2 and maintained there by intermittent addition of 3 N NaOH throughout the activation procedure. Titration can be stopped once the rate of fall in the pH of the reaction mixture (and hence the rate of addition of 3 N NaOH) has decreased markedly (usually 6–8 minutes after adding the CNBr). The "activated" Sepharose suspension is then filtered on a sintered-glass funnel and washed with 20–40 volumes of ice cold water. After the washing procedure is complete, the wet Sepharose cake is added to the IgG solution which is to be coupled. The ideal Sepharose to IgG ratio is approximately 30 to 1 (w/w). The protein (approximately 3 mg/ml) should be dissolved in 0.10 M sodium bicarbonate pH 8.0. The Sepharose-protein suspension is mixed at 4° for 18 hours, after which it is centrifuged and the supernatant fluid collected. The difference in the concentration of 280 nm of absorbing material in the initial uncoupled IgG solution, and this postcoupling fraction is a direct measure of the amount of IgG coupled to Sepharose. We have generally observed ≥90% IgG coupling under these conditions. The IgG-Sepharose is washed with whatever buffer is desired for subsequent column packing and operation, until no more 280 nm-absorbing material is present in the supernatant after centrifugation. The washed, coupled Sepharose can be stored as a suspension in buffer at neutral pH and at 4° without loss of activity for at least 4–6 months.

Column Preparation, Sample Application, and Elution

The column should be poured in any kind of convenient device. We have found that a plastic syringe with a gauze inner filter and polyethylene outflow tubing serves as an excellent chromatographic container. The adsorbent should be equilibrated in a buffer that (a) is not inhibitory to the function or antigenicity of the protein under consideration; (b) leads to optimal specific binding of the protein antigen; and, (c) maximally suppresses any potential nonspecific binding of the antigen and non-antigen protein to the adsorbent. Nonspecific binding is most easily detected by establishing a second column equivalent in size to the "antibody"

---

[4] M. Wilchek, V. Bocchini, M. Becker, and D. Givol, *Biochemistry* **10**, 2828 (1971).

column which contains the same solid matrix as the former; this material should be coupled to nonimmune IgG from the same animal species and in the same proportions as the "antibody" column. Equivalent aliquots of a crude extract or mixture, containing the desired antigen, are applied to both the "antibody" and "control" column, and fractions are collected. The amount of antigen in the wash through fractions is then determined.

Knowing how much total protein and specific antigen are initially applied to the column and how much of each washes through, one can calculate, by difference, the amount of protein which binds. In the absence of nonspecific binding, all the applied protein and antigen are present in the control wash fractions, and most of the nonantigen protein and none of the antigen are found in the "antibody" column wash fractions. Where there is nonspecific binding of antigen to the "control" adsorbent, column operation conditions must be altered to inhibit this effect. Nonspecific binding is often suppressed by increasing the ionic strength, altering the pH at which the column is equilibrated, adding to the column equilibration buffer a nonantigen protein such as purified bovine serum albumin (BSA), a nonionic detergent such as Triton X-100 or NP-40, or an organic solvent such as dimethylsulfoxide or ethylene glycol.

Elution of specific antigen protein from the immunoadsorbent can be achieved, in principal, by establishing conditions that will both promote disruption of the antigen–antibody complex and allow collection of the antigen in a form that is either biologically active or can be rendered so by simple manipulation. There are several means of disrupting antigen–antibody complexes. Among these are markedly raising or lowering the pH; addition of high concentrations of certain chaotropic ions, such as perchlorate, iodide, and thiocyanate; and addition of compounds, such as urea or guanidine, both of which are known to disrupt secondary and tertiary structure of proteins.[5] A detailed study of the effect of pH upon antigen-insoluble antibody complexes has been reported recently.[6]

Establishing successful chromatographic conditions is an empirical process. After developing a satisfactory system for sample application and for washing unbound protein from the column, elution conditions can be established. This last step is most easily accomplished by equilibrating the antigen with potential eluting solutions and then removing aliquots, either directly or after reequilibration with more physiological buffers, and testing for biological activity. By a systematic study, one can approach minimum and maximum changes in ionic strength, pH, and solvent conditions which the protein can withstand. At that point, direct elution of the

[5] P. L. Whitney and C. Tanford, *Proc. Nat. Acad. Sci. U.S.* **53**, 524 (1965).
[6] N. Welicky and H. H. Weetall, *Immunochemistry* **9**, 967 (1972).

protein can be attempted from the column, beginning with the minimum and then approaching the maximum tolerated disruption conditions. Clearly, the most desirable system for elution is that which is the least drastic and yet maximally efficient both in removing antigen and in preserving biological activity.

In the next section, methods for the purification of several different proteins will be described, each of which involves a unique set of elution conditions.

Specific Immunoaffinity Chromatographic Systems

Murine Type C RNA Tumor Virus Reverse Transcriptase.[7] All operations are at 4°. The Sepharose immunoadsorbent is prepared as noted earlier. It contains rabbit IgG purified from an antiserum raised against partially purified murine type C viral reverse transcriptase.[8] A 2–3-ml aliquot of this material (30–33 mg of protein per gram of adsorbent) is poured in a plastic syringe with plastic outflow tubing. The column is equilibrated in 50 mM Tris chloride, pH 7.8, 0.3 M KCl, 10 mg/ml bovine serum albumin, 2% Triton X-100 (buffer A). A sample containing the enzyme is equilibrated with the above buffer and then applied to the column by gravity. The volume of the sample is limited only by considerations of time and column flow rate; the latter tends to decrease markedly when viscous extracts are applied. In general, we have decreased viscosity of crude cell extracts by dilution with buffer A. The sample is applied, then the column is washed with 50 mM Tris chloride at pH 7.8, 0.3 M KCl (buffer B) until no further material absorbing at 280 nm washes through. The enzyme is eluted with 4–6 column volumes of 0.2 M NH$_4$OH, 0.3 M KCl, 10 mg of bovine serum albumin per milliliter (pH 10.8). NH$_4$OH, 10 mM, can be substituted for this eluting solution, with similar results. Fractions of 2 ml are collected at a flow rate of 0.4–0.5 ml/min. Up to 0.30-ml aliquots of enzyme can be assayed directly in a 0.5-ml reaction volume using poly(rA)·oligo(dT)(12–18) as a template–primer complex and [³H]TTP as substrate for poly(dT) synthesis.[9]

After completion of the run, the column is equilibrated with buffer B and stored in this buffer at 4°. Little or no loss in column binding capacity has been detected after 9 months, despite extensive use.

The procedure results in a high degree of purification of the enzyme

[7] D. M. Livingston, E. M. Scolnick, W. P. Parks, and G. J. Todaro, *Proc. Nat. Acad. Sci. U.S.* **69**, 393 (1972).

[8] W. P. Parks, E. M. Scolnick, J. Ross, G. J. Todaro, and S. A. Aaronson, *J. Virol.* **9**, 110 (1972).

[9] J. Ross, E. M. Scolnick, G. J. Todaro, and S. A. Aaronson, *Nature (London) New Biol.* **231**, 163 (1971).

although a definite specific activity cannot be presented since the protein concentration of the eluted material is so small that it cannot be measured with accuracy. The enzyme is regularly recovered in 20–40% yield. There is little or no binding of the enzyme to a control IgG-Sepharose column of identical size. The enzyme can be stored at pH 10.8 in liquid nitrogen for 3–4 days with retention of substantial activity when thawed and tested. When stored under these conditions for longer periods, it loses activity rapidly.

Avian Type C Viral Reverse Transcriptase.[10] The method is identical to that for the mouse enzyme with the following changes: (a) Rabbit antibody to avian type C viral reverse transcriptase is substituted for the antibody to murine polymerase.[8] The antimurine antibody does not significantly bind the avian polymerase and vice versa.[8] (b) After sample application, the column is washed with 50 mM Tris chloride at pH 7.8 containing 20% glycerol (v/v) and 1% BSA, until the absorbance 280 nm reaches a plateau in the wash fractions. (c) The column is then eluted with 50 mM NH_4OH containing 20% glycerol (v/v), and 1% BSA. (d) Volumes of both the column and the eluted fractions are kept to a minimum in order to suppress dilutional denaturation of this enzyme. (e) The assay is identical to that for the murine polymerase except that the final KCl concentration in the reaction mixture is 0.23 M.

Murine and Feline Type C Viral Group Specific (gs) Antigens.[11] A column of Sepharose-IgG (3–4 ml) is prepared at 4° in the same fashion as the anti-reverse transcriptase column. The major difference is that the IgG is from a goat antiserum raised against the group-specific protein from feline leukemia virus.[12] This antiserum reacts with both mouse and feline gs proteins, both single chain species of 30,000 daltons found in the inner core of the respective type C virions.[13,14]

The column is equilibrated in the same buffer used for initial equilibration of the anti-mouse reverse transcriptase IgG-Sepharose column, and sample application and column washing are exactly the same. However, the column is eluted with 12–15 ml of 1 N NH_4OH containing 0.3 M KCl, as one fraction. The eluate is dialyzed against three changes of 100 volumes each of 10 mM Tris chloride pH 7.8 containing 0.3 M KCl and

[10] D. M. Livingston, E. M. Scolnick, W. P. Parks, and J. Ross, *Virology* **50**, 388 (1972).
[11] W. P. Parks, D. M. Livingston, E. M. Scolnick, and G. J. Todaro, *J. Expl. Med.* **137**, 62 (1973).
[12] W. P. Parks and E. M. Scolnick, *Proc. Nat. Acad. Sci. U.S.* **69**, 1766 (1972).
[13] M. A. Fink, L. R. Sibal, N. A. Wivel, C. A. Cowles, and T. E. O'Conner, *Virology* **37**, 605 (1969).
[14] S. Oroszlan, C. Foreman, G. Kelloff, and R. V. Gilden, *Virology* **43**, 665 (1971).

then concentrated to between 0.5 and 1.0 ml by ultrafiltration through an Amicon PM-10 filter. This mouse or feline group specific protein may be assayed in a specific radioimmunoassay which has a minimum detection level of 0.1–1 ng.[12,15] Chromatography of a large quantity of crude mouse cell extract results in the isolation of group-specific protein with a 20-fold increase in specific activity compared to the starting material. Antigen recovery is 50%. The neutralized, concentrated protein can be stored at $-23°$ for several weeks with little or no loss in activity.

Mouse Interferon. According to the method of Sipe *et al.*[16] (a) Antibody to mouse L-cell interferon, induced by infection with Newcastle disease virus (NDV), is produced in sheep. The concentration of interferon used for immunization was 10^5 units/ml; 40 ml were injected weekly for 35 weeks. The antigen consisted of the supernatant culture fluid of L cells infected 24 hours previously with NDV virus. This suspension was acidified to pH 2.5 and maintained at that pH for 5 days, clarified by centrifugation at 4000 g for 15 minutes, brought to pH 7, and concentrated to 10^5 units/ml. The assay for interferon is a plaque reduction assay for vesicular stomatitis virus on L cells.[17] One unit is defined as the minimal amount necessary to reduce the plaque number by 50%. (b) Immunoglobulin, purified by ammonium sulfate fractionation, is coupled to Bio-Gel A, 1.5 M, after cyanogen bromide activation of the latter. After coupling, the immunoadsorbent is washed twice with 2 M urea, twice with 3 M NaCl, and equilibrated with 10 mM sodium phosphate at pH 7.4 containing 0.14 M NaCl (PBS), 100 μg of penicillin and 50 μg of streptomycin per milliliter. (c) Acidified and neutralized interferon containing culture fluid, 240 ml, is applied to a 20-ml column of the immunoadsorbent at 4°. The column is washed with three or more column volumes of PBS to remove all unbound material and then eluted at room temperature with 15 mM acetic acid–0.15 M NaCl at a flow rate of 0.5–0.8 ml per minute. The eluted material is neutralized by dialysis against PBS and assayed for interferon activity. (d) By this technique, the specific activity of crude mouse fibroblast interferon can be increased almost 2000-fold with a recovery of 55%.

Insulin. A Sepharose 2B immunoadsorbent containing covalently bound guinea pig antiporcine insulin IgG has been prepared by Akanuma *et al.*[18]

---

[15] E. M. Scolnick, W. P. Parks, and D. M. Livingston, *J. Immunol.* **109**, 570 (1972).

[16] J. D. Sipe, J. de Maeyer-Guignard, B. Fauconnier, and E. de Maeyer, *Proc. Nat. Acad. Sci. U.S.* **70**, 1037 (1973).

[17] R. R. Wagner, A. H. Levy, R. M. Snyder, G. A. Ratcliff, Jr., and D. F. Hyatt, *J. Immunol.* **91**, 112 (1963).

[18] Y. Akanuma, T. Kuzuya, M. Hayashi, T. Ide, and N. Kuzuya, *Biochem. Biophys. Res. Commun.* **38**, 947 (1970).

after immunizing guinea pigs with porcine insulin in complete Freund's adjuvant. Presumably, any high titered antiinsulin serum, containing reasonable levels of antiinsulin IgG, would be a useful source. This material is presented in detail in this volume [93].

Other Protein Purification Systems. Immunoaffinity chromatographic systems have also been developed for the SV40 T protein, in which the antigen is eluted with high concentrations of lithium chloride[19]; hypoxanthine-guanine phosphoribosyl transferase[20]; β-galactosidase, in which the enzyme is eluted from the column with 8 M urea and then reactivated by stepwise dialysis against nonurea-containing buffer[21]; and human placental lactogen.[22]

Thus, if specific, high-affinity antibody can be prepared and if reasonable operating and eluting conditions can be developed, the method is suitable for any protein. Its advantage lies in the potentially extreme specificity of immunoadsorbent binding of protein antigens, and, therefore, the high degree of protein purification rendered in one step.

[19] B. Del Villano and V. Defendi, *Virology* **51**, 34 (1973).
[20] A. L. Beaudet, D. L. Roufa, and C. T. Caskey, *Proc. Nat. Acad. Sci. U.S.* **70**, 320 (1973).
[21] F. Melchers and W. Messer, *Eur. J. Biochem.* **17**, 267 (1970).
[22] B. Weintraub, *Biochem. Biophys. Res. Commun.* 39, 83 (1970).

[92] Purification of the First Component of Human Complement and Its Subunit C1 Esterase[1]

By David H. Bing[2]

The classical serum complement system consists of 11 distinct proteins comprising 9 components. They are numbered C1 through C9 and collectively represent about 1% of the total serum proteins.[3] There is also evidence for an alternate complement pathway which involves at least 3 additional proteins; properdin, C3 Proactivator (glycine-rich β-glycoprotein, or GBG), and C3Pase (GBGase).[3-5] It is now clear that any

[1] Supported in part by Grant 1-RO1-AM 13697 from the National Institute of Health.
[2] Recipient of Research Career Development Award 1-KO4-AMF 0340-ALY.
[3] H. J. Müller-Eberhard, *in* "Progress in Immunology" (B. Amos, ed.), p. 553. Academic Press, New York, 1972.
[4] I. H. Lepow, *in* "Progress in Immunology" (B. Amos, ed.), p. 579. Academic Press, New York, 1972.
[5] C. A. Alper, I. Goodkofsky, and I. H. Lepow, *J. Exp. Med.* **137**, 424 (1973).

component may possess one, two, or all of the following properties: In serum it is a zymogen that can be activated to a functional protease; it is a specific substrate for another protease in the complement sequence, and cleavage results in release of a peptide that may have phlogistic properties; it can bind specifically to other complement proteins or immunoglobulins.[3,6]

The purification of these complement proteins is a formidable task. Most purification schemes have required a combination of ion exchange chromatography, gel filtration chromatography, and electrophoresis, and the result is usually a low yield of the enzyme and preparations which only approach homogeneity.[7] An example is C1s̄, a subunit of C1̄, which is present in serum at a concentration of 120 μg/ml. It can be obtained in a highly purified state by ion exchange chromatography and zonal electrophoresis in about 5% yield.[8] With repetition and recycling of various fractions, the yield can be increased to between 15 and 20%. However, the final product often remains heterogeneous with respect to other serum proteins, and there is evidence that C1s̄ purified by these procedures is not identical to the C1s̄ in whole serum.[9] Experiences such as these are not atypical in the purification of complement components and suggested that alternative approaches to purifying the complement proteins should be explored. It appeared logical, therefore, to develop an isolation procedure dependent on specific binding properties for a substrate or protein.

With the exception of C1̄ and C4, the naturally occurring substrates for the complement proteins are other complement proteins.[7] Synthetic amino acid ester substrates have been used to assay C1r, C1s, the C4̄,2̄ (C3 convertase) enzyme and C3b (C3 endopeptidase).[6] The substrates used are by no means specific, and, in fact, most have a large K_m and low V_{max}. Competitive inhibitors of C1q and C1s̄ have been described,[10-13] and a series of fluorosulfonyl benzamidines have been reported to be potent irreversible active site inhibitors of C1̄ and, perhaps, other complement components.[14,15] Because C1̄ and its subunits are the best char-

[6] N. R. Cooper, "Progress in Immunology" (B. Amos, ed.), p. 567. Academic Press, New York, 1972.
[7] H. J. Müller-Eberhard, *Advan. Immunol.* 8, 2 (1967).
[8] A. H. Haines and I. H. Lepow, *J. Immunol.* 93, 4561 (1964).
[9] K. Nagaki and R. M. Stroud, *J. Immunol.* 105, 162 (1970).
[10] D. H. Bing, *Biochemistry* 8, 4503 (1969).
[11] S. Westfall, G. H. Wirtz, and W. J. Canady, *Immunochemistry* 9, 1221 (1972).
[12] G. H. Wirtz, *Immunochemistry* 2, 95 (1965).
[13] C. R. Sledge and D. H. Bing, *J. Biol. Chem.* 248, 2818 (1973).
[14] B. R. Baker and M. Cory, *J. Med. Chem.* 14, 119 (1971).
[15] B. R. Baker and M. Cory, *J. Med. Chem.* 14, 805 (1971).

acterized complement proteins in terms of the known binding specificities for proteins and competitive inhibitors, it appeared that a logical starting point for applying affinity chromatography to complement proteins in general would be to develop an affinity chromatography purification strategy for $C\bar{1}$ and its subunits. Two kinds of resins were prepared. One series of resins were synthesized which contained benzamidine, a competitive inhibitor of $C1\bar{s}$. These resins were designed to isolate the $C1\bar{s}$ subunit of $C\bar{1}$. The other series of resins were agarose beads linked covalently to IgG.[16] This immunoglobulin can bind to $C\bar{1}$ via the C1q subunit.[3] Such a resin could be used to isolate either C1q or potentially the whole $C\bar{1}$ macromolecular complex. The preparation and use of these resins, including some resins which did not lend themselves to specific purification procedures for $C1\bar{s}$ are described.

1. Assays

$C1$[17] can be assayed either by its hydrolytic activity toward synthetic amino acid esters, or by its ability to form the $EAC\overline{1,4}$ intermediate in complement mediated hemolysis of sheep erythrocytes reacted with rabbit-anti-sheep red blood cell antibody. C1q, $C1\bar{r}$, and $C1\bar{s}$ can be assayed according to the ability of each to react with an excess of the other two to form $C\bar{1}$. The reconstituted $C\bar{1}$ is detected in turn in the assay in which $EAC\overline{1,4}$ formation depends on $C\bar{1}$ concentration. $C1\bar{s}$ can also be measured independently of C1q and C1r in terms of its ability to catalyze $EAC\overline{4,2}$ formation as well as its hydrolytic activity toward synthetic amino acid esters. C1q can be measured independent of C1r and C1s by its ability to agglutinate immunoglobulin coated latex particles. A list of these assays is given below.[18]

---

[16] IgG is the predominant type of immunoglobulin found in serum. It is the immunoglobulin which has a 7 S sedimentation coefficient and contains two kinds of polypeptide chains—the heavy chains (H chains) of the gamma type, and light chains (L chains) of the kappa or lambda type.

[17] Terminology for the complement system is that suggested in *World Health Organ. Bull.* 39, 935 (1968). Thus components of complement are designated C1 through C9. A bar across the top designates the active enzyme. C1q, C1r, and C1s are the subunits of C1.

[18] These assays are described in some detail in Chapter 4, "Kabat and Mayer's Experimental Immunochemistry" (E. A. Kabat, ed.), 2nd Ed. Thomas, Springfield, Illinois, 1961; in Chapter 7, "The Molecular Basis of Complement Action" by T. Borsos and H. J. Rapp, Appleton, New York, 1969; and in a section on complement in "Methods in Immunology and Immunochemistry" (R. F. Williams and M. W. Chase, eds.), Vol. 4, Academic Press, New York, 1974.

1.1 Hydrolysis of N-Carbobenzoxy-L-Tyrosine-paranitrophenyl Ester by $C\bar{1}$ and $C1\bar{s}$

This is a spectrophotometric assay which has been described.[10] Briefly, the enzyme is diluted to 1 ml in the cuvette in the presence of 0.005 ionic strength Tris-acetate at pH 8.0 containing 90 mM NaCl.[19] Thirty microliters of 1 mM acetone solution of N-carbobenzoxy-L-tyrosine-paranitrophenyl ester (N-Z-L-Tyr-Np) is added, the cuvette is mixed, and the reaction monitored for 5 minutes at 410 nm. The initial slope is taken as the rate. One unit of enzyme is defined as that amount which releases 1 nmole of nitrophenol in 5 minutes. The nitrophenol is determined by assuming a molar absorptivity of 1.66×10^4. N-Z-L-Tyr-Np is not a specific substrate; it is readily hydrolyzed by both trypsin and chymotrypsin.

1.2 $EAC\bar{1},4$ Formation

This is essentially the C1 transfer test of Borsos and Rapp[20] except that reaction volumes are reduced by one-fifth. C2 and CEDTA are of guinea pig origin. Titers are expressed as the reciprocal of the dilution which produces 63% lysis, or in terms of effective molecules.[21]

1.3 $EAC\overline{4,2}$ Formation

This is essentially the assay described by Nagaki and Stroud[22] except that the reaction volumes are reduced by one-fifth. Titers are expressed as the reciprocal of the dilution which gives 63% lysis.

1.4 Formation of $C\bar{1}$ from C1q, C1r and C1s

This is based on the method originally outlined by Lepow et al.[23] One-tenth milliliters of dilutions of C1q, C1r, and C1s are incubated at 0° for 30 minutes in the presence of 1 mM $CaCl_2$. The mixtures are assayed for $C\bar{1}$ activity with the $C\bar{1}$ transfer test.

1.5 C1q Latex Agglutination

A commercial preparation of IgG-coated latex particles (RA Test Kit, Hyland Laboratories) is used in this test. One drop of a C1q dilution is

[19] Prepared by dilution of 0.1 ionic strength Tris–acetate buffer (20.86 g of Tris and 2.89 ml glacial acetic acid diluted to 1 liter) and 3 M NaCl.

[20] T. Borsos and H. J. Rapp, J. Immunol. **95**, 559 (1965).

[21] The effective molecule is based on the one-hit theory for complement-mediated immune hemolysis. See M. M. Mayer, in "Kabat and Mayer's Experimental Immunochemistry" (E. A. Kabat, ed.), 2nd Ed., p. 187. Thomas, Springfield, Illinois, 1961.

[22] K. Nagaki and R. M. Stroud, J. Immunol. **102**, 421 (1969).

[23] I. H. Lepow, G. B. Naff, E. W. Todd, J. Pensky, and C. F. Hinz, Jr., J. Exp. Med. **117**, 933 (1963).

mixed on a microscope slide with 1 drop of the latex globulin reagent and 4 drops of the glycine EDTA buffer provided with the RA Test Kit. Agglutination is scored after mixing for 2–3 minutes, as strong (+ + +), moderate (+ +), weak (+), or doubtful (± or −).

2. Purification of C1s̄ on Benzamidine Resins

2.1 Background

Benzamidine and *p*-nitrobenzamidine have been shown to be a competitive inhibitor of human C1s̄ with $K_i = 0.1$ mM and 0.93 mM, respectively.[10] Baker and Cory described a series of meta-substituted benzamidines which were inhibitors of guinea pig C1̄.[14,15] One in that series, *m*-[*o*-(2-chloro-5-fluorosulfonylphenylureido)phenoxybutoxy]benzamidine, was found to inactivate specifically human C1̄ via the C1s̄ subunit.[24] Data of this kind led to the preparation of a series of *m*-benzamidine-substituted affinity chromatographic resins which have been tested for purification of human C1s̄. A list of these resins, with structural formulas, is given in Table I. All contain "arms" of varying length connecting the meta-substituted benzamidine to Sepharose, polyacrylamide, or poly-acrylamide agarose resins.

The methods for synthesizing these resins are based on previously reported procedures.[25-27] Resin I was prepared by coupling *m*-amino-

TABLE I

RESINS TESTED FOR PURIFICATION OF HUMAN C1s̄

Resin No.	B	R
I	—NH—	Sepharose 6B
II	—NHCOCH₂CH₂CONHCH₂CH₂NH—	Bio-Gel P60
III	—NHCOCH₂CH₂CONHCH₂CH₂CH₂NHCH₂CH₂CH₂NH—	Sepharose 6B
IV	—NHCOCH₂CH₂CONHC₆H₃(CH₃)NH—	Sepharose 6B
V	—OCH₂CH₂CH₂CH₂OC₆H₄(*m*—NH—)	Bio-Gel A5
VI	—OCH₂CH₂CH₂CH₂OC₆H₄(*o*—NH—)	Bio-Gel A5

[24] D. H. Bing, J. L. Mernitz, and S. Spurlock, *Biochemistry* 11, 4263 (1972).
[25] P. Cuatrecasas and C. B. Anfinson, this series, Vol. 22, p. 345.
[26] J. K. Inman and H. M. Dintzis, *Biochemistry* 8, 4074 (1969).
[27] P. Cuatrecasas and I. Parikh, *Biochemistry* 11, 2291 (1972).

benzamidine directly to cyanogen bromide-activated Sepharose 6B. Resins II, III, and IV were made by adding the bridge, B, to the resin, R (Table I) in one or more steps and then coupling *m*-aminobenzamidine with dicyclohexyl carbodiimide to a free carboxyl at the end of B group. Resins V and VI were prepared by coupling the entire benzamidine functional group, *m*-(*m*-aminophenoxybutoxy)benzamidine and *m*-(*o*-aminophenoxy-butoxy)benzamidine, to cyanogen bromide activated Bio-Gel A5.

2.2 Protein Preparation

The starting material for C1s̄ purification with these resins is the same, namely a euglobulin serum fraction obtained by low ionic strength-low pH precipitation of serum followed by extraction of the precipitate with a high ionic strength solution. It is prepared as follows: One liter of serum is diluted to 8 liters with ice-cold 20 mM sodium acetate at pH 5.5, and the pH of the final mixture is adjusted to 6.4 with 1 M NaOH. This mixture is stirred for 20 minutes at 0° and the precipitate is collected either by centrifugation for 30 minutes at 10,400 g at 4° or by allowing the precipitate to settle for 12–18 hours at 4°; the clear supernatant fluid is discarded and the precipitate is collected by centrifugation for 20 minutes at 10,400 g at 4°. The precipitate is washed twice with 0.02 ionic strength sodium acetate pH 5.5, is suspended in 100 ml of 0.3 M NaCl, and is stirred for 2–4 hours at 0°. This mixture is dialyzed against three changes of 0.15 ionic strength sodium phosphate at pH 7.4 over a period of 24 hours.[28] The C1 in this solution is activated from the zymogen form by incubation at 37° for 15 minutes,[29] and the entire mixture is then dialyzed at 4° against three changes of 0.02 ionic strength sodium phosphate at pH 6.3, 0.1 M NaCl and 1 mM Na$_2$H$_2$EDTA[30] for 36 hours. The ratio of volume of dialysis buffer to volume of protein is kept at about 100 to 1 throughout the dialysis. Just before chromatography, the protein is centrifuged for 1 hour at 48,200 g at 4°.

2.3 Resin Preparation

Resin I. *m*-Aminobenzamidine is prepared by low pressure reduction of *m*-nitrobenzamidine (Aldrich) and is recrystallized twice from 95% ethyl alcohol as the HCl salt (mp = 279–281). Twenty-seven mmoles

---

[28] Buffer *34* in Appendix II, "Methods in Immunology and Immunochemistry" (C. A. Williams and M. W. Chase, eds.), p. 401. Academic Press, New York, 1968.

[29] I. H. Lepow, G. B. Naff, and J. Pensky, *Complement, Ciba Found. Symp. 1964*, p. 74 (1965).

[30] Prepared by dilution of 0.1 ionic strength sodium phosphate, pH 6.3 buffer (buffer 33B[28]) 3 M NaCl and 0.1 M Na$_2$H$_2$EDTA.

(5 g) of *m*-aminobenzamidine are allowed to react at pH 9.0 in 0.1 *M* NaHCO$_3$ at 4° for 18 hours with 40 ml of settled Sepharose 6B which was previously activated with 4 g of CNBr. This procedure is repeated three times and results in a resin containing 0.6 mmole of benzamidine per milliliter of settled resin.[31]

The extent of modification is determined by titrating aliquots of washed *m*-aminobenzamidine-reacted resin to pH 12.6, i.e., the p*K* of benzamidine, and correcting to a blank which consisted of washed CNBr-activated Sepharose which had been treated identically except for omission of the amino compound from the reaction step at pH 9.0. The final preparation of resin has a slight yellowish tinge and has to be mixed with an equal volume of unmodified Sepharose in order to obtain flow rates of 10–15 ml per hour.

Resin II. *m*-Aminobenzamidine is coupled to succinyl aminoethyl aminopolyacrylamide resin.[26] The reaction is carried out with dicyclohexyl carbodiimide in a solvent consisting of dimethylformamide–pyridine (1:1, v/v) at 24–26° for 18 hours with stirring. Fifty milliliters of Bio-Gel P60 containing a total of 16.35 meq of carboxyl groups are allowed to react with 17 mmoles (3.11 g) of *m*-aminobenzamidine and 16.35 mmoles (3.37 g) of dicyclohexyl carbodiimide. About 50% of the benzamidine is covalently bound.

Resins III and IV. The details for synthesis of these resins are identical: Twenty milliliters of washed settled Sepharose 6B activated with 5 g of CNBr is reacted with 40 mmoles of 3,3'-diaminopropylamine (3.5 ml) for Resin III; or with 20 mmoles of 2,5-diaminotoluene·HCl (Aldrich) (3.9 g) for Resin IV. Both are allowed to react with the amine at pH 9.0 in 0.1 *M* sodium carbonate for 18 hours at 0°. Next, 20 mmoles of succinic anhydride (2 g) are allowed to react with the washed resin at pH 6 overnight at 0°. The resin is stirred with 100 ml of 0.1 *N* NaOH for 45 minutes at room temperature and washed with distilled water. Each resin is then allowed to react with 0.44 mmole (700 mg) of *N*-hydroxy-succinimide and 0.61 mmole (1.25 g) of dicyclohexyl carbodiimide in 40 ml of dioxane for 70 minutes at room temperature. They are washed sequentially with 200 ml of dioxane, 200 ml of methyl alcohol, and 200 ml of dioxane. Each resin is finally reacted overnight with 50 ml of 0.1 *M* *m*-aminobenzamidine (18.3 mg/ml) in 0.2 *M* sodium borate–0.16 *M* NaCl at pH 8.0. Glycine, 3.5 g, is added to each resin and the reaction continued 48 hours. Thereafter, the resins are washed with 500 ml of 0.5 *M* NaCl and 500 ml of 0.1 *M* acetic acid. Titration of the residues, as described above, indicated that Resin III contained 0.14

[31] D. H. Bing, *Immunochemistry* **8**, 539 (1971).

mmole of ligand per gram of resin (wet weight) and Resin IV contained 0.66 mmole per gram of resin (wet weight).

Resins V and VI. m-(o-Aminophenoxybutoxy)benzamidine, 1 mmole and m-(m-aminophenoxybutoxy)benzamidine, 1 mmole[32] are allowed to react, respectively, with 10 ml of settled Bio-Gel 5A activated with 2.5 g of CNBr. The coupling reaction is done in 0.1 M NaHCO$_3$ at pH 9.0, to which is added an equal volume of 2-methoxyethanol. The resins are washed successively with methanol, 0.15 M NaCl, and water, then titrated to pH 12 as described above. Resins V and VI contained 0.33 mmole per gram (wet weight) and 0.47 mmole per gram (wet weight), respectively. All resins are subjected to washing with all the elution buffers prior to equilibration against the 0.02 ionic strength sodium phosphate at pH 6.3–0.1 M NaCl containing 1 mM Na$_2$H$_2$EDTA.

2.4 Chromatography of C1$\overline{s}$

Between 10 and 20 ml of the euglobulin fraction is applied at 4° to about 5–10 ml of resin, equilibrated against 0.02 ionic strength sodium phosphate at pH 6.3, 0.1 M NaCl, and 1 mM Na$_2$H$_2$EDTA. Unadsorbed protein is washed through the resin, and elution is continued until the adsorbancy at 280 nm is less than 0.05. With certain resins, the column is eluted at this point with 0.4 M NaCl solution to remove nonspecifically adsorbed protein. In all cases, 0.2 M propionic acid is employed to elute the enzyme. These acid fractions are immediately neutralized with 0.5 M Tris, 0.4 M sodium phosphate, or by dialysis against 0.15 ionic strength sodium phosphate at pH 7.3.[28] With certain resins, a subsequent elution step with 0.1 M acetic acid is employed in an attempt to improve protein recoveries. Representative trials with the various resins described in Table I are presented in Table II.

Ia and Ib (Table II) represent the identical resin, but the euglobulin fraction of Ib was incubated for 30 minutes at 30° with 1 mM phenylmethanesulfonyl fluoride at pH 6.3, prior to chromatography. The inhibitor does not affect C1 esterase at this concentration.[24]

These experiments will be discussed in terms of the following context: affinity chromatography experiments usually are done in 2 steps. In the first step, the protein is adsorbed to the resin, and nonspecific protein removed; in the second step, the enzyme is eluted, preferably with a specifically binding substrate or inhibitor. If such a technique is not applicable, a solvent is used that can interrupt the bond between the immobilized ligand and the enzyme.

---

[32] The synthesis of these intermediates has been described. See M. Cory, Irreversible Inhibitors of Guinea Pig Complement, Ph.D. Dissertation, Univ. of California at Santa Barbara (1971).

TABLE II

PURIFICATION OF C1s̄ ON BENZAMIDINE RESINS

		Resin No.						
		Ia	Ib	II	III	IV	V	VI
Euglobulin fraction	Protein, OD$_{280}$	460	615	540	284	284	152	152
	Enzyme (units)[a]	18,750	24,800	20,300	3510	3510	6400	6400
	SA[b]	40.7	40.3	38.1	12.3	12.3	42.0	42.0
Nonadsorbed protein	Protein, OD$_{280}$	300	420	107	162	163	137	140
	Enzyme (units)	31,800	12,400	NT[c]	2610	6130	3770	0
	SA	106	29.6	—	16.1	37.5	27.5	—
0.1 M EDTA, pH 6.3 wash	Protein, OD$_{280}$	—	—	—	5.7	18.9	3.8	17.2
	Enzyme (units)	—	—	—	178	885	689	1300
	SA	—	—	—	31.1	47.8	81	75.7
0.1 M Propionic acid eluate	Protein, OD$_{280}$	46.5	90.7	0.4	2.45	10.6	9.5	5.8
	Enzyme (units)	10,950	8200	0	89	189	693	1480
	SA	236	90.5	—	36.3	17.8	12.3	256
0.1 M Acetic acid eluate	Protein, OD$_{280}$	13.2	21.4	0.4	—	—	—	—
	Enzyme (units)	2580	2960	0	—	—	—	—
	SA	195	138	—	—	—	—	—
Recovery (%)	Protein	78	76.6	20.6	60	68	98.7	107
Recovery (%)	Enzyme	241.2	95	0	81.8	200.3	80.7	43.5

[a] Activity measured with N-Z-L-Tyr-Np.
[b] Specific activity = units enzyme/OD$_{280}$.
[c] NT = not tested.

In regard to the first step, only Resin VI apparently adsorbed all the enzymatic activity detectable in the euglobulin fraction. In fact, with several resins (I and IV) more enzyme activity was detected in the nonadsorbed protein pool than in the starting material. It was speculated that this was due to an esterase or protease which would activate upon interaction with the resin; in the experiment summarized in Ib, phenylmethanesulfonyl fluoride, a nonspecific irreversible inhibitor of serine proteases, reduced the enzyme activity recovery from 241% to 95%, while the protein recovery remained the same.

In regard to the second step in affinity chromatography, elution of the C1s̄ was attempted with 10 mM benzamidine at pH 8.0 with all the resins. Benzamidine is a competitive inhibitor of C1s̄ ($K_i = 3.7 \times 10^{-4} M$). In every case, between 1 and 2% of the protein could be eluted by this compound, but no enzyme activity could be detected in the fraction even after 24–48 hours of dialysis. Dilute acid had been used with success to elute proteases from affinity columns,[33] and this was the only agent tested that would elute active C1s̄ from these resins, the exception being Resin II. Resin II was made with Bio-Gel P60, a polyacrylamide resin. It is similar to III, but the solid matrix excludes proteins above a molecular weight of 80,000. This resin adsorbed protein very well but did not allow its elution. Resin VI was the most effective. Apparently, activation of a nonspecific esterase did not occur during chromatography, all the initial enzyme activity was adsorbed, and the specific activity was 2.5 times better than in Ib, where activation of nonspecific esteratic activity was prevented with phenylmethanesulfonylfluoride. In Ia, Ib, and VI, it was proved that the propionic acid eluent had C1s̄ activity since these fractions had titers of 1000–5000 per OD_{280} unit in the assay in which the catalysis of the $\overline{EAC4,2}$ intermediate was measured. They also formed a precipitin arc with specific anti C1s̄ antiserum. The fractions were not, however, homogeneous. The propionic acid eluent from Ia and Ib contained 2 bands in 7% polyacrylamide gel electrophoresis[34]; the same fraction from VI contained 4 bands in the same analytical system. However, interpretation of these electrophoretic patterns was equivocal, because the acid eluents containing C1s̄ precipitated when they were concentrated or frozen and thawed.

These data indicate that the benzamidine resins probably do not represent a superior technique for obtaining highly active C1s̄ enzyme. It does appear that, wherever adsorption occurred, it was relatively specific since all the resins had similar concentrations of benzamidine co-

[33] G. K. Clina and W. Bushuk, *Biochem. Biophys. Res. Commun.* **37**, 545 (1969).
[34] O. Gabriel, this series, Vol. 22, p. 565.

valently bound; only one (Resin VI) really demonstrated good specificity for C1s̄. These results indicate the importance of the chemical nature of the side chain. For example, Resin IV is very similar to Resin VI, except that there is an amido linkage to the second benzene ring in Resin IV, while there is an ether linkage in the same position in Resin VI. With IV, no adsorption of enzyme was detected; with VI, all the enzyme was adsorbed. This observation agrees with inhibitor data in which it was found that m-[3-(m-nitrobenzamido)propoxy]benzamidine was not inhibitory toward C1̄, but m-[3-(m-nitrophenoxy)propoxy]benzamidine was.[35] Furthermore, the stereochemical orientation of the ligand appears to be important. Resins V and VI are identical except that the amino group linking it to the resin is ortho in Resin VI and meta in Resin V. Resin VI adsorbed enzyme, but Resin V did not. The conclusion at this stage is that it will be important to have fairly detailed knowledge about the interaction of a variety of benzamidine inhibitors with C1s̄ before a specific benzamidine affinity resin can be synthesized and successfully used to purify highly active, homogeneous C1s̄.

3. Purification of C1̄ and C1q on IgG Resins

The first component of complement is made up of three distinct subunits, C1q, C1r, and C1s. The association of these subunits is promoted by calcium ion and is enhanced at NaCl concentration less than 0.15 M. The C1q subunit can react with IgG via the Fc portion of the immunoglobulin; aggregation of the immunoglobulin by heat or formation of an antigen–antibody complex enhances C1q binding. Furthermore, the C1q-immunoglobulin interaction leads to the conversion of C1r and C1s from the proenzyme state.[3,6,7] This information provided the rationale for the development of IgG Sepharose resins for the purification of C1̄ and C1q, namely, cyanogen bromide-activated Sepharose linked to IgG immunoglobulin. Some experiments have been done with another IgG resin, IgG-p-azobenzamidoethyl-Sepharose. The preparation of this resin is also included as preliminary results indicate it may have higher capacity for C1̄,[36] and represents a method for attaching gram quantities of IgG to Sepharose.

3.1 Preparation of Resins

IgG is purified from serum by DEAE chromatography and ammonium sulfate precipitation.[37] It contains only IgG upon examination by urea

[35] M. Cory, personal communication.
[36] S. Assimeh and R. Painter, personal communication.
[37] J. L. Fahey, in "Methods in Immunology and Immunochemistry" (C. A. Williams and M. W. Chase, eds.), Vol. 1, p. 322. Academic Press, New York, 1967.

formate starch gel electrophoresis at pH 3.0. Four hundred milligrams are mixed with 40 ml of Sepharose 6B which had been activated previously at pH 11 with 4 g of CNBr. The activated resin is allowed to react with the IgG in 0.1 M NaHCO$_3$ pH 9.0 at 4° with stirring for 24 hours.[25] The resin is filtered and washed (see below). Between 7 and 9 mg of IgG are bound per milliliter of settled resin under these conditions. The extent of modification is followed by micro-Kjeldahl analysis of an aliquot of resin.[38]

IgG-p-azobenzamidoethyl-Sepharose is prepared by starting with 75 ml of Sepharose 6B which is reacted with 17.5 g of CNBr at pH 11. The activated resin is washed with 1 liter of ice-water and mixed with 15 g (220 mmole) of ethylenediamine (Eastman Kodak) diluted in 150 ml of water and adjusted to pH 11. The amine is allowed to react with the resin at 4° for 18 hours with stirring. The resin is washed on a Büchner funnel with water and then washed with a solution of dimethylformamide (DMF)–0.2 M sodium borate at pH 8.3 (6:4, v/v). The aminoethyl Sepharose is suspended in 200 ml of the same solution and allowed to react with 13.7 mmoles (2.7 g) of p-nitrobenzoyl azide (Eastman Kodak) for 4 hours at 23–24°. The resin is washed on a Büchner funnel with 500 ml of 50% aqueous DMF and then with 500 ml of 0.5 M NaHCO$_3$, pH 8.5. The resin is next mixed with 200 ml of 0.5 M NaHCO$_3$, pH 8.5, and 0.2 M sodium dithionite and reduced for 1 hour at 40° with stirring. The resin is washed sequentially with 200 ml of 0.5 M NaHCO$_3$ at pH 8.5, 200 ml of water, 200 ml of 0.5 N HCl, and, finally, suspended in 50 ml of 0.5 N HCl. This mixture is chilled to 0° and mixed with 100 ml of 0.1 M NaNO$_2$ to form the diazonium salt. After 7 minutes, the suspension is washed with ice-water on a Büchner funnel, and finally added at 0° to 100 ml of a 10 mg/ml solution of IgG in 0.1 M sodium borate at pH 8.0. The pH of the azo-resin protein mixture is usually about 3, and it is necessary to adjust to pH 8.0 with 1 N NaOH; the mixture immediately becomes orange after the pH is raised. The azo coupling of IgG to the Sepharose is allowed to proceed at pH 8.0 and 4° with stirring for 18 hours. The resin is then filtered and washed (see below), and the extent of modification is determined by analysis of unbound protein.[39] Usually 1 ml of resin binds 20–30 mg of IgG.

3.2 Equilibration and Regeneration of IgG-Agarose Resins

Prior to use, it is necessary to remove all noncovalently bound protein from the resins. Similarly, the resins after chromatography must be freed

---

[38] G. Schiffman, E. A. Kabat, and W. Thompson, *Biochemistry* **3**, 113 (1964).

[39] O. H. Lowry, N. J. Rosebrough, A. L. Farr, and R. J. Randall, *J. Biol. Chem.* **193**, 265 (1951).

of IgG and other proteins nonspecifically bound during chromatography. Although these proteins do not interfere with the chromatographic procedure, they will tend to slowly elute upon subsequent chromatography.

Therefore, after reaction of IgG with the CNBr activated Sepharose, the resin is filtered and successively washed on a Büchner funnel with 0.15 M NaCl, 0.1 M EDTA at pH 7.4, and 1 M NaCl–0.2 M 1,4-diaminobutane. The diaminobutane is redistilled prior to use. The volume of the washes is about 100 times the volume of the resin. The resin is then washed with water, and with 0.075 ionic strength Tris chloride at pH 8.0[40] and poured into a column. It is equilibrated against the starting buffer until the pH and conductivity of the effluent is identical with the input buffer. After chromatography, the resin is regenerated by stirring for 20 minutes in about 200 ml of 3 M NaCl adjusted to pH 9.0, or in 200 ml of 2 M guanidine·HCl adjusted to pH 7.4. The IgG-p-azobenzamidoethyl Sepharose is treated identically.

3.3 Purification of C$\bar{1}$ on IgG-Agarose

A column of the resin, 2.5 × 15 cm, is equilibrated with 0.075 ionic strength Tris chloride at pH 8.0 and 4°. The starting material is the euglobulin fraction prepared by precipitation with 20 mM sodium acetate, pH 5.5 (see Section 2.1). After extraction with 0.3 M NaCl, the protein is centrifuged at 48,400 g for 30 minutes at 4° and 10 ml of clear supernatant dialyzed against 2 changes of 1 liter each of 0.15 ionic strength Tris chloride, pH 8.0,[41] at 4° over a period of 24 hours. The protein is centrifuged at 48,400 g for 30 minutes, diluted with an equal volume of water and immediately applied to the resin at 4°. The unbound protein is washed from the resin until the absorbancy at 280 nm is less than 0.05. The C$\bar{1}$ is then eluted with 0.2 M 1,4-diaminobutane adjusted to pH 11.0 with 1 N HCl, and the eluate is dialyzed immediately against cold 0.15 ionic strength Tris chloride at pH 8.0, or 0.15 ionic strength Veronal-buffered saline, pH 7.4.[41] The pH of the eluting buffer is 11, but the pH of the eluted C$\bar{1}$ never rises above pH 9.5. The C$\bar{1}$ may be concentrated by dialysis against dry sucrose (commercial table sugar), or in an Amicon cell with a PM 30 membrane.

3.4 Purification of C1q on IgG-Agarose

A column of IgG-Sepharose, 2.5 × 15 cm, is equilibrated against 0.075 ionic strength Tris chloride at pH 8.0–10 mM Na$_2$H$_2$EDTA at pH 8.0 until the absorbancy at 280 nm is less than 0.05. The column

---

[40] Prepared by dilution of *Buffer 41* of reference cited in footnote 28.
[41] *Buffer 6B.*[28]

is eluted with 0.4 M NaCl–0.1 M Na$_2$H$_2$EDTA until the absorbancy is less than 0.05. Finally, C1q is eluted with 0.2 M 1,4-diaminobutane at pH 11.0, and is immediately dialyzed against cold 0.15 ionic strength Tris chloride at pH 8.0. The solution is concentrated by dialysis against dry sucrose or in an Amicon cell with a PM 30 membrane.

Table III contains a summary of representative experiments including protein and enzyme recoveries. The recovery of C1q is expressed only in terms of protein, since this is a structural protein which is without known catalytic activity. It does, however, participate in the complement sequence, and its activity in the conversion of EAC4 in the presence of C1r̄ and C1s̄, to the EAC̄1,4 intermediate and subsequent lysis by C2 and CEDTA, is expressed as dilution which produces 63% lysis of 1.5×10^7 EAC4 cells. The presence of C1q in various protein fractions is also monitored in a qualitative fashion by the latex agglutination test.

In the purification of C̄1 from euglobulin, it can be seen that more enzymatic activity was recovered in the purified protein than was applied to the column. Although not shown in the table, the total enzyme recovery becomes 95–100% when the euglobulin fraction is preincubated with phenylmethanesulfonyl fluoride. This treatment also results in a 10% or so loss in C̄1 activity in the assay measuring EAC̄1,4 formation and has therefore not been adopted as a standard procedure.

The purified C̄1 is concentrated to about 1 mg/ml and distributed into several tubes for storage at $-90°$. It maintains its activity for about 1 week at 0° in 0.15 M NaCl at pH 7.4. C̄1 purified by the resin reacts with specific anti-C1q and monospecific anti-C1s̄ antiserum, gives at least three precipitin arcs with anti-whole human serum antiserum, and gives four precipitin arcs with antiserum to the euglobulin fraction.[42] The purified C̄1 routinely has 4–5% IgM and IgG as determined by radial immunodiffusion analysis and contains 3 major staining bands in 7% polyacrylamide electrophoresis systems.[34] The data shows that the final yield from serum was about 169 μg/ml (Table III). If the molar ratio of the subunits of C̄1 are 1:2:4 (C1q:C1r:C1s) as indicated by Müller-Eberhard,[3] between 332 and 371 μg of C1 per milliliter should be present in serum. This gives a maximal overall yield of C̄1 of 46% in terms of protein recovered. It is difficult to assess the degree of purification in terms of enzyme activity: N-Z-L-Tyr-Np is not a specific substrate; enzyme activation occurs during the purification procedure; and C̄1 activity in serum and the euglobulin fraction is difficult to quantitate in the EAC̄1,4 assay because other contaminating complement components can interfere in subsequent steps of the lytic sequence. At worst, purification is 26-fold

[42] D. H. Bing, *J. Immunol.* **107**, 1243 (1971).

TABLE III
PURIFICATION OF $C\bar{1}$ AND C1q ON IgG-AGAROSE

	Resin	IgG-Sepharose	IgG-Sepharose
Serum	Volume (ml)	330	60
	Total protein	NT[a]	—
	Total enzyme	8.7×10^{14} em[b]	—
Euglobulin fraction	Volume (ml)	33	6
	Total protein (mg)	402	91.2
	Total enzyme		
	N-Z-L-Tyr-Np	14,520 units	NT (LA[d] ++)
	EAC4	6.45×10^{14} em	NT
	Specific activity		
	N-Z-L-Tyr-Np	33.1 units/mg	—
	EAC4	1.6×10^{12} em/mg	—
Nonadsorbed protein	Volume (ml)	350	65
	Total protein (mg)	296	350 (LA ±)
	Total enzyme		
	N-Z-L-Tyr-Np	13,600 units	NT
	Specific activity		
	N-Z-L-Tyr-Np	46 units/mg	—
EDTA eluent	Volume (ml)	—	15
	Total protein (mg)	—	1.5
	Total enzyme		
	N-Z-L-Tyr-Np	—	NT (LA −)
	EAC4	—	NT
0.2 M 1,4-Diaminobutane eluent	Volume (ml)	113	20
	Total protein	56.5 mg	2.4 mg
	Total enzyme		
	N-Z-L-Tyr-Np	49,000 units	0 (LA +++)
	EAC4	15.8×10^{14} em	950[c]
	Specific activity		
	N-Z-L-Tyr-Np	850 units/mg	—
	EAC4	2.8×10^{13} em/mg	—
% Recovery	Protein	86	89.8
	Enzyme	437	—
Protein purified		$C\bar{1}$	C1q

[a] NT = not tested.
[b] em = effective molecules; whole $C\bar{1}$ activity measured.
[c] Dilution of C1q in presence of excess of $C1\bar{r}$ and $C1\bar{s}$ which produced 63% lysis of 1.5×10^7 EAC4.
[d] LA = Latex agglutination test.

and at best 850-fold. The latter value is, in fact, an average value, since the specific activity of $C\bar{1}$ preparations in terms of N-Z-L-Tyr-Np activity has been observed as low as 400 and as high as 1400. The data presented here are derived from one representative experiment from a series of 50 such purifications.

C1q can also be resolved from the euglobulin fraction on IgG-Sepharose (Table III). In this instance the column is developed in the presence of EDTA in an attempt to prevent activation and/or binding of C1r and C1s. C1q is purified in about 26% yield from serum and is devoid of $C1\bar{r}$ and $C1\bar{s}$ activity.[43] C1q purified by this procedure is contaminated to 4–5% with IgM. The IgM protein can be removed by preparative sucrose gradient centrifugation and C1q be obtained in three steps: precipitation, chromatography, centrifugation. The product approaches 100% efficiency in the C1 transfer test which, incidentally, has also been observed for whole $C\bar{1}$.[43,44]

Whereas experience with the benzamidine resins indicated the difficulty of designing a resin with a specifically binding ligand, the experience with the IgG resins has presented another problem in terms of purifying complement components by affinity chromatography: even when there is a specific ligand, good purification is attained only with a specific eluting agent. Certainly EDTA and diamines do not fill this criteria; $C\bar{1}$ and C1q are contaminated with immunoglobulins and C1s and C1r still must be resolved in subsequent purification steps. However, these experiments do represent a start toward obtaining, in high yield, highly purified, fully active complement proteins with a minimal amount of time and effort. In this regard, affinity chromatography now does and probably will continue to represent a valuable tool for individual investigators interested in studying this interesting system of specifically interacting proteases.

[43] C. R. Sledge and D. H. Bing, *J. Immunol.* 111, 661 (1973).
[44] H. R. Colten, T. Borsos, and H. J. Rapp, *Science* 158, 1590 (1967).

[93] Insulin

By Yasuo Akanuma and Masaki Hayashi

There are several forms of specific binding between biological materials such as enzyme–substrate complexes, antigen–antibody reactions, and hormone–receptor bindings. Akanuma and Glomset[1] reported a method for coupling of lipoproteins to Sepharose by the method of Porath,[2]

[1] Y. Akanuma and J. Glomset, *Biochem. Biophys. Res. Commun.* 32, 639 (1968).

and they used it for the study of binding of the plasma enzyme, lecithin: cholesterol acyltransferase. Cuatrecasas has prepared a column of insulin-Sepharose and demonstrated the adsorption of insulin antibody to it at pH 8.8 and its elution with HCl.[3] Here, we describe an insulin-antibody-Sepharose with which we have concentrated insulin from biological fluids and have prepared an insulin-free plasma.[4]

Affixing the Antibody to Agarose[5]

The γ-globulin fraction is obtained from guinea pig antibovine insulin serum by ammonium sulfate precipitation at the final concentration of 40% of saturation. The precipitate is dialyzed against physiological saline and is subsequently used for the coupling with Sepharose 2B. Before coupling, Sepharose beads are washed extensively with distilled water to remove sodium azide. Ten milliliters of 25 mg cyanogen bromide per milliliter is added to 20 g, wet weight of Sepharose and stirred for 6 minutes at pH 11 with addition of 1 N NaOH to maintain the pH. The activated gel is washed with a large amount of ice-chilled distilled water, and then with 0.1 M NaHCO$_3$. The washed activated-Sepharose beads are mixed with anti-insulin globulin fraction (AIG) in 2 ml of 0.14 M NaCl–0.1 M NaHCO$_3$ and stirred gently for 20 hours at 4°. Finally, the gels are washed on a glass filter with 0.14 M NaCl–0.01 M Tris chloride (pH 8.6) to remove the unbound proteins. In the reaction of 650 mg of anti-insulin globulin with 60 g of cyanogen bromide-activated Sepharose, 92% of the anti-insulin globulin was coupled to Sepharose (AIG-Sepharose).

Guinea pig anti-HGH globulin is prepared in an entirely similar manner to anti-insulin globulin.

Insulin Binding to AIG-Agarose[5]

Maximum binding of insulin to AIG-Sepharose was studied by a batch technique. One gram of AIG-Sepharose (10 mg of protein per gram wet weight) was placed in a plastic vial with 1 ml of Tris-BSA buffer (0.14 M NaCl–10 mM Tris chloride–0.25% bovine serum albumin at pH 7.8). The amount of insulin in each vessel varied. Binding was found to become

---

[2] J. Porath, R. Axén, and S. Ernbäck, *Nature (London)* **215**, 1491 (1967).

[3] P. Cuatrecasas, *Biochem. Biophys. Res. Commun.* **35**, 531 (1969). See also Cuatrecasas and Parikh, this volume [85].

[4] Y. Akanuma, T. Kuzuya, M. Hayashi, Y. Ide, and N. Kuzuya, *Biochem. Biophys. Res. Commun.* **38**, 947 (1970).

[5] M. Hayashi, Y. Akanuma, T. Kuzuya, Y. Ide, and N. Kuzuya, *Tonyobyo* **14**, 63 (1971).

maximum and to plateau after 1 hour of incubation. Therefore, the insulin mixtures are incubated for 3 hours at 4°. After incubation, the gels are washed four times with ice-cold Tris-BSA buffer (8 ml per wash). The bound insulin is dissociated with 1 M acetic acid. Dissociation was found to occur beginning at pH 3.5; at pH 2.6, 90.4% of bound insulin is recovered. Insulin binding is maximum at an insulin concentration greater than 1000 mU/ml. One gram of AIG-Sepharose bound 420–460 mU of insulin at its maximum.

Anti-HGH globulin-Sepharose (4.4 mg protein per gram wet weight) bound only ^{125}I-HGH; no appreciable amount of ^{131}I-insulin was adsorbed. On the other hand, AIG-Sepharose bound only ^{131}I-insulin under these conditions. Thus, the specificity of antibody is preserved even after its coupling to the gel matrix. A control Sepharose, consisting of plasma high density lipoproteins-Sepharose (0.14 μmole of lecithin per gram wet weight) bound neither ^{131}I-insulin nor ^{125}I-HGH, thus demonstrating that nonspecific binding of antigens to the gel matrix is negligible.

Concentration of Insulin and Preparation of Insulin-Free Plasma

Column Chromatography. An AIG-Sepharose column (1.5 × 13 cm) is equilibrated with Tris-BSA buffer and is charged with 290 ml of the buffer, containing 109 μU/ml of pork insulin, at a flow rate of 60 ml per hour. Fractions of 9 ml were collected. The insulin contents in the eluent are less than 9 μU/ml. The column is washed further with the same buffer without obtaining insulin in the eluate. However, when the buffer is changed to Tris chloride, i.e., without albumin, the pH in the eluates increases slightly and small amounts of insulin are found to dissociate from the column. Finally, the column is washed with 1 M acetic acid thereby resulting in elution of the bound insulin. The eluate is dried by flash evaporation and the residue is dissolved in 2 ml of 0.01% HCl; the pH of the solution is brought to 8.0–8.6 with Tris-BSA buffer. The net recovery of insulin is about 90%; the highest concentration of insulin in the eluted solutions has been 2980 μU/ml (Fig. 1). After regeneration of the column by washing the gels with Tris-BSA buffer, there was no change in recovery of insulin from the column. Thus, the column can be used repeatedly without loss of its insulin-binding capacity.

For the experiments on plasma insulin extraction, an AIG-Sepharose column (1.5 × 11 cm) is equilibrated with the 0.2 M borate buffer containing 0.25% BSA. Ten milliliters of dog pancreatic vein plasma (1575 μU of insulin per milliliter) is charged, and the column is washed with 80 ml of the same buffer at a flow rate of 30 ml/hr. Bound insulin is eluted with 1 M acetic acid. The plasma which passes through the column contains between 0 and 3 μU/ml insulin; i.e., essentially insulin-free

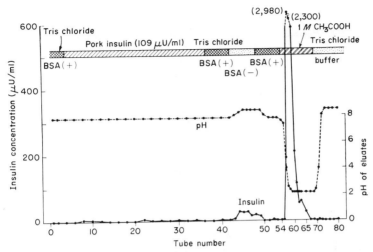

Fig. 1. Concentration of insulin with AIG-Sepharose. Pork insulin solution (290 ml; 109 μU/ml) was charged onto the column, which was washed with Tris-BSA buffer. Bound insulin was eluted with 1 M acetic acid.

plasma is obtained. The total insulin recovery after acetic acid elution is about 80% of the original plasma.

The effect of flow rate of the column procedure on the recovery of insulin has been studied. At a flow rate at 15, 30, and 60 ml per hour recoveries of 79%, 81%, and 84%, respectively, were obtained.

Batch Methods. The binding of insulin to AIG-Sepharose has also been studied by a batch method. One gram of AIG-Sepharose (5.4 mg of protein per gram wet weight) is incubated with 8 ml of dog pancreatic vein plasma (740 μU of insulin per milliliter) for 3 hours at 4°. The gel is collected by centrifugation, and the insulin content of the supernatant fluid is determined. In a typical experiment, 8 μU of insulin were present per milliliter of liquid. By acid treatment of the gel a total of 5920 μU of insulin were recovered, corresponding to 90% recovery.

The dissociation of antibody from the gels has been examined by the batch method. When 1 g of AIG-Sepharose was incubated with 8 ml of borate-BSA buffer, $6.2 \times 10^{-1}\%$ of the binding capacity was found in the supernatant fluid. When the gel was treated with 1 M acetic acid, $5.4 \times 10^{-3}\%$ of the binding capacity appeared in the supernatant fluid. When anti-insulin serum itself was coupled to Sepharose, instead of AIG, dissociation of the binding capacity from the gels was greatly reduced.[6]

---

[6] Y. Akanuma and M. Hayashi, unpublished observations.

Comments

AIG-Sepharose is able to bind insulin at between pH 7.8 and 8.6. The bound insulin is dissociated by 1 M acetic acid. Column chromatography and batch methods using this immunoadsorbent are useful techniques for both the extraction of insulin from biological fluids and for the preparation of essentially insulin-free plasma. The limitation of the methods is the dissociation of very small amounts of antibody from the gels during the treatment. The method could be improved by using anti-insulin serum-Sepharose instead of anti-insulin globulin-Sepharose.

[94] Thymocytes

By ZELIG ESHHAR, TOVA WAKS, and MICHAEL BUSTIN

Cells bearing specific surface components can be isolated from a heterogeneous mixture of cells with the aid of antibodies directed against these surface antigens. The lymphocytes constitute a heterogeneous population of cells. To elucidate the functions of the different lymphocyte subpopulations a number of fractionation methods have been devised. Some of these methods are based on fractionating the cells according to specific surface receptors or structural characteristics of the plasma membranes.[1-7] Thymocytes are considered to be a relatively homogeneous cell population toward which specific antibodies can be elicited.[8,9] The method presented affords quantitative fractionation of thymocytes by affinity chromatography on Sepharose conjugated with immunoglobulin derived from the sera elicited by either thymocytes or by fractions solubilized from thymocyte membranes.[10-12]

[1] S. Sell and T. An, *J. Immunol.* **107**, 1302 (1971).
[2] S. F. Schlossman and L. Hudson, *J. Immunol.* **110**, 313 (1973).
[3] H. Wigzell, P. Golstein, E. A. J. Svedmyr, and M. Jondahl, *Transplant. Proc.* **4**, 311 (1972).
[4] G. M. Edelman, U. Rutishauser, and C. F. Millette, *Proc. Nat. Acad. Sci. U.S.* **68**, 2153 (1971).
[5] P. Truffa-Bachi and L. Wofsy, *Proc. Nat. Acad. Sci. U.S.* **66**, 685 (1970).
[6] M. Sela, E. Mozes, G. M. Shearer, and Y. Karniely, *Proc. Nat. Acad. Sci. U.S.* **67**, 1288 (1970).
[7] A. Basten, J. Sprent, and J. F. A. P. Miller, *Nature* (*London*) *New Biol.* **235**, 178 (1972).
[8] M. C. Raff, *Transplant. Rev.* **6**, 52 (1971).
[9] B. H. Waksman, *Transplant. Rev.* **6**, 30 (1971).
[10] Z. Eshhar, *Eur. J. Immunol.* **3**, 668 (1973).
[11] Z. Eshhar, M. Gafni, D. Givol, and M. Sela, *Eur. J. Immunol.* **1**, 323 (1971).
[12] M. Bustin, Z. Eshhar, and M. Sela, *Eur. J. Biochem.* **31**, 541 (1972).

Preparation of Antisera and Immunoglobulins

To obtain antisera to rat thymocytes or soluble components derived from rat thymocyte membranes,[12] 10^9 thymocytes, or equivalent amounts of soluble material, emulsified in 2.0 ml of 50% complete Freund's adjuvant, are injected into rabbit at multiple intradermal sites. Booster injections are given intravenously with identical amount of material suspended in 1.0 ml of 0.14 M NaCl–10 mM sodium phosphate at pH 7.4 at 2-week intervals, and sera are collected 1 week after each booster injection. The potency and specificity of the antisera can be checked by the cytotoxic assay.[10,13] The immunoglobulin fraction of the various antisera are prepared by precipitation in 40% $(NH_4)_2SO_4$ and dialysis at 4° against the above-noted sodium phosphate buffer.[14]

Preparation of Antibody-Coated Agarose

One gram of Sepharose 4B activated by the cyanogen bromide procedure[15] is added to solutions containing 20 mg of protein in 1.0 ml of 0.1 N NaHCO$_3$ chilled to 4°. The suspension is mixed on a shaker at 4° overnight. The conjugated Sepharose is washed with the sodium phosphate buffer until the A_{280} is below 0.02. The amount of protein bound is determined by measuring the optical density at 280 nm (for globulins $\epsilon_{1\%}^{1 cm} = 14.0$) of the protein solution before addition to the activated beads and the optical density at 280 nm of the solutions used to wash the conjugated beads. By this procedure over 85% of the protein can be covalently linked to Sepharose. With bovine or rabbit serum albumin only about 60% of the protein is bound. The coated Sepharose beads can be stored at 4° in the presence of 0.02% NaN$_3$ and a few drops of toluene.

Thymocyte Suspensions and Labeling

Cells obtained from the thymus of rats, 6–10 weeks old, are gently teased through stainless steel mesh in Minimum Essential Medium (MEM, Microbiological Associates) made 0.1% in BSA (Eagle's-BSA). To remove aggregates, the cell suspensions are filtered through glass wool.

For labeling, 100 μCi of sodium chromate ^{51}Cr (Amersham) are added to 10^8 cells suspended in 1.0 ml of Eagle's-BSA, and the mixture is incubated for 1 hour at 37° with occasional shaking. Unbound isotope is removed by 4 washes in Eagle's-BSA at room temperature. To decrease

---

[13] A. R. Sanderson and K. I. Welsh, in "Transplantation Antigens" (B. D. Kahan and R. A. Reisfeld, eds.), p. 453. Academic Press, New York, 1972.

[14] P. Stelos, in "Handbook of Experimental Immunology" (P. Weir, ed.), p. 3. Blackwell, Oxford, 1967.

[15] J. Porath, R. Axén, and S. Ernbäck, Nature (London) 215, 1491 (1967).

the spontaneous release of the isotope, cells are incubated at 4° for 30 minutes in 20 ml of Eagle's-BSA medium and washed once again.

Microscopic Observation

To check whether the conjugated antibodies retain activity toward thymocytes, between 5×10^6 and 10×10^6 cells are added to 0.2 ml of a suspension of Sepharose conjugated with globulins derived from anti-thymocyte sera. The mixture is incubated at 37° for 10 minutes, and a drop from the bottom of the test tube is gently placed on a microscope plate. Particular care has to be exercised during these manipulations because the cells are very weakly attached to the Sepharose beads, and excessive shaking or rapid flow will remove cells from the conjugated Sepharose beads. Binding of thymocytes by the conjugated globulins is readily visualized with a microscope.[16]

Affinity Chromatography of Cells on Conjugated Agarose

Conjugated Sepharose is poured into columns made from the lower part of 10-ml disposable plastic pipettes (Falcon Plastics). A disk of stainless steel wire cloth (200-mesh) is wedged into the bottom of the pipettes. A short straight piece of polyethylene tubing is attached to the tip of the pipette so that the flow can be stopped when desired. (Use short, straight tubing because long, bent tubing will trap cells.) Routinely, columns with bead volumes of 1.2 ± 0.2 ml are used. The Sepharose beads are washed with Eagle's-BSA. Cells should be added to the columns in a volume of less than 20% of the total column volume. Routinely, 0.1 ml containing 1×10^6 to 10×10^6 cells are added to columns. The cells are washed into the column with a 0.1-ml portion of Eagle's-BSA. The outlet tube is closed to allow the cells to incubate on the columns for 15 minutes at room temperature. Columns are then washed with Eagle's-BSA. The fraction of unbound cells is defined as those cells which are collected in the first 4 ml of eluate. The number of cells eluted from, or bound to, the conjugated Sepharose is determined by ^{51}Cr counting. The radioactive counting is directly related to cell number since less than 6% of the counts invariably remain in the supernatant after centrifugation. The spontaneous release of ^{51}Cr from cells is not affected by the various Sepharose conjugates. The fraction of cells bound to the column is determined either by subtraction of the unbound cells from the total or by direct counting of the column. More than 90% of the thymocytes are bound to Sepharose columns conjugated with globulins derived from antithymocyte sera. In contrast, over 90% of cells do not

[16] Z. Eshhar, Ph.D. Thesis, Weizmann Institute, 1973.

TABLE I
BINDING OF THYMOCYTES TO VARIOUS AGAROSE CONJUGATES

Number of cells[a] $\times 10^6$	Agarose conjugate	Fraction of cells bound (%)
2	Anti-thymocyte globulin	95
10	Anti-thymocyte globulin	94
1	Anti-soluble globulin[b]	55
10	Anti-soluble globulin[b]	48
10	Anti-peak-I[b]	60
10	Anti-peak-III[b]	11
10	Normal rabbit globulin	5
2	Bovine serum albumin	6
10	Bovine serum albumin	5

[a] Specified number of cells in 0.1 Eagle's medium applied on columns of 1.2 ml total volume.

[b] Globulins isolated from antisera either against the whole mixture of components solubilized from rat thymocyte membrane or against the antigenically active peak-I or antigenically inactive peak-III obtained after gel chromatography of the soluble preparation [M. Bustin, Z. Eshhar, and M. Sela, *Eur. J. Biochem.* **31**, 541 (1972)].

bind to columns made of Sepharose conjugated either with globulins derived from normal rabbit sera or with bovine serum albumin (Table I). The number of cells bound is dependent on the amount of conjugated Sepharose used as well as on the number of cells added to a given column.

Inhibition of Binding by Soluble Antigenic Components

The binding of thymocytes to the Sepharose conjugates can be specifically inhibited by preincubation of the conjugated beads with solubilized thymocyte antigen. Prior to pouring of the columns, the Sepharose conjugates are incubated in test tubes at 25° with 1.0 ml containing per milliliter about 5 mg of antigen solubilized from thymocyte membrane by citraconylation.[9] The inhibition of thymocyte binding to Sepharose conjugates by antigen solubilized from thymus membrane is demonstrated in Table II. Proteins from rat spleen membranes solubilized by the same procedure do not inhibit the binding of thymocytes to Sepharose columns conjugated with globulins derived from serum elicited by thymocytes or by soluble preparations derived from thymocyte membranes.[11,12]

Release of Bound Cells from Sepharose Beads Conjugated with Globulins Derived from Antithymocyte Sera

The thymocytes which adhered to the columns can be quantitatively released by repeated resuspension of the resin with a Pasteur pipette.

TABLE II
INHIBITION OF CELL BINDING TO AGAROSE CONJUGATES BY ANTIGEN
SOLUBILIZED FROM THYMOCYTE MEMBRANES

Cell type[a]	Agarose conjugate[b]	Fraction of cells bound (%)	
		Without antigen	With antigen[c]
Thymocytes	Anti-thymocyte globulin	92	81
Thymocytes	Anti-thymocyte globulin	72	52
Thymocytes	Anti-soluble globulin	54	12
Thymocytes	Anti-peak-I	43	21
Thymocytes	Anti-peak-III ·	5	5

[a] One-tenth milliliter containing between 3×10^6 and 15×10^6 cells applied to columns of 1.2 ml.

[b] The Sepharose conjugated with various Ig are diluted with BSA-Sepharose to adjust the number of cells bound to desired values.

[c] Prior to pouring of the columns, the Sepharose conjugates were incubated in test tubes at 25° with 1.0 ml containing 5 mg/ml antigen solubilized from thymocyte membranes by citraconylation [Z. Eshhar, M. Gafni, D. Givol, and M. Sela, *Eur. J. Immunol.* **1**, 323 (1971)].

More than 30% of the released cells are viable as judged by vitality staining with a 0.3% solution of trypan-blue. Apparently resuspension of the cell–antibody–Sepharose complex exerts enough shearing force to break the bond between the antibodies and the cells.

Comments

1. Whereas red blood cells pass freely through the conjugate columns, nonthymic lymphocytes, especially B-lymphocytes, tend to adhere to Sepharose beads as well as to cotton wool or glass beads.[17,18] About 35% of lymphnode, spleen, blood or bone-marrow, lymphocytes bound to Sepharose conjugated to bovine serum albumin or normal rabbit globulins. Spleen cells which bind to these columns, when released, retain their capacity to produce hemolytic plaques on red blood cells.

To reduce this nonspecific adherence, the sticking cells can be eliminated by precycling the lymphocyte suspension on either Sepharose-BSA column (Table III), or on glass wool[2] or by sedimenting the cells that phagocytize carbonyl iron.[2] With immunoadsorbents made of Degalan beads, the nonspecific adsorption of lymphocytes may be smaller.[3]

2. When the binding forces between the cells and the antibody-con-

[17] N. M. Hogg and M. F. Greaves, *Immunology* **22**, 959 (1972).
[18] S. V. Hunt, *Immunology* **24**, 699 (1973).

TABLE III
RECYCLING OF BONE-MARROW CELLS ON BSA-AGAROSE COLUMN

	Fraction of bone-marrow cells (%)	
Step	Bound	Unbound
First application[a]	35	65
Release from column[b]	16	84
Second application[c] (eluted cells)	11	89
Second application[d] (bound cells)	47	53

[a] Four-tenths milliliter containing 40×10^6 bone marrow cells applied on 6 ml BSA-Sepharose columns.

[b] Cells bound on column released by gentle resuspending of the column content with a Pasteur pipette. In this step the total number of cells refer to the bound cells in first application, i.e., 35%. Note that most (over 80%) of the bound cells could be released from the columns.

[c] Three-tenths milliliter containing 15×10^6 unbound cells from first application reapplied to another column containing 4 ml BSA-Sepharose.

[d] Three-tenths milliliter containing 2.5×10^6 bound cells from first application (released by teasing) reapplied on column containing 4 ml BSA-Sepharose.

jugated beads are strong (e.g., with antibodies of high affinity), the method developed by Schlossman and Hudson[2] is recommended, whereby antibodies are conjugated to Sephadex rather than to Sepharose and the cells are released by digesting the resin with dextranase.

Author Index

Numbers in parentheses are reference numbers and indicate that an author's work is referred to although his name is not cited in the text.

G

Gaastra, W., 120, 492, 494(5)
Gabriel, O., 740, 744(34)
Gaertner, E., 254
Gafni, M., 750, 753(11), 754
Galbraith, W., 336
Gallop, P. M., 420
Gardner, E., 573
Garrett, L. R., 437
Garrison, O. R., 3(85), 6(85), 9, 339
Garvey, J. S., 621
Gauldie, J., 274, 282
Gazis, E., 508, 510(5)
Gell, P. G. H., 32, 49(10)
Gentry, P. W., 3(95), 6(95), 10
Germershausen, J., 3(81), 6(81), 9, 653
Gerster, J. F., 229
Gertler, A., 3(55), 5(55), 9
Gerwin, B. I., 3(65), 5(65), 9, 473
Ghuysen, J. M., 398(6, 7, 8), 399
Giannini, M., 559
Gibson, R., 390
Gielow, L., 373
Gilden, R. V., 67, 729
Gilham, P. T., 47, 48(31), 463, 469, 472, 473(26), 474(26), 478, 491
Ginsburg, S., 583
Giorgio, N., Jr., 700
Gitomer, W., 288
Givol, D., 183, 185(10), 186(5, 10, 12), 188(5, 11), 189(5), 190(18), 191 (18), 193, 703, 711, 712(10), 726, 750, 753(11), 754
Gladner, J. A., 449
Glaid, A. J., 598
Gleich, G. J., 709
Glenner, G. G., 103, 104(6b)
Glomset, J., 746
Godfrey, H. P., 32, 49(10)
Goebel, W., 394
Goetzl, E. J., 188
Goldemberg, S. H., 164
Goldman, R., 21, 71, 72(15), 103, 479, 703
Goldstein, A., 622
Goldstein, F. F., 173
Goldstein, I. J., 329, 336, 338, 502
Goldstein, L., 21, 71, 72(15), 103, 501, 502(14), 703

Golstein, P., 197, 750, 754(3)
Gomez, J. E., 592
Goncalves, J. M., 474
Good, R. A., 716
Goodkofsky, I., 731, 732(5)
Gordon, J. A., 317, 323(2), 328(2), 342 (9), 343, 347
Gordon, J. K., 3(88), 6(88), 10
Gordon, W. J., 359
Gorecki, M., 3(63, 76), 5(63), 6(76), 9, 104, 119, 138, 139(28), 183, 185 (6), 186(5), 188(5), 189(5), 294, 400, 492, 493(2), 494(2)
Gornall, A. G., 282
Gospodarowicz, D., 3(110), 7(110), 10
Göthe, P. O., 3(19), 4(19), 8, 595
Gotschlich, E. C., 715, 716
Gottesman, M., 3(100), 6(100), 10
Gottschalk, E., 166
Granner, D. K., 299
Grappel, S. F., 716
Gräsbeck, R., 305
Graves, D. J., 3(77), 6(77), 9
Graves, I. L., 499
Gray, E. G., 622
Greaves, M. F., 3(112), 7(112), 10, 754
Green, A. A., 367
Green, A. L., 583
Green, A. M., 388(7), 389
Green, D. M., 334
Green, I., 721
Greenaway, P. J., 333, 335(14), 346
Greenberg, D. M., 520
Greenblatt, J., 713
Greenfield, N. J., 274, 282
Gregoriadis, G., 688
Griffin, T., 14, 110, 113(8), 114(18), 115 (18), 117, 119(8, 18), 122(18), 123 (18), 125(18), 601
Griffiths, D., 347
Grimminger, H., 3(68), 5(68), 9
Grosjean, H., 471
Gross, J., 421
Gross, P. R., 624
Grossberg, A. L., 183
Grossman, S., 3(42), 4(42), 8
Grubner, D., 141
Grunberg-Manago, M., 485
Guilbert, B., 336, 338
Guilford, H., 121, 125(34), 163, 164(1),

Subject Index

Glycogen synthetase, 131–132, 164, 338
Glycolate oxidase, 302
Glycolipids, 85
Glycopeptide, 3, 332–341, *see also* specific glycopeptides
Glycoprotein, 6, 610, 688–691, *see also* Asialoglycoprotein receptors
desorption, 341
group-specific separation of, 331–341, *see also* specific glycoproteins
hydrazido-agarose derivatives, 85
insulin receptors, 668
lectin
covalent coupling, 337–338, 339
polymerization, 335, 337, 338
precipitation, 335, 337
purification, 335–338
β-Glycoprotein, 731
Glycoprotein receptors, 221
Glycosidase, 4
Glycosides, 4
β-Glycosides, 334
D-Glycosides, 336
β-L-Glycosides, 364–365
O-Glycosides, 318–323, *see also* specific glycosides
acetobromo derivatives, 321–323
O-Glycosyl polyacrylamide
affinity chromatography, 366–367
electrophoresis, 178–181
copolymers, 366
phytohemagglutinins, 361–367
preparation, 366
properties, 366
Glycylglycyl-(O-benzyl)-L-tyrosyl-arginine, 5
Glycylglycyltyrosine derivatives, 49
Glycyl-D-phenylalanine, 4
Glycyl-D-phenylalanine-agarose, 435
Glycyl-L-tyrosine, 395
GMP-agarose, 477
Gradient elution, 237–239
Griffonia simplicifolia, lectin from, 333
Group C anticarbohydrate antibodies, 712
Group densities, 52
Group-specific separation of glycoproteins, 331–341, *see also* specific compounds
GTP, *see* Guanosine triphosphate
GTP-agarose, 477

Guanidino group, 103
Guanine deaminase, 523–527
assay, 526
Guanosine diphosphate (GDP) as ligands, 482
Guanosine monophosphate, *see* GMP
Guanosine triphosphate (GTP), 4

H

HABA, *see* 4-(Hydroxyazobenzyl)-2'-carboxylic acid
Haemophilus influenza type b polysaccharide, 715–716
Hageman factor, 435
Haptens, immobilizing, 710
Haptoglobin, 6
Hemagglutinins, 328, 329–331
phytohemagglutinins, 367
Heme peptides, 6
of cytochrome *c,* 194
Hemoglobin, 5, 6, 706
Heparin, 6
Hepatitis β-antigen, 6, 338
Hexafluoroisopropanol, 656
Hexamethylenediamine, 4
as spacer, 80
Hexamethylenediamine-agarose, 510
as spacer, 72
Hexokinase, 246, 252, 328
Hindrance, steric, 118
Hirudin, 445
Histamine polyacrylamide, 47
Histidine, tRNA synthetase, 170
Histidine-binding protein, 133
Histone(s), 549–552, *see also* Cysteine-containing proteins
Histone F2al, 551
Histone F3, 551
Human chorion gonadotropin, 339
Hydrazide(s)
derivatives, 36–37, 50–51
group determination, 57
linkage, glass, 69
test for, 33
Hydrazide-agarose, 475–476, 490–491
Hydrazide gels, di- and trinitrophenylation of, 48, 49
Hydrazido-agarose, 84
Hydrazido-albumin, 92
Hydrazido-cellulose, 84–85